The Geodynamics Project is an international program of research on the dynamics and dynamic history of the earth with emphasis on deep-seated foundations of geological phenomena. This includes investigations related to movements and deformations, past and present, of the lithosphere, and all relevant properties of the earth's interior and especially any evidence for motions at depth. The program is an interdisciplinary one, coordinated by the Inter-Union Commission on Geodynamics (I.C.G.) established by I.C.S.U at the request of I.U.G.G., and I.U.G.S., with rules providing for the active participation of all interested I.C.S.U. Unions and Committees.

ISBN 3 510 65083 2

Design of the cover: Wolfgang M. Karrasch
Printed in Germany

Inter-Union Commission on Geodynamics
Scientific Report No. 38

Alps, Apennines Hellenides

Geodynamic Investigation along Geotraverses
by an International Group of Geoscientists

Editors
H. Closs, Hannover
D. Roeder, Tennessee
K. Schmidt, München

With 279 figures and 14 tables in the text
and on 5 foldouts

E. Schweizerbart'sche Verlagsbuchhandlung
(Nägele u. Obermiller) Stuttgart 1978

Preface

"Alps, Apennines, Hellenides" is the fourth contribution of the ICG National Committee for the Federal Republic of Germany to the ICG Scientific Report series. The previous contributions, ICG report numbers 8, 14, and 16, dealt with the rift belts that longitudinally split the Old World. This report, ICG number 38, addresses the collision fronts which traverse the lithosphere along the Mediterranean, normal to the rift system. Here between the rim of the Alps and Carpathians on the North and the South Atlas lineament and Cretic island arc on the South, instead of a stable front from classical plate tectonic *war of position*, there is a 1500 km wide zone of various fronts where the plate interactions take place. The orogenic belts as described in this volume are examples of how and where this *war of several fronts* between the African and Eurasian plates has shifted in geologic time.

The areas investigated are mostly beyond the boundaries of the Federal Republic. This volume represents a broad international collaboration between German scientists and their colleagues in the respective host countries. A mutual desire for understanding this European tectonic problem transcended the language barriers as well as the traditional limits between the scientific disciplines. Geoscientists of all persuasions and many nationalities worked side by side seeking a key which would open the way for a better understanding of this unique section of the Alpine system.

Both international and interdisciplinary collaboration required manpower to do the work, funds to pay for the work, and mentors to stimulate the work. All contributors to this book thank the authorities of the Deutsche Forschungsgemeinschaft (German Research Society) for funding, for coordinating, and for organizing the project in such an unbureaucratic manner. We all are grateful to those scientists and assistants who promoted the work by collecting new data, by creating new concepts, and by constructive criticism.

Still we are not yet in a position to define the driving mechanism of Alpine plate tectonics. However, for the model makers, the lot of new facts given in this book will place better limits on the realm of possibilities. Data collection and synthesizing is the best base for new concepts. Only those who ask nature will have a chance to learn more about the kinematics and dynamics of plate tectonics. Viewed in this light this volume will serve as a milestone on the long way towards the final comprehension of what has happened, and what is happening, geologically in the Mediterranean crust and lithosphere.

J. Henning Illies

Contents

CHAPTER 4
The Geotraverse Sicily - Southern Italy - Ionian Sea (IIIa+b)

CHAPTER 5
Aegean Region (Geotraverses IIIc and IV)

PART 4
Central Aegean Region

PART 5
Late Cenozoic Stratigraphy and Geodynamics

PART 6
Synthesis

CHAPTER 6
The Crusts of the Alps, Apennines, Hellenides: Comparisons and Correlations

CHAPTER 7
Plate Tectonic Conclusions

List of Contributors

Abbreviations (valid only for this book)

BGR	= Bundesanstalt für Geowissenschaften und Rohstoffe, Postfach 51 01 53, D-3000 Hannover 51
G	= Institut für Geologie or Geologisches Institut
P	= Institut für Paläontologie or Paläontologisches Institut
G + P	= Institut für Geologie und Paläontologie
GUMu	= Institut für Allgemeine und Angewandte Geologie, Universität München, Luisenstr. 37, D-8000 München 2
GTMu	= Lehrstuhl für Geologie der Technischen Universität, Lichtenbergstr. 4, D-8046 Garching bei München
Gp	= Institut für Geophysik or Geophysikalisches Institut
GpUMu	= Institut für Allgemeine und Angewandte Geophysik, Universität München, Theresienstr. 41, D-8000 München 2
M	= Institut für Mineralogie oder Mineralogisches Institut
Pe	= Institut für Petrologie or Petrographisches Institut
M + Pe	= Institut für Mineralogie und Petrologie or Petrographie
Me + Gp	= Institut für Meteorologie und Geophysik, Universität, Feldbergstr. 47, D-6000 Frankfurt 1
NLfB	= Niedersächsisches Landesamt für Bodenforschung, Postfach 51 01 53, D-3000 Hannover 51
U	= Universität
FU	= Freie Universität
TU	= Technische Universität

A = Austria, CH = Switzerland, D = Federal Republic of Germany (W.), F = France, G = Greece, I = Italy, USA = United States of America.

Ackermand, D. - M, U, Olshausenstr. 40/60, D-2300 Kiel

Agarwal, N.K. - Oil and Natural Gas Commission, 41 Chowringhee Road, Calcutta, India

Ahorner, L. - Erdbebenstation U Köln, Vinzenz Palottistr. 26, D-5060 Bensberg

Altherr, R. - M + Pe, TU Braunschweig, Gaußstr. 29, D-3300 Braunschweig

Angenheister, G. - GpUMu

Ansorge, J. - Gp, Schweizerischer Erdbebendienst, ETH-Hönggerberg, CH-8093 Zürich

Aric, K. - Lehrkanzel für Geophysik, U, Hohe Warte 38, A-1190 Wien

Bachmann, H. - c/o Preussag AG, Postfach 4829, D-3000 Hannover 1

Bannert, D. - BGR

Baumann, A. - M, U, Gievenbecker Weg 61, D-4400 Münster

Beblo, M. - GpUMu

Beccaluva, L. - Istituto di Petrografia, Università di Parma, Via Gramsci 9, I-Parma

Bechstädt, Th. - GUMu

Berckhemer, H. - Me + Gp

Berktold, A. - GpUMu
Besenecker, H. - NLfB
Best, G. - BGR
Bleil, U. - Gp, U, Postfach 10 21 48, D-4630 Bochum
Bodechtel, J. - Zentralstelle für Geo-Photogrammetrie und Fernerkundung, Luisenstr. 37, D-8000 München 2
Bögel, H. - GTMu
Böger, H. - G + P, U, Olshausenstr. 40/60, D-2300 Kiel
Borsi, S. - Laboratorio di Geocronologia e Geochimica Isotopica, Via Cardinale Maffi 36, I - Pisa
Brandner, R. - G + P, U, Universitätsstr. 4, A-6020 Innsbruck
Braune, K. - GTMu
Braunmühl, W.v. - Me + Gp
Büttner, D. - G, FU, Altensteinstr. 34A, 1000 Berlin 33, Germany
Buggisch, W. - G + P, TU, Schnittspahnstr. 9, D-6100 Darmstadt
Burger, H. - G, FU, Altensteinstr. 34A, 1000 Berlin 33, Germany
Butt, A.A. - G + P, U, Sigwartstr. 10, D-7400 Tübingen

Closs, H. - Gerlachstr. 9, D-3000 Hannover 1
Cortesogno, L. - Istituto di Petrografia, Università, I - Genova

Delibasis, N. - National Observatory, Seismological Institute, G - Athens
Del Moro, A. - Laboratorio di Geocronologia e Geochimica Isotopica, Via Cardinale Maffi 36, I - Pisa
Diersche, V. - Bogenstr. 37, D-2000 Hamburg-Schenefeld
Dietrich, V.J. - Institut für Kristallographie und Petrographie, ETH Zürich, Sonneggstr. 5, Ch-8092 Zürich
Doert, U. - G + M, U, Schloßgarten 5, D-8520 Erlangen
Dürr, St. - G + P, U, Lahnberge, D-3550 Marburg/Lahn

El Ali, H. - Gp, FU Berlin, Rheinbabenallee 49, 1000 Berlin 33, Germany
Ergenzinger, P. - G, FU, Altensteinstr. 34A, 1000 Berlin 33, Germany

Fabricius, F. - GTMu
Flügel, E. - P, U, Loewenichstr. 28, D-8520 Erlangen
Förtsch, O. - Geophysikalisches Observatorium, Ludwigshöhe 8, D-8080 Fürstenfeldbruck
Franz, U. - Kernforschungsanlage Jülich GmbH, Postfach 1913, D-5170 Jülich
Friedrichsen, H. - M + Pe, Abteilung Geochemie, U, Wilhelmstr. 56, D-7400 Tübingen 1
Frisch, W. - G, U, 1 Universitätsstr. 7, A - Wien
Froelich, R. - Department of Geology, University of Rhode Island, Kingston, R.I. 02881, USA
Fromm, K. - NLfB

Gardé, S. - BGR
Gebrande, H. - GpUMu
Germann, Kl. - Institut für Angewandte Geologie, FU, Wichernstr. 16, 1000 Berlin 33, Germany
Giese, P. - Gp, FU, Rheinbabenallee 49, 1000 Berlin 33, Germany
Gizycki, P.v. - GUMu
Görler, K. - G, FU Altensteinstraße 34A, 1000 Berlin 33, Germany.
Goerlich, F. - c/o Deutsche Forschungsgemeinschaft, Kennedyallee 40, D-5300 Bonn-Bad Godesberg

Götze, H.-J. - Gp, TU, Postfach 230, D-3392 Clausthal-Zellerfeld
Groscurth, J. - G, FU, Altensteinstr. 34A, 1000 Berlin 33, Germany
Günther, K. - G + P, TU, Callinstr. 30, D-3000 Hannover
Günther, R. - G, TU, D-3392 Clausthal-Zellerfeld
Gwosdz, W. - G + P, TU, Postfach 3329, D-3300 Braunschweig

Haessler, H. - Institut de Physique du Globe, Université, 5, Rue René Descartes, F-67084 Strasbourg Cedex
Heimann, K.O. - GTMu
Herm, D. - P, U, München, Richard-Wagner-Str. 10, D-8000 München 2
Hinz, K. - BGR
Höll, S. - GUMu
Hoernes, S. - M + P, Abteilung Geochemie, U, Wilhelmstr. 56, D-7400 Tübingen

Ibbeken, H. - G, FU, Altensteinstr. 34A, 1000 Berlin 33, Germany
Illies, J.H. - G, U, Postfach 6380, D-7500 Karlsruhe 1
Jacobshagen, V. - G, FU, Altensteinstr. 34A, 1000 Berlin 33, Germany
Jacoby, W. - Me + Gp
Jongsma, D. - Buro of Mineral Resources, POB 378, Canberra City 2601, Australia

Kauffmann, G. - G + P, U, Lahnberge, D-3550 Marburg/Lahn
Keller, J. - M, U, Hebelstr. 40, D-7800 Freiburg/Br.
Kelletat, D. - Geographisches Institut, TU, Schneiderberg 50, D-3000 Hannover
Kemmerle, K. - GpUMu
Klemm, D.D. - GUMu
Knoche, G. - G, FU, Altensteinstr. 34A, 1000 Berlin 33, Germany
Kockel, F. - BGR
Kopp, K.-O. - GTMu
Koschyk, K. - NLfB
Kreuzer, H. - BGR
Kronberg, P. - G, TU, D-3392 Clausthal-Zellerfeld
Kowalczyk, G. - G + P, U, Senckenberg-Anlage 32-34, D-6000 Frenkfurt

Lammerer, B. - GUMu
Leitz, F. - P, U, Loewenichstr. 28, D-8520 Erlangen
Lemcke, K. - c/o Gewerkschaften Brigitta und Elwerath, Fritz-Reuter-Str. 19a, D-8000 München 60
Leppig, U. - Peter-Thumbstr. 2, D-7800 Freiburg/Br.
Letz, H. - Gp, FU, Rheinbabenallee 49, 1000 Berlin 33, Germany
Leydecker, G. - Seismologisches Zentralobservatorium Gräfenberg, Krankenhausstr. 1, 8520 Erlangen
Loeschke, J. - G + P, U, Sigwartstr. 10, D-7400 Tübingen 1
Lohr, J. - c/o Gewerkschaften Brigitta und Elwerath, Riethorst 12, D-3000 Hannover 51
Losecke, W. - BGR

Makris, J. - Gp, U, Bundesstr. 55, D-2000 Hamburg 13
Marchetti, G. - Instituto di Geologia dell'Università, Strada Nuova 65, I - Pavia
Mariolakos, I. - Department of Geology, University, Akademias Str. 46, G - Athens

Maucher, A. - GUMu
Miller, Heinz - GPUMu
Miller, Hubert - G + P, U, Gievenbecker Weg 61, D-4400 Münster
Morelli, C. - Instituto di Miniere e Geofisica Applicata, Viale R. Gessi 4, I-34123 Trieste
Morteani, G. - Pe und Lagerstättenforschung, TU, Straße des 17. Juni, 1000 Berlin, Germany
Mostler, H. - G + P, U, Universitätsstr. 4, A-6020 Innsbruck
Müller, G. - Gp, U, Hertzstr. 16, Bau 42, D-7500 Karlsruhe-West (21)
Müller, M. - c/o Preussag AG, Geologisches Büro Süd, Christophstr. 20, D-8920 Schongau
Müller, P. - BGR
Müller, St. - Gp, Schweizerischer Erdbebendienst, ETH Hönggerberg, CH-8093 Zürich
Münzer, U. - Zentralstelle für Geo-Photogrammetrie und Fernerkundung, Luisenstr. 37, D-8000 München 2

Nagel, K. - G + P, TU, Hardenbergstr. 42, 1000 Berlin 12, Germany
Neugebauer, H.J. - Me + Gp
Nicolic, R. - Osservatorio Geofisico Sperimentale, I-34123 Trieste

Obenauf, P. - G, FU, Altensteinstr. 34A, 1000 Berlin 33, Germany
Okrusch, M. - M + Pe, TU, Gaußstr. 29, D-3300 Braunschweig
Otte, O. - Bonertstr. 42, D-4100 Duisburg 46

Papani, G. - Istituto di Geologia dell'Università, Via Massimo d'Azeglio 85, I - Parma
Perrier, G. - Laboratoire de Géophysique Interne, Université de Grenoble, Dom. Universitaire, F-38400 Saint-Martin-D'Heres
Picardo, G.B. - Istituto di Petrografia, Palazzo delle Scienze, Università, Corso Europa, I-16132 Genova
Pohl, J. - GpUMu
Puchelt, H. - Pe und Geochemie, U, Kaiserstr. 12, D-7500 Karlsruhe 1
Pucher, R. - NLfB

Raase, P. - M + Pe, U, Olshausenstr. 40-60, D-2300 Kiel
Raith, M. - M + Pe, U, Olshausenstr. 40-60, D-2300 Kiel
Reichert, C. - Gp, FU, Rheinbabenallee 49, 1000 Berlin 33, Germany
Reutter, K.-J. - G, FU, Altensteinstr. 34A, 1000 Berlin 33, Germany
Richter, D. - Laboratorium für Technische Gesteinskunde und Ingenieurgeologie, Fachhochschule, Bayernallee 9, D-5100 Aachen
Richter, D.K. - G, U, Postfach 2148, D-4630 Bochum-Querenberg
Richter, I. - Gp, U, Richard-Wagner-Str. 44, D-7000 Stuttgart 1
Risch, H. - Bayerisches Geologisches Landesamt, Prinzregentenstr. 28, D-8000 München 22
Roeder, D.H. - The University of Tennessee, Department of Geological Sciences, Knoxville, Tenn. 37916, USA
Roesler, G. - Schulstr. 3, D-6433 Philippstal (Werra), OT Röhrigshof
Rosenbach, O. - Gp, TU, Postfach 230, D-3392 Clausthal-Zellerfeld
Roßner, R. - G + M, U, Schloßgarten 5, D-8520 Erlangen
Rost, F. - Fachrichtung Mineralogie 17.2, Universität des Saarlandes, D-6600 Saarbrücken
Rumohr, J. - G, U, Olshausenstr. 40-60, D-2300 Kiel

Sassi, F.P. - Istituto di Mineralogia e Petrologia, Università, Corso Garibaldi 37, I-35100 Padova
Seidel, E. - M + Pe, TU, Gauss-Str. 29, D-3300 Braunschweig
Seidl, G. - GTMu
Sgavetti, N. - Istituto di Geologia dell'Università, Via Kennedy 1, I - Parma
Sijmons, K. - G, FU, Altensteinstr. 34A, 1000 Berlin 33, Germany
Skala, W. - G, FU, Altensteinstr. 34A, 1000 Berlin 33, Germany
Snoek, M. - Gp, U, Bundesstr. 55, D-2000 Hamburg 13
Soffel, H. - GpUMu
Spohn, T. - Me + Gp
Sprecher, Ch. - Gp, Schweizerischer Erdbebendienst, ETH Hönggerberg, CH-8093 Zürich
Scheelke, I. - Heppenheimer Str. 46, D-6090 Rüsselsheim-Königstädten
Schenk, V. - M, U, Postfach 2148, D-4630 Bochum-Querenberg
Schick, R. - Gp, U, Richard-Wagner-Str. 44, D-7000 Stuttgart 1
Schmedes, E. - Geophysikalisches Observatorium, Ludwigshöhe 8, D-8080 Fürstenfeldbruck
Schmidt, Kl. - GUMu
Schöler, W. - c/o Deutsche Texaco AG, Überseering 40, D-2000 Hamburg 60
Schreyer, W. - M, U, Postfach 2148, D-4630 Bochum-Querenburg
Schröder, B. - G, U, Postfach 2148, D-4630 Bochum-Querenburg
Schütte, K.-G. - Gp, FU, Rheinbabenallee 49, 1000 Berlin 33, Germany
Schult, A. - GpUMu
Schwan, W. - G + M, U, Schloßgarten 5, D-8520 Erlangen
Schwarz, G. - Gp, FU, Rheinbabenallee 49, 1000 Berlin 33, Germany
Stein, A. - NLfB
Steinwachs, M. - NLfB
Streif, H. - NLfB
Strobach, K. - Gp, U, Richard-Wagner-Str. 44, D-7000 Stuttgart 1

Teichmüller, M. - Geologisches Landesamt Nordrhein-Westfalen, Postfach 1080, D-4150 Krefeld
Teichmüller, R. - Geologisches Landesamt Nordrhein-Westfalen, Postfach 1080, D-4150 Krefeld
Tietz, G.F. - P, U, Loewenichstr. 28, D- 8520 Erlangen
Torge, W. - Lehrstuhl und Institut für Theoretische Geodäsie, TU, Nienburger Straße 6, D-3000 Hannover 1
Troll, G. - M + Pe, U, Theresienstr. 41, D-8000 München 2

Vecchia, O. - Istituto di vie e trasporti, Piazza Leonardo da Vinci 32 I - Milano
Venturelli, G. - Istituto di Mineralogia, Università di Parma, Via Gramsci 9, I-43100 Parma

chendorf, H. - G + P, TU, Postfach 3329, D-3300 Braunschweig
lbrecher, E. - G, FU, Altensteinstr. 34A, 1000 Berlin 33, Germany
emacher, J. - Fachrichtung Mineralogie 17.2, Universität des Saarndes, D-6600 Saarbrücken
Diefenbach, K. - GUMu
W. - Gp, U, Bundesstr. 55, D-2000 Hamburg 13
P. - Gp, FU, Rheinbabenallee 49, 1000 Berlin 33, Germany
- Gp, Herner Straße 45, D-4630 Bochum
-P. - G + P, U, Senckenberg-Anlage 32-34, D-6000 Frankfurt

Wissmann, G. - BGR
Wittlinger, G. - Institut de Physique du Globe, Université, 5, Rue René Descartes, F-67084 Strasbourg Cedex

Zacher, W. - GTMu
Zanferrari, A. - Centro di Studio per i Problemi dell' Orogeno delle Alpi Orientali, Corso Garibaldi 37, I - Padova
Zanzucchi, G. - Istituto di Geologia, Università, Via Kennedy 4, I - Parma
Zeil, W. - G + P, TU, Hardenbergstr. 42, 1000 Berlin 12, Germany
Zirpoli, G. - Centro di Studio per i Problemi dell' Orogeno delle Alpi Orientali, Corso Garibaldi 37, I - Padova

Introduction

H. Closs, F. Goerlich and Kl. Schmidt

An international group of geoscientists presents herewith the results of geodynamic research in three orogens. It has been known since long that these orogens are individual, though well interconnected, members of the Alpine family. They are special responses to events measurable on a worldwide scale: The old supercontinent Pangaea broke up in several phases into the big continental masses providing space for the young Atlantic Ocean which at present is still expanding. Contemporaneous with the opening of the Atlantic Ocean in the eastern continental mass a zone of instability developed in an eastwest direction, in the Tethyan area between Africa and Europe.

Perhaps this suture was predetermined by a late Paleozoic Variscan orogeny, remnants of which fringe the southern edge of Europe and the north edge of Africa and Asia minor. Even Caledonian thermal events are known there. Since late Triassic time all three geometric types of tectonic processes affected crust and mantle of this segment between Africa and Europe. Divergent decoupling of fragments had at least once, in the Jurassic, the character of real spreading with production of basic oceanic crust. Convergence affected the continental masses of Europe and Africa during several phases of different duration and rate. Also translative movements must have happened because microcontinents rotated counter-clockwise by up to 50°. We have had different periods of progradation of subduction zones in all orogens during the continent-continent collision. In the Alps a part of Africa or a microcontinent which formerly was a part of Africa has overridden Europe for several hundred kilometers. In other areas, such as the Southern Hellenides, Europe at present is overriding Africa. Tethys has disappeared by subduction and the largest part of the water masses of the Mediterranean marginal basins cover the northern edge of Africa or new ocean crust created since the end of Tertiary.

Generations of geologists not only from Europe have been attracted by this area and its wide spectrum of problems. We had little hope to decipher the secrets of the whole area, but many uf us critically reviewed, in the light of modern tectonic models, the data and ideas produced by former generations and added as many new data as possible in a short span of time. Due to the size of the study area and the financial and manpower limitations, we concentrated our research efforts on geotraverses (see fig. 1) although some complementary research was also done in adjacent areas. The following methods have been used: field geology, tectonics and fabric analysis, petrology, stratigraphy, sedimentology, paleoecology, inorganic geochemistry, kerogen maturation studies, paleotemperature studies by isotope geology, radiometric age dating, remote sensing, seismology, deep seismic sounding on land as well as on sea, magnetometry, paleomagnetics and magnetotellurics, gravity measurements and theoretical calculations with the method of finite elements.

This book is an attempt to present the results of specialists and to synthesize them. Nearly all contributions are short reviews of the relevant literature combined with recent research. Within the narrow confines of this book many contributions hardly exceed the character of an ab-

stract, and the length of an article should not be taken as a measure of its importance. Some matters, however, needed more space for an adequate presentation. We have tried to solicit, as much as possible, articles by small teams which not necessarily have unified ideas. Every author was free to present his views and to do his own speculation. In this way we hope to have avoided the danger of unification. At the end of every chapter there is a synthesis written upon invitation by one or several scientists. The best way to digest the book, we think, is to start with these syntheses and then to go into the details.

It was our desire to demonstrate that this book is more than just another attempt to apply plate tectonics to the three alpine orogens. The plate-tectonic approach here is based on new field work or has been tested in critical areas as shown by the single articles, but not all of the authors could accept pure plate tectonic solutions.

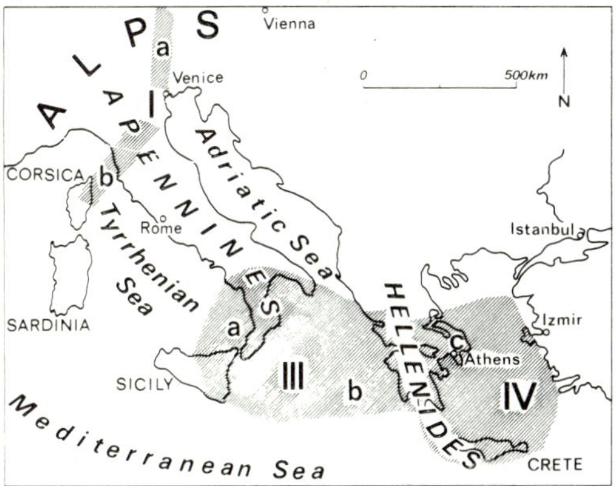

Fig. 1. Location of Geotraverses across the Alps, Apennines and Hellenides.

The authors form an international community of geoscientists which was based on many types of links. Of course, all worked under the umbrella of ICG, but more than that, there were meetings of different character in Austria, Germany, Greece, Italy, Jugoslavia and Switzerland which created official and personal contacts, served as forums of exchange of ideas and plans and were stimulating coordination.

Organisations or governmental agencies gave their support from Austria, France, Germany, Greece, Hungary, Italy and Switzerland. The *Commission International pour l'Exploration Scientifique de la Mer Méditerranée* supported the program.

The German group, guided by the national committee of ICG cooperated in the frame of a first special program of Geodynamics for five years duration.

The German group had a coordinator for each geotraverse. We thank these scientists for their fundamental share in the program. They are responsible for the syntheses.

From the beginning of the German part of the program two main coordinators (Closs and Wunderlich, later Schmidt) had the overall responsibility for implementation of the program. Prof. H. Wunderlich who had great experience in the Mediterranean area and gave us good advice died just two years after the start of the program. We also want to commemorize our colleague Prof. F. Karl who proposed the concept of the geotraverses and

who inspired leading principles of this program with his immense scientific and field experiences in the Alps. We regret deeply that both had to leave our group forever.

Prof. D. Roeder, University of Tennessee, joined the program for the last years. We thank him for his very stimulating cooperation and for editing the English texts.

German scientists could do their field work in Austria, Italy and Greece. We are extremely grateful that this was made possible and thank the relevant authorities, institutions and colleagues for their assistance and stimulation. It is impossible to list them all here.

The Geological Surveys in Hannover gave their support: We had fruitful discussions with Mr. H.D. Lang, scientific editor; Mr. E. Rode in collaboration with H. Stoll, H. Silber and Mrs. G. Schoebel of the Group Reproduction Techniques were very helpful; R. Leifheit as draftsman gave figures, where necessary, the final touch. The Reprocenter of the Institute of Anthropogeographic, Applied Geography and Cartography, Free University of Berlin-W under U. Gadewoltz was very cooperative. Mrs. Eva U. Becker completed with remarkable efficiency the final typewriting. Mrs. R. Closs was our allround assistant for about one year. We are very grateful to all of them.

Last not least the collaboration with so many geoscientists from all disciplines was a great experience and gave deep satisfaction.

Dr. E. Nägele of Schweizerbart publishers did his best to fulfill all our wishes regarding publication.

Chapter 1

Regional Reviews

Review of Earlier Geodynamic Concepts

Kl. Schmidt

When the German ICG program: Geodynamics of the Mediterranean Area started in 1970, the theory of lithospheric plates was already defined, but its principles had only partially been applied in the Alpine-Mediterranean complex.

Everywhere the modern ideas aroused enthusiasm during their triumphant advance around the globe, but they only met subdued optimism in Alpine geology, because here the effects of convective underflow and tectogenesis caused by colliding continents had already been discussed for more than half a century. The subfluence theory was already accepted and more or less modified by most Alpine geologists. After the establishment of the nappe theory in the Alps at the beginning of the 20th Century the way was paved for new geodynamic ideas and a critical evaluation of the traditional contraction theory.

In the following, therefore, the pre-plate tectonic elements of Alpine and Mediterranean geotectonics and the inventory of theories at the outset of the ICG program are reviewed.

Ampferer (1906) concluded that the obvious compression in fold and thrust belts cannot be explained by the shrinkage of the cooling crust, but only by underflow (*Unterströmungen*) of crustal and subcrustal material. In his sliding- and underflow hypothesis he already formulated one of the main principles of modern plate tectonics. The term *Verschlukkungszone* appeared in literature a few years later (Ampferer & Hammer, 1911). Ampferer (1911:33) emphasized that the splitting of continents,

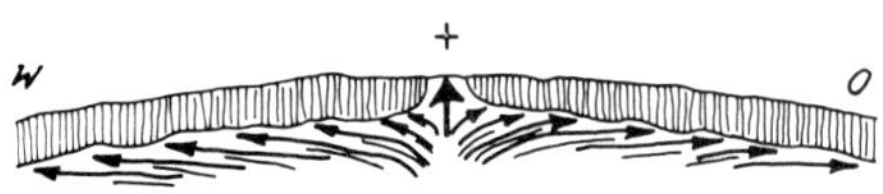

Fig. 1. Sea-floor spreading by viscous drag above a convective upwelling in the view of Ampferer (1941, fig. 2, p. 22). Original caption reads (translated) as follows: "Schematic view of the fragmentation of a continental mass by an upwelling subfluence and lateral removal of the fragments. The Atlantic median ridge, originating at + fixes the bisectrix of the fragmentation."

the formation of mountain arcs and adjacent sea trenches as well as related volcanic activities can easily be explained by underflow.

In his essay on the kinematics of the Atlantic ocean Ampferer (1941:

22) assumed a spreading mechanism which closely resembles modern models (fig. 1).

Argand (1924) and Staub (1924), however, developed their conceptions of the Alpine kinematics on the basis of the *Plattenramm-Theorie* in the frame of the contraction hypothesis (fig. 2). Schwinner (1920) supposed in his remarkable thermodynamic contribution, an about 120 km thick *tectonosphere* to which all orogenic and epeirogenic movements could be attached. This *tectonosphere* corresponds to the *lithosphere* of plate tectonics and to the rigid *sclerosphere* supposed by Kraus (1945/48) in contrast to the mobile asthenosphere.

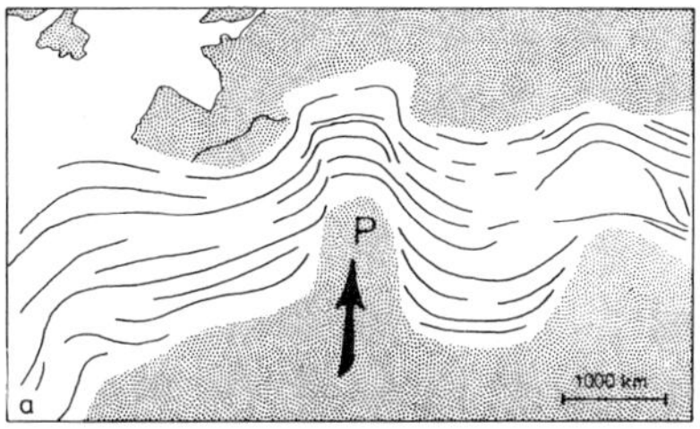

Fig. 2. Map showing, according to the plate ram theory (Argand 1924, fig. 20), the origin of the alpidic fold systems between European and the African plates. P, the African promontory, is equivalent of the later assumed Adria plate.

Ampferer's assumption of a mass compensation beneath the shortened nappe- and fold-profiles is, according to Schwinner's theory, satisfied by descending convection currents.

From 1930 on Kraus was especially engaged in underflow problems and improved the subfluence theory (Kraus 1931, 1935, 1936, 1945-1948, 1951) which at last, after decades of partly vehement controversy found wide approval. The southward underthrusting of both the North Alpine foothills and the volcanic-sedimentary sequences of the Penninic eugeosyncline underneath the Austroalpine complex represented, in the early sixties, the predominant tectonic model for the Eastern Alps (fig. 3).

Underthrusting of crustal segments beneath the Alps was called *subduction* since Amstutz (1951) and universally accepted 20 years later as a term describing a key element of the plate tectonic earth model (White et al. 1970). Although the implied mass balance and mantle-driven orogeny in Amstutz (1951) were the same as in the subfluence theory, *subduction* was neither considered as a translation of Ampferer's German *Verschlukkung* into French, nor did it necessarily imply the site of a convective downwelling.

In their ideas of the genesis of mid-ocean ridges, Ampferer and Kraus had come very close to the recent perception that there are direct relations between the opening of the Atlantic and the Alpine geodynamics of the Mediterranean area. They still lacked information, however, of the magnetic stripe patterns in the oceans, the Benioff- and low-velocity zones, to succeed in linking the chain of indications.

Furthermore, the earth model of crustal plates was already included in the *Plattenramm-Theory*. Kober (1925:169) wrote, though considering predominantly the continents: "The general regularity leads to a cognition

of eight such plates on the Earth, which are surrounded by mobile zones". In this concept besides the continents a North- and a South-Pacific mass were incorporated and interpreted as plates, although the concept of these crustal units was more supported by intuition and surface-geological inference than by data from crust and upper mantle.

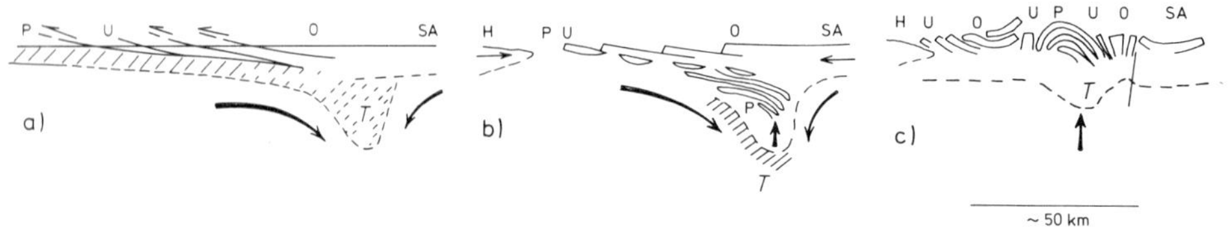

Fig. 3. Subfluence model of the Eastern Alps by Clar (1935). Stage (a) approximately Late Cretaceous, with Penninic unit (P) drifting southward while the upper plate comprising U, O and SA becomes imbricated. In stage (b) the earliest Tertiary subfluence zone is displayed southward, consuming the substratum of the upper Austroalpine unit (O). The southerly increase in crustal thickness delimits the range of influence of subcrustal corrosion. Stage (c) present state, contains the rise of the sialic tectogen (T), possibly with continuing subfluence.

Since oceanic zones play such an important part in the new global tectonics it may be recalled that the first oceanic rock assemblages (peridotites, pyroxenites, serpentinites, gabbros, pillow lavas and radiolarites) were described in the Alps and Apennines (Steinmann 1905, 1927). Steinmann's *oceanic trinity* is now confirmed by an increasing amount of deep sea drilling and is one of the principal items of plate tectonics.

In his epoch-making analysis of facies relations in the Swiss Helveticum Heim (1916) compared the Helvetic and Cretaceous with shelf-sediments and the Ultra-Helveticum with rock formations of continental slopes. The Penninic *Schistes Lustrés* were attributed to an oceanic environment. This picture has been repeatedly verified until today.

The Alpine *Wildflysch* was also a matter of special interest. Kraus (1932) gave a description and interpretation which corresponds to olistostromes, mélange formations and trench sediments and regarded the sediments as foreset beds of large underthrusting nappes, an interpretation favored by many studies of arc-trench assemblages.

Opinions vary on the dimensions of the originally existing oceanic zones in the Alps and in the Mediterranean. Within plate tectonic concepts the width of these vanished units represents the most important unknown factor, especially in paleogeographic reconstructions.

Steinmann (1925) for example, assumed a Late Jurassic deep sea zone of about 500 km width for the Alps and Apennines, which is similar to the values reported in recent plate reconstructions.

According to the subfluence theory as well as in plate tectonics, the approach of both continents induced or was accompanied by, rotations of interspersed microplates.

McPherson (1886) already realized the necessity of a rotation of the Iberian peninsula. Argand (1924) supposed a rotation of microplates in

the West-Mediterranean. Kraus (1951) thought that parts of the Austro-alpine nappe system were twisted and displayed irregularly, depending on local subfluence trends.

Mantle domes, presently discussed as geotectonic features, are well known basic principles of Haarmann's (1930) oscillation-theory and Van Bemmelen's (1933, 1952, 1960) undation-theory. Haarmann (1930) took the Alpine nappes as an example of his *secondary tectogenesis* and interpreted them as gravitational mass-displacements initiated by a *primary tectogenesis*, i.e. an ascending geotumour.

Van Bemmelen (1960) presumed the formation of a submarine geotumour (asthenolith) during the Middle Cretaceous in the Adriatic Sea from which the Alpine nappes are supposed to have slid northwards. In the Tertiary this asthenolith probably collapsed thus causing subsidence in the Adriatic basin.

These progresses were connected with crustal dilatations, fault tectonics, especially along the Periadriatic Line, and the intrusion of the Periadriatic granite-granodiorite plutons. Similar models were established later for the Betic-Rif arc, the Apennines, and the Carpathians (Van Bemmelen 1972, 1973).

Noteworthy contributions from the older generations of Alpine geologists to modern geotectonics make an impressive list. The earliest item on this list is nappism, furnishing evidence of large scale horizontal mass displacement. Next came the underflow or subfluence theory, followed by the term and the mechanical concept of *Subduction* and by the *Plattenramm*-theory of continent-continent collisions.

To conclude with some of the views gained during the ICG program, it is the author's view that plate tectonic models derived from the large rectilinear plate margins of the Pacific may have to be modified to fit the special geotectonic style of the Mediterranean. Differentiated and inhomogeneous margins of plates, continental microplates, and relatively narrow zones of oceanic crust form a kinematic pattern in which many subductions never gained lithospheric dimensions and thus failed to develop normal levels of arc vulcanism (Schmidt 1976).

The Alps and other Mediterranean orogens, however, may contribute to the problem of post-collisional tectonics involving the convergence and possible subduction of continental crust until the restored isostatic equilibrium terminates the subductive movements. Some of these orogens may supply critical data that show how far and how long horizontal stresses may maintain plate convergence after collision, against the isostatic buoyancy including the rheological impact of low velocity zones. The answer to this question will contribute further elements of solution to the puzzle of the new global tectonics.

References

Ampferer, O., 1906: Jb.geol.R.-Anst.Wien, 56: 539-622.

- , 1941: Sitzber. Akad. Wiss., Wien, math.-nat. Kl. Abt.I,150:19-35.

Ampferer, O. & Hammer, W., 1911: Jb. geol. R.-Anst. Wien, 61:531-710

Amstutz, A., 1951: Arch. Sci., 4,5: 323-329, Genf.

Dewey, J.F., Pitmann, W.C., Ryan, W.B.F.&Bonin, J., 1973: Bull. geol.Soc. Amer.,84:3137-3180.

Dietrich, V.J. & Franz, U., 1976: Geol. Rdsch.,65:361-374.

Haarmann, E., 1930: Die Oszillations-Theorie., 260, Stuttgart.

Heim, A., 1916: Z. Naturforsch. Ges.- 61:474-487, Zürich.
Kober, L., 1925: Gestaltungsgeschichte der Erde.- 200 p., Berlin, (Borntraeger).
Kraus, E.C., 1931: Geol. Rdsch.- 22:65-78.
- , 1932: Eclog. geol. Helv.- 25:39-129.
- , 1935: N. Jb. Min., Abh.- 73B:404-464.
- , 1936: Der Abbau der Gebirge I.- 352 p., Berlin (Borntraeger)
- , 1945-1948: N. Jb. Min.,Mh.Abt.- B:277-290.
- , 1951: Vergleichende Baugeschichte der Gebirge.- 587 p.,Berlin (Akad.Verl.).
Schwinner, R., 1920: Z.Vulkanologie. - 4:175-230, Berlin.
Staub, R., 1924: Beitr. geol. Kt. Schweiz. - 52:272 p.
Steinmann, G., 1905: Ber. Naturforsch.Ges.Freiburg/Br. - 16:18-67.
- , 1925: Geol. Rdsch. - 16:435-368.
- , 1927: Extr. C.-R. XIV.Congr.geol.intern.1926:3-33, Madrid.
Schmidt, Kl.,1976: Z.dt.geol.Ges. - 127:53-72.
Trümpy, R., 1975: Amer. J.Sci. - 275-A:209-238.
Van Bemmelen, R.W., 1933: Kon.Ak.v. Weetensch.Amsterdam, proc. - 36:686-694.
- , 1952: Geol.en Mijnbou. - 14:297-315.
- , 1960: Geol.Rdsch. - 50:474-499.
- , 1972: Geol.en Mijnbou. - 51:548-573.
- , 1973: Tectonophysics. - 18:33-79.
White, D.A., Roeder, D.H., Nelson, T.H. & Crowell, T.C., 1970: Bull. geol. Soc. Amer. - 81:3431-3432.

On the Kinematics of Global Plate Tectonics

W. v. Braunmühl

Introduction

Attempts to elucidate the dynamics of plate motion usually are based on two fundamentally different force models:
a. the lithosphere is dragged at its bottom by viscous coupling to a more rapid convective mantle flow (McKenzie 1969);
b. the concept of gravitational instability assumes that body forces in the lithosphere are responsible for the plate motion (Jacoby 1970).

It is possible to calculate the dynamics of plate motions without an a priori choice of one of the models. We assume that the direction of forces controlling the plate motions depend on observable or calculable parameters such as the length of the plate boundary, the surface area and plate velocity relative to an adjacent plate and to the deep mantle. By comparison of the computed forces on various plates with similar characteristics of motion one can determine the relative contribution of any force to the kinematics of a given plate. As a result of the calculation

we can thus decide which one of the two driving models should be correct.

The kinematics of plates carrying continents is significantly different from that of plates consisting only of oceanic lithosphere (e.g. the velocity, the relation of ridge length to length of subduction zones, the thickness the plate).

The Model

Our model is that of ideally rigid plates bounded laterally by ridges, transform faults and subduction zones and having a defined bottom boundary versus a viscous asthenosphere. According to the description of the motion as rotations of spherical caps we compute their torques as Cartesian vectors.

The following forces are calculated in our model (fig. 1):

1. the ridge push caused by gravity sliding perpendicular to the ridge axis;
2. the transform fault resistance acting parallel and proportional to the relative velocity of the adjacent plate;

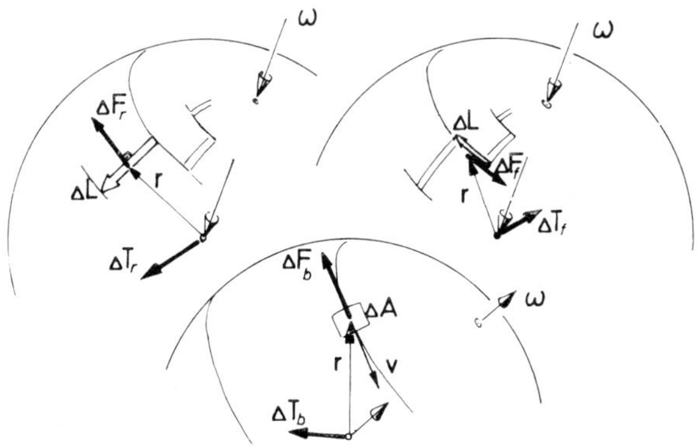

RIDGE PUSH $\Delta\vec{F}_r = C_r(\Delta\vec{L}\times\vec{r})$

SLAB PULL $\Delta\vec{F}_t = -C_t(\Delta\vec{L}\times\vec{r})$

GRAVITY SLIDING $\Delta\vec{F}_g = C_g\,\Delta A\, R(\vec{\omega}_{o,i}\times\vec{r})$

TRANSFORM FAULT RESISTANCE $\Delta\vec{F}_f = \pm C_f\,\Delta\vec{L}$

oder: $C_f\,\Delta L\, R(\vec{\omega}_{io}\times\vec{r})$

BOTTOM DRAG $\Delta\vec{F}_b = -C_b\,\Delta A\, R(\vec{\omega}_o\times\vec{r})$

SLAB DRAG $\Delta\vec{F}_s = C_s\,\Delta L\, R(\vec{\omega}_{io}\times\vec{r})$

COLLIDING RESISTANCE $\Delta\vec{F}_a = -C_a\,\Delta L\, R(\vec{\omega}_o\times\vec{r})$

TORQUES WITH RESPECT TO THE EARTH'S CENTER $\Delta\vec{T} = \vec{r}\times\Delta\vec{F}$

TOTAL TORQUE $\vec{T} = \Sigma\Delta\vec{T} = 0$

INDICES o OBSERVED PLATE
i NEIGHBOURING PLATE

n PLATES: 3 N EQUATIONS $T_i = \sum_{k=1}^{m} C_k f_{ik} = 0$

FOR M UNKNOWN PARAMETERS C_k

f_{ik} FUNCTION OF THE GEOMETRIE AND THE VELOCITY OF A PLATE

Fig. 1. Torques which according to our model determine the plate motions.

3. the slab pull caused by the high density of the cold descending plate modelled perpendicular to the trenches and proportional to the slab depth;
4. resistance against slab descent proportional to the absolute velocity acting as slab drag on the front and the sides of the slab, and as

5. colliding resistance acting across the interface to the neighbouring plate parallel and proportional to its relative velocity;

6. we assume that the oceanic lithosphere is thickening with increasing age, i.e. with increasing distance from the ridge axis. Correspondingly a force acts on surface elements which is caused by gravity sliding in the direction of plate thickening;

7. the different motions of lithosphere and asthenosphere imply at any case a viscous coupling dependent on the direction and magnitude of the absolute plate velocity. It does not matter if the motion is parallel or antiparallel. The sign of the force tells us whether the coupling acts as driving mechanism or as a resistance.

Results

A first important result for all plates is that the viscous coupling at the bottom of the lithosphere acts as a resistance against the plate motion and does not appear to be a driving force. Our model suggests that the concept of gravitational instability is the more important driving mechanism and not any kind of deep mantle convection.

Based on their magnitudes, we can compare the forces with physical properties determined by other geophysical methods (e.g. Forsyth & Press 1971; Press 1972). The order of magnitude of the resistance to the relative motion between lithosphere and asthenosphere has been calculated to correspond to a viscosity of 10^{20} to 10^{21} poises and the resistance to the downgoing slab relative to the mesosphere to 10^{22} to 10^{23} poises.

The body forces are in good agreement with a density contrast of the lithosphere with respect to the asthenosphere of about 0.05 g/ccm, a slab thickness of 100 km and a rise of the bottom of the lithosphere below the mid ocean of about 40 km.

A thickening of the Pacific plate between the ridge system and the subduction zones of about 40 - 60 km is confirmed by Leeds (1975). The calculated torques, acting as slab pull on the plates, require an average slab depth beneath the island arc chains in the western Pacific from 400-500 km and beneath the east Mediterranean belonging to the African plate from 150 to 200 km. This is in good agreement with Isacks & Molnar (1971). For a further interpretation of the results let us look at the Pacific plate as an oceanic one and the African plate as a typical continental one. The essential kinematic differences between them can be shown, if we take the net torques at the three parts of the plate (fig. 2): at the ridge zone (A), at the surface zone (B) and at the zone of descent (C).

In all three zones of the Pacific plate the net torques are of the same order of magnitude. The force vectors are nearly parallel or antiparallel to the direction of the plate velocity relative to the earth mantle. The most interesting result is that the ridge and the thickening of the plate are pushing the plate by gravity against the net resistance from the west Pacific trenches and thus compressing the plate approximately north west-southeastward (Richardson, Solomon & Sleep 1976).

The corresponding vector diagram for the African plate shows a much different and more complicated shape. The amounts of the differential torques are one order of magnitude less than those of the Pacific. This corresponds to the smaller velocity of the African plate while the surface areas are nearly equal. In addition, the African plate is to a large extend surrounded by spreading ridges and also the thickness increases

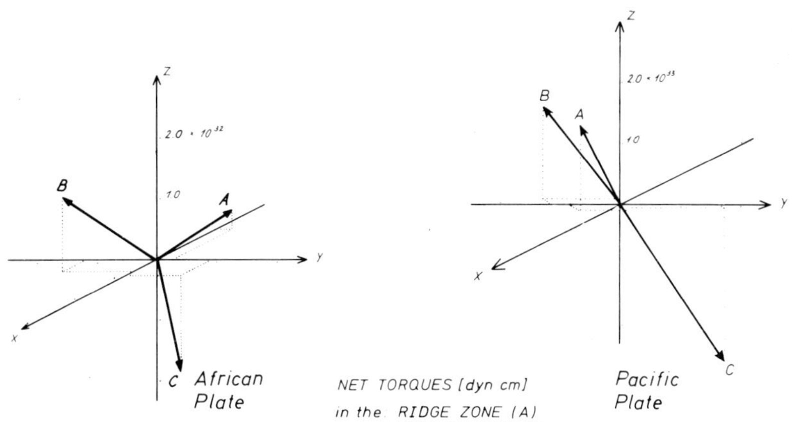

Fig. 2. Vectordiagram of the differential or net-torques in three areas of a plate. The forces are: in the ridge zone (A) the ridge push and the transform fault resistance; in the surface area: the gravity sliding and the viscous coupling of the lithosphere to the asthenosphere; and in the zone of descent (C): the slab pull, the deep slab resistance and the shallow colliding resistance.

from all of the borders except for the small subduction zone in the eastern Mediterranean. As a consequence, the net torques which are of great importance for the kinematic behaviour of a plate, are dominantly directed against each other. For calculations of the stress field and tectonic processes the scattered directions of the forces are of high importance (v. Braunmühl & Spohn 1977).

The role of the various forces for the kinematics of the African plate is not as clear as it is for the Pacific plate. The African plate as a typical continental one is driven by the push from the ridges and - in contrast to the oceanic plates - by a net pull from the subduction zone in the eastern Mediterranean.

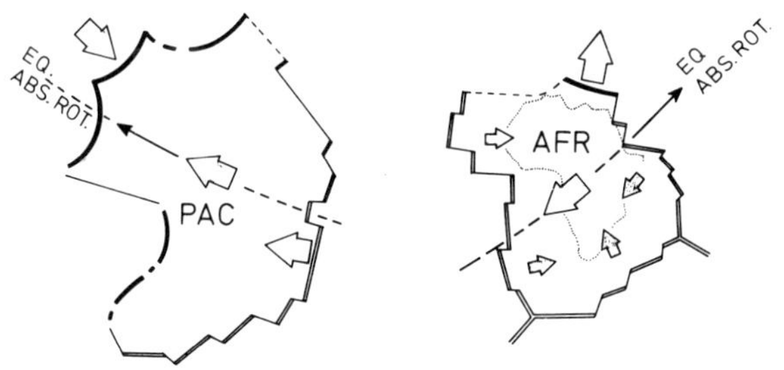

Fig. 3. Sketch map of the direction of forces acting on the Pacific plate as an oceanic plate and on the African plate as a typical continental one.

At smaller depths of the descending slab the ratio of the driving to resistance force is in general greater than unity, consequently the stresses in the slab are tensional (Leydecker 1975). The high importance of the downgoing slab for the kinematic of the African plate is also a consequence of the fact that the torques are widely scattered and dominantly directed against each other. This means that the net torques are

small. The main resistance comes from the viscous drag at the bottom of the plate although its absolute velocity is small. This probably means that the effective viscosity is greater under continental plates than under oceanic ones.

Two additional results are in a significant contrast to those for the Pacific plate. The amounts of the transform fault resistance and of the colliding resistance have relative high values for the African plate. We believe that the not well known boundary between the Eurasian and the African plate in the western Mediterranean (Isacks & Molnar 1971) acts dominantly as a transform fault. The absence of seismicity is here probably the result of the very low relative velocity between the two plates, so that the stresses can be released by creep.

The colliding resistance becomes more important because the crust and probably the lithosphere beneath the Mediterranean are not dominantly of oceanic type. So there might be no normal subduction of oceanic lithosphere but a mixed type of normal subduction and a continental-continental collision. This could explain the compressional stress field in the Aegean.

References

v.Braunmühl, W. & Spohn, T., 1977: Paper given at the meeting of German Geoph. Soc., Braunschweig.

Forsyth, D.W. & Press, F., 1971: J.Geophys. Res.-76:7963-7979.

Isacks, B. & Molnar, P., 1971: Rev. Geophys.Space Phys. - 9:103-174.

Jacoby, W., 1970: J. Geophys.Res. - 75:5671-5680.

Leeds, A.R., 1975: Physics of the Earth and Planetary Interiors. - 11:61-64.

Leydecker, G., 1975: Diss. publ. at: Institut für Meteorologie und Geophysik, Frankfurt/M.

McKenzie, D.P., 1969: Geophys. J.R.Astron.Soc. - 18:1-32.

Press, F., 1972: The Nature of the Solid Earth ed. by Robertson, E.C. - 147-171.

Richardson, R.M., Solomon, S.C. & Sleep, N.H., 1976: J. Geophys. Res. - 81:1847-1856.

On the Mechanism of Subsidence of Sedimentary Basins Adjacent to Passive Continental Margins

T. Spohn and H.J. Neugebauer

Seismic studies at passive continental margins give evidence that the crust beneath the continental slope and rise is subsiding (Sheridan 1974). Many mechanisms have been proposed to account for the subsidence of the crust and the formation of sedimentary basins there. A summary is given

by Wyllie (1971:228). Among these are two mechanisms that base on the gabbro-eclogite phase transition. According to the hypothesis that the Moho is a phase change boundary dividing a gabbroic crust from an eclogite mantle, a change in pressure, due to weight of the accumulated sediments, might cause the phase boundary to move and so initiate isostatic readjustments. On the other hand, if the Moho is a chemical boundary dividing a gabbroic crust from a mantle with peridotite as its major constituent a portion of the gabbroic crust might convert to denser eclogite. The eclogite tends to sink and dragging down a portion of the crust, it forms a depression at the surface.

Experimental work on the transition by Green & Ringwood (1967) give rise to the suggestion, that gabbroic material in the earth's crust should be in metastable state.

Under suitable conditions, that means sufficient supply of heat, the "frozen" gabbro could turn into the stable eclogite.

Beneath sedimentary troughs the thermal conditions may allow such a reaction. Detailed finite difference studies on the subject have been carried out by Spohn & Neugebauer (1977). The sediments accumulating in the trough act as a thermal blanket, because they are of lower thermal conductivity and richer in radioactive materials than the underlying rocks.

Fig. 1 shows a geotherm (Clark & Ringwood 1964) and the resulting geotherms after 50 my of sedimentation. The parameter of the curves is the sedimentation rate. The assumed thermal conductivities are $0.35 \cdot 10^{-2}$ cal/cm °C sec for the sediments and $0.5 \cdot 10^{-2}$ cal/cm °C sec for the upper crust. The conductivity is then varied with depth so, that if no sedimentation took place, the finite-difference scheme calculating the temperature would transform the original geotherm into itself.

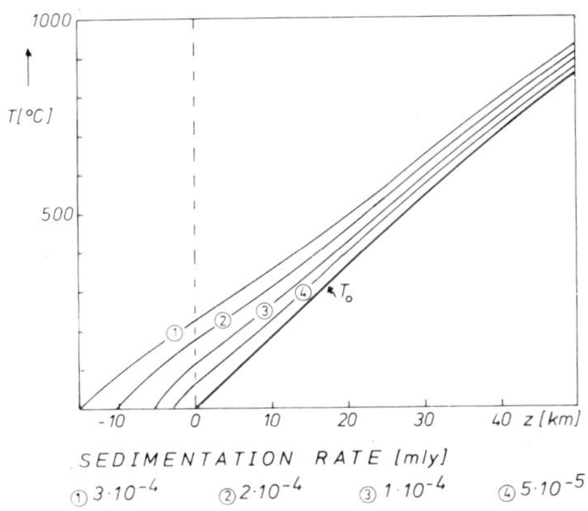

Fig. 1. Temperature depth relation after 50 my of sedimentation. Sedimentation rates as indicated.

Consequently radioactive heat sources in the crust and upper mantle are not considered separately. Heat production by sources in the sedimentary layer is assumed to be approximately 10^{-13}cal/cm³ sec. Deposition of 15 km sediments in 50 my causes the temperature in the crust to rise 100-200°C; the lowest sedimentation rate $5 \cdot 10^{-5}$m/y accumulates 5 km of sediments in 50 my and the temperature rises 10-50°C.

The transformation of a metastable gabbro or basalt requires that ions from various grains with different but homogeneous mineral assemblages react with each other. Thus an ion must be given enough activation energy

to surmount the barrier keeping it at its place and to pass through a grain boundary channel, usually 10^{-3} - 10^{-6} mm thick (Fisher & Elliot 1974), separating the different grains. Being brought together by diffusion the ions can react and build up new lattices. Both reaction and diffusion depend exponentially on temperature. According to Fisher & Elliot (1974) who established some criteria, we can assume that reaction takes place at least as fast as diffusion. So diffusion is the slowest mechanism and controls the transformation rate.

The process is therefore described by heat conduction and diffusion. As the temperature is highest at the base of crust, transformation starts there. The stored transformation heat is released and raises the temperature. This causes the adjacent grains to transform and the phase boundary slowly moves upward. Considering theoretical and experimental data for the diffusion coefficient gathered from the literature (Ahrens & Schubert 1975; Weertman & Weertman 1975; Stocker & Ashby 1973; Goetze & Kohlstedt 1973), we calculated the movement of the boundary between the two phases for different sedimentation rates. Fig. 2 shows an example.

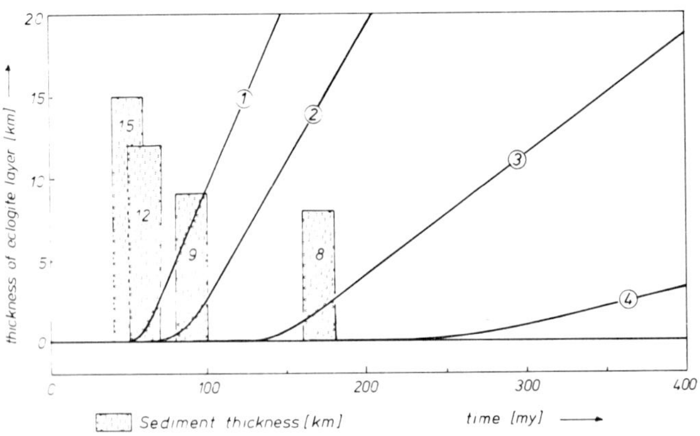

Fig. 2. Thickness of converted basalt as a function of time. The columns indicate the thickness of the sedimentary layer at the time when conversion starts. Sedimentation rates are the same as in fig. 1.

Assuming a sedimentation rate of $3 \cdot 10^{-4}$m/y, transformation starts after 50 my of sedimentation while 15 km sediments are deposited. The slope of the curve gets $2.7 \cdot 10^{-4}$ m/y comparable with the sedimentation rate, this fact is true for the other curves as well. With the slowest sedimentation rate which is taken into account, transformation starts after 170 my of deposition. The sedimentary layer is then 8.5 km thick. The curvature is very low and the curve reaches its linear part at t = 370 my, the slope is then $3.7 \cdot 10^{-5}$m/y.

Discussion

If sedimentation is rapid, the transformation of metastable gabbro or basalt will occur rather early but a relative thick sedimentary layer is

necessary. The transformation starts later for slower sedimentation rates but less thick sedimentary layers are required. In their paper, Spohn & Neugebauer (1977) have drawn further conclusions:

A linear relation is valid between the thickness of the sedimentary and the eclogite layer, but a minimum thickness of the former is required to start the transformation. The time delay between the beginning of sedimentation and the phase transition is a function of the activation energy for diffusion and the thermal properties of the crust.

The boundary between metastable gabbro or basalt and stable eclogite is not sharp but gradational. The width of a zone where the basalt is partly transformed can be calculated. This reaction zone may serve as an explanation of the layer of intermediate seismic velocity found beneath passive continental margins.

The process may perhaps play an important role for subsidence of sedimentary basins, especially because eclogite is supposed to be denser than the underlying peridotitic mantle. Numerical investigations on the causes of subsidence by Neugebauer & Spohn (1977) give strong support to this idea. It could be shown that the combined effect of the loading of sediments and the transformation to stable eclogite are sufficient to explain the observed rates of subsidence at passive continental margins being in a late stage of their development.

References

Ahrens, T.J. & Schubert, G., 1975: Rev. Geophys. - 13:383-400.
Clark, S.P. & Ringwood, A.E., 1964: Rev. Geophys. - 2:35-88.
Fisher, G.W. & Elliot, D., 1974: In: Hofmann et al. (ed.), "Geochemical Transport and Kinetics":231-241. Carnegie Inst., Publ. 634.
Goetze, C. & Kohlstedt, D.L., 1973: J. Geophys. Res. - 78:5961-5971.
Green, D.H. & Ringwood, A.E., 1967: Geochim. Cosmochim. Acta: 767-833.
Neugebauer, H.J. & Spohn, T., 1977: Submitted to Tectonophysics.
Spohn, T. & Neugebauer, H.J., 1977: Submitted to Tectonophysics.
Stocker, R.L. & Ashby, M.F., 1973: Rev. Geophys. Space Phys.- 11:391-426.
Weertman, J. & Weertman, J.R., 1975: An. Rev.Earth Planet.Sci. - 3:293-315.
Wyllie, P.J., 1971: The dynamic Earth: Textbook in Geosciences. Wiley, New York, 416 p.

Horizontal Compressive Crustal Stresses in Central Europe

L. Ahorner

There is strong evidence that large-scale geodynamic processes, which are still active in our time, like the interaction between ocean floor spreading in the North Atlantic ocean and the collision of the Eurasian plate and the African plate in the Mediterranean region, are generating a wide-spread tectonic stress field within the Earth's crust of Central Europe.

The existence and the orientation of the present-day stress field is known from fault-plane solutions of earthquakes (Schneider et al. 1966; Ahorner et al. 1972, 1974; Pavoni et al. 1974; Gangl 1974; Ahorner 1975; and others) and in-situ stress measurements using the hydrofracturing or the overcoring method (Dolcetta 1972; Greiner 1974; Rummel et al. 1975; Gysel 1975; Illies et al. 1976; Knoll et al. 1977; and others). Results of different methods of determination agree quite well and give a rather uniform picture for the direction of horizontal stresses within larger parts of the Alps and their northern foreland as far as the Lower Rhine area (fig. 1). The average stress field in this region is characterized by the following horizontal components of normal stresses:

- direction of maximum compressive stress (tectonic pressure axis) NW-SE
- direction of least compressive stress (tectonic tension axis) SW-NE

Local deviations from these mean directions are seldom greater than 30° (Ahorner 1975). If we look at Central Europe north of the Po valley, only in the structural bending zone of the Western Alps near Grenoble and in the easternmost parts of the Alps south of Vienna significant deviations from the average stress directions seem to exist. South of the Po valley, however, a quite different stress regime to that of the Alps is probably present. The few data in this area indicate an average axis of maximum compressive stress in SW-NE direction, that is nearly perpendicular to the general trend of the Apennin mountain chains. The sudden change of stress regime along the Po valley zone (line A-B in fig. 1) underlines the important role of this zone in the neotectonic development of the Alpidic orogen.

Seismicity distribution and seismotectonic block movements in Central Europe are clearly controlled by the present-day stress field (Ahorner 1975). Focal dislocations are predominantly of strike-slip type along major fracture zones striking NNE or WNW, because these zones of crustal weakness are arranged parallel to the direction of maximum shear stress of todays stress field. On the contrary, NW-trending fracture zones, which are arranged perpendicular to the direction of least compressive stress, show mostly dip-slip dislocations of tensional type (normal faulting). Fracture zones striking SW-NE, that is perpendicular to the direction of maximum compressive stress, consequently show dip-slip dislocations of compressional type (thrust faulting). The thrust-fault focal mechanism of Friuli earthquakes 1976 on the southeastern border line of the Alps is an actual example for this last type of seismotectonic movement (Müller 1977 this vol.).

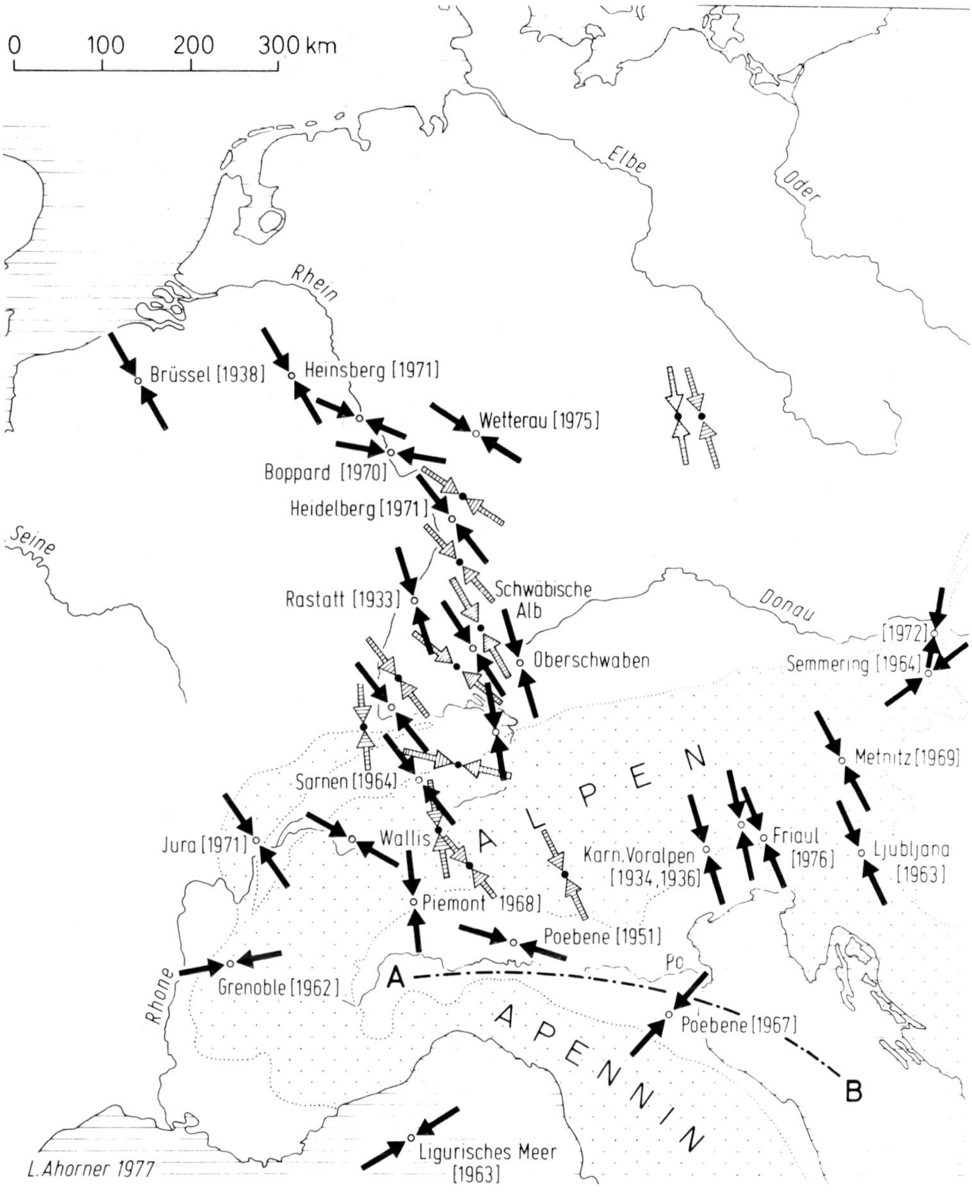

Fig. 1. Direction of horizontal compressive crustal stresses in Central Europe. From Ahorner (1975) with completions. Strike of the axis of maximum compressive stress is shown at each locality by a pair of arrows. Solid arrows are data of earthquake focal mechanism, hatched arrows are data of in-situ stress measurements.

References

Ahorner, L., 1975: Tectonophysics, 29:251-264.
Ahorner, L., Murawski, H.& Schneider, G., 1972: Geol. Rdsch.61:915-942.
Ahorner, L. & Schneider, G., 1974: Approaches to Taphrogenesis, Stuttgart:104-117.
Dolcetta, M., 1972: Ber.Int.Sympos.f.Untertagebau, Luzern:406-446.
Gangl, G., 1975: Proceed. XIV. General Assembl. Europ. Seismol. Comm. Berlin:409-410.
Greiner, G., 1974: Approaches to Taphrogenesis, Stuttgart:118-121.
Gysel, M., 1975: Tectonophysics, 29:301-314.
Illies, H. & Greiner, G., 1976: Oberrhein.Geol.Abhandlungen, 25:1-40.
Knoll, P., Bankwitz, P., Thurm, H., Schneider, M. & Thoma, K., 1977: Paper presented at the 6.Allunionsseminar für Spannungsmessungen im Gebirge, Novosibirsk.
Müller, G., 1977: J. Geophys. 42:343-349.
Pavoni, N. & Peterschmitt, E., 1974: Approaches to Taphrogenesis, Stuttgart:322-329.
Schneider, G., Schick, R. & Berckhemer, H., 1966: Zeitschr. f. Geophys., 32:383-393.

A Palaeomagnetic Guide for the Drift of the African and European Plates

K. Fromm and R. Pucher

Palaeomagnetic data from extra-Alpine Europe and Africa and from rocks dating between Late Carboniferous and Eocene have been computed to describe relative motion across Tethys and the Alpine orogenic belts, disregarding the Mediterranean microplates. The data have been uncritically taken from a paper of van der Voo & French (1974), in which gaps of the polar wandering curves (PWC) of Europe and Africa have been eliminated using the PWC of North America together with the pattern of the sea-floor spreading of the Atlantic ocean. Future revisions of some of the pole positional data will not touch the basic conclusions of that paper.

Transformation of the Polar Wandering Curves (PWC) into Components of Plate Movement

From each pole position of a PWC for a reference locality in the Alps near the Periadriatic Line (ϕ = 47°N; λ = 12°E near the village Mauls) 1) the *palaeomagnetic polar distance* has been calculated assuming that the magnetic field of the earth was always a central dipole and that the dipole axis was always parallel to the earth's spin axis. This allows the palaeomagnetic latitude to be expressed in palaeo-latitudes indicating North-South displacements of each plate. 2) The *angle of rotation* of each

plate is determined by the palaeomagnetic declination. This angle depends on the latitude, because it is measured against the meridian, which is not parallel to the next one. Therefore, the angles of rotation of the African and European plates had to be calculated for the *same* reference locality.

The results are summarized in table 1 and fig. 1:

1) Both continental plates have moved northward from Carboniferous up to present with a major reversal to the South during the Jurassic (tab. 1, col. 4,5).

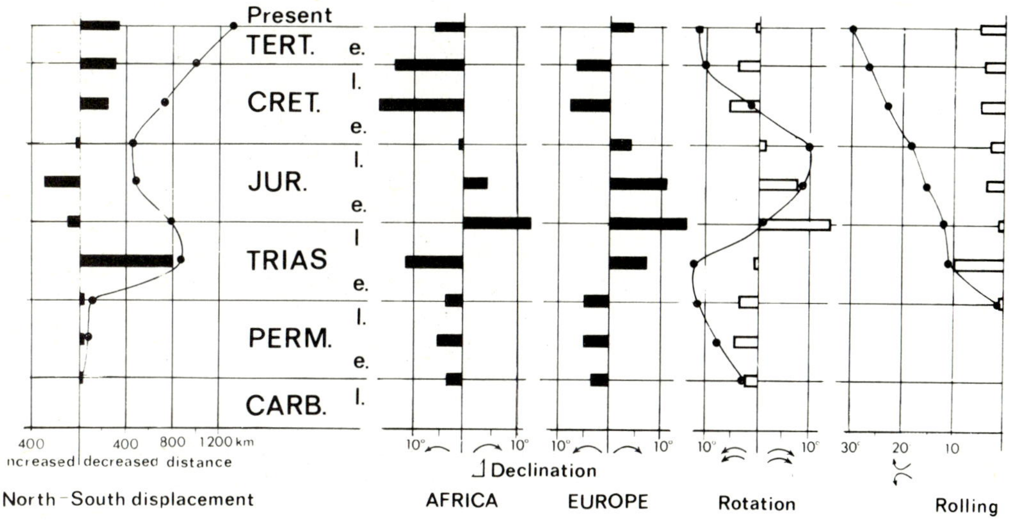

Fig. 1. Palaeomagnetic data for the locality Mauls (47.0°N/12.0°E) derived from the polar wandering curves of Africa and Europe (Van der Voo & French 1974) (see tab. 1).

2) The difference of the palaeolatitudes of the reference locality for each geological period, derived from the PWC of (A) and (E), give the latitudinal distance or difference of both continents (tab. 1, col. 6), the amount of the increased or decreased North-South distance (col. 7, and left part of fig. 1 in km) for each time interval.

3) The angle of rotation of Africa and Europe, effective at the reference locality, for each geological period (tab. 1, col. 11, 12 and fig. 1 middle part) was calculated from corresponding magnetic declination (tab. 1, col. 9, 10).

3a) The angles of rotation are remarkably small (highest value is 16°, col. 11).

3b) There are intervals in which for both the continents (plates) the

Period	Palaeomagnetic pole positions (°) Africa (A)	Europe (E)	Palaeo latitude (°) (A)	(E)	Difference of the palaeo-latitude (p E-p A) of each period	Increase (-)/decrease (+) of the latitudinal difference of neighboring periods	Cumulative latitudinal difference	Magnetic declination D(°) (A)	(E)	Difference of declination of subsequent periods ΔD(°) (A)	(E)	en-bloc rotation of (A) and (E) (°)	Cumulative en-bloc rotation	Rolling of (A) and (E) against each other (°)	Cumulative rolling
Present	90.0/0	90.0/0	47.0	47.0	0			0	0						
Tertiary early	75.4N/211.7E	78.6N/172.3E	33.1	36.1	3.0	3.0	12.1	355	4	- 5	4	- 0.5	-11.5	- 4.5	-30.5
Cretaceous Late	62.0/231.0	72.0/200.4	23.4	29.1	5.7	2.7	9.1	342	358	-13	-6	- 9.5	-11.0	- 3.5	-26.0
Cretaceous early	53.9/253.4	72.6/221.6	23.5	31.4	7.9	2.2	6.4	326	351	-16	-7	-11.5	- 1.5	- 4.5	-22.5
Jurassic late	42.0/242.0	59.9/201.8	9.4	17.2	7.8	-0.1	4.2	325	355	- 1	4	1.5	10.0	- 2.5	-18.0
Jurassic early	62.0/262.8	79.4/160.8	32.7	37.7	5.0	-2.8	4.3	329	6	4	11	7.5	8.5	- 3.5	-15.5
Triassic late	66.4/236.2	68.2/136.6	28.3	32.4	4.1	-0.9	7.1	342	21	13	15	14.0	1.0	- 1.0	-12.0
Triassic early	47.5/236.7	56.0/135.0	12.2	23.5	11.3	7.2	8.0	331	30	-11	9	- 1.0	-13.0	-10.0	-11.0
Permian Late	36.3/234.5	50.0/151.1	1.6	13.2	11.6	0.3	0.8	328	25	- 3	-5	- 4.0	-12.0	- 1.0	- 1.
Permian early	25.9/234.6	43.9/163.5	- 7.6	4.3	11.9	0.3	0.5	323	20	- 5	-5	- 5.0	- 8.0	0	0
Carboniferous Late	20.8/234.6	40.6/168.7	-12.1	0.0	12.1	0.2	0.2	320	17	- 3	-3	- 3.0	- 3.0	0	0
1	2	3	4	5	6	7	8	9	10	11	12	13	14	15	16

Table 1: Palaeomagnetic pole positions of Africa and Europe (VAN DER VOO & FRENCH, 1974) and derived data for the locality MAULS (47.0°N/12.0°E)

rotation is of nearly the same amount and direction and others with a rotation in opposite direction (col. 11, 12).
4) A condition of our dynamic model is to cause minimum friction energy during the movements of the plate. Therefore columns 11 and 12 of tab. 1 suggest the existence of *en-bloc rotations* of both continents in which they behave like a single plate, as well as other rotations in which *both continents roll* against each other. Since rotation is palaeomagnetically equivalent to East-West translation, a clockwise en-bloc rotation is equivalent to a translation to the East; a rolling component means that the plates are displaced in East-West direction against each other. The values of col 11 and 12 are mathematically devided into the two components of rotation, indicating the angle of each plate and not the angle difference of both plates (col. 13, 15).
4a) The rolling component of rotation turns out to be always anticlockwise for Africa (negative values of col. 15) and clockwise for Europe.

Conclusions Pertinent to Orogeny

Palaeomagnetic conclusions (see tab. 1 and fig. 1):
1) Between Late Carboniferous to Early Triassic (300 - 220 m.y.), Africa and Europe shifted to the North for approx. 22° latitude. The change of the relative distance of both plates adds up to only 0.8° lat., and the rotation was 12° anticlockwise for both continents (col. 14). This indicates that both continents had acted as one single plate during that time interval.
2) From Early Triassic to Late Triassic (220 - 180 m.y.), Europe moved for 9° lat. to the North (col. 5), but Africa even for 16° lat. (col. 4), resulting in an approach of both continents of 7.2° lat. (col. 7, fig. 1 left part) combined with a rotational component of 10° (col. 15) with opposite sense for both continents (tab. 1, col. 11, 12, 13; fig. 1 right part). The opposite rotation for (A) and (E) indicates a time interval of very strong compression of these continents including a gear-like rolling for 10°. The remarkable decrease of distance between both plates of about 800 km within 40 m.y. indicates the consumption of large parts of lithosphere. However, palaeomagnetic data indicate nothing about the proportions of oceanic and continental lithosphere involved in this consumption.

The opening of the Atlantic ocean as well as the separation of Australia + Antarctica from Africa + South America is put into Triassic (Heirtzler 1973; Smith & Hallam 1970).
3) During the continued drift to the North from Late Triassic to Early Jurassic, the two continents may have begun to separate, causing the origin of the Penninic ocean. The maximum spreading, of 2.8° lat. (col. 7) from Early to Late Jurassic, is accompanied by a reversed drift of both plates in the order of 20° lat. to the South (col. 4, 5). The latitudinal distance of both plates of 2.8° lat. is attributable to southward drift of both plates.

The Penninic ocean of possibly 400 km width, generated between Late Triassic and Early Cretaceous (190 - 120 m.y.), is comparable to the Red Sea, with a width of approx. 300 km and an age of 25 m.y. (Roeser 1975).
4) The last part of the drift of both continents is directed to the North from Late Jurassic up to the present with different rates. The

European displacement of 29.8° lat. (col. 5) has resulted in a reduction of the distance between both plates of 7.9° lat., or 800 km. This time span includes the alpine orogenic events. The remarkable differential South-North displacement to the two plates must have created lithospheric shortening and is a clear indication for the consumption of the Penninic ocean, for subduction in the Mediterranean area, especially for the Alps (for example Closs 1975; Dietrich & Franz 1976).

Plate tectonic models of the Mediterranean and the Alps often are based on the Atlantic opening model (Smith 1971; Dewey et al. 1973) resulting mostly in E-W movements. The result of the present consideration shows that from Triassic up to the present 1) Africa has shifted 35° lat. to the North, 2) Europe for 24° lat. to the North, 3) the distance of Africa from the Mid-Atlantic ridge is about 25° longitude. It becomes obvious that future geodynamic models should give more weight to the North-South displacements between Africa and Europe.

Acknowledgement: We are grateful to Prof. H. Closs who has encouraged us to carry out this study.

References

Closs, H., 1975: Rapp. Comm. int. Mer Médit., 23, 4a:55-57.
Dewey, J.F., Pitman III, W.C., Ryan, W.B.F. & Bonnin, J., 1973: Geol. Soc. Amer. Bull., 84:3137-3180.
Dietrich, V. & Franz, U., 1976: Geol. Rdsch., 65,2:361-374.
Dietz, R.S. & Holden, J.C., 1970: J. Geophys. Res., 75:4939-4956.
Heirtzler, J.R., 1973: in: "Implications of Continental Drift to the Earth Sciences. ed. by D.H. Tarling & S.K. Runcorn, London:191-196.
Roeser, H.A., 1975: Geol. Jb. D 13:131-153.
Smith, A. & Hallam, A.H., 1970: Nature, 225:139-144.
Van der Voo, R. & French, R.B., 1974: Earth Science Rev., 10:99-119.

Review of Palaeomagnetic Data from Italy and Adjacent Areas

H. C. Soffel

Introduction

The study of geodynamic processes in the Mediterranean area includes also palaeomagnetic measurements for the detection of movements (rotations) of smaller crustal blocks and for the improvement of the polar wander paths of the adjacent continents (Africa and Europe). Several review papers have been published in the last decade (Van der Voo & Zijderveld 1969; Zijderveld & Van der Voo 1969, 1973; McElhinny 1973; Manzoni 1976). A critical review of the latest data mainly with regard to the rock magne-

tic properties of the investigated rocks is presented in this paper. Such a discussion seems to be necessary in my view for a correct timing of the rotation of the Italian peninsula and other blocks in the Central Mediterranean.

Summary of Palaeomagnetic Data

The early palaeomagnetic measurements by Melloni* (1856), Folgheraiter* (1894), Chevallier* (1925) on volcanic rocks from Italy have been reviewed by Schmucker (1974) and shall not be reported here. Palaeomagnetic studies in Italy for the investigation of geological problems started in the early Sixties with the work of several Dutch palaeomagnetists on the late Palaeozoic (Permian) quartz porphyrites in the Southern Alps. Their results have been reviewd by Van der Voo & Zijderveld (1969). It could be shown that the declination of stable remanence is about 50° smaller than that of rocks from Central and Northern Europe of the same age. Later the same effect was shown by Roche & Westphal (1969), Westphal (1973), Henry (1975) for Permian and Triassic rocks of the Western Alps, by Manzoni (1970) for Triassic rocks of the Southern Alps, by Heller & Schmidt (1974) for Prepermian metamorphites and Tertiary andesites of the Ivrea-Verbano zone in Northern Italy and by Heller (1971) for the Middle Tertiary Bergell granite. Förster, Soffel & Zinsser (1975) found that also palaeozoic rocks from north of the Insubrian line (Ötztal Massif) have a declination which is about 50° too small compared with Stable Europe. Even Permian quartz porphyrites on the basis of the Northern Calcareous Alps in Vorarlberg (Austria) show the same effect (Soffel 1975c). The declination is here about 60° too small. However, an agreement with European Triassic directions was found for red sandstones of corresponding age on the basis of the Northern Calcareous Alps near Wörgl and Saalfeld (Austria) by the author. A similar result was obtained by Mauritzsch (1977) on Jurassic limestones of the Northern Calcareous Alps east of Salzburg (Austria). The same remanence directions as for the Bolzano rhyolites were also found for Permo-Triassic ignimbrites and tuffs from the easternmost part of the Southern Alps in Northern Slovenia (Yugoslavia) by Soffel & Miklic (1977).

In a palaeomagnetic study of Tertiary rocks from the Southern Alps (Colli Euganei, Soffel 1972, 1974) it could be shown that the palaeomagnetic data of Oligocene and younger rocks agree well with the corresponding data for Central Europe while the Eocene rocks have a declination of stable remanence which is about 50° too small. This result could be confirmed by a palaeomagnetic study on radionmetrically dated Tertiary rocks from the Monti Lessini in the Southern Alps (Soffel 1975a, b).

The palaeomagnetic investigations in Sardinia have been initiated by some Dutch palaeomagnetists in cooperation with Italian geologists (De Jong & Manzoni 1968; De Jong, Manzoni & Zijderveld 1969; Zijderveld, De Jong & Van der Voo 1970; Manzoni, De Jong, Carobene & Pasini 1972; Manzoni 1974, 1975a). They were later joined by French workers (Bobier & Coulon 1970; Bobier 1974; Coulon, Demant & Bobier 1974) and Americans (Alvarez 1972; Alvarez, Franks & Mairn 1973). It could be shown that all rocks formed earlier than Upper Oligocene-Lower Miocene have too small

* Not included in the list of references.

declinations compared with data from Central Europe while Middle Miocene and younger rocks have declinations which agree with the corresponding data from Central and Northern Europe. A similar result for Permian rocks from Corsica should also be mentioned here (Nairn & Westphal 1968). However, their result has been questioned by Storetvedt (1973) and Storetvedt & Petersen (1976) due to possible later remagnetization processes. A computer fit of Sardinia and Corsica to the coast of Southern France has been published by Westphal, Bardon, Bossert & Hamzeh (1973).

The palaeomagnetic investigations in Sicily were started by Schult (1973) who studied Upper Cretaceous volcanic rocks from the Ragusa platform in Southeast Sicily. His results were later confirmed by Barberi, Civetta, Gasparini, Innocenti, Scandone & Villari (1974) and Gregor, Nairn & Negendank (1975). In the same way as for the other areas the declination is about 50° too small with respect to Central Europe but in agreement with respect to Africa. A quite different result was obtained by Schult (1976) for Middle Jurassic and Upper Cretaceous rocks from Northwest Sicily. Although the data for Central Europe are not so well defined for this time interval there seems to be an agreement or even a tendency to larger values of declination for the rocks from Northwest Sicily. The declinations of Upper Cretaceous rocks from Northwest and Southeast Sicily are about 90° apart.

In contrast to all other areas, where most of the palaeomagnetic data have been obtained from authochthonous or parautochthonous magmatic rocks, those from the Italian peninsula originate mostly from allochthonous sediments. The difficulties for the interpretation of palaeomagnetic results due to this fact shall be discussed in the next section. The investigations in Central and Southern Italy were concentrated in Umbria and Toscana in the Central part of the peninsula, furthermore in Campania, Apulia and Calabria. Again American (Lowrie & Alvarez 1974, 1975; Alvarez & Lowerie 1974), British (Channell & Tarling 1975), Dutch (Klootwijk & Vandenberg 1975; Vandenberg, Klootwijk & Wonders 1975; Vandenberg & Wonders 1976) and Italian (Channell, D'Argento & Napoleone 1975; Manzoni 1975b) palaeomagnetists were active. Several limestones of the Ammonitico Rosso and Scaglia Rossa formations (Jurassic to Lower Tertiary) as well as bauxites (Apulia, Cretaceous) produced reliable results. All Jurassic rocks also show a declination of remanence which is about 50° too small compared with the yet poorly defined data from Central Europe. However, some of the Middle and Upper Cretaceous rocks have a tendency to smaller deviations of declination of stable remanence or even agree with the also poorly known Cretaceous directions for Central Europe. Preliminary measurements by the author on Jurassic and Cretaceous limestones in Central Italy (mainly Toscana) are in agreement with the above mentioned results.

A palaeomagnetic result from Greece should also be reported here. Pucher, Bannert & Fromm (1973) showed that Jurassic volcanic rocks of the Pindos mountain chains have also a too small declination of remanence (- 60°) compared with Central Europe.

Discussion of the Palaeomagnetic Results

A representation of all palaeomagnetic data from the Alps, Italy, Sicily, Sardinia, Corsica and schematically also Greece is shown in fig. 1. The large amount of data obtained in some cases in very small areas requires

Fig. 1. Generalized summary of palaeomagnetic data from the Alps, Italy, Sardinia, Corsica, Sicily and Greece. Arrows indicate direction (declination only) of remanence of rocks of different ages. Directions of remanence computed for a place in Central Italy (13° E, 42° N) from the European and African polar wander paths (McElhinny 1973) are presented separately for comparison. T_u, T_m, T_l = Upper, Middle and Lower Tertiary respectively. K, K_m, K_u = Cretaceous, Middle and Upper Cretaceous respectively. J = Jurassic. Tr = Triassic. P = Permian. C = Carboniferous. ? = preliminary result.

a restriction to the principle results. Therefore only representative data for larger geological units are shown. The position of the sampling sites within the geological units of the above mentioned areas should be taken from the original publications. The larger geological and tectonic units have also not been shown in fig. 1 in order to avoid a too complicated picture. The arrows in fig. 1 show the observed declination of remanence in the different areas. Inclinations are not shown because they do in general not differ significantly from the Central European data. The representative declinations of remanence computed from the polar wander paths for Stable Europe and Africa (McElhinny 1973) for the different formations are shown separately in fig. 1 as well, both for 13° E and 42° N (Central Italy). These reference directions should be compared with the observed directions in order to estimate the time and amount of relative rotation of the different crustal units.

Typical for the whole area (with the exception of Northwest Sicily) are the Northwestern directions of remanence. The declinations are in the average about 50° smaller than the corresponding directions for Central and Northern Europe and there is a tendency for an agreement with corresponding directions derived from the African polar wander path. With the usual assumptions in palaeomagnetism (axial, geocentric dipole, age of remanence equals age of rock formation) the data can be interpreted by postulating an anticlockwise rotation of the investigated rock formations with respect to Stable Europe. The time of rotation can be determined from the time of latest disagreement and earliest agreement with the Central European reference direction of the corresponding formation. The best and most conclusive result has been obtained from the volcanic rocks from Sardinia, where the anticlockwise rotation through 50° must have happened in Lower Miocene. The anticlockwise rotation of the (autochthonous or parautochthonous) Southern Alps probably occurred in Lower Oligocene. The same age can also be assumed for the anticlockwise rotation of at least part of the nappes in the Eastern Alps. The rotations in the Western Alps took probably place during the same time interval (Lower Oligocene). The time of rotation of the Italian peninsula (whether as a single unit or in different blocks) is still a matter of dispute. The correlation between age of rocks and age of remanence is not so well established for sediments, which can obtain their magnetization very much later in connection with diagenesis and later recrystallization. In most cases these secondary mineralizations cannot be detected by ore microscopic studies as is possible in magmatic rocks. Additional rock magnetic studies are helpful to determine the carriers of remanence but cannot decide in most cases which mineral is primary or secondary. Folding tests and tectonic corrections showed only that the stable remanence was formed prior to folding and tilting and must therefore be earlier than Miocene in Central Italy. The times of rotation which have been quoted for the Italian peninsula (Upper Cretaceous for the Northern Apennine, Middle Cretaceous for the Southern Apennine) have therefore to be regarded as upper limit, Miocene as lower limit.

The results from Calabria and Sicily are not yet wholly understood. Southeast Sicily is believed to be a part of the African plate. The palaeomagnetic data are in good agreement with this conception. The origin of Northwest Sicily and Calabria is still uncertain. Both areas must have carried out considerable amounts of rotation with respect to the Italian peninsula, Central Europe and Africa since Upper Cretaceous.

The few results from autochthonous or parautochthonous areas in Italy

indicate that also the basement of the peninsula has been rotated and not only the sedimentary cover or some nappes, as is probably the case in the Eastern Alps.

References

Alvarez, W., 1972: Nature, 235:103-105.
Alvarez, W., Franks, S.G. & Nairn, A.E.M., 1973: Nature, 243:10-11.
Alvarez, W. & Lowrie, W., 1974: Boll.Soc.Geol.Ital., 93:883-891.
Barberi, F., Civetta, L., Gasparini, P., Innocenti, F., Scandone, R. & Villari, L., 1974: Earth Planet. Sci. Letters, 22:123-132.
Bobier, C., 1974: Rend.Sem.Fac.Sci.Univ. Cagliari, 43:35-56.
Bobier, C. & Coulon, C., 1970: C.R.Acad.Sci. Paris, t. 270:1434-1437.
Channell, J., D'Argenio, B. & Napoleone, G., 1975:Soc.Geol.Ital.Seduta 12.12.
Channell, J.E.T. & Tarling, D.H., 1975: Earth Planet. Sci. Letters, 25: 177-188.
Coulon, C., Demant, A. & Bobier, C., 1974: Tectonophysics,22:59-82.
De Jong, K.A. & Manzoni, M., 1968: Boll.Soc.Geol.Ital.,78:713-171.
De Jong, K.A., Manzoni, M. & Zijderveld, J.D.A., 1969: Nature, 224:67-69.
Förster, H., Soffel, H. & Zinsser, H., 1975:N.Jb.Geol. Paläont. Abh., 149:112-127.
Gregor, C.B., Nairn, A.E.M. & Negendank, J.F.W., 1975: Geol.Rdsch., 64:948-958.
Heller, F., 1971: Z. Geophys., 37:557-571.
Heller, F. & Schmid, R., 1974: Schweiz.Min.u.Petr.Mitt., 54:229-242.
Henry, B., 1975: C.R. Somm. S.G.F.
Klootwijk, C.T. & Van den Berg, J., 1975: Earth Planet. Sci. Letters, 25:263-273.
Lowrie, W. & Alvarez, W., 1974: Nature, 254:285-288.
- , 1975: J. Geophys. Res., 80:1579-1592.
Manzoni, M., 1970: Tectonophysics, 10:411-424.
- , 1974: Rend.Sem.Fac.Sci.Univ. Cagliari, 43:238-295.
- , 1975a: Earth Planet.Sci.Letters, 27:275-282.
- , 1975b: Geophys.Res. Letters, 2:427-429.
Manzoni, M., De Jong, K.A., Carobene, L. & Pasini, G., 1972: Giorn. di Geologia, 28:5-9.
Manzoni, M., 1976: Giorn. Geol., 29:513-550.
Mauritzsch, F., 1977: Private Communication.
McElhinny, M.W., 1973: Palaeomagnetism and Plate Tectonics, Cambridge (Univ. Press).
Nairn, A.E.M. & Westphal, M., 1968: Palaeogeogr. Palaeoclimatol., Palaeocol. 5:179-204.
Pucher, R., Bannert, D. & Fromm, K., 1973: Bull. Geol. Soc. Greece, 10:169-171.
Roche, A. & Westphal, M., 1969: C.R. somm. S.G.F.:239-242.
Schmucker, U., 1974: in: Birett, H., Helbig, K., Kertz, W. & Schmucker,U. Zur Geschichte der Geophysik,Springer Berlin,Heidelberg,New York: 241-265.
Schult, A., 1973: Earth Planet. Sci. Letters, 19:97-100.
- , 1976: Earth Planet. Sci. Letters, 31:454-457.
Soffel, H., 1972: Earth Planet. Sci. Letters, 17:207-210.
- , 1974: Bull. di Geof. Teor. ed Appl. 16:333-355.

- , 1975a: J. Geophys. 41:385-400.
- , 1975b: J. Geophys. 41:401-411.
- , 1975c: N.Jb. Geol. Paläont. Mh.:375-384.
Soffel, H. & Miklic, F., 1977: Private Communication.
Storetvedt, K.M., 1973: Earth Planet. Sci. Letters, 21:22-28.
Storetvedt, K.M. & Petersen, N., 1976: J. Geophys. 42:59-71.
Van den Berg, J., Klootwijk, C.T. & Wonders, T., 1975: Progress in Geodynamics, Royal Netherlands Academy of Arts and Sciences, Amsterdam:165-175.
Van den Berg, J.& Wonders, A.A.H., 1976: Tectonophysics, 33:301-320.
Van der Voo, R. & Zijderveld, J.D.A., 1969: Verhandlingen Kon. Ned. Geol. Mijnbouwk. Gen., 26:121-138.
Westphal, M., 1973: Tectonophysics, 17:323-335.
Westphal, M., Bardon, C., Bossert, A. & Hamzeh, R., 1973: Earth Planet. Sci. Letters, 18:137-140.
Zijderveld, J.D.A. & De Jong, K.A., 1969: Geologie en Mijnbouw,48:559-564.
Zijderveld, J.D.A., De Jong, K.A. & Van der Voo, R., 1970: Nature, 226:933-934.
Zijderveld, J.D.A. & Van der Voo, R., 1972: In: Implications of Continental Drift to the Earth Sciences. Vol.1:133-161 In: Nato Advanced Study Institute, New York, Academic Press 1973.

Review of the Paleomagnetic Data of the Alps

R. Pucher and G. Morteani

A large-scale review of the paleomagnetic data of the whole Mediterranean area is given by Soffel (this vol.). The paleomagnetic data of the Alps are given in detail in this paper in fig. 1. It was the aim, a) to give for this area the paleomagnetic data as complete as possible, b) to correlate the data with the expected local directions of magnetization as deduced from polar wandering paths for Europe and Africa, and c) to correlate it with the large-scale tectonic units. In fig. 1 the direction of the magnetic declination is plotted as a bar. The value of the magnetic inclination is written at the end of the bar. The magnetic inclination is an important parameter although in most geodynamic applications of paleomagnetic data only the declination of the magnetization as an indication of rotational displacements is considered, and the magnetic inclination, which is correlated with the geomagnetic latitude, is disregarded. It should be mentioned that an inclination differing uniformly from the expected value in an entire region indicates a north-south displacement of this area. If the magnetic inclination, however, shows an irregular scatter it is a measure of the error which also applies to the magnetic declination.

For the Alpine region and for Northern Italy it is a matter of convention until now to describe the direction of the measured magnetization

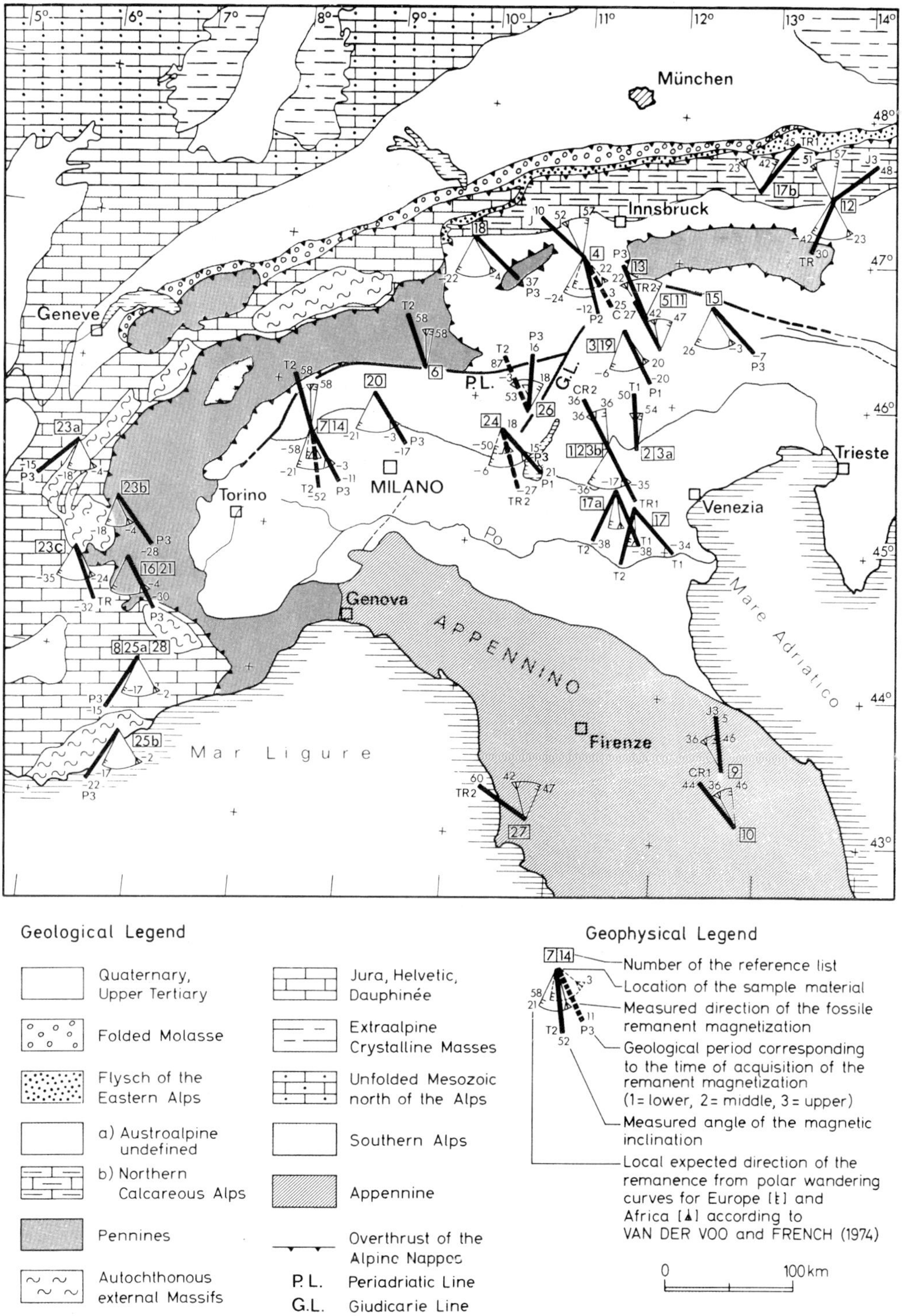
5°
6°
7°
8°
9°
10°
11°
12°
13°
14°
48°
47°
46°
45°
44°
43°
München
Innsbruck
Geneve
Torino
MILANO
Genova
Venezia
Trieste
Firenze
P.L.
G.L.
Po
APPENNINO
Mar Ligure
Mare Adriatico
Geological Legend
Quaternary, Upper Tertiary
Folded Molasse
Flysch of the Eastern Alps
a) Austroalpine undefined
b) Northern Calcareous Alps
Pennines
Autochthonous external Massifs
Jura, Helvetic, Dauphinée
Extraalpine Crystalline Masses
Unfolded Mesozoic north of the Alps
Southern Alps
Appennine
Overthrust of the Alpine Nappes
P.L. Periadriatic Line
G.L. Giudicarie Line
Geophysical Legend
Number of the reference list
Location of the sample material
Measured direction of the fossile remanent magnetization
Geological period corresponding to the time of acquisition of the remanent magnetization (1= lower, 2= middle, 3= upper)
Measured angle of the magnetic inclination
Local expected direction of the remanence from polar wandering curves for Europe [E] and Africa [A] according to VAN DER VOO and FRENCH (1974)
0
100 km

relative to the expected direction of "Stable Europe". Generally an anticlockwise rotation of the studied area is then postulated if the measured direction does not correspond to the expected direction of stable Europe. But if one compares the directions of the measured magnetization not only with the directions of "Stable Europe" but also with "Africa", a closer look to the data of fig. 1 shows that many data of Northern Italy and of the Alps agree well with the expected directions of the African plate. If this area belongs to Africa the assumption of a rotation seems to be not necessary. Particularly it can be seen that for the Southern Alps the paleomagnetic data from Permian to Tertiary time agree both in declination and inclination better with "Africa" than with "Stable Europe". Only the Adamello region, situated in the corner between the P.L. and the G.L., shows anomalous paleomagnetic directions with respect to the paleomagnetic data of the Southern Alps (site 26). This area appears to have undergone small-scale rotations with respect to the surrounding areas.

To the north, the paleomagnetic data of the Austroalpine units south of the Northern Calcareous Alps agree better with Africa than with Europe but they show a uniform difference of the direction to the East.

The data of the Northern Calcareous Alps agree better with Europe than with Africa.

Data of the Western Alps show for the Penninic units agreement with Africa (sites 23, 16 and 21) while the autochthonous or parautochthonous domains of Jura, Helvetic and Dauphinee, with one exception (site 23c) show agreement with stable Europe.

Comparing the data of the Apennines, the Po basin and the Southern Alps, Vandenberg & Wonders (1976) concluded that during Late Cretaceous and in Early Tertiary time the crustal blocks of the Southern Alps and of the Apennines must have been uncoupled to some degree. They propose a major fault zone in the area of the Po basin.

In the future, the different parts and tectonic units of the Alps should be studied by paleomagnetic methods in detail with the aim to define in time and space the rotations and translations of these units.

References

Channell, J.E.T. & Tarling, D.H., 1975: Earth Planet.Sci.Letters, 25:177-188.

Fig. 1. Geological sketch map of the Alps with paleomagnetic data. Symbols for geological period mean T = Tertiary, CR = Cretaceous, J = Jurassic, TR = Triassic, P = Permian. Paleomagnetic data refer to: 1 - Channell & Tarling 1975; 2 - De Boer 1963; 3 - De Boer 1965; 4 - Förster et al. 1975; 5 - Guicherit 1964; 6 - Heller 1973; 7 - Heller & Schmid 1974; 8 - Henry 1973; 9 - Klootwijk & Vandenberg 1975; 10 - Lowrie & Alvarez 1975: 11 - Manzoni 1970; 12 - Mauritsch 1976; 13 - Nienhuis 1958; 14 - Pucher et al. 1977; 15 - Pucher 1976; 16 - Roche & Westphal 1969; 17 - Soffel 1972; 17a - Soffel 1975b; 17b - Soffel & Milklic 1977; 18 - Soffel 1975c; 19 - Van Hilten 1960; 20 - Van Hilten & Zijderveld 1966; 21 - Van der Voo & Zijderveld 1969; 22 - Van der Voo & French 1974; 23 - Westphal 1973; 24 - Zijderveld & De Jong 1969; 25 - Zijderveld & Van der Voo 1973: 26 - Pucher et al. 1977; 27 - Vandenberg & Wonders 1976; 28 - Van den Ende 1977.

De Boer, J., 1963: Geologica Ultraiectina,11:178 p.
- , 1965: Journ. Geophys. Res., 70,4:931-944.
Förster, H., Soffel, H. & Zinsser, H., 1975: N.Jb.Paläont.Abh.,149:112-127.
Guicherit, R., 1964: Geologica Ultraiectina, 14:125 p.
Heller, F., 1973: Earth Planet. Sci. Letters, 20:180-188.
Heller, F. & Schmid, G., 1974: Schweiz.Min.u.Petr.Mitt.,54:229-242.
Henry, B., 1973: Earth Planet. Sci. Letters, 17:61-72.
Klootwijk, C.T. & Vandenberg, J., 1975: Earth Planet.Sci. Letters, 25:263-273.
Lowrie, W. & Alvarez, W., 1975: Journ. Geophys.Res.,80,11:1579-1592.
Manzoni, M., 1970: Tectonophysics, 10:411-424.
Mauritsch, F., 1977: Private Communication.
Nienhuis, G.H.W., 1958: Intern.Report.Geolog.Inst.No.25, 17 p.
Pucher, R., Fromm, K., Lensch, G. & Ahrendt, H., 1977: Priv.Communication
Pucher, R., 1976: Private Communication.
Pucher, R., Morteani, G. & Fromm, K., 1977: Geol. Jb. D: in press.
Roche, A. & Westphal, M., 1969: C.R. somm. S.G.F.:239-242.
Soffel, H., 1972: Earth Planet.Sci.Letters, 17:207-210.
- , 1975a: J. Geophys. 41:385-400.
- , 1975b: N. Jb. Geol. Paläont. Mh.:375-384.
Soffel, H. & Milklic, F., 1977: Private Communication.
Van Hilten, D., 1960: Geologica Ultraiectina, 5.
Van Hilten, D. & Zijderveld, J.D.H., 1966: Tectonophysics, 3:429-466.
Van der Voo, R. & Zijderveld, J.D.H., 1969: Verh.Kon.Ned.Geol. Mijnbouwk Gen. 26:121-138.
Van der Voo, R. & French, R.B., 1974: Earth Science Rev., 10:99-119.
Van den Ende, C., 1977: Dissertation Utrecht.
Vandenberg, J. & Wonders, A.A.H., 1976: Tectonophysics, 33:301-320.
Westphal, M., 1973: Tectonophysics, 17:323-335.
Zijderveld, J.D.H. & De Jong, K.A., 1969: Geologie en Mijnbouw, 48:559-564.
Zijderveld, J.D.H. & Van der Voo, R., 1972: In: Implications of Continental Drift to the Earth Sciences. Vol. 1:133-161 In: Nato Advanced Study Institute, New York, Academic Press.

Preliminary Results of the Lithospheric Seismic Alpine Longitudinal Profile, 1975, from France to Hungary*

Heinz Miller, J. Ansorge, K. Aric and G. Perrier

Since about 1956 refraction seismic measurements have been carried out throughout the Alps in international cooperation. The results of most of these earlier investigations have been summarized at various times (Closs & Labrouste 1963; Choudhury, Giese & De Visintini 1971; Giese & Morelli 1973; Angenheister et al. 1972; Giese & Prodehl 1976).

Most of the earlier refraction seismic observations, however, were carried out along profiles, the location of which was mainly determined by the more or less accidental position of available shotpoints. Thus the detailed interpretation was frequently handicapped due to the influence of strong horizontal velocity variations, which are introduced automatically when measurements are carried out at some angle to the strike of the geologic units.

As geodynamic considerations on the evolution of the Alpine system yielded a number of different models (Dewey & Bird 1970, Oxburgh 1972; Trümpy 1973; Closs 1975; Hawkesworth, Waters & Bickle 1975; Dietrich & Franz 1976), the need for a very detailed knowledge of crustal structure became apparent. Thus in September 1975 a long-range seismic experiment the Alpine Longitudinal Profile, 1975 (ALP 75), was carried out from France to Hungary in international cooperation with the aim to enlarge the resolving power with regard to crustal structure and to probe the lower lithosphere. A detailed description of the experiment and some very preliminary results are given by the ALPINE EXPLOSION SEISMOLOGY GROUP (1976).

The aim of this paper is to present some features of the crustal structure along the axis of the Alps as derived from the ALP experiment.

Fig. 1 shows the location of the observation lines and shotpoints of ALP 75 underlain by a geologic map after Frey et al. (1974). The main line between shotpoints A and F closely follows the general strike of the main geologic untis. As representative example of the data, figs. 2 and 3 show sections from the Western and Eastern Alps, that is profiles from shotpoints B and D to the East, respectively. The data for both sections have been processed identically. Therefore any differences in traveltime and wave patterns must be attributed to variations in the crustal structure. Details of the data processing procedure are given in ALPINE EXPLOSION SEISMOLOGY GROUP, 1976. The record section of fig. 2 covers part of the Lepontine and the westernmost part of the Eastern Alps yielding mainly information about the structure at depth below the Lepontine. A first interpretation including the reversed profile segment between shotpoints C and B has been given by Ottinger 1976. The record section of

* A Lithospheric Seismic Profile along the Axis of the Alps, 1975, Paper No. III.
Contribution Nr. 249, Institut für Allgemeine und Angewandte Geophysik, Universität München. Contribution Nr. 186, Institute of Geophysics, ETH-Hönggerberg, Zürich.

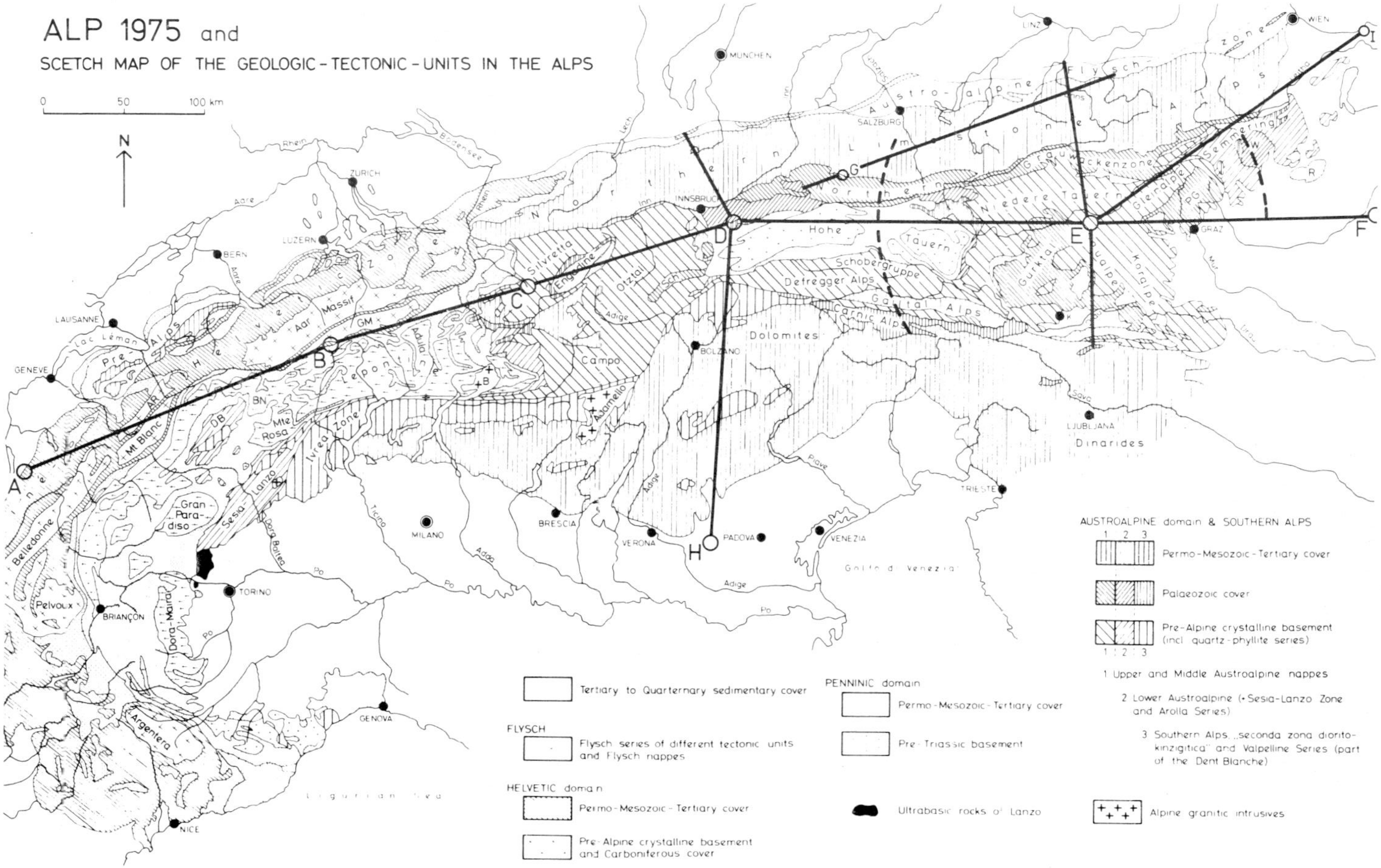

Fig. 1. Map of shotpoints and observation lines of the Alpine Longitudinal Profile, 1975. Circles indicate shotpoints with letters identifying them. The geologic map is taken from Frey et al. (1974).

fig. 3 gives information on the crustal structure below the Tauern window as it extends along the profile between shotpoints D and the Austrian-Hungarian border. The apparent velocities of the reversed profiles between shotpoints B and E indicate that horizontal velocity gradients - except for the top few kilometers influenced by sediments and weathering - are rather small within each profile segment. The assumption of plane layering is therefore reasonable, as long as the evaluation is restricted to the profile segments between adjacent shotpoints.

A comparison of both record sections clearly shows the following prominent differences: on the record section of fig. 2 (Lepontine) the appearance of seismograms is much more detailed in the distance range of 40 to 120 km and the interval of reduced travel times between 0.5 and 2 seconds than in the record section from the Eastern Alps (fig. 3). From this it clearly follows, that the structure of the upper crust is more differentiated in the Western than in the Eastern Alps. On the other hand the record section of fig. 3 shows more detailed later arrivals at distances larger than 100 km thus indicating a more complicated structure of the lower crust underneath the Tauern Window. This difference may be intensified in part by the better signal to noise ratio in the record section of fig. 3. From the record sections and the corresponding reversed profiles the velocity-depth distributions in fig. 4 were derived. Whereas the velocity-depth distribution for shotpoint D is characterized in its upper part by a very uniform velocity gradient (from 6 km/s at 2 km depth to 6.25 km/s at 19 km depth), the velocity-depth distribution for shotpoint B shows a multilayered upper crust including a low velocity zone with a p-wave velocity of 5.8 km/s. Both velocity-depth distributions have in common a rather marked low velocity zone which lies to the east at depths between 19 and 33 km and to the west between 26 and 43 km with *channel* velocities of 5.6 and 5.9 km/s, respectively.

The lower crust underneath the Tauern window is characterized by a thick transition zone with velocities around 6.8 km/s which is interrupted by a small inversion zone. This inversion zone apparently is a phenomenon closely associated with the Alpine orogeny as it is lacking north and south of the Eastern Central Alps and is most intense beneath them (Miller et al. 1977). In the Western Alps the transition zone is much thinner. Its velocity does not exceed 6.7 km/s, a value which is already found in an intermediate layer above the low velocity zone. The crust mantle boundary is found at approximately the same depth of 50 km under the Western and Eastern Alps.

The correlated and computed travel time curves for the models derived are eliminated from the record sections on purpose in order to avoid obscuring the record sections or biasing the reader. In addition to travel time computations the validity of the velocity depth distributions of fig. 4 has been checked by the computation of synthetic seismograms and rather satisfactory fits have been obtained. Still, we feel that these ralatively simple models will have to be changed in some details as more data like fans (fig. 1) and older data will be incorporated. For instance it is fairly obvious from fig. 3 that the very schematic upper crustal structure beneath the Tauern widow as represented by the uniform velocity gradient is a strong simplification, but the true structure may not be as complicated or differentiated as the structure of the upper crust in the Western Alps.

Fig. 5 shows four velocity depth distributions along the main line of ALP 75. The crustal models at shotpoints B and D are the ones described

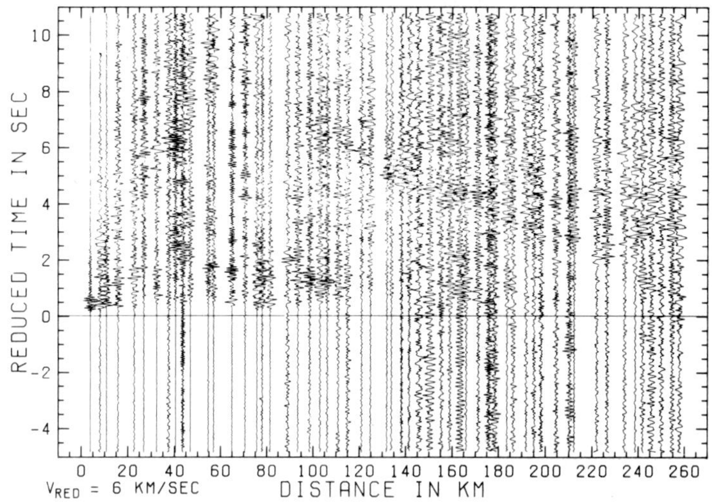

Fig. 2. Computer generated record section of the profile segment in the Western Alps from shotpoint B towards shotpoint D. All seismograms are band-pass filtered 1 - 12 Hz and normalized individually to the same maximum amplitude.

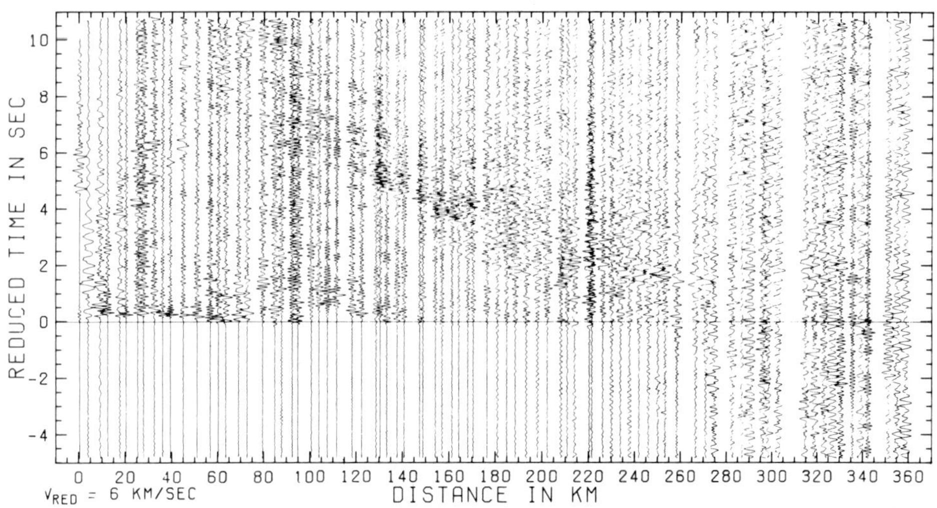

Fig. 3. Computer generated record section from the Eastern Alps between shotpoint D and the Austrian-Hungarian border. All seismograms are band-pass filtered 1 - 12 Hz and normalized individually to the same maximum amplitude.

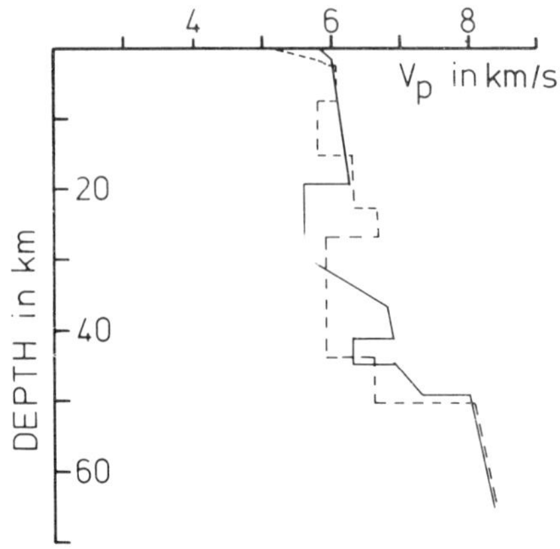

Fig. 4. Tentative velocity depth functions for the Eastern and Western Alps. Solid line: Velocity depth distribution beneath the Tauern Window. Dashed line: Velocity depth distribution below the Lepontine.

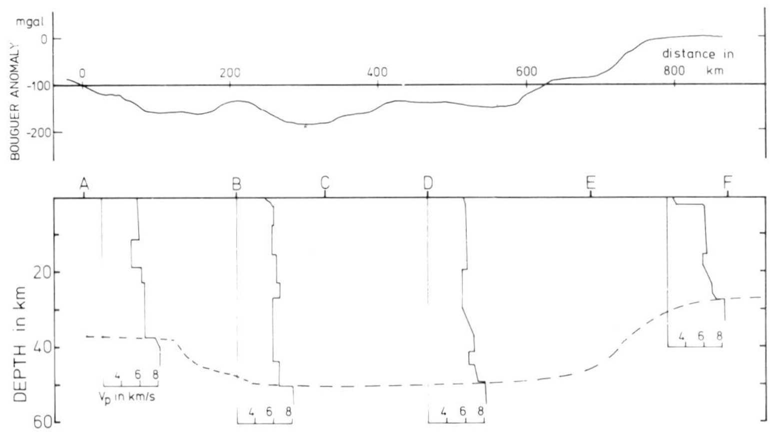

Fig. 5. Crustal structure along the main line of ALP 75 as given by selected velocity depth functions. The dashed line represents the Moho. Vertical exaggeration is five times. In the top part of the figure the Bouguer anomaly along the main profile is given after Makris (1971).

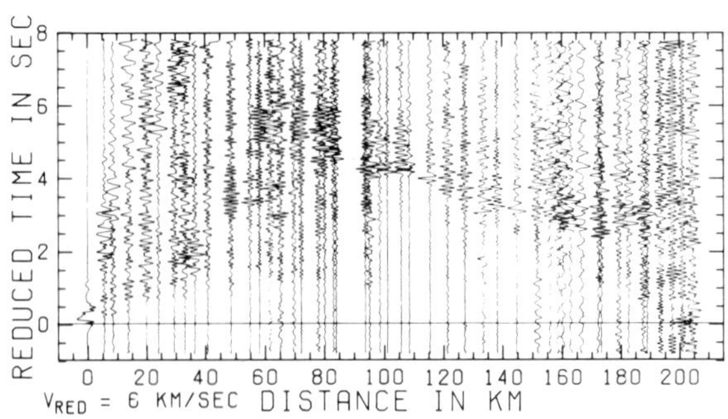

Fig. 6. Computer generated record section from shotpoint A towards shotpoint B in the westernmost part of ALP 75.

previously. The other two models do not represent truly Alpine structures. About 100 km to the West of shotpoint B the main line of ALP 75 does not continue parallel to the strike of the geologic units and the minimum of the gravity anomalies which also lies along the axis of the Alps, but rather starts to cross them at an oblique angle and shotpoint A is situated at the very margin of the Alps. The same is true at the eastern end where the main profile continues well into the Pannonian basin to shotpoint F. Therefore both of these velocity depth distributions should probably be considered as very preliminary since the effects of horizontal velocity variations cannot be disregarded. Nevertheless, the depth to the Moho and the gross structure of the crust are probably correct. The dashed line in fig. 5 gives the approximate Moho topography as derived from the ALP 75 data up to now. These Moho depths are in good agreement with the interpretations of previous data (see e.g. Giese & Prodehl 1976).

The rather steep dip of the Moho between shotpoints A and B in fig. 5 was derived by Thouvenot (1976) from the record section originating from shotpoint A (fig. 6). This record section differs strongly from those in the Western and Eastern Alps (figs. 2 and 3) by rather low p-wave velocities in the first kilometers, strong intermediate arrivals at a reduced travel time between 3 and 4 seconds at distances smaller than 100 km and strong variations of the apparent velocity of high energy arrivals, which are interpreted as overcritical reflections from the Moho. The dip of the Moho between shotpoints F and E has been derived by ray tracing technique using a program by Gebrande (1976) for two-dimensional inhomogeneous velocity distributions. Good agreement between the observed travel times of the P_n phase of shotpoint F and the calculated ones has been achieved for the particular Moho topography shown here. However, in order to attain greater confidence in the model and to further clarify the crustal structure in the transition regions between the Alpine orogen with its crustal thickness of about 50km and the areas outside with crustal thicknesses of as little as 25 km, many more model calculations will be necessary, especially since the fit for the data from shotpoint E is not yet satisfactory. There is some indication that in this region, the upper crust may also contain a thin low velocity zone (Aric & Gutdeutsch 1977). This however, is not shown in fig. 5, as it is not yet definitely identified.

In the top part of fig. 5 the Bouguer anomaly along the main line is given after Makris (1971). There is a good correlation between the Bouguer anomaly and the crustal thickness, although the mean values between the Eastern and Western Alps differ by about 40 mgal. The higher values for the Eastern Alps may easily be explained, even though the crustal thickness is about the same. The lower crust in the Eastern Alps has a higher mean p-wave velocity (fig. 4) and hence a higher density than in the Western Alps. The secondary maximum of the Bouguer values in the region of shotpoint B is most likely not caused by the crustal structure in that area. Rather it is a lateral effect of the Ivrea body (Kahle et al. 1976).

Whether the low-velocity zone forming the middle part of the crust may be viewed as the same structural element and therefore with the same implications for the alpine orogeny cannot yet be decided. Preliminary ray tracing calculations however, have shown that it would be consistent with the data if these low velocity zones were indeed the same in the Eastern as well as Western Alps, the only difference being the variation with

depth. Further calculations will be necessary to resolve this question.

With the aid of the new data from ALP 75 it has already been possible to further elucidate the crustal structure in parts of the Alps through the additional reinterpretation of earlier measurements. With respect to the International Geodynamics Project the structure of the crust along several geotraverses has been of special interest. Thus two crustal cross-sections have been derived, one in the Eastern (Miller et al.1977, see also Gebrande et al. this vol.) and one in the Western Alps (Müller et al. 1976).

In this manner and also with the aid of additional measurements especially in the Southern Alps it may be hoped that in the near future a well substantiated three dimensional model of the seismic structure of the Alps will be available serving as a strong bound on possible geodynamic models. In addition it will be possible to deduce the fine structure of the lower lithosphere from the ALP 75 long range data. No attempt however has been made up to now to do so, as a good knowledge of crustal structure is paramount to obtaining the true structure below the crust.

References

ALPINE EXPLOSION SEISMOLOGY GROUP, Reporter: Miller, H., 1976: Pageoph, 114,6:1109-1130.

Angenheister, G., Bögel, H., Gebrande, H., Giese, P., Schmidt-Thomé, P. & Zeil, W., 1972: Geol. Rdsch., 61,2:349-395.

Aric, K. & Gutdeutsch, R., 1977: Report Dtsch. Geophy. Ges.,Braunschweig.

Choudhury, M., Giese, P. & De Visintini, G., 1971: Boll. Geofis. teor. ed appl., 13:51-52, 211-240.

Closs, H. & Labrouste, Y., 1963: Mémoire Collectif, Année Géophysique Internationale, C.N.R.S., Sér. III, Fasc. 2.

Closs, H., 1975: Rapp. Comm. int. Mer Médit., 23,4a:55-57.

Dewey, J.F. & Bird, J.M., 1970: J. geophys. Res., 75:2625-2647.

Dietrich, V.J. & Franz, U., 1976: Geol. Rdsch., 65,2:361-374.

Frey, M., Hunziker, J.C., Frank, W., Bocquet, J., Dal Piaz, G.V., Jäger, E. & Niggli, E., 1974: Schweiz. Min.Petr. Mitt., 54:2-3.

Gebrande, H., 1976: In: Explosion Seismology in Central Europe, ed. P. Giese, C. Prodehl & A. Stein:162-167, Springer, Heidelberg.

Giese, P. & Morelli, C., 1973: Accad.nazion. dei Lincei, 183:317-362.

Giese, P. & Prodehl, C., 1976: In: Explosion Seismology in Central Europe, ed. P. Giese, C. Prodehl & A. Stein:347-375, Springer Heidelb.

Hawkesworth, C.J., Waters, D.J. & Bickle, M.J., 1975: Earth and Planet. Sci. Letters, 24:405-413.

Kahle, H.G., Klingele, E., Müller, St. & Egloff, E., 1976: Pageoph., 114:479-494.

Makris, J., 1971: Hamburger Geophys. Einzelschr.: 65 p.

Miller, H., Gebrande, H. & Schmedes, E., 1977: Geol.Rdsch.,66,2:289-308.

Müller, St., Egloff, R. & Ansorge, J., 1976: Schweiz. Mineral. und Petrograph. Mitt., 56:685-692.

Ottinger, Th., 1976: Diplomarbeit ETH Zürich:166 p.

Oxburgh, E.R., 1972: Nature, 239:202-204.

Thouvenot, F., 1976: M.S. Thesis, Institut du Physique du Globe, Straßbourg.

Upper Mantle Structure along a Profile through the Eastern Alps from Rayleigh Wave Dispersion*

St.Mueller and Ch. Sprecher

The Network of Stations

Over the past few years long-period vertical seismograph systems have been in operation at a number of stations in northern Italy in an effort to study the dispersion characteristics of Rayleigh waves. This special network was installed as part of a cooperative program of the University of Karlsruhe (Germany), the Swiss Federal Institute of Technology in Zürich (Switzerland), the University of California at Los Angeles (USA) and the University of Bologna (Italy). It comprised the following stations:
Bolzano (BLZ), Torino (TNO), Bologna (BOL), Grosseto (GSO).
In addition recordings of the long-period seismograph station in Fürstenfeldbruck (FUR) west of München (Germany) were used to obtain data along the north-south geotraverse (Ia & b) through the Eastern and Southern Alps and the Northern Apennines from FUR via BLZ and BOL to GSO. The station Trieste (TRI) of the World-Wide Standardized Seismograph Network (WWSSN) to the east like the stations Torino (TNO) and Zürich (ZUR) to the west of this profile served as the third station in tripartite configurations to determine the azimuth and phase velocity of Rayleigh waves from teleseismic events across the station network.

The Method of Investigation

Seismic surface waves which propagate along a layered medium, such as the earth, are subject to a geometric dispersion (see e.g. Seidl & Mueller 1977). In this case care must be exercised in distinguishing between phase velocity (c) and group velocity (U) of the spectral components contained in the surface wave train. Studies of the dispersion characteristics of seismic surface waves are a very powerful and economic tool for determining the gross structure of the crust and upper mantle. In regional investigations usually the phase velocity (c) of Rayleigh waves is measured as a function of wave period (T) by analysing the signal alteration of Rayleigh waves from a teleseismic event between two observation stations on the great-circle path through the epicenter. Azimuthal deviations from this path were detected by tripartite measurements based on records of suitably selected stations of the regional seismograph network. In this manner phase velocities of Rayleigh waves were obtained in the period range of 15 to over 100 sec for a total of 14 *profiles* in the Western and Central Alps (Sprecher 1976). The example presented here, the station line FUR - BLZ, covers the easternmost part of the region under investigation.

The dispersion analysis was carried out by a numerical data processing procedure introduced by Knopoff, Mueller & Pilant (1966) and supplemented by a program for the analysis of transient signals in the time-frequency

* Contribution No. 185, Institute of Geophysics, ETH Zürich, Switzerland.

domain as developed by Dziewonski, Bloch & Landisman (1969). As an example the time-frequency analysis is shown in fig. 1 (upper part) for the recording of an Alaskan earthquake at Bolzano in Northern Italy. This display is in the form of a so-called *Gabor Matrix*, i.e. group velocity (U) is plotted linearly as a function of the wave period (T) on a logarithmic scale. The amplitudes of the surface wave signal are represented as "contours of altitude" in this two-dimensional dispersion map. If the actual seismogram is compared to the Gabor Matrix the regular dispersion of the fundamental Rayleigh wave train becomes immediately apparent. It is interesting to note that one higher mode has also clearly been separated as seen in the upper left corner of fig. 1.

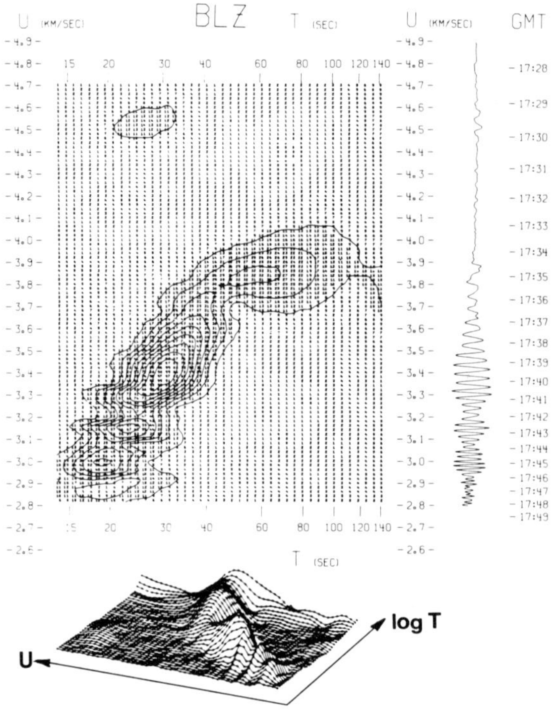

Fig. 1. Gabor Matrix of the seismic event 73-16 of 3 July 1973 (16:59:35.1 GMT) in South-East Alaska (58°.0 N; 138°.0 W) as recorded in Bolzano, Northern Italy, at a distance of 8150 km. U = group velocity, T = wave period. Upper part: 2 D map plot without instrumental corrections. Lower part: 3 D perspective display with instrumental corrections applied.

Instrumental Corrections and Smoothing of Phase Spectra

In a network of the dimensions used in this study particular attention must be paid to a careful correction for the phase distortion of the seismograph systems. Sprecher (1976) has described various methods for accurately determining the phase response of the recording systems used. In the lower part of fig. 1 the instrumentally corrected group velocity dispersion is shown in a three-dimensional perspective display. As is to be expected the correction mainly affects the long-period side of the recordings as is clearly visible in the uplifted amplitude *"mountain range"* corresponding to the fundamental Rayleigh mode. The phase spectra also have to be smoothed which in the present case was done very effectively by a new program of Sprecher (1976) using spline functions.

Results for a Traverse through the Eastern Alps

After all these preparatory steps in the data processing the phase velocity dispersion was computed for the various two-station lines. In fig.2 the resulting phase velocity curve for the profile Fürstenfeldbruck (FUR) - Bolzano (BLZ) is shown which covers the period range from 20 to 80 seconds. For the interpretation of these observational data it is necessary to go through an inversion process which allows to determine the shear wave velocity (V_S) as a function of depth (z) in the earth from the phase velocity measurements. After the average structure of the crust (essentially based on the model of Mueller & Talwani 1971) and of the

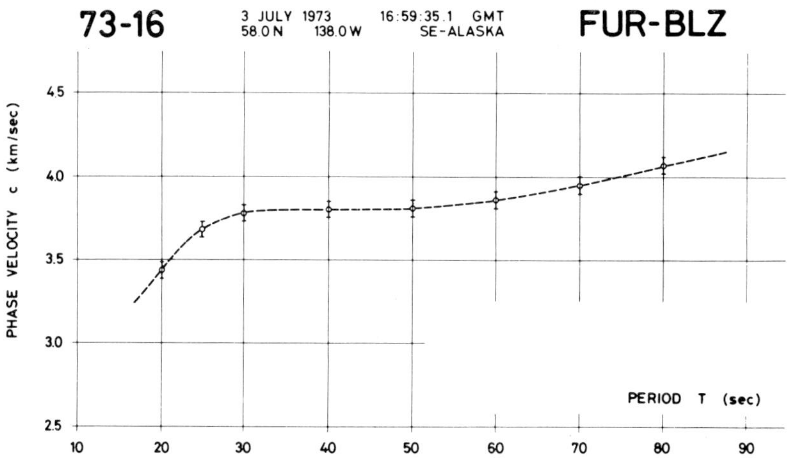

Fig. 2. Phase velocity (c) dispersion curve as a function of wave period (T) of the fundamental Rayleigh mode for the profile Fürstenfeldbruck (FUR) - Bolzano (BLZ). Event 73-16 (SE-Alaska) as in fig. 1.

deeper mantle was fixed in the form of *boundary conditions* a systematic search was performed in a four-parameter space for the set of permissible solutions which were compatible with the measured phase velocities allowing for the experimental uncertainties as indicated in fig. 2. The four variable model parameters were the average shear velocities and thicknesses of the lower lithosphere and of the asthenosphere, whose lower boundary could vary between a depth of 200 to 250 km. Each two-station line was inverted following a procedure first described and termed the "*hedgehog*"method by Keilis-Borok & Yanovskaya (1967). The solution thus arrived at for the profile FUR - BLZ traversing the Eastern Alps yields a lithospheric thickness of 120-140 km. For the lower boundary of the asthenosphere under that part of the Alps a depth of 230-250 km was obtained.

In fig. 3 all available recent results for the gross structure of the crust and upper mantle have been summarized for a traverse through the Eastern Alps extending from Ingolstadt (Germany) in the north to Lago Lagorai near Bolzano in Northern Italy. This line roughly coincides with the profile discussed by Mueller & Talwani (1971). The crust-mantle

boundary (M) has been adopted from a recent paper by Miller, Gebrande & Schmedes (1977), in which evidence is presented for a P velocity inversion in the uppermost mantle about 15 km below the M discontinuity.

As can be seen in fig. 3 the lithosphere-asthenosphere boundary dips from north to south reaching a depth of about 140 km under the Southern Alps. The corresponding shear velocities in the lower lithosphere and in the asthenosphere are indicated in fig. 3. In this model the lower boundary of the asthenosphere has been kept fixed at a depth of 240 km.

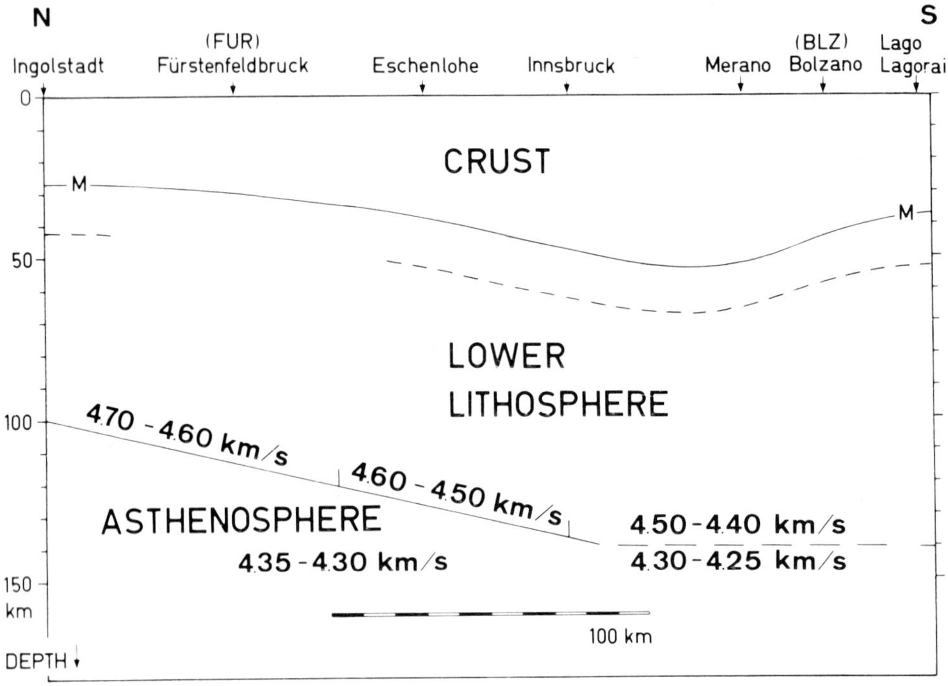

Fig. 3. Crust-mantle section across the Eastern Alps from seismic refraction and surface-wave dispersion data. M = crust-mantle boundary. - - - = P velocity inversion in the uppermost mantle. The numbers in the lower part of the section are the average shear velocities in the lithosphere and asthenosphere, respectively.

Regional Results for the Central and Western Alps

In the regional survey carried out in the Central and Western Alps by Sprecher (1976), and which is limited in the east by the station line FUR - BLZ, an attempt was made to interpret the multiple *hedgehog* solutions in terms of a comprehensive shear velocity model of maximum smoothness. The preferred model shows the following structural characteristics: The shear velocity of the lower lithosphere averaged over the entire region of investigation has a value of 4.45 $\pm$ 0.05 km/sec. It is lower (4.35 $\pm$ 0.05 km/sec) in the southwest and south and higher (4.65 $\pm$ 0.05 km/sec) in the northeast, i.e. in the neighborhood of the profile FUR -

BLZ (see fig. 3, lower left). The total thickness of the lithosphere is about 70-80 km in the northwest under the Rhinegraben area in agreement with the basic model deduced for southwestern Germany by Seidl, Mueller & Knopoff (1966) and thickens substantially towards the southeast. Under the central part of the Swiss Alps the observational data can best be explained by a lithosphere about 120-130 km thick with an average shear velocity of 4.35 ± 0.05 km/sec in the asthenosphere underneath. The data available so far are consistent with a depth of 200-250 km to the base of the asthenosphere. For the preferred regional model presented here the lower boundary of the asthenosphere was found to lie at a depth of 240 km.

Conclusions

Although the recent results of Sprecher (1976) have clearly demonstrated that there exists an asymmetric structure in the uppermost mantle under the Alps which may be explained as the remnant of an early southward dipping subduction zone, further work will be required - particularly in the Eastern Alps - in order to explore the details of this important structural feature as well as a suspected extension towards the south.

References

Dziewonski, A., Bloch, S. & Landisman, M., 1969: Bull. Seism. Soc. Amer., 59:427-444.
Keilis-Borok, V.I. & Yanovskaya, T.B., 1967: Geophys.J.R.A.S.,13:223-234.
Knopoff, L., Mueller, St. & Pilant, W., 1966: Bull. Seism.Soc. Amer., 56:1009-1044.
Miller, H., Gebrande, H. & Schmedes, E., 1977: Geol. Rdsch.,66:289-308.
Mueller, St. & Talwani, M., 1971:Pure and Appl.Geophys.,85:226-239.
Seidl, D., Mueller, St. & Knopoff, L., 1966:Z.Geophys.,32:472-481.
Seidl, D. & Mueller, St., 1977: J. Geophys., 42:283-328.
Sprecher, Ch., 1976: Diss. ETH Zürich, Nr.5864:156 p.

Gravimetric Measurements on three N-S-Profiles through the Eastern Alps - Observational Results and Preliminary Modelling

H.-J. Götze, O. Rosenbach and W. Schöler

Introduction

The measurements were planned in cooperation with the Geophysics Department of the University of Vienna and carried out by the Institute of Geophysics of the Technical University of Clausthal in 1973, 1974 and 1975.

The lines of the profiles are
Profile A: Landshut-Wasserburg-Reit im Winkl-Kitzbühel-Mittelsill-Lienz-Plöckenpaß
Profile B: Altötting-Zell am See-Großglockner-Lienz
Profile C: Freistadt-Linz-Liezen-Neumarkt-Klagenfurt.
They are shown in fig. 1 together with the surface geology.

Methods of Observation and Reduction

On the three profiles the measurements are spaced at about 3 km in the alpine foreland, at about 2.5 km in the alpine margins, and at 1.5 - 2.0 km in the central Alps.

The techniques are as described by Ehrismann et al. (1973) for gravity measurements in mountainous terrain. Terrain corrections for the innermost zone were made by the method developed by Schöler (1976). Götze et al. (1976) describe the procedure for measuring the vertical gradient, with special emphasis on the reproductibility of the readings.

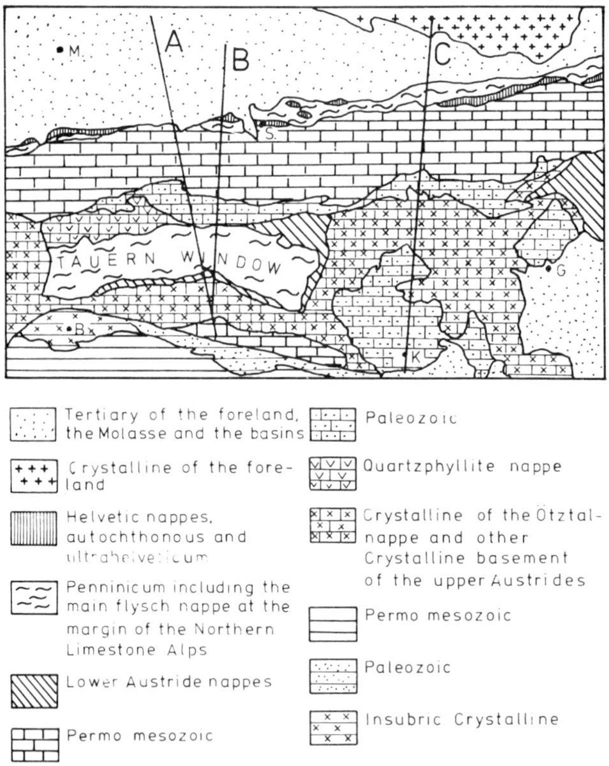

Fig. 1. Position of the gravity profiles A, B and C (Eastern Alps). M = München, S = Salzburg, G = Graz, B = Brixen, K = Klagenfurt.

The Gravity and Gradient Observations

The gravity differences between the individual stations were measured with a Worden Master gravity meter using a stepwise procedure. The differences between the field stations were each measured twice and these

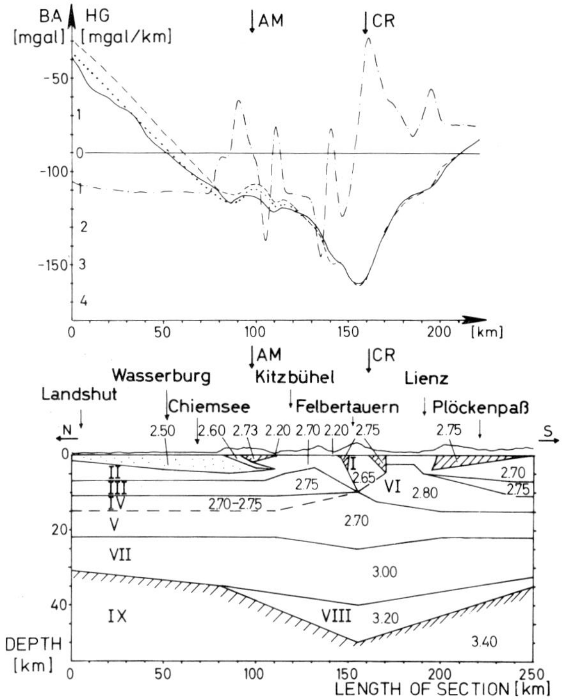

Fig. 2. Profile A
For crustal models figs.2-4:
AM = Alpine northern margin
CR = Central ridge
BA = Bouguer anom. in mgal.
HS = Horizontal grad
Depth in km, N left side of fig.

________: modified Bouguer anom.

--------: comp.Bouguer anom. (for A layer IV = 2.75 gr/cm³)

........: comp.Bouguer anom. (for A layer IV = 2.70 gr/cm³)

-.-.-.-.: comp.horizontal grad. curve (for A layer IV = 2.70 gr/cm³).

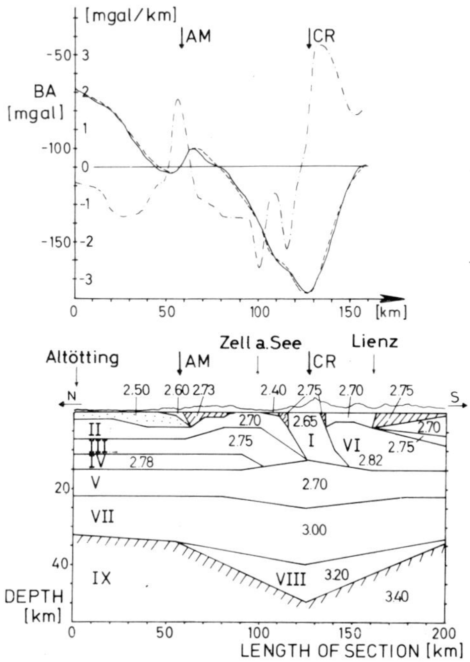

Fig. 3. Profile B.

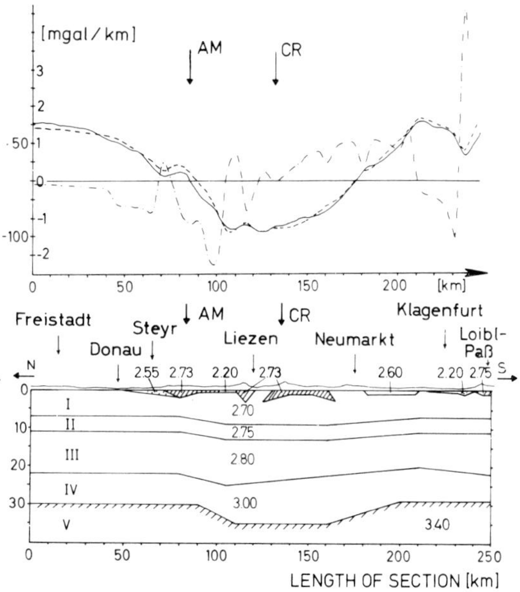

Fig. 4. Profile C.

between base stations four times.

The profiles are tied to the Austrian national gravity network. The gradients were measured with a LaCoste and Romberg Type D micro gravity meter.

Results

The following anomalies are shown in figs. 2 and 4:

modified Bouguer anomaly:

$$(\Delta g_o'')^+ = g + \delta g^+_{Niv} + \delta g_{Top} + \delta g_{Bpl} - \gamma_o$$

The symbols have the following meanings: g = measured gravity; γ_o = normal gravity at sea level according to the International Formula of 1930; δg_{Top} = terrain correction; δg_{Bpl} = infinite slab Bouguer correction; δg^+_{Niv} = Free air correction using mass-corrected vertical gradient.

Remarks

1) Note that the modified Bouguer anomaly is computed using the mass-corrected vertical gradient at the station: see Götze (1976).
2) The terrain correction for the most distant zones is made by the method of Ehrismann et al. (1966, 1971).
3) The accuracy of the gravity anomalies is about $\pm$ 0.5 mgal. This error is larger than in flat lands owing to the difficulties of levelling and of estimating the terrain corrections in rough terrain. The standard density of 2.67 gr/cm^3 is used in the Bouguer slab and terrain corrections. In addition to this standard method we have calculated geological corrections based on three-dimensional models of the investigated area and its surface geology.

Model Calculations

Model calculations in connection with the results of earlier field measurements are in progress. Three of these models concerning the profiles A, B and C are shown in the figs. 2, 3 and 4 (at the top). The concept of these two-dimensional models is based especially on the following publications: in geophysics: Closs (1957), Harcke (1972), Angenheister et al. (1972); in geology: Geologische Karte von Österreich nach Exner (1964), Tollmann (1963).

The gravity profiles, computed using the method of Talwani (1959), and the densities shown in figs. 2, 3 and 4 are given at the bottom of these figures. They are in good agreement with the modified Bouguer anomaly profiles.

Profile A (fig. 2)

In contrast with earlier work the model of fig. 2 contains much detail in the crustal structure down to 10 km.

For example the block of low density (see layer I; = 2.65 gr/cm^3) within the region of the Tauern window between 150-165 km on the profile

is placed near the surface. The lower side of the structure is located in the depth of ca. 10 km. In contrast with earlier model-calculations (Makris 1971) the low density block (layer I) has only local signification.

The low velocity layer between the depth of 10 and 25 km has a density of 2.70 gr/cm³ (see layer V). The model of fig. 2 contains two variations:

a) low velocity layer between the depth of 15 and 25 km

b) low velocity layer between the depth of 11 and 22 km (together with layer IV).

The second variation (b) is in good agreement with the modified Bouguer anomaly.

The transition layer (layer VIII; density = 2.30 gr/cm³) is placed above the Moho-surface, which is located appr. 30 km under the alpine foreland and appr. 50 km below the central Alps.

Profile B (fig. 3)

The distance runs up to appr. 60 km between profile A and profile B. This explains why both models have the same deep structure (see layers V, VII-IX). This is especially true for the transition layer VIII. The depth of the Moho under the foreland and under the central Alps is still the same as in model A.

However, contrast to the position of the low velocity layer of model A the layer V of model B is symmetrically to the central Alps.

Relative to the upper crust structures (down to 15 km) model A and B contain many local variations. See for example the position and depth of the block with the density = 2.65 gr/cm³.

Profile C (fig. 4)

The gravimetric density model of the last profile C shows a crustal structure without any geological disturbance. The model reveals a more horizontal system of layers in comparison with profiles A and B, down to appr. 25 km. The low velocity layer and transition layer are absent in the very east Alps. The Moho-surface dips from the depth of appr. 30 km under the alpine foreland to the depth of only 35 km under the central part of the profile (km 110 - 160).

Remarks

The proposed model calculations of the east Alps are preliminary working models. Further conclusions will be possible when the model calculations, now in progress for the adjacent areas, have been completed. These results will enable us to revise the density models as shown in figs. 2 - 4 in the near future taking also into account the seismic results of the *Alpine Longitudinal Profile 1975*.

Acknowledgement

We are indebted to the following institutions for their help and co-operation:

- Bayerisches Landesvermessungsamt, München, Präsident Dr.E.Messerschmidt
- Bundesamt für Eich- und Vermessungswesen, Wien, Hofrat Dr.E. Senftl

- Lehrkanzel für Geophysik der Universität Wien (Vorstand: Prof. Dr. R. Gutdeutsch), a.o. Univ. Prof. Dr. P. Steinhauser
- Geologisches Institut der Universität Wien, Prof. Dr. A. Tollmann
- Felbertauern Straße AG, Lienz.

References

Angenheister, G., Bögel, H., Gebrande, H., Giese, P., Schmidt-Thomé, P. & Zeil, W., 1972: Geol. Rdsch., 61:349-395.
Closs, H. & Hahn, A., 1972: Geol. Jb., 72:503-528.
Ehrismann, W., Müller, G., Rosenbach, O. & Sperlich, N., 1966: Boll. geofs. teor. ed appl., 8,29:3-20.
Ehrismann, W. & Lettau, O., 1971: Arch. Meteor. Geophys.Bioklim. Ser. A 20:383-396.
Götze, H.-J., 1976: Arch. Met. Geophys. Bioklim. Ser. A 67-77.
Harcke, H., 1972: Dissertation Inst. Geophysik Univ.Karlsruhe.
Schöler, W., 1976: Arch. Met. Geophys. Bioklim. Ser. A :79-88.
Talwani, M., Worzel, J.L. & Landismann, M., 1959: J. Geophys. Res., 64:49-59.
Tollmann, A., 1963a: Geol. Rdsch., 52:226-237.

Coalification Studies in the Alps

M. Teichmüller and R. Teichmüller

Plant remains, in particular wood and the exines of spores and pollen, react very sensitively when heated during diagenesis and anchi-metamorphism. The degree of alteration of wood or of its coalification product, vitrinite, is measured microscopically on the basis of the optical reflectivity (Rm_o and $Rmax_o$)* (cf. fig. 1). The alteration of spore exines is determined by their colour in normal transmitted light or by their fluorescence colour under blue, violett or ultraviolett irradiation. Such microscopic determinations of the degree of alteration or of coalification (= rank) can be carried out also on finely divided organic matter in rocks (Stach et al. 1975). The degree of coalification depends directly upon the heat exposure, whereas pressure, beyond the early compaction of lignites, retards the coalification process, - at least before the stage of semi-anthracite is reached. The heat exposure includes the time factor (*cooking time*). Therefore it is rarely possible to deduce

*Rm_o = average or random reflectance under oil immersion (measured in unpolarized light). $Rmax_o$, $Rmin_o$ = maximum reflectance, resp. minimum reflectance,under oil immersion (measured with polarizer in linearly polarized light, rotating the microscope stage).

exact temperature data from coalification data alone. This holds true also for paleotemperatures determined with the electron spin resonance of kerogene. Nevertheless, the determination of the main coalification stages (fig. 1) permits rather detailed statements about the past heating of a sample up to the stage of anchi-metamorphism or up to about 300°C. The progress of coalification is not influenced by the chemical environ-

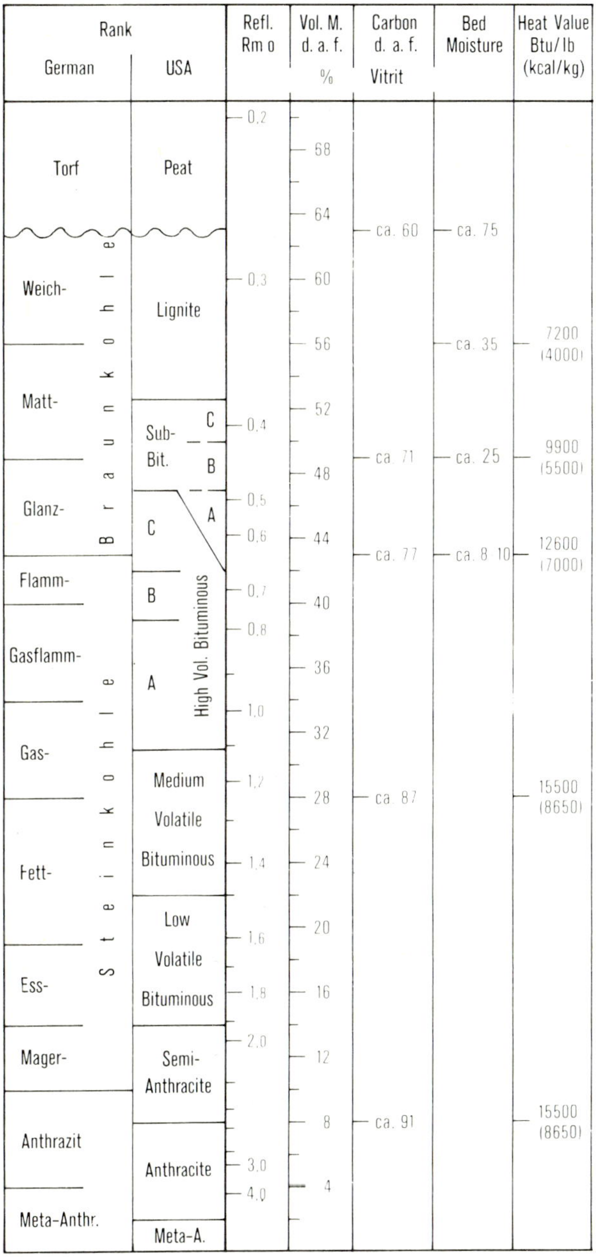

Fig. 1. Correlation between the German and the North American coal classification on the basis of different rank parameters (after Stach et al. 1975, p. 42).

ment. Beginning with the anthracite stage also shear stress can promote the orientation and the extension of graphite-like crystallites, thus increasing maximum reflectance and reflectance anisotropy.

In the Alps microscopical coalification studies have covered only a few areas, but our data demonstrate that the method permits certain conclusion about the geothermic history and the tectonic development.

Subduction Zone at the Northern Margin of the Alps

The coalification conditions were studied in two cross sections (fig.2). In the section north of the Tegernsee Alps, the rank of coal in the autochthonous Molasse increases very slowly with depth. The boundary Braunkohle/Steinkohle (0.6 % Rm_o) is reached not until a depth of ca. 4500 m. This low temperature gradient is also reflected in unusually deep petroleum occurrences, such as near Darching at 4410 m. Within the Helveticum, likewise, the rank of coaly matter remained low: in the deep-

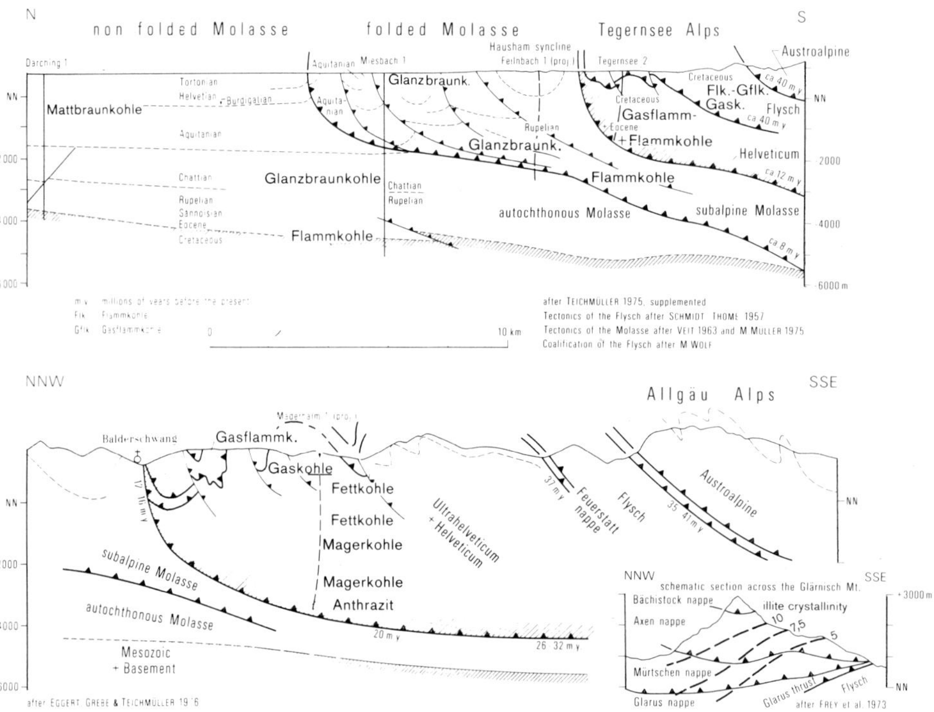

Fig. 2. The coalification conditions in the subduction zone at the northern margin of the Tegernsee Alps and the Allgäu Alps. The cross section through the Glärnisch Mt. was added for comparison. The value 7,5 of the illite crystallinity probably corresponds to the top of the meta-anthracite zone. Note the increase of coalification from the East to the West at the base of the Helvetic pile of nappes (i.e. in the direction Tegernsee Alps - Allgäu Alps - Glärnisch Mt.).

est part of the imbricated Helveticum the coalification did not proceed beyond the stage of ca. 0.7 % Rm_o (Flammkohle). In the Flysch nappes the rank of coal is only a little higher.

In the borehole Maderhalm 1 (fig. 2, lower part) the rank of coaly matter increases with depth to near the base of the Helvetic pile of nappes where 2.8 % Rm_o (anthracite stage) is attained. Near Maderhalm, coalification correlates poorly with stratigraphic or imbricate burial. This suggests that coalification in the Helvetic nappe complex postdates its internal structure but predates the emplacement of the complex upon the Molasse (as a matter of fact, boreholes in the Austrian Calcareous Alps encountered a relatively weakly coalified Oligocene Molasse (0.5 - 0.7 % Rm_o) below the Alpine nappe complex). In the Allgäu Alps the conditions are similar to those at the Glärnisch Mt. (fig. 2, lower right), where, according to Frey et al. (1973), the illite crystallinity increases with depth, independent of the stratigraphic age of the strata. Here, the diagenesis is slightly younger than the nappe imbrication (Frey et al. 1973). The rise of the lines of equal illite crystallinity towards the south (fig. 2, lower right) is related to the Oligocene heating of the Alps as may be deduced from the rejuvenation of the K/Ar-ages of glauconites.

The strikingly slow increase of coalification with depth at the northern margin of the Alps lets us assume that the heat flow was relatively low for a long time, perhaps related to the subsidence of the upper mantle within this subduction zone.

Northern Calcareous Alps

Very few coalification data exist from the Northern Calcareous Alps. In the Eocene of Häring the coalification did not develop beyond the stage of Glanzbraunkohle (0.51 % Rm_o). As expected, the same is valid for the Gosau coals of the Upper Cretaceous in the Salzkammergut (0.46 % Rm_o). Near Grünbach in Lower Austria the Gosau coals just reached the Flammkohle stage (vitrite: 5 % moisture, 40 % volatile matter d.a.f., 78,5 %C d.a.f.). In the Lunz series of the Upper Triassic near Gaming, the coalification only reached 0.95 % Rm_o and 1.03 % Rm_o. although, here, the coal seams are folded, imbricated and mylonitisized. Moreover, the Lunz nappe was subduced below some sub-nappes of the Ötscher nappe and is believed to have been buried about 3000 m deep. In the uppermost nappe of the Northern Calcareous Alps (Inntal nappe) the Upper Triassic oil shale of Seefeld (Tirol) hardly reached the maturity stage ($<$ 0.5 Rm_o).

Northern Grauwacke Zone

In contrast to the strikingly low degree of coalification in the Northern Calcareous Alps, coniferous wood embedded in Permoskythian rocks near Mittenberg was transformed to meta-anthracite. The high reflectance of more than 9 % $Rmax_o$ points to the stage of almost semi-graphite. The anthracitization was caused by tectonic burial and intense movement, possibly by an additional thermal event suggested by unpublished microprobe data. In the Veitsch nappe near Trieben and Kaisersberg latest Carboniferous strata have reached the meta-anthracite stage; in local shear zones, semi-graphite and microcrystalline graphite. Within the strongly

dislocated Stephanian of the Steinach nappe of the Austroalpine near Nösslach, the coalification has reached only the stage of low rank meta-anthracites at 5.4% $Rmax_o$, i.e. anchi-metamorphism.

Central Zone of the Eastern Alps

In Styria, the coal-bearing Middle Miocene overlies the nappes of the central zone of the eastern Alps. The coals of the Leoben district, deposited along the Mur/Mürz line, in spite of their young age and their shallow burial of only a few hundred meters, were transformed into Glanzbraunkohle (vitrite:7-9% moisture, 75-77% C d.a.f.,ca.7000 kcal/kg a.f.). Similar conditions are valid for the Glanzbraunkohle of Fohnsdorf, although, here, the burial was deeper (ca.1500 m). The coals of both deposits have almost reached the Steinkohle stage (cf. Patteisky & M.Teichmüller 1960, fig.4-9). This suggests a high geothermic gradient, similar to the values of 5 to 6°C/100 m locally observed in the Styrian and south Burgenland Tertiary (Janoscheck 1976). Post-tectonic thermal domes may have promoted the coalification, in a similar fashion as the Tauern crystallization between early Miocene and 8 m.y. B.P.(Frey et al. 1974).

Swiss Alps

Only a few coalification samples have been studied so far. In the deepest Chattian, near Weissbach at the north front of the Alps, the Steinkohle stage is just reached (0.56-0.61 % Rm_o). In the Eocene of the Gemmen-Alphorn (Wildhorn nappe of the uppermost Helveticum) the coalification has already reached the stage of 0.79 % Rm_o. In the autochthonous and the parautochthonous Mesozoic of the Aar Massif, such as in the Rhaetian of the Weissgandstöckli, strongly anisotropic meta-anthracites occur with 5.09 % Rm_o, 6,86 % $Rmax_o$. This rank is similar to that of Carboniferous meta-anthracites within the Helvetic massifs where an average of 17 localities is 5.04 % Rm_o, 6,83 % $Rmax_o$. This suggests that alterations during the Alpine orogeny were considerably stronger than those during the late Variscan orogeny.

Metamorphism (= graphitization after Schüller 1961) is widespread in the Penninic zone, but in the front of the Penninic St. Bernhard nappe near Valdigna di Aosta, Carboniferous coals only reached the stage of strongly anisotropic meta-anthracites (Petrascheck 1937). Also, the Prealpine Klippen of Penninic provenance show lower coalification. In the middle Jurassic of Chateau d'Oex and of Klus in the Simmental, only 0.98 % Rm_o and 1.14 Rm_o are attained.

Sesia Zone of the Western Alps

The unconformable post-orogenic Oligocene cover (Scheuring et al. 1974) contains detrital plant remains altered to meta-anthracite with 7.5% $Rmax_o$ and semi-graphite with 8.8 % $Rmax_o$, 0.4 % $Rmin_o$ (fig. 3). This anchi-metamorphism may be in genetic connection with the intrusion of tonalites near Biella and Miaglione (Stadler & Teichmüller 1976).

In the Westphalian C-age beds of Manno, within the South alpine lithosphere plate, the coalification only reached the Gasflammkohle stage

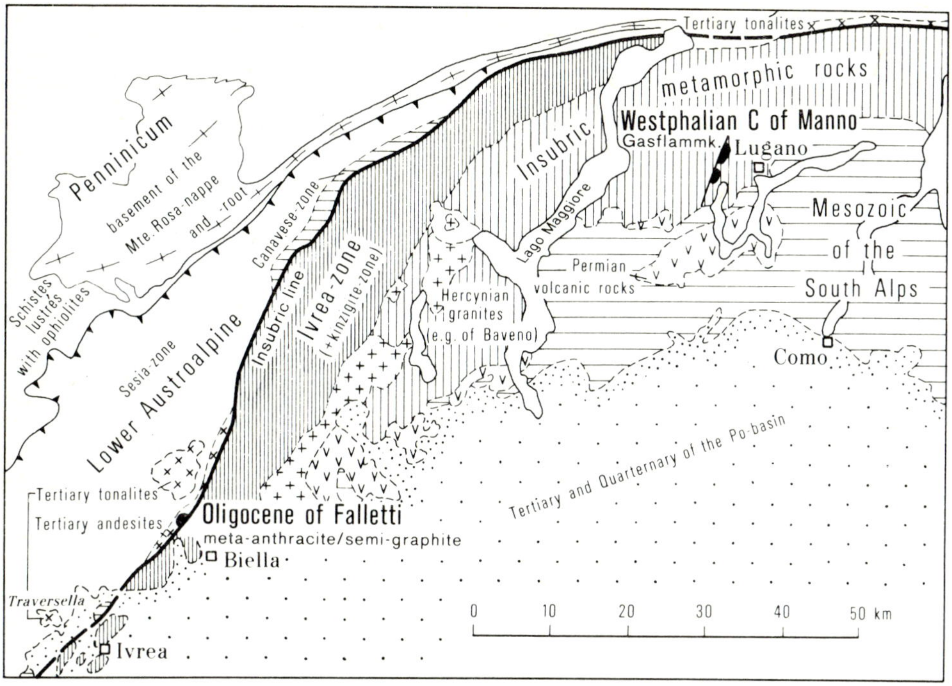

Fig. 3. The tectonic position of the Oligocene of Falletti and the Carboniferous of Manno.

with 0.80 % Rm_o (fig. 3). The Carboniferous of Manno belongs to the large geanticline of Lugano which to the east borders the Bergamask trough.

These few examples show that coalification data provide relatively precise geothermal information within the range of diagenesis and anchimetamorphism.

Acknowledgements

Samples for coalification studies were kindly provided by Dr. Agiorgitis, Dr. Bögel, Dr. Dietrich, Prof. Feys, Prof. Hantke, Dr. Hunziker, Mr. Leistikow, Prof. Patteisky, Prof. Ruttner, Prof. Trümpy, Dr. M. Wolf, Mr. Zingg and the oil companies BRIGITTA-ELWERATH, PREUSSAG and DEUTSCHE TEXACO. We are very grateful to all of them. We thank also Dr. N. Bostick and Prof. Roeder for their kindness to revise the English text.

References

Eggert, P., Grebe, H., Teichmüller, M. & Teichmüller, R., 1976: N.Jb. Geol. Paläont. Abh., 152,1:112-136.

Frey, M., Hunziker, J.C., Roggwiller, P. & Schindler, C., 1973: Contr. Mineral. and Petrol., 39:185-218.

Frey, M., Hunziker, J.C., Frank, W., Bocquet, J., Dal Piaz, G.V., Jäger, E. & Niggeli, E., 1974: Schweiz. Mineral.Petrogr.Mitt.54,2/3:247-290.

Janoscheck, W., 1976: Vortrag vor der Geol.Ver.Hannover, 28.2.
Müller, M., 1975: Erdöl und Kohle, Ergänzungsband-Compendium 75/76:63-76.
Patteisky, K. & Teichmüller, M., 1960: Brennstoff-Chemie,41:79-84, 97-104, 133-137.
Petrascheck, W.E., 1937: Z.dtsch. geol.Ges., 87:622-632.
Schüller, A., 1961: N.Jb. Mineral.Abh., 96:250-290.
Stach, E., Mackowsky, M. Th., Teichmüller, M., Taylor, G.H., Chandra, C. & Teichmüller, R., 1975: Stach's Textbook of coal petrology. 2nd ed. 428 p., Gebr. Borntraeger, Stuttgart.
Stadler, G., Teichmüller, M. & Teichmüller, R., 1976: N. Jb. Geol. Paläont. Abh., 152,2:177-198.
Teichmüller, M. & Teichmüller, R., 1975: Geol. Bavarica,73:123-142.

For additional references see Teichmüller & Teichmüller (1975), Eggert et al. (1976) and Stadler et al. (1976).

Chapter 2

THE GEOTRAVERSE EASTERN ALPS (Ia)

The Eastern Alps - An Introduction

H. Bögel and D. Roeder

The Alps are a stack of thrust sheets composed of a basement complex of Paleozoic metasediments and crystalline rocks, Mesozoic and minor pre-Mesozoic ophiolites, and Paleozoic to Neogene sedimentary rocks, transported northward and northwestward in Cretaceous to Neogene time, strongly warped, metamorphosed, intruded by minor granitoid plutons after the nappe emplacement, and presently rising fast to form one of the highest orographic belts in Europe and the circum-Mediterranean.

The Western Alps, west of the Rhine and Adda rivers, are a complex and polyphase imbrication with many of the essential elements exposed. In the Eastern Alps, however, most of these units are largely covered by the structurally highest and southernmost unit, the Austro-Alpine complex of nappes. This vast complex is rooted in the circum-Adriatic hinterland of the Alps. It is articulated by imbrications, warps, domed metamorphic isograds, late Alpine steeply dipping faults. It continues (paleogeographically) in the Southern Alps which show minor deformation of Alpine age but the beginning of a south vergent thrust belt (Miocene to Recent).

Within the structurally highest warps or domes of the Austro-Alpine complex, the Engadin and Tauern windows expose strongly deformed units composed of pre-Alpine basement and associated sediments, middle European miogeosynclinal deposits and eugeosynclinal ophiolite-associated metasediments. The correlation of these units with their equivalents in the Western Alps and the geotectonic interpretation of the nappe stack and its causes are discussed in the context of plate tectonic models of orogeny.

The geology of the Eastern Alps is summarized in a geologic-tectonic map (fig.1, foldout), a not exaggerated synoptical section (fig. 2) and a diagrammatic cross section of the nappe units (fig. 3; both backside of foldout).

The northern edge of the Alps (fig. 1) between Bregenz and Wien is a multiple thrust front in which foredeep clastic (legend 2), pre-foredeep shelf rocks (15), and a synorogenic flysch unit (16, 17) are stacked beneath the Austro-Alpine thrust complex (3 to 9) which occupies most of the map area.

The northern foreland is a wedge-shaped basin with its pre-Mesozoic basement rising northward and outcropping north of Linz in the Bohemian massif (26). At the foot of the Alps, this basement is estimated at up to 5 km below sea level. Beneath the southern foreland the basement is estimated at 10 to 15 km of depth but the lobate mountain front near Brescia and Udine is a strong southfacing homocline without or with minor southward thrusting but considerable seismotectonics. Thick Neogene molasse (2) overlies directly or with minor intercalated flysch (10) the Mesozoic platform carbonates(11). The west edge of the Austro-Alpine thrust complex

near Bregenz and Chur is a serrated west-facing thrust contact with north- to southwesterly sense of displacement, exposing in cross section the nappe complex with some of its articulations, several intermediate slices of geotectonic significance (16, 17, 20, 21), strongly deformed oceanic suites (18), crystalline (22), and late orogenic plutons (24).

The east edge of the Austro-Alpine thrust complex is an unconformable overlap of the post-orogenic Neogene (2) fill of the Vienna and Pannonian basin with neotectonic rifting and subsidence. The thrust front continues to the NE into the Carpathians of Czechoslovakia, Poland, and Rumania, but the eastward extension of the nappe character is poorly known.

The nappe complex is pierced by two windows, the smaller Engadin window to the W and the Tauern window at the center of the map. The Engadin window exposes ophiolites and shelf carbonates of intermediate slices (20,21), flysch (17), and fine clastics with oceanic affinites (18), but no basement. In the Tauern window the section is more complete, with shelf carbonates and pre-Alpine crystalline basement, formerly part of the foreland (see fig. 2,3), beneath oceanic suites. However, high-rank metamorphism has obliterated the difference between basement and Alpine sediments.

A major system of east-west faults, offset near Bolzano, articulates the thrust complex, the Periadriatic lineament. Its function as a subduction suture, repeatedly suggested in the past, is rather unlikely because it dissects continental crust with a common geological history from the late Paleozoic into the Neogene. Late Paleozoic and Tertiary plutons are aligned on this fault system, however.

Fig. 3 is a correlation chart which shows the present superposition and topological arrangement of the principal nappe units within the area of the geotraverse. The inset, lower right, shows the geotectonic disposition of these units, with material of the Eurasian plate (E), overlain by (G) eugeosynclinal material in black and flysch with dot pattern, and in turn overlain by (A) Africa-associated small plate material, split at the vertical Peri-Adriatic lineament (PN) into an Austro-Alpine and South-Alpine part.

Legend fig. 1. For further explanation see text.
1: Area of main investigations of the Geotraverse Ia, Eastern and Southern Alps; 2: Molasse, Tertiary undivided, Quaternary, f: folded Molasse, w: Waschberg belt; 3-5: Lower Austro-Alpine nappe system: 3: sediments; 4: quartz phyllites, 5: basement; 6-9: Upper (and Middle-) Austro-Alpine nappe system: 6: sediments (Northern Calcareous Alps); 7: Paleozoics undivided; 8: Central Alpine Mesozoics; 9: basement (Altkristallin, partly Paleozoics and quartz phyllites included); 10-14: Southern Alps: 10: flysches, 11: Permo-Mesozoics, 12: quartz porphyrites, early Paleozoics; 14: basement; 15: Helveticum, Ultrahelveticum and Klippen belts; 16-22: Flysch, Penninicum: 16: Rhenodanubian Flysch, 17: Penninic flysches, 18: Bündner Schiefer and equivalents, 19: Untere Schieferhülle, 20: Arosa, Platta nappes, 21: Falknis, Sulzfluh, Tasna nappes, 22: crystalline rocks, Central Gneiss (Tauern Window); 23: Tertiary volcanics, 24: Tertiary plutons; 25: Permian plutons (included small Triassic plutons near Predazzo); 26: Bohemian Massif; 27 above: normal and wrench faults, below (barbed): thrust faults.

to p. 58

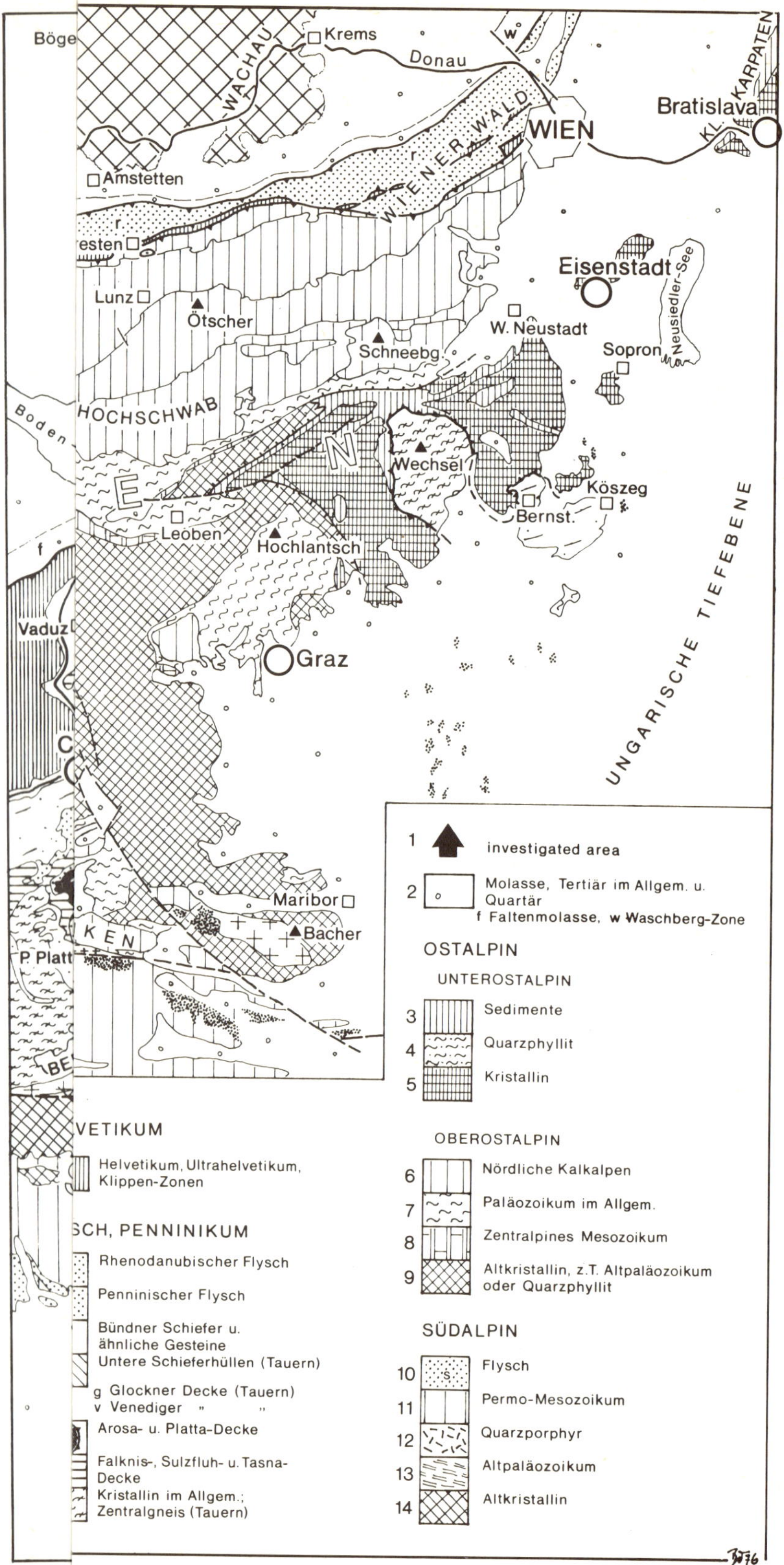

Aus Bög

:acking order of thrust sheets of the Eastern and

: the main elements: E former Eurasian Plate, G
. former Adria-Microplate.

ıeral, P Penninicum, V Foreland. O Upper Austroal-
stroalpine after Tollmann, NKA Northern Calcareous
ruck quartzphyllites, GD Glockner nappe, RT Rhe-
Klippenbelt, H_2 allochthonous Helveticum, H_1 au-
:, z Central-Gneiss Core + Paleozoic + Untere

· Triassic, j Jurassic, cr Cretaceous, te Tertiary;
quartzphyllites, 3 Lower Paleozoics, 4 Upper Pa-
asin and seamount facies, 7 mainly pelagic facies
ıden schist), 9 Flysch, 10 black triangles: oli-
n: ophiolites, 11 zone with retrograde metamor-
area with Hercynian metamorphism, 15 area with
area with meso-Alpine metamorphism (36-40 my).

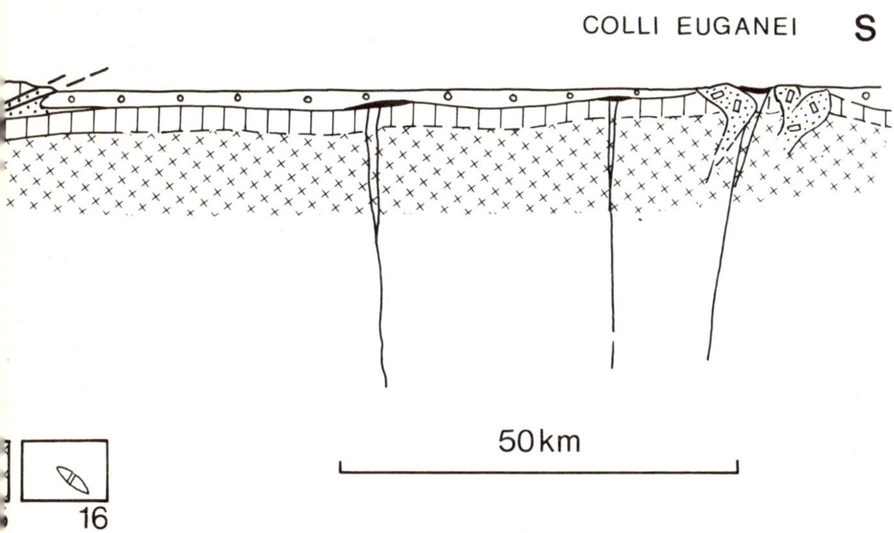

ıt (so-called Altkristallin) including quartzphyllites, Central Al-
ic of the Drauzug (d); 9 Paleozoics and quartzphyllites (Southern
'riassic) plutons; 11 Permian plutons; 12 quartzporphyrites (below)
mites; 13 flysches of the Southern Alps; 14 Tertiary volcanics
basement; 16 ophiolites; PN Periadriatic (Insubric) lineament.

Pre-Alpine crystalline and sedimentary basement (legend 1) occurs in the foreland (V), and also as a gneiss complex emerging in the Tauern window such as the Venediger nappe (VD), in slices (M) at the base of the Austro-Alpine complex, referred to as Mittelostalpin by Tollmann (1963), which consists of Altkristallin mainly (see Oxburgh 1968 for instance). Basement furthermore occurs south of the Peri-Adriatic lineament (PN).

The nappe stack beneath the Austro-Alpine consists near its north front, of miogeosynclinal Mesozoics (7) overlain by Neogene Molasse (12). Three kinematic units are distinguishable. The base, the autochthonous Helveticum (H_1) is overlain by the dislocated Helveticum (H_2), in turn overlain by the Ultrahelveticum (UH). The palinspastic position of the next following Rhenodanubian Flysch (RF, see Dietrich & Franz 1976), presently located between the foreland and the Austro-Alpine and therefore correlatable with the Penninic group of thrust plates (P), depends on the bias of the worker. Fixistic models assume adjacent and parallel depositional troughs for at least the late Cretaceous parts of the Helveticum, the Ultrahelveticum, the Rhenodanubian Flysch, and the late Cretaceous parts (Gosau) of the Austro-Alpine (Bögel 1976). Very slow convergence of a subduction zone below the Flysch belt is compatible with this fixistic view. Mobilistic models assume deposition of the Rhenodanubian Flysch at the subducting north front of the Austro-Alpine complex, either as an accretionary wedge (Dietrich & Franz 1976), or as a forearc basin fill (Roeder 1976).

The internal parts of the lower (Eurasian) continental mass outcrop in the Tauern window. They consist of a group of polyphase (Roeder 1976) thick-skinned gneiss cores, probably dislocated recumbent folds (Bögel 1976), grouped as the Venediger nappe (VD) with an imbricated miogeosynclinal sedimentary cover (7), and overlain by a complex of fine clastic, olistostromal, and ophiolitiferous sediments and metasediments grouped as Bündnerschiefer (8) and considered as the fill of the Alpine geosyncline proper. In the Tauern window, this complex constitutes the Glockner nappe (GD); its base may be a major transport surface or a transgressive contact with minor dislocation. At any rate, the internal structure of the Glockner nappe suggests very strong deformation. All internal units are overprinted by a more or less postkinematic intermediate-t/low-p metamorphism. The lower units of the Glockner nappe show remnants of high-p/low-t metamorphism.

The Austro-Alpine complex, probably not always a solid plate, has the shape of a north-pointing wedge, modified by polyphase imbrication and by late Alpine distortion. It has one or several shear planes at its base and the present erosional surface at its top.

An external part ot the north, grouped as the Northern Calcareous Alps (NKA), consists of two or three thrust masses or nappe complexes. The Lower Austro-Alpine (U) consists of early Paleozoic metasediments (2) with a retrograde metamorphism of Alpine (unpubl. data by Morteani & Satir) age (11). The Upper Austro-Alpine (O) contains, with tectonic or unconformable basal contact, a Paleozoic graywacke belt (3), a partly nonmarine clastic unit of Permo-Pennsylvanian and early Triassic age (4), a major carbonate complex with reefs, platforms, and offreef basins of Triassic age (5) grading into a deeper-water Juro-Cretaceous series (6 and 7), and unconformably overlain by Molasse-type Tertiary clastics.

An internal part, the Mittelostalpin (M) surrounding the Tauern window, contains crystalline basement overlain by Mesozoic carbonates.

The Peri-Adriatic lineament (PN) is a straight, steeply dipping fault

parallel to the strike of the Alps. It is not a subduction suture, and its one-time function as an arc-associated wrench fault (Fitch 1970) is less well documented than is its last and major function. Dip-slip uplifts the central-Alpine block to the north, dislocating late Alpine metamorphic isograds and post-Alpine morphological surfaces.

South of this fault zone, the Southern Alpine complex consists of crystalline basement (1) and quartz-phyllitic early Paleozoic metasediments intruded by late Hercynian (13) and late Alpine granitoid plutons. Above clastic to calcareous late Paleozoic transgressive series (5 and 7) lies a Mesozoic to Tertiary series which are very similar to that of the Northern Calcareous Alps, but somewhat thicker. Relatively minor thrusting toward the south involves the Tertiary and repeats the entire shelf section. Recent activity, however, appears to be favoring this thrust belt, as suggested by the seismotectonics and high seismicity of the Friule region. Corroborating the neotectonics is the stratigraphy of the Alpine hinterland in the Padan-(Po)-Plain and the Venice basin which contains a Southern Alpine section buried abruptly under several kilometers of clastic Neogene.

The Alpine contributions to the present volume are grouped according to the Alpine framework as discussed. Papers about the Northern foreland contain data in support of extensive thrust imbrication and a deeply buried pre-Alpine basement and miogeosynclinal cover. The papers of part 2 are focussing on the complex synsedimentary rifting that affected Tethyan small-plate material prior to its convergence and on the pre-sollisional stratigraphy of the Austro-Alpine thrust complex. Part 3 deals with the geology of the Tauern window and the crystalline rocks of its frame. In part 4, a rear deep and a neotectonic retrocharriage on the backside of the Alpine subduction system are reviewed.

References

Bögel, H., 1975: Geodynamics Project, Nat.Comm.Fed.Republic of Germany, Progress Rept.:42-50, Bonn.

Dietrich, V.J. & Franz, U., 1976: Geol. Rdsch., 65/2:361-374.

Fitch, T.J., 1972: J-Geophys.Res. 77(23):4432-4460.

Oxburgh, E.R., 1968: Proceed. Geol. Assoc., 79/1:1-46, London.

Roeder, D., 1976: Z. Dtsch. Geol. Ges., 127:87-103.

Tollmann, A., 1963: Ostalpensynthese: 256 p., Wien (Deuticke).

Part 1

The Northern Foreland and Geophysical Researches in the Northern Alps

Summary of Post-Permian History of the Northern Alpine Foreland

K. Lemcke

The northern Alpine foreland covers an area of about 760 km in length from Geneva (Switzerland) and St. Pölten (Austria), with the biggest width of about 140 km (Regensburg) and smallest of 7 km (Amstetten/Upper Austria) (fig. 1).

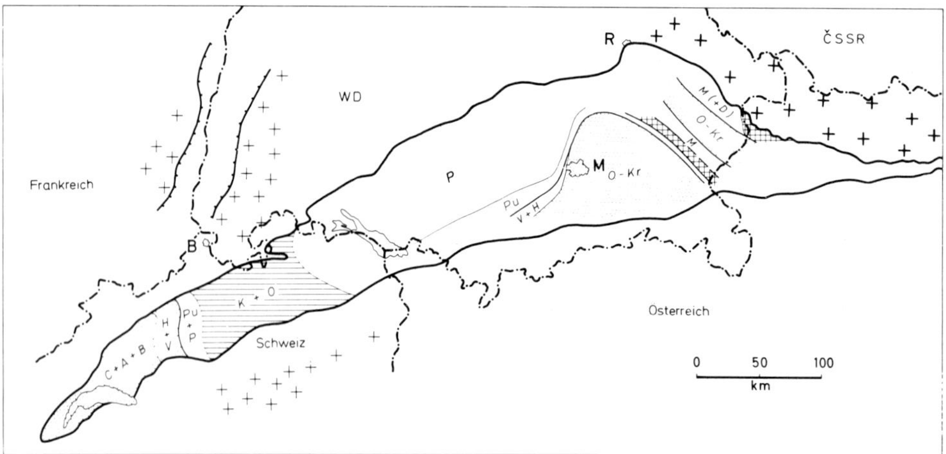

Fig. 1. Map showing substratum of the Molasse (Tertiary uncovered). Narrow crosshatching = basement; dots = Cretaceous; horizontal hatching = swell region between Black Forest and Aar Massif; D = Dogger; M = Malm in general, O = Oxfordian, K = Kimmeridgian, P = Portlandian, Pu = Purbeckian; V = Valanginian, H = Hauterivian, B = Barremian, A = Aptian + Albian (Gault); C = Cenomanian, Ca = Campanian; thick crosses = recent basement areas, at the beginning of the Molasse time without sediment cover; thin crosses = recent basement areas, at the same time with sediment cover; fault symbols = recent border faults of the Upper Rhine Graben. B = Basel, M = München, R = Regensburg. - From: Lemcke 1975.

The post-paleozoic history of this area mainly derived from numerous wells drilled for oil and gas has confirmed that it belongs to the southern border of the Variscan consolidated Mesoeurope.

The evolution took place in three epeirogenetic stages separated by 2 major transgressions in Malm (late Jurassic) and Helvetian (middle Mio-

cene) time. During the first stage the Vindelician land mass was covered by transgressive onlapping sediments. The second stage comprises the paleogeographic influence of the uplift of the Rhenish Shield and the foundering of an East Bavarian marginal trough with movements that continue well into the period of Molasse deposition. The third stage after Helvetian time comprises the elimination of all mesozoic paleogeographic trends and the integration of the area into the Alpine realm. Epeirogenetic movements however, persisted as evident in two reversals of sediment transport at the beginning (E-W) and after the end of upper Miocene time (W-E).

Since the late Eocene, the present Alpine foreland became gradually incorporated as northern flank into the Molasse trough which developed far south, with the trough axis located 10-20 km south of the present alpine front. In Austria, south of the salient Bohemian massif, the trough axis may have been located even farther south. In the center of the trough Flyschmolasse was formed in the early Oligocene gradually changing over into Molasse facies. Subsequently the trough axis shifted northward into the present foothill region (Folded Molasse belt), perhaps interrupted by some retrogradations.

Sediments of the former southern flank of the Molasse trough have been preserved as intra-alpine "Unterinntal-Tertiär" (Lower Inn Valley Tertiary) covering parts of the Calcareous Alps. It was thrusted over the axis of the autochthonous Upper Eocene-Oligocene Molasse trough in the Late Chattian time with the vehicle of its underlying Austro-Alpine formations. The northward transport might have reached a range of 50 km.

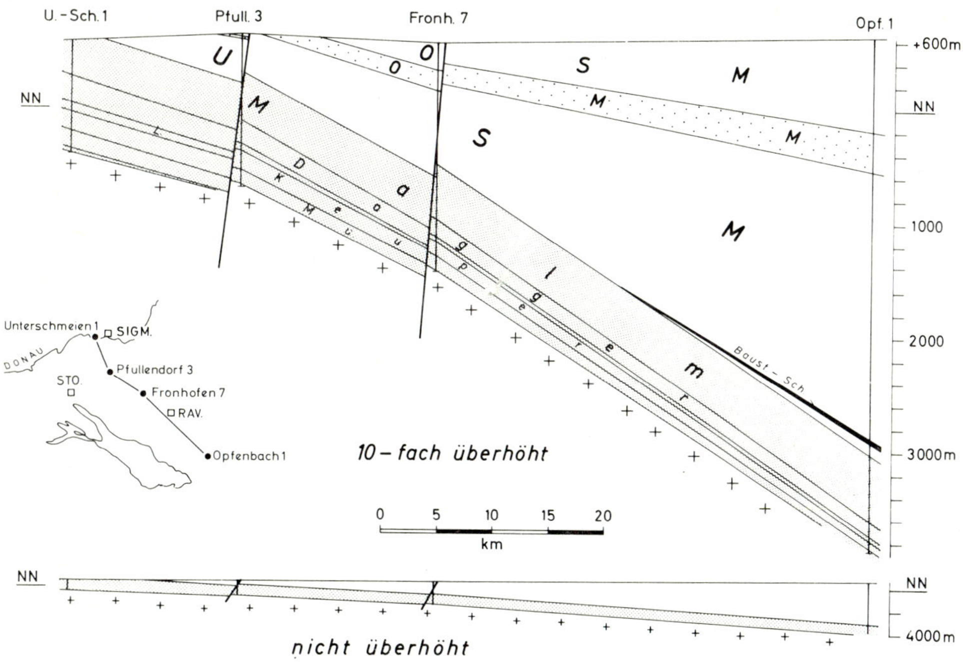

Fig. 2. A cross section through the Molasse trough north of the lake Bodensee. - From: Lemcke 1975.

Southward tilting of the northern foreland continued until early Pontian with sediments south of a zero edge increasing to >5250 m in thickness, sourced mainly in the rising Alps and deposited in marine, brackish and non marine environments.

Outside of the Foothills belt, the foreland is deformed by numerous E-W striking sometimes synsedimentary normal faults, more frequently in the south, and antithetic preferably in the northern part of the basin (fig. 2). Caused by the final emplacement of the Austro-Alpine nappes upon the former axial region of the Molasse trough in post-Sarmatian time, the folds of the Foothills belt were initiated. By this event also the whole unfolded Molasse area was compressed and forms now a flat syncline often with steep flanks at the south.

Subsidence diagrams prepared from well data show not only the change from mesozoic continental shelf to Tertiary orogenetic foredeep but also a post-Molasse rebound and uplift of the foreland.

In Western Switzerland, events of unusually rapid subsidence took place in late Triassic and Jurassic time and peristed through Molasse

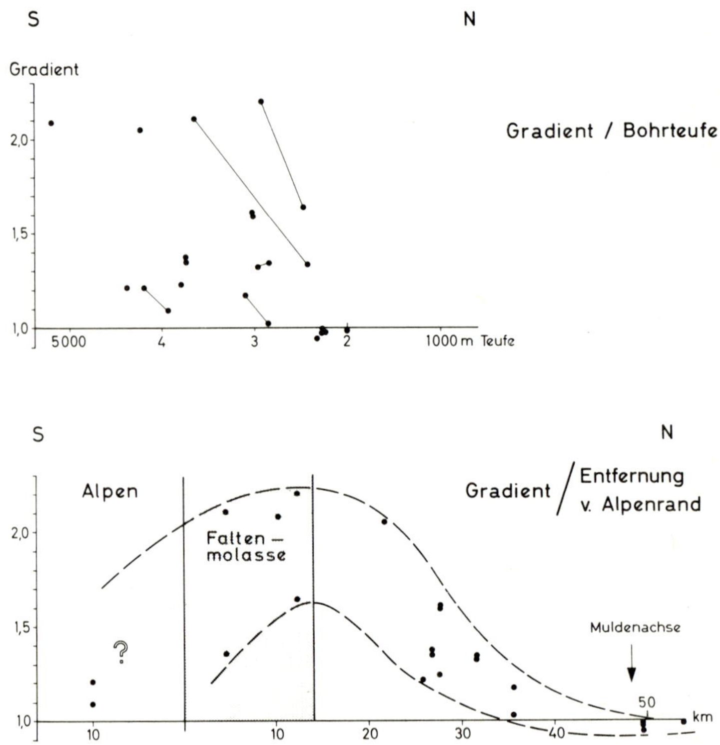

Fig. 3. Upper Part: Hydrostatic pressure gradients versus depth: No correlation can be recognized. - Lower Part: Hydrostatic pressure gradients versus distance from the Alpine orogen: The pressure gradient increases in the direction to the Alps and reaches its maximum in the northern part of the Folded Molasse (Faltenmolasse). It is smallest in the present axis of the Molasse syncline (Muldenachse) and north of it. - From: Lemcke 1973.

time, but are presently replaced by unusually rapid uplift, of 0.3 to 0.4 mm/y.

Formation pressures in the sediments of the basin demonstrate the role of the Alpine foreland as the link between the central European craton and the orogene. Suprahydrostatic pressures originating from Alpine stress (fig. 3), coexist laterally and vertically with subhydrostatic pressures indigenous to the non-Alpine basement. The seismic velocities especially in the Upper Molasse increase towards the orogene and demonstrate tangential stress in the foreland (Lohr 1969).

References

Lemcke, K., 1973: Geologica Bavarica. - 69:5-48.
- , 1974: Ecl. geol. Helv. - 67:121-133.
- , 1975: Jber. u. Mitt.oberrh.geol.Ver., N.F. - 57:87-115, Stuttgart.
Lohr, J., 1969: Geoph. Prosp. XVII. - 2:111-125.
Müller, M., 1976: Compendium 75/76, Erg.-Bd.z.Erdöl u.Kohle:63-76.
Richter, M., 1951: Z.Dtsch.Geol.Ges. - 102:177-180.

Miesbach 1 and Staffelsee 1 - two Basement Tests below the Folded Molasse

M. Müller

Introduction

The deep wells Staffelsee 1 (S 1) and Miesbach 1 (M 1), drilled in 1964-1966 and in 1974-1975 (Müller 1970, 1975) in the Alpine foothills of Bavaria (fig. 1), penetrated the northernmost part of the Folded Molasse and the autochthonous Molasse. This report summarizes the stratigraphic (fig. 2) and structural results of these wells based on the results of drilling, reflection seismic profiles, coal mining data, and surface geology.

Alpine North Front

The structural north front of the Eastern Alps is located within the Molasse foreland basin. A north-dipping homocline involving the uppermost layers of the autochthonous Molasse separates the Unfolded Molasse to the north from the Folded Molasse and its series of W-E striking synclines to the south. The mountainous nappes of the Helveticum, the Flysch, and the Calcareous Alps adjoin further south, as shown in the two cross sections of fig. 3 and 4.

The age of the northernmost thrust fault is post-Sarmatian (< 5 m.y.) (Schmidt-Thome 1962). All thrusts of the Folded Molasse dip steeply near the surface and sole out within the autochthonous Molasse series, as shown by reflection seismic results (Veit 1963). S 1 and M 1 drilled

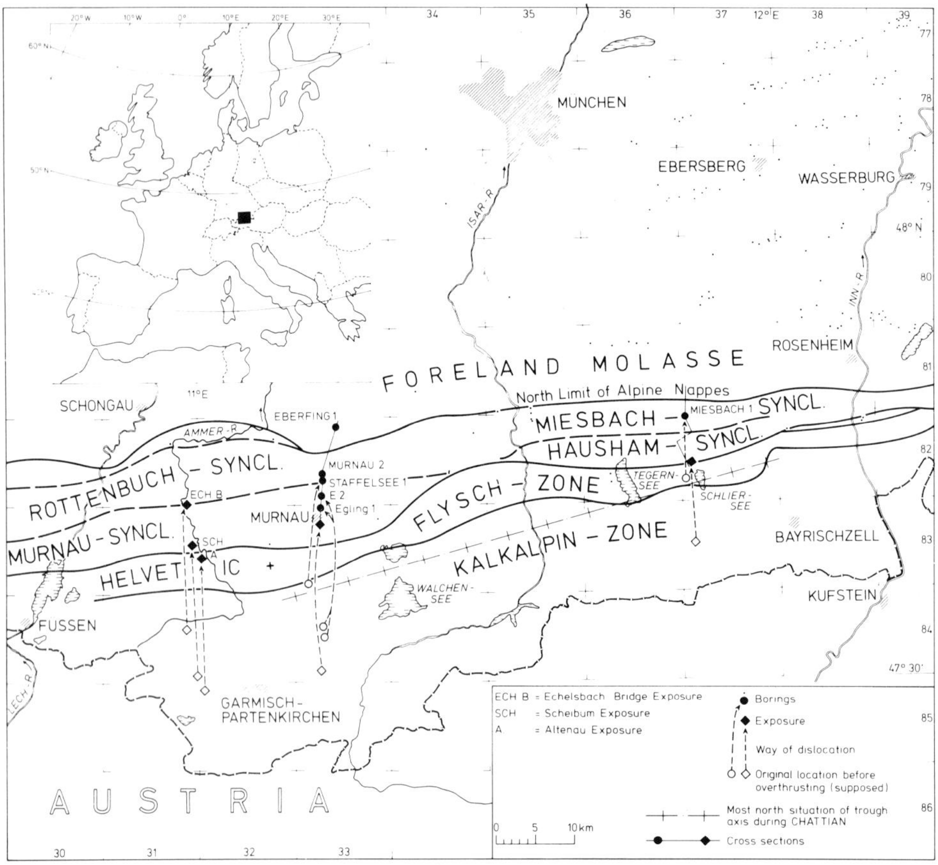

Fig. 1. Tectonic sketch map of the transition of the Alps to the unfolded (Foreland-)Molasse between the rivers Lech and Inn. Directions and amounts of horizontal minimum dislocations of some reference points in the Molasse are shown.

through the thrust planes without encountering shreds or dislocated blocks, but in Murnau 2, blocks of Chattian Coloured Molasse and Chattian/Rupelian Baustein beds occurred beneath the thrust plane.

At S 1, the thrust is indicated by a lithological change, but at M 1 it is located within marine shales of different stratigraphic position and evident in a distinct increase in calcite veining and slickensided fibrous and layered calcite material. Such material occurs in outcrop in the gorge of the Ammer river (fig. 1) and is inferred to form the base of the Murnau syncline 20 km farther east, based on the results of the well Egling 1 and from reflection seismic profiles.

There are no facies differences above and below the thrusts. In M 1 the average seismic velocity in the upper tectonic unit is somewhat higher than in the lower. The maturation of the organic substance is

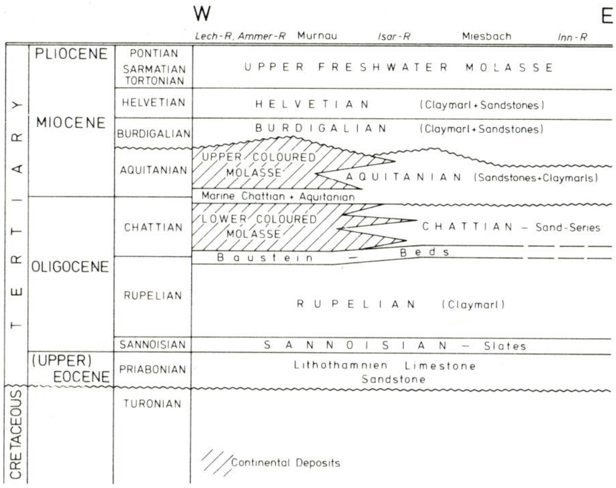

Fig. 2. Stratigraphic sequence and correlation of the Molasse in the area of the Wildcats M 1 and S 1.

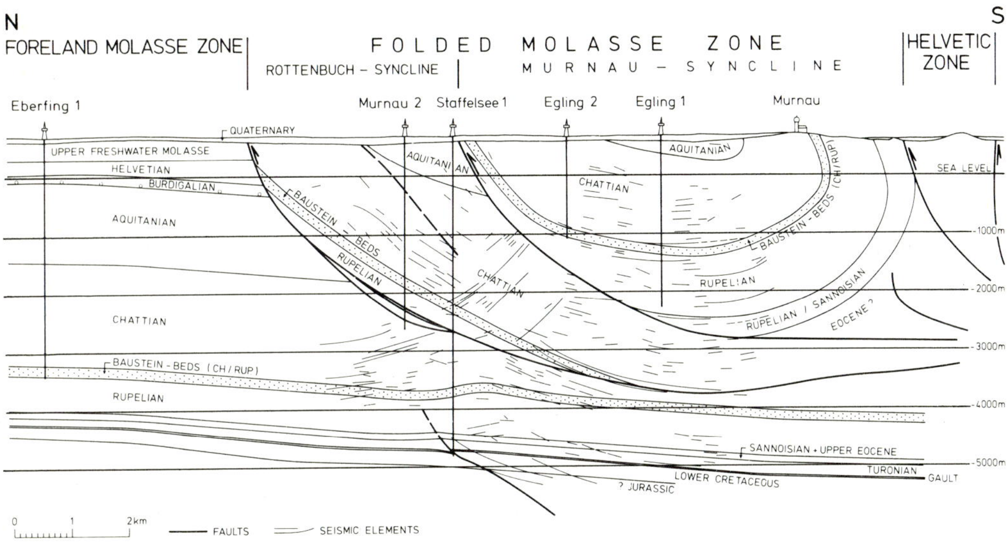

Fig. 3. N-S cross-section through the Molasse near Murnau after Müller 1970. The reflection seismic elements are only partly migrated.

lower than in other mid-European basins (Teichmüller 1975) and the maturation gradient (Rm/100 m) is even lower than in wells about 40 km north of M 1 (Jacob et al. 1977). In the upper tectonic unit of M 1 the maturation gradient of 0.006% is somewhat smaller than in the lower unit where

it is 0.008%. Details of the maturation with depth in M 1 indicate that the maturation took place before the dislocation of the Folded Molasse. From these observations a further increase of thickness of the autochthonous Molasse to south before the dislocation is inferred.

With these and other geological observations, both tectonic models, subduction or obduction are compatible. From regional tectogenetic investigations of the Molasse Fuchs (1976) postulated subduction as the fundamental tectonic process, also Frisch (this volume).

No indications could be found that the Folded Molasse had ever been covered by other nappes.

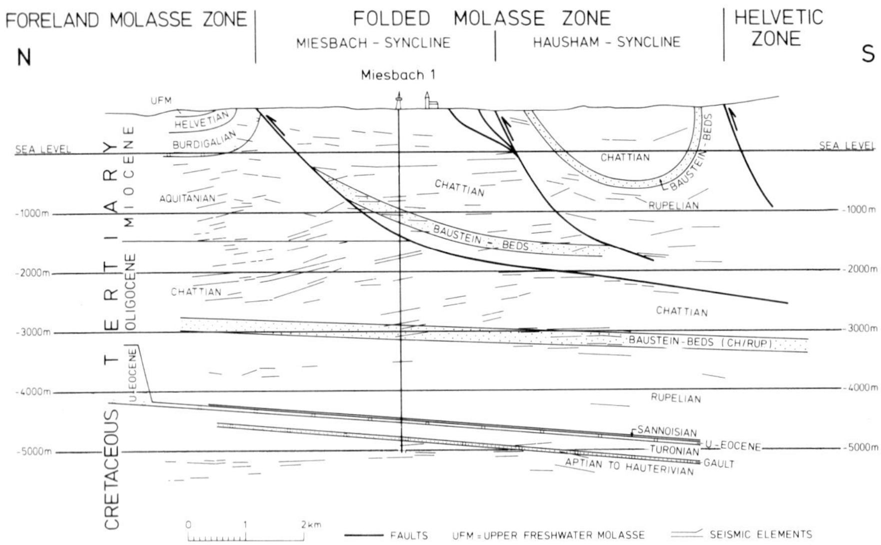

Fig. 4. N-S cross-section through the Molasse near Miesbach. After: Müller 1975; Paulus et al. 1968; Stephan et al. 1966; Veit 1973.

The Autochthonous Molasse and the Basement

The predominating tectonic features in the autochthonous Molasse are a gentle dip to south, increasing with depth, and north-dipping antithetic faults causing monoclinal structures. Below the thrustplane the extensional tectonic features of the Foreland-Molasse continue to south.

Under the northern front of the thrust, the base of the Molasse dips with 6,9% at the Miesbach meridian. There are no seismic indications showing an increase of dips under the Folded Molasse or in the mesozoic basement. This means that the Tertiary movements were the essential post-Jurassic tectonic events. The seismic reflections allow to extend this also for the adjacent southern 5 km.

At S 1 a tectonic breccia with a thickness of 30 m in the lower Cretaceous was found. Therefore small dislocations caused by the thrustings also in the mesozoic basement are probable.

A considerable increase in thickness of the Chattian (M 1) resp. the

Lower Coloured Molasse (S 1) as well as the Lower Marine Molasse (Rupelian) occurs from north in the direction to both wildcats. That means they are located still north of the axis of the Oligocene trough. In other words, the Molasse will extend considerably more to the south than presently explored. In fig. 3 and 4 the unfolded Molasse was presumed at least to extend as far as under the southern border of the Murnau- respectively of the Hausham syncline (see fig. 1). From all conclusions concerning the oligocene trough axis it follows that the original thickness of the Molasse underneath the Calcareous Alps of this day will have been some thousands of meters on the line Garmisch-Partenkirchen--Bayrischzell.

Horizontal Dislocations

The original area of deposition of the Folded Molasse may have been located at the present location of the Helvetic zone. If some allowance is made for erosion above the Folded Molasse, the Chattian could be restored to a minimum value of 13.5 km south of the well S 1, respectively 7 km south of the well M 1. Accordingly, the Chattian depocenter was located south of the Walchensee and Schliersee. The horizontal dislocation of the present southern border of the Folded Molasse amounts to a minimum of 18 km for S 1 and of 10 km for M 1 (fig. 1).

References

Fuchs, W., 1976: Jb.Geol.B.A. - 119, H. 2, Wien.
Jacob, H. & Kuckelkorn, K., 1977: Erdöl-Erdgas-Ztschr. - 93:115-124.
Kraus, L. & Paulus, B., 1962: Erdöl u.Kohle. Erdgas.Petrochem. - 15:783-790.
Lemcke, K., 1973: Geologica Bavarica. - 69:5-48.
Müller, M., 1970: Geologica Bavarica. - 63:86-106.
- , 1975: Geologica Bavarica. - 73:113-121.
- , 1975/76: Compendium:63-76, Leinfelden-Echterdingen.
Paulus, B. & Stephan, W., 1968: Erl. Geol. Karte Bayern 1 : 25 000, Bl. Nr. 8237 Miesbach:202-286, München.
Schmidt-Thomé, P., 1962: Ztschr.Dtsch.Geol.Ges. - 113:231-260.
Stephan, W. & Hesse, R., 1966: Erl. Geol. Karte Bayern 1 : 25 000, Bl. Nr. 8236 Tegernsee, 304 p., München.
Teichmüller, M. & Teichmüller, R., 1975: Geologica Bavarica. - 73:123-142.
Veit, E., 1963: Bull.Ver.Schweizer.Petrol.-Geol.u.-Ing. - 30:15-52.

Alpine Stress Documented by Anomalous Seismic Velocities in the Molasse Trough

J. Lohr

In the three Tertiary basins of Germany, very detailed determinations of seismic velocities have been made by velocity measurements in deep oil and gas wells.

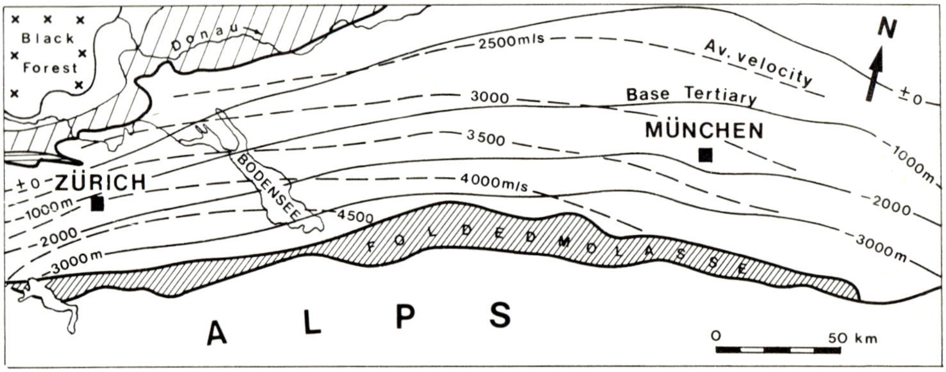

Fig. 1. Lines of equal depth of base Tertiary and lines of equal average velocity down to base Tertiary.

For the Molasse basin, fig. 1 shows that the lines of equal depth and the lines of equal average velocity are not parallel, suggesting that the normal velocity-depth function is not applicable in this region.

The velocity-depth functions for the average velocities are given in fig. 2 for the German basins. For the Tertiary of the Rheingraben and for Northwest Germany, a nearly linear function (I, II) describes the situation. However, the velocity-depth functions of the three principal areas of the Molasse basin north of the Alps (III-V) are different, although the lithology in all basins (I-V) can be taken as approximately uniform

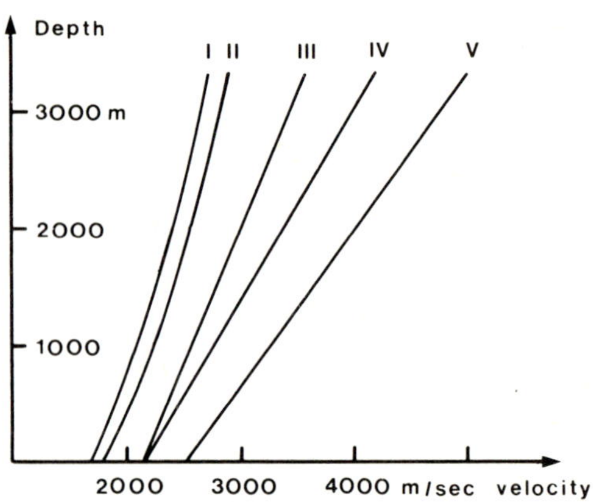

Fig. 2. Average velocities for comparable Tertiary-basins in Germany. For further explanation see text.

for these considerations (sands, clays, claystones etc.). Therefore, strong influences in addition to normal compaction must have been effective to produce these differences in the velocity-depth-functions.

The velocity behaviour of the layers of the Tertiary sediments in areas I and II is apparently caused by normal compaction. The other three curves are abnormal: The average velocity in the western Molasse-basin (V) starts near the surface with values which are normal for depths between 2500-3000 m in the basins I and II. The near-surface values for the seismic velocity in the areas III and IV are reached in the area I at a depth of only 1000 m.

From fig. 2 it can be seen that we have not only a depth gradient of the velocity and a gradient in an east-west direction (III-V) but also a gradient in the north-south direction perpendicular to the border of the Alps.

From fig. 2 it follows that e.g. for a depth of 500 m in area IV of the Molasse trough, the average velocity is of the same order as the velocity of layers at about 2500 m depth in the areas I and II. An additional lateral compression originating from the Alps corresponding to the weight of 2000 m of Tertiary sediments must be assumed, that is, a pressure of 460 bar at a specific weight of 2,3. In the same fig. 2 for 1000 m depth in area IV, there is a velocity equivalent to a weight of 3300 m sediment or 760 bar on the average.

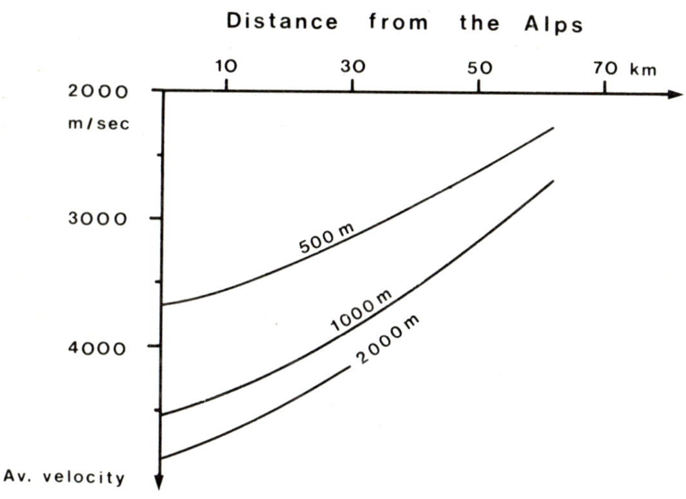

Fig. 3. Relation between velocity and distance to the Alps (for selected depths) near the Bodensee (Area IV).

These complex relations become still more plausible if one compares the velocity-depth function of area IV in fig. 2 and its variations in dependence of the distances from the Alps as given in fig. 3. The stress field was relatively strong. Probably the water in the pores of the unconsolidated sediments played an important role for the propagation of the stress field in a south-north direction. From Lemcke (this volume) we know that an overpressure in Tertiary strata had been found in wells,

with pressure increasing toward the Alps. Although this pressure is too small to produce the observed velocity anomalies, the fact of the over-pressure in the area confirms the effectual lateral compressions inferred from seismic velocities.

The lateral velocity changes in the Molasse basin III to V seem to correlate to some extent with the structure of the foothills of the Alps. In the eastern Molasse (III) the abnormality in the velocities is smallest where the Folded Molasse pinches out. On the other hand, the highest velocities are found in area V where the Molasse trough is relatively narrow and where a Variscian massif, the Black Forest, is situated close to the Alps.

The velocity anomaly due to the lateral compression was already taken into consideration by Reich, when discussing refraction measurements (1945, 1946, 1949), as well as by Brix, Götzinger, Kröll, Logigan (1963) who did not exclude the influence of the Alpine tectonics on seismic velocities in Austria. The observation of the strong influence of lateral effects is supported also by geological observations along the boundary between the Folded and the Unfolded Molasse (Albrecht & Furtak 1965; Voll 1953). In particular Lemcke (1972, 1973) commented on lateral compression in the Unfolded Foreland Molasse. According to Davis (1975) an increase of seismic velocity of Cretaceous sediments, which was observed in the neighbourhood of the disturbed belt of the margin of the Rocky Mountains in Southern Alberta, Canada, is also due to lateral compression, and it is still present in the plains at a distance of up to 100 km from the eastern limit of the folded and thrust-faulted mountains.

In conclusion, it can be said that, in spite of the great evidence of dilatational forces, the Alpine Foreland Molasse was also exposed to compressional stress having its source in the Alps. They influenced in particular the younger Molasse sediments and found their strongest expression close to the Folded Molasse.

References

Albrecht, K. & Furtak, H., 1965: Geol.Mitt. - 5:227-248, Aachen.

Brix, Götzinger, Kröll & Logigan, 1963: 6.World petrol.congr.Sect. 1 paper. - 3:247-265, Frankfurt.

Davis, Th.L., 1975: Paper 45.Intern.SEG meeting. -

Elberskirch, W. & Lemcke, K., 1955: Z.Dtsch.Geol.Ges. - 105:307-323.

John, H., 1956: Erdöl u. Kohle. - 9:290-299, Hamburg.

- , 1957: Erdöl U. Kohle. - 10:493-496, 570-573, 661-664.

Lemcke, K., 1973: Geol.Bav. - 69:5-48.

- , 1974: Ecl.geol.Helv. - 67:121-133.

Lohr, J., 1967: Geophys.Prospect. , XVII. - 2:111-125.

- , 1967: Bull.Ver.Schweiz.Petr.-Geol.u.Ing. - 34:29-38.

Lohr, J. & Nodop, I., 1955: Erdöl u.Kohle. - 8:461-464, Hamburg.

Reich, H., 1945: Verh.d.Geol.Bundesanst:55-66.

- , 1946: Die Naturwissenschaften. - 33:345.

- , 1949: Erdöl u.Kohle. - 2:81-87, Hamburg.

Voll, G., 1953: Beol.Bavarica. - 17:135-143.

Electrical Conductivity below the Eastern Alps and their Northern Foreland

A. Berktold, M. Beblo and K. Kemmerle

Along a NS-profile Chiemsee-Drautal magnetotelluric measurements have been made at 65 and geomagnetic depth soundings at 21 sites. Additional geomagnetic depth soundings have been performed at 6 sites along an EW-profile Zell am See - frontier Austria/Hungary.

Preference Direction of the Induced Electric Field

Most of the stations show a "preference direction" of the induced electric field $\tilde{E}$. In this direction the electric field has its strongest amplitudes. In fig. 1 the preference directions of $\tilde{E}$ (short lines) at most of the stations are plotted in a simplified geological map.

Along the profile Chiemsee-Drautal the major rock units have different electrical conductivities. The average resistivity $\bar{\rho}$ of the rocks is:

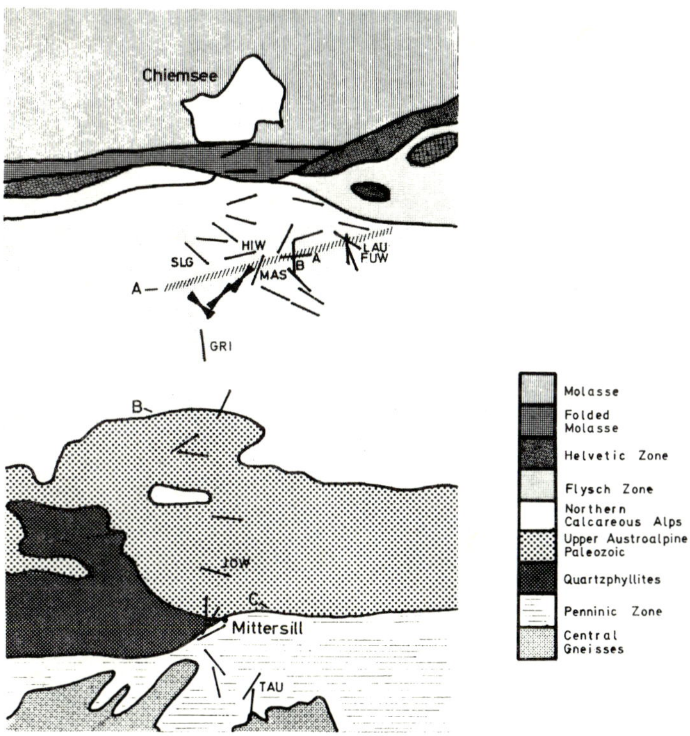

Fig. 1. The preference directions of the induced electric field in the northern part of the Eastern Alps. (For fig. 1 - 5 more information in the text.)

$\tilde{\rho} \approx 5$ - 20 Ωm in the Molasse Basin, $\tilde{\rho} \approx$ 200 - 1000 Ω m in the Calcareous Alps, $\tilde{\rho} \approx$ 300 Ωm in the Upper Austro-Alpine Paleozoic and $\tilde{\rho} \approx$ 600 Ω m in the Penninic Zone. In the southern part of the Molasse Basin the preference directions of $\tilde{E}$ are more or less parallel to the strike of the Alps (fig. 1). At the morphological border of the less conducting rocks of the Calcareous Alps the preference directions of $\tilde{E}$ remain about parallel to the average strike of the Alps as far as about 10 km south of the morphological border of the Alps (in fig. 1 marked by the hatched line "A"). From there to the south the preference directions rotate towards a NS-direction. This indicates that the southern border of the Molasse Basin is situated about 10 km south of the front range of the Alps below the nappe pile of the Calcareous Alps. The local rotations of the preference directions are caused by the lateral inhomogeneities of the electrical conductivity.

Transition Zone Molasse Basin to Calcareous Alps

For all stations the apparent resistivity ρ_a (T) has been calculated as a function of the period T. For the interpretation of the field measure-

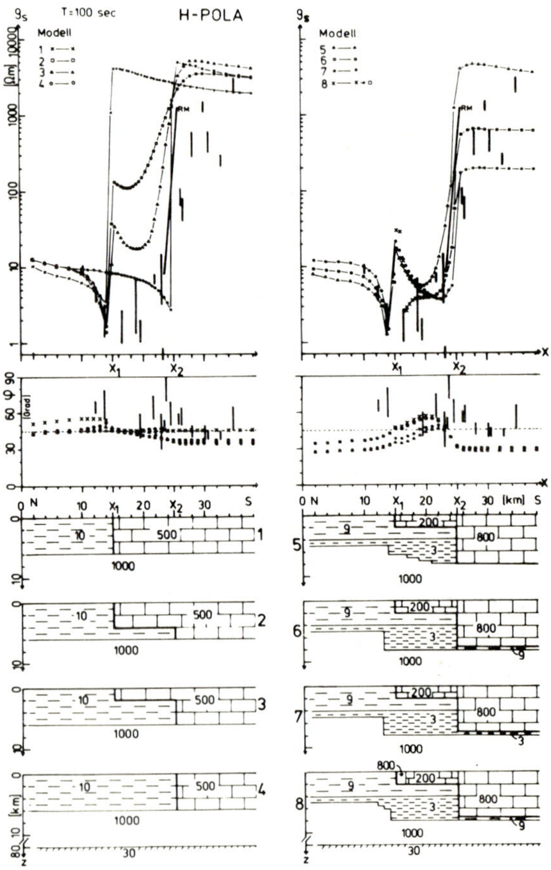

Fig. 2. Results of the H-polarization.

ments by 2-dimensional models the observed horizontal components of the electric and magnetic field were split into two components, which are parallel ($E_{\parallel}$, $H_{\parallel}$) respectively perpendicular ($E_{\perp}$, $H_{\perp}$) to the average strike of the Alps. From the ratios $E_{\parallel}/H_{\perp}$(E-polarization) respectively $E_{\perp}/H_{\parallel}$(H-polarization) the apparent resistivities $\rho_{a\parallel}(T)$ resp. $\rho_{a\perp}(T)$ were obtained as a function of the period T as well as the phase retardations $\phi_{\parallel}$ (T) and $\phi_{\perp}$ (T). The results for the H-polarization are shown in fig. 2. In the upper part 4 diagrams are shown, in which the apparent resistivity ρ_a resp. the phase retardation ϕ is plotted for the period T = 100 sec along the northern section of the NS-profile. The results of the field measurements are plotted as short vertical lines. The position X_1 on the X-axis means the morphological border of the Alps, X_2 means a position 10 km south of X_1. (In fig. 1 X_2 would be within the hatched line "A"). The results of the model calculations are plotted as different symbols connected with lines. In the lower part of fig. 2 four simple models of the conductivity distribution along the northern section of the profile are shown on the left side and four more differentiated models are shown on the right side. The numbers in the models are the resistivities of the different rock units in Ω m (rocks of the Molasse Basin including Flysch and Helvetikum: left side 10 Ω m, right side 9 and 3 Ω m; rocks of the Calcareous Alps: left side 500 Ω m, right side 200 and 800 Ω m; rocks of the cristalline basement: 1000 Ω m). Below the depth of 80 km a resistivity of 30 Ω m has been assumed in all models. The observed phase retardations are about 20° smaller than the calculated phase retardations, but the spatial variations along the profile are rather similar. The models 3 and 4 on the left side and the model 8 on the right side best fit the results of the field measurements. This is also confirmed by the results of the E-polarization.

In a special field program the distortion of the induced electric field by local conductivity anomalies near the earth's surface has been investigated. As far as possible the local distortions have been eliminated.

In fig. 3 the results of geomagnetic depth sounding along the northern section of the profile (Chiemsee - Reit im Winkl) are compared with results of model calculations. The models are similar to those of fig. 2. Above basins filled with well conducting sediments, such as the Molasse Basin, the amplitudes of the geomagnetic variations are increased in comparison with the amplitudes of the same variations in the surrounding area as additional currents are induced in the well conducting sediments. Since these currents are flowing parallel to the strike of the basin, that means nearly in East-West-direction in the Molasse Basin, the superimposed magnetic field has been measured in the NS-component $\tilde{H}$ which is perpendicular to the strike. The EW-component $\tilde{D}$ is not affected by the additional electric currents flowing in the Molasse Basin. So for the interpretation of the geomagnetic depth sounding the amplitude ratios $\tilde{H}/\tilde{H}_n$ and $\tilde{Z}/\tilde{H}_n$ were calculated as a function of the period T. $\tilde{H}$ is the NS-component of the magnetic variations at the respective site, $\tilde{H}_n$ that at a base station, $\tilde{Z}$ is the vertical component at the respective site.

From the magnetotelluric measurements and the geomagnetic depth sounding we conclude that

a) the Calcareous Alps are underlain in the northern section of the profile by well conducting sediments (Molasse, Flysch, Helvetikum) up to about 10 km south of the morphological northern border of the Alps,

b) the well conducting sediments below the Calcareous Alps are thicker

than the Calcareous Alps themselves,
c) the southern border of the well conducting sediments is rather abrupt.

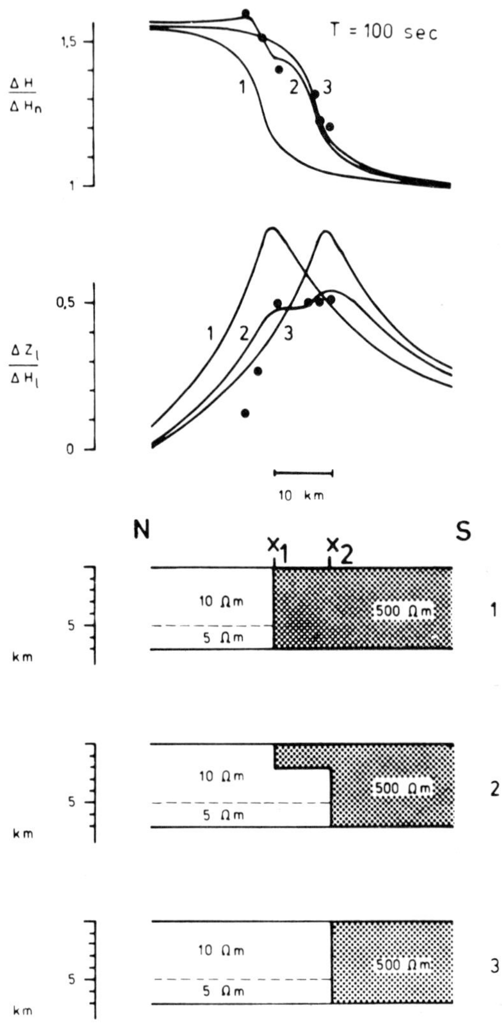

Fig. 3. Results of geomagnetic depth sounding along the northern section of the profile.

The Conductivity Distribution below the Eastern Alps

Below the Eastern Alps the distribution of the electrical conductivity was determined to get information about the distribution of temperature as the electrical conductivity of rocks in greater depth is a function of temperature. From the magnetotelluric measurements on the line Chiemsee - Hohe Tauern the apparent resistivity $\rho_a(T)$ and the phase retardation ϕ (T) were determined for each station as described above. Some

2-dimensional models were calculated. No clear indication could be found for an increased electrical conductivity which would indicate a temperature anomaly in the crust below the Eastern Alps.

Geomagnetic depth sounding has been made along the NS-profile Chiemsee - Drautal, along a short NS-profile crossing the Zillertal Alps and along an EW-profile Zell am See - frontier Austria/Hungary. The location

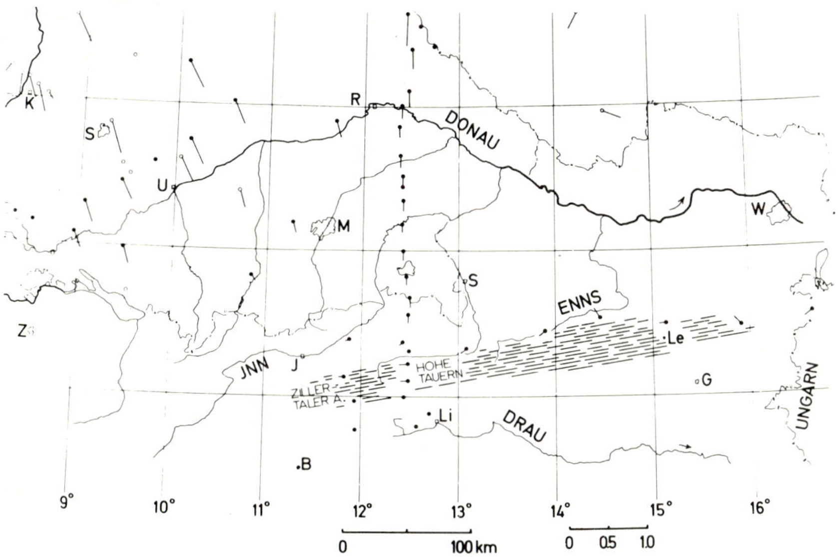

Fig. 4. Induction arrows for periods of about 20 Min.

of the stations is shown in fig. 4. In this figure *Wiese* induction arrows are plotted for periods of about T = 20 min. These arrows point away from zones of increased electrical conductivity. The length of the arrows represents the amplitude ratio $\tilde{Z}/\tilde{H}_c$, where $\tilde{H}_c$ is the horizontal component which is maximally correlated with $\tilde{Z}$. The length of the shown induction arrows is rather small and constant in the whole Eastern Alps compared for instance with the induction arrows north of the river Donau. The direction of the arrows varies from site to site.

In fig. 5 the amplitude ratio $\tilde{H}/\tilde{H}_n$ is plotted along a NS-profile Donau - Chiemsee - Drautal for the periods 100, 500, 1000 and 5000 sec. The amplitudes of the magnetic variations are clearly increased in the area of the well conducting Molasse sediments but there are only small local increases within the Eastern Alps such as in the Hohe Tauern and at the Periadriatic Line near Toblach. Figs. 4 and 5, corroborating the results of the magnetotelluric measurements, show that below the Eastern Alps there are only some weak and local conductivity anomalies but no indication of a significantly increased electrical conductivity.

Acknowledgements

We are grateful to Prof. Dr. G. Angenheister for his interest and support and we have to thank Dr. V. Haak for using his computer programs. The computing has been made at the Leibniz-Rechenzentrum of the Bavarian Academy of Sciences.

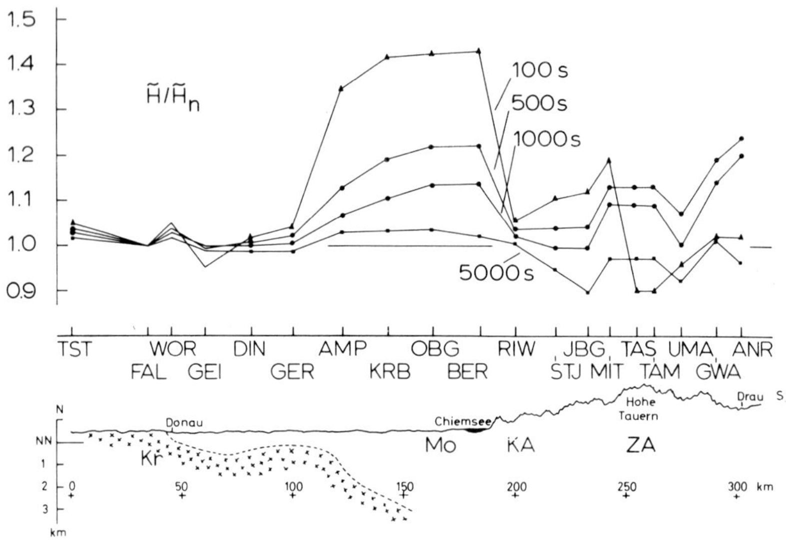

Fig. 5. The amplitude ratio $\tilde{H}/\tilde{H}_n$ on a N-S profile crossing the Molasse trough and the Eastern Alps for T = 100, 500, 1000, and 5000 sec. Kr = crystalline basement of the northern foreland of the Alps, Mo = Molasse, KA = Calcareous Alps, ZA = Central Alps.

References

Beblo, M., 1974: Protokoll Erdmagnetische Tiefensondierung:201-219, Grafrath.

Berktold, A., 1974: Protokoll Erdmagn. Tiefensondierung:58-70, Grafrath.

Berktold, A., Beblo, M., Kemmerle, K., 1976: Geol.Rundschau, 65:715-732.

Kemmerle, K., 1974: Protokoll Erdmagn.Tiefensondierung:220-240,Grafrath.

Interpretation of the Berchtesgaden Magnetic Anomaly

U. Bleil and J. Pohl

A large positive magnetic anomaly with maximum values of 120 gamma exists in the area of Berchtesgaden (Bleil & Pohl 1976). The strike direction is about 120°, the half-width normal to the strike about 40 km. A charasteristic feature of the anomaly is the absence of a pronounced negative minimum in the north. A southern minimum may at least in part be due to a negative anomaly in the Hohe Tauern. From regional surveys in the Bavarian and Austrian Alps it appears that anomalies similar to the Berchtesgaden anomaly can be found both to the east and west of the surveyed area. The total EW extension of these anomalies, which are aligned approximately parallel to the northern margin of the Eastern Alps, is about 150 km.

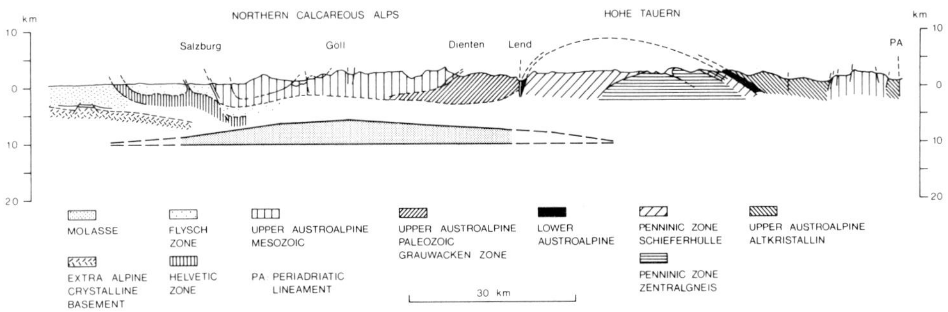

Fig. 1. Cross-section through the Eastern Alps after Bögel (fig. 7, Profile 1 in Angenheister et al. 1972, simplified) showing the location of a possible magnetic model body with respect to tectonic units.

Two-dimensional models were calculated for a profile normal to the strike. A maximum depth of the magnetic body is given by the Curie depth for magnetite at about 20 km if a normal temperature gradient of 30°C/km is assumed. An upper limit is probably given by the thickness of sediments in the surveyed area, estimated to be about 5-6 km.

In models assuming a magnetization in the direction of the present geomagnetic field (D = 0°, I = 60°), the magnetic body would have to extend 100 to 150 km to the north of the center of the anomaly in order to suppress the strong minimum on the north edge of the anomaly in the model curves. Such an extent of the body is geologically improbable.

With a south-dipping magnetization (D = 180°, I = 45°) good fits with the measured anomaly can be obtained. Models calculated with this direction of magnetization and a homogeneous magnetization intensity of 10^{-3}G have approximately trapezoidal cross-sections, upper boundaries between 5 and 9 km and lower boundaries between 13 and 20 km. Higher magnetization intensities result in similar model bodies with reduced thicknesses. With a magnetization of $2 \cdot 10^{-3}$ G it is possible to raise the lower boundary to about 10 km. The maximum elevation of the magnetized body is then about 5.7 km below the surface. In all cases the model bodies have a large

horizontal extension compared with their thickness. The NS extension along the profile is about 70 km, with the northern boundary of the magnetic body located at about 10 km NNE of Salzburg. The south-dipping direction of magnetization used in model calculations could be interpreted as resulting from a reverse late Paleozoic or early Mesozoic component and an additional component of similar intensity in the direction of the present field.

From the magnetization required for a satisfactory fit of computed and measured anomalies some conclusions about the magnetized material can be drawn. A comparison with mean values for susceptibility, remanent magnetization and Königsberger ratios suggests mafic and possibly in part ultramafic rocks as the most probable magnetic source layer.

A layer of mafic and perhaps ultramafic material in this area appears to be confirmed by the similarity between the gravity effect of the magnetic models and the Bouguer gravity data (Makris 1971; Angenheister et al. 1972). For calculation of the gravity effect a density contrast of 0.1 and 0.2 g/cm^3 with the less magnetic and less dense surrounding rocks was assumed. The comparison indicates that models with a thickness of the order of 5 km are more probable than models with greater thickness (fig. 1).

Two interpretations are possible:
1) The magnetized layers could belong to the metamorphic complex of the Bohemian Massif extending below the Northern Calcareous Alps. In this case, irrespective of the origin of these layers, there would be no relation between the magnetic anomaly and Alpine orogeny. However, anomalies of similar extension are not known elsewhere in the Bohemian Massif.
2) Another tentative interpretation is to consider the magnetic sources as an ophiolite suite and thus possibly as part of an ancient ocean floor at the northern boundary of the Alps. In this case plate tectonic models proposed for the geodynamical evolution of the Eastern Alps should be considered (e.g. Hawkesworth, Waters & Bickle 1975). The mean magnetic properties of ophiolites as measured e.g. by Vine & Moores (1972) on the Troodos Igneous Massif would be in agreement with the magnetization adopted in the model calculations. It has to be noticed, however, that the seismic velocities are lower in this part of the Alps than in the average oceanic crust (Angenheister et al. 1972; Fox, Schreiber & Peterson 1973).

References

Angenheister, G., Bögel, H., Gebrande, H., Giese, P., Schmidt-Thomé, P. & Zeil, W., 1972: Geol. Rdsch. - 61:349-395.

Bleil, U. & Pohl, J., 1976: Geol. Rdsch. - 65:756-767.

Fox, P.J., Schreiber, E. & Peterson, J.J., 1973: J. Geophys. Res. - 78:5155-5172.

Hawkesworth, C.J., Waters, D.J. & Bickle, M.J., 1975: Earth Planet. Sci. Lett. - 24:405-413.

Makris, J., 1971: Hamburger Geophys. Einzelschr., H. 15, 65 p.

Vine, F.J. & Moores, E.M., 1972: Geol. Soc. America Mem. - 132:195-205.

Upper Crustal Structure in the Northern Alps Derived from Seismic Refraction Measurements

M. Will, H. Miller and H. Gebrande

From earlier seismic measurements the existence of Flysch, Helvetic, and Molasse units underneath the Northern Calcareous Alps was known (Angenheister et al. 1972). To determine the thickness and southward extension of these units the northern part of the Geotraverse Ia has been explored by detailed seismic refraction measurements. Fig. 1 shows the study area and the location of refraction lines.

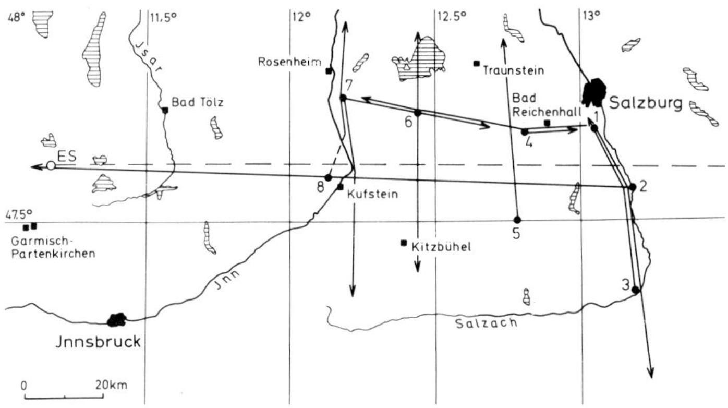

Fig. 1. Location of refraction lines in the northern part of "Geotraverse Ia". The numbered heavy dots are shot points: 1 Gartenau, 2 Golling, 3 St. Johann i.P., 4 Antoniberg, 5 Lofer, 6 Marquartstein, 7 Rohrdorf, 8 Kiefersfelden. The arrow-heads indicate the most distant recording sites for the different refraction lines. The dashed line represents the profile Eschenlohe-East which has been measured within an earlier project.

Measurements were carried out along 11 profiles originating at 8 different quarries and using quarry blast as a seismic source. The profiles lie mostly within the Northern Calcareous Alps but some of them extend northward into the Flysch and Helvetian zone and into the Molasse-basin (Vorland-Molasse) and some extend southward into the Central Alps (Penninic zone). Generally the refraction lines are about 60 km long. One profile (Golling-W) attains 160 km in length.

Many of the observed travel time curves are markedly affected by horizontal variations of seismic velocity. Therefore results obtained with the conventional assumption of horizontally layered media are only first approximations. Improvements involved iterative model calculations using a ray tracing method (Will 1975) developed for this project. With this method ray paths and travel times for two-dimensional inhomogeneous media may be calculated. The models may consist of any number of layers, the boundaries of which are arbitrarily formed polygons. Within a layer the seismic velocity is either constant or is defined by a velocity at the top and a constant gradient of arbitrary direction. Thus models with

nearly any kind of two-dimensional velocity distribution can be constructed. By comparing observed and computed travel times one improves the model with appropriate alterations. This process is repeated until a reasonably good fit between calculated travel times and the significant features of the observed travel time segments is reached. The confidence limits of the final model not only depend on the fit between calculated and observed travel times but even more so on the quality and amount of the original data. A system of well documented reverse and overlapping travel time curves will inspire more confidence in the model than data from a single shotpoint.

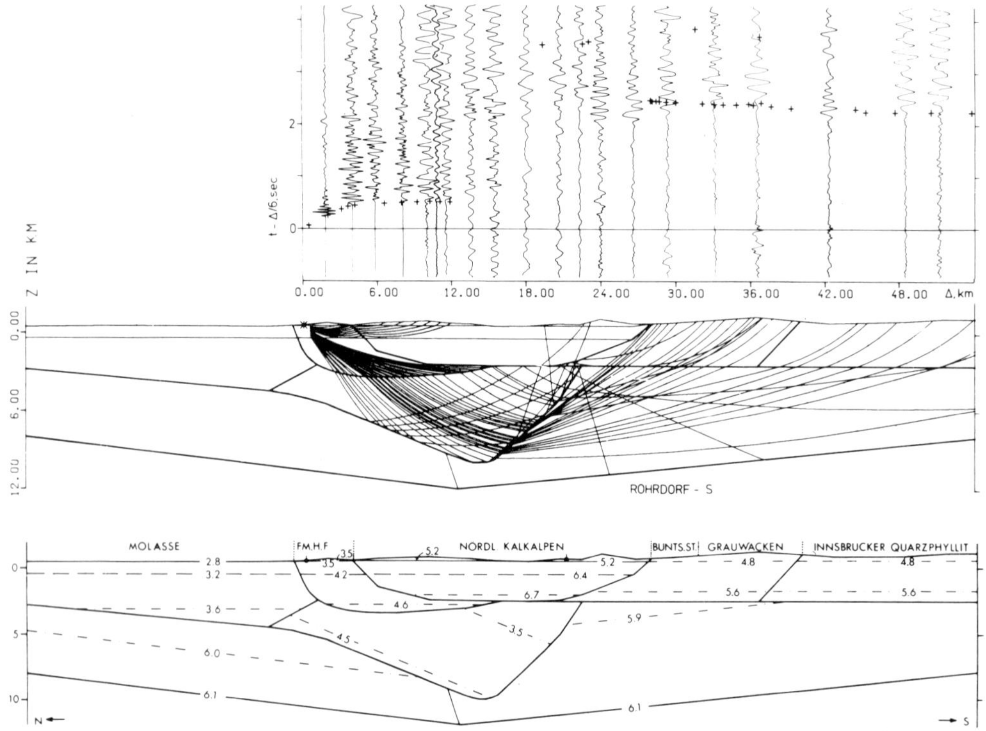

Fig. 2. Results for the profiles Rohrdorf-South and Kiefersfelden-North. Lower part: Final two-dimensional model. Dashed lines are lines of equal p-wave velocity with values in km/sec. Solid lines may partially be interpreted as boundaries of lithologic units. FM = Folded Molasse, H = Helvetian, F = Flysch. Middle part: computed raypaths for the model shown below and a shotpoint at the location of the quarry near Rohrdorf. Upper part: Record section of the refraction line Rohrdorf-South with computed travel times (crosses).

As an example fig. 2 shows the results for the profile Rohrdorf-S. The lower part shows the final model in detail. Solid lines separate zones which in part may be interpreted as lithologic units. P-wave velocities within these zones are defined by dashed and dot-dashed lines of equal velocity. The velocity gradients are constant between adjacent lines of

equal velocity and change at dashed lines within any particular zone. The middle part shows the calculated ray paths whose travel times are marked by crosses in the record section shown in the top part of the figure.

The essential features of this model are considered as representative for most parts of the study area. In particular the surprisingly high velocities of p-waves (up to 7 km/sec) within the Northern Calcareous Alps have been established on most of the other profiles. On profiles measured within the Northern Calcareous Alps these high velocities cause a shadow zone and usually prevent the gathering of information about layers with velocities less than 7 km/sec below the Northern Calcareous Alps. Applying the Herglotz-Wiechert method to the travel time curves of these profiles up to the beginning of the shadow zone, we can infer a general thickness of 3 km for the Northern Calcareous Alps. This thickness has been incorporated in the model presented in fig. 2 as a fixed boundary condition.

The record section of fig. 2 shows the sudden disappearance of energy in the first arrivals at a distance of some 12 km. This is due to the high velocity of the limestones and due to the steep dip of their northern boundary acting as a reflector for rays with certain angles of incidence. For reasons of clarity these particular rays are not drawn in fig. 2. A new travel time segment can be observed at greater distance with a marked time lag (reduced travel time 2.2 sec). These arrivals can be explained by rays penetrating the low velocity zone beneath the Northern Calcareous Alps. A satisfactory agreement between calculated and observed travel times could only be reached by assuming a rather steep dip for the wedging out of the low velocity zone. This would also be in good agreement with results from magnetotelluric measurements (Kemmerle 1973; Berktold et al. this volume).

The rather low mean p-wave velocity of this layer of about 4 km/sec is a strong indication for poorly consolidated sediments. They may consist of Flysch and/or Helvetian and/or Molasse series. The maximum thickness of these sediments under the Northern Calcareous Alps is about 7 km and they extend southward for at least 15 km from their northern margin.

The model presented will not necessarily be the last one. The record section shows some arrivals in the distance range of 12 to 26 km which may possibly be explained by converted waves (p-s waves, which could penetrate the limestones as s-waves). They could also be caused by p-waves incident from below which could penetrate the Northern Calcareous Alps because their base may not be an ideally reflecting planar boundary but a somewhat corrugated interface.

Although further refinements may be possible in the future we feel that the model presented here will not have to be changed very much with respect to its main features. The existence of a rather thick sedimentary layer beneath the Northern Calcareous Alps is a boundary condition which must be satisfied by any successful geodynamic model of Alpine orogeny. Furthermore the low seismic velocity of these sediments suggests that they may have retained enough porosity to be prospective for hydrocarbon accumulations.

References

Angenheister,G., Bögel,H., Gebrande, H., Giese,P., Schmidt-Thomé, P. & Zeil, W., 1972: Geol.Rdsch. - 61:349-395.

Kemmerle, K., 1973:Dipl.-Thesis,Inst.f.Allg.u.Angew.Geoph.Univ. München.

Will, M., 1975: Diss.Universität München, 145 p.

Seismicity Studies in the Northern Part of the Eastern Alps

O. Förtsch, K. Koschyk and E. Schmedes

In the northern part of the Alps and the folded molasse only a small number of earthquakes is recorded and located by the international seismological surveys. However, the seismic activity in this region consists of a high number of microearthquakes, which may be recorded by mobile stations set up in the area under investigation. These microearthquakes can only be caused by the still active stress field of the Alpine orogeny. Two seismicity studies were carried out in the folded molasse (region of Peißenberg) and in the northern part of the Eastern Alps.

The Earthquakes in the Region around Peißenberg (fig. 1)

Between 1962 and 1971 about 100 earthquakes were recorded. Most of them were microearthquakes which were locally perceptible but were not destructive. Nearly all earthquakes were recorded by the seismological station Fürstenfeldbruck (FUR) and the stronger ones by most of the Middle-European stations. In order to localize the shocks, mobile seismological stations were set up in the region of Peißenberg.

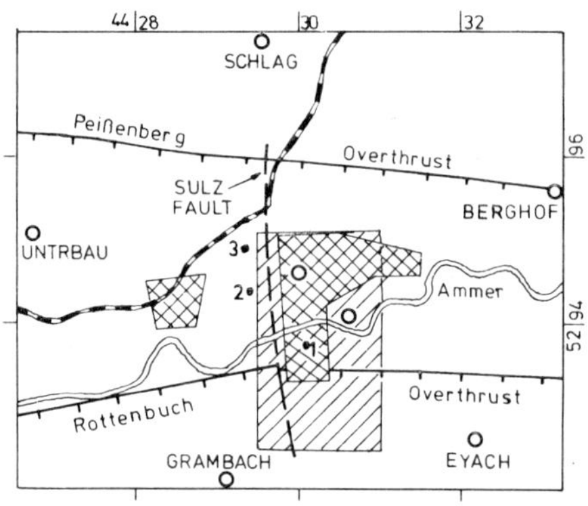

Fig. 1. Region of Peißenberg with epicenter area (hatched area) and mining district (cross-hatched area) of the coal mine from 1961 to 1970. Points 1 - 3 give epicenters located with 4 stations. (After Koschyk 1973).

Peißenberg is situated on the northern part of the Peißenberg-fault. Coal was mined at Peißenberg mine until 1971. The earthquakes were felt most strongly in that part of Peißenberg which is situated over the eastern mining field of the coal mine, but even the stronger shocks did not cause any destructions or deformations in the mine. Calculations showed that all hypocenters lie at a depth between 1000 and 2000 meters and that depth increases with increasing energy. The epicenters of the located earthquakes generally lie in the region of the eastern mining field, where coal was exploited at that time (see fig. 1).

By subsidence above the extracted coal seams rock bursts were caused in the mine. These rock bursts, however, were never recorded in Fürstenfeldbruck (FUR). The vibrations of these rock bursts are of such high frequencies that they are soon attenuated by absorption.

Thus we can conclude that the shocks in Peißenberg are tectonic earth-

quakes with hypocenters below the coal mine in the neighborhood of the Peißenberg thrust. They must be caused by the pressure of the stress field of the Alps. However, the coincidence with the mining activity at Peißenberg suggests a certain triggering effect of the mine. The stress field must be deformed in the region around the coal mine and therefore fracturing may be facilitated. By the small earthquakes the existing stress was lowered, possibly preventing a destructive earthquake in this region.

Seismicity Studies in the Northern Part of the Eastern Alps (fig. 2)

Between January 1972 and December 1975, about 310 local earthquakes were recorded by the seismological station Schlegeis (SCE). Most of them were microearthquakes, which were not localized by the international seismological surveys. Therefore four other stations were set up in Bad Reichenhall (BHG), Garmisch-Partenkirchen (GAP), Innsbruck (IBK) and Obergurgel (OGA). Since March 1973 about 30 earthquake epicenters could be localized. These epicenters show about the same distribution as the epicenters of the larger earthquakes which were reported in this century by the in-

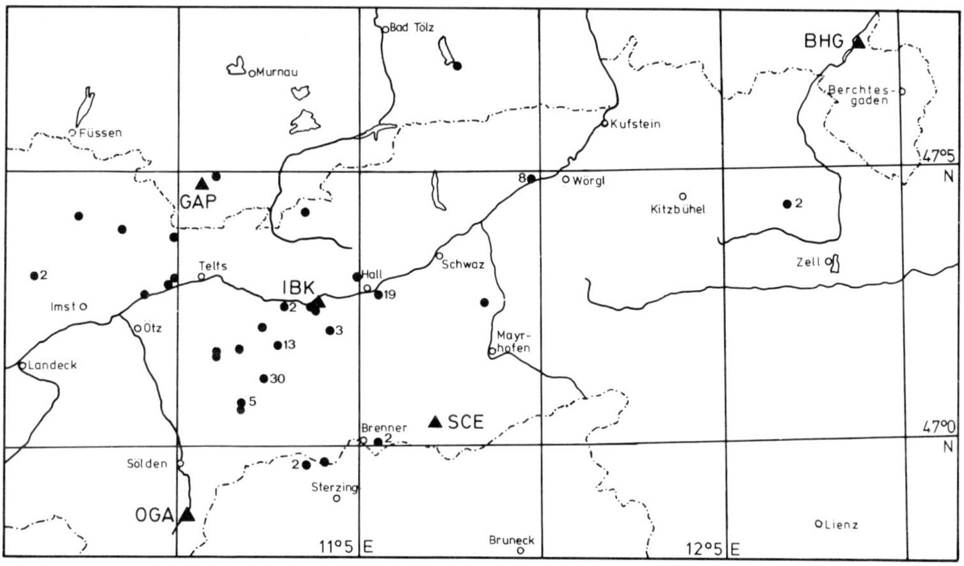

Fig. 2. Location of seismological stations BHG, GAP, IBK, OGA and SCE and epicenters of earthquakes localized from March 1973 to December 1975. Numbers indicate earthquake swarms and give the number of events for each swarm.

ternational surveys. From the epicenter map (see fig. 2) we can see the following active regions: the region of Innsbruck, the Mieminger Plateau, the Lechtaler Alps, the region Brenner - Sterzing and the Northern Calcareous Alps between Kufstein and Berchtesgaden.

Very often swarms of microearthquakes (up to 30 microshocks were re-

corded within a few hours) are observed, which shows that in this region tectonic stresses are preferably reduced by small shocks rather than by large ones.

For two events, macroseismic investigations were carried out. In both cases maximum intensity IV on the MSK 1964 scale was reached. The areas of perceptibility had a seize of 8 and 4 square kilometers only, which proves that the hypocenters must have been near the surface.

We conclude that the investigated region is still tectonically active and that by microearthquake-studies active areas can be determined within a relatively short time.

References

Förtsch, O. & Schmedes, E., 1976: Geol. Rdsch., 65:748-755.

Koschyk, K., 1973: Diss. Univ. München.

Part 2

The Northern Calcareous Alps and the Southern Alps

Paleo-oceanographic Aspects of the Upper Cretaceous Geosynclinal Sediments of the Eastern Alps

A. Butt and D. Herm

Introduction

Syndepositional lateral and vertical movements affected the Alpine areas already before the uplift of dislocated nappes dated by the Paleogene beginning of Molasse sedimentation. Crustal shortening during the late Cretaceous created narrow and steep-flanked basins, such as deep-sea trenches with Flysch sedimentation, narrow shelves, and submarine platforms merging laterally with marginal seas. This paleogeographic setting is inferred from northeast-Alpine data, but is typical of the entire northern Tethys during the Upper Cretaceous.

The present paper represents a review of a paleoenvironmental study carried out by Butt (1976). A reconstruction of the paleoenvironment is based on biostratigraphic correlation, the ecologic significance of some foraminifera and on the correlation of lithologic and geochemical data, mainly on calcium carbonate.

In the recent oceans, the biogenic components of the sediment on the sea-floor reflect the pelagic productivity of the overlying water masses. This relation of productivity and sedimentation to depth is complicated in near-shore basins by dilution of pelagic sediments due to terrigenous influx, in offshore oceanic regions by dissolution of the pelagic carbonate below calcite compensation depth (CCD), and in the trenches and abyssal plains by tectonically controlled turbidity currents.

By comparing the results of the fossil sediments of the Alps with that of recent analogous oceanographic data of the sediment in cores and in suspension (DSDP 1968-1976; Lisitzin 1972; Hsü & Jenkyns 1974; Sliter et al. 1975) arguments are presented that the litho- and biofacies of the Eastern Alpine Upper Cretaceous basins were mainly controlled by the tectonic evolution in their relation to the open sea.

A paleobathymetric analysis of the Flysch Zone (Hesse & Butt 1976) is instrumental in reconstructing the paleoenvironments in the northern part of the Eastern Alps.

Paleontological and lithologic characteristics of the sediments and their depositional environments

Coniacian to Maestrichtian sediments are present, from north to south, in the paleogeographic zones of the Helvetikum, Ultrahelvetikum, Marginal Flysch, Flysch, and Gosau, but exposed sediments represent only relicts of the ancient sedimentary basins.

Data gathered from the typical sections of the zones are used to interpret the local depositional environments of the basins as shown in fig. 1, whereas their lateral paleobathymetric and syndepositional tectonic settings are demonstrated in fig. 2.

Helvetikum Zone

Santonian to Maestrichtian are represented by a sequence of marlstones (Amdener Formation), glauconitic sandstones (Stallauer Formation), calcareous marlstones (Prinswanger & Pattenauer Formations), silty marl-

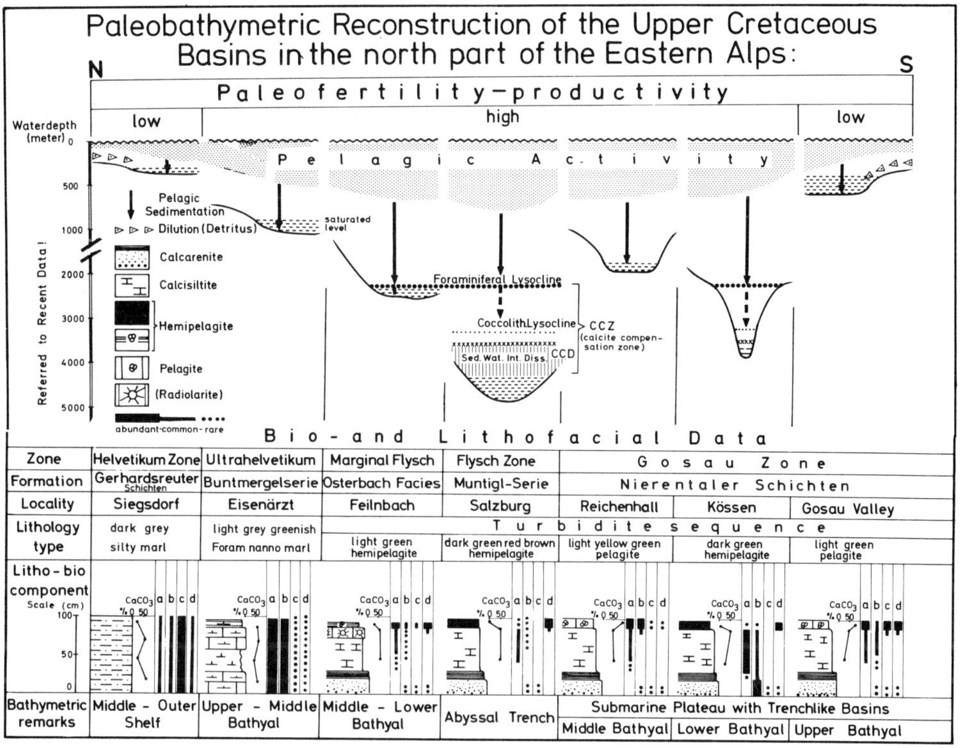

Fig. 1. Schematic paleobathymetric setting and pelagic productivity of the overlying water mass. Comparing the bio- and lithologic data with analogous oceanographic data from the Recent, dilution and dissolution relation of the Upper Cretaceous basins are shown. The Helvetikum and Gosau-Valley basins show dilution of the carbonate due to detritus influx. The Ultrahelvetikum and Reichenhall basins show saturation levels with regard to calcium carbonate. But the Marginal Flysch, Flysch, and Gosau of Kössen basin indicate different levels of calcium carbonate dissolution.
a) Coccoliths; b) Planktonic foraminifera; c) Calcareous benthonic foraminifera; d) Agglutinant benthonic foraminifera.
Note: The Kössen basin was originally situated North of the Reichenhall basin!

stones (Gerhardsreuter Formation), and mica-rich sandstones (Haçhauer Formation). Lithofacies and biofacies indicate open shelf conditions.

Fluctuating ratios of planktonic to benthonic foraminifera and of calcium carbonate content suggest an instability of the water depth and of the associated sedimentary regimes. Lower Campanian and Upper Maestrichtian glauconitic and mica-bearing sandstones suggest an extreme shallowing of the shelf, with shallow water mollusks remaining as the sole indicators of the depositional environments, reflecting near-shore shelf to inner-shelf conditions (25-50 m). During the Middle Campanian to Lower Maestrichtian and the Middle Maestrichtian calcareous marl- and silty-marlstones were deposited, in which the ratio of planktonic to benthonic foraminifera, and the relative abundance of calcareous benthonic foraminifera (nodosariids, rotaliids, buliminids) suggest middle to outershelf conditions (100-300 m).

Ultrahelvetikum Zone

Santonian to Maestrichtian are represented by a series of varicolored pelagic marlstones (*Buntmergelserie*). Planktonic foraminifera confirm that ultrahelvetic sedimentation continued during the Upper Cretaceous without hiatus, although the sediments at present are exposed in small discontinuous outcrops. In contrast to the Helvetic sediments, the Ultrahelvetic sediments are free of coarse-grained detrital material, and the bulk of the sediment is made up of foraminifera-bearing biomicrite in which the fine-grained matrix is mainly composed of coccolith plates and calcareous mud. The foraminifera consist of 98% planktonics and 2% benthonics, including agglutinants such as *Glomospira* and calcareous forms such as *Gyroidinoides* and *Osangularia*. Biofacies and lithologic data indicate an upper to middle bathyal depth (700-1000 m), suggesting a marked southward deepening of the Helvetic shelf. These bathyal basins of the Ultrahelvetikum Zone are interpreted as the upper portion of the northern continental slope in this region of the Tethys.

Marginal Flysch Zone (Butt 1976)

The sediments of the Marginal Flysch Zone display an abrupt facies contrast with the Helvetic and Ultrahelvetic rocks. They are mainly composed of *wildflysch*, that is pebbly mudstone and coarse- to medium-grained turbidite sediments, which were deposited on a paleoslope, deeper than and south of the Ultrahelvetic bathyal areas. This paleoslope presumably was part of the northern continental slope of Tethys. The wildflysch and conglomeratic turbidite sediments indicate the tectonic fragmentation of the continental slope, with chaotic debris being transported into the marginal area of the Flysch basins through submarine canyons of the slope region. Medium- to fine-grained turbidite deposits (Osterbach section) consist of interbedded calcarenites, calcisiltites, calcilutites, and hemipelagites indicating a midslope environment. On the basis of calcium and microfossil contents of the hemipelagite (hemipelagic claystone), Hesse & Butt (1976) suggested that this midslope basin was located above CCD. The hemipelagic claystone contains a moderate amount of calcium carbonate, rare and poorly preserved planktonic foraminifera, and abundant coccoliths (fig. 1). Although the turbidite basin was located above CCD.

scarce and poorly preserved planktonic foraminifera apparently indicate that the basin was located below the foraminiferal lysocline (Butt 1976). Other bicomponents of the hemipelagic claystone are radiolarians, fish debris, and benthonic foraminiferas including agglutinants and calcareous species (fig. 2). Faunal and floral composition of the hemipelagic claystones are indicative of middle to lower bathyal depth levels of the basin (1000-2000 m).

The interpretation of the coarse grained turbidites in the Marginal Flysch zone as lower slope deposits contradicts the previous interpretation by Hagn (1960) who inferred a conglomerate-furnishing *Cetic Ridge* separating the Ultrahelvetic trough and the Flysch trough.

Flysch Zone

This zone represents an oceanic-trench environment in front of northward advancing Austroalpine nappes. This trench which was filled entirely by turbidite sediments resulted from subduction of the Penninic oceanic basin during the Cretaceous (fig. 2). Santonian to Maestrichtian sediments (Piesenkopf, Kalkgraben, Oberste Bunte Mergel, Bleicherhorn Formations in the north; Zementmergelserie in the south) indicate turbidity-current environments, revealing differences in the supply of material, rate of sedimentation, and proximal versus distal relation of the sediment source and the Flysch basin (Butt 1976). The turbidites consist of interbedded calcarenite, calcisiltite, calcilutite, and hemipelagite litho-units. The calcarenite layers had neritic sources, whereas the fine-grained calcisiltites (light gray mudstone), and the calcilutites (light green claystones) containing pelagic ooze are indicative of bathyal sources. The hemipelagites (olive to dark green and redbrown hemipelagic claystones) on the other hand, are indicative of sea-floor environments of the Flysch basin itself. In contrast to the turbidite litho-units (such as the calcisiltite: gray mudstones), which display a relatively high calcium carbonate content contain abundance calcareous fossils, the hemipelagites are typically calcite-free and consist almost exclusively of deep-sea agglutinating foraminifera (fig. 1). Calcareous fossils such as planktonic and benthonic foraminifera are almost absent, but occasionally rare and poorly preserved specimens of coccoliths may be preserved. Fish debris and some pyritized radiolarian specimens are also occasionally present. The biologic constituents of the hemipelagite correspond to those found in the recent deep-sea, calcite-free abyssal clays. Comparing the results of the Upper Cretaceous turbidite sediments of the Eastern Alps with the recent oceanographic data (DSDP), Hesse & Butt (1976), suggested that the Flysch basin of the Eastern Alps was located below the calcite compensation depth (CCD). Accordingly, syndepositional structural position, presence of fine-grained pelagic ooze, calcite-free hemipelagic clay, relative abundant agglutinant foraminifera, and presence of coiled and meandering trace fossils, Butt (1976) suggests that the Flysch basin of the Eastern Alps indicates abyssal trench environments (3500-5000 m). Hesse (1974) presented similar views on the Early Cretaceous turbidite sediments of the Eastern Alpine Flysch Zone. Reviewing the paleobathymetric aspects of the Monte Antola Flysch sediments, Scholle (1971) also arrived at similar conclusions.

Gosau Zone of the Northern Calcareous Alps

This zone comprises biostratigraphic stages (Coniacian to Maestrichtian) with the same facies types locally continuing into the Early Tertiary. Exposures of the Gosau Zone are small, sporadic, located on the higher nappes of the Lechtal-, Inntal-, Staufen- and Berchtesgaden-units. The unit unconformably overlies the Triassic and Jurassic carbonate complexes in the north. In the south, however, the Gosau sediments have transgressed on the rocks of the *Altkristallin* (Paleozoic, metamorphosed crystalline rocks). Rocks of the Gosau Zone represent a spectrum of synorogenic environments confined to several sedimentary basins. On basis of a facial comparison of the Gosau and Flysch Zones of the Eastern Alps, earlier workers (Kockel 1922; Hagn 1960) believed that the Gosau and Penninic (Flysch Zone) seas were isolated by a W-E extending ridge called *Rumunischer Rücken*, while others (Tollmann 1963; Oberhauser 1968) considered this ridge partly equivalent to the *Ultrapienidic Ridge*, which possibly supplied ophiolitic detritus to the Gosau basins in the south. On the contrary, we suggest that the Gosau and Flysch Zones were not paleogeographically isolated, in fact, their paleobathymetric relation indicates the existence of an oceanic trench with subduction of the Penninic realm to the north, and an arc to the south (Dietrich & Franz 1976) may thus indicate reworking of oceanic crust breccia of Penninic origin. The arc-trench, representing sediments of the Gosau Zone, corresponds to a submarine plateau environment in the north (Triassic-Jurassic calcareous platform basement) and a marginal sea environment in the south (Kainach Gosau on *Altkristallin* basement).

Gosau Basins of the Submarine Calcareous Plateau

Three selected Gosau basins i.e. Reichenhall (Herm 1962), Kössen (Franz 1967) and Gosau-Valley (Wille-Janoschek 1966) and their sedimentary record show three sedimentary phases suggesting a strong crustal subsidence. During the first sedimentary phase (Coniacian-Santonian) shallow marine shelf environments prevailed. Rudist colonized littoral ridges interspersed with deeper basins with inner- to outershelf conditions.During the second sedimentary phase (Campanian) the shallow marine environments of the first phase subsided to bathyal depths at which both pelagic and fluxoturbidite sediments were deposited. During the third sedimentary Phase (Maestrichtian) the three basins underwent still deeper subsidence with turbiditic sedimentation predominating. The carbonate platform evolved into a submarine plateau with trough-shaped basins (fig 1 & 2). Supporting the Campanian-Maestrichtian syndepositional subsidence of the Gosau basins Hesse & Butt (1976) suggested on the basis of calcium carbonate and microfossil contents of the turbidite sediments, that the Kössen basin subsided below local CCD or to lower bathyal depth, whereas the Reichenhall and Gosau-Valley basins (Butt 1976) remained above CCD, or at middle and upper bathyal depth.

Gosau Basins of the Marginal Sea

The Kainach and Krappfeld Gosau (Hinte 1963, 1965) basins represent a marginal environment with poor connection to the open sea. The graben-shaped

basins were separated by island ranges. The absence of Middle to Upper Maestrichtian marine sediments in the marginal sea, however, may indicate an overabundance of clastic supply or an isolation from the open sea by barriers (comp. Brandenberg area: Herm 1977).

Pelagic Productivity and Lateral Water Mass Relation of the Upper Cretaceous basins

Analogous to recent sediments, the pelagic Upper Cretaceous sediments of the North Eastern Alps are made of coccolith,planktonic foraminifera and radiolarian ooze. The nature and distribution of the pelagic content, however, reflects surface biologic productivity and fertility of the overlying water mass, indicating distance from the shore, water depth and current circulation patterns (fig. 1).

The Helvetic sediments consist of lower ratios of planktonic to benthonic foraminifera. Furthermore, the composition of planktonic foraminifera shows a relative low abundance of globotruncanids compared to heterohelicids and hedbergellids. This association of planktonic foraminifera is typical of shelf-biomass, suggesting a low productivity region. Abundant detritus indicates near-shore conditions. The sediments of the Ultrahelvetic and Reichenhall-Kössen Gosau basins consist of abundant planktonic foraminifera and coccolith ooze. In contrast to the Helvetic sediment, the globotruncanids are relatively abundant. The composition of the pelagic fauna suggests a region with relatively high productivity. These bathyal basins were farther from shore and formed an open sea environment with upwelling currents. In the Kössen Gosau basins, the planktonic foraminifera are concentrated in the thin turbidite sequences, suggesting a high level of plankton productivity of the water-mass, although calcite-free hemipelagic sediments indicate that the basin was located below CCD during the Upper Campanian - Lower Maestrichtian. The Gosau-Valley basin was shallow with a relatively low productivity of planktonic foraminifera. The pelagic sediments (interbedded pelagic layers of turbidites) were diluted due to a higher rate of terrigenous influx, suggesting a proximity to shore.

The hemipelagic layers of the Flysch basin of the Eastern Alps are barren of planktonic foraminifera, coccoliths, and calcareous benthonics, though turbiditic layers contain relatively abundant pelagic fossils. The absence of planktonic foraminifera and coccoliths in the hemipelagic claystones neither indicates a lack of pelagic activity nor a land-locked or paleogeographic isolation of the Flysch basin, as it was earlier believed (Ksiazkiewicz 1961; Pflaumann 1967), but pre-sedimentary dissolution below the calcite compensation zone (CCZ), representing several stages of dissolution; the foraminiferal and coccolith lysocline, calcite compensation depth (CCD), and sediment-water interface dissolution (fig. 2). The calcareous pelagic ooze was redeposited below the CCZ by the turbidity currents in the Flysch basin, and its calcareous components were rapidly buried and hence not dissolved. The generic composition and diversity pattern of the planktonic foraminifera (Herm 1974; Butt 1976) in the Ultrahelvetic and Gosau Zones indicate that the Flysch basin was not a paleobiogeographic barrier with regard to pelagic floral and faunal migration laterally in the seas of the Helvetic Zone in the north, and Gosau in the south.

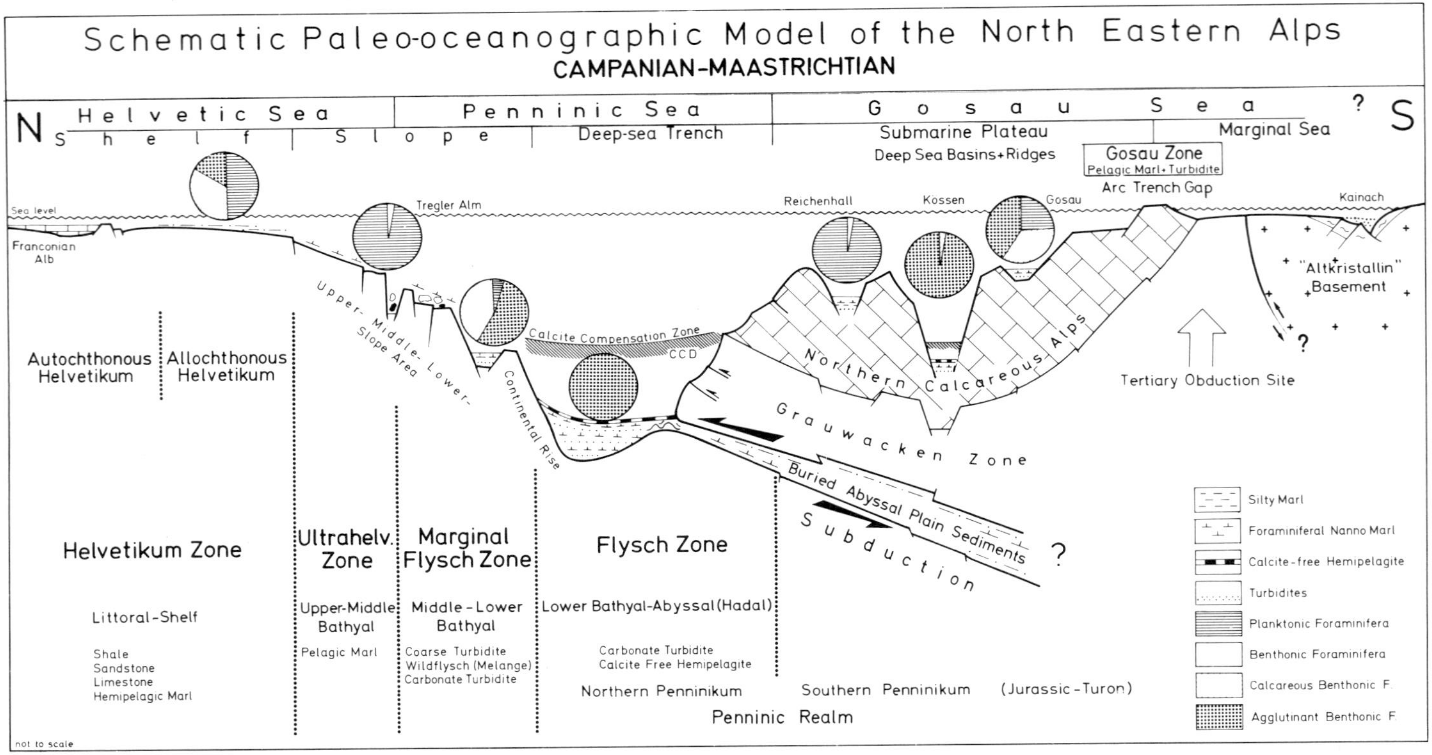

Fig. 2. Schematic paleobathymetric cross-section representing a reconstruction of the Upper Cretaceous paleo-oceanographic situation. Foraminiferal associations and lithologic data of the sediments are used to interpret the paleoenvironments of the paleogeographic zones (Helvetikum, Ultrahelvetikum, Marginal Flysch, Flysch and Gosau). Note: The Kössen basin was originally situated North of the Reichenhall basin!

Concluding Remarks

Fig. 2 shows that the depositional environments of Upper Cretaceous basins representing from N to S the zones of the Helvetikum, Ultrahelvetikum, Marginal Flysch, Flysch and Gosau of the Northern Calcareous Alps were not paleogeographically isolated. Although the dislocated zones display an extreme facies divergence of their sediments, the genetic properties of the sediments suggest an interconnected paleobathymetric setting, consisting of a shelf, slope, abyssal-trench, submarine plateau, and a marginal sea environment along a N-S cross-section of the Eastern Alps. On the basis of this paleooceanographic reconstruction three main paleogeographic realms can be subdivided. The Helvetic sea in the north, and the Gosau sea in the south represent the lateral flank areas of the Eastern Alpine geosyncline, whereas the Penninic sea (Flysch Zone) represents the central deep part of the geosyncline.

We are grateful to Drs. H. Bögel and R. Hesse for stimulating discussions and encouragements.

References

Butt, A., 1976: Unpubl., Ph.D., Thesis, Tübingen University, 128 p., 40 plates, Tübingen.

Dewey, F.J., Pitman, C.W., Ryan, F.B.W. & Bonnin, J., 1973: Bull. Geol. Soc. Amer. - 84:3180, Boulder.

Dietrich, V. & Franz, U., 1976: Geotekt. Forsch. - 50:85-109, Stuttgart.

Franz, U., 1967: Gosau - in: Erläuterungen zur geologischen Karte von Bayern, 1:25.000, Blatt 8240, Marquartstein, München:117-125.

Hagn, H., 1960: Geologica Bavaria. - 44:1-208, München.

Herm, D., 1962: Abh. Bayer. Akad. Wiss., Math.-Nat. Kl., N.F. 104:119 p., München.

Herm, D., 1974: Third Plantonic Conference Abstract:33, Kiel.

- , 1977: Mitt. Bayer. Staatsslg. Paläont. hist. Geol. - 17:257-277, 5 Abb., München.

Herm, D. & Butt, A., 1975: International Symposium "Geodynamik der Ostalpen". (DFG) Abstract. - :11, Salzburg.

Hesse, R., 1974: Bull. Geol. Soc. Amer. - 85 (6):859-870, Boulder.

Hesse, R. & Butt, A., 1976: J. Geol. - 84:505-533, Chicago.

Hinte, J.E., van, 1963: Jb. Geol. B.A. - 8:1-147, Wien.

- , 1965: Kooninkl. Nederl. Akad. Wetenschappen, Ser.B - 68(2):72-92, Amsterdam.

Hsü, J.K., 1972: Repts. 24th Inter. Geol. Congr. - 6:67-74, Montreal.

Hsü, J.K. & Jenkyns, H.C., 1974: (edit.): Int. Ass. Sed., Spec. Pub. - 1:1-447, Oxford.

Hsü, J.K. & Schlanger, S.O., 1971: Bull. Geol. Soc. Amer. - 82:1207-1218, Boulder.

Kockel, C.W., 1922: Mitt.Geol.Ges.Wien - 15:63-168, Wien.

Ksiazkiewicz, M., 1961: Soc.Geol.Pologne. - 31:3121, Kraków.

Laubscher, H.P., 1975: Amer.J.Sci. - 275:865-876, New Haven, Conn.

Lisitzin, A.P., 1972: Soc.Econ.Paleontol.Mineral., Spec.Publ. - 17 (edit. Rodolfo, K.S.), 218 p., Tulsa.

Oberhauser, R., 1968: Jb.Geol.B.A. Wien. - 111:115-145, Wien.

Pflaumann, U., 1967: Geol. Rundsch. - 56:200-227, Stuttgart.

Scholle, P.A., 1971: Bull. Geol. Soc. Amer. - 82:629-658, Boulder.

Sliter, W.V., Bé, A.W.H. & Berger, W.H., (editors), 1975: Cush. Found. Foram. Res., Spec. Publ., 13:1-159, Kansas.
Tollmann, A., 1963: Ostalpensynthese, 256 p., Wien (Deuticke).
- , 1969: Geologie, 18:1131-1155, Berlin.
Wille-Janoscheck, U., 1966: Jb. Geol. B.-A., 109:91-172, Wien.

The Gosau Formation (Northern Calcareous Alps), an Example of an Alpine Trench-Slope-Continental Margin

J. Dietrich and U. Franz

Extensive work dealing with ophiolitic detritus in the Upper Cretaceous and Tertiary Gosau deposits of the Northern Calcareous Alps has been done (Dietrich & Franz 1976). The conglomerates, breccias, sandstones, marls, clays, and minor amounts of rudist limestones represent clastic sedimentation within a system of Upper Cretaceous to Eocene basins.

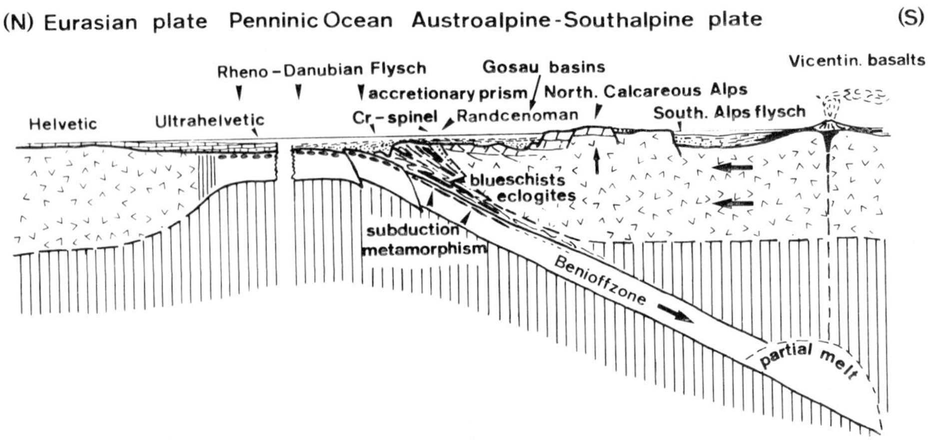

Fig. 1. Palinspastic cross-section through the Austro-Alpine area during the period: Cenomanian to Eocene (100-45 m.y.).

These sediments are transgressive on already deformed Mesozoic sequences of the Calcareous Alps. The basal conglomerates and breccias as well as several other clastic horizons of the Gosau formation have been regarded as tectonic markers for different phases of the alpine orogeny.

Whereas the coarse-grained and poorly rounded detritus of the Calcareous Alps appears in all clastic Gosau deposits, the Upper Santonian/ Lower Campanian clastics contain the exotic ophiolitic detritus:

Sandstones with meta-basaltic, serpentinite and ophicalcite components, in many cases with considerable amounts of chromium spinel;
Sandstones consisting principally of serpentine and ophicalcite grains, and conglomerates with meta-basaltic (pillos- and hyaloclastite fragments), gabbroic, serpentine and ophicalcite pebbles.

Ophiolitic components are well-rounded, polished an comparatively small in size (not exceeding 50 mm in diameter). Their undeformed structures and metamorphic mineral assemblages (chrysotile/lizardite serpentinites and ophicalcites as well as the meta-basalts in zeolite- and lower greenschist facies) are comparable to the low grade metamorphic rocks of the oceanic crust.

The characteristics of the Upper Santonian/Lower Campanian clastic Gosau sediments suggest that the ophiolitic detritus experienced long-distance transport into the Gosau basins by mass flow and turbidite activity. Coarse carbonate clasts (in some instances greater than house-size blocks) may have been derived from nearby scarps within the shelf and slope of the Calcareous Alps.

Therefore the Gosau basins or troughs are regarded as locally confined offshore basins within the accretionary prism between the continental shelf (Northern Calcareous Alps, fig. 1) and a tilted flysch wedge, which formed the highest part of the subduction zone, (similar to contemporary arc-trench gap and trench-slope situations in several parts of the Pacific; Seely et al. 1974; Karig & Sharman 1975). Ophiolitic detritus was shed from downgoing slices and sediments of the Penninic oceanic crust (Dietrich 1976). Southward-directed subduction thrust the Penninic oceanic lithosphere under the Austroalpine-South-alpine sialic crust and caused the internal thrust sheets of the Northern Calcareous Alps as well as the uplift of the North-alpine Flysch deposits.

References

Dietrich, V., 1976: Geotekt. Forsch. - 50:1-84.
Dietrich, V. & Franz, U., 1976: Geotekt. Forsch. - 50:85-109.
Karig, D.E. & Sharmann III. G.F., 1975: Geol. Soc.Amer. Bull.- 86:377-389.
Seely, D.R., Vail, P.R. & Walton, G.C., 1974: The Geology of Continental Margins (ed. Burk, C.A. & Drake, C.L.) - 249-260, Springer, New York.

Vulcanism and Manganese Ore Deposition in the Liassic of the Northern Calcareous Alps

K. Germann

In the Northern Calcareous Alps manganese minerals are particularly concentrated in early to middle Jurassic bituminous shales and marls, red nodular limestones and radiolarites (Germann 1971, 1972, 1973). Fe-Mn-

carbonate (Ca-rhodochrosites, oligonite) deposits (fig. 1) with some million tons of reserves, but lacking actual economic importance (Gudden 1969; Germann & Waldvogel 1971), are concentrated in thinly laminated upper Liassic shales and marls (Manganschiefer) reflecting sedimentation under prevailing stagnant conditions. Facies evolution from poorly oxygenated to anaerobic sediments is suggested in some of these deposits by the vertical sequence of the iron minerals, from siderite to chamosite to pyrite. In the Berchtesgaden deposits laminated braunite-rhodochrosite-hematite ores indicate, however, that Liassic manganese accumulation also occurred in oxidizing environments.

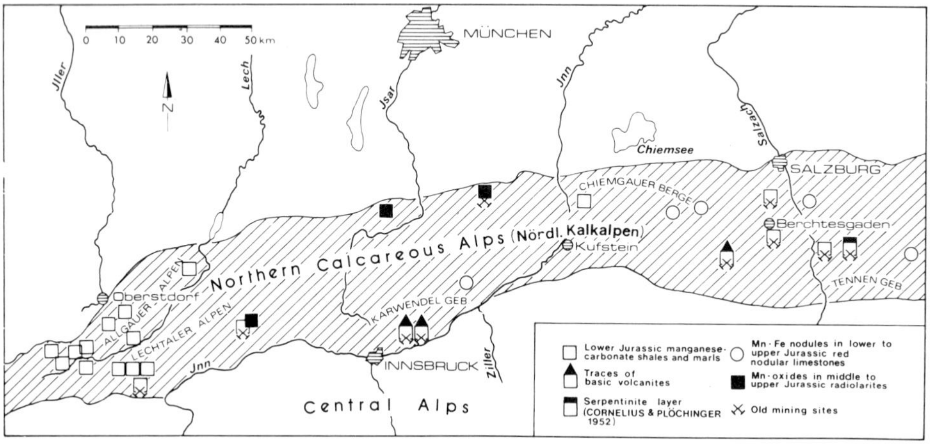

Fig. 1. Jurassic manganese occurrences in the Northern Calcareous Alps.

Near the southern margin of the Northern Calcareous Alps (Karwendel and Berchtesgaden Mts., see fig. 1) at two manganese shale occurrences, celadonite-bearing rhodochrosite rocks have been detected Germann 1972). Celadonite, a common alteration product of Mg-Fe-rich eruptive rocks, is indicative of volcanic activity during early Jurassic deposition of manganese sediments. Co- and Ni-anomalies corroborate the interpretation of celadonite as remnants of probably tuffaceous basaltic material. With their mineralogy (e.g. braunite) and geochemistry (elevated contents of Ba, Cu, Zn, low Mn/Fe ratio) distinctly different from normal marine-sedimentary ores, the Liassic manganese accumulations should be grouped with the *greenstone* type (Shatskiy 1964) of volcanogenic-sedimentary manganese deposits. Probably the heavy metals have been supplied by locally ascending magmatic solutions as proposed for similar manganese-iron accumulations on the East Pacific Rise (Boström et al. 1974) or the northern Afar Rift (Bonatti et al. 1972), minor contributions being due to leaching by seawater from basaltic material (Bischoff & Dickson 1975).

Volcanogenic manganese deposition in the Alpine Liassic coincides with the major block-faulting events and initial stages of rifting in the Alpine-Mediterranean Tethys (Bernoulli & Jenkyns 1974; Evans 1975) and most probably is related to earliest ophiolitic activities. Reconstruction of lower Jurassic plate margins for the Eastern Alps by Smith (1971) was based on a late Triassic or Jurassic sinistral wrench fault in the Salz-

burg Alps (Fischer 1965). Cornelius & Plöchinger (1952) reported serpentine streaks in manganese shales in the vicinity of this structure, producing additional evidence of lower Jurassic uprising of mantle material in the Northern Calcareous Alps. This region of the Tethys remained marginal, however, to the centers of magmatic activity. Early Jurassic Alpine-Mediterranean metallogenesis reached its maximum development in the Lower Austroalpine (*Unterostalpin*) trough and the Dinarides, where sulfide- and Cr-deposits are related to ophiolites (Petrascheck 1966).

References

Bernoulli, D. & Jenkyns, H.C., 1974: Soc. Econ.Paleontologists and Mineralogists Spec.Publ. - 19:129-160.
Bischoff, J.L. & Dickson, F.W., 1975: Earth Planet.Sci.Lett.- 25:385-397.
Bonatti, E., Fischer, D.E., Joensuu, O., Rydell, H.S. & Beyth, M., 1972: Econ. Geol. - 67:717-730.
Boström, K. & Joensuu, O., Kraemer, T., Rydell, H., Valdes, S., Gartner, S., Taylor, G., 1974: Geol.Fören.Stockh.Förh. - 96:53-60.
Cornelius, H.P. & Plöchinger, B., 1952: Jb. geol.B.A.- 95:145-225, Wien.
Evans, A.M., 1975: Mineral.Deposita. - 10:254-260.
Fischer, A.G., 1965: Verh.geol.B.A.- 1965:20-33, Wien.
Germann, K., 1971:N.Jb.Paläont.Mh. - 1971, 3:133-156.
 - , 1972: Tschermaks miner.petrogr.Mitt. - 17:123-150.
 - , 1973: In: Amstutz, G.C. & Bernard, A.J. (Edits.),Ores in sediments: 129-138, Berlin.
Germann, K. & Waldvogel, F., 1971: N.Jb.Paläont.Abh. - 139:316-345.
Gudden, H., 1969: Erzmetall. - 22:482-488.
Petrascheck, W.E., 1966: Sitzber.österr.Akad.Wiss., math.-nat.Kl.Abt.I. - 175:57-74.
Shatskiy, N.S., 1964: Intern.Geol.Rev. - 6:1030-1056.
Smith, A.G., 1971: Bull.Geol.Soc.Amer. - 82:2039-2070.

Middle Triassic Block Faulting in the Eastern and Southern Alps

T. Bechstädt, R. Brandner, H. Mostler and Kl. Schmidt

The breakup of the Paleoeuropean shelf is evident in the Middle Triassic; probably it started as early as Upper Permian and not, as was formerly believed, in the early Jurassic. The Middle Triassic disrupture is supported primarily by sudden vertical facies changes, fissures, and the reappearance of volcanic facies.

In the Alps the Mid-Triassic period is a time of very strong facies differentiations. Possible reasons are:
1) eustatic fluctuations of the sea level (glacio-eustatic, tectono-eustatic or geoidal-eustatic);

2) selective bioconstruction and erosional activities by currents, and
3) synsedimentary tectonics.

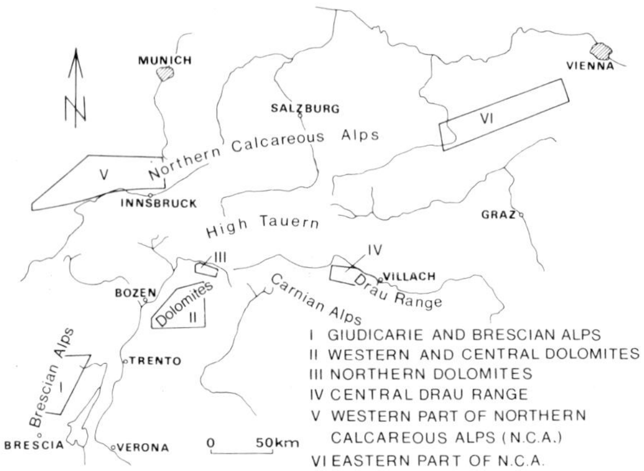

Fig. 1. Sketch-map of the Eastern and Southern Alps showing the locations of the facies-patterns.

A relative rise of sea-level must be antecedent to the start of bioconstructions on pre-existing platforms. Apart from climatic changes, differences in the rate of sea-floor spreading (Forney 1975) and geoid-changes (Mörner 1976) are a possible cause for sea-level fluctuations. Even a few decameters of sea-level rise would be enough to start reef building, a rate far below that taken into consideration by Forney or Mörner. The final reef-basin association may then be, as in the Bahamas, the result of selective bioconstruction on the platforms and erosion in the troughs (Andrews et al. 1970; Schlager et al. 1976).

Is this model valid also for the Middle Triassic of the Alps? Here the reef-basin association is situated on top of huge, uniformly developed platforms (Upper Sarldolomite of the Southern Alps, *Zwischendolomit* of the Drau Range, middle Alpine Muschelkalk resp. Steinalm limestone of the Northern Calcareous Alps). Here we often observe an immediate superposition of shallow water sediments by basinal deposits. Which water depths are to be considered for these deposits?

Water depths of basinal sediments are still problematic (contrary to platform sediments, which correspond closely to present day models of shallow-water sedimentation, such as the Bahamas). They can, however, be inferred:

a) by geometrical construction, as for example from the difference in altitude between synchronous sediments of the platform-rim and the basin (800-1000 m water depth, Bosellini & Rossi 1974). Because of strong tectonics especially in these transitional areas this method can only rarely be used.

b) by the faunal associations. Water depths of 500 m and deeper are indicated by psychrospheric ostracodes (Benson & Sylvester-Bradley 1971; Gründel & Kozur 1975); special siliceous sponges and deep-water holothurians (Mostler, in press) are evidence for minimum depths of 200 m.

In the upper Anisian/Ladinian, these faunal elements appeared for the first time in the post-Hercynian history of the Eastern and Southern Alps. The Triassic sediments of the E show basin characteristics earlier than those of the W (Bechstädt & Mostler 1976). The immigration of the faunas

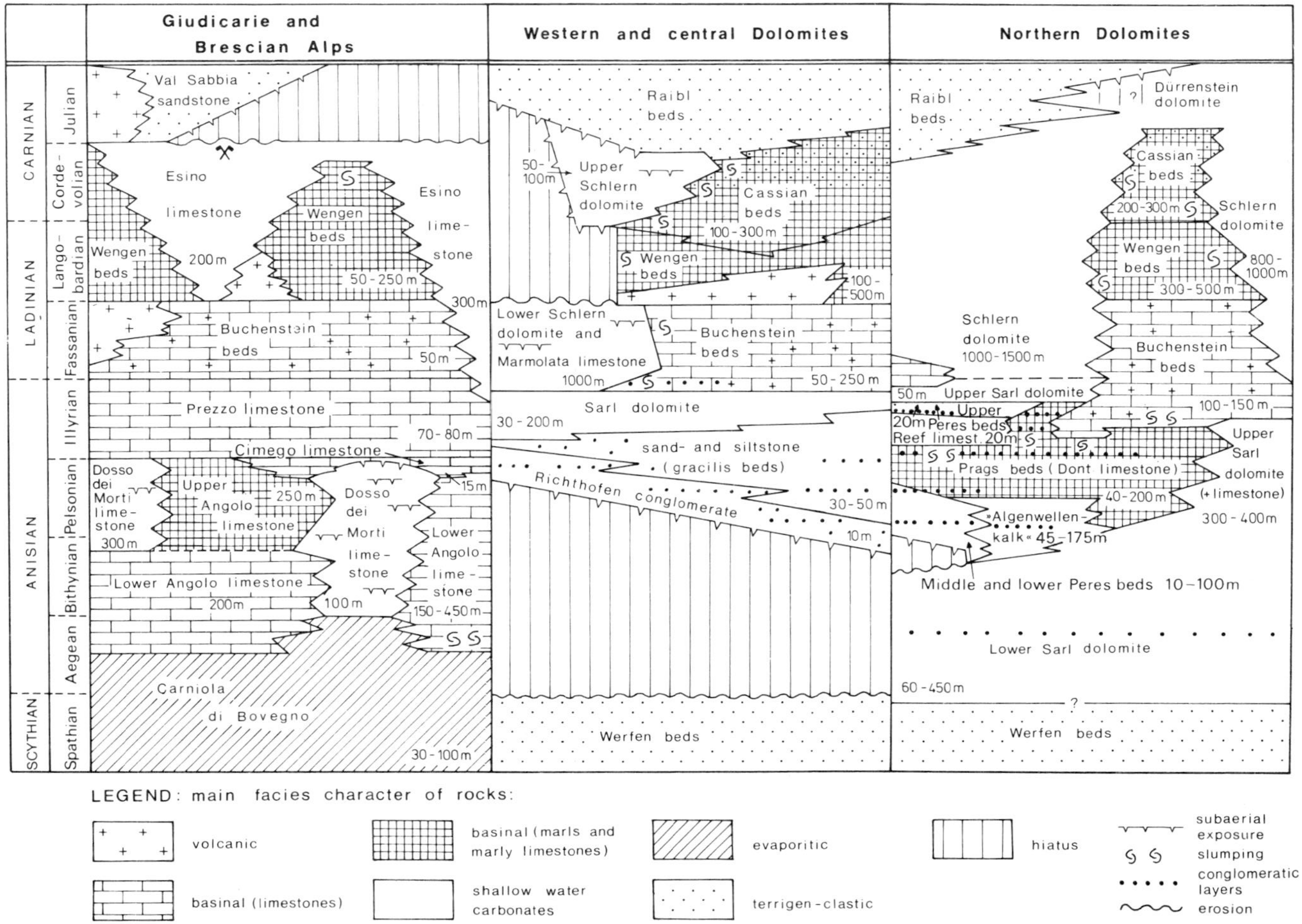

Fig. 2. Facies patterns from selected areas of the Southern Alps.

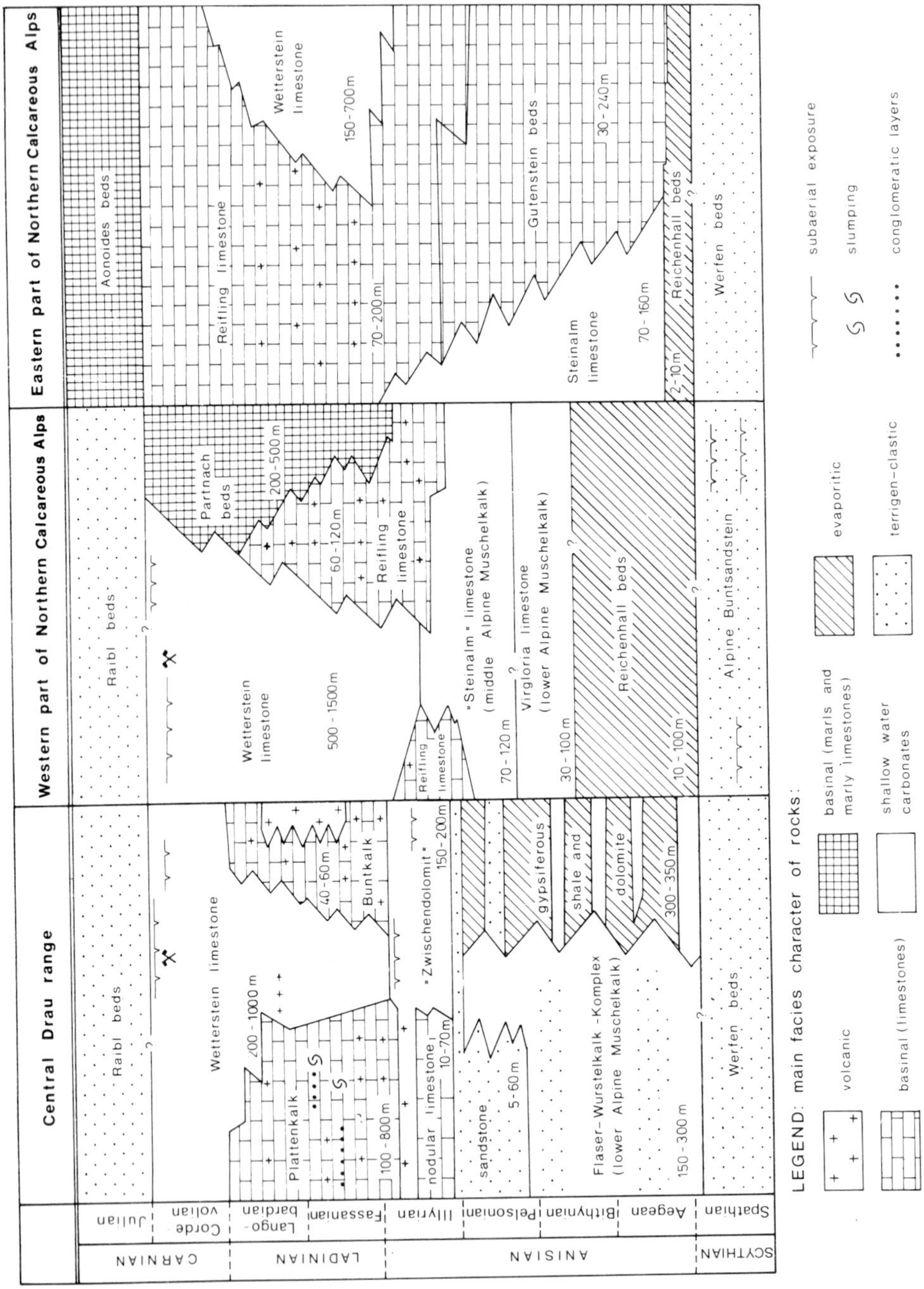

Fig. 3. Facies patterns from selected areas of the Drau Range and the Northern Calcareous Alps.

indicate the connection of the newly developed basins with the open ocean. The rich radiolarian content further supports pelagic conditions for that time. The faunal content shows a stepwise deepening of the basins of 100-200 m in the middle Anisian, 500 m and more in the upper Anisian/Ladinian, and to depths below the carbonate compensation level in the Jurassic.

Such sudden changes from platform-sediments to these deeper marine deposits indicate changes in water depth of several hundred meters. Sudden eustatic sea-level changes within this scope are considered improbable. But also other indications for tectonic movements in the Triassic exist: Neptunian dykes (Fischer 1964) and fissures, angular discordances, fine textured graded flows, slumping and olistoliths, lateral facies variety following a nearly peneplained relief, and vulcanism. Of course only a combination of several of these factors is convincing as a syn-sedimentary tectonic process. Fissures and evidence of vulcanism are of special importance in pinpointing a disrupture.

Aborted Rifting in the Triassic

The accelerated tectonism causes many stratigraphic changes, especially in middle and late Anisian and in the Rhetian to lower Liassic time. The tectonic processes are probably caused by an initial rifting stage, a disruption of the continental crust in connection with graben tectonics (see also Schlager & Schöllnberger 1974; Dietrich 1976; Bechstädt et al. 1976). Hallstatt-type limestones, volcanics, reefal sediments and ophiolite sequences indicate an ocean that is expanding from the SE (Asia minor, Greece) to the Alps in the NW.

The post-Hercynian tectonic disintegration started already in Upper Permian time, earlier in the E than in the W. After the large scale tectonic subsidence in the Upper Anisian, basins surrounded the few remnants of the former huge shelf area. Here frequently reef-building started or reef-like sediments accumulated. Reef-building was possible for the first time now due to the connection to the open sea.

The facies pattern follows preferredly NE-SW and NW-SE directions in the Southern Alps, E-W directions in the Austro-Alpine unit. These are possibly post-Hercynian structures (Flügel 1975) which can be found from Permian to the Jurassic.

The tectonism is linked with volcanic activity of an early *Red-sea-type* (Hoffman et al. 1974). The development of sodium-rich basic magmatites (Upper Permian, Lower Triassic) to alcalibasalts (Middle Triassic) coincides with characteristics of rifting.

The first phases of tectonic disintegration were followed by a long period of relative inactivity. From the upper Ladinian to the Norian the carbonate facies spread into the marginal regions of the opening ocean (the limestone-shale alternation in the middle and upper Carnian are probably related to eustatic processes). Especially the platform-period of the Hauptdolomite-Dachstein-limestone marks a distinct tectonic quiescence.

At the end of the Norian substage rifting reappeared, but the main spreading-zone is now to be found in the N, in the Penninic region, a formerly epicontinental area. Jurassic plate-disintegration affected much larger areas than Mid-Triassic rifting. In most of them the deep subsidence could not be compensated by sedimentation, basin facies was preserved for a long period.

References

Andrews, J.E., Shepard, F.P. & Hurley, R.J., 1970: Geol. Soc. Amer. Bull. - 81:1061-1078.
Bechstädt, T., Brandner, R. & Mostler, H., 1976: Geol. Rdsch.- 65:616-648.
Bechstädt, T. & Mostler, H. 1976: Z.dtsch.geol.Ges. - 127:271-289.
Benson, R.H. & Sylvester-Bradley, P.C., 1971: Bull. Centre Rech. Pau-SNPA, 5 (suppl.):63-91, Pau.
Bosellini, A. & Rossi, D., 1974: Soc. Econ. Paleont. Miner., Spec.Publ.-18:209-233.
Dietrich, V., 1976: Geotekton.Forschg. - 50:1-84.
Fischer, A.G., in: Symposium on cyclic sedimentation, Kansas Geol. Surv. Bull. - 169:107-149.
Flügel, H.W., 1975: Geol.Rdsch. - 64,1:1-62.
Forney, G.G., 1975: J.Geol. - 83:773-779.
Gründel, J. & Kozur, H., 1975: Freiberger Forschungsh. - C 304:39-49.
Hoffmann, P., Dewey, J.F. & Burke, K., 1974: Soc.Econ.Paleont.Miner., Spec. Publ. - 19:38-55.
Mörner, N.-A., 1976: J.Geol. - 84:123-151.
Mostler, H., in press: Ber.naturw.-med.Ver.Innsbruck, 64.
Schlager, W., Hooke, R.L. & James, N.P., 1976: Geol. Soc. Amer. Bull. - 87:1115-1118.
Schlager, W. & Schöllnberger, W., 1974: Mitt.geol.Ges.Wien.- 66/67:165-198.

Sedimentation and Tectonics of the Thiersee-Karwendel Synclines (Northern Tyrol, Austria)

W. Zeil and K.-H. Nagel

In the Thiersee-Karwendel Synclines (Northern Tyrol, Austria) a sequence of Triassic platform carbonate rocks approx. 3000 m in thickness, is overlain by Jurassic sedimentary rocks of 200 to 1000 m thickness. Due to incipient relief development the early to Middle Liassic beds show considerable diversity in facies and in thickness over short distances. The area deepened in early Jurassic and Middle Jurassic time and in Late Jurassic time sediments of the *Radiolarite group* (Radiolarite and Tauglboden beds) were deposited under bathyal conditions.

The main part of the Jurassic sequence consists of pelagic carbonates. Their facies and thickness indicate two distinct sedimentary basins and synsedimentary tectonics. On submarine plateaus thin-bedded 60 to 200 m of *Biancone* limestones were deposited. In regions of strong subsidence episodic debris flows and turbidites (clastic sediments) resulted in a ten times higher rate of sedimentation with 600 m of thickness. Sedimentation of *Biancone* limestones persisted into the Lower Cretaceous. They are topped by 200 to 500 m of marls and marly limestones locally attain-

ing Barremian to Aptian age.

Structurally the area forms part of the southern margin of the Lechtal unit. It is delimited by the following complexes:

To the east: Kaisergebirge thrust mass

To the south: Wetterstein front of the Pendling range and front of the Inntal unit

To the west: Wetterstein mountains

To the north: Wamberger anticline.

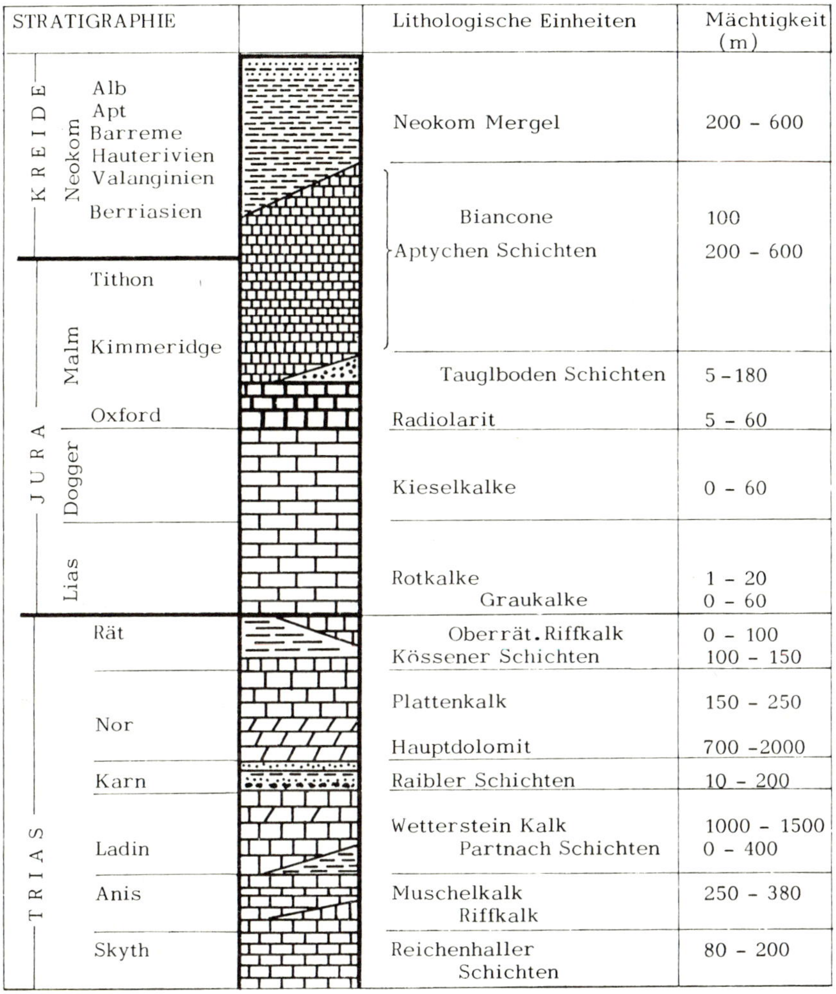

Fig. 1. Stratigraphic Standard profiles peripheral to the Thiersee and Karwendel synclines.

A remarkable tectonic element, a double kink, apparently joins the Thiersee and Karwendel synclines. Some authors, e.g. Quenstedt (1933) and Ampferer (1941) have argued for steeply plunging folds. Originally this

was supposed to be a syncline trending 70 km E-W from Kiefersfelden to Mittenwald.

Tectonic analysis has emphasized plotting of tectonic axes. Slickensides and fractures were considered of subordinate importance. Joint Diagrams reveal two intersecting joint maxima: One with NE strikes and steep SE dips, the other with NW strikes and NE dips. Both attitudes have been observed repeatedly in the calcareous Alps; they are almost certainly genetically related. These diagonal joints vary only with variations of material properties of the lithological units.

The fold axes reveal a recognizable pattern. In the center of the area the E-W axes are rotated about 50°, and steepened by about 30°. Axes trending between 218 and 238 are clearly concentrated in the southern part of the area. This is borne out by both observation in the field and by mapping. Since these axes are superimposed on the E-W axes, the question arises whether the latter can be related to the original stress field.

Mechanically we have an elastic plate which can carry out rotations, but not translations. Transverse extension is inhibited; hence any attempt at N-S shortening will result in N-S stresses as well as compressive stresses. These compressive stresses increase with progressive N-S deformation. Hence N-S shortening and inhibition of E-W extension result in biaxial compression (actually triaxial if overburden pressure is included). In this situation the line of flexure is no longer curved

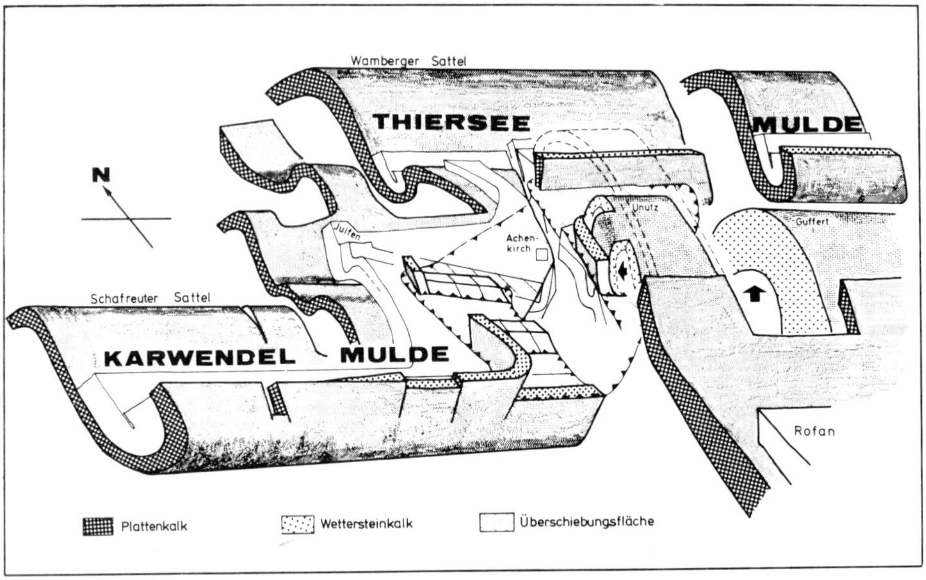

Fig. 2. Structural model of the surroundings of Achenkirch.

cylindrically (the axes trend only E-W), but doubly curved. This applies particularly where the extent of the plate is considerable.

Indeed, on a large scale superposition of E-W fold axes is conspicuous. Four large tectonic units can be recognized:

the 3 E-W trending major folds: Thiersee syncline, Schafreuter anticline, Karwendel syncline
the Achentaler thrust mass.

The *Thiersee syncline* can be followed as a coherent element from the Inn valley cut near Kufstein to the east, to Achenwald, south of the Achen pass to the west. Its north limb is the south limb of the Wamberg anticline. The syncline is isoclinal and faces north. Its south limb is thrust over its north limb and in some areas overrides the Neocomian marls of its core. The tectonic plan is further complicated by diagonal faults. The first glance upon the geological map even evokes the impression that these faults have only affected the north limb. In reality, the fault system continues in the south limb, but there it is differently developed in parts.

South of the Thiersee syncline, the adjoining Schulterberg anticline forms part of the strongly differentiated *Schafreuter anticline*. The anticlinal hinge of this fold, which is up to 12 km wide, plunges gently east. West of Fall the Schafreuter anticline merges immediately with the Wamberg anticline. Only the equally gently east plunging Thiersee syncline subdivides this major structure into two independent anticlines with opposing axial plunge. To the east of the Bächental the Schafreuter anticline is overridden by the Achental thrust mass. This unit corresponds to the sheared off south limb of the adjoining *Karwendel syncline*. It is in the area in which the Achental thrust sheet passes into the south limb of the Karwendel syncline that we find the only proven bend which connects generally E-W striking beds with generally N-S striking beds. Steeply plunging folds, however, would require at least four such localities.

Both the north limb and the south limb of the Karwendel syncline are further deformed by doming. Between Rissbach and Plumsbach the syncline is isoclinal and faces north.

The characteristic fold pattern - narrow synclines, wide anticlines - is due to variations in lithology in the fold cores. The ductile synclinal core consisting of Neocomian marls permits north facing isoclinal synclines. The south limb of the syncline can rotate until resisted by the north flank of the adjacent anticline. Where synclinal core material is highly mobile, the limbs of the syncline become isoclinal. Tensional stresses in synclinal hinges generate glide surfaces in the synclinal core, especially since these are favored by the highly ductile material there.

In contrast, anticlinal cores consist of incompetent limestones and dolomites. Hence they are much wider.

Fuchs (1944) interprets the variability in axial direction as an effect of different external stresses, approximately orthogonal to the axes. Since this relationship exists only statistically, in no case deterministically, this model represents a considerable simplification. In contrast to this model of a discontinuous external stress field with variable orientation we propose the model of a continuous stress field with constant orientation.

In the analysis of solid body deformation, four parameters must be considered: 1. External forces; 2. Material properties; 3. Mechanical constraints; 4. Dimensions of the body.

Pure fabric analysis deals only with the sum of the effects of these parameters. Changes in the fabric diagram do not correspond solely to changes in the externally applied forces.

Further, in the case of large-scale deformation - the only kind of concern in Alpine tectonics - preceding deformation should not be neglected, since its influence on subsequent deformation is decisive. At the same time progressive deformation causes local confining stresses which vary with time, side by side with variations with time of material properties.

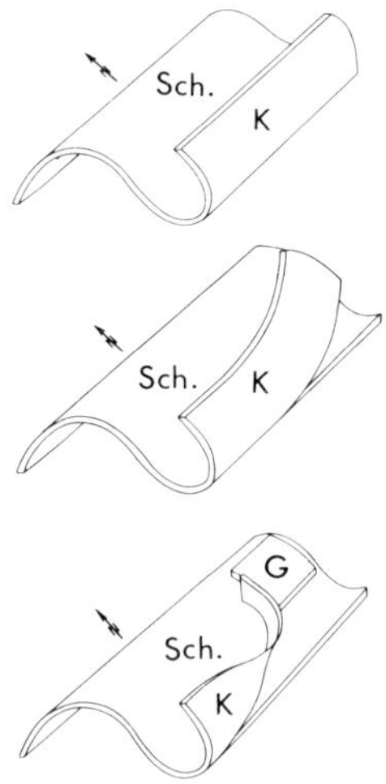

Fig. 3. Stages of deformation of the Karwendel syncline from middle Cretaceous (above) and Oligocene (below), schematic. Sch = Schafreuter anticline, K = Karwendel syncline, G = Gröbner Hals fault.

In the area under consideration the wide Hauptdolomit anticlines could serve as local counterforts to the narrow synclines only after their formation in Malmian time. Only the narrow isoclinal synclines permit tensile stresses in their south limbs and hence lead to overthrusting (Achental thrust mass). Constant orientation of externally applied forces, therefore, does not result in constant orientation of the internal stresses which are the immediate cause of deformation. For instance, inhibition of lateral escape in an E-W direction results in high E-W stresses which do not require externally applied forces (E-W compression).

This picture is complicated by the fact that the ductile marls suffer intense deformation, but take up the smallest stress. Hence, a homogenous stress field is not possible, even with constant applied forces, because stress trajectories are deflected into the more competent layers (limestone, dolomite). The proposed model postulates applied forces active over a long time interval (Malm to Pliocene): N-S compression.

Consideration of the above mechanical critera shows that the proposed model does not conflict with the known geological observations of this area.

References

Ampferer,O.,1941: Sitzungsb.d.Akad.d.Wiss.Wien,Math.-natwiss.Kl.Abt.I., 150:183-199.

Fuchs, A., 1944: N.Jb.Miner.Geol.Paläont., Beil.,88:337-373.

Nagel,K.H., Schütz, K.I., Schütz, S., Wilmers, W. & Zeil, W., 1976: Geol.Rdsch., 65:536-557.

Nagel, K.H., 1975: Geotekt. Forsch., 48:139 p.

Quenstedt, W., 1933: Z.Dtsch.Geol.Ges.,85:459-461.

Schütz, K.I., 1975: Dissertation TU Berlin:182 p.

Paleogeography of the Permian in the Southern Alps

W. Buggisch, E. Flügel, F. Leitz and G.-F. Tietz

Sedimentological, paleontological and geochemical studies of the Permian sediments in South Tyrol, in the Carnic Alps and in the Karawanken (Buggisch 1976; Buggisch et al. 1976; Flügel 1975, 1977) suggest the following paleogeographical evolution:

The *basement* of the Permian rocks are Variscan quartz phyllites in the Dolomites. In the area between the western Carnic Alps and Slovenia Upper Carboniferous Auernig beds are exposed.

These cyclic deposits are overlain by the *Lower Permian* Rattendorf formation. The boundary between Carboniferous and Permian corresponds to the change from mainly clastic to predominately carbonate sedimentation; the first Lower Permian index fossils (fusulinids) are found approximately 40 m above the uppermost clastic of the Auernig formation.

In the western area (Dolomites) locally continual coarse-grained clastics (Waidbruck conglomerate resp. Verrucano alpino) represent the transition from the Carboniferous to the Permian. Thick volcanic rocks (Bozener Quarzporphyr) were formed during the Lower Permian in South Tyrol. According to the *caldera model* (Bosellini 1965) no volcaniclastics have been found in the contemporaneous sediments of the Carnic Alps.

The Lower Permian Rattendorf and Trogkofel formations of the Carnic Alps and the Karawanken are characterized by a shelf- and shelf-edge sedimentation. During the Rattendorf stage (Asselian) three lithological units can be recognized: The *Lower Pseudoschwagerina limestones* were deposited cyclically in a near-shore inner-shelf area by alternating transgressive and regressive phases. Four lithological cycles with basal clastics and overlying carbonate are known (Homann 1969). The carbonate members consist of irregularly bedded biomicrites and algal buildups. Low-diverse invertebrate communities and monomict algal associations indicate restricted, shallow-marine and subtidal environments.

During the uppermost parts of the Lower Pseudoschwagerina limestones the source area of the clastics and its erosive activities were changing. The erosion of metamorphic and acid volcanic rocks and the increasing sedimentation of clastics in a near-shore high-energy environment with intertidal to subtidal conditions is responsible for the origin of the sandstones and silty shales of the *Grenzland group*. For the upper part of this group an additional source area with acid plutonites is suggested by heavy mineral analysis (Tietz 1975).

The Upper Pseudoschwagerina limestones show an even repetition of microfacies sequences, which, together with the medium to high biotic diversity, are evidence of a repeated shifting of ecological zones from very shallow water areas to off-shore environments in an open-marine shelf lagoon with normal water circulation.

This outer-shelf platform facies continued during the Trogkofel stage (Sakmarian and Artinskian) in the western parts of the Carnic Alps (Forni Avoltri) and in the western Karawanken (Dolzanova soteska near Trzic). In contrast to these well-bedded *Trogkofel limestones* organic buildups with encrusting algae (Tubiphytes Maslov) and with fenestellid bryozoans, brachiopods and crinoids are predominant in the Sexten Dolomites, in the

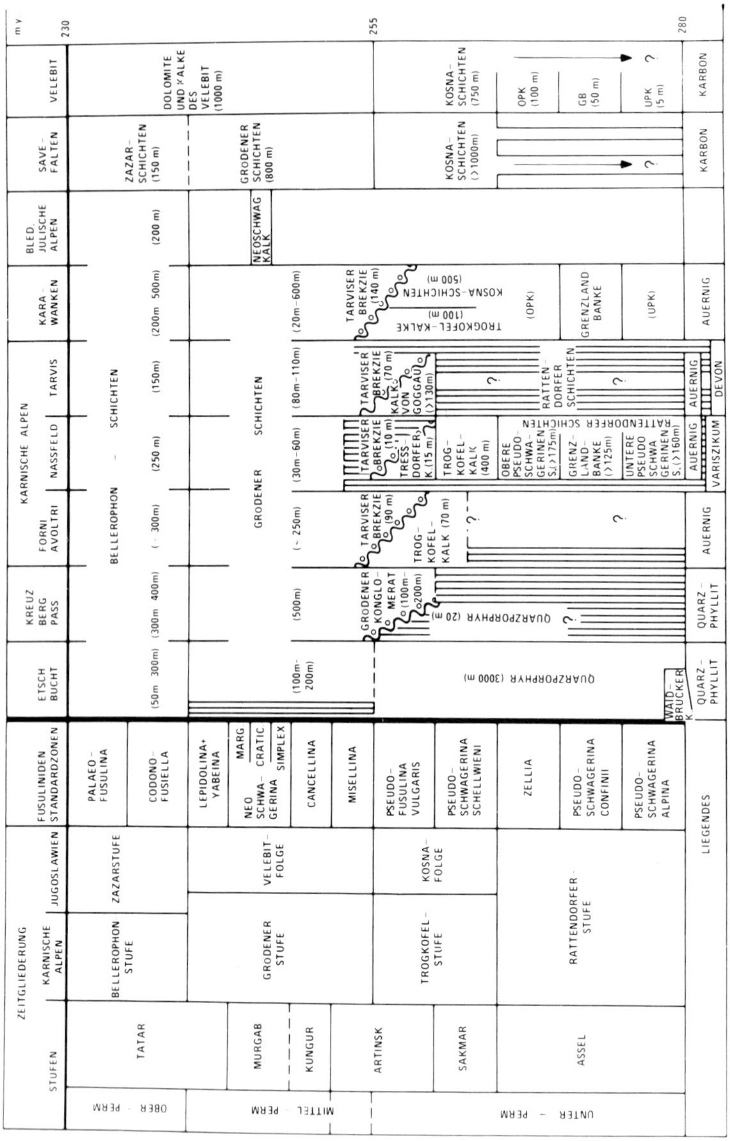

Fig. 1. Correlation table. Time units after Kahler (1974).

Carnic Alps (Trogkofel; Tarvis) and in the Karawanken.

Besides the Trogkofel area all the Tubiphytes limestones are known only from resedimental Trogkofel components within the Tarvis breccia. A shelf-edge position of some of these *Trogkofel reefs* is probable.

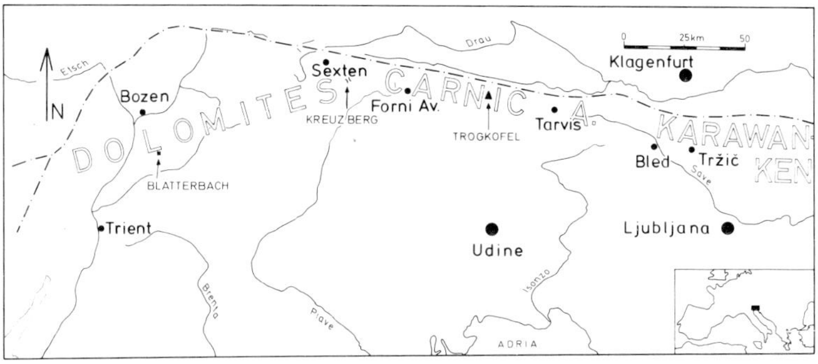

Fig. 2. Location Map.

The mobile inner shelf, developed in the Carnic Alps and in Slovenia and Croatia, is characterized by the "Clastic Trogkofel formation", consisting of up to 2000 m quartz-limestones, conglomerates, and some limestones. Lithoclastic limestones suggest a long period of synsedimentary re-sedimentation (Kochansky-Devidé et al. 1973).

A widespread regression at the end of the Lower Permian is manifested by breccias and conglomerates (*Tarvis breccia*, *Gröden conglomerate*). These sediments have been formed in areas of intensive intra-Permian block-faults. A tectonic uplift of some sedimentary regions seems to be responsible for widespread submarine and subaerial destruction of the Trogkofel limestones in high-energy environments.

The red bed facies of the *Gröden formation* (Middle Permian), known from South Tyrol to the Karawanken, unconformably overlies different Lower Permian or pre-Variscan rocks.

Within the sedimentation area of the Gröden formation two facies units have been observed:

a) Coarse-grained and ill-sorted conglomerates and sandstones consist predominantly of reworked material from the basement. In the Carnic Alps and in the Western Karawanken this facies appears only at the base of the section, but in South Tyrol and in the Western Karawanken it clearly predominates and shows additional characteristics, in particular authigene kaolinite formed by intensive in-situ weathering of feldspars and a lack or a low content of clacite and no dolomite.

b) Fine-grained, medium- to well-sorted, interbedded silt- and sandstones, overlying the facies unit A in the area between the Karawanken and the river Etsch (Adige) in South Tyrol. Characteristics of this are the high feldspar contents (20-30%), high carbonate content (predominately dolomite), and the occurrence of chlorite together with the lack of kaolinite.

The transition between these two facies is characterized by gypsum, coal, enrichment of Pb, and by typical clay minerals (montmorillonite, mixed layers). Representative section are located in the Gröden valley and in the Bletterbach gorge near Radein.

A particular Middle Permian facies is exposed in the area of Bled, Julian Alps, consisting of high-energy reef-limestones (Ramovs 1955).

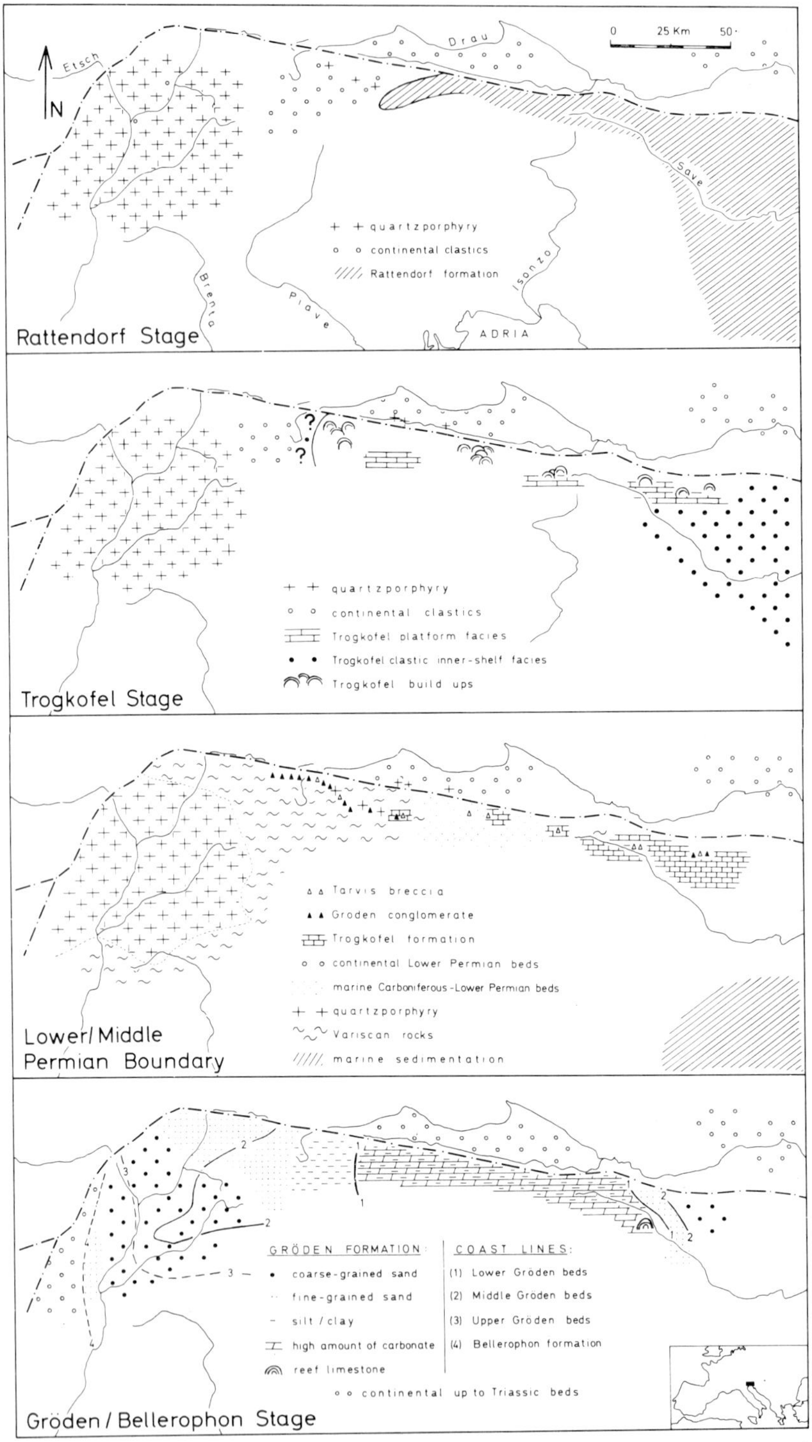

Fig. 3. Reconstruction of Permian paleogeography in the Southern Alps.

The attributes of facies A corroborate a deposition in a continental environment; fresh-water algae (Kreuzberg pass, Sexten Dolomites) supports this interpretation.

The transition facies shows characteristics of a coastal region with marginal lagoons. Paleontological data (tetrapod traces, driftes cephalopods and foraminifera) are in accordance with this model.

Facies type B is deposited in a marine environment as suggestes by the supply of Mg for dolomitization and formation of the chlorites, enrichment of Mn, scarse foraminifera, ostracods and gastropods.

In summary, the paleogeographical development during the Middle Permian is as follows: Continental sedimentation at the beginning, marine transgression in the western Carnic Alps and in the western Karawanken during the early Gröden stage; progressive transgression during the middle Gröden stage to the west, with a coast line near the Gröden valley and the Bletterbach gorge and a coastline near the river Etsch at the end of the Middle Permian.

The increasing transgression during the Upper Permian resulted in the despotion of bituminous sediments and basal evaporites of the *Bellerophon formation*. In the south-western near-shore area evaporites are dominating (facies fiammaza, Accordi 1958, sabhka environment whereas in northeastern South Tyrol gypsum with open marine carbonates prevails (facies badiotica). The Bellerophon formation of the Carnic Alps indicates connections between the marine Zazar formation in the Save Folds area and with the limestones and dolomites of the Velebit Mountains.

References

Accordi, B., 1959: Estud.Geol. - 15:1-17, Madrid.

Bosellini, A., 1965: Mem.Mus.Storia Nat.Ven.Trid.- Anno 28,3:1-72,Trento.

Buggisch, W., 1976: Habil-Schrift TH Darmstadt, 84 p.

Buggisch, W., Flügel, E., Leitz, F. & Tietz, G.-F., 1976: Geol. Rundschau, 65/2:649-690.

Flügel, E., 1975: Carinthia II, 164/84:43-62, Klagenfurt.

- , 1977: in: Flügel, E. (Ed.): Fossil Algae. Recent Research and Developments. - Berlin-Heidelberg-New York (Springer).

Homann, W., 1969: N.Jb.Geol.Paläont.Mh. - 1969:265-280.

Kochansky-Devidé, V., Buser, St., Cajhen, J. & Ramovs, A., 1973: Razprave, Diss. XVI/4, 19 p., Ljubljana.

Ramovs, A., 1955: N.Jb.Geol.Paläont.Mh. - 1955:455-457.

Tietz, G.F., 1975: Carinthia II, 164/84:115-124, Klagenfurt.

Upper Jurassic Radiolarites in the Northern Calcareous Alps (Upper Austroalpine Unit)

V. Diersche

The distribution, thickness and facies of Jurassic sediments in the Northern Calcareous Alps - investigated mainly between the longitudes of Salzburg and Kufstein - have been influenced since the breakdown of the carbonate platforms at the end of the Triassic times by accentuated selective crustal movements. These movements resulting from a differentiated interplay between embryonic folding, sedimentation and vertical faults formed a system of E-W trending troughs and ridges parallel to the orogen.

The *troughs* were filled during the Lias and Dogger with grey basinal marls and marly limestones of mainly pelagic character with a distinct terrigenous influx. Thickness of these sediments locally exceeds 1000 m (Jacobshagen 1964). The *ridges* were covered with thin red pelagic limestones ("ammonitico rosso"). This sedimentation pattern is quite similar to that of the Southern Calcareous Alps (Aubouin 1964; Bernoulli 1964) and of the Klippen belt in Poland (Birkenmaier 1960).

Sedimentation of the overlying *Upper Jurassic radiolarites* continues within the same tectonical and sedimentological pattern. At the turn Dogger/Malm the radiolarite starts in the *troughs* with a grey, evenly bedded facies. Submarine *ridges* are covered on the flanks by grey radiolarite, and on top by red radiolarite of reduced thickness. Locally the radiolarite is found on the ridges not before the Upper Oxfordian (Wendt 1969; Krystin 1971). In the upper part of the radiolarite a red knobby bedded facies is widely distributed (except in troughs with extremely high sedimentation rates, fig.2). With the beginning of the radiolarite sedimentation the terrigenous influx stops abruptly. At the same time, there is a marked change in bio-microfacies from sediments rich in calcite filaments to sediments rich in radiolarians.

The southernmost troughs show a maximum thickness of more than 350 m and may contain sedimentary breccies and slumps (Tauglboden-Berchtes-

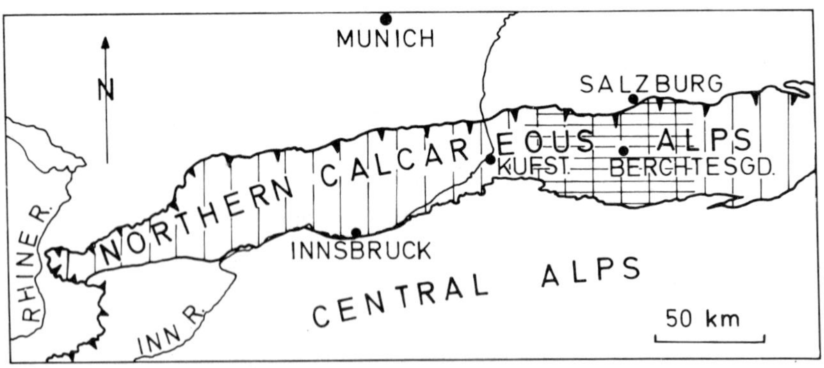

Fig. 1. Location map. Cross hatched = main working area.

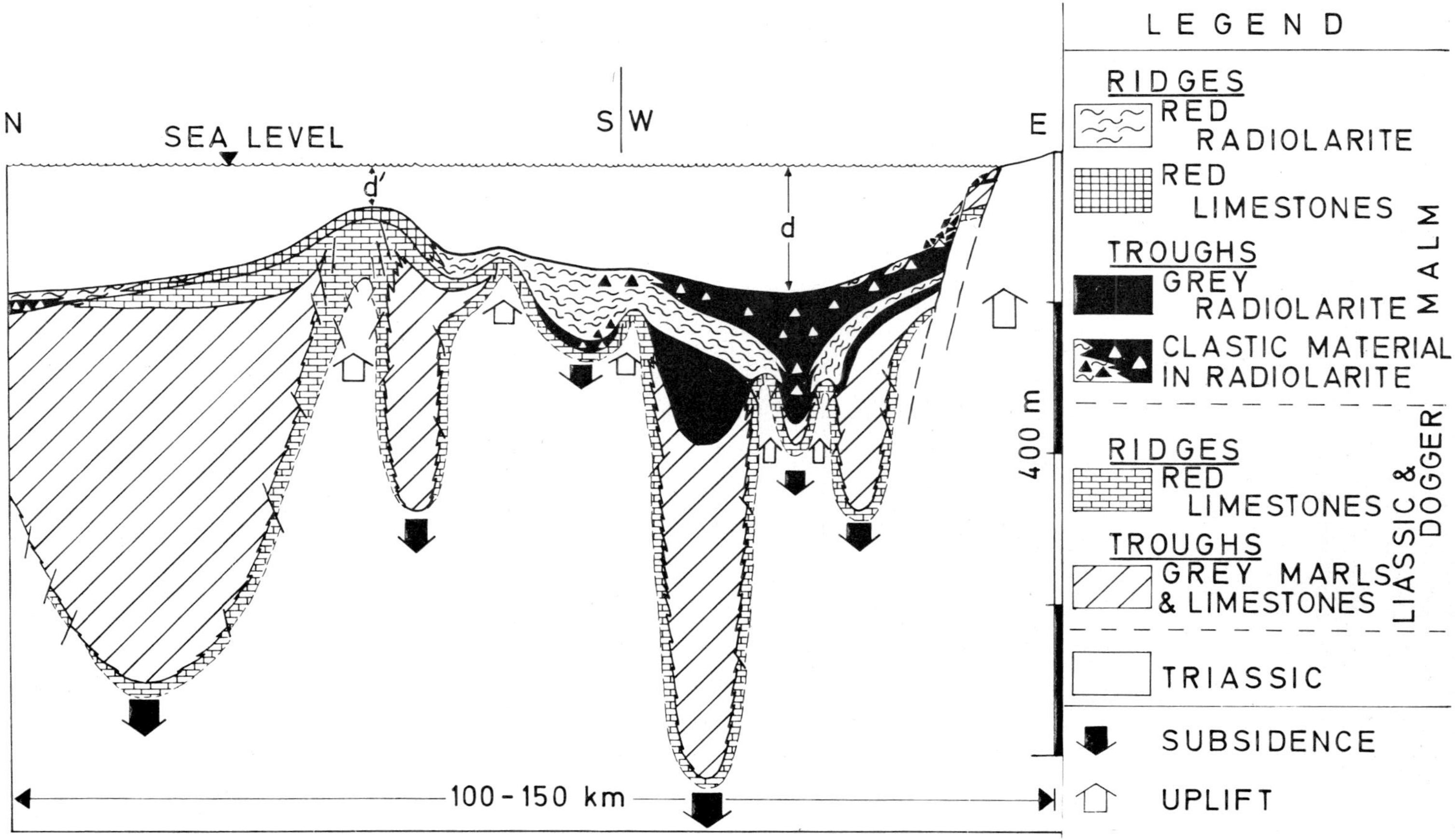

Fig. 2. Palinspastic facies section across the Austroalpine Unit for the Late Radiolarite sedimentation stage (Upper Oxfordian/Kimmeridgian). Section along the line Zinnkopf - Saalach - Berchtesgaden. Sediment thicknesses only approximately. Water depth not to scale.

gaden area south of Salzburg, and Achensee area north of Innsbruck), whereas the northern troughs show reduced maximum thicknesses of 10 to 20 m. On the northernmost ridge extending over several hundreds of kilometers, the radiolarites are laterally substituted by thin red pelagic limestones ("ammonitico rosso"). These limestones with the foraminifer *Protoglobigerina* represent a sediment of similar biogenic and geochemical composition as the recent Globigerina ooze; their distribution in the Tethys is restricted to the submarine ridges (fig. 2), however. Further ridges with red pelagic limestones partially or completely substituting the radiolarites occur also in the middle and southern part of the Northern Calcareous Alps (Zürs area, Sonnwend Mountains, Dachstein Mountains). The radiolarites at the outermost northern rim of the Northern Calcareous Alps show a conspicuous amount of terrigenous material (? proximity of the Lower Austroalpine Unit, comp. Finger 1975).

For the first time we could prove *syngenetic volcanism* for the Austroalpine radiolarites. The volcanics are at least partly rhyolithic-dazitic in mineralogy and geochemistry and are represented by crystal tuffs and heavily altered scoria.

The *accumulation rates* in the radiolarites vary between 1 - 2 mm/1000 y (submarine ridges) and maximum 23 mm/1000 y (troughs). Some sections, e.g. in the Tauglboden - Berchtesgaden special trough, show accumulation rates of up to 110 mm/1000 y (see also W.&M.Schlager 1973). As an overall mean accumulation rate 12 mm/1000 y is obtained for the Austroalpine radiolarites for the compacted sediment (40% compaction). This accumulation rate is higher than that of W.Schlager (1974), Garrison & Fischer (1969), and Bosellini & Winterer (1975). The red pelagic limestones on the ridges show minimum accumulation rates of 0.4 mm/1000 y (comp. with the sedimentation rates in Schwab 1976).

Up to 80% of the *radiolarite* is composed of radiolarians; coccolith fragments are very rare. Benthos is extremely scarce (sponge spicules) and nekton occurs in the form of fish teeth, aptychi, belemnites and rhyncholiths. The *red pelagic limestones* are rich in coccoliths, *Globochaete* and *Protoglobigerina*, mixed with radiolarians. Besides there occur the relics of a rich benthos, e.g. up to 10% crinoids, echioids, foraminifers (*Lenticulina*, *Spirillina*, and agglutinating forms), silicisponges and the nowadays psychrospheric ostracode *Bythocypris*. Nekton is represented by fish teeth, belemnites, *ammonites*, aptychi and rhyncholiths.

The *carbonate content* of the radioclarite of 22% (mean from 76 representative samples) is very low compared to 98% $CaCO_3$ of the red pelagic limestones on the ridges (in the whole the radiolarite carbonate content is 7 times higher than that of the Recent oceanic radiolarian ooze (mean about 3% $CaCO_3$)). This low carbonate content indicates *carbonate dissolution*, proven in the radiolarite by the lack of the pelagic calcitic fossils, e.g. *Globochaete* and *Protoglobigerina* (Colom 1967). The increase in silicisponges, *Spirillina* and the psychrospheric ostracode *Bythocypris* towards the flanks of the limestone ridge is interpreted as a reaction to decreasing temperatures - supposed Tethyan surface temperatures were about 20-25°C, bottom water in the troughs about 10-15°C - affecting the carbonate dissolution in the radiolarite troughs. In addition, an increased differentiation in carbonate dissolution in the radiolarite troughs is marked by the *complete lack* of aragonitic fossils - only one ammonite was ever found in the Tethyan radiolarites - indicating sedimentation below the aragonite compensation depth.

As to the *sedimentation depth* of the radiolarites we may not equate the compensation depths for calcite or aragonite of the warm Upper Jurassic Tethys with those of the Recent tropical oceans. One of the reasons is the increase in biogenic carbonate production due to calcareous planktonic organisms (e.g. coccolithophorids , globigerinids) since the Upper Jurassic. The much lower biogenic carbonate input in the Upper Jurassic and the higher water temperatures also of the deep ocean water may have shifted them downwards. On the other hand cold currents and/or the existence of small troughs with bad circulation may have caused the described differentiated carbonate dissolution at much shallower depths than expected.

The *water depth* can be concluded from *biogenic* (a) and *sedimentologic-tectonic* (b) facts:

a) The maximum water depth for the *red pelagic limestones* - corresponding to the upper limit for the *radiolarite* (see fig. 2) - may range between 150 and 400 m (Ziegler 1967; Wendt 1969, 1970; Bernoulli & Renz 1970; Krystin 1972). The lack of any calcareous algae as well as the occurrence of boring (? blue-green) algae in organic carbonate substrate of the pelagic limestones points to a similar range of sedimentation depth (comp.Wendt 1970; Golubic & Lukas 1975). This hypothesis is supported by the fact that no intact coccospheres are found within these pelagic limestones. In the Recent tropical oceans the majority of coccospheres disintegrates to water depths of 200 m.

b) Assuming a slope gradient of 3° to 4° from ridge to trough (mean gradient of the Recent Continental Slope) and 50 km as the largest distance between the submarine ridges with pelagic limestones - distance between the Wendelstein and the Sonnwend Mountain area - we calculate a relief of 1300 m (3° angle) to 1800 m (4° angle) between these ridges and the trough floor between. Adding the water depth (fig. 2, d) above the ridges to the paleorelief (ridge/basin) we get a maximum water depth of the radiolarite troughs of about 2200 meters (fig. 2, d). Nevertheless smaller basins may have had shallower depth ranges. Garrison & Fischer (1969) calculated for the same area among others a maximum water depth of more than 5000 m for the radiolarite time. This calculation needs steep slopes of up to 10°. Such steep slopes were never observed over a long distance along the red limestone ridges. Slumping structures along these ridges are an exception. Steep slopes of smaller relief are found along the southernmost ridges, connected with slumping structure and turbidites.

The *trace element concentration* of the radiolarites reaches only 1/5 to 1/10 of that of recent radiolarian oozes. This may have been caused by a restricted and/or by a relatively small outcrop of oceanic crust in the central part of the Tethyan Sea (Penninic belt, Dietrich & Franz 1976).

Summarizing we may state that the Upper Jurassic radiolarites of the Northern Calcareous Alps are *deep sea sediments*. Their sedimentation starts together with an increased subsidence in the area of the Upper Austroalpine Unit proved by the decline in terrigenous influx, an increase in the pelagic character of the sediments, a shift in the trace fossil pattern, a uniform spreading of the radiolarite facies over a wide area, and by a restricted carbonate sedimentation.

The differentiation of the sedimentation area into small troughs and ridges points to a tectonic setting at a *continental margin*. This margin may have belonged to the disintegrating African plate. A similar tectonic setting may be represented by the Recent California Borderland.

Acknowledgements

For valuable discussion and help the author wishes to thank Prof. Dr. H. Zankl (University of Marburg), Prof. Dr. W. Zeil and Dipl.-geol. H.-R. Walter (Technical University of Berlin-West), and Prof. Dr. K. Germann (Free University of Berlin-West),and Mrs. Hoenig for assistance in the translation.

References

Angenheister, G., Bögel, H., Gebrande, H. & Giese, P., 1972: Geol. Rdsch. 61:349-395.
Angenheister, G., Bögel, H. & Morteani, G., 1975: N. Jb. Geol. Paläont., Abh.,148:50-137.
Aubouin, J., 1964: Bull. Soc. géol. France,75:730-766.
Bernoulli, D., 1964: Beitr.Geol.Karte d.Schweiz, N.F.,118.Lfrg.:1-134.
Bernoulli, D. & Jenkyns,H.C.,1974:In:Modern and Ancient geosynclinal sedimentation (R.H.Dott & R.H. Shaver, eds.). Spec. Publ., 19.Soc. Econ. Paleont.Mineralog.:129-160.
Bernoulli, D. & Renz, O., 1970: Eclog.geol.Helv.,63:573-607.
Birkenmajer, K., 1960:Jb.Geol.B.-A., 103:1-36, Wien.
Bosellini, A. & Winterer, E.L., 1975: Geology, 3:279-282.
Colom, G., 1967: Palaeogeogr.,Palaeoclim.,Palaeooecol.,3:229-310.
Dietrich, V.J. & Franz, U., 1976: Geol. Rdsch., 65:361-374.
Fenninger, A. & Holzer, H.-L., 1974: Mitt.Geol.Ges., 63:1-141, Wien.
Finger, W., 1975: IXth Intern.Congr. of sedimentology, Thème 4 (Tectonique et sédimentation), 1:119-125, Nice.
Garrison, R.E. & Fischer, A.G., 1969: In: Depositional environments in Carbonate rocks etc.(G.M.Friedman, ed.). Spec.Publ.Soc.écon. Paleont. Mineral., 14:20-56.
Golubic, St. & Lukas, K.J., 1975: In:The study of trace fossils (R.W.Frey, ed.). Ch.12:229-254, Berlin.
Jacobshagen, V., 1964: Geol. Romana, 3:303-318, Roma.
Krystin, L., 1972: Ann. Nat. hist. Mus. Wien, 76:195-310.
- , 1971: Verh. geol. B.-A.:486-509, Wien.
Nagel, K.-H., Schütz, K.-J., Schütz, S., Wilmers, W. & Zeil, W., 1976: Geol. Rdsch., 65:536-557.
Schlager, W., 1974: In: K.J. Hsü & H.C. Jenkyns, eds.: Pelagic sediments, on Land and under the Sea,Spec.Publ.Int.Ass.Sedimentol.,1:37-58,Oxford.
Schlager, W. & Schlager, M., 1973: Sedimentology, 20:65-89.
Schlager, W. & Schöllnberger, W., 1974: Mitt.Geol.Ges.Wien, 66-67:165-193.
Schwab, F.L., 1976: Geology, 4:723-727.
Trauth, F., 1950: Verh.Geol.B.-A.:145-218, Wien.
Trümpy, R., 1960:Bull.Geol.Soc.America.,71:843-908.
Wendt, J., 1969:N.Jb.Geol.Paläont.,Abh., 132:219-230.
- , 1970: N. Jb.Geol. Paläont., Mh.:433-448.
Ziegler, B., 1967: Geol.Rdsch.,56:439-464.

Paleomagnetic Study of the Adamello Region (Southern Alps)

R. Pucher, G. Morteani, K. Fromm

In the Southern Alps the Periadriatic Line (P.L.) and the Giudicarie Line (G.L.) form a triangle (see fig. 1). The southern part of this triangle is formed by sedimentary and volcanic rocks of Paleozoic to Tertiary age. Its northern part consists mainly of the granodiorites of the Adamello pluton of Tertiary age. Granodiorites, xenoliths included in the granodiorites, porphyry dykes, sandstones metamorphosed at the contact with the magmatic rocks of the Adamello pluton as well as Permian sandstones and rhyolites are investigated paleomagnetically and petrologically. In this paper only a short summary of the paleomagnetic results is given. For further details see Pucher, Morteani & Fromm (in press). The data given here are included also in the review paper of Pucher & Morteani (this vol.).

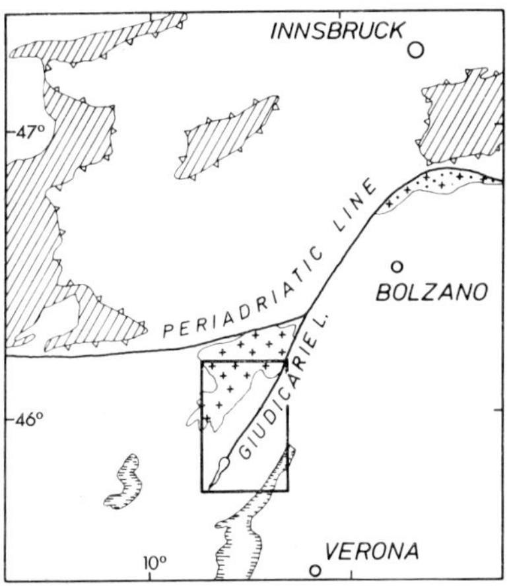

Fig. 1. Geological sketch map of a part of the Central Alps locating the sampling area.

Alpidic Samples

Granodiorites and xenoliths of the Adamello pluton do not show a stable remanent magnetization, whereas the porphyry dyke, genetically connected with Adamello intrusion (site 160, tab.1), the sandstone tectonized by the G.L. (site 244, tab. 6) and the sandstone metamorphosed to hornfels at the contact with the Adamello pluton (site 233, tab.1) show a stable mean direction of remanence of D/I = 331°/87° (see upper part of tab.1). The magnetic declination D of the remanence is comparable to that of the Bergell intrusion (Heller 1973) with D/I = 337°/58°, whereas the magnetic inclination is higher than that of the Bergell. This high inclination can be the effect of a short-time excursion of the earth's magnetic field (secular variation) or due to a tilting of the wedge interposed between the P.L. and the G.L. According to radiometric age determinations of Borsi et al. (1977) the magmatic rocks of the Adamello intrusion show from south to north younger cooling ages. This systematic trend can be

Table 1: Petrological and magnetic data of the samples

site Nr	number of samples	rock type	optimum of demagnetization (kA/m) +)	remanent magnetization: intensity (mA/m) +)	remanent magnetization: direction: declination/inclination (°) with angle of confidence (95%) in situ	dip corrected	dip of the layering direction/amount (°)	CURIE temperature (°C)
alpidic samples								
233	8	hornfels	-64	0.7	122/85(15°)	267/28	270/66	530
244	7	sandstone	-80	1.3	325/78(14°)	172/75	160/26	
160	6	porphyry (intrusive)	-80	1.1	271/88(29°)	255/74	253/14	535
varistic samples								
183	10	sandstone	-64	32.7	346/57	5/32 (3°)	34/32	
188	9	sandstone	-64	9.2	355/60	12/33 (6°)	34/31	none
249	9	sandstone	-64	1.1	342/34	360/38(9°)	70/30	
227	9	sandstone	-32	0.7	33/31	13/38 (10°)	280/30	
194	10	sandstone	-64	1.7	157/12	154/20(9°)	32/14	545
199	9	sandstone	-40	3.0	165/16	161/32(13°)	10/18	
164	6	porphyry	-80	2.3	178/-24	170/-26(6°)	253/14	537

+) magnetic field: 100 Oersted ≅ 8 kA/m in S.I. units
magnetization: 1 Gamma ≙ 10 mA/m

interpreted as (a) due to increasing younger intrusion from south to north or (b) due to a postintrusive tilting of that area to the south. Such a tilting cannot explain the high inclination value which would result in even lower inclination values. At present, the authors interpret the magnetic inclination as the result of secular variation.

Varistic Samples

The samples of Upper Permian age are normally magnetized, those of Lower Permian are inversely magnetized (see lower part of tab.1 and tab.2). The remanence of both sample groups shows a clear directional difference. This difference can be interpreted in terms of tectonism.

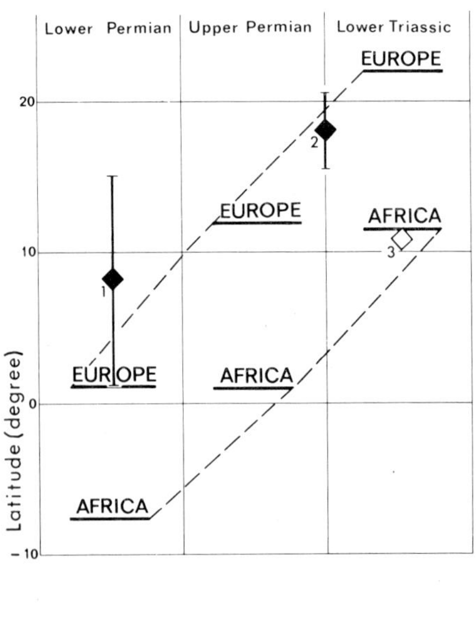

Fig. 2. Comparison of the paleolatitudes of the Adamello region for the Permian and Lower Triassic deduced from polar wandering curves of Van der Voo & French (1974) for Europe and Africa with data from the samples with error of measurement $\alpha_{95}°$ and data from Zijderveld & De Jong (1969) (open square).

Table 2: Paleomagnetic interpretation of the data

geological period	magnetic polarity	number of sites	samples	mean values of magnetic declination/inclination (°) with angle of confidence (95%) in situ	dipcorrected	virtual pole position
Tertiary	normal	3		327/87 (13°)	252/64 (56°)	
			21	331/87 (10°)	253/61 (16°)	
Lower Permian	normal	4		1/48 (27°)	7/33 (4°)	
			37	0/48 (7°)	7/33 (4°)	177°E/61°N
Early Lower Triassic	inverse	3		166/2 (38°)	161/9 (52°)	
			25	163/7 (11°)	159/16 (13°)	144°W/33°N

In fig. 2 the paleolatitudes of the Adamello region for Permian and Early Triassic time is computed from the polar wandering curves of Van der Voo & French (1974) for Europe and Africa. Paleolatitudes derived from the data of the present study are marked with black squares (1 = reversed, 2 = normal). The value of Zijderveld & De Jong (1969) is in fig. 2 represented by the open square. The samples of Lower Permian with reversed direction of remanence (symbol 1 in fig.1) show a large scatter and are in better agreement with Stable Europe than with Africa. The samples of Upper Permian age have normal direction of remanence and a smaller scatter than the sample of Lower Permian. They show a direction of remanence more similar to that of European Early Triassic than Permian. They show furthermore the same increase of latitude as it can be deduced for Europe and Africa from the polar wandering curves (Fromm et al.this vol.).

Data of Zijderveld & De Jong (1969) measured on rocks originating from the southernmost part of the triangle between the P.L. and the G.L. south of the studied area, on the contrary, agree with the paleolatitude of Africa. Comparing the present data with that of Zijderveld & De Jong (1969) one has to ask for a tectonic line between both areas. Until now there is no indication for such a line.

Two points one has to keep in mind: a) With the assumption of horizontal sedimentation the inclination of the dip corrected remanence of sedimentary rocks indicates the polar distance. Differences of the measured and the expected magnetic inclination therefore are a strong indication for north-south displacements. b) Van der Voo & French (1974) have included the results of the sea-floor spreading of the Atlantic Ocean to construct their polar wandering curves since the real polar wandering curve for Africa still is very poor. The final shape of the polar wandering curve of Africa may differ from the derived one.

The present investigation indicates with the direction of the remanence of varistic samples that the region between the Periadriatic Line and the Giudicarie Line behaved as an independent tectonic unit later than Lower Triassic times. The high inclination of the alpidic samples, probably an effect of secular variation, allows no tectonic conclusions.

References

Borsi, S., Callegari,E., Del Moro,A., Ferrara,G., Fratta,M.,Giuliani,O., Macera,P., Pardini,G., Pescia,P.,Tonar,S.& Villa,I.,1977:Econ.V,Pisa.

Heller, F., 1973: Earth Planet. Sci. Letters, 20:180-188.

Pucher, R., Morteani, G. & Fromm, K., 1977: Geol. Jb., D, in press.

Van der Voo, R. & French, R.B., 1974: Earth Science Review, 10:99-119.

Zijderveld, J.D.H. & De Jong, K.A., 1969: Geologie en Mijnbouw, 48: 559-564.

Part 3

The Central Area of the Eastern Alps

Metamorphism of the Penninic Series in the Western Tauern Window (Austria/Italy)

D. Ackermand, G. Morteani and P. Raase

The Penninic series of the western Tauern Window comprise the *Zentralgneis* (central gneisses), the rocks of the *Untere Schieferhülle* (lower schist mantle), and the rocks of the *Obere Schieferhülle* (upper schist mantle). The Zentralgneis is composed of granitic, granodioritic, and tonalitic orthogneisses and migmatites, the Palaeozoic Untere Schieferhülle mainly consists of paragneisses, mica schists, quartzites, and amphibolites, and the rocks of the Permo-Mesozoic Obere Schieferhülle essentially are calc mica schists with intercalated greenschists. A recent review of the petrostratigraphic and tectonic setting, metamorphism, and age of the Penninic series in the western Tauern Window is given by Morteani (1974).

The texture and composition of the rocks in the western Tauern Window are mainly the result of the Alpidic metamorphism (*Tauernkristallisation* according to Sander 1911) although some relics of a pre-Alpidic metamorphism are preserved.

The migmatic structure of the gneisses in the central area of the Tauern Window is ascribed to a high-grade pre-Alpidic metamorphism clearly related to the Variscan magmatism. The granodioritic and tonalitic rocks often exhibit diatectic structures. Probably, they are of anatectic origin. From field relations it may be seen that the youngest rocks generally are of granitic composition. Radiometric dating on granitic gneisses from the western Hohe Tauern yielded two whole rock isochrons of about 240 m.y. (Jäger et al. 1969; Satir 1974) and one isochron of 282 m.y. on aplitic rocks (Besang et al. 1968). Two more whole rock isochrons of about 245 m.y. have been determined in the eastern Tauern Window (Lambert 1964; Cliff 1968, 1971). Because the youngest granitic rocks intruded the migmatic gneisses, the high-grade Variscan metamorphism and anatexis must be older than the granites, i.e. older than about 240 m.y.

The regional variation of the metamorphic grade may be derived from the anorthite isograds in epidote-bearing gneisses (Morteani & Raase 1974), the biotite-isograd (Hoernes 1973), the variation of the optic angle of the K-feldspar (Raase & Morteani 1976), and from the isotherms determined by $^{18}O/^{16}O$-isotope studies (Hoernes & Friedrichsen 1974) (see fig. 1).

The highest metamorphic grade occurs in the central part of the Zillertaler Alps and decreases towards the margin of the Tauern Window. In the central area, the amphibolite facies was reached as shown by the occurrence of garnet-bearing plagioclase amphibolites, by the appearance of staurolite and kyanite and the breakdown of margarite and paragonite + quartz in rocks of suitable composition (Morteani 1971, 1974; Raith 1971;

Raase 1972; Hoernes 1973; Ackermand & Morteani 1973). The latter reaction suggests a maximum metamorphic temperature of about 630°C. Near the margin of the Tauern Window typical parageneses of the greenschist facies are observed. By oxygen isotope thermometry, Hoernes & Friedrichsen (1974) determined maximum temperatures of about 650°C in the western part of the central Zillertaler Alps decreasing to about 400°C towards the margin of the Tauern Window.

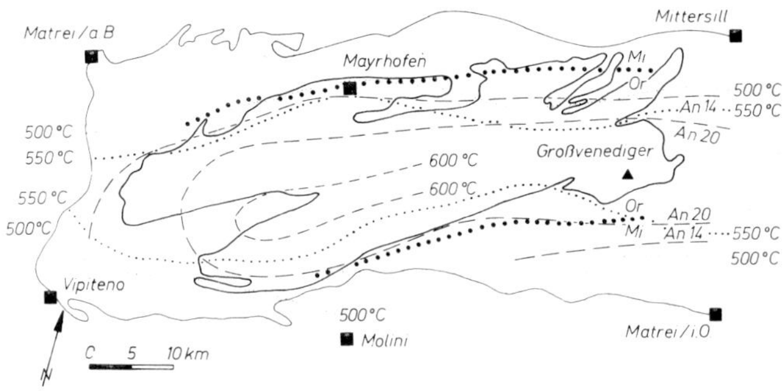

Fig. 1. Isograds in the western Tauern Window. Broken line: An-contents of plagioclase rims in epidote-bearing gneisses and schists (Morteani & Raase 1974). Thick dotted line: boundary between maximum microcline (Mi) and orthoclase to intermediate microcline (Or) (Raase & Morteani 1976). Thin dotted line: Isotherms based on $^{18}O/^{16}O$-thermometry (Hoernes & Friedrichsen this volume).

The metamorphic pressure in the western Tauern Window must have exceeded 5 kbars as suggested by the occurrence of kyanite, by the Al-rich composition of the amphiboles, by the phengitic composition of the white micas, and by the occurrence of eclogitic and glaucophanitic assemblages including the paragenesis zoisite + kyanite + quartz in the southern Venediger region (Morteani 1971; Raith 1971; Ackermand & Morteani 1973; Hoernes 1973; Abraham, Hörmann & Raith 1974; Miller 1974).

The crystallization sequence, in particular the reaction textures and the mineral zoning, suggest that the Alpidic metamorphism was essentially a two-stage process. Morteani & Raase (1974) studied the compositional zoning of plagioclase in epidote-bearing gneisses. In the central part of the Zillertaler Alps plagioclases are usually reversely zoned with about 15% An in the core and 20 - 30% An at the rim of the grains. Towards the margin of the Tauern Window the An-contents decrease to about 0 - 5% An in the core and 15 - 20% An at the rim and finally, unrimmed albite is found in the outermost zone of the Tauern Window. The reverse zoning of the plagioclase is generally discontinuous, suggesting a discontinuous increase of metamorphic temperatures during the final stage of the Alpidic crystallization.

The amphiboles may also show discontinuous zoning with an actinolite core and a rim of barroisitic Al-rich hornblende indicating an increase of temperature (Abraham et al. 1974; Raith et al. in this volume). These zoned amphiboles commonly occur in basic rocks of the Mesozoic Obere Schieferhülle, thus indicating two Alpidic crystallization stages.

Ackermand & Morteani (1976) have shown by microprobe analyses that the garnets of the Zillertaler Alps usually are continuously zoned. The Mg- and Fe-contents of these garnets increase and the Mn- and Ca-contents decrease from core to rim following a bell-shaped curve. Such zoning pattern, although not conclusive, may be taken as an additional argument for a late temperature increase. The intensity of garnet zoning significantly decreases towards the center of the Zillertaler Alps, indicating a homogenization in consequence of higher temperatures. In a few rocks garnets were found in which the rims show continuous zoning (as described above) but which, in addition, contain a sharply defined core which is considerably more Mn-rich and Ca-poor. This core may be interpreted as a relic of an earlier metamorphic phase.

Zonal structure may also be found in K-feldspar (Raase & Morteani 1976) and in epidote (Hörmann & Raith 1973; Raith 1976).

The Alpidic metamorphism was syn- to posttectonic (Sander 1911). This follows from microscopic studies of the crystallization-deformation relations and from the isograd pattern shown in fig. 1 which demonstrates that the isograds follow the major tectonic structures, although in detail, they may be discordant to the rock boundaries.

The age of the Alpidic metamorphism has been derived from K/Ar-determination on amphiboles, biotites, and white micas (Raith et al. this volume). An early Alpidic phase at about 70 m.y. is suggested from K/Ar-dates of glaucophane from the southern Venediger area. The main Alpidic event probably took place in the time-span between 50 and 30 m.y., and the latest post-kinematic crystallization representing the last climax of the Alpidic metamorphism occurred at about 20 m.y.. The time of the last cooling suggested from the lowest biotite ages was earlier in the Venediger region (18 - 19 m.y.) than in the western Zillertaler Alps (13 - 14 m.y.), indicating an earlier uplift and erosion in the middle part of the Tauern Window than in the west part.

The ages of the metamorphic phases can be correlated to the last phases of the geodynamic evolution of the Penninic complex. The glaucophane dates of about 70 m.y. indicate an early Alpidic high pressure/low temperature metamorphism which may be generated by the process of subduction of the Penninic series under the Austroalpine continental plate (Oxburgh & Turcotte 1974; Hawkesworth et al. 1975; Frisch 1976). The main Alpidic metamorphism (50 - 30 m.y.) of greenschist to amphibolite facies then was induced by rising temperatures within the Penninic series during thermal equilibration towards a normal geothermal gradient at the final stage of overthrusting. The last metamorphic crystallization (20 m.y.) was post-kinematic and was followed by an updoming and erosion of the Tauern Window.

References

Abraham, K., Hörmann, P.K. & Raith, M., 1974: N. Jb.Min.Abh., 122,1:1-35.
Ackermand, D. & Morteani, G., 1973: Contr.Mineral.Petrol. - 40:293-304.
- , 1976: Tschermaks Min.Petr.Mitt. - 23:117-136.
Besang, G., Harre, W., Karl, F., Kreuzer, H., Lenz, H., Müller, P. & Wendt, T., 1968: Geol.Jahrb. - 86:835-844.
Cliff, R.A., 1968: N.Jb.Geol.Paläont.Mh.II:655-663.
- , 1971: Contr.Min.Petrol. - 32:274-288.
Frisch, W., 1976: Geol.Rundschau 65:375-393.

Hawkesworth, O.J., Waters, D.J. & Bickle, M.J., 1975: Earth Planet.Sci. Lett. - 24:405-413.
Hoernes, S., 1973: Tschermaks Min.Petr.Mitt. - 20:81-106.
Hoernes, S. & Friedrichsen, H., 1974: Schw.Min.Petr.Mitt.54:769-788.
Hörmann, P.K. & Raith, M., 1973: Contr.Miner.Petrol. - 38:307-320.
Jäger, E., Karl, F. & Schmidegg, O., 1969: Tschermaks Min. Petr. Mitt. 13:251-273.
Lambert, R.St.J., 1964: Verh.Geol.B.A. Wien:16-26.
Miller, C., 1974: Schweiz.Min.Petr.Mitt. - 54:371-384.
Morteani, G., 1971: Verh.Geol.B.A. Wien: 287-314.
- , 1974: Fortschr.Miner.52, Beiheft 1:195-220.
Morteani, G. & Raase, P., 1974: Lithos 7:101-111.
Oxburgh,E.R. & Turcotte, D.L., 1974: Schweiz.Min.Petr.Mitt.,54:641-662.
Raase, P., 1972: Unpubl.Doctor's Thesis,Univ.Kiel, 164 p.
Raase, P. & Morteani, G., 1976: Geol.Rundschau. - 65:422-436.
Raith, M., 1971: Verh.Geol. B.-A., Wien:163-207.
- , 1976: Contr.Mineral.Petrol. - 57:99-117.
Sander, B., 1911: Denkschr.Akad. Wiss. Wien, math.nat.Kl. - 82:257-319.
Satir, M., 1974: Schweiz.Min.Petr.Mitt. - 54:213-228.

Early Palaeozoic Geodynamics in the Eastern and Southern Alps

R. Höll, J. Loeschke, A. Maucher and Kl. Schmidt

Radiometric data and their different interpretation have stimulated renewed discussions about Caledonian mountain building in the evolution of polymetamorphic basement rocks in the Eastern and Southern Alps. The results are compiled in the Metamorphic Map of the Alps (1 : 1 Mio.) (gen. coord. E.Niggli 1973).

Sassi & Zanferrari (1972) postulated an unconformity separating the crystalline basement of the Deferegger Alps (Cornetto di Confine area) from the overlying Thurntaler Quarzphyllit complex, as evidence for Caledonian orogenic processes in the strict sense. Sassi et al. (1974 a, b) assumed that the gneiss-mica schist basement received its amphibolite-facies metamorphism during the Caledonian orogeny, while the quartzphyllite obtained its low-grade (greenschist facies) character during the Hercynian orogeny. They take for granted that the postulated unconformity and the corresponding metamorphic hiatus are due to late Caledonian basement elevations ranging from 10 to 30 km.

Heinisch & Schmidt (1976) and Schönlaub & Daurer (1977), however, could not agree with these theories.

Nevertheless, an intense pre-Upper Ordovician, i.e. *pre-Sardinian*, metamorphism is proved in parts of the Northern Graywacke Zone by Schönlaub & Daurer (1977), and in the Gleinalpe by Frank et al. (1976). In the Ötztal region, anatexis did not begin earlier than 450 m.y. B.P. (Satir 1976). Brack et al. (1977) obtained 440 m.y. for a metamorphism in the

Schobergruppe. However, as no regional unconformity is evident in Upper Ordovician to Devonian rock series of the Alps, terms such as *Caledonian event*, *thermal event* etc. are used by some authors (Schmidt 1976).

Moreover, there are no reliable indications of a tectogenesis between Upper Ordovician and Devonian or Carboniferous time in nonmetamorphic Early Palaeozoic sedimentary-volcanic complexes. Local facies changes and intercalations of clastic rocks can be interpreted as epirogenic movements or eustatic sealevel oscillations in a locally mobile sedimentary basin.

An increasing number of biostratigraphic and radiometric data indicate intense basaltic and calc-alkaline volcanism and deep-seated granitic-granodioritic magmatism in the Eastern Alps lasting from the Ordovician (Caradocian) to Silurian time (Mostler 1970; Loeschke 1973, 1975). Basic submarine lavas and tuffs (spilites, greenschists, amphibolites) are the oldest volcanic rocks, followed by partly subaerial acid volcanic rocks (ignimbrites, keratophyres, porphyroids).

The widespread, felsic Blasseneck Porphyroid can be interpreted as a surface equivalent of the deep-seated magmatism. As it was not coupled with incisive regional alterations of palaeogeographic patterns and trends, this magmatism is classified as *anorogenic*. Comparisons of the Early Palaeozoic evolution of the Alps and Central Europe display striking similarities. Deep-seated magmatism and calc-alkaline volcanism have the same age, and the palaeogeography reveals analogous trends. Therefore, one may suppose that both areas are part of the same West European *marginal sea system* which was bordered by Caledonian subduction systems in the NW (Mitchell 1976) and by the European Platform in the NE (Schmidt, in press).

Although we have no evidence for spreading centers or stripes of oceanic crust in Central Europe, there are well-known domains of volcanism with long lasting, but interrupted activity (Proterozoic-Lower Carboniferous) and shifting eruption centers.

The original basement of the volcanic rock bearing series in the Eastern Alps is unknown. The local existence of underlying gneisses or granites, however, cannot be excluded, since Schönlaub (in press) discovered granite- or granite-gneiss-conglomerates in Ordovician sequences below *porphyroid* complexes.

A comparison of Cenozoic and Recent volcanic rocks of different geotectonic settings (Carmichael et al. 1974) encourages the view that the Early Palaeozoic volcanics of the Austrian Alps are similar to volcanic rocks of marginal basins and volcanic chains within island arc systems. Volume distribution, the bathymetric development, chemical composition and the stratigraphic succession favour this interpretation (Mitchell & Reading 1971; Miyashiro 1974; Ringwood 1975). Furthermore, geochemical data of *immobile* elements (Ti, Zr, Y, P) might be interpreted as indications for the existence of a subduction zone of Ordovician age. Loeschke (1973, 1975) considers the Early Palaeozoic volcanics as differentiation products generated above a possibly south-dipping subduction zone. Volcanic processes would reflect an island arc evolution from submarine to local subaerial stages whereby the thickness of the crust gradually increased.

Höll (1975) and Höll & Maucher (1976) stimulated speculations concerning the Early Palaeozoic metallogenesis of the Eastern Alps. They have pointed to the ore mineralization and the arrangement of the Early Palaeozoic stratiform ore deposits on palinspastic schemes of the Eastern Alps.

Most of these ore deposits show a close spatial, temporal and genetic relationship to the Early Palaeozoic submarine volcanism. Such ore assemblages and ore belts are also known from North America (Sutherland Brown & Cathro 1976), from the Andes (Bolivia), the Kuriles, Japan (Ei Korikoshi 1976), and the northern part of New Zealand (Weissberg 1969) where they are bound to subduction systems (Clarke et al. 1976; Sillitoe 1976; Mitchell 1976). Höll (1975) and Höll & Maucher (1976) design a north-dipping Benioff-Zone, deduced from a comparison of the transverse variations in the relative abundance of the ore metals within the Early Palaeozoic ore deposits of the Eastern Alps and the Mesozoic and Cenozoic ore deposits in the Circum-Pacific areas. Thereby they assume that the major stratabound ore deposits of the Eastern Alps containing W-Mo-Bi-Sn were resulted from the furthest down-dip ore generation above the Benioff-Zone.

The Early Palaeozoic subduction processes of the Eastern Alps provided favourable conditions for magma emplacement and metallogenesis without transforming the geosyncline into a Caledonian mountain chain.

The final consolidation of the crust was completed much later by the Variscan orogeny, which comprised a complete orogenic cycle including a flysch stage, tectogenesis with folding and gravity tectonics, magmatism, metamorphism, uplift and a molasse stage. It is to be considered in context with the formation of the Late Palaeozoic Pangea.

References

Brack, W., Köhler, H. & Müller-Sohnius, D., 1976: Geol.Rdsch. 65,H. 2: 483-511.

Carmichael, I., Turner, F. & Verhoogen, J., 1974: Igneous petrology. New York (McGraw Hill), 739 p.

Clarke, A.H., Caelles, J.C., Farrar, E., Haynes, S.J., Lortie, R.B., McBride, S.L., Quirt, S.G., Robertson, R.C.R. & Zentilli, M., 1976: The Geological Association of Canada, Spec. Paper. - 14:23-58.

Ei Horikoshi, 1976: The Geological Association of Canada, Spec. Paper. - 14:121-142.

Frank, W., Klein, P., Nowy, W. & Scharbert, S., 1976: Tscherm.Mineral. Petr. Mitt. - 23:191-203.

Heinisch, H. & Schmidt, K., 1976: Geol. Rdsch. - 65:459-482.

Höll, R., 1975: Habil.Schrift.Bayer.Akad.Wiss., Math.-naturw. Kl. Abh., N.F., H. 157 A, 114 p.

Höll, R. & Maucher, A., 1976: Handbook of Strata-Bound and Stratiform Ore Deposits, Vol. 5:1-36, Amsterdam (Elsevier).

Karig, D.E., 1974: Ann.Rev.Earth Planet.Sci. - 2:51-75.

Loeschke, J., 1973: N. Jb.Mineral.Abh. - 119:20-56.

- , 1975: Geol.Rdsch. - 64:62-74.

Mitchell, A. & Reading, H., 1971: J. Geol. - 79:253-284.

Mitchell, A.H.G., 1976: The Geological Association of Canada, Spec. Paper, 14:3-21.

Miyashiro, A. 1974: Amer.J.Sci. - 274:321-355.

Mostler, H., 1970: Festbd. Geol.Inst. 300-Jahr-Feier Univ.Innsbruck: 507-522.

Ringwood, A., 1975: Composition and petrology of the earth's mantle. New York (McGraw Hill) 618 p.

Sassi, F.P. & Zanferrari, A. 1972:Boll.Soc.Geol.Ital.,91:533-557,Roma.

Sassi, F.P., Zanferrari, A. & Zirpoli, G., 1974a: N.Jb.Geol.Paläont. Mh. 1974:609-624.
- , 1974 b: Mem. Mus.Tridentino di Sci.Nat., 20:3-32.
Satir, M., 1976: Geol.Rdsch. - 65:394-410.
Schmidt, Kl., 1976: KOSSMAT-Symposium 1974, Nova acta Leopoldina,224, Halle.
Schönlaub, H.P. & Daurer, A., 1977: Geodynamics and geotraverses around the Alps (Abstracts). Meeting 28.2.-1.3.1977 in Salzburg:47-49.
Sillitoe, R.H., 1976: The Geological Association of Canada, Spec. Paper. 14:59-100.
Sutherland Brown, A. & Cathro, R.J., 1976: Canadian Institute of Mining and Metallurgy, Spec. Vol. - 15:7-16.
Weissberg, B.G., 1969, Econ. Geol. 64:95-108.

Oxygen Isotope Studies of the Austroalpine and Penninic Units of the Eastern Alps

S. Hoernes and H. Friedrichsen

Oxygen isotope data were collected on rocks and minerals from the Penninic Tauern window and from Austroalpine rocks of the Ötztal - Stubai Alps and associated overlying Brennermesozoikum, in order to obtain information on a) the scale of oxygen isotope exchange and b) the grade and extent of Alpine metamorphism.

The pre-Mesozoic rocks of the area investigated were metamorphosed during several events. Radiometric age determinations (Besang et al.1968; Satir 1974; Schmidt et al. 1967; Raith et al. this volume) indicate thermal events from Ordovician up to Neogene. The degree of Alpine metamorhpism is usually lower than the pre-Alpine events.

Experimental procedure

Oxygen was extracted from minerals using standard methods employing BrF_5. Details are described in Friedrichsen et al. (1973), Hoernes & Friedrichsen (1974).

Oxygen isotope compositions are reported using the common δ-notation (relative to SMOW), with an analytical uncertainty of ± 0.1‰. Temperatures were derived from oxygen isotope fractionations between mineral pairs using fractionation curves described in Bertenrath et al. (1973) and Hoernes & Friedrichsen (1974) and Hoernes & Friedrichsen (in prep.).

Results

a) Whole rock data: Unaltered granitic ortho-gneisses typically have

$\delta^{18}O$ values of +5 to +9‰, paragneisses +10 to +20‰ and marbles +25 to +30‰.

Histograms showing calculated $\delta^{18}O$ values for whole rock samples of the Penninic units and of the Austroalpine units are given in fig. 1. These calculated oxygen isotope compositions cover the whole range of metasedimentary to metavolcanic rocks. From the scatter of $\delta^{18}O$ values of the Mesozoic rocks (i.e. peripheral Schieferhülle and Brennermesozoikum), we conclude that no large scale transport of oxygen took place during Alpine metamorphism.

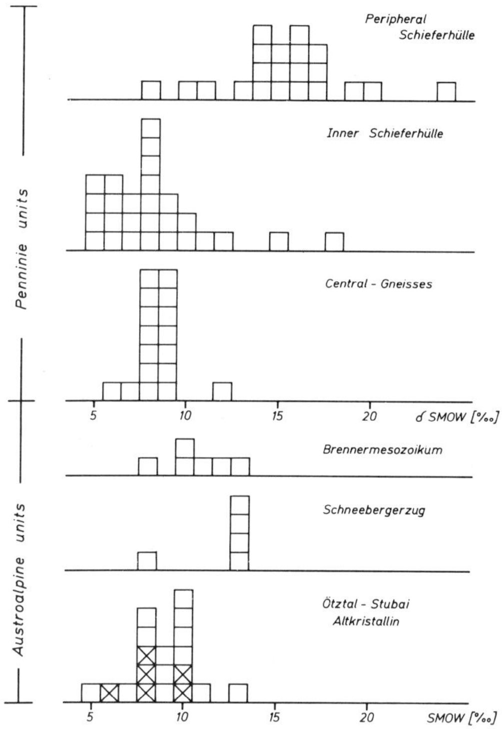

Fig. 1. Histograms showing calculated whole rock compositions. Each square represents one sample. The squares with the cross represent orthogneisses from the Ötztal-Stubai Altkristallin, the squares without the cross are the para-gneisses of this area.

In contrast to the Mesozoic rocks, metasediments of the Inner Schieferhülle show significant large-scale homogenization of their oxygen isotopes with the central gneisses. Similarly, large scale oxygen isotope exchange took place between the para-gneisses and ortho-gneisses of the Ötztal-Stubai Altkristallin. This large-scale homogenisation of oxygen isotopes must be related to pre-Alpine events.

B) Oxygen isotope data for cogenetic minerals

1) Penninic units: During Alpine metamorphism, oxygen isotope equilibration between minerals was attained in most of the rocks from the Tauern window.

The criteria for this argument were discussed in Hoernes & Friedrichsen (1974).

Paleo-isotherms for Alpine metamorphism, based on oxygen isotope frac-

tionations between cogenetic minerals indicate a temperature maximum of 650°C at the Alpenhauptkamm (geographically highest parts of the Alps) near Schlegeis. North and south of the Alpenhauptkamm, the temperatures of metamorphism were distinctly lower. Near Gerlos, in the northern part of the Tauern window and in the Dorfertal to the south, the temperatures were as low as 415°C (fig. 2).

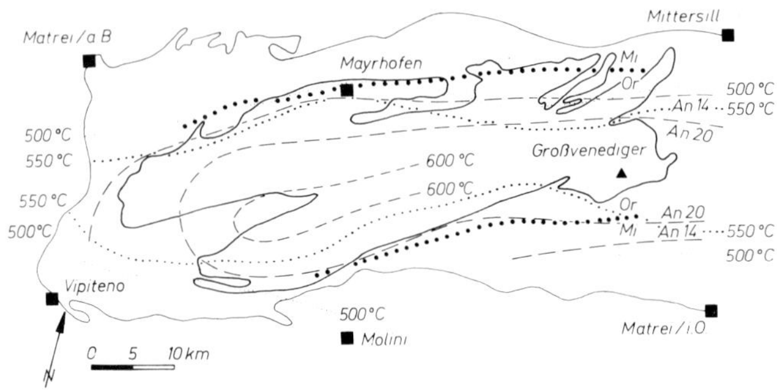

Fig. 2. Paleo-temperatures in the Tauern window (see also Ackermand et al. this volume).

When plotted on a P - T diagram (fig. 3), temperatures and pressures (interpreted from published experimental data on mineral parageneses present and an estimation of the thickness of the overlying tectonic units) describe a normal, continental geotherm. This geotherm could be produced by a heat flow of 2 HFU. (Richardson 1970). Recent measurements in the Tauern area yielded a heat flow of 2.5 HFU. (Oxburgh & Turcotte 1974).

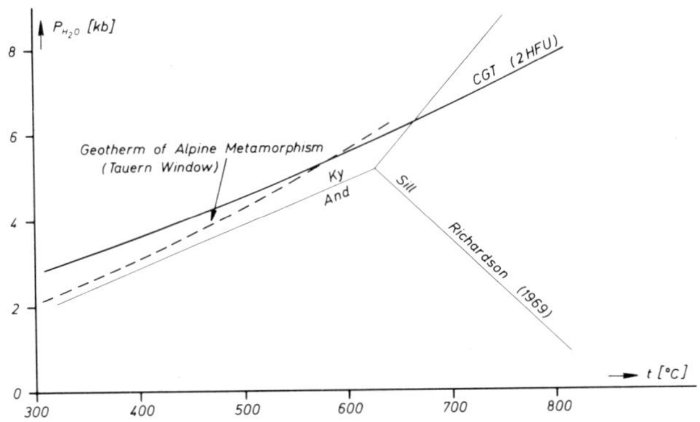

Fig. 3. P-T diagram illustrating the conditions of Alpine metamorphism in the Tauern window. The temperatures are derived from the oxygen isotope fractionations, pressures from an estimation of the thickness of the overlying units and from mineral parageneses found in this area. (Kyanite is the only found aluminium silicate).

Only a few cases of oxygen isotope disequilibrium were found. Silicate phases in eclogites from the peripheral Schieferhülle south of the Großvenediger, and in some coexisting carbonate and silicate phases of the

peripheral Schieferhülle, yielded nonconcordant mineral fractionations. In these rocks, oxygen isotope equilibrium, if it exists at all, must be restricted to a volume of less than a few ccm, and may occur only at grain boundaries.

2) Austroalpine units: The basal Permo-Carboniferous conglomerates of the Brennermesozoikum are isotopically inhomogenous on a mm scale, as shown by different $\delta^{18}O$ values for detrital and matrix quartz.

Detrital quartz has a $\delta^{18}O$ of + 8 to + 9‰, in contrast to + 12‰ for matrix quartz. Micas and the finest-grained portion of the matrix quartz did reach oxygen isotope equilibrium, however, during Alpine metamorphism. The detrital quartz pebbles of these conglomerates, on the other hand, did not recrystallize during Alpine metamorphism, and hence it seems that recrystallization is required to promote extensive oxygen isotope exchange.

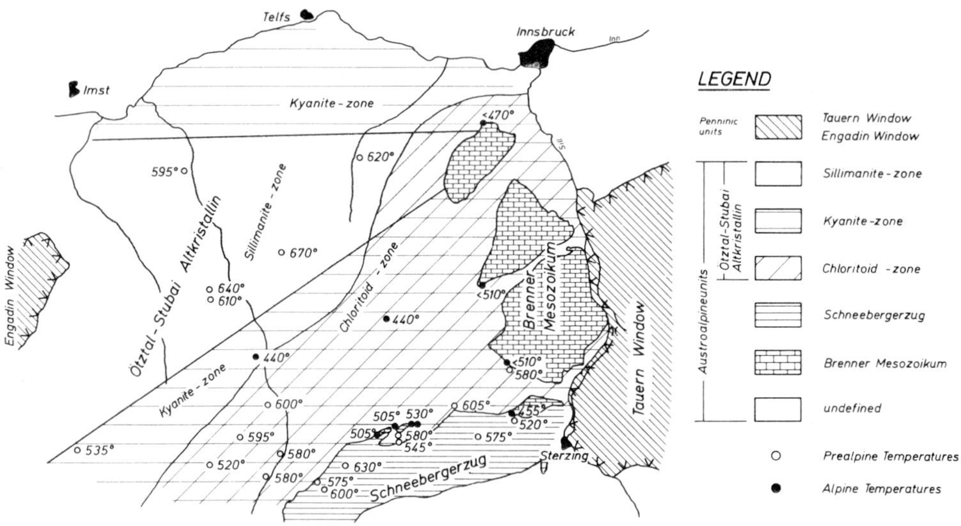

Fig. 4. Simplified geological map of the northern Ötztal-Stubai Alps, showing the mineral zones (Purtscheller 1968) and the temperatures derived from oxygen isotope fractionations. Open circles represent pre-Alpine - closed circles the Alpine temperatures. The temperatures found in the Schneebergerzug may be Alpine, but the preservation of pre-Alpine *relict temperatures* is also possible.

The temperatures of Alpine metamorphism in the Austroalpine units of the Ötztal-Stubai Alps were as high as 500°C in rocks of the Brennermesozoikum. Higher temperatures found in rocks of the Schneebergerzug up to 600°C, may also be due to the Alpine metamorphism, but the preservation of some *relict temperatures* of pre-Alpine metamorphism can not be excluded. In the sillimanite zone of the northern part of the Ötztal-Stubai Alps, a pre-Alpine origin for temperatures above 600°C is compatible with metamorphic mineral assemblages (Purtscheller 1968) (fig. 4).

Conclusions

During the Alpine orogeny the Austroalpine units of the Ötztal-Stubai Alps and the Penninic units of the Tauern window reached metamorphic conditions at different times. Assuming a cooling rate of 20°C/1 m.y. the biotites of the Brennermesozoikum (with a cooling age of 80 m.y. according to Schmidt et al. (1967) and a temperature of formation of 500°C) must have been crystallized at least 90 m.y. ago. For this calculation we assume a blocking temperature of 300°C for the radiogenic isotopes in the biotites (Jäger 1967). During that time the conditions of a low temperature/high pressure metamorphism in the Penninic units are assumed. Petrographic evidence for these conditions based on some relict mineral parageneses are reported by Miller (1974) and Fry (1973). (See also Ernst 1973). The recrystallization of the minerals in the Tauern window during the later high temperature phase of Alpine metamorphism yielded a complete reequilibration of the oxygen isotopes. Isotope equilibrium between the minerals and the fluid phases has been obtained on a small scale (i. e. within a hand specimen).

References

Bertenrath, R., Friedrichsen, H. & Hellner, E.E., 1973: Fortschr. Mineral. 50, Beih. 3:32-33.

Besang, C., Harre, W., Karl, F., Kreuzer, H., Lenz, H., Müller, P. & Wendt, I., 1968: Geol. Jb. - 86:835-844.

Ernst, W.G., 1973: Bull. Geol. Soc. Amer. - 84:2053-2078.

Friedrichsen, H., Müller, G. & Stahl, W., 1973: Contr. Mineral. Petrol.- 40:159-168.

Fry, N., 1973: Min. Mag. - 39:121-122.

Hoernes, S. & Friedrichsen, H., 1974: Schweiz. Min. Petr. Mitt.-54:769-788.

- , in prep.: Oxygen isotope data of the polymetamorphic area of the northern Ötztal-Stubai Alps (Tirol).

Jaeger, E., Niggli, E. & Wenk, E., 1967: Beitr. Geol. Karte Schweiz N.F. 134, 67 p.

Miller, Ch., 1974: Schweiz. Min. Petr. Mitt. - 54:371-384.

Oxburgh, E.R. & Turcotte, D.L., 1974: Schweiz. Min. Petr. Mitt. 54:641-662.

Purtscheller, F., 1968: Tschermaks Miner. u. Petrogr. Mitt. - 13:35-54.

Richardson, S.W., 1970: Fortschr. Mineral. - 47:65-76.

Richardson, S.W., Gilbert, M.C. & Bell, P.M., 1969: Amer. I. Sci. - 267:259-272.

Satir, M., 1974: Schweiz. Min. Petrogr. Mitt. - 54:213-228.

Schmidt, K., Jäger, E., Grünenfelder, M. & Grögler, N., 1967: Eclogae geol. Helv. - 60:529-536.

Geodynamics of the Western Tauern Window

D.D. Klemm, B. Lammerer and K. Weber-Diefenbach

The area of interest is situated in the centre of geotraverse I A. It contains the deepest tectonic units which have undergone the highest degree of Alpine metamorphism as well as considerable tectonic deformation.

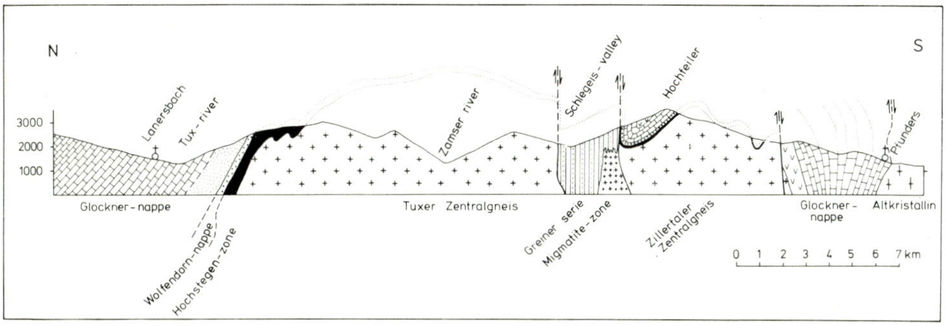

Fig. 1. Geologic section of the Western Zillertal Alps partly after Baggio et al. (1969) and Frisch (1974).

The deepest tectonic elements are represented by Variscan gneisses (fig. 1), which are separated by the Paleozoic Greiner schist series and a migmatic zone, probably of Varistic age, too. A Mesozoic cover of mainly carbonate sediments is partly dated by fossils (Hochstegen Zone and Wolfendorn nappe). The basal gneiss complex consists of meta-granodiorites, - granites, - quartzdiorites, amphibolites and acidic and basic dikes. Geochemical considerations (Reissinger 1971; Prosser 1975) lead to the concept of widespread palingenesis during the late Variscan orogeny. Restites of the unassimilated old rock series are found as dark xenolithic inclusions within the magmatites.

Petrographical and geochemical studies show that the Greiner series is composed of the metamorphic equivalents of a sedimentary sequence with bituminous clays, marls, arkoses or greywackes, conglomerates and breccias, sandstones and subordinate limestones, and a sequence of greenschists, hornblende gneisses, amphibolites and serpentinites, which must be considered to have been mainly vulcanites (Lammerer 1972; Haydn 1970; Weber-Diefenbach 1976; Reissinger 1976; Lammerer et al. 1976).

The rapid change from one rock type to another, the commonly low thickness of the layers, their intense lateral fluctuation and their petrographic composition mediate the concept of a periterrestrial marine sedimentary environment characterized by tectonic disturbance and clastic material derived from a metamorphic basement. A coastal landscape or a local graben or trough with terrestrial sedimentation and marine ingressions accompanied by vulcanic events might best describe the environmental situation.

A major thrust separated this part, which is considered to belong to the European plate, form the ophiolitic Glockner nappe, which represents the oceanic or oceanized realm of the Penninic zone. Sandy and argillaceous carbonates from a pile of sediments, about 1000 m in thickness which underwent high pressure metamorphism (lawsonite, glaucophan) followed bv high temperature metamorphism (Morteani 1971; Hoernes & Friedrichsen 1974; Miller 1974).

With a nearly vertical dipping tectonic contact representing the former subduction plane, the Glockner nappe borders on to the Altkristallin in the south, whose Variscan metamorphic gneisses and amphibolites underwent only low-grade Alpine metamorphism. Post-tectonic granites and dikes appear to be of Alpidic age, but have not yet been dated. The Altkristallin and the series to the south form the base of the Austroalpine Unit which is part of the African plate (resp. Carnian or Adriatic subplate). Considerable vertical movements took place after the nappes were emplaced in the study area. This is testified not only by the still continuing orographic rise (Senftl & Exner 1973), but also by differential movements along E-W (NE-SW) striking vertical faults which have produced extensive drag folds (e.g. in the Pfitsch valley). The northern fault blocks are relatively raised so that the intensity of the Alpine metamorphism decreases rapidly toward the south. About 15 km to the south the Periadriatic Lineament separates the metamorphic central Alps from the non-metamorphic southern Alps.

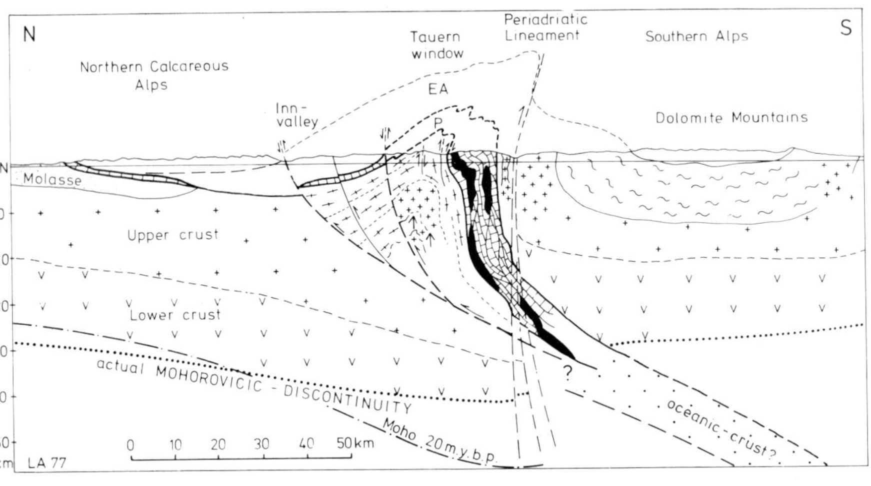

Fig. 2. The western Tauern Window and its tectonic setting in a section through the Alps.

Fig. 2 shows a geodynamic interpretation of the present surface geology and geophysical data based on the supposition of a southward-dipping subduction zone in the late Cretaceous and the Paleogene.

Penninic oceanic or oceanized crust was subducted under the African plate. Part of the oceanic sedimentary cover was imbricated between both plates. In the subducted European plate part of the upper crust may have been sheared off from the lower crust (Closs 1975), forming an initial

anticlinal structure and a thickening of the upper crust. Frictional heat (Roeder 1976) in addition to updoming of isogrades may be responsible for late thermal metamorphism and may have produced sialic melts forming diapiric granitoids which forced local updoming, but also not exposed in the present erosion level.

The Periadriatic lineament is not considered as a former subduction plane, but as a part of the fault system, which was established to re-adjust buojancy between the plates, after subduction had come to rest.

Uplift of the Tauern window of the Alps was followed by subsidence in the foreland where the Molasse basin formed. The final gravity-gliding of the Austro-alpine nappes into the sediments of the Molasse trough might have been possible as a result of this opposing sense of movement, that is an uplift of 10 - 20 km behind the Austro-Alpine nappes, and contemporaneous subsidence of about 5 km in front of the nappes.

References

Angenheister, G., Bögel, H., Gebrande, H., Giese, P. Schmidt-Thomé, P. & Zeil W., 1972: Geol. Rdsch. - 61,2:349-395.

Closs, H., 1975: Rapp. Comm. int. Mer Médit. - 23, 4a:55-57.

Frisch, W., 1974: Mitt.Geol.Ges.Wien 66-67, E.Clar Festschrift:9-20.

- , 1975: Verh. Geol.B.-A., 2-3:201-221.

- , 1976: Geol. Rdsch. - 65,2:in press.

Hawkesworth, C.J., Waters, D.J. & Bickle, M.J., 1975: Earth and Planetary Science Letters. - 24:405-413.

Hoernes, S. & Friedrichsen, H., 1974: SMPM. - 54:769-788.

Klemm, D.D. & Prosser, E., 1975: Nachr. dt. geol.Ges. - 13:25-26.

Lammerer, B., 1975: AV-Jahrbuch 100:16-25.

Lammerer, B., Fruth, I., Klemm, D.D., Prosser, E.& Weber-Diefenbach, K., 1976: Geol. Rdsch. - 65:in press.

Miller, Ch., 1974: Schweiz. Min. Petr. Mitt. - 54:371-384.

Morteani, G., 1971: Verh. Geol. B.-A.:287-314, Wien.

Prosser, E., 1975: Diss. Univ. München. - 70 p.

Reissinger, M., 1971: Unveröffentl.Dipl.-Arb. Univ. München.

Roeder, D., Continental Convergence in the Alps. - in press.

Senftl, E. & Exner, Ch., 1973: Verh.Geol.B.-A. - 2:209-234, Wien.

Thiele, O., 1974: Jahrb. Geol. B.-A. - 117:55-74.

Tollmann, A., 1975: N.Jb. Geol. Paläont. Abh. - 148,3:286-319.

Weber-Diefenbach, K., 1976: Tschermaks Min. Petrl. Mitt. - 23:1-22.

The Northern Margin of the Tauern Window North of the Tuxbach River, Tirol

Hubert Miller

Between the relatively autochthonous *Tuxer Gneis* in the south and the Lower Austroalpine *Innsbrucker Quarzphyllit* in the north, on both sides of the Tuxbach river, there is a narrow zone of stratigraphically and tectonically very variable rock complexes. On the southern side of the Tuxbach river Frisch (1968) and Thiele (1970) have contributed to the knowledge of the parautochthonous and Penninic units. North of the Tuxbach river, the author and collaborator (Lagally & Miller 1974) have recently confirmed an allochthonous tectonic style.

The classic *Tuxer Phyllite* (Sander 1921) can be devided into a Paleozoic and a Mesozoic series.

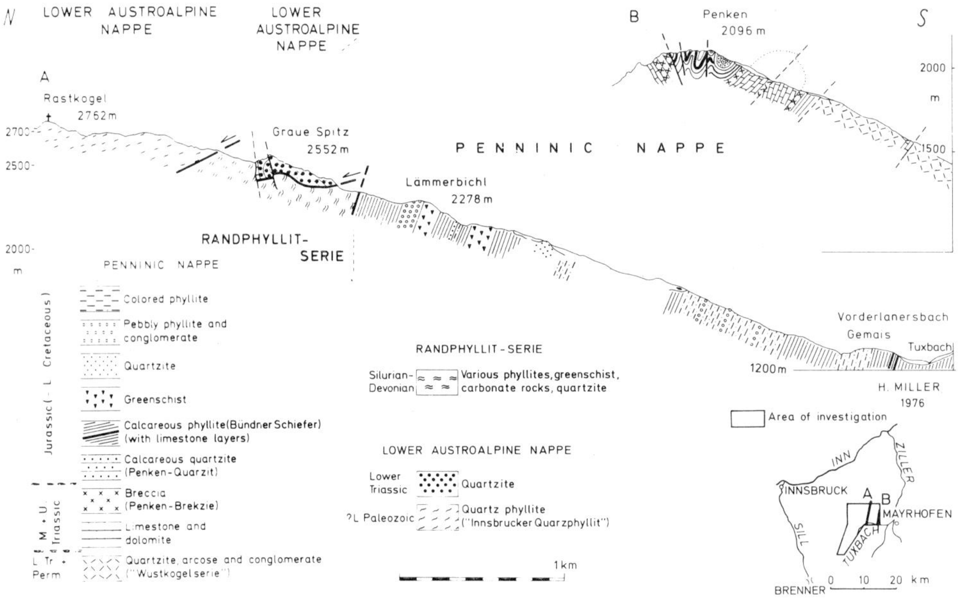

Fig. 1. Two profiles showing the structural units of the northern margin of the Tauern window north of the Tuxbach river.

The Paleozoic phyllites with intercalated greenschists, marbles, quartzites, dolomites and magnesites (*Randphyllit-Serie*) have been placed (e.g. Mostler 1973; Frisch 1974) into the *Innsbrucker Quarzphyllit* of the Lower Austroalpine unit.

By detailed mapping and analysing the mesoscopic structures, it was found that this paleozoic group of metasediments and metavolcanic rocks (*Hoarberger-Lage* in Lagally & Miller 1974) has to be separated from the *Innsbrucker Quarzphyllit* complex. Syngenetic scheelite concentrations of economic interest are known from the *Habach-Serie* (Penninic unit) and the paleozoic series described here as *Randphyllit-Serie*. The Randphyllit-

Serie may form a kind of transition zone between the Austroalpine Innsbrucker Quarzphyllit and the Penninic zone. It is overlain by some tectonic klippen of Lower Austroalpine rocks (e.g. Graue Spitz and Hippold-Spitz).

The Mesozoic *Bündner-Schiefer-Serie* can be divided into three members: The lower one is composed of phyllites of low carbonate content, calcareous quartzites and some limestone beds; the intermediate one is characterized by multicoloured phyllites, calcareous phyllites, greenschists, pebbly phyllites and conglomerates; the uppermost one consists mainly of typical calcareous phyllites (*Bündner Schiefer*).

The Permo-Triassic sediments of the Penken and Gschößwand area in the eastern continuation of the study area onto Mayrhofen had been previously attributed to the Lower Austroalpine nappe (e.g. Kristan-Tollmann 1962). Later Thiele (1970) and Frisch (1974) have shown that the biggest part of the Penken-Gschößwand area belongs to the Penninic nappe. Recent mapping (Vels 1975) in the area between the Penken summit region and the village of Vorderlanersbach proves that the Penken summit region (Penken-Gschößberg- and Knorn-Schuppe; Kristan-Tollmann 1962) is also forming part of the Penninic nappe. The phyllites of the Penken summit area are typical Bündner Schiefer; the Penkenquarzit and the Penkenbrekzie are identical with calcareous quartzites and breccias of the area NE of Vorderlanersbach, which are stratigraphically related to typical Bündner Schiefer of the Penninic nappe.

References

Frisch, W., 1968: Mitt.Ges.Geol.Bergbaustud. - 1967, 18:287-336.
- , 1974: Mitt. Geol. Ges. Wien, 66-67:9-20.
Kristan-Tollmann, E., 1962: Mitt.Geol.Ges. Wien, 54, 1961:201-227.
Lagally, U. & Miller, H., 1974: N.Jb. Geol. Paläont.Mh., 1974:374-384.
Mostler, H., 1973: Veröff. Univ. Innsbruck, 86:237-266.
Sander, B., 1921: Jb. Geol.St.-Anstalt. - 70, 1920:273-296.
Thiele, O., 1970: Verh. Geol. B.-A. - 1970:230-244, Wien.
Vels, B., 1975: Unveröffentl. Dipl.-Arbeit Univ. Münster: 106 p.

For a more extensive bibliography see Lagally, U. & Miller, H. (1974).

Stratigraphy and Structure of Mesozoic Complexes in the Quartz Phyllite Nappe (Radstädter Tauern, Austria)

R. Roßner

The sedimentary series of the Radstädter Tauern with an unique facies in its Mesozoic part, belong structurally to the lower Austroalpine "Lungauridenring" (Kober e.g. 1955, p. 292). This ring of petrographically similar rocks surrounds the Tauern cupola on all sides. This fact has been interpreted by Kober with the existence of a tectonic window. The internal

structure of the Radstädter Tauern Mountains confirms the window interpretation.

The structure of the Radstadt zone consists of a system of imbricate thrust sheets dipping to the North. Their extent can be seen in fig. 1.

The highest of these thrust units is the Quartz phyllite nappe. In its northern part, extensive Permo-Mesozoic beds can be found only at the

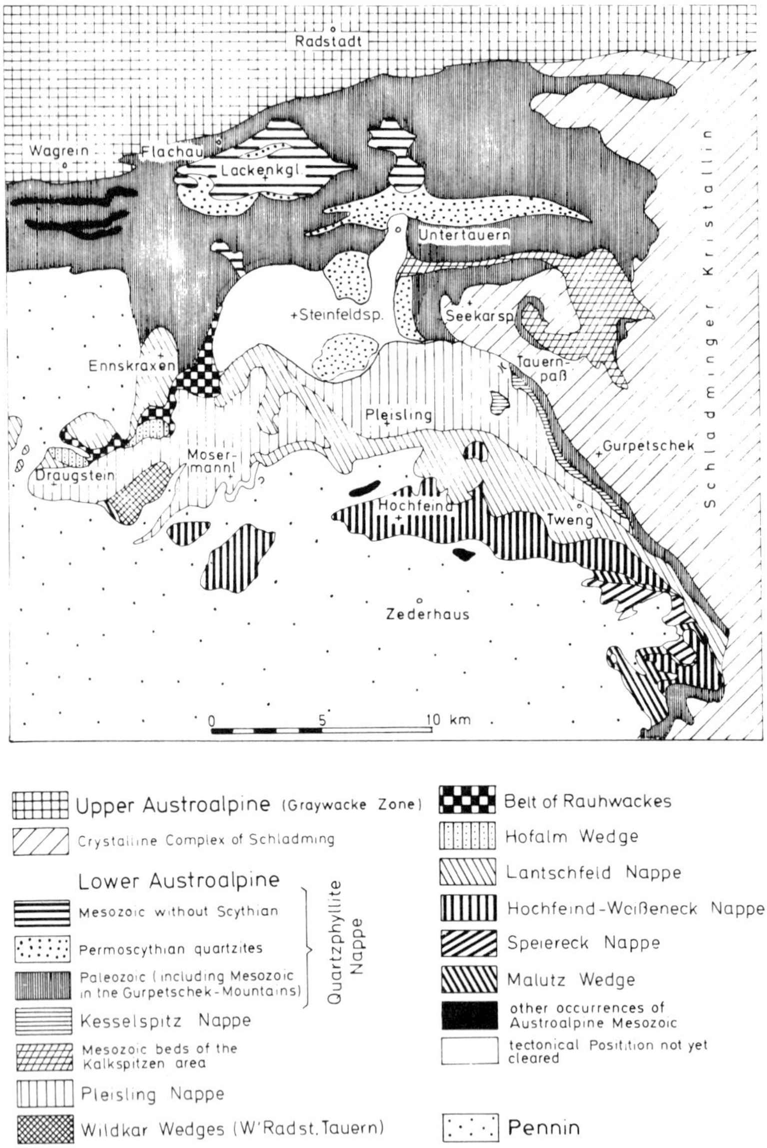

Fig. 1. General map of the tectonic structure of the Radstädter Tauern Mountains (according to Demmer 1962; Frech 1901; Tollmann 1963a and b, 1964 and investigations of the author).

Lackenkogel and along the Taurach valley north of Untertauern. The remainder of this unit consists of older Paleozoic, possibly Silurian or Devonian quartz phyllite (Schönlaub, unpublished ms).

Mapping and stratigraphic studies of the Lackenkogel area show that the Permo-Mesozoic sequence is in upright position. It includes a sediment suite ranging from Permo-Scythian to Raetian. The Permo-Scythian part consists of quartzites which are at some places underlain by quartzitic schists, possibly equivalents of the Alpine Verrucano (Tollmann 1960). Anisian beds show abnormally thick limestones and shales and an abundance of dolomite in the upper part of the column. The massive and stratified Ladinian dolomites are generally devoid of internal sedimentary fabric, but in the south part of the Lackenkogel zone, the microfacies indicates a transition from reef to back reef lagoon similar to facies in the Wetterstein limestone (Ott 1972; Roßner 1976). This transitional facies differs from the uniform lagoonal facies with diplopores and gastropods (Ott 1972) distributed uniformly through the central Radstatt zone. Tectonic deformation has greatly reduced the thickening of the slaty and arenaceous part of the Carnian beds. Dark grey dolomites overlying these rocks can probably also be dated as Carnian. They gradually pass upward into the Hauptdolomit of the Norian stage.

The Raetian suite can be subdivided into a lower part with Kössen development and an overlying part consisting of Raeto-Liassic reef limestone with corals and sponges (Fabricius 1966).

Micropaleontological studies yielded conodonts and sclerites of holothurians, for the first time in Triassic rocks of the Central Alps; Anisian beds of the Lackenkogel area contain: *Theelia sp.*, *Schizotheelia spiniperforata* (Zawidzka) and Norian beds: *Priscopedatus triassicus* (Mostler). In Ladinian beds from the southern rim of the Quartz phyllite nappe (Steinfeldspitz area) there could be discovered an association of sclerites of holothurians (*Theelia anisica* Mostler, *Priscopedatus heisseli* Mostler) and conodonts (*Gondolella mombergensis* Tatge, *Gladigondolella tethydis* Huckriede ME) (all fossils det. Mostler).

The series of Mesozoic rocks along the Taurach valley north of Untertauern is overturned throughout. The rock suite extends from Scythian quartzites to Ladinian dolomites and resembles the Mesozoic of the Lakkenkogel area.

The facies of the Anisian beds in both complexes suggests strong affinities with Tollmann's (1958) *northerly facies* in parts of the Lantschfeld nappe.

Furthermore, the facies of the Anisian shows only little similarity to the Anisian of the Northern Calcareous Alps that would be expected if palinspastically the sedimentation areas of the highest Radstadt nappe and the lower Calcareous Alp nappes had been adjacent.

According to the minor structures the Lackenkogel complex is a large syncline overturned to the North and lying upon the Quartz phyllite mass (see fig. 2). The Middle Triassic substrata of the central Hauptdolomit-Raetian complex of the syncline was strongly deformed and imbricated. Secondarily the northern rim of the syncline was pressed to the South and produced southerly overturned structures. During a later phase the eastern part, east of the Zauch valley was tilted clockwise along N-S-striking faults.

On both flanks of the Taurach valley, north of Untertauern, the Mesozoic beds and fold axes dip and plunge underneath the Quartz phyllite mass suggesting a tectonic window in this region, if this term is used to

describe the stratigraphically younger inverted middle limb of a nappe fold appearing beneath its older core.

Existing map data suggest an extended syncline-anticline structure of the Quartz phyllite nappe with its subjacent limb and remains of an upper limb both composed of Mesozoic rocks (fig. 2). The outlines of this nappe crosscut and limit the internal structure. Similar conditions are known from some parts of the central Radstädter Tauern Mountains.

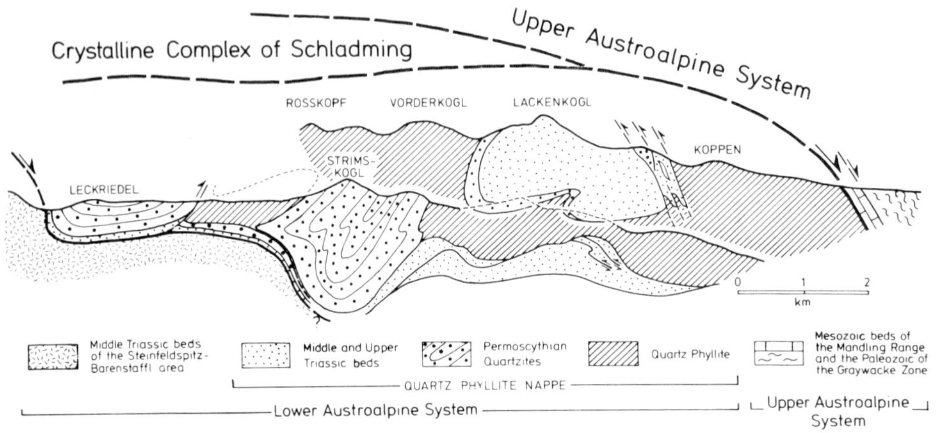

Fig. 2. Schematic section across the northern part of the Radstädter Tauern Mountains along the meridians of Altenmarkt/Pongau and Radstadt illustrating the structure and position of the Quartz phyllite nappe.

New data from the southern rim of the Quartz phyllite nappe confirm the highest position of this unit among the other parts of the Radstadt nappe system.

Nevertheless the facies connection of Mesozoic strata in the Quartz phyllite nappe with those belonging to the northerly, dark facies of the Lantschfeld nappe may be an indication of a close connection between the areas of sedimentation belonging to both sequences.

References

Demmer, W., 1962: Unpubl. thesis. University of Vienna, 196 p.

Fabricius, F., 1966: Intern Sed. Petr. Ser. - 9:143 p.

Frech, F., 1901: Geol.u.paläont. Abh. (ed. by v. Koken) . - 5:1-66.

Kober, L., 1955: Bau und Entstehung der Alpen. - 379 p., Wien.

Ott, E., 1972: Mitt. Ges. Geol. Bergbaustud.- 21:251-276.

Roßner, R., 1976: N. Jb. Geol. Paläont. Mh. - 1976:541-557.

Schönlaub, H.P., 1972: Unpublished manuscript, 10 p.

Tollmann, A., 1958: Jb. Geol. B.-A. - 101:79-115.

- , 1960: Mitt. Geol. Ges. Wien, 52 (1959):117-131.

- , 1963a: Ostalpensynthese. - 256 p., Wien.

- , 1963b: Mitt. Geol. Ges. Wien, 55(1962):85-125.

- , 1964: Mitt. Geol. Ges. Wien, 57:49-56.

The Age of the Alpidic Metamorphism in the Western Tauern Window, Austrian Alps, according to Radiometric Dating

M. Raith, P. Raase, H. Kreuzer and P. Müller

Introduction

The Tauern Window in the Eastern Alps, since Termier (1903), is considered to be a tectonic window exposing the Penninic rock series which are the lowest structural unit of the East Alpine nappe system. The frame of the Tauern window is formed by the structurally higher Austroalpine units.

The Penninic complex consists of two main rock units:

1) A pre-Alpidic crystalline basement mainly composed of Palaeozoic mica schists, paragneisses, and amphibolites (lower Schieferhülle) which are in part magmatized and intruded by Variscan granodiorites and tonalites (Zentralgneis);

2) and a cover of Permo-Mesozoic metasediments and metavolcanics (Upper Schieferhülle).

During the Alpidic orogeny the entire complex underwent intense deformation and metamorphism, ranging from greenschist facies in the peripheral part to low-grade amphibolite facies in the central part of the window. The processes of deformation and metamorphism were induced by the thrusting of the Austroalpine nappe system onto the Penninic rock complex. Recently, the geodynamic evolution of the Penninic and Austroalpine rock series has been discusses by Oxburgh and Turcotte (1974), Bickle et al. (1975), Hawkesworth et al. (1975), Frisch (1976), Dietrich (1976) applying the concepts of plate tectonics. In these models, the timing of the deformational and metamorphic events depends heavily on radiometric dating and its interpretation. The interpretation of the radiometric data, especially in the Zentralgneis and the lower Schieferhülle, is highly complicated by the complex thermal and deformational processes which have transformed the Penninic rock series during the Variscan and Alpidic orogeneses.

Rb/Sr whole-rock age determinations on granitic to tonalitic rocks of the Zentralgneis gave pre-Alpidic intrusion ages in the range of 243-282 m.y. (Cliff 1968, 1971; Lambert 1964; Jäger et al. 1969; Satir 1974, 1975; Besang et al. 1968; Hawkesworth 1976).

The Alpidic metamorphic events which affected the Penninic rock series were investigated by Rb/Sr and K/Ar mineral dating (Oxburgh et al. 1966; Besang et al. 1968; Jäger et al. 1969; Lambert 1970; Cliff 1971; Cliff et al. 1971; Hawkesworth 1973, 1976; Kreuzer et al. 1973; Satir 1973, 1974, 1975). According to these studies the climax of Alpidic metamorphism within the Penninic rock series probably was at 40 - 35 m.y. The last thermal event ended in early to middle Miocene time, but generally the scatter of the dates was not small enough to define clear cooling ages.

The existence of an early Alpidic high pressure / low temperature metamorphism is suggested from the occurrence of eclogites and glaucophane schists in the southern Venediger area (Abraham et al. 1974; Miller 1974). This metamorphic phase could be contemporaneous with the Eoalpine blueschist metamorphism recognized in the Penninic of the Western Alps (c.f. Frey et al. 1974), but radiometric confirmation is still needed.

In the Austroalpine Altkristallin sheet near the border of the Tauern

Window an Upper Cretaceous metamorphism of amphibolite facies has been ascertained by petrographic and radiometric investigations (c.f. Hawkesworth 1976).

The present research largely extends the studies of Besang et al. (1968) and gives the results of K/Ar analyses carried out on 136 mineral fractions of samples which were collected from the Zentralgneis and the Schieferhülle of the Großvenediger area, the Zillertaler Alps, and the Aurina valley (Ahrntal). These new dates indicate the existence of three thermal Alpidic events within the western Tauern window:
1) an early Alpidic crystallization related to the high pressure eclogite-glaucophane schist metamorphism,
2) the main Alpidic event related to the greenschist to amphibolite facies metamorphism, and
3) a late post-kinematic crystallization stage.

The final cooling history is deduced from the K/Ar mica-dates.

The areal distribution of the K/Ar dates of biotites, white micas, and amphiboles is shown in the geological map of the western Tauern window (fig. 1). In addition, the variation of the dates is presented in four N-S profiles and in one E-W profile across the Tauern window (fig. 2 and fig. 3).

Results and Discussion

Amphiboles: In the Obersulzbach valley, isolated massive tonalite bodies of Variscan age were found which, according to microscopic studies, proved to be nearly unrecrystallized. Oxygen isotope studies on these rocks yielded magmatic crystallization temperatures of about 670°C, in contrast to about 500°C for the deformed and recrystallized adjacent gneisses (Friedrichsen et al. 1973). Obviously, the coexisting minerals of the massive tonalites were not reequilibrated during the Alpidic metamorphism. K/Ar determinations on hornblendes and biotites of one of these occurrences yielded Variscan apparent ages between 360 and 290 m.y. However, two Rb/Sr dates on biotites, e.g. 32± 2 and 36 ±4 m.y., uncertainty in initial ratio taken into account, proved the Alpidic influence in these rocks. The occurrence of Variscan K/Ar apparent ages may be interpreted by inherited argon: in spite of diffusion in the mm-scale as proved by the Rb/Sr dates, the rocks as a whole retained argon if they did not recrystallize.

On amphiboles of metabasites from the Venediger region K/Ar dates covering a wide range between Variscan and Alpidic apparent ages, i.e. 210 to 50 m.y., have been determined. Especially in the Venediger Massif the dates seem to be related to the degree of recrystallization of the Variscan brown-green hornblende rather than to variable losses of argon by diffusion. The more abundant the microlites within the amphibole or the greater the proportion of a young blue-green amphibole the lower are the apparent ages. Such dates (210 to 50 m.y.) obviously are to be interpreted as mixed ages. From the lowest date an upper limit of 50 m.y. is suggested for the age of the main stage of Alpidic crystallization in the orthogneiss massif of the Venediger.

In the eastern Zillertaler Alps amphiboles yielded apparent ages of 61 m.y. for a greenschist sample from the lower-grade Schieferhülle and 42 and 39 m.y. for amphibolites from the higher-grade Zentralgneis. In the

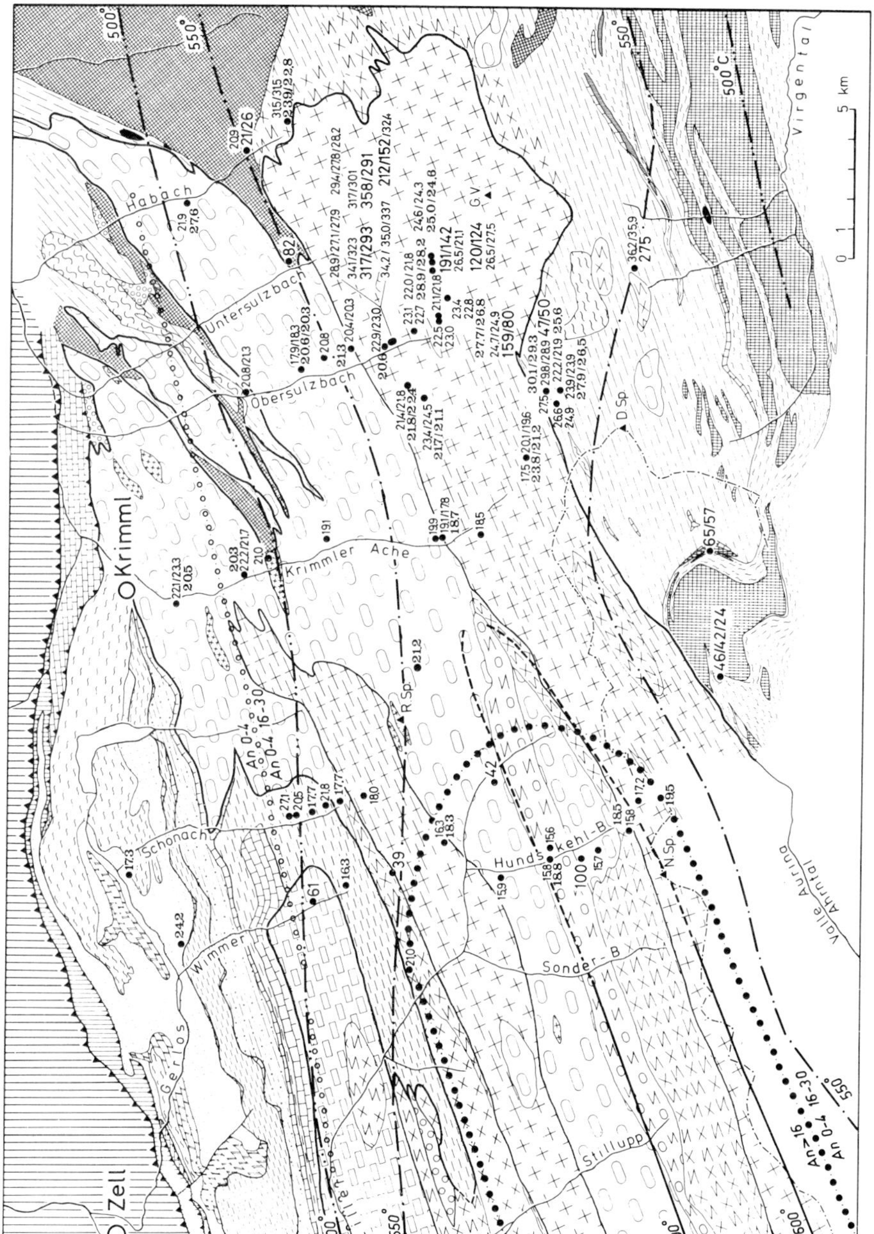
Zell
Krimml
Habach
Untersulzbach
Obersulzbach
Krimmler Ache
Schonach
Wimmer
Gerlos
Stillupp
Sonder-B
Hunds kehl-B
Valle Aurina
Ahrntal
Virgental
500°
550°
600°
500°C
0 1 5 km
R.Sp.
N.Sp.
D.Sp.
G.V.
An 0-4
An 0-4 16-30
An≥16
16-30

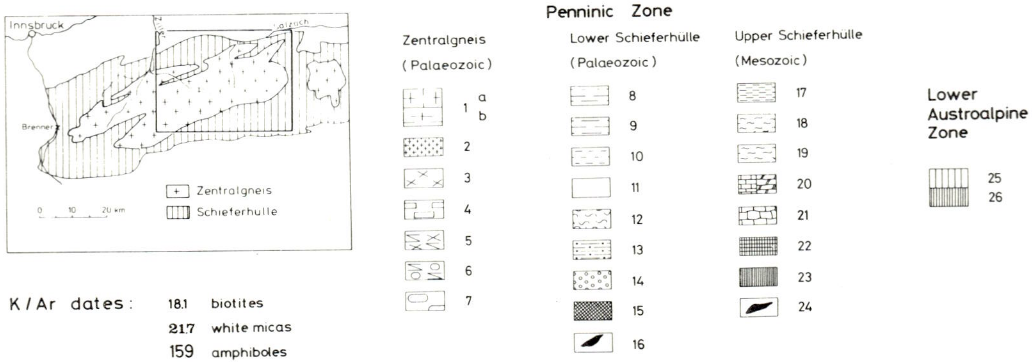

Zentralgneis

1a Unfoliated metatonalite to metagranodiorite
1b foliated metatonalite to metagranodiorite
2 Metagranite to -grandodiorite
3 Leucocratic metagranite
4 Porphyritic metagranite
5 Migmatic and banded gneiss
6 Diatectic augen gneiss
7 Augen and flaser gneiss

Lower Schieferhülle (Paleozoic)

8 Mica schist and paragneiss
9 Epidote-biotite schist
10 Phyllitic to quartzitic acid metatuffite and lava
11 Metaarkose, metaquartzite, and metaporphyroid (Wustkogelserie)
12 Graphitic phyllite
13 Metaarkose (Porphyrmaterialschiefer)
14 Acid metatuffite and metaagglomerate
15 Chlorite schist to amphibolite
16 Serpentinite

Upper Schieferhülle (Mesozoic)

17 Phyllite to meta-arkose
18 Phyllite, calc-poor
19 Calc-mica schist
20 Calcitic and dolomitic marble (Triassic, Jurassic)
21 Marble (Hochstegenkalk, Jurassic)
22 Chlorite schist to amphibolite
23 Eclogitic rocks
24 Serpentinite

Lower Austroalpine Zone

25 Quartzphyllite of Innsbruck
26 Phyllite, calc phyllite and quartzite (Richbergkogelserie)

17–19 Bündner Schiefer

GV Großvenediger

D Sp Dreiherrn Spitze

N Sp Napf Spitze

R Sp Reichen Spitze

Fig. 1. Geological map of the Venediger region and the eastern Zillertaler Alps with apparent K/Ar ages of micas and amphiboles. An-isogrades according to Morteani & Raase (1974); $^{16}O/^{18}O$-isotherms according to Hoernes & Friedrichsen (1974).

Greiner series of the western Zillertaler Alps the K/Ar data of actinolites range from Variscan to 35 m.y. (Morteani et al., in prep.). From the lowest of these dates an age of about 40 m.y. may be assumed for the main Alpidic crystallization. Yet, further proof is needed as the dates might reflect the opposite influences of excess argon and argon loss.

Two amphibole samples from the low-grade Mesozoic Upper Schieferhülle of the Aurina valley (Ahrntal) were studied. Apparent ages of about 60 m.y. are determined for two fractions of a homogeneous coarse prismatic actinolite sample and apparent ages of about 44 m.y. for the two fractions of the core actinolite from a zoned amphibole sample. The bluegreen hornblende rim of the latter amphibole yielded the significantly lower age of 24 m.y. In both samples the analysed fractions of actinolite differ in the concentration of atmospheric argon. For both samples atmospheric argon increases with decreasing apparent age and there is no linear correlation. Therefore, no meaningful isochron interpretation can be given (c. f. Shafiqullah & Damon 1974).

From the petrological investigations, it follows that the actinolite cores crystallized at temperatures below 500°C and the hornblende rims slightly above 500°C. According to the diffusion experiments of Gerling et al. (1965) amphiboles which are low in iron may retain argon even considerably above 500°C. Therefore, it may be assumed that the actinolites have not suffered significant argon loss, although this possibility cannot be excluded. If there really was no argon loss, and if excess argon is taken into account, the minimum date of 42,5 m.y. could be interpreted as the upper limit of the main Alpidic crystallization.

The K/Ar date of 24 m.y. of the hornblende rim of the zoned amphibole will give a rather good approximation of the time of the late Alpidic crystallization stage because argon losses during cooling from slightly above 500°C will only be insignificant. A considerable contribution of excess argon is improbable because of the low age, not essentially higher than most of the mica dates. Further indication for a crystallization at the transition from Oligocene to Miocene is given by the date of 21 m.y. of one of the two fractions of a weakly zoned clear bluegreen hornblende from the Habach valley, which is concordant with the date of the coexisting biotite.

Preliminary results on minerals of eclogites and glaucophane schists from the southern Venediger region indicate an upper limit of 70 m.y. for the low temperature / high pressure stage of the Alpidic metamorphism (Raith et al., in prep.).

Micas: Alpidic ages varying between 18 and 30 m.y. were determined on micas from deformed and recrystallized gneisses and tonalites of the Venediger Massif (fig. 1 and 2). The white mica ages are generally higher than the biotite ages, but the difference may vary up to 7 m.y. and some of the biotites have an even higher age than the coexisting white micas. Furthermore, there are differences in the ages of biotites as well as white micas on the meter-scale in the field and often also between different size-fractions of the same sample.

Mica dates higher than about 25 m.y. were only found in massive and slightly deformed and recrystallized tonalites and amphibolites. It appears that these higher ages are due to only partial degassing of the rocks, varying from place to place. The mica dates of the Venediger Massif obviously cannot be interpreted as clear cooling ages.

Partly, the distinctly different K/Ar dates determined on the different size-fractions may be caused by the occurrence of more than one Alpidic mica generation.

Judging from the highest Alpidic mica dates from the Venediger massif, the micas might have formed before about 30 m.y. during the main Alpidic crystallization at about 40 m.y. suggested from the actinolite K/Ar data. This notion is further supported by white mica dates close to 30 m.y. from samples of a NS-section through the Mesozoic Schieferhülle south of the Venediger Massif (Raith et al., in prep.).

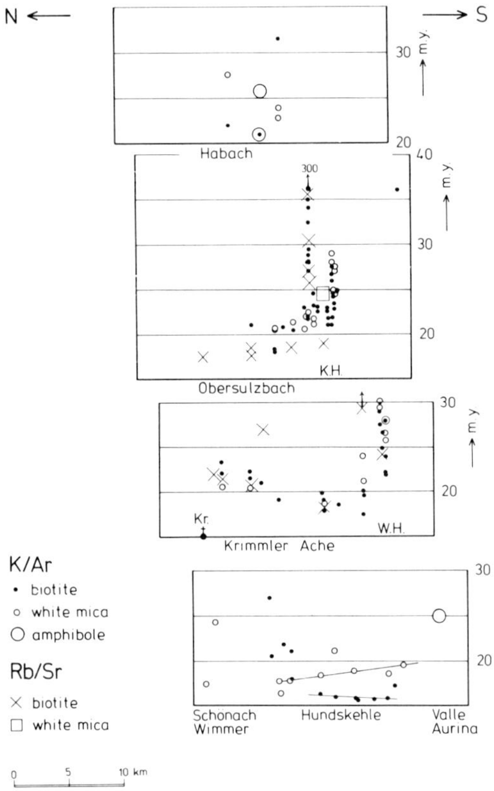

Fig. 2. Apparent ages of the Venediger region and the eastern Zillertaler Alps in N-S profiles. The Rb/Sr-dates are mineral-whole rock values from Besang et al. (1968) and Jäger et al. (1969). Kr.= Krimml, W.H. = Warnsdorfer hut, K.H. = Kürsinger hut.

In the region north of the Venediger Massif and in the eastern Zillertaler Alps the K/Ar dates of micas are much more uniform than in the Venediger Massif (fig. 1 and 2). White mica ages are mainly in the range of 20 to 17 m.y. and biotite dates in the range of 23 to 16 m.y. The lowest and most uniform ages of biotites were found in the higher-grade metamorphic gneiss series of the Hundskehl valley, i.e. 16 m.y. In this region the white mica ages are systematically higher by about 3 m.y. This age pattern is regarded to be caused by cooling, and a cooling rate of about 30°C/m.y. is obtained assuming closing temperatures of 300°C for biotite and 380°C for white mica (Purdy & Jäger cit. by Hunziker 1974).

In the northern region of lower metamorphic grade the mica dates show a significant spread. The biotite ages increase up to 23 m.y. with a

single value of 27 m.y. (fig. 2).

In the western Zillertaler Alps, the lowest biotite dates (13 - 14.5 m.y.) were determined which are significantly lower than those in the Hundskehl valley (fig. 3 and Satir 1975).

On the whole the local spread and the regional variation of the mica dates cannot be interpreted as a clear cooling pattern. Also, environmental excess argon should be excluded as the main cause of high biotite dates because, in the northern Krimmler Achental, the same elevated biotite dates are determined by the Rb/Sr-method (Besang et al.1968; Jäger et al. 1969) as well as by the K/Ar-method both slightly higher than the K/Ar ages of white mica (fig. 2). We may explain this age pattern by assuming that the phengitic white micas crystallized at a late postkinematic stage (20 to 17 m.y.) at a temperature which was too low to allow complete degassing of preexisting Variscan and early Alpidic biotites.

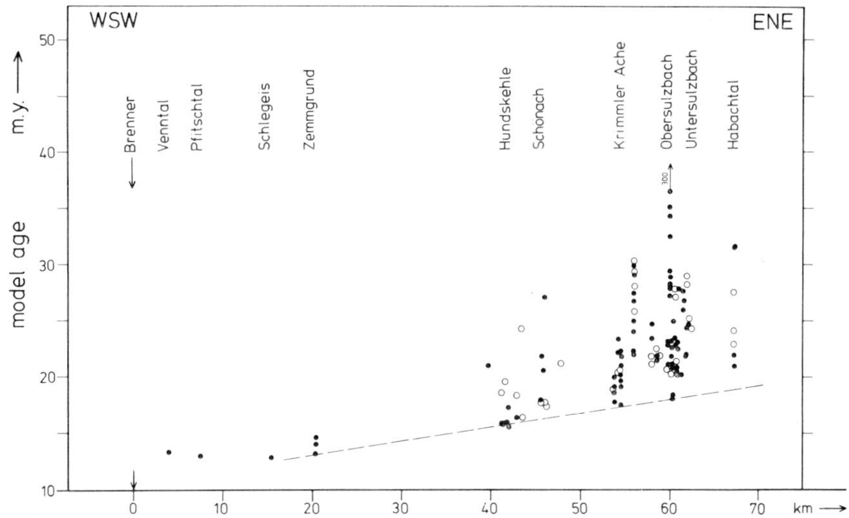

Fig. 3. K/Ar mica ages from the western Tauern window, projected to an E-W profile. • = biotite, o = white mica. Lowest biotite ages decrease with a gradient of -1.3 m.y./10 km from E to W.

If we regard the lowest biotite dates as cooling ages a steady decrease is observed from the Venediger region (18 m.y.) towards the western Zillertaler Alps (13 m.y.) indicating that the biotites closed later with respect to argon in the western area, i.e. that cooling was delayed by 5 m.y. in the west. From these dates a gradient of about -1.3 m.y./10 km may be derived (fig. 3). A similar gradient of about -1 m.y./10 km from south to north is suggested by the Rb/Sr biotite ages of the augen and flaser gneisses from the Obersulzbach valley and by the K/Ar white mica ages from the valleys of Hundskehle and Schönach (fig.2). The regional age pattern can be attributed to an updoming along an E-W axis, subsequent to the cooling caused by uplift and ersosion, and by a proceeding of uplift and denudation with amounts increasing to the west.

Conclusions

The radiometric mineral data of this investigation show that the Penninic rock series in the western Tauern window experienced a thermal history similar to that described by Cliff et al. (1971), Hawkesworth (1976) for the eastern Tauern window, and by Hunziker (1974), Frey et al. (1974) for the Western Alps. Although they do not permit a precise dating of the different phases of Alpidic metamorphism, the mineral dates give an indication of three distinct thermal events:

1) An early Alpidic phase of high pressure/low temperature conditions is suggested from K/Ar dates ranging from 70-50 m.y. determined on glaucophane from the southern Venediger area.

2) The main Alpidic event probably took place in the time-span between 50 and 30 m.y. A minimum age of about 30 m.y. can be derived from the highest Alpidic mica ages, whereas an upper limit of 50 m.y. is indicated by some amphibole dates from the Venediger region.

3) The latest post-kinematic crystallization representing the last climax of the Alpidic metamorphism, took place at about 20 m.y. This is inferred from the white mica dates of the northern Venediger region which are interpreted as formation ages and from the occurrence of post-kinematic rim-hornblendes with ages of abot 22 m.y. in the Habach and Aurina valleys.

The final cooling (300°C) of the Penninic rock complex caused by uplift and erosion began at about 19-18 m.y. in the Venediger region and proceeded towards the west (14-13 m.y.) and the north (17 m.y.) with a gradient of about -1.3 m.y./10 km. This is suggested from the lowest biotite K/Ar dates.

A cooling rate of about 30°/m.y. between 380 and 300°C has been derived from the age difference between the white mica and biotite dates from the highest grade region of the eastern Zillertaler Alps. The distribution of the mica dates along the N-S profiles indicates a pronounced updoming of the central Tauern window along an E-W axis after the time of closing of the micas.

Finally, the structural and metamorphic development of the Penninic series in the Western Tauern window may be discussed applying the concepts of plate tectonics (Clar 1965; Oxburgh & Turcotte 1971; Frisch 1976; Hawkesworth et al. 1975). The post-Variscan geodynamic evolution of the Penninic complex, according to these models, is characterized by three phases:

a) Formation of the Penninic eugeosyncline with sediments of the Bündner Schiefer facies and ophiolites during the Jurassic and Lower Cretaceous.

b) Subduction of the Penninic rock series under the Austroalpine continental plate in a southerly direction during the Upper Cretaceous to Eocene. By the process of subduction a metamorphism of low temperature/high pressure facies is generated. Indication for this early Alpidic metamorphism in the western Tauern window is given by eclogites and blueschists. The highest glaucophane date of 70 ± 12 m.y. is suggested to be a maximum age for this metamorphism, because excess argon is proved in the coexisting garnet.

c) Completion of the subduction process by collision of the Penninic and Austroalpine continental plates and final overthrusting of the Austroalpine sheet until Late Eocene time. Thermal equilibration of the East alpine nappe pile towards a normal geothermal gradient during these processes and subsequent updoming and erosion. Within the Penninic rock

series, rising temperatures generate a Barrow-type metamorphism in the time-span between 50 - 30 m.y. A post-kinematic crystallization stage at about 20 m.y. represents the latest stage of the Alpidic metamorphism. (Constants used for age calculation are those of the "Phanerozoic Time-scale", Harland et al. 1964).

References

Abraham, K., Hörmann, P.K. & Raith, M., 1974: N. Jb. Mineral. Abh., 122:1-35.

Besang, C., Harre, W., Karl, F., Kreuzer, H., Lenz, H., Müller, P. & Wendt, I., 1968: Geol. Jb., 86:835-844, Hannover.

Bickle, H.J., Hawkesworth, C.J., England, P.C. & Athey, D.R., 1975: Earth Planet. Sci. Letters, 26:13-28.

Clar, E., 1965: Z. dtsch. geol. Ges., 116:T2:267-291.

Cliff, R.A., 1971: Contr. Mineral. Petrol., 32:274-288.

Cliff, R.A., Norris, R.J., Oxburgh, E.R. & Wright, R.C., 1971: Jb. geol. B.-A., 114:121.172, Wien.

Dietrich, V.J., 1976: Geotekton. Forsch.: 1-84, Stuttgart.

- , 1976: Geol. Soc. Amer., 4,3:147-152.

Frey, M., Hunziker, J.C., Frank, W., Bocquet, J., Dal Piaz, G.V., Jäger, E. & Niggli, E., 1974: Schweiz. Mineral.Petrogr. Mitt.,54:247-290.

Friedrichsen, H., Müller, G. & Stahl, W., 1973: Contr.Mineral.Petrol., 40:159-168.

Frisch, W., 1976: Geol. Rdsch., 65:375-393.

Gerling, E.K., Kol'tsova, T.V., Petrov, B.V. & Zul'fikarova, Z.K., 1965: Geochemistry International 2,1:148-154.

Harland, W.B., Smith, A.G. & Wilcock, B. (Eds.), 1964: Quart.J. geol. Soc. London, 120s:450 p.

Hart, S.R., 1964: J. Geol., 72:493-525.

Hawkesworth, C.J., 1973: ECOG-3, Oxford, Abstracts of Geochron. and Isotope Geol. 9, Tervuren.

- , 1976: Contr. Mineral. Petrol., 54:225-244.

Hoernes, S. & Friedrichsen, H., 1974: Schweiz.Min.Petrogr.Mitt.54:769-788.

Hunziker, J.C., 1970: Eclogae geol. Helv., 63:151-161.

- , 1974: Geol. Min. Univ. Padova, 31:54 p.

Jäger, E., Karl, F. & Schmidegg, O., 1969: Tschermaks Mineral. Petrogr. Mitt., 13:251-272.

Kreuzer, H., Harre,W., Müller,P., Raase, P. & Raith,M., 1973: ECOG-3, Oxford, Abstracts of Geochron. and Isotope Geol., 9, Tervuren.

Lambert, R.St.J., 1970: Eclogae geol. Helv., 63:197-205.

Morteani, G. & Raase, P., 1974: Lithos 7:101-111, Oslo.

Miller, Ch., 1974: Schweiz. Mineral. Petrogr. Mitt., 54:371-348.

Oxburgh, E.R., Lambert, R.St.J., Baadsgaard, H. & Simons, J.G., 1966: Verh. Geol. B.-A.:17-33, Wien.

Oxburgh, E.R. & Turcotte, D.L., 1974: Schweiz. Mineral. Petrogr. Mitt., 54:641-662.

Satir, M., 1973: ECOG-3, Oxford, Abstracts of Geochron.and Isotope Geol. 9, Tervuren.

- , 1974: Schweiz. Mineral. Petrogr. Mitt., 54:213-228.

- , 1975: Mem. Ist. Geol. Mineral. Univ. Padova, Vol. XXX: 82 p.

Shafiqullah, M. & Damon, P.E., 1974: Geochim. cosmochim. Acta, 38:1341-1358.

Termier, M.P., 1903: Bull. Soc. géol. France:711-766.

The "Altkristallin" of Eastern Tyrol between Tauern Window and Periadriatic Lineament.

G. Troll

Austroalpine Altkristallin surrounds the Tauern Window from the Oetztal Mass in the west along to the south and to the Mur Alps in the east. Rb-Sr *age determinations* of leucocratic gneisses yield similar results of Caledonian events (see table 1) interpreted for the studied area as ages of homogenization during metamorphism. The gneisses commonly intercalate with tholeiitic amphibolites and amphibole-bearing gneisses.

Table 1: Comparison of data of geological events in gneisses around the Tauern Window, Eastern Alps (decay constant $Rb^{87} = 1.47 \cdot 10^{-11} \cdot a^{-1}$)

Author	Rock type	Age of isochron	Initial ratio of Sr^{87}/Sr^{86}	locality
BORSI*) et al. 1973	orthogneiss	434 ± 12 (2)	0.7086 ± 0.0034	Antholz Southern Tyrol
BRACK in: TROLL et al. 1976	microcline augengneiss	439 ± 20 (2)	0.7107 ± 0.0016	Wangenitzsee, Schobergruppe
FRANK et al. 1976	plagioclase gneiss	500 ± 45 (2)	0.7044 ± 0.0012	Gleinalm
GRAUERT 1966	orthogneiss	428 ± 17 (2)	0.7121 ± 0.006	Silvretta
SATIR 1975	orthogneiss	436 ± 17 (2)	0.7102 ± 0.0014	Oetztal Alps, Tyrol**)

*) The data given by BORSI et al. have been newly calculated by the method of WENDT (in BROOKS 1968). Hence slight deviations of error and initial ratio result.

**) Further comparable data from the Oetztal mass have been yielded by HARRE et al. (1968), MILLER et al. (1967) and SCHMIDT et al. (1967).

The *sub-sequence of metamorphic rocks* in the Schobergruppe contains clastic metasediments covering the largest area, the mentioned amphibolitic metavolcanites including amphibole eclogites (so-called eclogite-amphibolites), and, moreover, layers of several meters consisting of augengneisses or mylonitic gneisses with similar modal composition. Tectonically, the Schobergruppe plays a distinctive role because it contains two different complexes with gently inclined boundaries (Troll & Hölzl 1974, fig. 1).

The lower complex (Liegend complex, see fig. 1) is monotonous with a psammitic character, the upper complex (Hangend complex), however, is characterized by a higher amount of metabasites. The boundary in-between cannot be attributed to any event. Since the metamorphic rocks have partly undergone a strong mylonitization along the low-angle faults, the thrusting must post-date the *Caledonian* metamorphism, but whether its age is Hercynian or Alpidic, is uncertain.

In general, the tectonic pattern of the metamorphic rocks of the Schobergruppe is mainly formed by the following folding events. B_1 ca. per-

pendicular to B_2, appears in open flexures and folds (slight amplitudes large wavelengths, plunging to NE or SW). The strongest compression is directed from NE to SW (isoclinal B_2 plunging 10 - 25° to SE). This behaviour of the upper complex of mainly metabasites is modified in the lower complex consisting of schists and gneisses; there two fold axes are found (B_1 plunging 25 - 35° to ENE; first formation of faults in Okl and ac directions). B_1 and B_2 are simular. B_1 and B_2 enclose an acute angle. B_3 may be of alpidic age (plunging gently to E), but that has yet to be verified (cf. Troll et al. 1976).

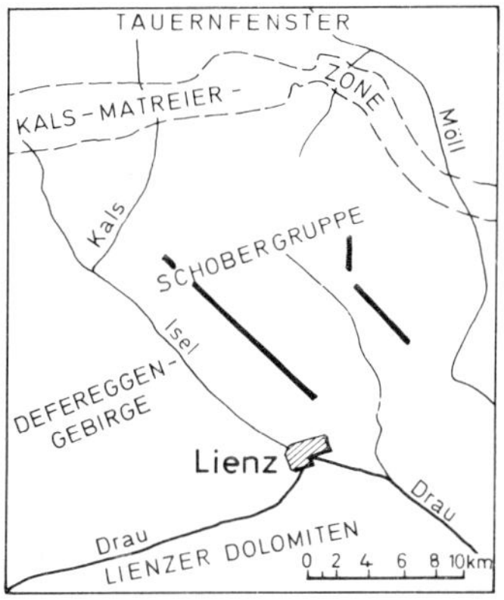

Fig. 1. Index map for the profile of fig. 2.

A comparison with the area of Kalkstein in the Deferegger Alpen sheds some more light on the succession and on the effect of the tectonic events. There, a Triassic series has been wedged in the Altkristallin basement so that pre-alpidic deformations can be recognized (Guhl & Troll 1977). In the Kalkstein area also four folding phases occur: B_1 to B_4 are present in the metasedimentary rocks of the Altkristallin, B_2 to B_4 in the metamagmatites, B_3 to B_4 in the Permo-Triassic metasediments. The orientations of the folding structures, however, are different from those in the Schobergruppe. The Triassic beds show flexural folding in the incompetent layers and fracture cleavage in the competent layers of the Altkristallin and of the Triassic dolomites. B_4 indicates the last alpidic deformation consisting of synclinal and anticlinal structures, with wavelengths in the order of 1 km, and of a few faults.

The main deformations (B_1 to B_2 in both areas) have been outlasted by crystallization of mineral parageneses in low- to highrank amphibolite facies. The alpidic history is difficult to reveal. Retrograde metamophism is everywhere present and can be detected by regeneration of eo-alpidic mineral ages (cf. Oxburgh et al. 1966).

In garnet-mica schists of the western Schobergruppe (in the Liegend complex) a 0.5 to 1 m thick band of conglomerate was discovered. The pebbles in the *conglomerate layer*, ellipsoidic and up to 10 cm in size, are

Fig. 2. Cross sections through central parts of the Schobergruppe. The upper section representing the area of the Wangenitzsee where the Rb-Sr isochron age has been obtained, has been mapped by Schwarzböck (1968), the lower section 1969 to 1974 by the author and co-workers.

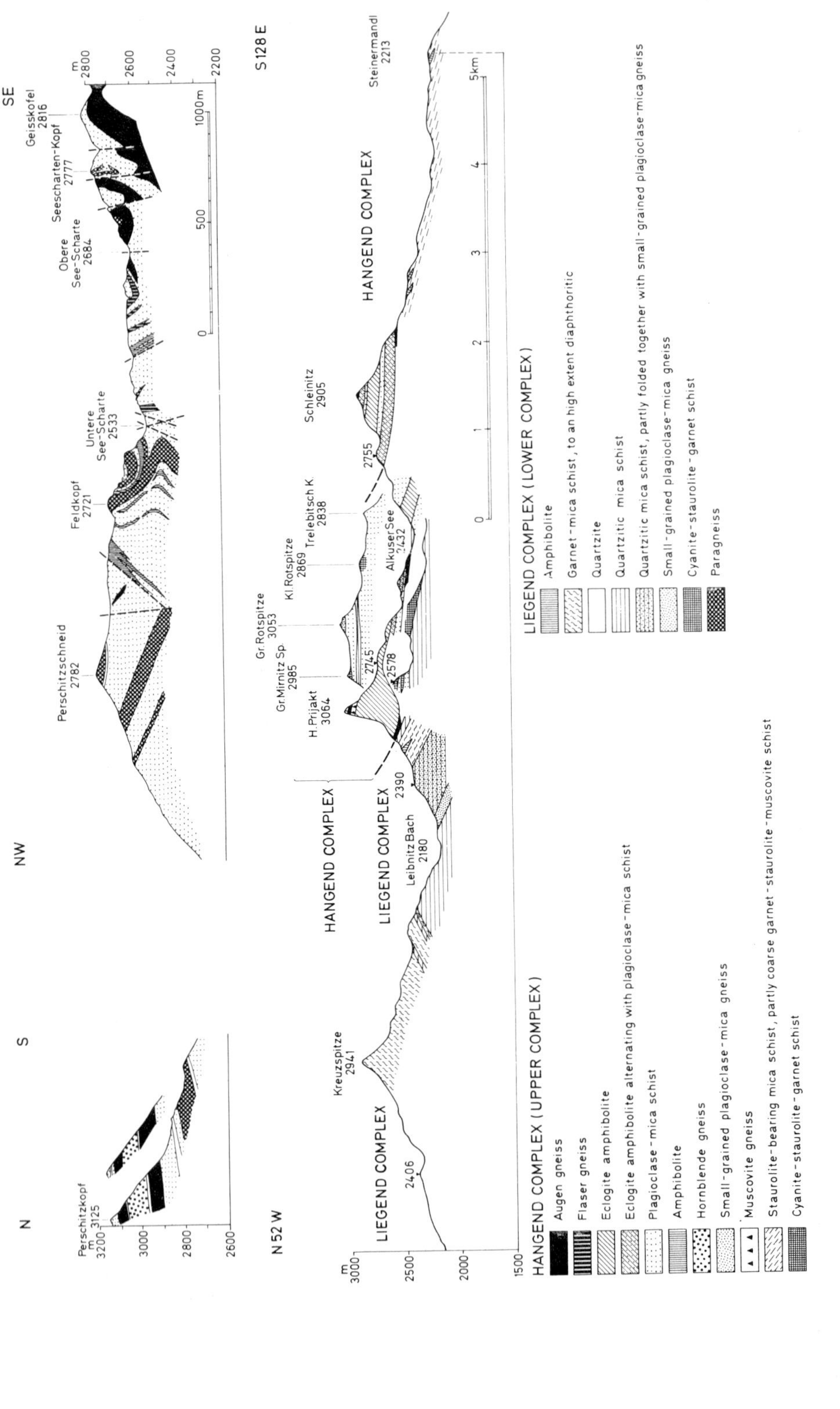
N
S
NW
SE
N 52 W
S 128 E
Perschitzkopf 3125
Perschitzschneid 2782
Feldkopf 2721
Untere See-Scharte 2533
Obere See-Scharte 2684
Seescharten-Kopf 2777
Geisskofel 2816
0
500
1000m
Kreuzspitze 2941
2406
Leibnitz Bach 2180
2390
H.Prijakt 3064
Gr.Mirnitz Sp 2985
Gr.Rotspitze 3053
Kl.Rotspitze 2869
Trelebitsch K. 2838
Alkuser See 2432
2745
2578
2755
Schleinitz 2905
Steinermandl 2213
5km
LIEGEND COMPLEX
HANGEND COMPLEX
HANGEND COMPLEX (UPPER COMPLEX)
Augen gneiss
Flaser gneiss
Eclogite amphibolite
Eclogite amphibolite alternating with plagioclase-mica schist
Plagioclase-mica schist
Amphibolite
Hornblende gneiss
Small-grained plagioclase-mica gneiss
Muscovite gneiss
Staurolite-bearing mica schist, partly coarse garnet-staurolite-muscovite schist
Cyanite-staurolite-garnet schist
LIEGEND COMPLEX (LOWER COMPLEX)
Amphibolite
Garnet-mica schist, to an high extent diaphthoritic
Quartzite
Quartzitic mica schist
Quartzitic mica schist, partly folded together with small-grained plagioclase-mica gneiss
Small-grained plagioclase-mica gneiss
Cyanite-staurolite-garnet schist
Paragneiss

without exception fine-grained garnet-quartzites that are slightly altered (table 2). Such garnet-quartzites are unknown from other places in the Altkristallin. They show the existence of an older sedimentary basement from which the pebbles are derived. Another, however less probable interpretation of the occurrence would be an original manganese quartzite layer deformed by boudinage and folding.

Table 2: Modal analyses of pebbles of garnet quartzites in garnet-bearing mica schists from Außersteiner N Ainet, Schobergruppe

Section No.	10479/1	10479/2	13171/1	13171/2	Average
garnet	43.2	41.3	34.8	37.8	39.3
quartz	23.2	27.6	35.5	32.6	29.7
biotite	8.0	2.6	3.1	3.3	4.3
chlorite (Pennine)	19.5	18.5	18.0	19.0	18.7
opaques	5.0	7.9	5.8	5.6	6.1
accessories (rutile, calcite, zircon, sericite, apatite, titanite)	1.1	2.1	2.8	1.7	1.9
Points	1304	1000	1500	1500	5304

The *amphibole eclogites* alternate with variable amphibolites on the base of the Hangend complex in a series of 400 m in thickness (e.g. in the northern flanc of Prijakt Mountain, Schobergruppe). This series is interpreted as geosynclinal volcanism, essentially as tholeiitic flows and tuffs. This interpretation is based on chemical analyses of the amphibole eclogites (Troll et al. 1976, p. 493-497). The mineralogy of these eclogites is characterized by primary and secondary hornblendes. The primary hornblendes are in contact with the coexisting clinopyroxenes, are of pargasitic-tschermakitic composition, rich in sodium, and lie on the boundary between Ca-hornblendes and subcalcic amphiboles. The lack of cyanite in the eclogitic rocks is surprising, especially in respect to similar rocks in the Koralpe and Saualpe, Carinthia.

The *grade of metamorphism* can be determined only roughly by the mineral assemblages of the amphibole eclogites as well as of the metapelitic schists. The distribution coefficients between coexisting minerals show a genesis of the amphibole eclogites in a temperature range of 350 to 550°C under pressures of 4.5 to 8 kb. The crystallization of cyanite partially replacing staurolite, suggests a Barrovian type of metamorphism, in which the stability field of staurolite is still compatible with the maximum temperatures of eclogite formation in the pressure field from 4.5 to 8 kb.

In a part of the Schobergruppe augengneisses (with conspicuous microcline porphyroblasts) frequently occur (Troll et al. 1976, fig. 12). Thin sections of these rocks show the presence of calcite so that only few locations were appropriate for *Rb-Sr isotopic age determinations* for example the area of Lake Wangenitz (geographic coordinates: 12°48'E; 47°56'N), 10 km south of the margin of the Tauern Window. Orthogneisses (augengneisses), paragneisses, two amphibolite samples and one eclogite sample were measured.

The obtained data give an *isochron for orthogneisses as well as for paragneisses* indicating an age of 440 $\pm$ 13 (2 σ) m.y. (Ordovician/Si-

lurian). This is interpreted as an age of homogenization caused by a *Caledonian* metamorphis (Brack 1977). The strontium initial isotope ratio is 0.7102 ± 0.0008 (2 σ). This isochron is a revised and completed version of the preliminary isochron already published (Troll et al. 1976, p. 509-510).

The *mineral isochrons* (apatite-potassium feldspar-biotite and apatite-potassium feldspar-muscovite, respectively) yield an age of 70 ± 4 m.y. for biotite and of 79 ± 5 m.y. for muscovite in the area of Lake Wangenitz. Further muscovite ages from the Schobergruppe have also values of about 80 m.y. These ages are considered to be cooling ages of a Cretaceous metamorphism.

Close to the margin of the Altkristallin along to the zone of Kals-Matrei (ca. 1 km S Matrei) a muscovite age of ca. 44 m.y. was measured and is interpreted as a mixed age between both Cretaceous and Tertiary alpidic events.

The mineral isochron of Lake Wangenitz area demonstrates that about 80 m.y. ago apatite, potassium feldspar, and phengite have been an open system for strontium causing a rehomogenization of Sr isotopes between the minerals. The age of minerals dates the cooling after metamorphism, and the termination of the Sr isotope exchange. The measured biotites show errors of 2 σ still touching each other. This is possibly an effect of different behaviour of diffusion for Sr isotopes.

Certainly, the thermal influence of the Cretaceous metamorphism was strong enough that all minerals were affected by the change of the Sr isotope ratio. Eventual traces of Pre-alpidic metamorphic events between the total rock age of 440 m.y. and the Cretaceous metamorphism were completely extinguished. Post-Cretaceous metamorphism or thermal events can only be found in the marginal areas of the Schobergruppe close to the Tauern Window (cf. Hawkesworth 1976, for the region SE Tauern Window).

This means with respect to the Alpidic geodynamic history, that the Altkristallin mass to the south of the Tauern Window belongs to an intermediate structural level. The distinct metamorphic episode at the end of the Carboniferous, as it is established by intrusions of Penninic orthogneisses within the Tauern Window, has not been observed in the Schobergruppe.

References

Borsi, S., Del Moro, A., Sassi, F.P. & Zirpoli, G., 1973: Mem. Soc. Geol. Ital., 12:549-571, Pisa.
Brooks, C., Wendt, I. & Harre, W., 1968: J. Geophys. Res., 73:18p.
Brack, W., 1977: Diss. Fachbereich Geowiss. Univ. München:97 p.
Frank, W., Klein, P., Nowy, W. & Scharbert, S., 1976: Tschermaks Miner. Petr. Mitt., 23:191-203.
Grauert, B., 1966: Earth Planet. Sci. Lett., 1:139-147, Amsterdam.
Guhl, M. & Troll, G., 1977: Verh. Geol. B.-A., 45-52, Wien.
Hawkesworth, C.J., 1976: Contrib. Mineral. Petrol., 54:225-244.
Heinisch, H. & Schmidt, K., 1976: Geol. Rdsch., 65:459-482.
Oxburgh, E.R., Lambert, R.S., Baadsgaard, H.B. & Simons, J.G., 1966: Verh.Geol. B.-A.:17-33, Wien.
Purtscheller, F. & Sassi, F.P., 1975: Tschermaks Miner. Petr. Mitt., 22:175-199, Wien.
Sassi, F.P., Zanferrari, A. & Zirpoli, G., 1974: Mem. Museo Tridentino Sci. Nat.,37/38,20:5-30, Trento.

Satir, M., 1975: Mem.Ist.Geol.e Mineral. dell'Universitá di Padova,30: 84 p.
Schwarzböck, H., 1968: Diss. phil. Fakultät Univ. Wien: 214 p.
Troll, G. & Hölzl, E., 1974: Jahrb. Geol. B.-A., 117:1-16, Wien.
Troll, G., Forst, R. & Söllner, F., 1976: Geol. Rdsch.,65:483-511.

Contribution to the Geodynamic Interpretations in the Eastern Alps

F. P. Sassi, S. Borsi, A. Del Moro, A. Zanferrari and G. Zirpoli

This paper summarizes the results of a multi-disciplinary research carried out within a test-area. Our aim was that of obtaining a large amount and variety of analytical data useful for a geodynamic approach.

As test-area we chose the Austridic region between Mules (Mauls) and the Iseltal, in which our previous research had shown a wide variety of Alpine and/or probably Alpine effects of a magmatic, metamorphic and dynamic nature (Borsi et al. 1973; Sassi et al. 1974).

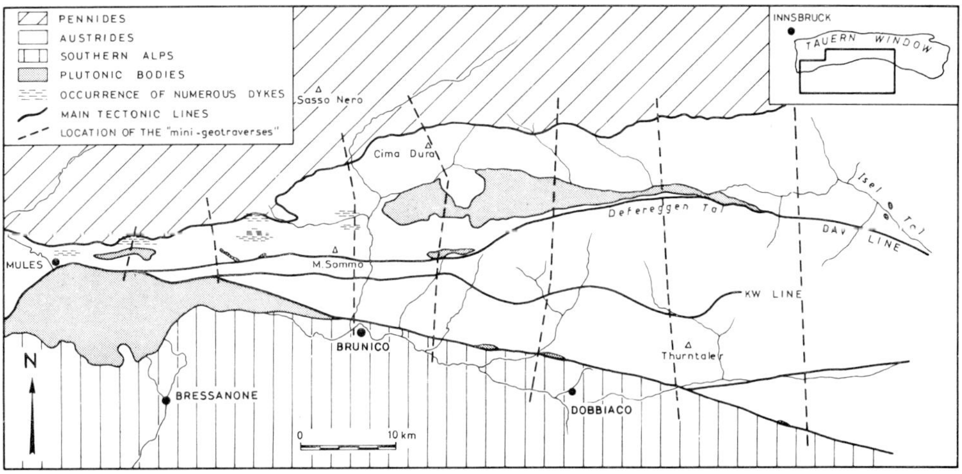

Fig. 1. Structural sketch showing the location of the investigated area.

This wide variety of Alpine effects required a detailes study, with the effort concentrated along a series of selected *mini-geotraverses*. Fig. 1 shows the trace of these sections, along which a detailed field survey, new petrographic studies and many radiometric age determinations were carried out. The main results of these new studies are only summarized here; they are extensively reported and discussed in separate pa-

pers (Borsi et al. 1977 a, b, c ; Zanferrari 1977).

Geological data

In the structural sketch in fig. 1 the existence of three Austridic blocks is quite clear, due to the occurrence and trend of the two main infra-Austridic tectonic lines: the *Defereggental-Anterselva (Antholz)-Valles (Vals)* line (DAV), and the *Kalkstein-Vallarga (Weitental)* line (KW) (Dal Piaz 1934; Sassi et al. 1974). These lines consist of cataclastic and mylonitic belts, ranging in thickness between a few dozen to several hundred meters. Subhorizontal striae and drag folds with subvertical axes show that strike-slip movements occurred along these belts, at least during very late times.

The three blocks are different from the structural point of view, as can partly be deduced merely from the new geological sections we constructed (Borsi et al. 1977 c):

The southern block: is characterized by Alpine post-crystalline meso- and macroscopic folds; however, these deformations did not eliminate the pre-Alpine mesoscopic structural pattern;

The intermediate block: has kept its Variscan structure (Schlingen Bau) more completely, and the Alpine deformation is essentially confined to block faulting;

The northern block: is characterized by the fact that the Alpine structural pattern is substantially synmetamorphic, at least outside the phyllonitic belts, whereas its pre-Alpine structures are relics and recognizable only locally.

These important differences in the Alpine structural pattern imply that the three blocks underwent different fates during the Alpine tectogenesis, and that they were involved in different environmental conditions, especially from the thermic point of view. As can be seen later on, these conclusions are strictly consistens with those independently reached using petrographic and radiometric data.

Petrographic data

The rocks outcropping along the sections were submitted to new petrographic analysis, particularly detailed in those samples utilized for geochronological research. The main results are as follows:

a) In the block south of the KW line, metamorphic or important magmatic Alpine products are not present, so that only the Variscan and pre-Variscan metamorphic situations can be studied. By the term *pre-Variscan* we mean an intermediate to high-pressure metamorphism in the amphibolite facies, certainly older than the low-pressure metamorphism representing the Variscan thermal climax. The pre-Variscan metamorphism thoroughly affects the whole Austridic continental margin in the Eastern Alps (Sassi & Zanferrari 1972; Borsi et al. 1973; Purtscheller & Sassi 1975; Kleinschmidt et al. 1976; Kleinschmidt & Ritter 1976).

This old metamorphism could also represent an early Variscan metamor-

phic phase. However, we have disregarded this interpretation mainly because:
1) the idea of a two-phase development of the Variscan metamorphism requires demonstration which, for the time being, has no support;
2) in every geodynamic model implying the development of two distinct metamorphic events during a single orogenic cycle, the effects of the older metamorphism should be found within a narrow belt; they should not occur widespread over a given large region, as could do the effects of the younger metamorphism;
3) the widerspread, important, acid to intermediate magmatism (plutonism and vulcanism) which developed approximately during the latest Ordovician, is strong evidence for an important thermic event operating, during the Orodovician, on a regional scale.
b) In the intermediate block too, rock-producing Alpine processes are not detectable, so that there is no difference between the intermediate block and the southern one. From this point of view, this fact may suggest that the structural differences existing between the two blocks could not imply a difference in the Alpine dynamic events, but only the different kind of reaction of each block to an essentially similar dynamic activity.
c) In the block north of the DAV line, however, the Alpine effects are very abundant.
c_1) First of all, we mention the occurrence of *metamorphic crystallizations of certain Alpine age*. Among these, the following are to be emphasized as typical products, because they are systematically lacking in the rocks which preserve Variscan radiometric ages of the biotites:
1) *crystallization of celadonite-rich muscovites:* fig. 2 shows the histogram of the b_0 values of the K-white micas from the northern block (phyllonitic belts excluded) in counterposition to that of the muscovites from

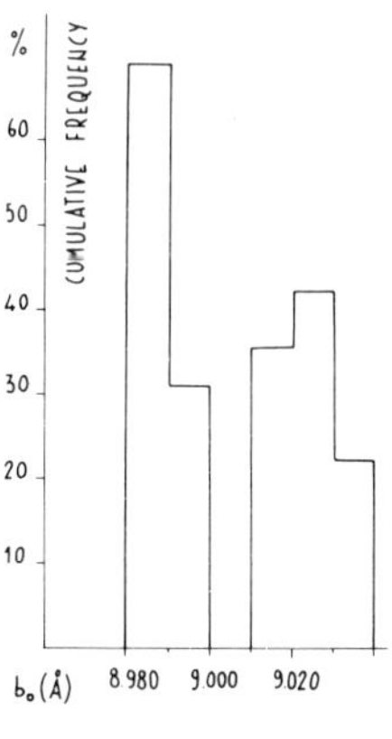

Fig. 2. b_0 values of the potassic white micas from the *Altkristallin* metapelites. The histogram with the higher values concerns rocks from the Northern block, affected by the Alpine recrystallization. The lower values histogram concerns rocks lacking in Alpine overprints.

other Austridic areas which are lacking in Alpine overprints;
2) *recrystallization effects in the phyllonitic belts*. The significance of these recrystallized phyllonites is not yet completely clear; we believe that their study can supply very important data;
3) *fine-grained kyanite aggregates*, similar to the Variscan *Disthen-Flasern* which are well known, for example, in the Saualpe. Within our area, these aggregates only occur in rocks in which biotites have Alpine cooling ages.

The crystallization of a lavender-blue amphibole, liable to be con-

fused with glaucophane even in thin section, has also been referred to this metamorphic phase (Sassi et al. 1974). However, we stress here that this amphibole is not a glaucophane: chemical and X-ray analyses have ascertained that it is a somewhat rare lithium amphibole, the holmquistite (Borsi et al. 1977 c).

c_2) We also mention the development of an *acid to intermediate plutonism*. The whole infra-Austridic plutonic alignment (Rensen, Monte Alto-Altenberg, Rio Vena-Winnebach; Vedrette di Ries-Riesenferner, Cima di Vila-Zinsnock) can be attributed to this process only speculatively. However, the emplacement age of the following bodies is certainly Alpine:

1) the younger dykes outcropping in this area (see locations in fig. 1), because some of them cut the Penninic Bündnerschiefer (Scolari & Zirpoli 1972);

2) the Riesenferner massif, from which Rb/Sr whole-rock isochrons around 30 m.y. have been obtained (Borsi et al. 1977 b). The values of the initial Sr isotopic ratio which we found could indicate a crustal genesis, with minor contributions of materials having low Sr isotopic initial ratios;

3) the Rensen massif, in which a Rb/Sr wholerock isochron of 41 ± 19 was obtained; there are reasons for believing that the true age of this massif is better represented by the Rb/Sr age of muscovite, 29 m.y. (Borsi et al. 1977 a).

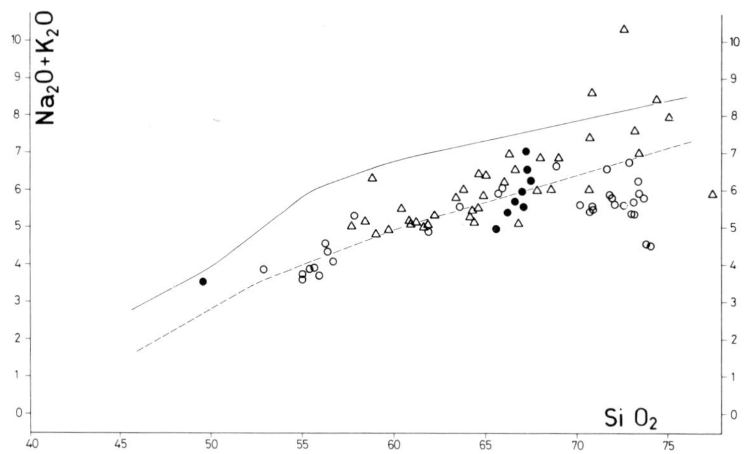

Fig. 3. SiO_2 versus (Na_2O + K_2O) diagram concerning rock samples from the Vedrette di Ries massif (triangles), dykes of the Valles and Rio Vena valleys (open circles), M. Alto and Rensen massif (closed circles).

To conclude, the Oligocene age of the whole infra-Austridic plutonic alignment, however partly speculative, can be reasonably assumed on the basis of our new data. This implies the production of abundant calc-alcaline melts (fig. 3) and their emplacement along an E-W tectonic lineament during the Oligocene.

The other two blocks were also affected by this magmatic activity, but to a much lesser extent: for example, the small dioritic body of Polland (near Lienz) has the same Alpine age (Borsi et al. 1977 c).

It should be amphasized that the value of approx. 30 m.y. for the Alpine plutonic rocks is becoming more and more recurrent in the whole of the Alps (see the review in Exner 1976). It is worthy of emphasis that this age value is very near the age of the Alpine thermic climax (Lepontine and Tauern metamorphisms).

Radiometric data

We have mentioned here our radiometric data concerning the plutonic rocks; we can thus discuss the results we obtained on the biotites along the *mini-geotraverses*. Borsi et al. (1977 c) report all these Rb/Sr cooling age values, arranged according to the localization of the samples north or south of the DAV and KW lines.

A striking result is that the biotites from the south of the DAV line give cooling ages of around 300 m.y., while those from the north of it give age values lower than 30 m.y. This situation is very evident in fig. 4. The Periadriatic Lineament and the Austrides/Pennides boundary do not imply any sharp change of the biotite cooling ages. On the contrary, it is the DAV line which represents, in the now outcropping levels, the barrier north of which the basement rocks were involved in the Alpine

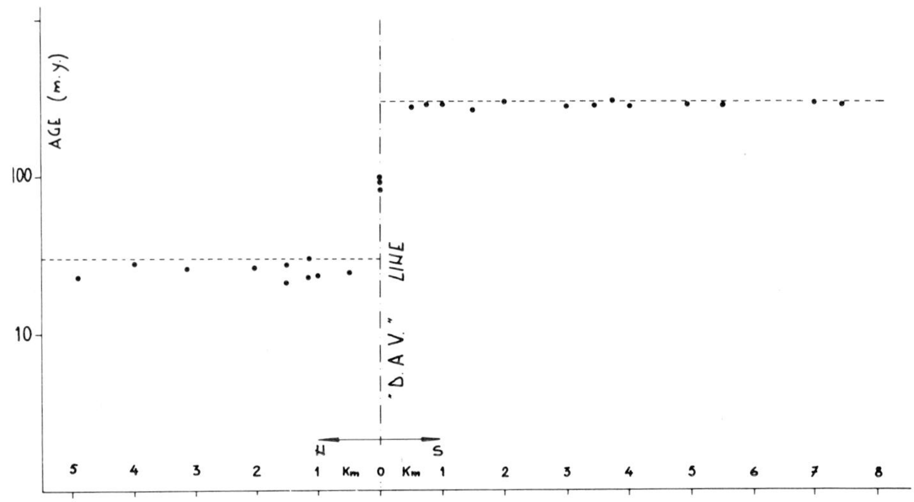

Fig. 4. Sharp change of the Rb/Sr biotite cooling age values at the DAV line; Hercynian values occur in the Southern block, Alpine values in the northern one.

regional heating. Therefore, it is not merely accidental that Alpine granitoids and important Alpine metamorphic and structural overprints occur only in the northern block, while in the two more southerly blocks the metamorphic situation is pre-Alpine, and the Alpine deformations, although important, are post-crystalline.

The northern block also seems to have some pre-Alpine pecularities by comparison with the southern ones. In fact, the Ordovician metagranitoids outcrop largely in the two southern blocks, while well-foliated

aplites and pegmatites prevail in the northern one. A new important datum is that some of these metamorphosed aplites have a certain middle-Permian age (258 ± 18 m.y.; Rb/Sr whole-rock isochron).

Concluding Remarks

By linking the geological, petrographic and radiometric data concerning the Austridic block we studied, and setting them in the regional context taken from the literature, a geodynamic model for the Eastern Alps could be attempted. However, we prefer to give here a picture strictly linked to unquestionable data, renouncing the speculative approach for the time being. Consequently, the following results are given as well-established facts:

1. The continental margin we studied was involved in a heating process subcontemporaneously with the development of the Tauern metamorphism in the Pennides. The Austridic rocks of the northern block reached temperatures slightly lower than 500°C, in the currently outcropping levels, under intermediate pressure conditions. A similar situation, perhaps, can also be found in the deeper levels of the two more southerly blocks. During this metamorphism, a fold-producing, dynamic activity developed, at least on micro-and mesoscopic scales.
2. It was probably this thermic event which determined the deep melting of crustal materials containing some amounts of basic rocks. These melts were emplaced 30 m.y. ago approximately, making up variously-sized massifs and dyke swarms. Most of the melts were injected along a *weak belt* whose E-W trend runs subparallel to the border of the continental margin, i.e. to the buried Benioff zone, and subparallel to the Periadriatic Lineament. The injection of the dykes is younger than the *nappe emplacement*, in view of some dykes cutting the Austrides/Pennides boundary. Moreover, the structural control detected in the dyke swarms implies that, at least locally, block-faulting tectonics developed before their injection.
3. Another metamorphic phase developed, at least locally, under lower temperature and probably lower pressure, more than 18 m.y. ago. Therefore, if we also take into account the Cretaceous metamorphism linked to subduction (demonstrated outside our area, e.g. Frey et al. 1974), the low-temperature, low-pressure metamorphism quoted here under Point 3 would represent a third metamorphic event linked to the Alpine orogenesis.
4. The phyllonitic belt occurring in the northern block formed before the injection of the Riesenferner pluton and before the regional heating linked to the *Tauernkristallisation*. This phyllonitization is probably related to the *nappe emplacement*.
5. After the intrusion of the Oligocene granitoids, the DAV line was active. By means of paleothermic considerations concerning the situation under point 1, a vertical displacement of more than 5 km can be estimated.

References

Borsi, S., Del Moro, A., Sassi, F.P. & Zirpoli, G., 1973: Mem. Soc. Geol. Ital. - 12:549-571, Pisa.

- , 1977 a (in print): On the Radiometric Age of the Rensen Massif

(Eastern Alps).
- , 1977 b (in print): On the Age of Vedrette di Ries (Riesenferner) Periadriatic Massif.
Borsi, S., Del Moro, A., Sassi, F.P., Zanferrari, A. & Zirpoli, G., 1977 c: Mem. Sci. Geol. Padova, 1.
Dal Piaz, Gb., 1934: Mem. Ist. Geol. Univ. Padova. - 10:1-242.
Exner, Ch., 1976: Verh. Geol. B.-A. - 1976, 2:3-64, Wien.
Frey, M., Hunziker, J.C., Frank, W., Bocquet, J., Dal Piaz, G.V., Jaäger, E. & Niggli, E., 1974: Schweiz. Min. Petr. Mitt. - 54:248-290, Zürich.
Kleinschmidt, G. & Ritter, V., 1976: Carinthia II, 166/86:57-91, Klagenfurt.
Kleinschmidt, G., Sassi, F.P. & Zanferrari, A., 1976: N. Jb. Geol. Paläont. Mh.:421-434, Stuttgart.
Purtscheller, F. & Sassi, F.P., 1975: Tschermaks Min. Petr. Mitt. - 22:175-199, Wien.
Sassi, F.P. & Zanferrari, A., 1972: Boll. Sco. Geol. Ital. - 91:533-557, Roma.
Sassi, F.P., Zanferrari, A., Zirpoli, G., Borsi, S. & Del Moro, A., 1974: N. Jb. Geol. Paläont. Mh.:421-434, Stuttgart.
Scolari, A. & Zirpoli, G., 1972: Mem. Ist. Geol. Min. Univ. Padova. - 29:1-32.
Zanferrari, A., 1977: Mem. Sci. Geol. Padova., 1.

Granites and Granodiorites at the Periadriatic Line in SW of the Tauern Window

P. v. Gizycki and Kl. Schmidt

The granitic and granodioritic igneous rocks at the Periadriatic Line (P.L.) are partly of Hercynian, partly of Alpidic age, thus indicating the repeated magmatic activity of that prominent lineament separating the Southern Alps from the Austroalpine and the Penninic units further North (Bögel 1975; Exner 1976). At present the surface areas of exposure of Hercynian and Alpidic plutons along and near the P.L. are in a proportion of 1 : 1,7.

In the cross section of the Vals Valley (fig. 1) the Hercynian granodiorite of Brixen and the adjacent Rensen Granodiorite, though separated by the P.L., approach one another within a range of 1,4 km. If radiometric dates would not exist one might imagine close genetic relations and consanguinity of the rather similar rocks. Petrographic and tectonic indications however, suggest an independent magmatic and tectonic position of the Rensen Pluton (Nollau 1974). On the other hand, the northern con tact of the Rensen granodiorite is in a position only 600 m south of the

fault zone between the Austroalpine and the *Obere Schieferhülle* of the Tauern Window. Right at the fault-zone, aplitic and lamprophyric veins cut the Austroalpine crystalline as well as the Penninic schists affected by the Tauern crystallization (Scolari & Zirpoli 1973).

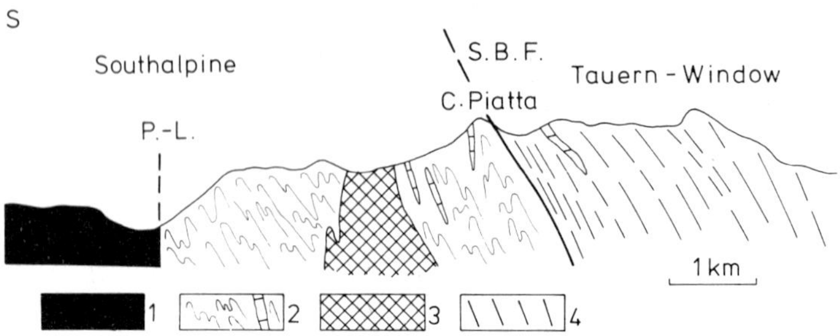

Fig. 1. Crosscut through the Vals Valley. 1. Brixen-granodiorite, 2. Hercynian crystalline with aplitic and lamphrophyric veins near the S.B.F. 3. Rensen-granodiorite, 4. Penninic rocks of the Glockner-Nappe. P.L. Periadriatic Line, S.B.F. Southern border fault of the Tauern-Window.

A direct spatial connection of these veins with the Rensen granodiorite could not be ascertained by recent mapping (Gizycki). Geochemical data,however, are in favour of a close genetic relation. Besides muscovite- and biotite ages of 29 respectively 17 $\pm$ 4 m.y. there had been no

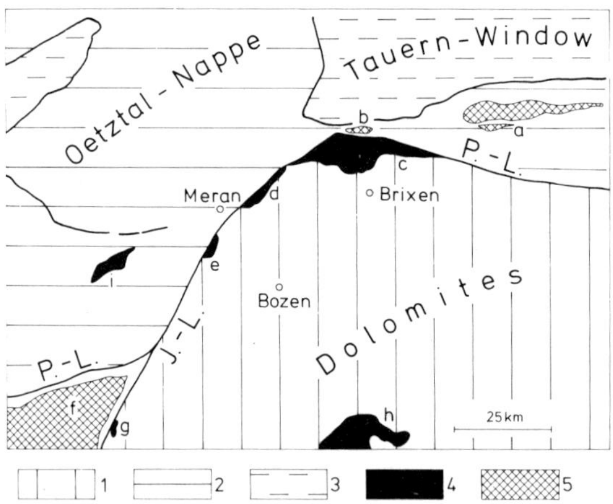

Fig. 2. Hercynian and Alpidic plutons at the Periadriatic Line (P.-L.). a. Rieserferner, b. Rensen-, c. Brixen-, d. Iffinger-, e. Kreuzberg-, f. Adamello, g. Mte. Sabion-, h. Cima d'Asta, i. Martell-pluton I.-L. Iudicarian-Line. 1. Southalpine-, 2. Austroalpine-, Penninic units, 3. Hercynian plutons, 4. Alpidic plutons.

reliable indications for the intrusion age of the granodiorite (Borsi et al. 1973; Sassi et al. this volume). The pluton contains neither mineral assemblages nor petrofabrics attributable to the Tauern crystallization. The veins, however, exhibit a slight schistosity and an *epimetamorphism* due to a peripheral or late effect of the postkinematic (polyphase) crystallization of the Tauern Window.

A geochemical comparison of the Periadriatic intrusions (fig. 2) has proved very useful. The Hercynian (Brixen, Iffinger, Kreuzberg, Mte. Sabion, Cima d'Asta) and the Alpidic (Adamello, Rieserferner) plutons, dated radiometrically, also differ significantly with regard to their contents of trace elements (Ca, K, Ti, Zr, Sb, Sr, Ba).

On two component diagrams the Rensen granodiorite plots clearly in the field of the Alpidic igneous rocks.

With respect to the main deformation of the Tauern Window and the surrounding Austroalpine, the Rensen granodiorite is classified petrologically and tectonically as a late to posttectonic intrusion.

The pluton is affected, however, by postcrystalline cataclastic deformations caused by late activities of the P.L., in the same way as the Brixen granodiorite.

Considering all the above data and arguments, we conclude that the Rensen and Rieserferner granodiorites are better explained as post-collisional melting-products than as arc magmatism related to a Cretaceous subduction (Schmidt 1976).

The thickening of continental crust by underthrusting of sialic crustal segments, high compression and the generation of friction heat are most likely processes which probably initiated low-velocity channels and the formation of granitic melts.

The late tectonic relaxationslead to dilatating fault tectonics, especially along the P.L., and to crustal elevations, thus providing space for the rise and intrusion of granitic plutons.

References

Bögel, H., 1975: Verh. Geol. B.-A. - 2,3:163-199, Wien.

Borsi, S., Del Moro, A., Sassi, F.P. & Zirpoli, G., 1973: Mem. Soc. geol. Ital. - 12:549-571, Pisa.

Exner, Ch., 1976: Verh. Geol. B.-A. - 2:3-64, Wien.

Nollau, G., 1974: Erlanger geol. Abh. - 98:1-92, Erlangen.

Schmidt, Kl., 1976: Z.dt. geol. Ges. - 127:53-72, Hannover.

Scolari, A. & Zirpoli, G., 1972: Mem. Ist. Geol. Univ. Padova,29:1-32.

Scolari, A., 1973: Mem. Acc. Pat.Sc.Lett.Art.,85:19-48, Padova.

Origin and Metamorphism of Ultramafic Rocks along the Periadriatic Line in the Western Alps

F. Rost and J. Wannemacher

Ultramafic rock bodies in various sizes occur in crystalline zones on both sides of the contact between the Penninic and Southern Alps, extending from the Ivrea zone to the Etsch (Adige) valley. Their mantle origin has been demonstrated geophysically near the Ivrea zone; a similar origin of the other bodies is beyond doubt. Petrologically, they fall into two types: *Magmatic* ultramafites with spinel and two pyroxenes (Lensch 1971) occur in the southern part of the Ivrea zone. Ultramafites containing pyrope-rich garnets or the kelyphitisation products of retrograded garnets occur only between the middle Ivrea zone and the Ultental (fig. 1).

The ultramafic rocks disumed in this paper are pre-Alpine, possibly Hercynian, in age and differ genetically from the Jurassic or Cretaceous *serpentinites* found in ophiolite suites. The latter are not treated in the present paper.

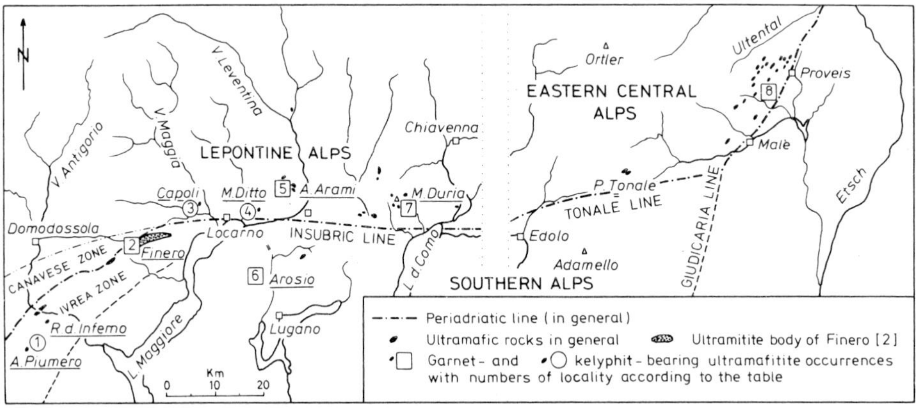

Fig. 1. Sketch map of ultramafic rock occurrences in crystalline zones along the boundary between Southern and Central Alps.

T a b l e

Occurrences of garnet and kelyphite bearing ultramafitites along the Insubric and Tonale line in the western Alps according to fig. 1.

1) Alpe Piumero and Rio del Inferno, middle Ivrea zone (Lensch & Rost 1972);
2) Finero (Lensch 1975);
3) Capoli and Monte Salmone;
4) Monti di Ditto;
5) Alpe Arami and Croveggio (Rost, Wannemacher & Grigel 1974);
6) Arosio/Sotto Ceneri;
7) Region of the Monte Duria (Fumasoli 1974);
8) Mountain range S of the Ultental (Brenneis 1971; Amthauer, Brenneis, Grigel & Rost 1971, 1972).

The rock bodies studied are arranged in fig. 1 and the following table. Our research on ultramafites along the Periadriatic line used emission spectrography, X-ray vacuum spectrometry and powder diffractometry on the kelyphites especially in the localities 3-7. It refines previous results and may promote a discussion of geodynamic processes at the northern boundary of the Mediterranean.

Analytical results about chemistry and mineralogical composition of ultramafic rocks from occurrences 3, 4, 6 and 7 show a similar distribution to ultramafites of Alpe Arami and Ultental. Most of the rocks are peridotites; a trend to harzburgites and enstatite-rich lherzolites is evident; websteritic pyroxenites seldom appear as small intercalations.

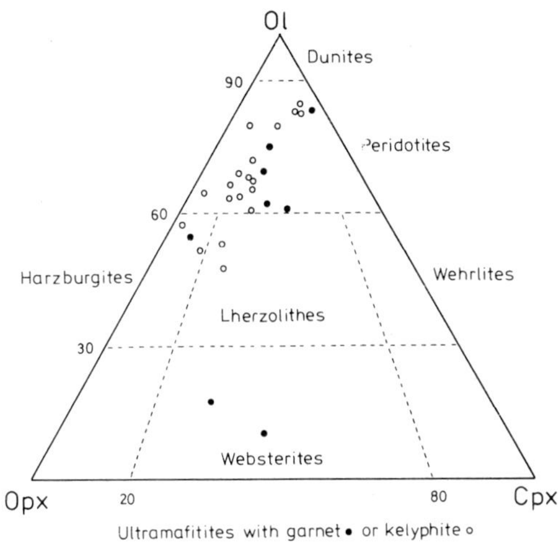

Fig. 2. Diagram of normative mineral composition of the ultramafitites from the occurrences 3, 4, 6 and 7 in the table and fig. 1.

Conclusions

Garnet ultramafites may originate under pressure conditions higher than 10 kbar. The occurrence of such rocks demonstrates, along the Periadriatic line, a pre-Alpine deep burial of crystalline basement rocks, in which ultramafites intruded tectonically from the Upper Mantle. It is significant that the burial happened on both sides of the Periadriatic line.

For the origin of garnets in the ultramafites there are two interpretations:

1) Preexisting garnet ultramafites from deeper levels of the Mantle - more than 40 km - have been emplaced into deeply buried crystalline basement rocks of the prealpine orogen.

2) Spinel-pyroxene-ultramafites of magmatic facies (facies 1 in the diagram fig. 3) tectonically have been emplaced into basement gneisses of different zones and after the emplacement both were buried to deeper levels. Here the pressure conditions caused a secondary metamorphic growth of pyrope-rich garnets in the ultramafic rocks (facies 2 in fig.3). During this process the gneisses of the environment partly were converted into granulitic rocks (Ultental) respectively *stronalithes* (Middle Ivrea zone).

For the garnet peridotites of Alpe Arami, enclosed in Lepontine gneisses of amphibolite facies, Den Tex (1971) concludes that fabric, mineralogy and geochemistry indicate an origin, most probably from deeper parts of a Benioff zone. In the ultramafic rocks of Ultental (Amthauer et

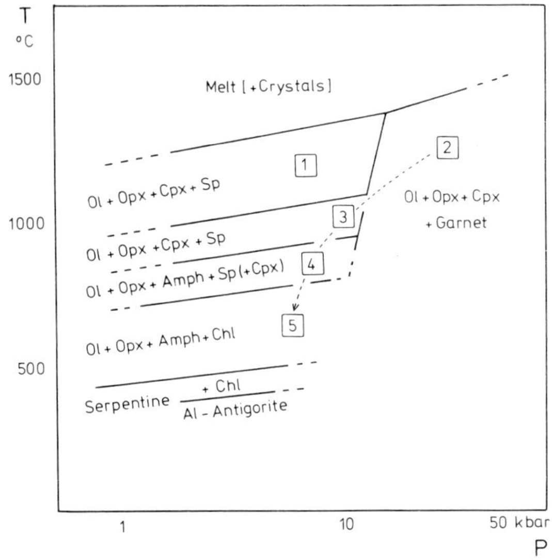

Fig. 3. Simplified facies diagram of "orogenotype" ultramafitites in crystalline environments.

al. 1971) of the area around the Monte Duria and of Arosio the pyrope garnets appear as separate grains as well as exsolution lamellae in both pyroxenes caused by the high pressure conditions (fig. 4). This exsolution may be explained better according to the interpretation 2.

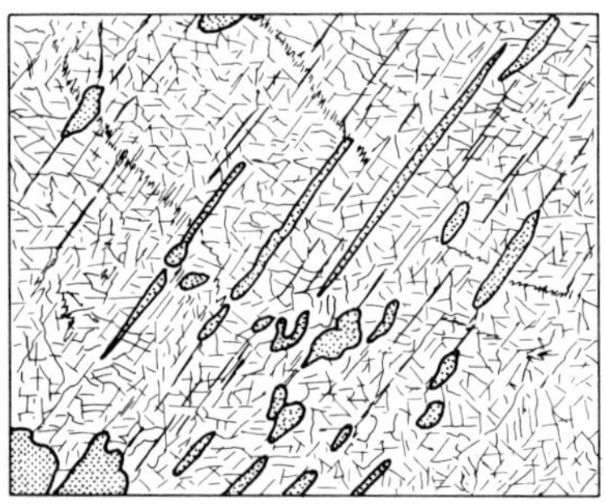

Fig. 4. Exsolution lamellae of garnet (dotted) in orthopyroxene of garnet lherzolithe from Alpe di Brez/Ultental. (Section ⊥ c, lamellae // (100) of orthopyroxene; figure length=2mm).

After this primary phase of deep burial an Hercynian or older orogeny in the Alps, uplift connected with decreasing conditions of pressure and temperature caused a retrograde metamorphism in the ultramafic rocks. In reversion of the garnet equation

enstatite + diopside + spinel — pyrope garnet + olivine

the garnet reacted with surrounding olivine grains to *kelyphite*, a radially fibrous aggregate of enstatite, diopside and spinel corresponding to

the rock facies 3 in fig. 3. The chemical composition of these secondary kelyphite minerals differs from the primary mineral conposition, especially in the content of isomorphic elements, such as Al in both pyroxenes, Ca in enstatite, Fe and Cr in spinel. In the study area this first stage of kelyphitisation is missing. The minerals composition of the kelyphite here is a third spinel, enstatite and Al-rich amphibole according to the minerals of the rock facies 4 in fig. 3. The estimated conditions of ca. 800 - 900°C and less than 10 kbar appear to have remained stable during longer times, since the kelyphitisation of this second stage is very common in the studied peridotites. This may be interpreted as an interruption of uplift.

The next phase of retrograde metamorphism is characterized again by decreasing pressure and temperature connected with an increasing influence of aqueous fluids in supercritical conditions. Spinel and the Al-rich amphibole became unstable and altered to chromium bearing clinochlore and Al-poor actinolite, whereas olivine and orthopyroxene remained unaltered (facies 5 in fig. 3). This phase of retrograde metamorphism was accompanied by tectonic deformation. The chlorite bearing ultramafites partly underwent foliation and changed into chlorite amphibole schists. Serpentinization at temperatures of less than 500°C may remain out of consideration.

Relics of all the different metamorphic stages with their typical minerals are preserved in most of the ultramafic rocks along the Insubric line and Tonale line.

References

Amthauer, G., Brenneis, P., Grigel, W. & Rost, F., 1971: Fortschr. Miner. 49, Beiheft 1:11-13.

- , 1972: Ber. Deutsch. Ges. geol. Wiss. B. Miner. Lagerstättenf. 16:227-240.

Brenneis, P., 1971: Dissertation Saarbrücken.

Fumasoli, M.W., 1974: Dissertation Zürich.

Lensch, G., 1971: Ann. Univ. Sarav. - 9:5-146.

- , 1975: N. Jb. Miner. Mh.:333-336.

Lensch, G. & Rost, F., 1972: Schweiz. Min. Petr. Mitt. - 52:237-250.

Rost, F., 1971: Verh. Geol. Bundesanstalt Wien:266-286.

Rost, F., Wannemacher, J. & Grigel, W., 1974: Schweiz. Min. Petr. Mitt. - 54:353-369.

Den Tex, E., 1971: Fortschr. Miner. - 48:69-74.

A Plate Tectonics Model of the Eastern Alps*

W. Frisch

The prevailing tectonic models of the Eastern Alps (e.g. Ernst 1973; Closs 1975; Hawkesworth et al. 1975; Dietrich 1976) are based on one southward-dipping subduction zone which passes through several stages of evolution by the successive advance of the overriding Adriatic plate.

The model presented here explains the tectonic and facies relations of the Eastern Alps in terms of three southward-dipping subduction or underthrusting zones active in succession during the Alpidic orogeny. Two of these three mobile zones, the South Penninic and North Penninic zones, are attributed to subduction of oceanic basement, but the third and most external one is an underthrusting foredeep on the late European continental margin. The three mobile zones which relieve each other in time govern the orogenic evolution.

Up to the beginning of the Jurassic, the Helvetic, Penninic and Austroalpine realms of the Alps are part of the Eurasian continental plate. Concerning their facies, the Triassic sediments of these three domains are continuous and of shallow-water or platform type indicating an increasing rate of subsidence towards the south or southeast. Basin sediments (e.g. the Hallstatt facies) are found approaching the continental margin of the Tethys ocean southeast of the Austroalpine - South Alpine domain.

In early Jurassic time, the *South Penninic* (Piémontais) ocean begins to open giving rise to the deposition of the Schistes lustrés (Bündner Schiefer) of the later Glockner nappe (Tauern window) which are sediments mainly deposited on oceanic crust or in a trench (Frisch 1974, 1976; Tollmann 1975). With the opening of the South Penninic ocean the Austroalpine basement, now being part of the newly formed Adriatic plate, split off the Eurasian plate. The South Penninic ocean is considered to be the continuation of, and to be connected by transform faults with, the central Atlantic which opened at the same time (Roeder 1976). Thus, applying the geometric rules of plate tectonics and taking into account that subduction in the South Penninic ocean commenced at some time in the early Cretaceous, long enough before the closure of the ocean at about 80 m.y., the maximum width of this ocean is estimated at 1000 to 500 km in the early Cretaceous. With the formation of a trench the predominantly calcareous Schistes lustrés sedimentation (Glockner facies of the Tauern window; Frasl & Frank 1966) of the late Jurassic to early Cretaceous changes to a predominantly lime-free sedimentation (Fusch facies; loc. cit.) which is partly developed as flysch (Prey 1975). The Schistes lustrés of the Glockner nappe are interpreted as a tectonic mélange formed prior to the closure of the South Penninic ocean by the continent-continent collision in the Upper Cretaceous (Hawkesworth et al. 1975). During subduction, serpentinites were scraped off remnant oceanic crust belonging to the overriding plate; these subducted serpentinites are now part of the Matrei

* Publication No. 33 of the academic research program "Geologic Deep Structure of the Eastern Alps" financed by the Austrian Fund for the Support of Scientific Research.

zone which represents the uppermost part of the mélange series. Obducted parts of ophiolites at the base of the Glockner nappe are less frequent; in this position are common highly contorted Mesozoic shelf sediments scraped off the northern continental margin during collision (fig. 1).

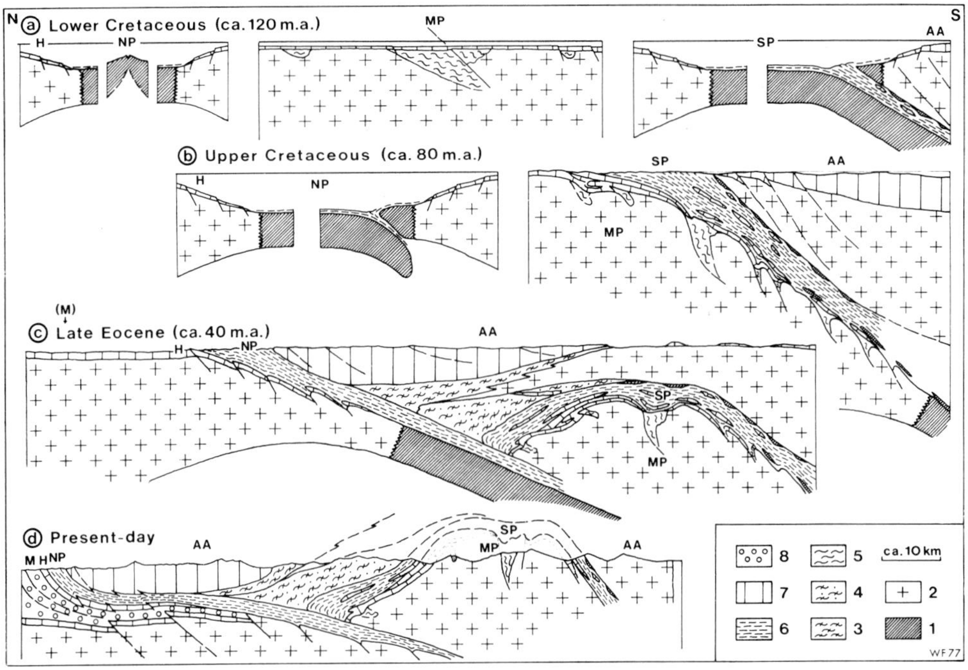

Fig. 1. Profiles showing evolution of the Eastern Alps during Alpidic orogeny. No exaggeration. Middle Penninic zone unrolled accurately, other zones are more schematic. Present-day section runs from Tegernsee in the north to Bruneck in the south. Profiles are cutting the western Tauern Window.
M = molasse; (M) in Profile c indicates site of succeeding Molasse foredeep; - H = Helvetic zone; - NP, MP, SP = North, Middle and South Penninic zones; - AA = Austroalpine zone; 1: oceanic crust; - 2: continental crust (including Hercynian sedimentary series where not shown separately); - 3: Paleozoic (only Grauwacken zone) of Upper Austroalpine (Profiles c and d); - 4: Paleozoic schists of Lower Austroalpine (Profiles c and d); - 5: Paleozoic schists of Middle Penninic zone; - 6: oceanic sediments (Mesozoic to Cenozoic); - 7: Mesozoic to Cenozoic cover of continental crust; - 8: molasse (Tertiary).

Generally, the sediments of the *Austroalpine* realm in the Jurassic and Lower Cretaceous are of increasingly deeper water environment compared with the Triassic. This is due to the separation of the Adriatic plate which is now largely surrounded by oceanic areas. In the Lower Cretaceous, trench sediments of flysch type, such as the Rossfeld beds of late Valanginian to early Aptian age, at the fronts of internal décollement nappes indicate early compressive movements in this southernmost part of the

Eastern Alps (Faupl & Tollmann in press).

North of the South Penninic ocean, the *Middle Penninic* zone of the Eastern Alps is in the same position as the Briançonnais in the Western Alps (Clar 1965; Tollmann 1965). Today we find this area in the Venediger nappe of the Tauern window, the basement of which is represented by the Zentralgneisses. A palinspastic reconstruction of the Venediger nappe is presented in fig. 1 (a) (labelled "MP"). It is shown that there existed gneiss cupolas separated by Paleozoic schist synclines due to Hercynian orogeny, as well as a Hercynian thrust nappe (Frisch 1977).

In the Middle Penninic zone of the Tauern window, the sediments are of shallow-water type or missing due to non-deposition in the Triassic, and those of a subsiding shelf in the Jurassic (Upper Jurassic Hochstegen marbles). In the southern part, there are Lower Cretaceous clastic sediments near the continental margin. Parts of the Middle Penninic zone may have been emergent even in Jurassic and Lower Cretaceous times.

In the Lower Cretaceous, a trough or basin begins to develop between the Middle Penninic zone and the Helvetic domain which is still adherent to the European continent. This basin is considered to be the continuation of the *North Penninic* (Valais) trough of the Swiss Alps, and is the zone where the Rhenodanubian flysch is deposited. The basement of the Rhenodanubian flysch is not preserved. For this reason, and because of the depositional environment of the flysch which is a deep-water basin and slope sediment (Faupl 1976 and personal communication), it is likely that the flysch basin was floored by oceanic crust. Remnants of continental crust complicate the paleogeographic pattern at least near the eastern end of the Rhenodanubian flysch basin (Faupl 1975).

According to the model presented here, the Rhenodanubian flysch basin has spread since the Lower Cretaceous and commenced to be consumed along a subduction zone at its southern margin in the Upper Cretaceous (ca. 80 m.y., figs. 1 and 3). At this time, the Middle Penninic zone is considered to have collided with the Austroalpine basement squeezing the oceanic South Penninic Glockner nappe and some "Lower Austroalpine" slices in between. With the collision, the subduction activity shifted from the South Penninic to the North Penninic ocean. Thus, the Middle Penninic Venediger nappe, from this time welded onto the overriding plate, was ready to become an allochthonous feature (figs. 1 and 2).

The North Penninic (Valais) trough ends towards the west in the northern part of the French Alps. It is suggested here that this basin has had a considerable width due to spreading over a period of at least 50 m.y. and was underlain by oceanic crust. I speculate therefore, that it did not end in the Western Alps but was offset to the south by a prominent fault. The suggested continuation is the trough of the Pyrenees and the Bay of Biscay which has a history consistent with the North Penninic basin (Williams 1975). The suggested continuation of the Middle Penninic (Briançonnais) zone is Corsica-Sardegna (excluding the South Penninic nappe in northeastern Corsica) and the Iberian peninsula north of the Betic cordillera. Thus, the development of the Alpine "geosyncline" is intimately related to the successive break-through of the Atlantic ocean to the north.

The North Penninic basin was closed in the Eocene with the collision of the Helvetic and Middle Penninic zones and the gravitative gliding of the Northern Calcareous Alps off the updoming Austroalpine basement (fig. 1). The *Helvetic* domain, prominent in the Western Alps, probably ends somewhere west of the southward protrusion of the Bohemian massif (Angen-

heister et al. 1975). Rudimentary décollement slices of mainly post-Jurassic sediments are exposed west of Salzburg, the basement being entirely obscured at depth. The "klippen" zones to the east (Gresten, St. Veit) are defined as "Ultrahelvetic".

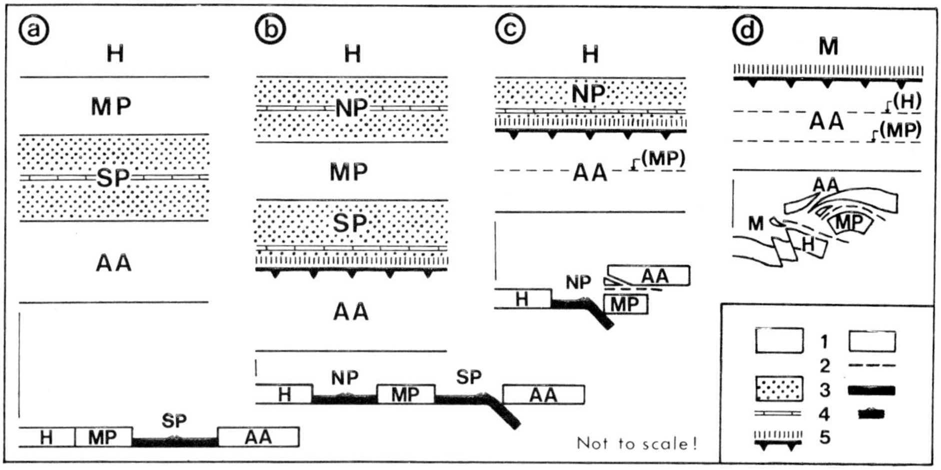

Fig. 2. Schematic sketch of the crustal zones of the Eastern Alps during different periods of Alpidic orogeny. Maps (above) and north-south profiles (below) during (a) Jurassic and early Lower Cretaceous (ca. 180 - 130 m.y.) - (b) late Lower Cretaceous and early Upper Cretaceous (ca. 130 - 80 m.y.) - (c) late Upper Cretaceous to Eocene (ca. 80 - 40 m.y.) - (d) Oligocene and Miocene (ca. 40 - 5 m.y.).
M = Molasse foredeep; - H = Helvetic zone; - NP, MP, SP = North, Middle and South Penninic zones; - AA = Austroalpine zone. 1: continental crust and sedimentary cover; - 2: scraped-off oceanic sediments; - 3: oceanic crust; - 4: spreading ridge; - 5: trench (foredeep) with subduction (underthrusting) zone. Not to scale; oceanic areas are to be thought much more extensive.

With the Eocene event, the axis of the depositional trough shifts again to the north, forming the external *Molasse* basin which is a foredeep on the margin of the underthrusting Eurasian continental plate. Underthrusting continues during deposition; however, this is not a "subduction" proper as the material was no more carried down to the asthenosphere but rather contributed to the thickening of the sialic crust. This process is quite similar to that in other orogens with large foredeeps, e.g. the Zagros fold belt or the Himalayas.

The *metamorphic history* of the Eastern Alps reflects the shifting subduction and underthrusting activity sketched above. In the Cretaceous we find a paired metamorphic belt (Hawkesworth et al. 1975) due to the subduction of the South Penninic ocean: a thermal event affects the Austroalpine domain, while probably simultaneous high-P/T metamorphism is preserved in the mélange series at the base of the Glockner nappe. The existence of this paired belt has been terminated with the collision of the Middle Penninic and Austroalpine zones. This event is indicated by

the radiometric cooling ages of micas clustering near 80 m.y. in parts of the overriding plate (Miller et al. 1967; Schmidt et al. 1967; Lambert 1970; Cliff et al. 1971).

With its climax in the Eocene, there is regional metamorphism with elevated temperatures in the units of the Tauern window which are at this time part of the overriding plate. High-P/T metamorphism is unknown for this period; the series possibly affected by such metamorphism - the southern continuation of the Rhenodanubian flysch - are still obscured beneath the higher tectonic units.

A number of radiometric mica ages cluster near 18 - 25 m.y. (Besang et al. 1968; Cliff et al. 1971) in the Penninic zones of the Tauern window. The ages, commonly attributed to cooling, may well indicate a slight thermal event at this time which coincides with the main period of subsidence and underthrusting in the foredeep (Molasse basin).

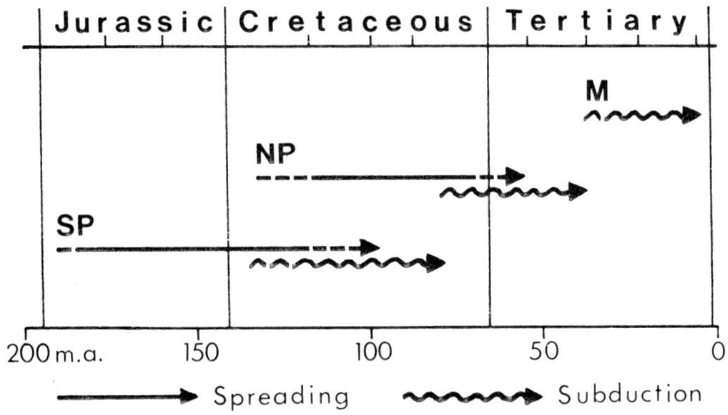

Fig. 3. Sketch showing active periods of the mobile zones of the Eastern Alps. M = Molasse foredeep; - NP = North Penninic ocean; SP = South Penninic ocean.

Fig. 1 reflects the palinspastic situation along a profile cutting the western Tauern window during several stages of the evolution of the Alpine orogen. Fig. 2 is a schematic sketch of the evolution of the East-Alpine orogen, and fig. 3 indicates the active periods of the mobile zones in the Eastern Alps. The figures illustrate the shifting of the orogenic activity from internal to external domains. Not shown in the figures are the internal structures of the Austroalpine domain, such as the internal nappe movements and the formation of the Gosau basins in the Upper Cretaceous.

This model shall be a working hypothesis, and emphasis is placed on the Penninic zones of the Tauern window which are shown in palinspastic reconstructions drawn to scale (fig. 1). A similar model, based on the same fundamental idea, i.e. several subduction zones relieving each other with time, is concisely discussed by Bögel & Schmidt (1976)(see also Frisch 1976). The mechanism of shifting subduction activity is termed "suture progradation" by Roeder & Bögel (this volume). The paleo-, meso- and neo-Alpine deformation phases of Trümpy (1973) correspond to the three events of underthrusting (collision) of continental crust described above; however, Trümpy's concept provides only one Penninic basin for the Eastern Alps.

The author thanks H. Bögel and P. Faupl for many worthful discussions. P. Faupl critically read the manuscript.

References

Angenheister, A., Bögel, H. & Morteani, G., 1975: N. Jb. Geol. Paläont. Abh. - 148/1:50-137.
Besang, C., Harre, W., Karl, F., Kreuzer, H., Lenz, H., Müller, P. & Wendt, I., 1968: Geol. Jb. - 86:835-844.
Bögel, H. & Schmidt, K., 1976: Kleine Geologie der Ostalpen. - 231 p., Ott Verlag, Thun.
Clar, E., 1965a: Z. Dtsch. Geol. Ges. - 116/2:267-291.
- , 1965b: Verh. Geol. B.-A., Sonderh. G:11-35, Wien.
Cliff, R.A., Norris, R.J., Oxburgh, E.R. & Wright, R.C., 1971: Jahrb. Geol. B.-A. - 114:121.272, Wien.
Closs, H., 1975: Rapp. Comm. int. Mer Médit. - 23/4a:55-57.
Dietrich, V.J., 1976: Geotekt. Forsch. - 50:1-84.
Ernst, W.G., 1973: Geol. Soc. Amer. Bull. - 84:2053-2078.
Faupl, P., 1975: N. Jb. Geol. Paläont. Mh. - 1975:528-540.
- , 1976: N. Jb. Geol. Paläont. Mh. - 1976/8:449-470.
Faupl, P. & Tollmann, A., in press: Geol. Rundschau.
Frasl, G. & Frank, W., 1966: Der Aufschluß, Sonderh. - 15:30-58, Heidelberg.
Frisch, W., 1974: Mitt. Geol. Ges. Wien. - 66-67:9-20.
- , 1976: Geol. Rundschau. - 65/2:375-393.
- , 1977: N. Jb. Geol. Paläont. Mh. 11:675-696.
Hawkesworth, C.J., Waters, D.J. & Bickle, M.J., 1975: Earth Planet Sci. Lett. - 24:405-413.
Lambert, R.S.J., 1970: Eclogae geol. Helv. - 63/1:197-205.
Miller, D.S., Jäger, E. & Schmidt, K., 1967: Eclogae geol. Helv. - 62/2:537-541.
Prey, S., 1975: Verh. Geol. B.-A. - 1975/4:291-295, Wien.
Roeder, D., 1976: Z. Dtsch. Geol. Ges. - 127:87-103.
Schmidt, K., Jäger, E., Grünenfelder, M. & Grögler, N., 1967: Eclogae geol. Helv. - 60/2:529-536.
Tollmann, A., 1965: Mitt. Geol. Ges. Wien. - 57/2:469-478.
- , 1975: N. Jb. Geol. Paläont. Abh. - 148/3:286-319.
Trümpy, R., 1973: In: De Jong, K.E. & Scholten, R., Gravity and tectonics:229-251, J.Wiley & Sons, New York-London-Sydney-Toronto 1973.
Williams, C.A., 1975: Earth Planet. Sci. Lett. - 24:440-456.

Part 4

The Southern Foreland of the Alps and the Po Plain

Aftershock Investigations and Fault-Plane Solutions of the Friuli Earthquakes 1976

H. Gebrande, H. Häge, Heinz Miller, G. Müller and E. Schmedes

The Friuli earthquake of May 6, 1976, at 20h 00m 12.6s GMT, LAT.: 46,23°N, LONG.: 13,20°E, h = 7 km (EMSC working group, 1976) was the largest seismic event in Central Europe ever recorded by seismic stations. Although the area was known to be seismically active, an event of such disastrous intensity was not expected. The question is how this event, and the two other large shocks which followed on September 15, 1976, may be incorporated in geodynamical models of Alpine orogeny. Investigations of aftershocks and focal mechanisms may contribute to answer this question.

Aftershock Investigations

Preliminary hypocenter determinations for the main shock of May 6, performed by different seismological centers, gave rather inconsistent results because of lacking nearby stations. For an improvement of the accuracy an extension and closer spacing of the existing network seemed necessary. Immediately after the main shock two automatic, digitally recording stations (Gebrande et al. 1976) were installed by the Institut für Allgemeine und Angewandte Geophysik und Geophysikalisches Observatorium, Universität München, near the Austrian-Italian border at Naßfeld and Arnoldstein (fig. 3). A third station of the same type was running at Sylvenstein in the Northern Calcareous Alps about 200 km NW of the epicenter area. Normalized histograms of S-P traveltime differences for these stations (fig. 1) indicate an extension of the aftershock area of about 20 km in a North-South and 40 km in an East-West direction. This is the order of magnitude expected for the dimensions of the focal plane of a magnitude 6 earthquake.

For more accurate determinations of hypocenter and origin time, a fourth station of the same type was installed at Malemaseria within the epicenter area (fig. 3).

To calculate origin times from S-P values the velocity ratio v_p/v_s must be known. At first the standard value 1.73 was used, but a consistency check with the aid of record sections clearly showed that a higher value was more appropriate. With the aid of record sections in the form of Wadati diagrams - two examples are shown in fig. 2 - a mean value of 1.9 for the v_p/v_s ratio was obtained. Taking this high ratio into account, origin times changed by up to one second and consequently locations of hypocenters shifted by several kilometers.

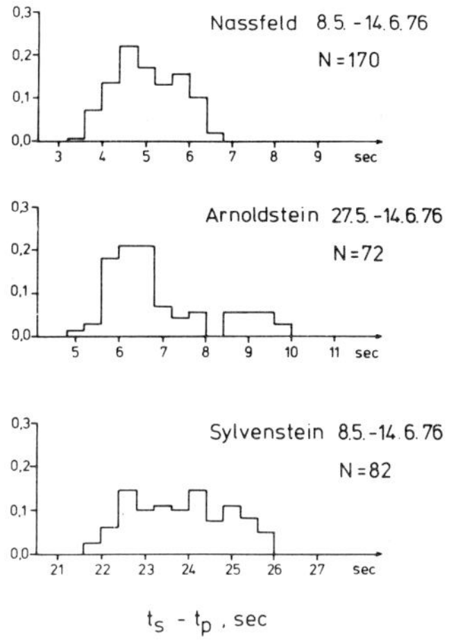

Fig. 1. Normalized histogram of S-P traveltime differences of Friulian aftershocks for the seismic stations Naßfeld, Arnoldstein und Sylvenstein.

The ratio of 1.9 seemed surprisingly high, and at first it was suspected that this value might be time dependent and a possible indicator of dilatancy. However, according to Finetti et al. (1976) this value was still valid after the September shocks, and it rather seems to be a property of limestones. This last interpretation is supported by ultrasonic measurements on different specimens of Alpine limestones (Städtler 1973).

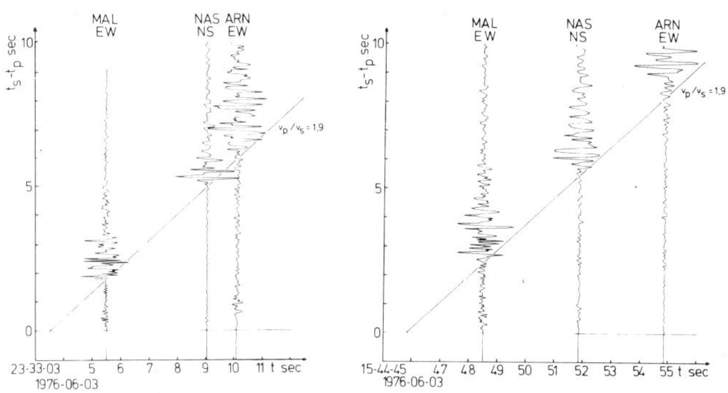

Fig. 2. Seismograms of two different events arranged similar to a Wadati diagram. A v_p/v_s ratio of 1.9 is deduced from these plots.

Fig. 3 shows the location of epicenters for some selected events for the period of May 29 to June 14 as determined with the station triangle. Numbers indicate focal depth, which varies between 0 and 17 km with a mean value around 8 km.

The accuracy of these determinations has been checked by two-dimensional model calculations using a realistic model of crustal structure which was derived from refraction seismic measurements (Miller et al. 1976). An example of such a model calculation is given in fig. 4 for an aftershock with a depth of 6 km. There seems to be satisfactory agreement between calculated and observed traveltimes. This supports not only the

hypocenter determination, but also the validity of the crustal model.

During the time span of our observations no time dependent trend was observed in the hypocenter distribution. Yet when looking at the complete aftershock series (Finetti et al. 1976; EMSC Working Group 1976; Schmedes et al. 1976) one notices a definite trend. From a rather narrow clus-

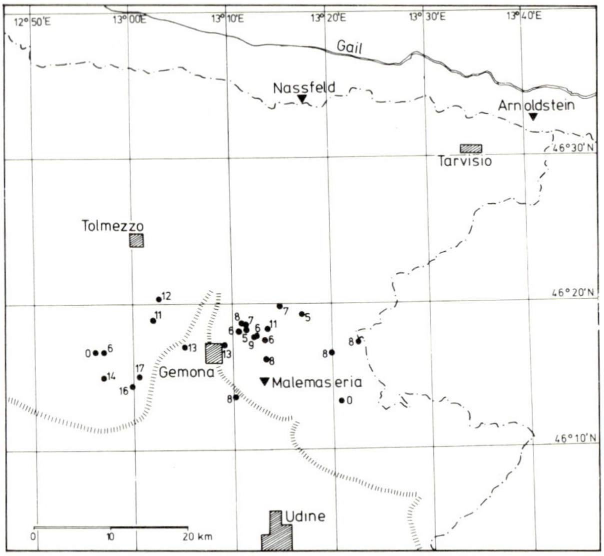

Fig. 3. Outline map showing location of seismic stations and epicenters. Triangles: stations, closed circles: epicenters. Numbers indicate focal depth.

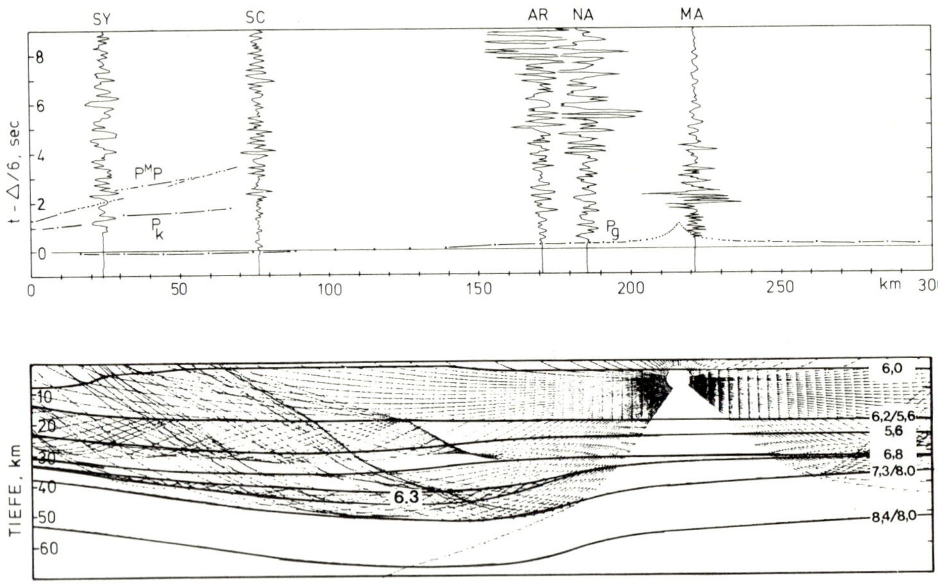

Fig. 4. Ray paths and traveltimes calculated for a realistic model of crustal structure for a Friulian aftershock with a focal depth of 6 km. Calculated traveltimes are compared to seismograms observed at the seismic stations Sylvenstein, Schlegeis, Arnoldstein, Naßfeld and Malemaseria.

ter around Gemona hypocenters first spread out in East-West direction and then more or less discontinuously moved northwards in the September series of shocks. Simultaneously, hypocenter depths became larger. This northward tendency fits well the fault-plane solutions for the three large shocks of May and September.

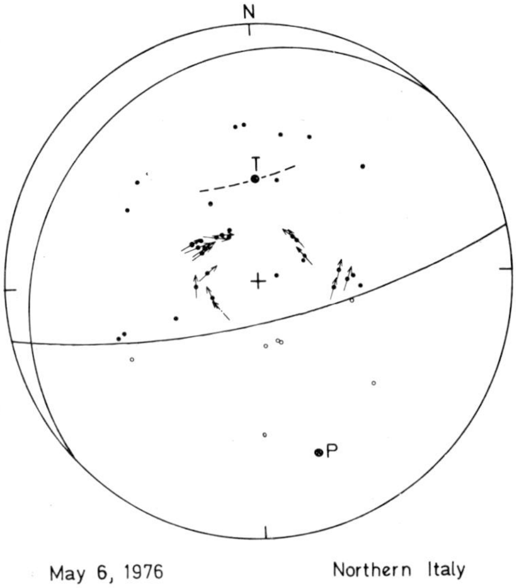

Fig. 5. Fault-plane solution for the Friuli earthquake of May 6, 1976. Equal-area projection of the lower focal hemisphere. Closed circles: P-wave compressions, open circles: P-wave dilatations, arrows: S-wave polarisations. T is the tensional axis, P the compressional axis. Nodal lines are indicated by continuous lines. The data are from long-period seismograms.

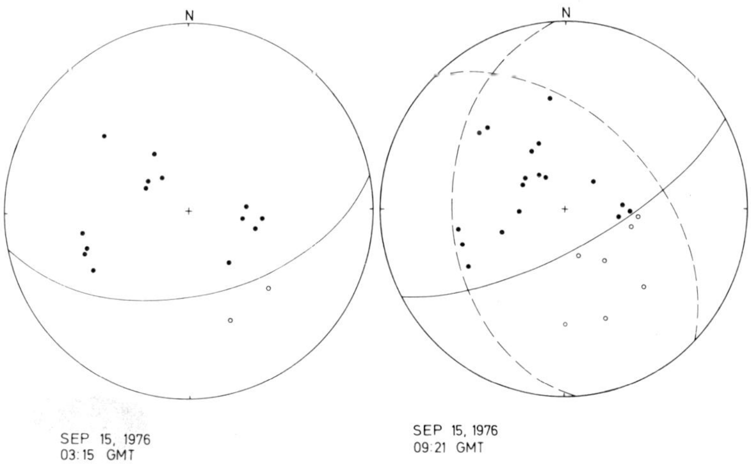

Fig. 6. Fault-plane solutions for the Friuli earthquakes of September 15, 1976. Only one nodal plane can be drawn. It is well determined for the shock at 09.21 GMT. In this case limiting orientations for the second nodal plane are given (dashed lines).

Fault-plane Solutions

The construction of the fault-plane solution of the main shock (fig.5) has been described in detail elsewhere (Müller 1977). For the September shocks (fig. 6) much fewer data were available. Therefore only one of the two nodal planes for P waves can be derived. In both cases it is close to the steeply dipping nodal plane of the May shock. This indicates that the fracture has probably occurred in a similar way in all three cases.

In the case of the May shock the gently dipping nodal plane must be assumed as fracture plane. The dislocation on this plane corresponds to a shallow underthrust from South to North with an azimuth of 348° and a dip of 13°. The same holds probably for the September shocks. This interpretation is in agreement with many geological observations in this region of the Southern Limestone Alps, which indicate extensive subduction of Paleocene-Eocene sediments below the Triassic Hauptdolomit (Martinis 1975).

The P-axis of the fault-plane solution of the May shock corresponds to a tectonic stress from NNW to SSE. The same tectonic stress is found for earthquakes in South-West Germany (Schneider et al. 1966; Ahorner et al. 1972 and this vol.). The regional tectonic stress therefore seems to be relatively homogeneous from North-Eastern Italy to South-West Germany. The orientation of the stress direction perpendicular to the strike of the Eastern Alps suggests that the present stress pattern has the same orientation as the one that formed the Alps. In the light of plate tectonics the active subduction from the South in Friuli is most easily explained by convergent movements between the Eurasian and the African lithospheric plate. The extensive underthrust of Molasse sediments under the Northern Limestone Alps which is geophysically well documented also fits into this picture of plate interaction. The alternative geodynamical hypothesis of updoming of the Central Alps and gravitational sliding of nappes to the North and South, which has been renewed by Van Bemmelen (1977), requires a much more complicated explanation of the Friuli earthquakes.

References

Ahorner, L., Murawski, H.& Schneider, G., 1972: Geol. Rdsch., 61:915-942.

Gebrande, H., Miller, H. & Schmedes, E., 1976: Proc.15th ESC Gen.Ass. in Krakau, in press.

EMSC Working Group on the Friuli Earthquakes, 1976: Paper at the Meeting on Friuli Earthquake, Udine.

Finetti, I., Giorgetti, F., Haessler, H., Hoang, T.P., Slejko, D. & Wittlinger, G., 1976: Paper at the Meeting on Friuli Earthquake, Udine.

Martinis, B., 1975: In: Structural model of Italy (L.Ogniben, M. Parotto, A. Praturlon (edt.), Consiglio Nazionale delle Richerche, Roma.

Miller, H., Gebrande, H. & Schmedes, E., 1977: GeolRdsch.66: in press.

Müller, G., 1977: J. Geophys., 42:343-349.

Schmedes, E., Miller, H. & Gebrande, H., 1976: Paper at the Meeting on Friuli Earthquake, Udine.

Schneider, G., Schick, R. & Berckhemer, H., 1966: Z.Geophys.32:383-393.

Städtler, G., 1973: Diplomarbeit, Universität München.

Van Bemmelen, R.W., 1977: Tectonophys.39, T13-T19.

Aftershocks of the Friuli 1976 Earthquakes and Tectonic Implications

G. Wittlinger and H. Haessler

The epicentral region of the Friuli earthquakes of 1976 located in the north eastern point of Italy, occupies the mid-valley of Tagliamento on the border between the Carnic and the Julian Alps. The major tectonic feature of this region is the predominance of large east-west oriented thrust faults (Martinis 1977). These thrust faults are traversed by transverse faults of NE-SW orientation, particularly on the right bank of Tagliamento river.

The Friuli region has frequently been hit by important earthquakes, some of which catastrophic. Gemona, situated at the middle of the epicentral zone of 1976, was already destroyed in March 1511 by an earthquake of intensity IX. The maps of maximum seismic intensity show that the epiceneral zone of 1976 is located on the alignment of great intensity which extends from Verona in Italy to Vienna in Austria (Karnik 1969, 1971).

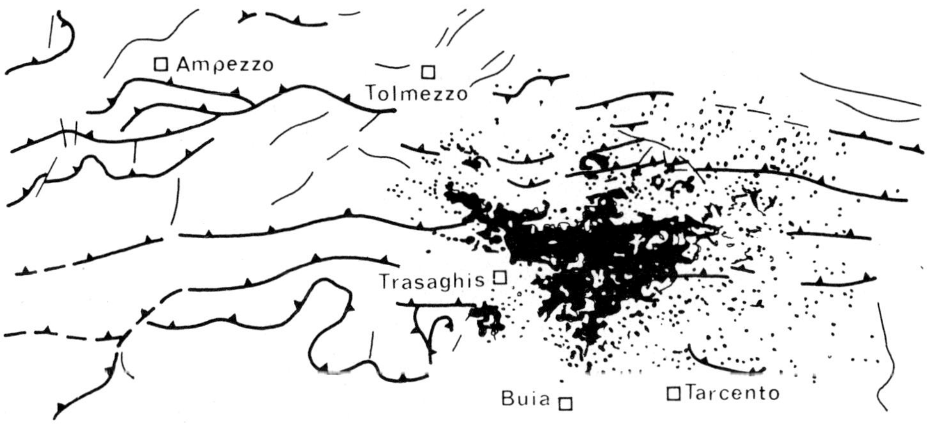

Fig. 1. Simplified tectonic map of the 1976 Friuli earthquakes area (after Martinis, Milano) and epicenters of more than 3000 aftershocks.

The seismic activity which affected Friuli 1976 and still continuing in autumn 1977, is characterized by its intense aftershock sequence, and 100 days after the primar shock by the occurrence of a second shock as important as the main.

The study of the two major shocks of which we recall the parameters (C.S.E.M. Working Group 1976):

6 MAY	1976	20h.00	46,23 N	13,20 E	h = 7 km	M = 6,5
15 SEPT	1976	9h.21	46,30 N	13,18 E	h = 11 km	M = 6,3

cannot, for the moment, be a definitive contribution to our knowledge of the regional tectonic. Fault plane solutions obtained for the May 6,1976 event and the interpretation deduced, differ considerably from author to author. G. Mueller (1976) proposes an underthrusting of the Adriatic

block under the Alps striking almost E-W and slightly north. On the contrary Van Bemmelen (1977) suggests a normal fault striking E-W and G. Weber et al. (1977) proposes a N-S strike-slip fault of sinistral movement.

The considerable number of events we recorded in Friuli allows us to investigate the tectonic problem with the help of a large number of perfectly homogeneous data. Recording with a telemetered 7 station network was carried out for 4 days just after the May event and for 9 months from September 29, 1976.

The aftershocks are concentrated on a relatively small region bounded by the isoseismal IX of the May 6 shock. A tendency of E-W alignment appears underlying the general direction of some thrust faults, particularly that of periadriatic fault.

The vertical section (fig. 2) along N-S shows, for the sake of clarity, only the shocks of $M_L > 2.5$. The aftershocks seem to lie on a northerly dipping surface having a dip of 60° near the surface and 15° at depths. The composite focal mechanism of aftershocks which define the curve A of fig.2 is a thrust fault striking E-W and dipping 40° towards North, agreeing with the mean dip direction of the curve. We also remark a good agreement between the position of the focus and the geological trends of the tectonic scheme of Roeder and Bögel (this vol.). We also observe that the aftershocks of May 6 event are shallower and situated more towards south than those of Sept. 15 event. This shift of aftershocks zones is in conformity with the foci of the two major shocks, thus indicating that two nearby blocks were successively affected.

All the aftershocks do not show thrusting mechanism alone; some composite fault plane solutions show strike-slip movements particularly

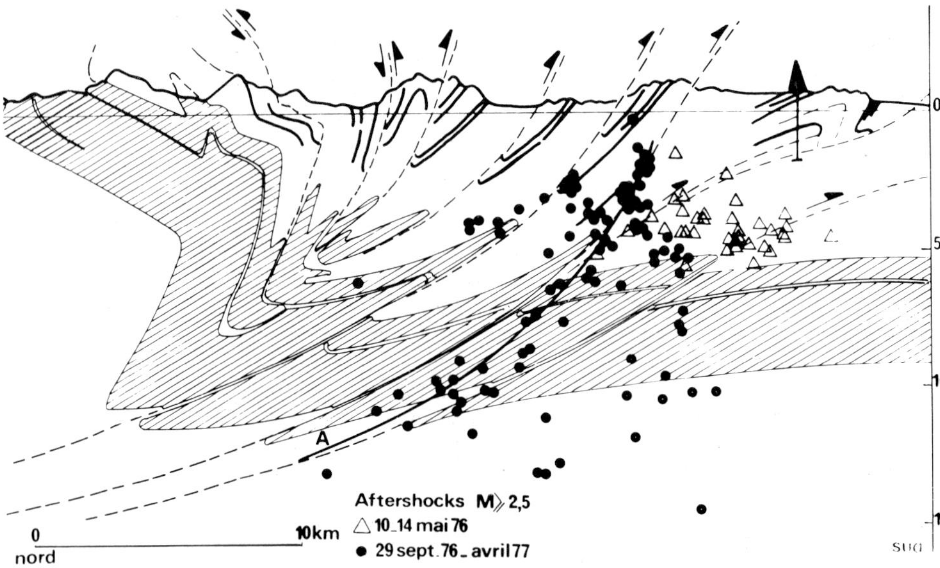

Fig. 2. North-South section of the 1976 Friuli earthquakes area. The tectonic section is taken after Roeder & Bögel (this vol.). Only the biggest of $M_L > 2.5$ aftershocks are plotted on this section.

south of Tolmezzo and on the west bank of Tagliamento river, precisely where transverse faults have been detected.

However, the major aftershocks ($M_L > 3$) as a whole, show thrust movements striking E-W and dipping north at less than 45°. We shall, therefore, retain this general tendency of thrusting which seems to charaterize the main tectonic feature of this part of Friuli. This view is in good agreement with the general picture of plate tectonics which suggests an overthrusting of the Alps on the Adriatic block.

References

Martinis, B., 1977: National Research council. Geological and Mining Sciences Committee Italian "Geodynamics" Project. Milano.

Karnik, 1969: Seismicity of the European area, Part 1:361 p., Praha.

 - , 1971: Seismicity of the European area, Part 2:218 p., Praha.

C.S.E.M. Working Group, 1976: Friuli earthquakes meeting, Udine.

Mueller, G., 1977: Journal Geophys., 42:343-349.

Van Bemmelen, R.W., 1977: Tectonophysics, 39:T13-19.

Weber, C. & Courtot, P., 1977: Le séisme du Frioul dans son contexte séismotectonique. O.C.D.E. Meeting, Rome.

A Seismic Model of a Refraction Profile Across the Western Po Valley

A. Stein, O. Vecchia* and R. Froelich

Contents

* Supported by Consiglio Nazionale delle Ricerche, Italy.

Introduction and Technical Data

Seismic recordings were made along a NE-profile across the Po Valley in 1966 by 15 groups from West Germany and Italy following nine large explosions set off in a high mountain lake in the Mercantour Massif (Stein et al. 1967). This profile connects the Western Alps with the Southern and Central Alps across the Po Valley. Recordings were made on an arc at a shotpoint distance of 120 km connecting the profile with the Ivrea Zone in the Western Alps. This was done to facilitate the identification of the seismic phases along the profile. Fig. 1 shows the seismic refraction lines in the western part of the Alps and the Po Valley. The thicker lines in the Alps are the profiles of the "ALP 1975" project (Miller et

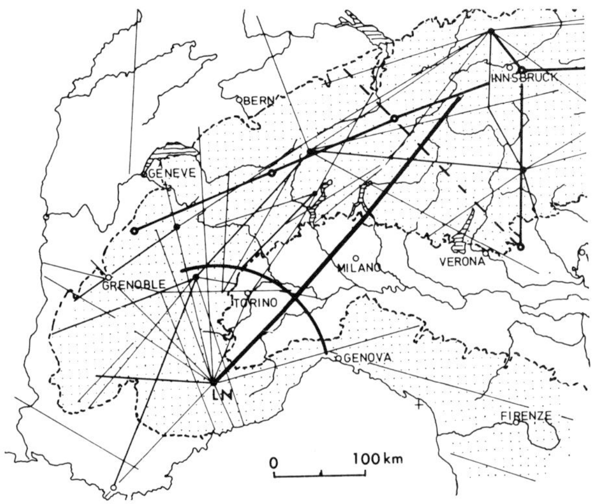

Fig. 1. Position of seismic refraction lines in the Western Alps, the Po-Plain and the Central Alps; further explanation in the text.

al. this vol.). Our NE-profile, starting from the Lac Nègre shotpoint (LN in fig. 1), and the above-mentioned arc are marked by the thickest lines. The profile ends in the Central Alps at a shot-point distance of nearly 400 km. The seismic arc covers an interval between N 25°W and N 75°E. Seismogram sections of the arc and the NE-profile are presented in fig.2 and the upper part of fig. 3, respectively.

In this paper, a two-dimensional seismic model of the profile will be discussed, which is based on ray tracing methods.

Correlation of the Seismic Phases

1) Correlation along the 120 km Arc

The results of explosion seismology in the Alps have been reviewed recently by Giese & Prodehl (1976). An extensive compilation of the original literature is given there. A strong positive gravity anomaly occurs along the middle of the Ivrea Zone at the eastern margin of the Western Alps. The anomaly is caused by upper mantle material of high density and seismic velocity which has been thrust in a northwesterly direction onto sialic material of the Western Alps, which is in agreement with the concept of plate tectonics. A seismic refraction profile from Lac Nègre along the axis of the anomaly has shown small but precise forerunners of 7.4 km/sec at distances beyond 70 km. This P_i wave is refracted at

depths of only 5-10 km. It is followed by the corresponding overcritical reflection P^I and the P^M phase, which is overcritically reflected from the upper mantle. The P^M wave is strongly retarded by a low velocity channel below the Ivrea body, which consists of the sialic rock mentioned above.

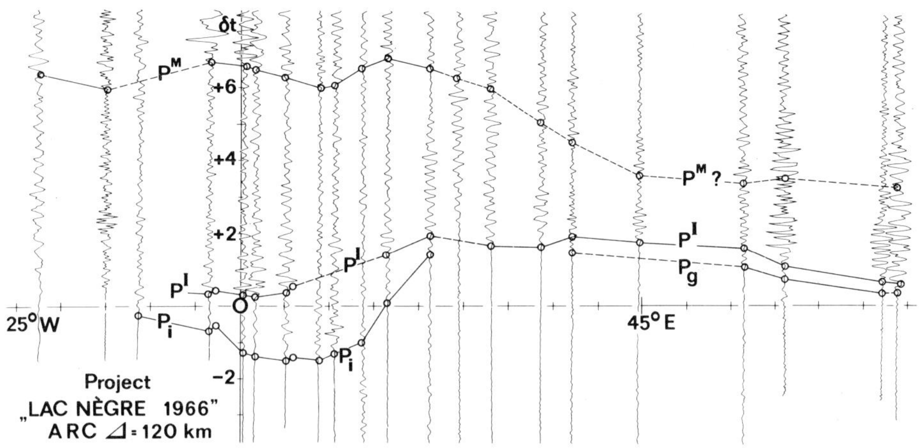

Fig. 2. Project "Lac Nègre 1966"; seimogram section of the arc in 120 km distance from the shot point, reduced time scale $\delta t = t - \Delta/6.0$ sec; (the time-domain $\delta t > 8$ sec is omitted, because it showed seismic noise only); further explanation in the text.

In order to identify the seismic phases along the NE-profile, the P_i, P^I and P^M waves were correlated along the arc, starting at the axis of the gravity anomaly (about N 10°E in fig. 2). The quality of the correlation is determined by the spacing of the observation points. The most probable correlation is plotted in fig. 2. From west to east along the arc, the P_i phase is progressively more retarded by the young sedimentary fill of the Po Valley in the region of Turin and by the easterly dip of the Ivrea body. The two arrivals on the trace at N 45°E, which is on both the arc and the NE profile, can be identified as P^I and P^M.

2) Correlation along the NE-Profile

The correlation on the NE-profile is plotted in fig. 3. The P_g wave was recorded as a first arrival up to distances of 115 km. The P^I wave, identified from the data on the arc, was correlated at a distance interval of 43 to 139 km. At 118 to 260 km from the shotpoint, the P^M wave arrived with a high signal-to-noise ratio, despite the seismic noise level, which

Fig. 3. Project "Lac Nègre 1966"; upper part: seismogram section of the NE-profile, reduced time scale $\delta t = t - \Delta /6.0$ sec; (the time-domain $\delta t > 5$ sec is omitted, because it showed seismic noise only); lower part: attempt of a seismic model;further explanation in the text.

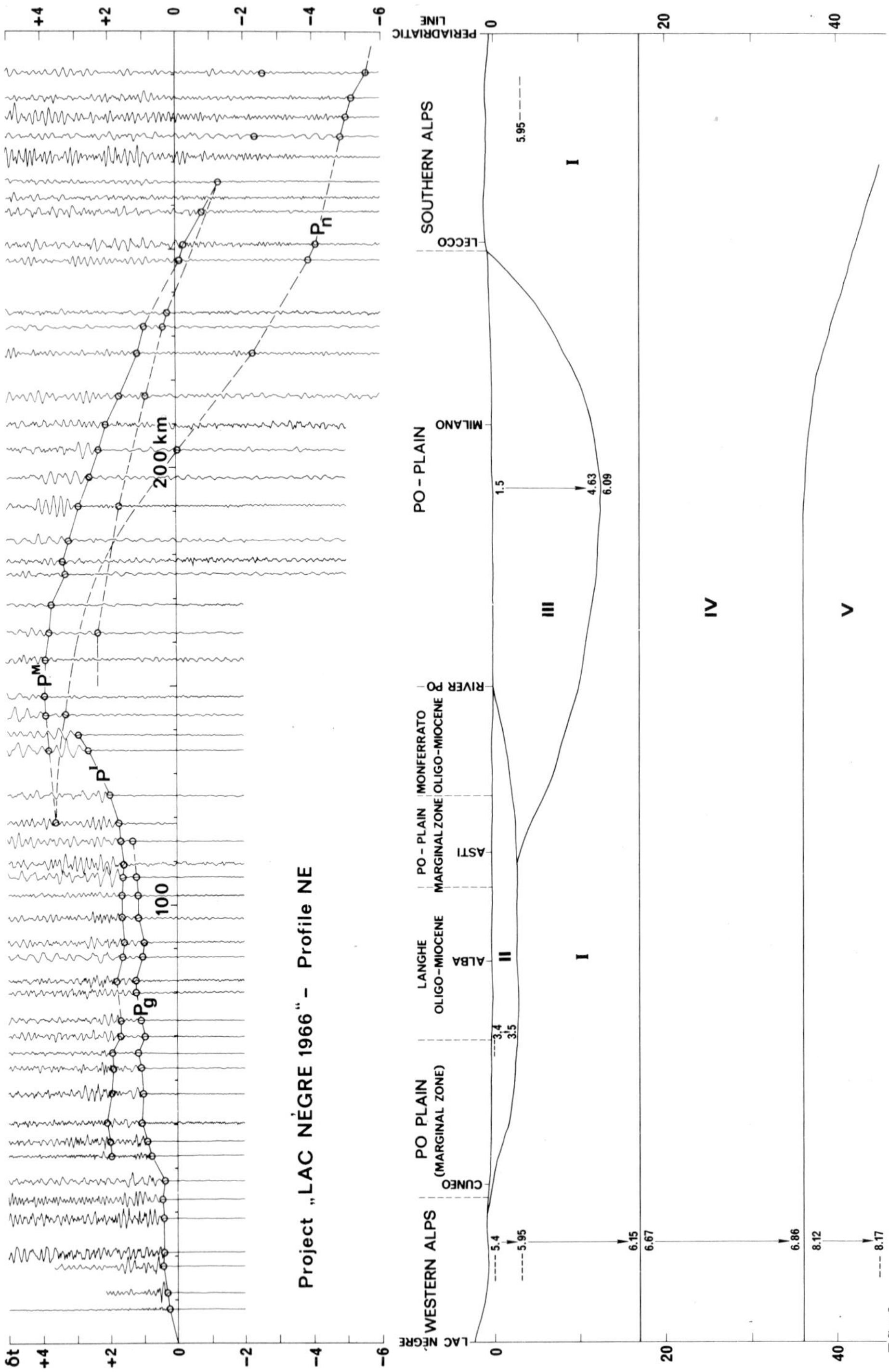
Project „LAC NÈGRE 1966" – Profile NE
δt
P_g
P^I
P^M
P_n
100
200 km
LAC NÈGRE
WESTERN ALPS
CUNEO
PO PLAIN (MARGINAL ZONE)
LANGHE OLIGO-MIOCENE
ALBA
PO - PLAIN MARGINAL ZONE
ASTI
MONFERRATO OLIGO-MIOCENE
RIVER PO
PO - PLAIN
MILANO
LECCO
SOUTHERN ALPS
PERIADRIATIC LINE
I
II
III
IV
V
5.4
5.95
6.15
6.67
6.86
8.12
8.17
3.4
3.5
1.5
4.63
6.09
Z [km]

proved to be 25 times higher in the region of Milan than in the Central Alps. The curvature of the P^M travel-time curve even changes its direction because of the strong retardations due to the sediments of the Po Valley.

Giese et al. (1970) have discussed the areal distribution of the P^I, P_i, P^M, and P_n phases (called c(i), d(i), c, and d in their paper). According to their concept, the surface of the Ivrea body changes continuously into the surface of the upper mantle below the Po Valley, where the upper mantle of the Western Alps no longer exists. Accordingly, the P^I curve on our NE profile should change continuously into the phase which is marked by P^M in fig. 3. Such a correlation would be an important alternative to the one used for our model. Indeed, if the doubtful P^M correlation along the arc for azimuths of about 45° and on the NE-profile between 120 and 140 km (fig. 2 and the dashed part of the P^M line in fig. 3) is not accepted, P^I can be linked to P^M on the profile by a smooth curve. The main argument against such a correlation are the strong P^M arrivals at distances of up to 250 km, which then must be overcritical reflections from the Ivrea surface near the shotpoint (depth about 17 km). Overcritical reflections normally die out at distances which are large compared with the depth of the reflection point. Therefore, this alternative is not considered further in this paper.

At the southern end of the arc, the correlations of P^M and P^I are compatible with the data from the seismic profile Lac Nègre-Genoa (fig. 1). A depth of 28 km is estimated for the surface of the Ivrea body on the basis of this profile and a depth of 50 km for the upper mantle of the Western Alps (Giese, person. comm.).

The absence of the P_n wave at most of the stations in the Po Valley may be explained by the small amplitudes of the P_n wave, which corresponds to a low velocity gradient in the upper mantle. These are masked by the seismic noise level. In the Southern Alps, however, ($\Delta > 250$ km) P_n was recorded as a clear first arrival at many stations.

Calculation and Discussion of a Seismic Model

Because of the difference in orientation between the Ivrea body and the NE-profile, the descending seismic rays near the shotpoint do not pass through a vertical plane, so that three-dimensional ray tracing is necessary. However, the density of seismic data is only sufficient for a simplified two-dimensional model. Such a model is presented in the lower part of fig. 3 and is discussed in the following.

1) Mathematical Basis of the Computer Program

The travel-time curves of the seismic model are calculated by ray tracing. A constant vertical velocity gradient is allowed in every layer. This means a velocity/depth funtion of the type

$$v_i(Z) = v_{io} + g_i Z,$$

where i is the index of the layer, v_{io} the velocity extrapolated to sea level, Z the depth in km, $v_i(Z)$ the relevant velocity in km/sec, and g_i the vertical velocity gradient.

The curved interfaces between the layers are approximated by connecting points of equal horizontal spacing by straight lines. Refraction or overcritical reflection of the seismic ray at any interface is provided in

the computer program (Stein 1969). Every ray is composed of a series of circular arcs. The horizontal and vertical projections of these arcs and the corresponding travel times are then given by simple mathematical expressions. In the case of our model, the deviations between measured and calculated travel times do not normally exceed 0.4 sec.

2) Alpine Body and Granitic Layer

The metamorphic rocks of the Alps and the crystalline basement below the Po Valley correspond seismically to the P_g wave. At greater depths, the velocity of this seismic layer slightly exceeds a value of 6.0 km/sec, independent of the petrographic character of the rock material. Near the suface, however, such material has velocities of about 5 km/sec. For this reason, the granitic layer "I" in the lower part of fig. 3 is schematized by two velocity/depth dunctions:

$$V_I(Z) = 5.4 + 0.174\ Z \qquad \text{for } Z \leqslant 3.16 \text{ km}$$

and $$V_I(Z) = 5.9 + 0.015\ Z \qquad \text{for } Z > 3.16 \text{ km.}$$

For a depth of 3.16 km the two expressions yield the same value of 5.95 km/sec.

The curvature of the P_g travel times near Lac Nègre corresponds to the higher velocity gradient of the first expression.

Where the profile enters the forelands near Cuneo, P_g shows a systematic retardation, which increases the reduced travel times in a northeasterly direction to values of about 1 sec. In our model, the retardations are explained by sediments of about 3 km thickness. The P_g wave dies out at a distance of 115 km from the shotpoint.

The layer I velocity functions are also used for the Southern Alps (see lower part of fig. 3, right side).

3) Transition Zone of Northern Apennines

The great variety of geological structures along the profile between Cuneo and the River Po is schematized by layer II (fig. 3). The velocities may be somewhat too high according to the function assigned to this layer:

$$V_{II}(Z) = 3.4 + 0.05\ Z$$

Sufficiently precise geological information can be taken from the structural model of Italy by Ogniben et al. (1973). According to it, the regions of Cuneo and Asti contain Plio-Pleistocene sediments. Near Asti, the Pliocene base reaches its maximum depth of about 600 m below sea level. In the region of Alba, the profile cuts the northwesterly part of the Langhe with its Oligo-Miocene sediments. Oligo-Miocene and older sediments also outcrop in the Monferrato Hills, which is of special interest for our considerations.

4) The Monferrato Overthrust

Along the profile beyond the River Po, the flat part of the Po Valley begins with Plio-Pleistocene sediments of low velocity. The thickness of this layer is more than 3 km in the structural model. It was assumed to be separated from the Monferrato Hills by a normal fault. Thus one would have expected retardations of the seismic arrivals to begin at the River Po (km 150 of the profile) and increase in a northeasterly direction. The seismogram section, however, shows very strong retardations of the P^I phase (more than 1 sec) beginning already about 10 km northeast of Asti (km 120 on the profile, fig. 3). This must be interpreted by a major

overthrust of the Oligo-Miocene and older sediments of the Monferrato region onto the Plio-Pleistocene sediments of the Po Valley. The horizontal displacement of this overthrust is therefore about 30 km at least. Fig. 4 shows the velocity/depth function of our model for this zone at a shotpoint distance of 130 km. The overthrust is indicated by the low-velocity channel in the depth range from 2 km to 7 km.

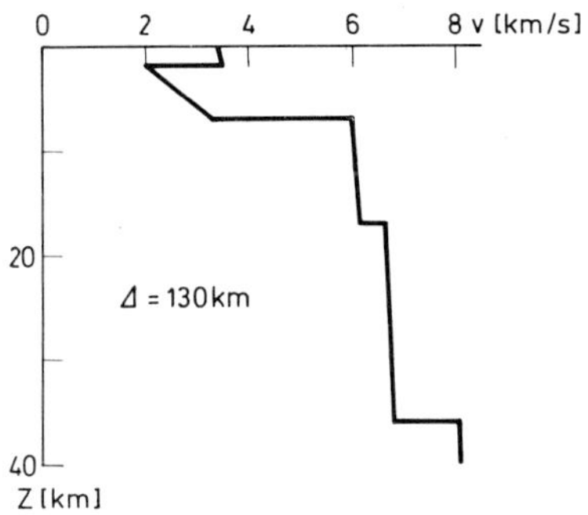

Fig. 4. Velocity/depth function in the area of the Monferrato Overthrust.

The identification of the Monferrato fault as an overthrust is an example of the close relationship between deep seismic sounding and seismic prospecting for oil and gas. Seismic reflection measurements are less suitable for exploring overthrusts, because the interfaces are disturbed in such a zone and do not reflect seismic energy well. Any deep waves which come up at a sufficiently small angle of incidence are suitable for exploring overthrusts similar to the case discussed here.

As a result of the seismic anomaly, magnetotelluric investigations were proposed which confirmed the exceptional structure of the Monferrato region (Losecke this vol.).

5) Sediments in the Po Valley between the River Po and the Southern Alps

Low velocity sediments must be present down to very deep depths in the center of the basin between the River Po and the Southern Alps to explain the unequal travel time retardations of the P^M wave. These retardations lead to a negative curvature for this part of the travel-time curve which is normally positively curved.

The velocity function of sediments in the Po Valley, zone III in our seismic model (fig. 3), is

$$V_{III}(Z) = 1.5 + 0.25\ Z.$$

Near the surface, this expression is in good agreement with deep-well data which had been made available by courtesy of AGIP. But for greater depths, the calculated values are lower than the measured ones. At a depth of 3 km the formula yields 2.25 km/sec as opposed to deep-well data of 2.5 km/sec for the Messiniano Formation and 3.2 km/sec for the Lower Jurassic. A Mesozoic dolomite was met at a depth of 5.4 km in the Malossa oil field 20 km east of Milan (Groppi et al. 1977). For this depth, the formula yields 2.85 km/sec, which is certrainly too low for a dolomite. In our model, the Po Valley sediments are present down to 12.5 km in the center of the basin. If the assumed velocities are too low, the base of the sediments in the model must really be considerably

deeper than 12.5 km to fit the measured P^M travel-time curve. This conclusion requires the upper mantle to have little or no relief at a shotpoint distance of 60-120 km. If the "sediments" reach down to 15 km and more, the degree of metamorphosis may increase continuously with depth without any distinct interface between sediments and cristalline basement.

6) The Intermediate Layer
The P^I reflections (fig. 3) need a reflector at distances half of the recording distance. This means that only at a shotpoint distance of 20-70 km is it necessary to postulate the existence of the surface of layer IV. At this distance this interface corresponds to the surface of the Ivrea body. This has been proven by arc shooting. Naturally, the interface is not a horizontal plane at a depth of 17 km as assumed for the model. It is, however, advisable to keep the model simple. The velocity of the intermediate layer is assumed to be

$$V_{IV}(Z) = 6.5 + 0.01\ Z.$$

This leads to a velocity of 6.67 km/sec at a constant depth of 17 km for the surface of the layer. In this case, the P^I arrivals at distances less than 70 km are taken as undercritical.

At distances beyond 70 km, the intermediate layer cannot be derived from the seismic data and is only an equivalent for the increase in velocity with depth.

7) The Upper Mantle
The reliability of the seismic model is increased by the P_n arrivals from the Southern and Central Alps, which are not influenced by the Po Valley sediments. The velocity function of the upper mantle was assumed to be

$$V_V(Z) = 7.9 + 0.006\ Z.$$

This means a velocity of 8.12 km/sec at the top of the upper mantle below the Po Valley. This velocity occurs at a depth of 36 km in our model. The average apparent velocity of P_n in the Southern and Central Alps was determined to be 7.4 km/sec, which requires the surface of the upper mantle in our model to dip 8.5° towards the Central Alps and reach a depth of 49 km below the Periadriatic line.

The graph of our model in fig. 3 ends 300 km from the shotpoint, where the Periadriatic Line separates the Southern Alps from the Central Alps. The maximum recording distance, however, was 396 km and the profile ends on the profile along the axis of the Alps reported by Miller (Alpine Explosion Seismology Group 1976). For the point of intersection, about 65 km west of shotpoint D, the preliminary cross-section by Miller shows a maximum depth of 50 km for the upper mantle. Our model would yield a value of 62 km by extrapolation of the upper mantle slope. According to Müller et al. (1976) the greatest depth of crust/mantle boundary on the Swiss geotraverse is 58 km below the Lepontin.

8) The Central Alps
Although the Po Valley is the main object of this paper, some remarks should be made about the Central Alps. Between the Periadriatic Line and the end of the profile, some poor later arrivals with P_n-like velocities were recorded (P_{n2} in fig. 5). Fig. 5 shows a part of the seismogram section for this part of the profile. The weak arrivals can be determined a bit better than visible in the seismogram section, since only one of the

traces of a station could be presented in the section.

The retardation of 1.5 sec for the travel time (fig.5) indicates that the upper mantle in the Central Alps is 15 km deeper than in the Southern Alps. The fault plane related to the Periadriatic Line could form the boundary where this depth difference occurs. This depth difference corresponds qualitatively to the results on the Swiss Geotraverse (Müller et al. 1976) and to the subduction model for the Eastern Alps given by Closs (1975). The value of 15 km for this difference in the depth to the upper mantle seems high to these authors. On the other hand, the depth difference between the surface of the Ivrea Zone and the upper mantle in the Western Alps even exceeds 35 km.

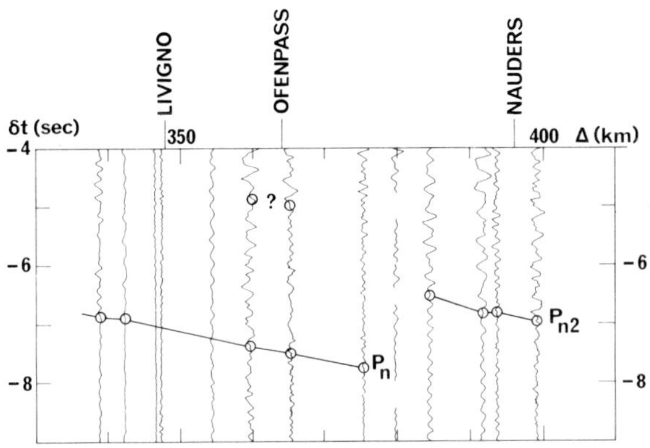

Fig. 5. Project "Lac Nègre 1966"; part of the seismogram section of the NE-profile with the phases P_n and P_{n2}; distance range 330 - 400 km.

Final Remark

The seismic model presented here is one of the probable solutions which are compatible with the widely spaced, deep-seismic sounding data. However, we hope that the essential features of the model will stimulate further discussion of the geodynamics of the investigated region.
For further information see also Losecke et al. and Günther et al.this vol.

References

Closs, H., 1975: Rapp. Comm.intern.Mer Médit.,23,4a:55-57.

Giese, P. & Prodehl, C., 1976: in: Explosion seismology in Central Europe, ed.P.Giese, C.Prodehl,A. Stein:347-375,Berlin/Heidelberg/New York.

Giese, P., Morelli, C., Prodehl, C. & Vecchia, O., 1971: XIIe Assemblée Générale de la Commission Séismologique Européenne Luxembourg 21-29 Sept.; Série Géophysique N°101,Observatoire Royal de Belgique.

Groppi, Muzzin & Vaghi, 1977: Nachr. Dtsch. Geol. Ges., 16:116.
Miller, H., 1976: Pageogh., 114, Birkhäuser Verlag, Basel.
Müller, St., Egloff, R. & Ansorge, J., 1976: Schweiz. mineral. petrogr. Mitt. 56:685-692.
Ogniben, L. (ed.), 1973: Structural model of Italy, sheet north, scale 1:1 000 000. Consiglio Nazionale della Recherce.
Stein, A., Vecchia, O., Steinbeck, J. & Fröhlich, R., 1967: Report XIV. IUGG General Assembly, Vol. II: IASPEI.
Stein, A., 1969: Transactions Internat. Conference of Experts on Explosion Seismology (Leningrad):67-78, Kiev: Naukova dumka (in Russian).

Part 5

Synthesis

Geodynamic Interpretation of the Alps

D. Roeder and H. Bögel

Contents

Introduction

Earlier geodynamic models of the Alps showed surprising similarities to modern plate tectonic views and helped significantly to bring about their world-wide acceptance (Schmidt this vol.), but most of the modern, rigorously plate tectonic interpretations of the Alps (Closs 1975; Dewey et al. 1973; Dietrich & Franz 1976; Ernst 1973; Hsü 1971; Roeder 1977b; Toksöz & Bird 1977) are based on criteria which have been developed elsewhere in the world.

If one admits that it would be unreasonable to assume special tectonic processes exclusively for the formation of the Alps (Trümpy 1960), a plate tectonic interpretation of the data presented in this volume is the only acceptable form of a synthesis. However, to the same extent as everywhere else in the world, there is no single criterion which by itself would support an entire plate model: the weight of the evidence in favor of a plate tectonics interpretation of the Alps must come from the combined list of circumstances, as follows:

1) Paleomagnetic data show Africa and Europe separating during the Jurassic over several 100 km, approaching and finally colliding (Fromm et al. this vol.).

2) Breakup of a peri-Tethyan carbonate platform during early to middle

Jurassic is suggested by faulting, anomalous thicknesses, and geochemical suggestions of basaltic vulcanism (Germann this vol.). The appearance of complete and incomplete ophiolite suites in the Western Alps during the Cretaceous (Dietrich & Franz 1976) and of deep-sea sediments in the Eastern Alps during the Upper Cretaceous (Butt & Herm this vol.) is interpreted as sea-floor spreading and the formation of Atlantic-type margins within the future Alpine realm.

3) This peri-Tethyan fragmentation is complicated by an earlier, late Permian to early Triassic, episode of breakup and ensuing sea floor spreading (Roeder 1976), inferred from lithofacies and thickness distribution and possibly related to Triassic calcalkaline vulcanism in the southern Alps (Sassi et al. this vol.).

4) The asymmetric structure of the present Moho discontinuity beneath the Alps suggests an extinct subduction zone, with a gradual southward descent and with a higher and horizontal position beneath the southern Alps abutting less gradually against the lowest position (Gebrande et al. this vol.).

5) In a discontinuous metamorphic belt along the central axis of the Alps, high p/t products occur externally adjacent to low p/t products of similar, probably Cretaceous age (Miller 1974). Both types are separated by major thrust faults which from independent evidence are parts of a boundary between two major geotectonic provinces. Therefore, the juxtaposed types of metamorphism can be interpreted as a paired metamorphic belt. The paired event is post-dated and in part superimposed by a post-collisional low p/t event in the deeper structural units (Raith et al. this vol.).

6) Palinspastics of the visible structural units of the Alps and their seismically determined crustal thicknesses demonstrate the disappearance of several 100 km of crust and upper mantle which cannot possibly be achieved by *Verschluckung* as in the pre-plate tectonic models.

7) In the Ivrea zone of the Western Alps, the appearance at surface and above a thrust contact, of subsialic mantle material can only be achieved by lithospheric subduction with a suture below the Ivrea zone (Giese 1968; Nicolas & Jackson 1974).

8) The unusually deep subsidence of the Po basin on the back side of the Alps suggests a backarc mechanism as part of the evolution of the Alpine upper plate.

9) The Alpine zone of convergence or subduction continues eastward into the Carpathians where intermediate earthquakes (Barazangi & Dorman 1969), shallowest seismotectonics (Birkenmajer 1974), and Pliocene and younger arc vulcanism (Stille 1953) suggest an active subduction system. The abundant Cretaceous to Neogene arc magmatism was related to marginal underthrusting or subduction by Stille (1953) and to plate tectonics by Radulescu & Sandulescu (1973) and by Burchfiel (1976).

The following Alpine model has been developed by matching available data with plate tectonic models of orogeny (Dewey & Bird 1970; Roeder 1973), the models of arc-trench gaps (Karig & Sharman 1975; Seely et al. 1974; Hamilton 1977), the models of continent-continent collision (Bird et al. 1975), and models of low-angle thrusting and gravity spreading (Elliott 1976; Chapple 1975).

Based on this approach, the Alps would appear as a north-vergent stack of thrust sheets and recumbent folds above and south of uninvolved parts of a European foreland. The structures would be composed of the material of either plate. They would have resulted from south-dipping subduction

of Mesozoic oceanic crust during the Cretaceous, followed in the Eocene by a continent-continent collision. The Eurasian craton and associated or interspersed basins floored by oceanic or quasi-oceanic crust would have functioned as a lower plate, and intra-Tethyan or Gondwana-associated material as an upper plate.

The beginning of subduction is not directly documented. It must have postdated the Jurassic breakup and spreading and must have predated the crystallization of high p/t metasediments. Their late Cretaceous age would have been fixed by an uplift of at least 24 km, which is considerably more than can be documented for that time in the structure of the present upper plate. The uplift, therefore, must have happened before the emplacement of the upper plate. This inference implies the erosive and subductive removal of an intercalated, possibly oceanic plate of which the South Penninic ophiolites of the Western Alps may be remnants. The present (Austro-Alpine) upper plate, devoid of Cretaceous arc vulcanism, would not have become an upper plate until Tertiary time.

After collision, persistent plate convergence caused a major change in tectonic style. The original suture was abandoned and movement shifted to shear surfaces deeper in the crust of the lower plate. The upper plate and higher parts of the lower plate were folded into an asymmetric anticlinorium with a stack of gravity-driven thrust sheets on the north flank and a kinked root zone with *retrocharriage* on the south flank. Uplift and thrusting continue at present, but plate convergence across the Alps, while likely, is not established.

Structural Location of Suture

A collision suture is defined as the original thrust contact between collided continental masses, and it is a basic requirement of any collisional plate model. In the Alps it is best identified at the base of the Austro-Alpine thrust unit, because this unit is conterminous with undeformed central parts of the Adria plate south of the Alps and east of the Apennines, because it consists of shelf sediments above cratonic basement, and because it overrides eugeosynclinal sediments associated with ophiolite complexes.

These remnants of oceanic crust probably are parts of the lower or upper plate of a subduction. Any other choice of a collision suture in the Alps would lead to paleogeographic difficulties. However, the juxtaposition of two plates at a suture is complicated by an intervening wedge of sediments deposited on and moving with either or both plates. Such sediments are referred to as *interplate sediments*. Present knowledge of such deposits is confined to the interpretation of recent arc-trench models (Karig & Sharman 1975) where they appear as tensional forearc basin fill, accretionary wedges, tectonic friction carpets, and in foreland troughs.

Interplate sediments with uncertain or changing plate attribution and timing of deformation occur below the Austro-Alpine thrust mass and its basal suture. The most internal outcrops of the suture are located at the fringes of the Tauern and Engadin windows and at the west edge of the Austro-Alpine nappe system. The upper plate and its sialic basement of Hercynian gross radiometric age overlie deep-water metasediments, marine basalts, and ultramafic fragments or bodies (Trümpy 1960). The youngest documented ages of the sediments range from late Cretaceous to

early Eocene; they date the collision if the interplate sediments are considered as part of the lower plate, but they record only local complications if they are considered as part of the arc-trench system.

Types of Subduction Models

It is tempting, but not conclusively possible, to identify some or all of the sedimentary interplate units with the sedimentary units of the recent arc-trench model. The interplate sediments and crustal complications in the lower plate give rise to several kinematic models.

In the Western Alps and in the Tauern window, the sialic lower plate is strongly deformed into north-vergent polyphase recumbant folds. Ophiolite suites and deep-marine metasediments alternating in paleogeographic space with shallow-water sediment suites suggest that the lower plate consisted of a field of small sialic blocks or remnant arcs separated by rifts or ocean basins and possibly arc-trench systems. Subduction of this heterogeneous material, therefore, may have occurred with frequent changes in rate and azimuth of subduction, and in structural and metamorphic style.

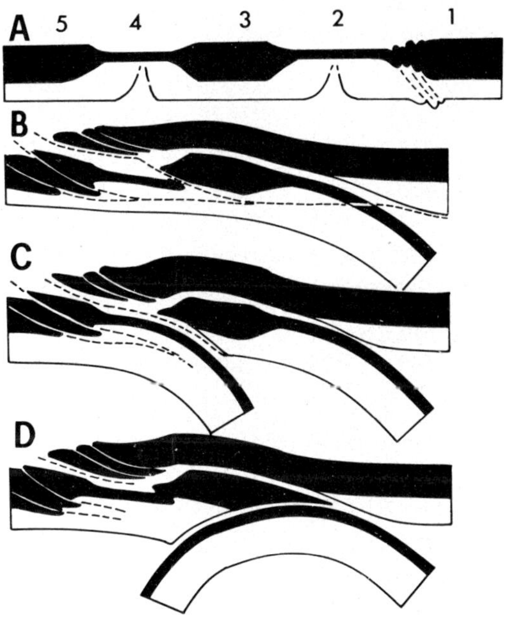

Fig. 1. Cross sections of three possible plate tectonic models of the Alps. Schematic. Black, crust. Outlined shapes, lithosphere. A, pre-orogenic stage, with 1, Adriatic craton, 2, south Penninic ocean, 3, middle Penninic craton, 4, north Penninic ocean, 5, Eurasian or Helvetic craton. B, two-plate model with continuous subduction and intracrustal suture progradation in dashed lines. C, two-plate model with discontinuous subduction. D, three-plate model with oceanic intervening south Penninic plate.

The visible effects of such irregularities have been combined in three simple types of plate models proposed in the literature for the pre-collisional stages of the Alps (fig. 1):

1) Two-plate model with continuous subduction.
2) Two-plate model with discontinuous subduction.
3) Three-plate model.

Authors of the first of these choices (Closs 1975; Dietrich & Franz 1976; Hawkesworth et al. 1975; Hsü 1971) emphasize the similarities of the Alpine orogen with classical arc-trench models. They recognize the subdivision of the arc-trench gap into an accretionary wedge, a trench-slope structural high, and a forearc basin. They also explain the distribution of detached ophiolite bodies and melanges as products of scraped oceanic crust, and they point out the typical subduction-related metamorphism.

The second type of model (Frisch this vol.; Bögel 1974; Dietrich & Franz 1976, in part) provides plausible explanations for the unusually long duration of the entire process of convergence and for inconstistencies in the ages of collision. The lack of evidence for arc vulcanism in normal quantities (Bögel 1976) may suggest that the total convergence was absorbed by several consecutive shallow subductions instead of one deep-reaching persistently downgoing slab.

The third type of model is suggested by the necessity of an additional heat contribution to explain the combined thermorogenic events between 40 and 7 m.y. (Trümpy 1973; Roeder 1977b). The contribution is suggested to originate in the Eurasian plate as upper-plate arc magmatism in the intermediate of three plates of an arc-arc collision.

Other complete or partial models have been published (Trümpy 1976), are being suggested in several papers of this volume, or have been discussed in letters (Dennis, personal to Roeder 1977). Until the perfect Alpine model has been developed, none of the present efforts should be discarded entirely.

Paleogeographic Location of Alpine Subduction

Paleogeographic data from Alpine Mesozoic shelf sediments suggest that the Alpine subduction originated in a marginal oceanic trough within a field of disintegrating small plates northwest of the open oceanic part of Tethys. This trough was created during late Triassic and early Jurassic by sea-floor spreading mechanically linked to the spreading of the north-central Atlantic (Roeder 1976; Frisch this vol.). Precise plate reconstructions are obscured by later small plate rotations, but the palinspastic arrangement and the age of deep-water seaways and microcontinents are evident from subsidence analyses of Triassic to Cretaceous mio- and eugeosynclinal sediments in southern Europe and North Africa. The lower plate of the Alpine subduction shows remnants of two oceanic regions separated by a cratonic high, known as the south, middle, and north Penninic realm. The troughs and cratonic high areas are exposed in the Western Alps. In the Tauern window, an incomplete stacked series suggests part of a trough and part of a cratonic high (Frisch this vol.). Sedimentation rates compiled for all of the Mesozoic shallow-water deposits do not suggest any oceanic crust of Triassic age within the Alpine realm. However, they clearly suggest a cycle of breakup and decelerating subsidence starting in early or middle Triassic (Roeder 1976; Bechstädt et al.

this vol.). Steady increases in thickness toward the east and southeast in the Alps, and toward the north in Tunisia, Egypt and Lebanon (Roeder this vol.) suggest that this sea floor spreading took place within Tethys itself. Vulcanogenic-sedimentary deposits of Mn ore associated with anomalously high metal contents, suggest basaltic explosive and submarine mid-ocean ridge vulcanism in early to middle Jurassic time within the Austro-Alpine thrust complex (Germann this vol.). The same type of data also suggests that a new cycle of rifting and plate fragmentation started either in early to middle Triassic (Roeder 1976) or in late Triassic to early Jurassic (Bechstädt et al. this vol.; Frisch this vol.) in the middle of the shelf involved in the subsidence of the previous cycle. This new event created the Penninic troughs and possibly other sites of divergence within the subsiding peri-Tethyan shelf, as transformed extensions of the spreading north-central Atlantic (Dewey et al. 1973; Roeder 1976; Frisch this vol.), or as an RRR triple junction, with the North Penninic trough being a failed arm (Dietrich & Franz 1976). It is in the southern one of the two Late Triassic or Jurassic troughs that the Alpine subduction originated.

Evidence for these inferences is contained in rates of shallow-water sedimentation and their regional distribution, in paleobathymetric data (Bernoulli & Jenkyns 1972; Bosellini & Winterer 1975) and in scattered and sparse suggestions of vulcanism. The Penninic ophiolites were probably generated during these spreading events, but their dislocation has so far prevented their age dating.

Structure of Lower Plate

The articulation of the lower plate surface into high-riding micro-continents and deep ocean basins is accompanied by recumbent basement involving folds (Milnes 1974; Frisch 1974; Bögel 1974). Such structures are not compatible with low-angle thrusting and transport (Roeder 1967; Elliott 1976), but their formation requires a transport overprinting of formerly upright folds. The original upright folds were produced by subhorizontal compression at less than 2 to 5 km of overburden, an unlikely condition after burial beneath the upper plate in its present or restored thickness. The folds are therefore likely pre-collisional in age. Within the late Cretaceous Alpine scenario, these conditions are met in three different interpretations. The folds could belong to an antithetic fold-thrust belt in the upper plate of a north-dipping subduction, such as in a three-plate model (Roeder 1977b). Secondly, the folds could have accompanied major wrench faults at the edges of the microcontinents. Finally, discontinuous subduction could create waves of folds ahead of the arc-trench system. Arrival of a high-riding microcontinent in the trench would eliminate the topographic slope, a prerequisite for sledrunner thrusting (Elliott 1976). Continued convergence would cause the movement to prograde into the foreland, creating a belt of thick-skinned folds until subduction resumed within the next-following small ocean basin.

Thick-skinned folding in West Alpine parts of the lower plate appears to have occurred over a wide range of kinematic and/or geologic ages, but subsequent to initial nappe-forming movements (Milnes 1974). This suggests that the interplay between plate movements, folding, and low-angle thrusting may be much more complex than depicted above.

Structure of Upper Plate

The upper plate of the Alpine subduction is represented by the Austro-Alpine nappe complex consisting of a cratonic basement, a miogeoclinal sediment cover, and a variety of synorogenic to postorogenic clastic rocks. In any plate tectonic model, this complex should contain older structures of rifting related to the south-Penninic breakup, and younger elements of a subduction system, that is, an arc-trench gap, a magmatic arc, and a backarc basin or stable craton. The original palinspastic arrangement of these units can be expected to be severely disturbed by imbrication and folding.

The Austro-Alpine complex consists of interleaved thrust sheets with polyphase internal deformation received at any time since the beginning of subduction, with post-collisional warping, kinking, and with faulting on the Periadriatic fault system. The present gross thickness of the upper plate increases from a pointed edge at the north front of the Austro-Apline to a crustal thickness of more than 40 km (Angenheister et al. 1975; Giese 1968), exposed in the western part of the Alpine south edge above the basal thrust identified as the suture. In detail, the thickness of this complex is not well known. Near the Tauern window, Oxburgh & Turcotte (1974) estimate the thickness at 10 to 15 km, which in our opinion is somewhat too high, but more accurate data are not available.

The northward thinning of the upper plate is in accordance with the expected cross-sectional shape of a thrust mass with its sole fault cutting up-section in the direction of movement. Intra-crustal thrusting at the Austro-Alpine north front is consistent with the two stages of subduction assumed in the present Alpine plate model. Within other plate models, however, this wedge shape has been explained as inherited from the Penninic plate fragmentation and its localized stretching and thinning to form a north-facing Atlantic-type margin (Roeder 1977b). Alternatively, the present crustal thickness at the Alpine south edge has been proposed to be thickened by Alpine thrust imbrication (Giese, pers. comm. to Roeder).

The magmatic arc of the Alpine system is poorly documented. Dietrich & Franz (1976) find no Upper Cretaceous arc products except a few undiagnostic basic to intermediate dikes. Basaltic arc vulcanism of Eocene age is known in limited quantities from the area NW of Vicenza in the Southern Alps, and in large quantities of andesitic detritus that makes up most of the *Tavayannaz* sandstone of the Helvetic nappe units of Switzerland. The south-Alpine occurrence of arc vulcanism is an apparent point source located on the narrow Vicenza arch (Rocco & Jaboli 1960). Its lateral extensions along Alpine strike would be buried beneath several kilometers of post-collisional Tertiary of the Po basin. However, there are no available subsurface data supporting these lateral extensions. The Tavayannaz volcanics may not have been sourced within the Austro-Alpine plate.

The lack of Cretaceous vulcanism despite evidence for Cretaceous subduction suggests that the Austro-Alpine complex did not become the upper plate of the subduction until the Tertiary, and that the Cretaceous magmatic arc is not contained in the imbricate series exposed in the Alps at present. Since the Austro-Alpine complex is conterminous to the south with undeformed plate parts, the Cretaceous arc can have been located only externally to the Austro-Alpine complex, north or east of it and later overridden by it. The basic dikes could represent rifting on the cratonic edge of a backarc basin.

Arc-Trench Model

It is useful to identify the clastic rock units related to the Austro-Alpine complex, in particular the sedimentary interplate units, with elements of the recent arc-trench systems, because they are datable and because their paleogeography can be reconstructed. They furnish data with which one can chart the details of the Alpine subduction and its changing geometry.

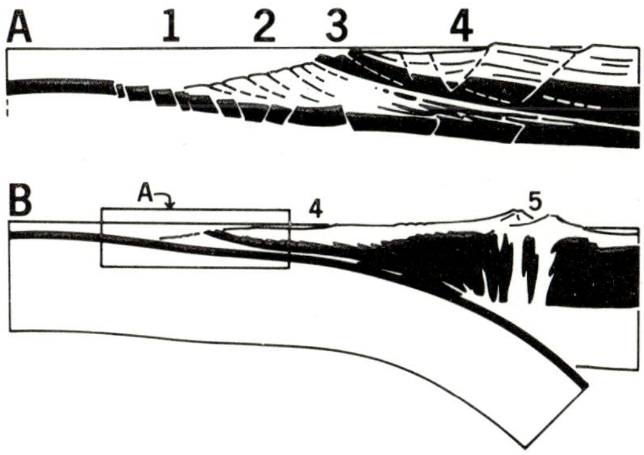

Legend

1 trench axis,
2 trench innerwall and accretionary wedge,
3 Outer non-volcanic arc or median high,
4 fore arc basin,
5 volcanic arc.

Fig. 2. Arc-trench model by Karig & Sharman (1975) illustrated in two cross-sections. Black, crust, compressed and metamorphic parts of accretionary wedge, and material of median high. A, detailed section. B, small-scale section of entire subduction zone.

The classical tripartite subdivision of arc-trench gaps (Seely et al. 1974; Karig & Sharman 1975) consists (fig. 2) of an *accretionary wedge* of clastic sediments derived from either side of the trench, a *median high* or stack of thrust plates involving metamorphosed trench fill or older rocks, and a gently extensional *forearc basin* on pre-subductional basement or on the tectonized trench fill in the rear extension of the median high. Volume and stratigraphic age span of these units depend on the availability of sediment and on the depth of submergence of the lower plate. Sediments arriving in the accretionary wedge from the trench side are progressively deformed. Their imbrication uplifts the median high and provides a wall retaining the forearc basin fill. Bouldery or fragmented deposits (olistostromes and melanges) originate at the submarine escarpment of the median high and are shed over the accretionary wedge, in particular during a steepening of the lower slope, during increased thrusting or during low rates of sediment supply. Fragments of oceanic crust and mantle found as olistoliths in melanges (Gansser 1974) or outcropping in outer non-volcanic arcs (Van Bemmelen 1949), or tentatively identified on reflection seismic profiles (Beck & Lehner 1974), have likely originated as fragments torn loose from the lower plate of the subduction and were incorporated into the upper plate whenever promontories of lower-plate material surpassed the level of the suture. Small normal faults, common in many recent trench outer walls, could provide sufficient relief to generate small shallow ophiolite fragments. However, for major

ophiolite thrust units, this explanation may not suffice. They may represent either lower-plate material incorporated into the upper plate by a major event of suture progradation (Roeder 1977a), or oceanic upper-plate material obducted to reach the surface. Both alternatives suggest collision events.

Interplate Clastic Units

Sediments of this type include the Cretaceous and Paleogene Rheno-Danubian Flysch at and below the front of the Austro-Alpine system, furthermore the finer clastic *Bündnerschiefer* of about the same age and presently located below the central and south part of the Austro-Alpine system. Bouldery deposits adjacent to, or part of, the clastic units are key features in interpreting their geodynamic setting.

The Rheno-Danubian Flysch (Oberhauser 1968), with 1000 m or more of turbidite sequences, spans the maximum possible period of Alpine subduction of oceanic crust (Barremian to late Eocene). In a mobilistic interpretation, therefore, it could represent the frontal part of an accretionary wedge (Dietrich & Franz 1976) which was attached to the front of the upper plate and moved with it during the entire process of subduction, collision, and intraplate convergence. However, its internal deformation is not well known and its original top and bottom are not preserved. Polyphase deformation can be expected for internal parts of an accretionary wedge, but erosional unconformities between folding stages are not required in this environment. An alternative, more fixistic interpretation (Angenheister et al. 1975; Frisch this vol.) would place the Rheno-Danubian Flysch into the lower plate, that is, into the rapidly subsiding North Penninic trough above early and mid-Mesozoic shallow-water deposits. In this interpretation, its present association with the Austro-Alpine complex is fortuitous; it was reached and overrun by this thrust mass from early Eocene on. This would explain the apparent lack of polyphase deformation and metamorphism in the Rheno-Danubian Flysch.

Marine sediments carrying sedimentary boulders or tectonic inclusions occur palinspastically south and north of the Flysch depositional area. Some deposits, sliced into the Flysch body, are of doubtful palinspastic position without a firmly established kinematic sequence of tectonic events. Although they are not critical for the paleogeography of the Rheno-Danubian Flysch and for a decision between the two interpretations, other bouldery deposits very clearly originated at a median high and thereby lend essential support to the model of discontinuous subduction proposed in this article.

On the external side and overridden by the Rheno-Danubian Flysch complex, the Ultrahelvetic *Wildflysch* of Cretaceous to lower Oligocene age contains foreland-derived clasts implying active fault scarps in the lower plate long before the collision (Butt & Herm this vol.). Therefore they are not critical for the palinspastic location of the depositional area of the Rheno-Danubian Flysch. Spilitic flows and clasts in the *Wildflysch* and related formations, dated as Maestrichtian to Eocene, may record a tensile downbending of the lower plate in front of the approaching arc-trench system, but the timing of this vulcanism is judged as somewhat too early for this mechanism.

Foreland-derived fragments of the *Gresten Klippenzone* in the foreland of the easternmost Alps, embedded in Ultrahelvetic shales of Upper Creta-

ceous to Paleocene age, are interpreted (Gwinner 1971) as tectonically detached and incorporated in the Flysch complex in Neogene time.

Stratigraphic ties of the Rheno-Danubian Flysch body to the mobile upper plate are doubtful to non-existant, but tentative paleogeographic ties can be established with bouldery deposits. The *Pienide Klippenzone* (see Andrusov 1972; Mahel 1968) of the Carpathians and easternmost Alps contains fragments of Jurassic and lower Cretaceous carbonate, serpentinites, and basalts embedded in a flysch matrix of Upper Cretaceous age. This zone is strongly suggestive of an ophiolitic melange (Gansser 1974). Its pre-collisional age implies a palinspastic position south of the site of collision, although it is closely associated with the Rheno-Danubian Flysch, probably by polyphase thrusting in Neogene time.

The *Aroser Zone* is an ophiolitic melange (Gansser 1974) extending southward beneath the Austro-Alpine complex for about 100 km where it merges with the South Penninic Platta ophiolite body, with the contact being overprinted by later thrust faulting. The Aroser Zone contains both olistoliths and shear fragments likely derived from all units in contact with the Aroser Zone and embedded in a sedimentary matrix of Cretaceous age. It overlies tectonically the Rheno-Danubian Flysch as well as all identified crustal elements of the lower plate. At its top, this unit has been smeared along the sole of the Austro-Alpine complex during the Tertiary emplacement of the latter, and therefore it does not require a palinspastic position of its own. Its olistoliths and their embedding, however, suggest a major tectonic event during the Cretaceous. This event took place south of the combined depositional areas of the Rheno-Danubian Flysch and the southerly adjacent *Bündnerschiefer* of the North Penninic trough, but north of the Austro-Alpine complex. It involved part of the South Penninic ocean floor as shown by the connection between Aroser Zone and Platta ophiolite. The plate tectonic nature of this event becomes more apparent when also data on metamorphism are considered.

Upper-Plate Clastic Units

Clastic units that are part of the Austro-Alpine complex include the lower Cretaceous olistostromal *Rossfeld* formation, the upper Cretaceous conglomeratic to olistostromal *Rand Cenoman* formation, and the upper Cretaceous to Paleogene marine clastics, conglomerates, and reefs of the *Gosau* group. These units are deposited in rifted or perched basins conformable above continuously subsiding shelf or unconformably on prestructured shelf sediments and basement. In plate models assuming the Austro-Alpine complex as the site of the arc during the entire period of subduction (Dietrich & Franz this vol.; Roeder 1976, 1977b), these deposits would be remnants of forearc basins north of the magmatic arc. However, in the model proposed in the present article, they would be located palinspastically south of the arc, possibly on the south slope of an oceanic backarc basin. Clastic components in the Gosau group include ophiolite detritus which has been ascribed to lower-plate material scooped up and redeposited in the Gosau forearc basins (Dietrich & Franz 1976). Chromitiferous ultramafites outcropping subaerially in the median high could easily furnish this detritus. As an alternative source of chromite, the Eohellenic obductites of the Dinarides have been proposed (Roeder 1976), possibly with northward fluviatile or shallow marine transport.

Metamorphism

Radiometric work on Alpine metamorphic and plutonic rocks has evolved into a key instrument in understanding the orogenic mechanism. Despite uncertainties in the dynamic interpretation of the crystallization processes involved, the data body now contains more details than can be resolved by the present plate tectonic approximation. Data not only from the East Alpine geotraverse, but also from the Western Alps of Switzerland, Italy, and France, are consistent with nearly all of the plate tectonic models proposed to date. Although this demonstrates the usefulness of the combined radiometric, petrologic, and tectonic approach, it suggests a shortage of plate tectonic speculation.

Caledonian and latest Paleozoic thermorogenic events (Satir 1976; Höll et al. this vol.) affected the cratonic basement of the Austro-Alpine complex and, possibly, the lower plate. They have generated the building blocks of the Alpine plate system and require analytical attention because they are constituents of the mixed ages which are common in both plates.

The Alpine thermorogenic history is documented by Mesozoic and Tertiary radiometric dates which demonstrate four gross groups of events (Bickle et al. 1975; Frey et al. 1974; Hawkesworth 1976; Jäger 1973; Miller 1974; Raith et al. this vol.; Satir 1976). The dates of two of these events, formerly called Paleo-Alpine (Trümpy 1973; Roeder 1977b) but now uniformly referred to as Eo-Alpine (Dal Piaz et al. 1972), are scattered between 130 and 60 m.y. One of these, dated at 105 to 65 m.y. by Rb/Sr and K/Ar ages on micas, has low p/t assemblages characterized by staurolite and andalusite or sillimanite. It occurs at the base of the Austro-Alpine complex around the Tauern window and farther west in equivalent upper-plate rocks. The other of these events, with widespread high p/t assemblages of blueschists and eclogites, occurs in lower-plate or interplate rocks of the southwestern and western Alps and is dated at 130 to 60 m.y., with a peak at around 80 m.y. Within the Tauern window, similar parageneses are correlated with the Eo-Alpine event but not dated (Miller 1974; Raith et al. this vol.). The third and fourth groups of events range in age between 38 and 7 m.y., are of a Barrovian low p/t type attaining amphibolite grade with kyanite, and occur in domed or rising areas of the lower plate. A peak in dates between 40 and 36 m.y. is termed Meso-Alpine, and another between 28 and 7 m.y. is called Neo-Alpine. In the following, these radiometric dates and their groupings are interpreted as fixed by paroxysmal tectonic events during Alpine subduction and after collision.

Elements of a Paired Belt and their Preservation

In all plate tectonic models of the Alps, the two Eo-Alpine events are interpreted as a paired metamorphic belt (Miyashiro 1961) and a major argument for a subduction model of orogeny (Ernst 1975). In such a pair, the low p/t type with high t reflects the heating associated with arc magmatism and is therefore diagnostic of the upper plate. The high p/t type with high p reflects the anomalously cool conditions and rapid burial associated with subduction; it is therefore diagnostic for the lower plate or the deepest parts of the accretionary wedge belonging to the upper plate.

The high p/t metamorphic suites of the Tauern window, consisting of eclogites and minor glaucophane schists without lawsonite, crystallized or closed during uplift through 8 to 11 kb of overburden pressure and through 400 to 500°C (Miller 1974). At an average crustal density of 2.65 this pressure range translates to 20 to 41 km, and the temperatures indicate geothermic gradients of 10 to 17°C/km which are abnormally low for sedimentary basins on continental crust (Levorsen 1967) but very similar to those found in the Shimanto belt and the East Honshu basin, that is, Mesozoic to Neogene trench fill on the Pacific side of Japan (Uyeda & Horai 1964). This geodynamic environment suggests that the protolith of the eclogite-bearing formation, possibly late Jurassic or early Cretaceous trench fill, reached this depth range by subduction, but as part of an accretionary wedge belonging to the interplate unit.

Retrogradation to greenschist facies during the Meso- or Neo-Alpine events took place at 3.6 to 4 kb (Miller 1974) or 13 to 15 km of depth, estimated to be the restored thickness of the overburden by Frank (1969), composed of the Bündnerschiefer of the window content, a thin melange unit, and the Austro-Alpine thrust complex. Between both metamorphic events, therefore, the host rocks of the eclogites became exhumed in reversal of the effects of subduction and rose through at least 14 to 28 km of overburden, most likely by a process of upwarping and erosion. The central parts of the Austro-Alpine complex and its substratum are affected by major upwarping which kinked the thrust complex into a steeply dipping or overturned root zone (Roeder 1973). However, the amount of uplift achieved during this upwarping is only 10 to 15 km (Frey et al. 1974) and the structural event is of Oligocene to Recent age and therefore cannot explain the preservation of the Paleo-Alpine metamorphism and its peak of cooling ages at about 60 m.y.

A more appropriate mechanism would have to be timed properly to coincide with the late Cretaceous radiometric dates, and would have to uplift the metamorphics into a depth range of sub-metamorphic pressures and temperatures, and shallow enough to remain in a low rank of retrogression during the reburial beneath the Austro-Alpine thrust sheet.

These requirements imply an event of upwarping and erosion which can have taken place only before the emplacement of the Austro-Alpine thrust complex. Since the eclogites are located at less than 5 km below the level of the Austro-Alpine basal thrust, the amount required for the earlier uplift is more likely 24 to 38 km.

Buried and pressure-metamorphosed rocks could re-emerge as part of the median high by transport on the basal thrust of this structure, if the underlying accretionary wedge became temporarily attached to the lower plate. The bulge of the subducted sediment wedge would uplift the median high and the adjacent backarc by about the thickness of the wedge, that is, a maximum of 5 km. Higher amounts of uplift can be achieved only if the wedge, parts of the arc-trench system, and the lower plate are distorted, kinked, and upturned, as in the South-Alpine root zone (Roeder 1973), or in a flip or polarity change of the subduction zone.

There are no structural records of a late Cretaceous event of root-zone kinking, but a number of stratigraphic observations are consistent with such an event. The Cretaceous melanges at the tip of the interplate series, in particular the Aroser zone, the Matreier zone, and the Pienide zone suggest a pre-tectonic, essentially olistostromal, stage of development. One may think of mass waste, that is, a rapid and extensive disintegration of once coherent thrust sheets, in the form of slump blocks

pushed off a very active submarine thrust fault scarp. The Cretaceous or Upper Cretaceous timing of this group of events coincides with the Eo-Alpine high p/t metamorphism. The palinspastic position of the olistostromal units just internal to the eclogite-bearing units fits their probable point of origin at a thrust-faulted median high above a distorted and fast rising accretionary wedge. We therefore consider the mass waste documented in the melanges as local surface indications of the same process of uplift that led to the cooling and preservation of the Eo-Alpine blueschists and eclogites (fig. 3). This fault scarp and the site of melange deposition at its foot cannot yet have been covered by the Austro-Alpine thrust mass although internal deformation may have

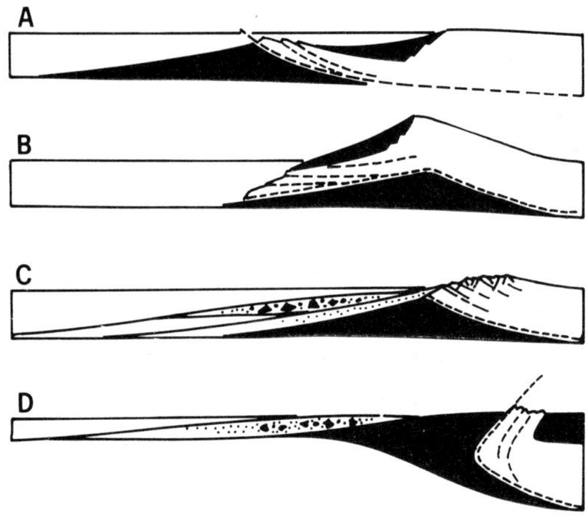

Fig. 3. Cross sections showing two mechanisms of preserving blueschists in trench innerwalls of subduction zones, schematic, Black, accretionary wedge and forearc basin fill. Dotted: melanges. A, state of accumulating an accretionary wedge and beginning of thrusting at median high (dashed). B, accretionary wedge underthrust beneath median high and forearc basin, without erosion. C, same as B, but with structural high eroded and melanges emplaced. D, Alternative mechanism with rotation-zone kinking of median high and blueschist bearing rocks in the vicinity of the high.

existed already within the latter. Kinking of the median high could have attained the full rate of subduction, measured in cm/y, if the accretionary wedge was fully coupled to the lower plate. Depending on the rate of subduction and on the dip of the sole fault, the re-emergence of the blueschists and eclogites could have taken a reasonable 5 to 15 m.y.

Arc and upper plate of this Cretaceous accretionary wedge were not the Austro-Alpine thrust complex, but probably an oceanic arc with south-Penninic ocean floor in both plates (fig. 4). This interpretation is consistent with the extensive South-Penninic ophiolites and their role as forearc region of a Cretaceous arc system, with the complete disappearance of the arc, and with the nearly total lack of Cretaceous-age arc magmatism in the Austro-Alpine thrust complex. This view is also consistent with the northern provenience of chromite within the Gosau sediments.

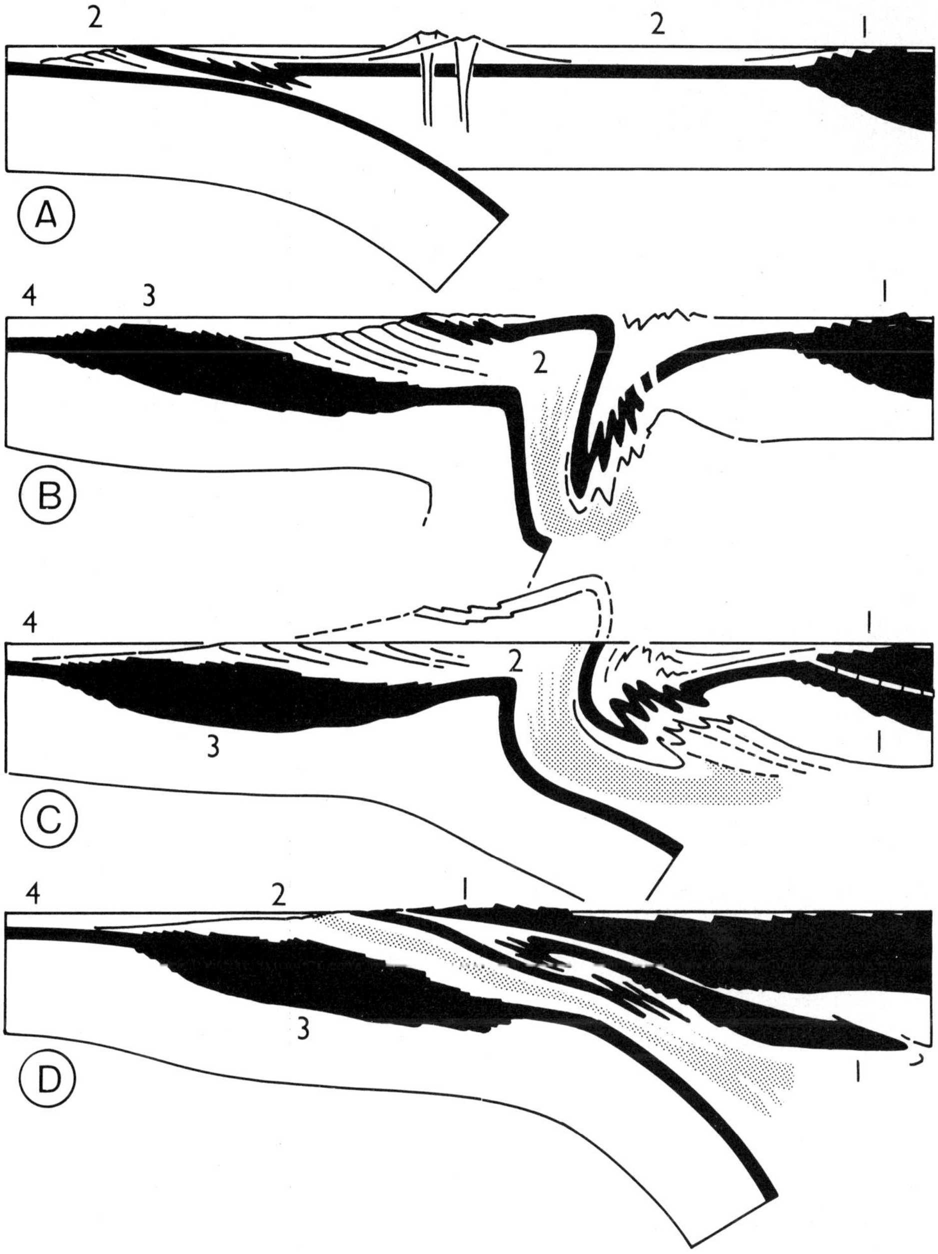

Fig. 4. Four time-successive cross sections suggesting a mechanism for providing producing and preserving the Eo-Alpine high p/t metamorphism (stippled), schematic, not to scale. 1, Calcareous Alps. 2, South Penninic oceanic area with Bündnerschiefer and lateral equivalents. 3,Middle Penninic (Briançonnais) continental fragment and lateral equivalents. 4, North Penninic trough and lateral equivalents. A, about 120 m.y., South

This heavy mineral suggests rapid erosion of an oceanic arc and its crust during the late Cretaceous kinking, with sediment transport across a backarc basin onto the Austro-Alpine region.

The paleobathymetric model by Butt & Herm (this vol.) requires exposed sources of carbonate detritus. Such sources must be assumed available at the oceanic arc. Possibly subduction started at the edge of yet another series of plate fragments carrying early Mesozoic carbonate banks.

The low p/t metamorphism is found in Austro-Alpine rocks. Therefore, the present view holds that the two elements of the paired belt, at present closely juxtaposed across a thrust fault, originated independently and separated possibly by several 100 km, and probably were caused by unrelated tectonic mechanisms. Substantial age differences between the high p/t and the low p/t events are therefore conceivable, although the very wide ranges of both types match closely. The oldest of the low p/t dates of around 100 m.y. for example, could record the beginning of backarc spreading, a prominent heat source in arc-trench systems (Karig 1970; Dennis 1976).

The final item in the present view is an uncertainty. The hypothetical south-Penninic arc collided with the mid-Penninic craton in middle or late Cretaceous time, touching off a major event of root-zone kinking and erosion. The collision transferred the plate convergence either forward into the North Penninic trough or backward to the front of the Austro-Alpine complex which had been an Atlantic-type margin until then. The choice is determined by the provenience of carbonate boulders in the interplay melanges of the Arosa and Pienide zones. If these boulders are sourced in the Austro-Alpine during the sedimentary stage of the melange, the source complex must have collided with the back side of the obducted south Penninic crust before the (upper Cretaceous) sedimentary stage of the melange. If these boulders are derived, similar to the carbonate detritus in the Rheno-Danubian Flysch, from fragments of carbonate banks associated with the hypothetical arc, the conversion of the inactive Austro-Alpine complex into a subduction front and a volcanic arc may be deferred until the Eocene when the only known arc vulcanism becomes apparent.

Post-Collisional Strain Heat

The Meso- and Neo-Alpine metamorphic events (Trümpy 1973), lumped together in some syntheses (Frey et al. 1974; Roeder 1977b), appear as se-

Fig. 4 (cont.) Penninic seaway showing mid-oceanic subduction zone, volcanic arc, and Atlantic-type margin along north edge of Austro-Alpine continent (1). B, about 100 m.y., collision of South Penninic area arc with Mid-Penninic continent (3), kinking of both plates, and high p/t metamorphism in Bündnerschiefer. C, about 90 m.y., buoyant rise of Mid-Penninic south edge and root-zone system, exposing blueschists. Incipient subduction or crustal thrusting at north edge of Austro-Alpine. D, about 60 m.y., with Austro-Alpine plate overriding stacked Middle Penninic, South Penninic ocean fill with blueschists, South Penninic ophiolites, and frontal Austro-Alpine craton.

parate events with peaks of uplift-related cooling ages around 38 and 20 m.y. (Raith et al. this vol.). Metamorphism related to these age groups has been explained by burial beneath a stack of nappes (Bickle et al. 1975). However, heat transfer, estimates of the available overburden, and the timing of structural events require an additional heat contribution (Oxburgh & Turcotte 1974; Raith et al. this vol.). Such additional heat could have been furnished by arc magmatism in the lower plate if it had functioned as an upper plate prior to an arc-arc collision and the subductive elimination of a third intervening plate (Roeder 1977b, fig. 3).

Alternatively, the heat contribution could come from friction on intracrustal shear surfaces (Bird et al. 1975; Roeder 1977b). At plate tectonic strain rates, this heat is sufficient to create metamorphic halos, anatectic melts, and small plutons. In the present Alpine model, strain heat rather than arc heat is assumed to have caused the peri-Adriatic plutons (Sassi et al. this vol.). Also the radiometric dates from metamorphic rocks in the domes areas of the Tauern and the Pepontine Alps of Switzerland are ascribed to strain heat; they would reflect erosion and cooling in the centers of updoming, but there is as yet no convincing structural cause to explain two separate radiometric events.

Suture Progradation

After collision of the Austro-Alpine complex with the Eurasian craton, plate convergence in the Alps continued at least through the late Miocene and possibly through the present. Movements during this phase consisted of strong uplift and weak compression in the central parts and northward thrusting in the external parts, with waves of transport spreading northward and downward into the foreland sediments. This forward movement of orogenic activity is an old observation (Trümpy 1960; Wunderlich 1966). Its plate tectonic appearance is called suture progradation (Bird et al. 1975; Roeder 1977a, b), and its significance is in the gain of more detailed geophysical knowledge of surbsurface structure, and in relating metamorphism and plutonism to post-collisional plate deformation.

The basic mechanism during this stage is interpreted either as prograding subduction of limited (crustal) extent on individual scars or sutures (Trümpy 1973; Bögel 1974; Frisch this vol.) or progradation by the development of intraplate branches of the original suture (Roeder 1977b). Within the fragmented and mixed crust of the Alpine lower plate, both interpretations may be end members of a spectrum of possible combinations, although geophysical data appear to favor the second interpretation. Subhorizontal intracrustal low-velocity layers (Giese 1968, this vol.), are interpreted as the base of the tectonically active Alpine body. This base is modelled as a décollement with transport and deformation above and without deformation below, reaching surface in the marginal thrust belts. As the core area rises, the décollement is abandoned and replaced by new thrust surfaces deeper in the crust and involving larger peripheral areas, while the older décollement surfaces become uplifted, deformed, and finally eroded.

Evidence in support of this model is the kinking and southward overturning of the entire Austro-Alpine complex and parts of the substratum to form the South Alpine root zone. Folding and uplift are also documented in the central Alps and north of the south Alpine root zone by the Meso- and Neo-Alpine radiometric cooling ages (Frey et al. 1974; Raith et

al this vol.). Its amount can be estimated at about 15 km, if a 10 km thick Austro-Alpine composite plate (Oxburgh & Turcotte 1974) is assumed to have had its top at sea level prior to the collision, and if its base is assumed, by restoring central-Alpine erosion, at a present elevation of 5 km.

The early stages of this uplift were accompanied by subhorizontal compression, as is evident in the overturning of the root zone and in mesoscopic tectonic fabrics of the Tauern window (Roeder 1976) where two post-collisional kinematic stages of folding show a decreasing influence of subhorizontal compression.

Thrusting affected the northern foreland since the collision which is apparent as a strong surge in depositional rates during the late Eocene in the Molasse foreland (Lemcke this vol.). Structural details in the frontal belt of the Austro-Alpine complex suggest at least two phases or age groups of thrusting. The older phase created the imbrication or stacking of the upper Austro-Alpine, the Flysch, and possibly the Helveticum, probably during the collision.

The younger group may represent a wide time span. An early pulse in late Oligocene time (Lemcke this vol.) and another larger one in early Miocene time (Angenheister et al. 1975) are suggested by stratigraphic analysis. Later thrusting postdates the formation of subalpine synclines which involve late Miocene deposits (M. Müller this vol.). During one or more of these late pulses, the Alpine thrust units down to a depth of 3 or 4 km overrode flat-lying Molasse for about 10 km (Berktold et al. ; Will et al.; Müller all this vol.).

Outward migration of thrust activity and associated strain, amply documented stratigraphically, is illustrated by deformed fossils. Located at present within the folded part of the Molasse area of deposition, these fossils indicate that their host rocks had received up to 25% strain prior to the development of the folds (Albrecht & Furtak 1966). Similar waves of compressional strain ahead of the outward migrating thrust fronts have been found in the southern Appalachians (Dean & Kulander 1977).

Gravity Spreading

For interrelating quantitatively the two components of late to recent Alpine tectonism, central uplift and marginal thrusting, we are applying the dynamic model of gravity spreading (Bucher 1956; Elliott 1976). This model describes the motion of thrust sheets with equations developed for the motion of glaciers. A forward directed slope of the rock-air interface maintains gently rising forward directed transport at the base of the thrust unit. The movement tends to eliminate the slope, whereas the compressional uplift in the Central Alps tends to maintain it. Although north-dipping thrust plates occur quite commonly in the northern Alps, gravitational downhill gliding is clearly not an essential part of the convergent movements (fig. 5). North-dipping sole faults on which such gliding could have taken place, predate the central Alpine uplift which tilted them into their present-day attitude. They are crosscut and blocked by faults which offset Neogene structures. Extending clearly across the Northern Calcareous Alps, these faults offset the Austro-Alpine basal thrust and the northdipping thrust plates, create thrust windows of flysch and Aroser zone, and must be inferred to sole out in the central Alps far below the present and recently uplifted outcrops.

In the Elliott (1976) version of the gravity-spreading model, the origin of the basal décollement is not explained, and its depth range is only loosely controlled by the shape of the detached and spreading body. For orogenic systems of a wide range of shapes in cross section, décollements at any depth greater than 8 to 10 km are stable. Therefore, the depth to the basal décollement is determined not by the gravity-spreading mechanism, but by crustal processes. In the Alps, these processes are attributed to collision.

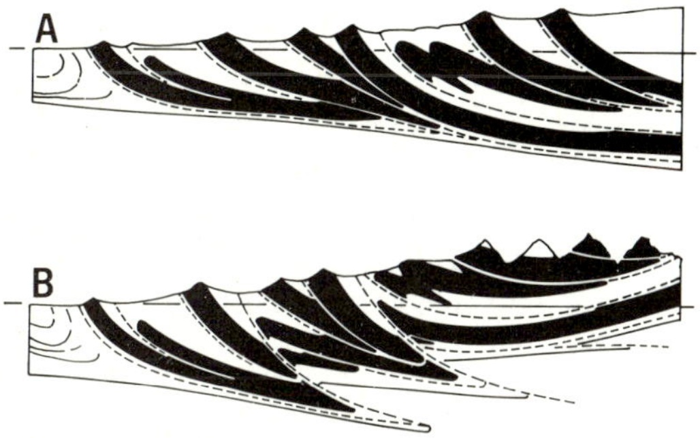

Fig. 5. Mechanism of Late-Alpine polyphase sledrunner thrusting, schematic. A, structure during late Eocene collision, with Neogene to Recent uplift of Central Alps restored. B, present structure with thrusts intersecting older tilted thrusts.

In a cross section that extends through the south Alpine root zone, for example, it is geometrically possible (Roeder 1977b, fig. 4) to attribute transport on the intracrustal low-velocity layer to the kinking of both plates in an interpretation compatible with the Elliott model. In the vicinity of the geotraverse, the root zone is less pronounced and the crustal velocities are more complex (Gebrande et al. this vol.) suggesting a more diffuse crustal deformation. Gravity spreading over a stable basement should produce a fairly regular thrust belt. In the Alps, however, the progradation of the suture may also affect the basement. Therefore, the kinematic details of the north Alpine nappe tectonics, with crosscut thrusts, folded-fault structures, gentle warps, and recumbent, possibly mushroom-shaped folds, would reveal further details of the interplay between gravity spreading in the suprastructure and convergence within the crust.

Neotectonics

Present Alpine movements consist of uplift in the central parts and supracrustal thrusting at the flanks, in directions similar to or enclosing angles of 20 to 30 degrees with, the directions of European crustal compression. This suggests either that crustal convergence beneath the Alps is continuing unchanged since the collision, or that the Alpine neotectonic rise and spreading reflects the isostatic adjustment of a once

thickened and heated portion of the crust which at present is nearly free of horizontal compressional stress. In the following, therefore, the significant evidence is listed, followed by the arguments for each of these views.

Shallow micro-earthquakes in the subalpine Molasse (Förtsch et al. this vol.) suggest continued Molasse folding above uninvolved basement with little or no change in style with respect to Miocene and older folding. A focal plane solution of the Friuli shock of 1976, its aftershocks (Gebrande et al. this vol.), and the deep well Agip Bernadia #1 have outlined a southvergent thrust belt superimposed on originally north-vergent thrusts involved in the south Alpine root zone (fig. 6). Instantaneous differential uplift of parts of the central uplift is measurable in two places in the Tauern window as uplift of a central area relative to marginal areas (Angenheister et al. 1975). At present this uplift amounts to about 1 km/m.y., whereas the finite uplift during the past 40 m.y. attained an estimated rate of only 0.3 km/m.y.

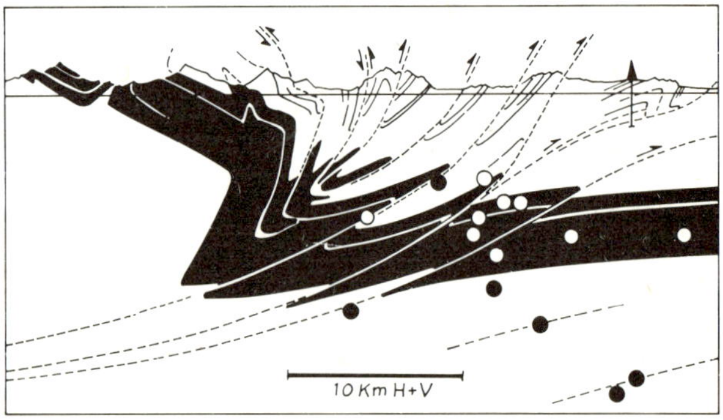

Fig. 6. Cross section of Alpine south front in Friuli–Gemona, showing originally north-vergent thrust belt kinked and rethrust with south vergence. Black, lower and middle Triassic. White and black dots, hypocenters of aftershocks of Friuli earthquake 1976, after Gebrande et al. this vol. Surface geology after Carta Geologica delle tre Venezie 1 : 100 000.

Arguments for a neotectonic model without horizontal compression (Illies et al. 1977) include the mechanism of gravity spreading itself: the documented recent differential uplift of central Alpine parts, such as the Tauern, is sufficient to maintain a multiple of this amount of spreading and marginal thrusting, depending on the cross-sectional area and shape of the Alps. Even intracrustal (thick-skinned) thrusting does not, by itself, desprove a fixistic mechanism. The last kinematic stage in the Tauern window (Roeder 1976) is dip-slip movement on vertical faults causing central Alpine uplift but not compression. Instantaneous convergence between the Eurasian and African plates, of 2.3 cm/y (LePichon et al. 1973) may be completely absorbed at the Calabrian subduction. The B-axes of the youngest Molasse folds and the first motions on the Friuli earthquake (Gebrande et al. this vol.) indicate a direction of transport normal to the average slope of the Alps but divergent from the northwesterly striking intraplate p-axis in continental Europe between the English Channel and the Adriatic Sea (Ahorner this vol.). Finally, heat-

flow data in the areas of central Alpine updoming are normal (Angenheister et al. 1975).

In this model, therefore, neotectonic uplift would be powered by buoyant isostatic rise of thickened and heated crustal material in the classical fashion of orogenic history (Stille 1924).

Arguments for a neotectonic model with crustal convergence are based on the refraction-seismic structure of the crust. The neotectonic centers of Alpine uplift are localized anticlines and therefore too small to reflect an isostatic rise of the entire lithosphere given the normal range of flexural-rigidity values (Walcott 1972). The narrow centers of uplift are underlain by less distorted or flat low-velocity layers suggesting anomalously high temperatures and very young or recent strain heat released by subhorizontal shear. The coupling of Alpine rise with equally fast subsidence in the Po valley basin is better explained by kinking of the lithosphere under subhorizontal pressure.

In the second model, the Alpine neotectonics is interpreted assuming equilibrium state of the stresses involved. This approach makes the neotectonic crustal processes in the Alps similar to those studied in other belts of continental collision, in particular the Himalayas and the Zagros range (Bird et al. 1975).

Arguments acceptable to both views include the contrast between the low and decreasing topographic gradient toward the high and rising central European massifs to the north, and the very steep topographic and structural gradient toward the deep and probably sinking Po plain. Since a topographic gradient is required for continued transport of sled-runner-shaped thrust sheets, a tectonic style change can be expected to have taken place beneath the Alpine north slope. Possibly, the late Alpine centers of uplift are folds replacing nappe transport and thereby expressing this style change. Southward thrusting on the southern mountain front can also be considered as an adaptation to the increasing structural and topographic slope, but the origin of the south Alpine root zone, largely responsible for the evolution of the south Alpine topography, is much older than the present Friuli thrust belt.

The contrast in average topography between both sides of the Alpine orogen is not repeated in the depth of the M discontinuity, which is about the same on both flanks of the Alps. Its rapid sinking and the 16 km of sediment fill (Stein et al. Losecke et al., all this vol.), however, makes the crust of the Po basin likely quasi-oceanic rather than purely continental. A tensional regime and backarc dilation may have thinned this crust prior to the flip of thrust vergence.

References

Albrecht, K. & Furtak, H., 1966: Geol. Mitt., 5,3:227-248.

Andrusov, D., 1972: Tectonics, Sect. 3, Int. Geol. Congr. Paroc., No. 24: 273-281.

Angenheister, G., Bögel, H. & Morteani, G., 1975: N. Jahrb. Geol. Paläont. Abh., 148, 1:50-137.

Barazangi, M. & Dorman, J., 1969: Bull. Seismol. Soc. Amer., 59:369-380.

Beck, R.H. & Lehner, P., 1974: Am.Assoc. Petrol.Geol.Bull.,58,3:376-395.

Bernoulli, D. & Jenkyns, H.C., 1974: Soc. Econ.Pal.Mineral.,Spec. Publ., 19:129-160.

Bickle, M.J., Hawkesworth, C.J. & England, P.C., 1975: Earth Planet. Sci.

Lett., 26:13-18.
Bird, P., Toksöz, M.N. & Sleep, N.H., 1975: J. Geophys. Res.,80:4405-4416.
Birkenmajer, K., 1974: Geol. Soc. London Spec. Pub., 4:127-157.
Bosellini, A. & Winterer, E.L., 1975: Geology, 3,5:279-282.
Bögel, H., 1974: Sitz. Ber. DFG Schwerpunkts-Progr. (abstr.).
- , 1975: Geodynamics Project, Nat.Comm.Fed.Republ.Germany Progr. Rept.: 42-50.
Bucher, W.H., 1956: Geol. Soc. Amer. Bull., 67:624-649.
Burchfiel, B.C., 1976: Geol. Soc. Amer. Spec. Pap., 158:82 p.
Chapple, W.M., 1975: NE Sect. 10th Ann.Mtg.,Geol.Sco.Amer. Abstr.7,1:38-39.
Closs, H., 1975: Rapp.Comm.Int.Mer Médit., 23,4a:55-57.
Dal Piaz, G.V., Hunziker, J.C. & Martinotti, G., 1972: Mem. Soc. Geol. Ital., 77,4:433-466.
Dean, S.L. & Kulander, B.R., 1977: Abstr.with Progr.26 Ann Mtg. S.E. Sect. Geol. Soc. Amer., 9,2:132-133.
Dennis, J.G., 1976: Z. dtsch. geol. Ges., 127:73-85.
Dewey, J.F. & Bird, J.M., 1970: J. Geophys.Res.,75:2625-2647.
Dewey, J.F., Pitman III, W.C. & Bonnin, J., 1973: Bull.geol.Soc. Amer., 84:3137-3180.
Dietrich, V.J. & Franz, U., 1976: Geol.Rdsch.,65,2:361-374.
Elliott, D., 1976: J. Geophys. Res., 81:949-963.
Ernst, W.G., 1973: Geol. Soc. Amer. Bull., 84:2053-2078.
- , 1975: Benchmark Papers in Geology, 19:450 p.
Frey, M., Hunziker, J.C., Frank, W., Bocquet, J., Dal Piaz, G.V., Jaeger, E. & Niggli, E., 1974: Schweiz.Mineral.Petrogr.Mitt., 54:247-290.
Gansser, A., 1974: Eclogae geol. Helv., 67:479-508.
Giese, P., 1968: Schweiz.Mineral.Petrogr. Mitt., 48:261-284.
Gwinner, M.P., 1971: Geologie der Alpen: 394 p., Stuttgart.
Hamilton, W., 1977: Amer.Geophys. Union. M.Ewing ser. 1.
Hawkesworth, C.J., Waters, D.J. & Bickle, M.J., 1975: Tectonophysics, 24:405-413.
Hawkesworth, C.J., 1977: Contr. Min.Petr. 84:225-244.
Hsü, K.J., 1971: Nature, 233, 5314:44-48.
Illies, J.H. & Greiner, G., 1977: Tectonophysics, in press.
Jäger, E., 1973: Eclogae geol. Helvet., 66:11-21.
Karig, D.E. & Sharman III, G.F., 1975: Geol. Soc. Amer. Bull.,86:377-389.
Levorsen, A.I., 1967: Geology of Petroleum, 2nd ed.,724 p.,San Francisco, London.
Mahel, M., 1968: Int. Geol. Congr., 23rd Sess., Sect.3:77-86.
Miller, C., 1974: Schweiz.Mineral.petrogr.Mitt.,54:371-384.
Milnes, A.G., 1974: Geol. Soc. Amer. Bull. 85:1727-1732.
Miyashiro, A., 1961: J. Petrol., 2:277-311.
Nicolas, A. & Jackson, E.D., 1972:Schweiz.Mineral.Petrogr.Mitt.,52:479-495.
Oberhauser, R., 1968: Jb. Geol. Bundesanst.,111:115-145.
Oxburgh, E.R. & Turcotte, D.L., 1974: Schweiz.Mineral.Petrogr.Mitt.,54: 641-662.
Radulescu, D.P. & Sandulescu, M., 1973: Tectonophysics, 16,3/3:155-161.
Rocco, T. & Jaboli, D., 1958: Habitat of oil and some factors that control it, Amer.Assoc.Petrol.Geol.Symposium:1153-1167.
Roeder, D.H., 1973: J.Geophys.Res.,78,23:5005-5024.
- , 1976: Z.dtsch.Geol.Ges., 127:87-103.
- , 1977a: Geology 5:203-206.
- , 1977b: Tectonophysics, 40:339-350.
Satir, M., 1976: Geol. Rdsch.,65,2:394-409.

Seely, D.R., Vail, P.R. & Walton, G.G., 1974: The geology of continental margins:249-260.
Stille, H., 1953: Beihefte Geol. Jahrb., 8:239 p.
- , 1924: Grundfragen der vergleichenden Tektonik:443 p.,Berlin.
Toksöz, M.N. & Bird, P., 1977: Tectonophysics, 41:181-193.
Trümpy, R., 1960: Geol. Soc. Amer.Bull., 71:843-908.
- , 1973: Gravity & Tectonics: 520 p., New York,London,Sidney,Toronto.
- , 1976: Eclogae geol. Helv.,69,2:249-264.
Uyeda, S. & Horai, K., 1964: J.Geophys.Res., 69:2121-2141.
Van Bemmelen, R.W., 1949: The Geology of Indonesia I: 732 p., The Hague.
Walcott, R.I., 1970: J. Geophys. Res. 75:3941-3954.
Wunderlich, H.G., 1966: Wesen und Ursachen der Gebirgsbildung. B.I. Hochschultaschenbücher, V:339-367.

Chapter 3

The Geotraverse Northern Apennines - Corsica (Ib)

Introductory Remarks on the Double Orogene of Northern Apennines and Corsica (Western Alps)

K.-J. Reutter

Wherever the different segments of the Mediterranean mountain belt system with their special lithological and structural features meet or pass into each other, problems concerning the palinspastic arrangement and the orogenic movement pattern become much more intricate than in the segments themselves. Nevertheless it seems that just these transitional zones may yield the most important informations about the geodynamic evolution of the whole system.

One of these important zones is the passage from the *Europe-vergent* Alps into the *Africa-vergent* Apennines. Its significance was already recognized at the beginning of this century when P.Termier and G.Steinmann studied the mutual relations of the Alps and the Apennines. In the following decades a lot of contrasting hypotheses were set forth about the possible connections of the Alpine chains which culminated in the well-known loops drawn through the Mediterranean region by L. Kober, R.Staub, H. Stille and others . Most of these theories proved untenable in the course of time as geological knowledge was improved by thorough geological mapping, precise stratigraphic dating and detailed sedimentological and structural studies.

Thus it could be shown that the main orogenic movements of the Western Alps leading to the westward displacement of eugeosynclinal sediments and ophiolites onto the pre-Alpine European foreland occurred in the late Eocene (*Ligurian phase*), while the northeastward directed movement of eugeosynclinal nappes upon the Italian crustal block (the later *Adria microplate*) occurred in the late Miocene (*Tuscan phase*). The flip of the tectonic transport direction corresponds to a change of the foreland and hinterland functions since during the *Tuscan phase* the Western Alps and Corsica together with the former European foreland acted as a hinterland, while the Adria microplate became a foreland. Very similar or even identical eugeosynclinal sequences were found in allochthonous position in the Western Alps and in Corsica as well as in the Northern Apennines. Thus it became clear that both mountain chains represented a double orogene with a common eugeosyncline.

The stepwise consumption of this Penninic-Ligurian eugeosyncline must be considered responsible for the expulsion of its sediments and ophiolites towards both sides.

With the discovery of the plate tectonic mechanism, a new dimension was introduced into the geotectonical considerations of Mediterranean orogeny. At the same time, the progress in the knowledge of metamorphic processes, the advancement of marine geology, the development of marine and terrestrial seismic methods, the disponibility of satellite imagery

and other new techniques offered new possibilities to get further insight into the orogenetic problems. These new possibilities stimulated the research work in the Western Alps and in the Northern Apennines. National and international geoscientific research programs were started and new theories based on the new global tectonics were advanced.

The present chapter - also partly chapter 2, part 3 - is dedicated to a geotraverse passing through the Northern Apennines and Corsica as a prolongation of the Western Alps. An attempt was made to obtain new data and elaborate new aspects of the double orogene by special investigations using different branches of geoscience. First of all, the crustal structure had to be examined by an explosion seismic program in order to ascertain if and to what extent the oceanic Penninic-Ligurian eugeosyncline had been consumed and collision between the Corso-Sardinian continental crustal fragment of pre-alpidic Europe and the Adria microplate had occurred. Some light could be shed upon the origin and consumption of the eugeosyncline by petrological studies of the ophiolites and their metamorphism. Structural investigations of the nappe tectonics in Corsica and the Northern Apennines were necessary for a better understanding of the overthrust processes. Nappe transport in the Northern Apennines was preceded by the installation and migration of graywacke flysch sedimentary basins, whose depositional patterns were detected by stratimetric methods. As heat transfer and tectonics are intimately related, it was important to obtain information about the thermal history of the Apennines. Coalification studies were found to be very useful in this respect. The interpretation of satellite imagery helped illuminate fossil and actual stress patterns and their directional changes. Finally, photogeologic and field observations revealed subrecent tectonic features.

By the combination of all these different research techniques, we intended to contribute to the solution of the problems posed by the Northern Apennines - Western Alps double orogene and to the understanding of the geodynamics governing its formation.

Part 1

Geophysical Research

Seismic Refraction Measurements in the Ligurian Sea and in the Northern Apennines

H. Letz, C. Reichert, P. Wigger and P. Giese

In the Alps greatest crustal thickness coincides with the central zone as evidenced by gravimetric and seismic data. The Apenninic crustal structure, however, must be quite different because the Apennines are situated between the gravity high of the Ligurian Sea and the minimum of the Po-Plain.

The first seismic refraction data of the area under discussion were given by a profile running along the Ligurian coast. Observed in 1966 from the Western Alps, it reveals an average crustal thickness of about 30 km for the region E and W of Genova (Giese & Morelli 1975), but the record section obtained results in a very complex pattern of travel-time curves, indicating a more complicated structure. Their interpretation, in the form of a structure model of the earth's crust in the Ligurian Apennines, required further data.

In 1974, in close cooperation between French, German, and Italian geophysical research institutions, a large-scale crustal seismic project, based on previous experience gathered in Southern Italy, was carried out in the triangle Corsica - Tuscany - Liguria (fig. 1). In order to detect

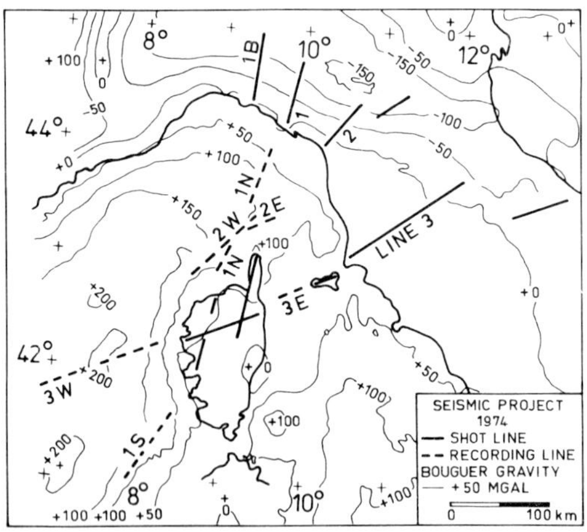

Fig. 1. Position map showing the seismic refraction lines observed in 1974.

lateral variations of crustal structure in detail, the spacing of shot-points and stations was kept at a constant 2.5 to 5 km. Shooting was possible only on sea, performed by the Italian research vessel "Marsili".

A total of 40 seismic recording stations was operating on land. Measurements were taken on four profiles with the object to investigate the crustal structure of the Northern Apennines between Genova and Elba. Besides, by the same shots, the crustal structure of Corsica was to be investigated (fig. 1).

This paper deals with the two profiles evaluated by the German group. Profile 1 begins 10 km W of La Spezia and extends up to the border of the Po-Plain near Parma. On this line, blasts were observed from shooting sections 1S, 1M, and 1N. The gaps are due to crossing telephone cables. The evaluation of the data of this profile is being made by Letz (1977). Profile 3 runs from Elba in NE direction via Siena into the Apennines. Here, shots were recorded, fired in the channel between Corsica and Elba (3 E) and W of Corsica (3 W). Data of this line were interpreted by Reichert (1977) and Wigger (1977). It should be noted that the results of the other profiles fit into the general picture of crustal structure presented here.

Due to the large number of shots, a great amount of records has been obtained. The individual seismograms are compiled to record sections in two ways: one type of presentation uses a fixed station and moving shotpoints. The reversed display is obtained by keeping the shotpoints fixed and moving the recording point. Owing to the great density of the material, the variations of the picture and the arrangement of the typical wave groups from section to section are mostly weak. This fact enables the identification and the correlation of the various wave groups from section to section.

In the next step of data evaluation, the correlated travel-time branches were displayed in an intercept-time-distance section (Pavlenkova 1973). This kind of presentation gives first qualitative information on the main lateral changes of crustal and upper mantle structure and allows the determination of true velocity at the refracting boundaries. In the final step, this time section is inverted into a depth cross section. This model, obtained as a first approximation, is controlled by the ray-tracing method. Discrepancies between the observed and calculated travel-time curves induce further improvements of the cross section.

Line 1

On line 1, 20 recording stations were in operation which recorded 45 shots. The shots fired SW of Corsica gave no readable records due to the large distance. But from the portions 1M and 1N, good to excellent records are available. About 40 record sections have been compiled. As an example, the record section of recording station 110 is presented in fig. 2, showing some of the most important wave groups.

In general, the following main wave groups can be distinguished (fig. 2):

1) The P_g-wave shows clear first arrivals and has been recorded up to a distance of 120 km. The apparent velocity increases from 5.5 to a remarkably high 6.4 km/s.

2) In the distance range from 55 to 115 km, a reversed branch, called here P_1^M can be recognized. The apparent velocities are ranging from 7.0 to 8.1 km/s. In some sections, this branch is split into two parts (P_1^M and P_2^M).

3) A wave group with clear arrivals appears beyond distances of 120 km.

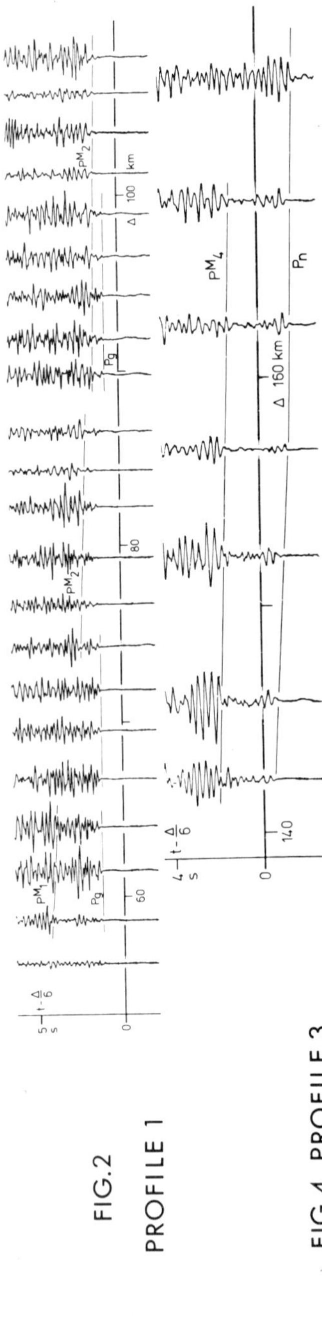

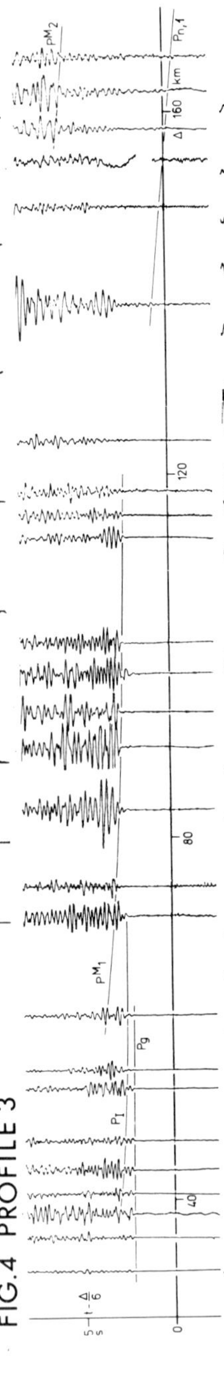

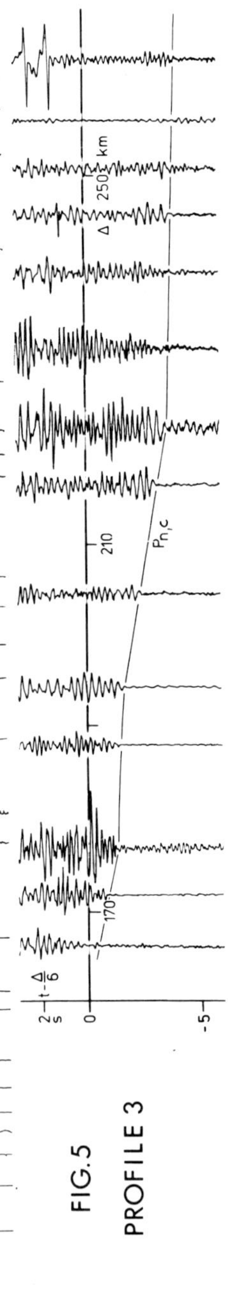

Fig. 2. Record section with reduced time axis from station 110 on line 1.

Fig. 4. Record section with reduced time axis of line 3. The records are obtained on Elba and in Tuscany from a shot fired in the Corsica-Elba channel.

Fig. 5. Record section with reduced time axis of line 3. The shot observed was fired W of Corsica.

The apparent velocities vary between 7 and 8 km/s. The true velocity is 7.4 km/s. Although this value is distinctly reduced in respect to the normal upper mantle value of 8.0 to 8.2 km/s, this branch is regarded as a P_n branch and thus, the corresponding boundary is seen as separating the crust from the upper mantle.
4) Behind the P_n-group, a further wave group with large amplitudes has been recorded. It is interpreted as reversed segment P_3^M showing apparent velocities between 7.3 and 8.2 km/s. In some cases, even a further group P_4^M, separated from the P_3^M, is visible in the record sections. It is practically parallel to the P_3^M branch.

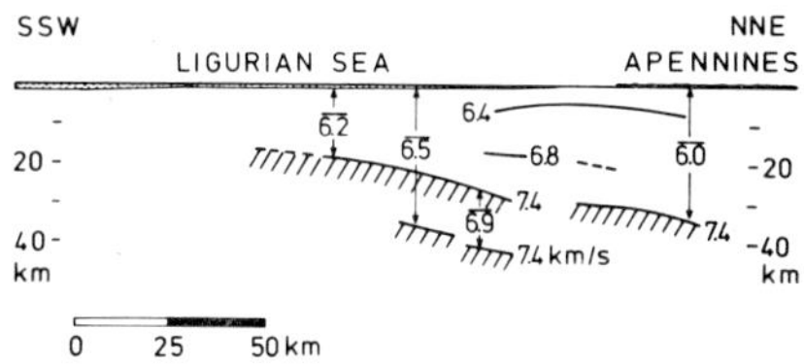

Fig. 3. Cross section along line 1, showing lines of equal velocity.

Fig. 3 shows the cross section derived from the wave groups described above. The top of the basement, associated with the velocity line 6.4 km/s, could be traced only in the central portion of the whole line owing to the given positions of shot- and recording points and the limited observation range of the P_g-wave. The depth of the basement varies between 4 and 6 km. A weak culmination could be detected 10 - 20 km offshore.

From the P_1^M wave, crustal thickness under the southern flank of the Northern Apennines is well established. The crust/mantle boundary, lying under the coast at a depth of about 30 km, dips northward down to 35 km and more. The uppermost mantle shows a distinctly reduced velocity of about 7.4 ± 0.2 km/s.

Under the Ligurian Sea, the crust/mantle boundary is evidenced and its depth determined by the P_n-wave with a reduced velocity of 7.4 km/s. Half-way between La Spezia and Cap Corse, crustal thickness is about 20 km, whereas 30 km were measured 30 km offshore La Spezia. The cross section of fig. 3 shows just offshore an interruption between both parts of the crust/mantle boundary. This gap is caused by the fact that, using the two groups P_1^M and P_n, the given positions of the shooting and observing lines do not allow an overlapping of both groups in respect to the boundary under discussion. Nevertheless, a displacement of the crust/mantle boundary in this part of the profile may exist, because just here, the thickness of the crust changes distinctly within some 10 km along the profile.

The P_3^M and P_4^M branches, parallel to the P_n branch, are interpreted as reflected waves. Using the z_{max}-relation one obtains a reflector depth of 36 to 43 km. The interval velocity between this deep reflector and the crust/mantle boundary amounts to 6.9 km/s. The significance of this outstanding reflector is understandable only within the pattern of the general structure of the region (Giese et al. this volume).

Line 3

The records of line 3 have been treated in the same way as those of line 1. From the seismograms of this line, over 50 record sections were compiled. In detail, the following wave groups can be recognized (fig. 4).

a) Shots in the channel Corsica - Elba
   1) P_g-wave, clear first arrivals up to a distance of about 50 km.
   2) P_1^M-wave, showing large amplitudes. The corresponding travel-time curve is of retrograde nature.
   3) $P_{n,1}$-wave with weak amplitudes observed between 140 and 200 km. It can be associated with teh P_1^M wave.
   4) P_2^M-wave recorded between 150-200 km with large amplitudes, showing a time delay of about 6-7s in respect to the P_n-wave.

b) Shots W of Corsica (fig. 5)
   1) $P_{n,c}$-wave, penetrating the upper mantle under Corsica and Elba.

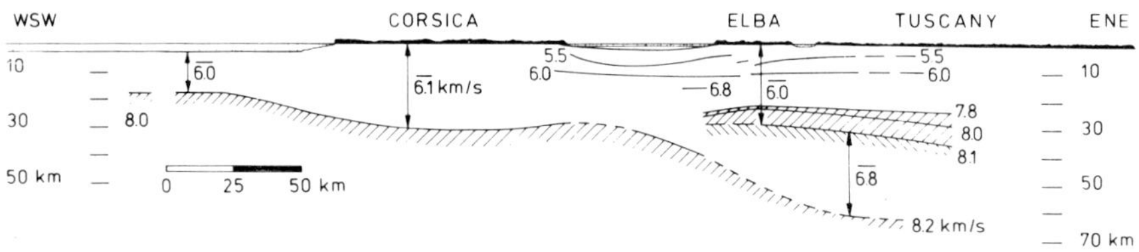

Fig. 6. Cross section along line 3, showing lines of equal velocity.

With these wave groups the crustal structure between the channel of Corsica-Elba and Tuscany can be explored, that is the key region containing the transition between the block of Corsica-Sardinia and the northern Apennines, beginning on Elba (fig. 6).

In accordance with seismic reflection measurements, far more than 5000 m postorogenic sediments must be expected between Corsica and Elba (Finetti & Morelli 1974).

Under Elba and the coastal region of Tuscany, the 6.0 km/s iso-velocity line has been determined at an average depth of 9-11 km. A clear identification and determination of the top of the basement is not possible here due to the very complex geological structure.

Special importance has to be attached to the P_1^M wave group which suggests the existence of a discontinuity at a depth of 20-26 km, where velocity increases to about 7.5 to 8 km/s, as proved by the $P_{n,1}$-wave. This discontinuity shows a weak culmination under Elba. Remarkable is the observation that the $P_{n,1}$-wave disappears at a distance of 200 km. Thus the discontinuity either breaks off towards NE or/and a velocity inversion follows below this boundary.

The wave described as P_2^M-wave may be interpreted either as multiple of the P_1^M-wave or as primary reflection of an interface at about 60 km depth. The position of the corresponding reflection element would be found under the coastal region of Tuscany.

The shots fired W of Corsica were observed successfully from 165 km on up to a distance of 370 km. In such a distance interval, only the P_n phase can be recorded. Apparent velocity along this travel-time curve

varies between 7 and 11 km/s, due to strong lateral inhomogeneities. To transform this travel-time curve into depth values, knowledge of the crustal structure of Corsica and that of the sea W of Corsica is needed. According to Hirn et al. (1976), crustal thickness - in agreement with other authors (Fahlquist & Hersey 1969; Finetti & Morelli 1973) - amounts to about 15 km oceanic crust, and it increases towards Corsica to 30 km. This depth has to be associated with th $P_{n,c}$-wave observed on Elba and the continent. By aid of the wave front method, the course of the corresponding interface has been constructed. The crust/mantle boundary shows a gentle updoming under the Corsica-Elba channel adjacent to a steep dip towards the NE under Elba. The already mentioned reflection P_2^M from 60 km depth can then be easily associated with this crust/mantle boundary plunging towards NE. Because similar very late reflections have been observed also on profile 1 and in Southern Italy (Schütte this volume), this wave is regarded as a primary one. Furthermore, this deeper crust/mantle boundary has no connection with the shallow crust/mantle boundary at a depth of 20-25 km under Elba described above.

In the region between these two discontinuities, both meeting the criteria of a crust/mantle boundary, interval velocity amounts to $\leq$ 6.8 km/s. It has to be mentioned that this is an average value and that in zones that can easily be several km thick, velocity may clearly be smaller than 6 km/s.

A gravity model derived by Carozzo (personal communication) from this crustal section produces a Bouguer anomaly which agrees with that observed to a satisfactory extent.

References

Fahlquist, D.A. & Hersey, J.B., 1969: Inst. Océnographique, Monaco, 67:1386.

Finetti, J. & Morelli, C., 1973: Boll. Geofis. teor. ed appl.,15:263-341.

- , 1974: Rend.del Sem. Fac.Scienze Univ.Cagliari.

Giese, P. & Morelli, C., 1975: Quaderni de "La ricerca scientifica" n. 90 CNR:453-489.

Hirn, A. & Sapin, M., 1976: Bull.Soc.géol.France, 5:1195-1199.

Letz, H., 1977: Dipl.-Arbeit, Freie Universität Berlin, 60 p.

Pavlenkova, N.I., 1973: Physics of the Solid Earth. - 8:544-550.

Reichert, C., 1977: Dipl.-Arbeit, Freie Universität Berlin, 55 p.

Wigger, P., 1977: Dipl.-Arbeit, Freie Universität Berlin, 69 p.

Review of the Crustal Structure of the Northern Apennines, the Ligurian Sea, and Corsica

P. Giese, C. Morelli and R. Nicolic

In this contribution the results obtained by seismic reflection and refraction measurements as well as by gravimetric studies have been compiled in order to obtain an overall picture of crustal and upper mantle structure of the Northern Apennines and their adjacent areas. The elaborated model may serve as a basis for further geotectonic considerations.

Geological development and crustal structure suggest the subdivision into the following regions:

- the Balearic Sea and the deep-sea part of the Ligurian Sea,
- the Corsica-Sardinia block,
- the border region between the European continental fragment Corsica and the Adria microplate,
- the Adria microplate,
- the border region between the Apennines and the Western Alps W of Genova.

The Balearic Sea

The structure of crust and upper mantle in the Balearic Sea and the deep-sea part of the Ligurian Sea has been investigated by Fahlquist & Hersey (1969), Finetti & Morelli (1973), and Hirn & Sapin (1976). This basin is of oceanic nature with sediments up to 7 km thick. While Fahlquist et al. (1969) determine the depth of the crust/mantle boundary at 12 km, Finetti et al. (1973) gave 15,8 km as crustal thickness. A reinterpretation of these wide-angle reflection data by Nicolic (in press) reveals an increase of velocity from 6 to 8 km/s in the depth range between 10 - 14 km, thus indicating a transition from the oceanic crust to the upper mantle.

The Corsica-Sardinia block

The block of Corsica and probably also the Variscan and Prevariscan part of Sardinia are of purely continental structure as evidenced by the crustal thickness of 30 km and the average crustal velocity of 6.1 km/s (Hirn et al. 1976). The transition from the continental crust of Corsica to the oceanic crust of the Balearic Sea takes place within a distance of 30-40 km, mainly by a thinning of the upper crust, the sialic part of the Corsican continental crust disappears practically completely towards W.

Fig. 1 shows a tentative isochrone contour map of the basement around northern Corsica. The mapped horizon does not necessarily represent the same geological formation at all points. This map demonstrates quite clearly the northern extension of Corsica up to the latitude of Gorgona. Here the Corsican crust seems to touch directly the crust of the Adria microplate.

On the E and NE side of Corsica there must be the contact to the crust of the Adria microplate which shows surely a complicated structure.

Therefore, before this contact zone is discussed, the crustal structure of the Adria microplate is described in the following.

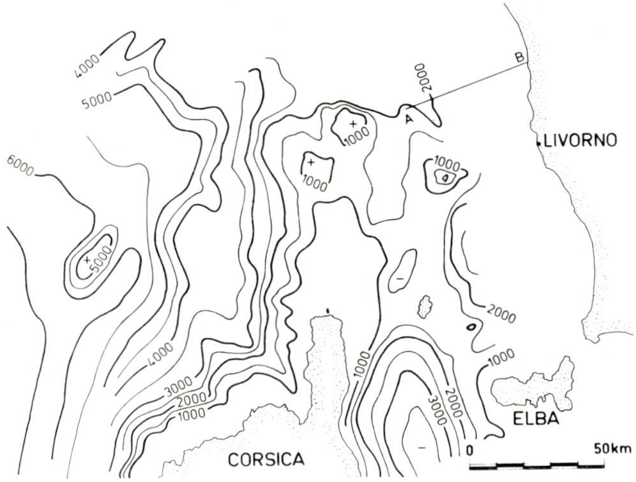

Fig. 1. Contour map showing the two-way traveltime between the surface and the crystalline basement (contour interval 500 ms) (Nicolic in press)

The Adria Microplate

The channel between Corsica nad Elba shows a typical graben structure. The total thickness of the sediments, being of Pliocene and Miocene age, is about 8000 m. Within the graben, there is no indication of compressional structures, only normal faults and flexures can be recognized. The nature of the acoustic basement is open. N of Gorgona this Graben disappears.

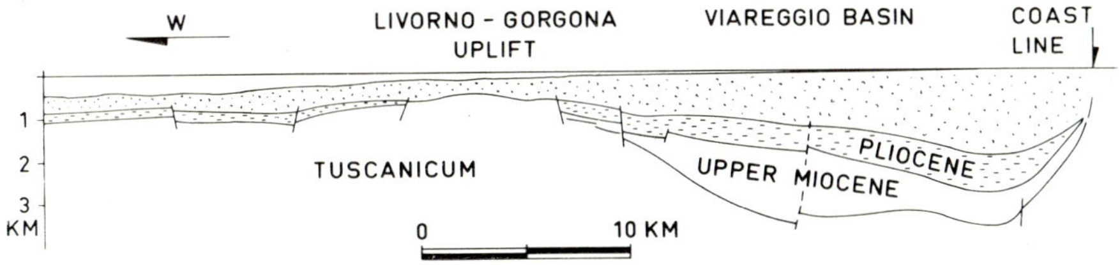

Fig. 2. Cross section obtained from seismic reflection measurements through the Viareggio basin. The position of this profile is shown in fig. 1 by the line A - B (Nicolic in press).

The uplift of Elba with its granitic batholit extends northward to the Livorno-Gorgona uplift (fig. 2). Between this uplift and the coast the Viareggio basin is intercalated, filled with 3000 m sediments of Quaternary and Messinian age. In a smoothed form this uplift continues northwards and enters the Northern Apennines E of Genova.

The structure of the deeper crust is revealed by seismic refraction measurements (Morelli et al. 1977; Letz et al. this volume). Under the Ligurian Sea and the island of Elba, a discontinuity at a depth between 15 and 25 km could be clearly detected. It is characterized by a velocity between 7.4 and 8.0 km/s. Although these values are somewhat smaller than normal, this boundary is regarded as crust/mantle boundary of the SW margin of the Adria microplate. From offshore towards Liguria and western Tuscany the crust/mantle boundary dips strongly towards the Apennines' crest, reaching here a depth of about 35 km. The question whether this represents the greatest crustal thickness in the Northern Apennines cannot be answered up to the present. Under the Po-Plain a depth of the crust/mantle boundary of about 35 km can be expected (Giese 1972; Giese & Morelli 1975). Supposing that the basement of the Po-Plain extends without great displacements as far as under the main crest of the Apennines, crustal thicknesses larger than 40 km can hardly be expected in the Northern Apennines.

In Tuscany the crust/mantle boundary could be detected only as far as under the coastal region. Its further course towards E is still open because the upper mantle wave disappeared here.

Special structures within the Apenninic crust/mantle boundary are possible without, however, having been proved by observations from, after all, rather widely spaced seismic refraction stations. The transition from the thin crust of the Ligurian Sea to the thicker one under the Apennines may by all means occur along deep-reaching fault zones. These flexures or fault zones may become more intensive in southward direction.

Border between the Corsica Block and the Adria Microplate

Which are the relations between the Corsican and the Apenninic crusts? The reply to this question is given by the results of the seismic refraction line Corsica-Elba. It has been outlined by Morelli et al. (1977) and Letz et al.(this volume) that a deep discontinuity exists under Elba, steeply downdipping towards NE which seems to be linked up with the crust/mantle boundary under Corsica. The question whether this deep-seated crust/mantle boundary represents the direct continuation of the Corsican crust/mantle boundary or whether, by tectonic processes, crustal portions were welded on to the Corsica block at that depth cannot be answered here. In the following, however, this deep boundary under Elba and under the Ligurian Sea as well is considered to be related to the Corsica block. On the other hand, the shallow Apenninic crust/mantle boundary disappears on the W side of Elba so that the western border of the Adria microplate has to be expected between Elba and Corsica, as is already known from geological considerations. Under the shelf and coastal region of the Ligurian Sea, at a depth of 35 to 50 km, the existence of a discontinuity could be proved, at least in sections, which is regarded as belonging to a deeper second crust/mantle boundary connected with the Corsican crust/mantle boundary.

The western and southwestern margins respectively of the Adria microplate must be assumed W of Elba, running northwards, passing half-way between Cap Corse and La Spezia, and entering the Apennines near Genova. The islands of Pianosa and Capraia may be situated on this border line.

In order to check, by aid of gravity measurements, the pattern of

crustal structure as obtained by seismic data, a number of Bouguer profiles has been computed for the area under discussion (Carozzo & Nicolic in press). It can be stated in general that the structure derived from the seismic records is in agreement with the gravimetric data. In some cases minor modifications of the seismic results must be made. One of the most important improvements by gravity data is the statement that, under the Corsica-Elba channel, an uplift of the upper mantle must exist, having an amplitude of about 5 km. Due to the arrangement of the seismic measurements on this profile, this uplift cannot be detected by refraction seismics.

Western Alps - Northern Apennines Border

Western Alps and Northern Apennines developed from the same Penninic-Ligurian eugeosyncline. Near Genova the Adria microplate changes its behaviour in respect to this eugeosyncline; this plate is backland for the Western Alps, whereas in the SW-corner of the Po-Plain it begins to act as foreland for the Apennines. What does the crustal structure look like in this complicated zone?

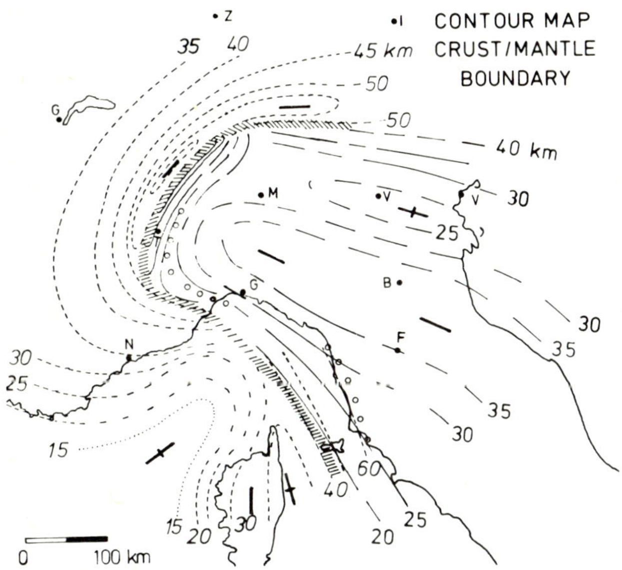

Fig. 3. Contour map showing the depth of the crust/mantle boundary.
Adria microplate: ———— evidenced by seismic refraction data
—— —— — extra/interpolated from gravity data
European plate: ---------- evidenced by seismic refraction data
- - - - - extra/interpolated from gravity data
Ligurian Sea .... oceanic crust/mantle boundary
////////// western border of the Adria microplate
oooooooooo as far as eastern extension of the European/Corsica crust goes at least.
N = Nice, G = Genova, T = Torino, M = Milano, V = Verona, V = Venezia
B = Bologna, F = Firenze, G = Genève, Z = Zürich, I = Innsbruck.

Giese et al. (1971) studied the western border of the Ivrea body and the eastern extension of the European crust/mantle boundary under the Po-Plain. By a reinterpretation of the profile Lac Nègre - Genova (Roewer et al. unpubl. report), this study can be completed.

There is not only an overlapping crustal structure in the Ivrea zone, but also betwen Cuneo and Savona. The European crust/mantle boundary extends at least to W of Torino and up to Savona, reaching here a depth of 45 - 50 km. On the other side, the western and southwestern margins of the Adria microplate are hidden under units of the Western Alps. As near as between S of Ivrea and Cuneo, the gravity high of Ivrea is covered by formations of the Western Alps. S of Cuneo the crust/mantle boundary of the Adria microplate could be detected 10 km NE of the Massif Mercantour, having here a depth of 25 km.

Summarizing these results, the following can be stated: In the southwestern part of the Po-Plain, the Adria microplate has been overthrust onto the European continent but, in the same region, it has itself been overthrust by the Apenninic units. By backward folding the marginal zones of this microplate are covered by Western Alpine units.

Fig. 3 comprises the results previously discussed as contour map showing the depth of the crust/mantle boundary of the Adria microplate as well as that of the European continent. The western border of the Adria microplate is indicated by a hatched zone.

References

Carozzo, M.T. & Nicolic, R., 1978: Boll. Geofis. teor. ed appl., in press.

Fahlquist, D.A. & Hersey, J.B., 1969: Bull.Inst.Océanographique, Monaco, 67:1386.

Finetti, J. & Morelli, C., 1973: Boll. Geofis. teor. ed appl. 15:263-341.

Giese, P., 1972: Unternehmen Erdmantel, Franz Steiner Verlag:75-100.

Giese, P., Morelli, C., Prodehl, C. & Vecchia, O., 1971: Proc. 12th Gen. Ass. Europ. Seismol. Comm. (Luxembourg 1970), Obs. Royal de Belgique, Comm. Sér. A-N° 13, Sér. Geophys. 101:182-183.

Giese, P. & Morelli, C., 1975: Quaderni de "La ricerca scientifica" n.90. CNR:453-489.

Hirn, A. & Sapin, M., 1976: Bull. Soc. géol. France, no. 5:1195-1199.

Morelli, C., Giese, P., Hirn, A., Colombi, B., Eva, C., Guerra, I., Letz, H., Nicolic, R., Reichert, C. & Wigger, P., 1977: International Symposium on the Structural History of the Mediterranean Basins. Split (Yugoslavia), Edition Technip, Paris:281-286.

Nicolic, R., 1978: Boll. Geofis. teor. ed. appl., in press.

Roewer, P. & Giese, P., 1977: The profile Lac Nègre - Northern Apennines, unpubl. report.

A Microearthquake Study along the Sestri - Voltaggio - Line

M. Steinwachs

This paper presents evidence to suggest that the Sestri-Voltaggio-Line may not be an active fault at present although it is located within a tectonically active region.

Seismological Situation of the Study Area

The distribution of epicentres (Bossolasco et al. 1970) from recorded seismograms and seismological maps (Giorgetti 1973) reconstructed from historical data shows a comparatively low frequency of epicentres in the region between the Western Alps and the Northern Apennines. A correlation with the course of the Sestri-Voltaggio-Line is not possible.

There are two permanent seismological recording stations in the vicinity of the Sestri-Voltaggio-Line. One of them is on the Monte Capellino, approximately 10 km east of the Sestri-Voltaggio-Line and the other one in Genova. Both stations work on comparatively low sensitivity and are not well suited for microearthquake investigations.

Table 1

Station	Longitude (Greenwich)	Latitude
1	8° 47' 47" E	44° 36' 47" N
2	8° 52' 13" E	44° 33' 53" N
3	8° 55' 03" E	44° 37' 10" N

Epicentres located

Date	Focal time	Longitude (Greenwich)	Latitude
a) Jul.4.73	00:02:03,15	8° 41' 36" E	44° 32' 30" N
b) Jul.4.73	12:49:15,17	8° 46' 00" E	44° 25' 30" N
c) Jul.5.73	07:07:12,36	8° 07' 24" E	44° 07' 54" N
d) Jul.7.73	11:58:40,39	8° 31' 25" E	44° 20' 36" N
e) Jul.7.73	09:46:40,28	8° 36' 24" E	44° 29' 00" N

Seismological Recordings and Results

During the period of 25 June 1973 to 19 July 1973 recordings were made in the Voltaggio area, using three mobile high-sensitivity seismological stations. The stations operated with radiotelemetric data transmission and magnetic tape recordings. Stations 1 and 2 each used a vertical seismometer whereas at station 3 a vertical seismometer and 2 horizontal seismometers were used. Apart from many quarry blastings, 42 microearth-

quakes were recorded, whose difference in travel time from the P- and S-wave at station 3 was between 3 and 14 seconds. Five of these micro-earthquakes showed onsets good enough to determine the epicentres using the spherical section procedure. Table 1 gives the data of the station points and the located epicentres.

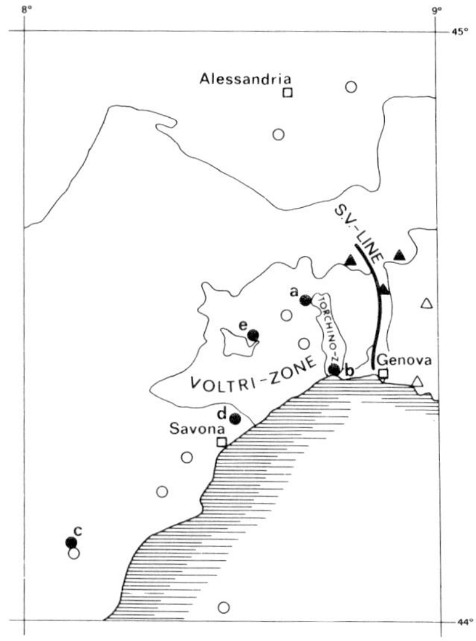

Fig. 1. Map of epicentres. Full dots: epicentres calculated from microearthquakes; circles: epicentres of earthquakes after Bossolasco et al. (1970); full triangles:sites of the mobile stations; triangles: sites of the permanent stations.

Fig. 1 shows the distribution of the calculated epicentres. In addition, the epicentres of the earthquakes observed and located in the investigation area during the period 1967 - 1970 (Bossolasco et al. 1970) have been inserted on the map. The map shows that there are no epicentres immediately on the Sestri-Voltaggio-Line, whereas three of the five microearthquakes which were located are in and border on the Torchino Zone which is situated west of the Sestri-Voltaggio-Line. After Bussolasco et al. (1970), two epicentres of earthquakes were located in this area between 1967 and 1970.

Discussion

The results presented should be regarded as the beginning of such investigations in the Northern Apennine region. Due to the relatively short duration of recordings, only 3 1/2 weeks, a comprehensive statement cannot be made as yet.

An epicentre pattern along the Sestri-Voltaggio-Line could not, however, be determined. The result appears to indicate that the Sestri-Voltaggio-Line is not a permanently active fault. Nevertheless, the results clearly show that there is microearthquake activity in the Voltri Zone west of the Sestri-Voltaggio-Line. If one takes into consideration that of a total of 42 recorded events, only 5 were located, it can then be assumed that in the Voltri Zone, there are still a considerable number in addition to the 3 epicentres located (fig. 1). From this it can be concluded that in the western marginal zone of the Western Alps recent tectonic movements occur.

References

Bossolasco, M. & Eva, C., 1970: XIIe Assemble Generale de la Commission Seismologique Europeene, Luxembourg:1-7.
Giogetti, F. & Jaccarino, E., 1973: Carta degli epicentri dei terremoti della anno 0 al 1893. - Comitatio Nationale Energia Nucleare.
Görler, K. & Ibbeken, H., 1964: Geol. Rundschau, 53,1:73-84.

Results of Magnetotelluric Measurements in the South-Western Po Plain, Italy

W. Losecke and I. Scheelke, with a Contribution of H. Closs

In 1972 magnetotelluric measurements (MT) were carried out in the Monferrato hills and in the Molasse areas north (Po-plain) and south of it (area around Asti). A discontinuity was expected there which should separate the thick sedimentary layers of the Po basin from the adjacent hills which are the northwestern continuation of the Pedeapennine (Ogniben et al. 1973).

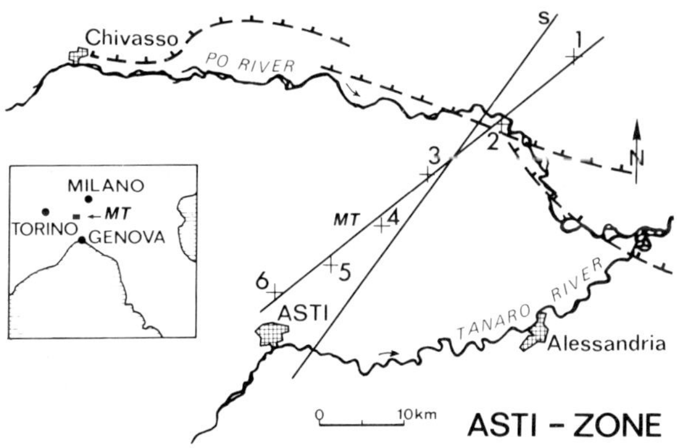

Fig. 1. Index map of the magnetotelluric measurements (MT). Numbers indicate MT stations; S = part of a deep seismic sounding profile (see Stein et al. this vol.).

The sedimentary layers of the Po basin were expected to have low resistivity and, therefore, probably could be separated from a basement. Fig. 1 shows the location of the stations and of a refraction profile (Stein et al. this vol.). The MT measurements were carried out at 6 stations with a spacing of 8-10 km. Station 1 (Zeme) is situated in the Po basin, station 2 (Frassincto) in the area of the expected geological discontinuity, station 3 and 4 (Casale and Fons Satera) in an area where older material was expected to be overthrusted over younger sediments. The sta-

tions 5 (Montemagno) and 6 (Asti) have a more internal position in respect to the northern front of the Apennine.

One and two-dimensional calculations for tensor rotated apparent resistivities were carried out (Vozoff 1972). As a result one model is given in fig. 2. Because of the small number of MT stations and the wide

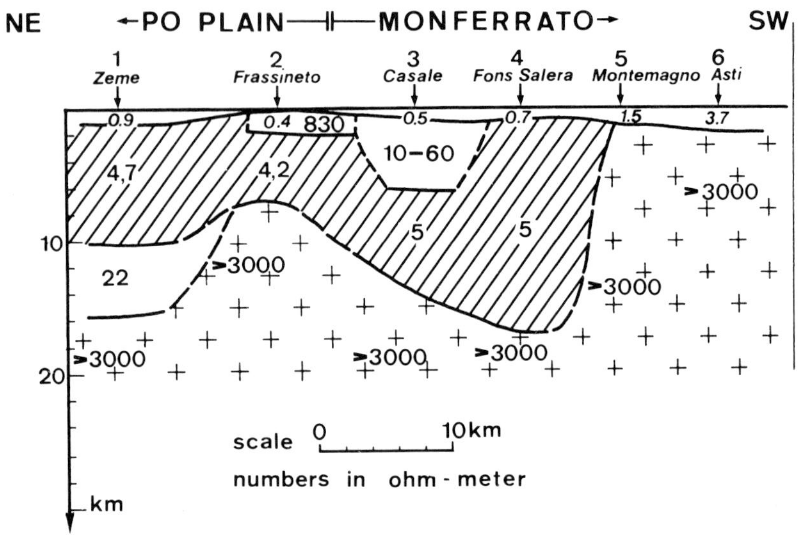

Fig. 2. Resistivity cross section derived from magnetotelluric measurements.

spacing, the model can only be a very rough one, but we are convinced that the main structure of the underground down to about 20 km can be detected.

In the Po-basin low resistivity layers occur down to an approximate depth of 10 km, underlain by an intermediate layer of slightly higher resistivity, in turn underlain by a highly resistive MT basement. The intermediate layer likely represents sedimentary rocks with their base at about 16 $\pm$ 2 km of depth. The occasionally very low resistivities (below 1 Ohm·m) of the near surface layer are probably caused by a high Ion-content of the water in the pores of the youngest sediments. At stations 2 and 3 the low resistivity layer (4-5 Ohm·m) is sloping downwards to the south under a more resistive partly thick stack of sediments. Under station 4 it comes again near to the surface and its thickness reaches here about 16 km. From station 2 to station 4 it is in direct contact to the basement. At stations 5 and 6 the highly resistive MT basement is in the model rising up and comes near to the surface.

It should be emphasized that the presented model is only one possible model to explain the measured data - it is, however, the model which we prefer after having calculated a number of other configurations of layers with somewhat different resistivities.

We interpret our model as an indication of a remarkable overthrust of the Apennines over the foreland of the Po-basin, and we come qualitatively

to the same conclusion as the interpretation of the deep seismic sounding (Stein et al. this vol.). Furthermore, it seems to us that the detection of a high resistivity layer in the region of Asti ($>$ 3000 Ohm·m) in a depth between 2 and 3 km from the surface is also an important result. It excludes that a thick stack of Apennine nappes exists in this region. We assume that in this relatively small depth the crystalline basement begins. There is no low resistivity layer in this crystalline of the upper crust which could be interpreted as an indication of a nappe structure with intercalations of sedimentary sequences of a thickness comparable with the depth.

Apennine nappe material occurs below Frassineto (station 2) and Casale (station 3). Under Frassineto the thickness of the sedimentary nappes is in the order of 2 km, whereas in the area of Casale a nappe stack of about 6 km thickness and a width up to 10 km is compatible with the measured data.

It may be mentioned that fig. 2 is not exaggerated. The southern border of the intermediate layer (22 Ohm·m) seems to be very steeply dipping, also the southern border of the low resistivity (5 Ohm·m) layer between Fons Salera and Montemagno. It is a matter of geotectonic speculation to prefer faultzones or frontparts of crystalline nappes thrusted in NE-direction as interpretation of these strongly inclined resistivity boundaries. Some shallow earthquakes in this region demonstrate neotectonic processes.

For further geotectonic interpretation see Günther & Reutter this vol.

References

AGIP-Mineraria 1959: I giacimenti dell'Europa occidentale. Atti Conv. Milano 1957, 2, Acc. naz. Lincei, Roma.

Ogniben, L., ed., 1975: Structural model of Italy, Cons. Naz. Ric.

Scheelke, I. & Losecke, W., 1974: Kolloquium erdmagnetische Tiefensondierung, Grafrath, unpubl. report.

Part 2

Geological Contributions

Structure and Geotectonic Evolution of Northeastern Corsica

G. Seidl and W. Zacher

The island of Corsica consists of the western foreland of an Alpine orogenic belt of which small remnants are exposed in the east and northeast parts of the island. Accordingly, the largest part of Corsica is underlain by Hercynian granites of Mississippian to Permo-Triassic age. Small volumes of tectonized roof pendants consist of high-grade metamorphic rocks of unknown age and of nonmetamorphic sedimentary rocks of lower Ordovician (?), lower Silurian, late Devonian and Carboniferous age. These roof pendants suggest a Caledonian or older and a pre-Pennsylvanian orogeny, unconformably followed by Pennsylvanian, Permian, or Tertiary sedimentation. Triassic, Jurassic, and Cretaceous sediments form part of the parautochthonous series near Ponte Leccia and Corte. Eocene conglomerates, graywackes, sandstones and limestones overlap the crystalline basement toward the west. Alpine-age westward thrusting and regional metamorphism affected the parautochthonous rocks as well as the marginal parts of the Hercynian granite complex in the Tenda massif.

The following is a brief listing of the recognized tectonic units and their stratigraphic-lithologic content.

The Geology of Northeast Corsica

The *Parautochthonous* unit is an essential part of the Schuppen zone at the east margin of the crystalline basement between Ponte Leccia and Corte. Above an involved strongly sheared basement, it contains Permian rhyolites and arcosic conglomerates, Triassic and Jurassic carbonates, and Eocene unconformable clastics and carbonates. Near Corte this series contains low-grade metamorphism.

The *Caporalino-St. Angelo nappe* consists of Jurassic to Eocene clastics and the Caporalino limestone (Malm).

The *St. Lucia-Padule nappe* involves pre-Permian granites and older metamorphites, overlain by a Permian to Upper Cretaceous (?) clastics grading into flysch.

The *Cima Pedani Klippe* has a Triassic to Jurassic sequence similar to, and is tectonically part of, the Parautochthonous unit. This klippe must have reached its present emplacement, in part above Schistes Lustrés, by polyphase thrusting.

The Schistes Lustrés form a thick pelitic, eugeosynclinal complex of Jurassic and Cretaceous, possibly Triassic, age with intercalations of ophiolitic rocks and epizonal metamorphism, deposited possibly in part on oceanic crust but in places also on Permian metavulcanites or Triassic (?) quartzites and dolomites (Delcey & Meunier 1966). Besides extensive

nappe transport, this unit shows at least three generations of folding in the north part (Zacher 1975), but only one folding phase to the south.

The *Ophiolite Unit* forms exposures of the typical ophiolite suite near Ponte Leccia and in the Balagne, comprising serpentinites at the base, followed by spilites, pillow lavas, and radiolarites. The magmatic rocks, of assumed Jurassic crustal age and of similar composition as the rocks of the Mid-Atlantic ridge, were metamorphosed successively to greenschist grade, to a high-pressure facies, and again to greenschist grade (Ohnenstetter et al. 1976). Ophiolites are found in two trends of different tectonic position. Low-grade metamorphosed bodies occur on top of the Schistes Lustrés and are overlain by the *Schistes de Aiti* (Balagne, Ponte Leccia, Inzecca). A second thick, more highly metamorphosed complex with sheared metagabbros, greenschists with abundant glaucophane and partly garnet, and more complex and stronger folding, but only few pillow lavas and cherts, occurs within the Schistes Lustrés.

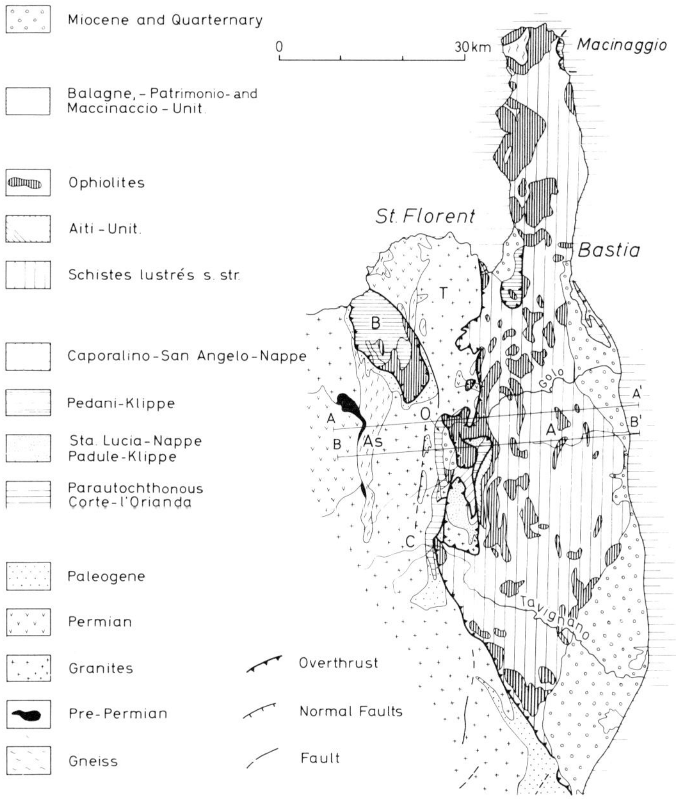

Fig. 1. Tectonic sketch map of NE-Corsica. As = Asco; B = Balagne; C = Corte; O = Crête de l'Orianda; T = Tenda; A = Monte San Angelo/Casinca.

The *Aiti Schists* above the ophiolites and Schistes Lustrés, are, toward their base, lithologically similar to the Schistes Lustrés. Stratigraphically higher in their section, however, they involve Jurassic or

Cretaceous conglomerates, flysch-like sediments, phyllites, limestones, cherts, and graywackes. They are unconformably overlain by Eocene. The nature of the contact between the *Aiti Schists* (above) and the ophiolite-related radiolarites (below) may be sedimentary or tectonic.

The Alpine Orogeny of Corsica

The following chronological account of geotectonic events is suggested by the known stratigraphic column. Other workers, however, have extensively applied plate tectonics and subduction theory. Alvarez (1972), Westphal et al. (1976) and others have demonstrated the rotation of Corsica by 30° and Sardinia by 60° with tectonic and paleomagnetic methods. High pressure metamorphism in the Schistes Lustrés suggest a mechanism of an extinct subduction zone.

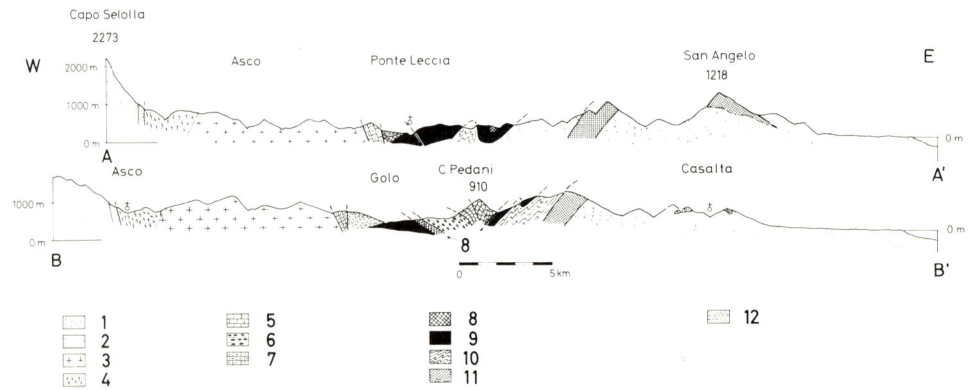

Fig. 2. Schematic cross sections through the middle part of NE-Corsica. (Location see fig. 1). 1 = Permian; 2 = Upper Carboniferous; 3 = Granite; 4 = Gneiss; 5 = Pedani Klippe; 6 = Padule Klippe; 7 = Parautochthonous of Crête de l'Orianda; 8 = Aiti Schists; 9 = Ophiolites, weekly metamorphosed 10 = Schistes Lustrés nappe; 11 = Ophiolites, metamorphosed; 12 = Miocene.

Permian: Rhyolitic vulcanism above granitic plutonism affected and covered a truncated post-Variscan peneplain, with porphyrites ignimbrites, tuffes, conglomerates, and arcoses.

Triassic and Liassic: Subsidence and transgression created a shelf series to the west and the Penninic ocean basin to the east.

Dogger: Within the Penninic basin, a mid-oceanic ridge with basic vulcanism formed while parts of the shelf were truncated down to the basement.

Malm: The ophiolite vulcanism ended, the ridge submerged beneath radiolarian cherts, schistes, and Calpionella limestones. Shallow-water carbonates and crystalline conglomerates were deposited on the shelf.

Lower Cretaceous: Ophiolite complexes slid gravitationally westward over the subsiding Schistes Lustrés complex, with post-nappe deposition of flysch. Renewed truncation on the shelf created more unconformable conglomerates.

Upper Cretaceous: The Schistes Lustrés and the autochthonous parts of the ophiolites were buried to great depth, affected by high/low t metamorphism and strong folding. The allochthonous ophiolites and the overlying flysch were also folded but metamorphosed to a lesser degree.

? *Paleocene - Eocene:* By subduction of the remaining oceanic crust the Schistes Lustrés were pressed against the craton and imbricated its sedimentary cover. In a later phase parts of this imbrication slid over the Schistes Lustrés.

? *Upper Eocene - Oligocene:* Isostatic adjustments elevated parts of the area, most strongly the Schistes Lustrés complex. Parts of the unit and their overlying sediments slid gravitationally to the west and overthrusted the Eocene sediments lately deposited there.

Miocene: Thrusting and folding ceased, and Molasse transgressed unconformably the Alpine structures.

References

Alvarez, W., 1972: Nat. Phys. Sci. - 235/58:103-105.

Baudelot, S., Doubinger, J., Durand-Delga, M. & Vellutini, P., 1976: Bull. Soc. géol. France, (7) XVIII, 5:1221-1228.

Delcey, R. & Meunier, A., 1966: Bull. Serv.Carte Geol.France,61,Nr.278.

Denizot, G., 1952: Rev. Scient. - 90(2), Nr. 3316.

Durand-Delga, M., 1975: In: Géologie de la France, Bd.II:465-478.

Fallot, P.M., 1958: Annuaire du Collège de France, 58e année.

Nardi, R., 1968: Mem.Soc.Geol. Ital., 7:323-344.

Ohnenstetter, M., Ohnenstetter, D. & Rocci, G., 1976: B.S.G.F., 7, t. XVIII, 1:115-134.

Prösl, K.H., 1973: Unpubl. Manuscr., TU München.

Staub, R., 1928: Naturf. Ges. Zürich, 73:298-348.

Westphal, M., Orsini, J. & Vellutini, P., 1976: Tectonophysics,30:141-157.

Zacher, W., 1975:Rapp.Comm.int.Mer Médit. - 23,4a:71-72.

The Pile of Nappes in the Northern Apennines, its Unravelment and Emplacement

K.-J. Reutter and J. Groscurth

Introduction

In the Northern Apennines essential problems of tectonics, stratigraphy and dating of geological events have been resolved (see Sestini 1970; Elter 1973; Dallan Nardi & Nardi 1975). There is no question that a pile of nappes consisting mostly of eugeosynclinal (oceanic) Mesozoic and Cenozoic sedimentary rocks is overlying miogeosynclinal (deeply submerged continental platform) deposits of the same age. It is also no matter of

discussion that an important mechanism of nappe transport was gravitational gliding and that nappe movement started in the Lower Miocene and was completed in the Upper Miocene ("Tuscan phase").

Ambiguities are remaining concerning the individuality of nappes, their mutual tectonic relationships and, especially, their palaeogeographic arrangement. These questions are rendered difficult not only by the complicated Apenninic nappe movement pattern but, above all, because some of the eugeosynclinal units had already participated in earlier (Upper Eocene) alpidic nappe tectonics with different transport directions ("Ligurian phase").

Iaccarino et al. (1974) showed that the scattered slabs of the eugeosynclinal sequences cannot be assembled to a satisfying palaeogeographic picture by means of stratigraphic and sedimentologic comparisons. In the present paper more significance is attributed to the stacking order of the nappes than to lithological similarities. If it is presumed that the actual superposition of nappes was esentially acquired during the Apenninic orogeny (Lower Miocene to Upper Miocene) and not already before it, the palaeogeographic arrangement, which results from an unravelment of the nappes according to this stacking order, is not worse than those resulting from other considerations. For the emplacement of such unravelled nappes, models of modern active plate margins can be used.

The Pile of Nappes

Autochthonous Units: The Mesozoic and early Cenozoic sequences of the Po Plain, the Umbro-Marches and Tuscany are very similar to those of the Southern Alps. They can be interpreted as sediments on a subsiding carbonate platform which from Middle Jurassic to Tertiary times was rather deeply submerged.

During Apenninic orogeny, the southwestern border of this platform was tectonically deformed. Only the Po Plain, the Umbro-Marches and corresponding sequences, which underly the allochthonous materials of the northeastern flank of the Apennines, can be considered to be really autochthonous, although the sedimentary cover is slightly folded and faulted as a consequence of its detachment at the Upper Triassic evaporites. The basement, however, is supposed to be rather rigid and to extend from the Po Plain far beneath the Apennines.

In the so-called Tuscan Autochthon, situated in the southwestern (internal) parts of the Northern Apennines, the sedimentary cover and its palaeozoic basement are low-grade metamorphic and strongly folded. The contact between the rigid external and folded internal basement is hidden beneath allochthonous sediments (fig. 1). There are palaeogeothermal (Reutter et al. this volume) and palaeomagnetic (reinterpretation of Van den Berg & Wonders 1976) indications that the mobile internal parts of the platform an their sediments are thrusted upon the external rigid basement complex.

Allochthonous Units: The autochthon is overlain by several nappes which are entirely composed of sedimentary rocks except for some ophiolites. The sediments are all Mesozoic or Cenozoic in age but they considerably differ in facies, thickness and stratigraphic range. The sequence of nappes is most complete in the internal parts of the Apennines (fig. 2).

Tuscan Nappe: The lowermost nappe, which directly or with intermediate parautochthonous slices overlies the "Tuscan Autochthon", is a non metamorphic sequence of Triassic to Lower Miocene sediments. It is derived from the most internal part of the carbonate platform, where it was detached at an evaporitic Upper Triassic member during the orogenic compression. The front of the nappe is considered to be the huge overturned M. Orsaro - M. Cetona anticline in the middle of the Northern Apennines.

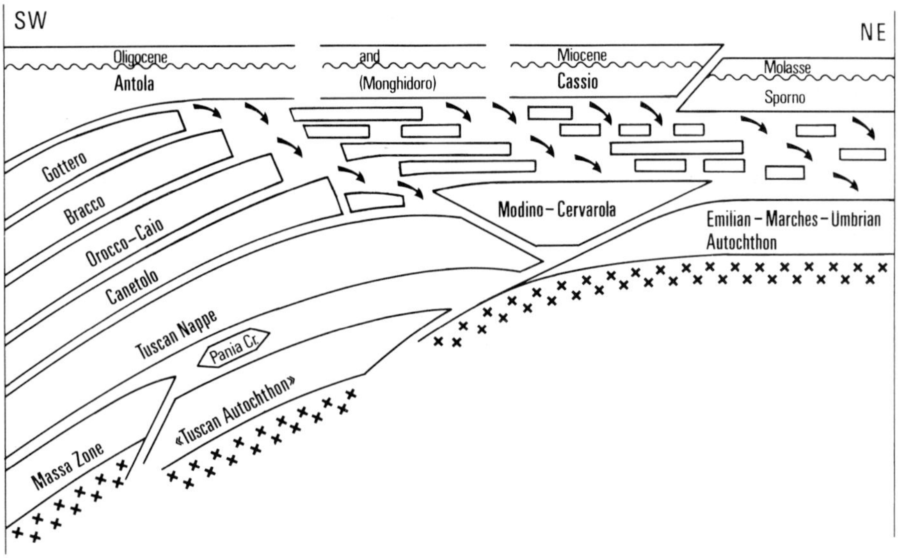

Fig. 1. Schematic illustration of the stack of nappes in the Northern Apennines.

Ligurid Nappes: The allochthonous terranes, which overlie the Tuscan and Umbro-Marchean sediments, considerably differ from these in facies and thicknesses. As some of them bear ophiolites, they must be considered to be of oceanic origin (Zanzucchi this vol.).

Canetolo Nappe: The Macigno graywacke flysch of the Tuscan Nappe is overlain by mostly disordered ophiolite-free flysch formations consisting of calcareous microbreccias, detritical limestones, marls, shales and graywackes of Paleocene to Eocene age and of Oligocene (Lower Miocene?) andesitic tufts and conglomerates. Since in the uppermost parts of the Macigno the same material forms olistostromes, gravitational gliding results to be the transport mechanism of the nappe. Possibly this lowermost Ligurid nappe represents the Tertiary stratigraphic sequence sheared-off from the mostly Cretaceous Orocco-Caio unit (Zanzucchi this vol.).

Fig. 2 (foldout). Areal distribution of the tectonic units in the Northern Apennines.

Orocco-Caio Nappe: The following nappe on top of the Canetolo unit is composed of a dark calcareous Helminthoidea flysch (Santonian-Paleocene) of deep water origin (Scholle 1972), and a basal chaotic complex made up of some arenacous flysch, but mainly of olistostromes and reworked masses of ophiolites and hercynian granitoids. It reflects strong synsedimentary tectonic interaction between oceanic and continental crust during Cenomanian and Turonian times (Elter & Raggi 1965).

The exact delimitation of the Orocco-Caio nappe is difficult because similar sediments of transitional aspect are also found in the overlying Ligurid nappes, especially in the uppermost one. But wherever the Orocco-Caio unit appears in its defined tectonical position it is not overlain by Oligocene or younger sediments thus allowing its classification as an independent nappe.

Bracco Nappe: North of La Spezia the Orocco-Caio nappe disappears towards the west under a sequence composed of an ophiolitic basement, cherts, siliceous limestones and shales (Pagani et al. 1972). Decandia & Elter (1972) showed that this Upper Jurassic to Lower Cretaceous sedimentary sequence of oceanic origin, together with a relatively thin veneer of oceanic basement (ophiolites, Galli et al. 1972), sheared off and was deformed to a stack of large recumbent isoclinal folds. Corresponding to their Apenninic transport direction, this folding probably is younger than the Upper Eocene (Ligurian Phase).

Gottero Nappe: The next higher Gottero unit is separated by tectonic contacts frcm the Bracco unit, but according to its facies and stratigraphic extent ranging from the lower Cretaceous up to the Paleocene, it may represent the younger deposits of the Bracco unit. Its characteristic formation is the Gottero-Graywacke-Flysch of Cenomanian-Turonian (Paleocene?) age, with tectonized palaeocene sediments at its top.

Antola - Cassio Nappe: The uppermost nappe of the Northern Apennines, which here is called Antola-Cassio nappe, is composed of different subunits (Antola, Monghidoro, Cassio and Sporno complexes; Zanzucchi, this volume). It was tectonically assembled in the course of the late Eocene "Ligurian phase", which involved these subunits, except for the Sporno complex, in folding and thrusting with Alpine, i.e. westerly, directions. An unconforming sedimentary cover of molasse type rocks (e.g. Ranzano sandstone) of Upper Eocene to Upper Miocene age is common to all of these subunits and connects them.

The Antola complex comprises an Upper Cretaceous to Paleocene sequence with a thick Helminthoidea flysch of deep water origin (Scholle 1972). The Cassio sequence is similar, but its basal complex contains pebbles and blocks of continental, probably insubric origin (Sames 1970; Elter et al. 1966). It is noteworthy that the Cassio complex, with the exception of its Tertiary molasse sediments, is also found in an allochthonous position in the Western Alps (Simme flysch, Elter et al. 1966). The Monghidoro Maestrichtian to Paleocene sequence is dominated by clastics. The marly Sporno flysch (Paleocene to Middle Eocene) has a Cretaceous base which resembles that of the Orocco-Caio nappe. Southeast of the transverse Prato - Sillaro line, the Antola-Cassio unit essentially consists of the calcareous Morello flysch (possible equivalent of the Sporno flysch), its Upper Cretaceous base and syntectonic M. Miocene sediments.

North of Genoa the hetrogeneous Antola-Cassio nappe is firmly connected by the Upper Eocene to Miocene molasse sediments to the Sestri-Voltaggio, the Piemont and the Briançonnais zones of the Western Alps. During the Apenninic orogeny (Oligocene to Upper Miocene) the Western Alps acted as a hinterland while the Antola-Cassio unit at their rim was thrust over the lower Apenninic nappes as far as the external zones. Thus it overlies in the SW the Gottero nappe, in the central zone the Orocco-Caio nappe, and in the NE the autochthonous Umbro-Marches (figs. 1, 2). During the nappe transport the sedimentation of molasse type formations must have continued on the back of the Antola-Cassio nappe in such a way that in the external zones these sediments pass almost continuously from the Oligocene through the whole Miocene into the neoautochthonous Pliocene of the Po Plain (e.g. Loiano, Merla 1950).

Syntectonic Modino-Cervarola Unit: Nappe tectonics of the Northern Apennines are puzzled by the presence of the Modino-Cervarola unit, which holds an intermediate position between the allochthonous Tuscan unit and the Umbro-Marches zone. It is mainly composed of graywacke sequences similar in facies to the Tuscan Macigno and the Umbro-Marches Marnoso-arenacea Formation. Its age (Oligocene to M. Miocene) seems to overlap that of the other two formations or at least of parts of them. It differs, however, by the great amount of repeated intercalations of olistostromes and gravity slide masses of Ligurid origin, especially at its internal border, by innumerable examples of synsedimentary tectonics and by intense folding. The Modino-Cervarola unit overlies the Macigno along the overturned frontal anticline of the Tuscan nappe, and at its external border it is thrusted towards the NE over the Marnoso-arenacea (Umbro-Marches).

There are several contradictory features which do not allow to consider the Modino-Cervarola unit in the same way as the other nappes. For example, it seems to be a higher nappe than the Tuscan nappe (Rentz 1971; Plesi 1975), but it can also be considered as external equivalent of it (Bortolotti et al. 1970). It is here concluded that the Modino-Cervarola graywackes were deposited at the front of the advancing Ligurid nappes as an internal border of the different levels of the Macigno, lateron of intermediate graywackes (external Cervarola graywackes and Castel Guerrino sandstone, Groscurth 1971), and finally of the Marnoso-arenacea formation. Together with the front of the Ligurids the just deposited marginal sediments were pushed to the NE over the sediments of more central parts of the basin, until in a final phase they were overwhelmed by the Ligurid nappes and buried under them (fig. 3, sections B, C).

Unravelment

The provenance of the nappes from the southwestern zones, now almost totally hidden beneath the Ligurian sea, results from the direction of folds, the reduction of the number of nappes (fig. 1), and their increasing loss of coherence towards the Po Plain. In the same direction the Tertiary graywacke flysches (Macigno, Cervarola and Marnoso-Arenacea formations) are of progressively younger age. On the assumption that the individuality of the units and the succession of nappes, as it was outlined, is correct, and that no reciprocal overtaking of nappes took place, an unravelment of the nappes (without consideration of the syntectonic Modino-Cer-

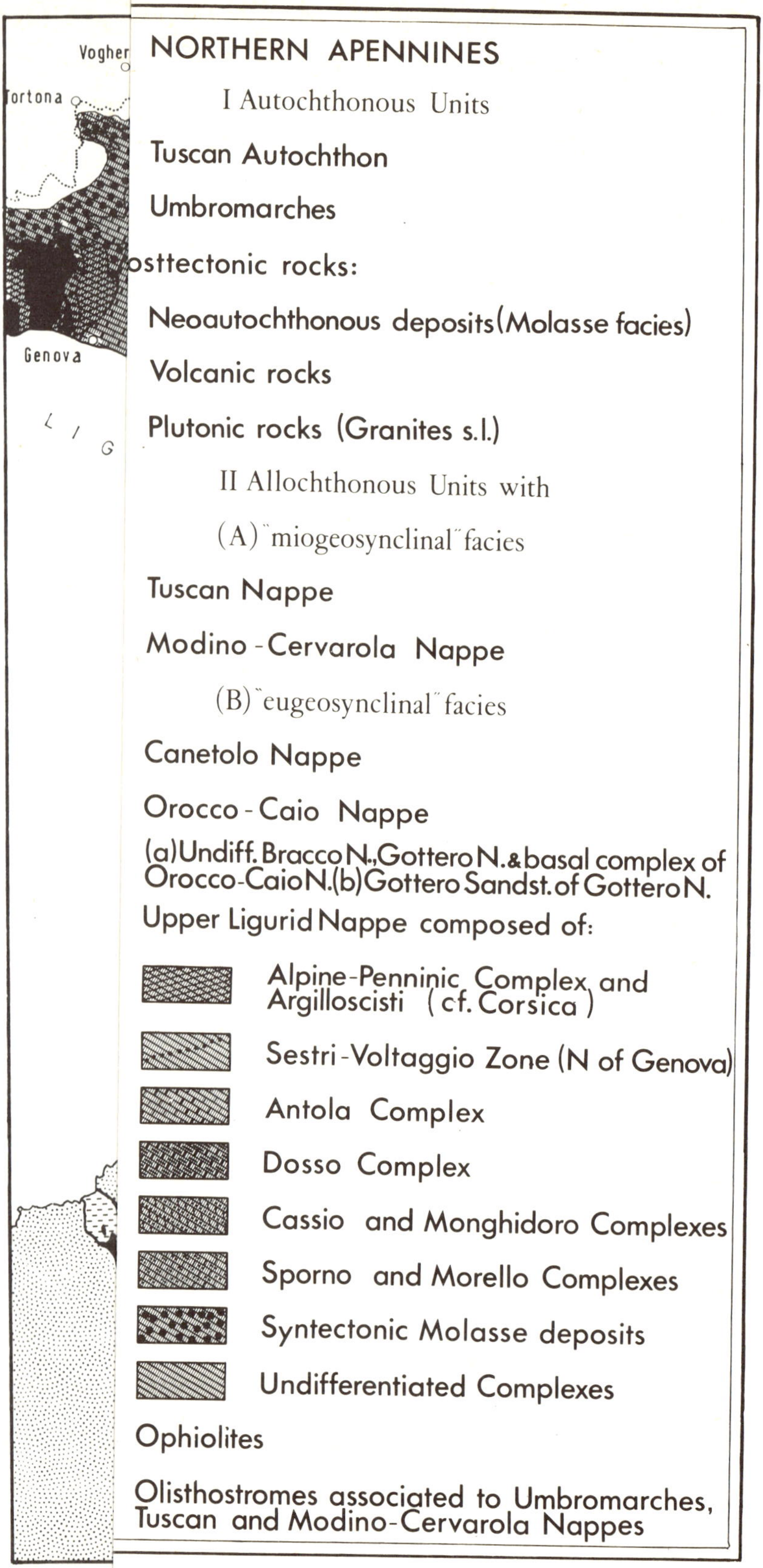

Fig. 2. Alps, Apennines, Hellenides

to p. 238

varola unit), according to the stacking order argument, leads to the following pre-Apenninic paleogeographic arrangement (from SW to NE):

Antola-Cassio-Sporno nappe Gottero nappe Bracco nappe Orocco-Caio nappe Canetolo nappe	oceanic ("eugeosynclinal") facies
Tuscan nappe "Tuscan Autochthon" Po-Plain - Marches Autochthon	deeply submerged platform ("miogeosynclinal") facies

The eugeosynclinal units must not all have been arranged in a lateral sense. The Gottero unit probably overlay the Bracco unit. As the Gottero unit stratigraphically only reaches the Paleocene it may have been overlain by younger formations (e.g. the Sporno flysch) which by diverticulation lateron formed a nappe of their own. Through the unravelment and by means of such considerations a pre-Apenninic palaeogeography results as it is shown in fig. 3, section A. In it a distinction is made between internal Ligurids (Antola-Cassio unit) and external Ligurids (Orocco-Caio unit and Canetolo unit; Braga 1972), based on the presence (internal Ligurids) or absence (external Ligurids) of synorogenic molasse deposits of Oligocene and Miocene age on top of the units.

This palinspastic arrangement of the eugeosynclinal units does not satisfy all the relationships indicated by stratigraphy, facies and clastic components. Isopic formations as Antola and Orocco-Caio Helminthoidea flysches or the Canetolo and Sporno calcareous flysches are separated from each other. In the same way the Cassio sequence with its insubric components would be situated in the middle of the eugeosyncline, far away from the insubric continental slope. Some of the palaeogeographic problems may be due to the pre-Apenninic Alpine orogeny during the Upper Eocene, which already involved the Antola and Cassio Flysches in nappe movements, but it is also possible that the configuration of the eugeosyncline was more intricate than generally supposed.

Emplacement of the Nappes

It is supposed that during the Ligurian phase only a western part of the oceanic Penninic-Ligurian eugeosyncline was destructed, while the part belonging to the external Ligurids, the Gottero and Bracco units and the Sporno complex was preserved. Also the Antola, Monghidoro and Cassio complexes, with the exception of their allochthonous counterparts in the Western Alps, may not have left the oceanic basement, although they were affected by the late Eocene movements. The Apenninic orogeny can be considered to be an effect of the consumption of this remaining oceanic lithosphere, which must have started in the Oligocene with a southwestward dipping subduction zone (Giese et al., this volume).

Along almost flat lying branches of the subduction plane, similar to models of recent subduction presented for the coast of Oregon by Seely, Vail & Walton (1974) and for the Java trench by Beck & Lehner (1974), the thick eugeosynclinal sediments were sheared off their basement together with a thin veneer of oceanic crust, now preserved in the Bracco

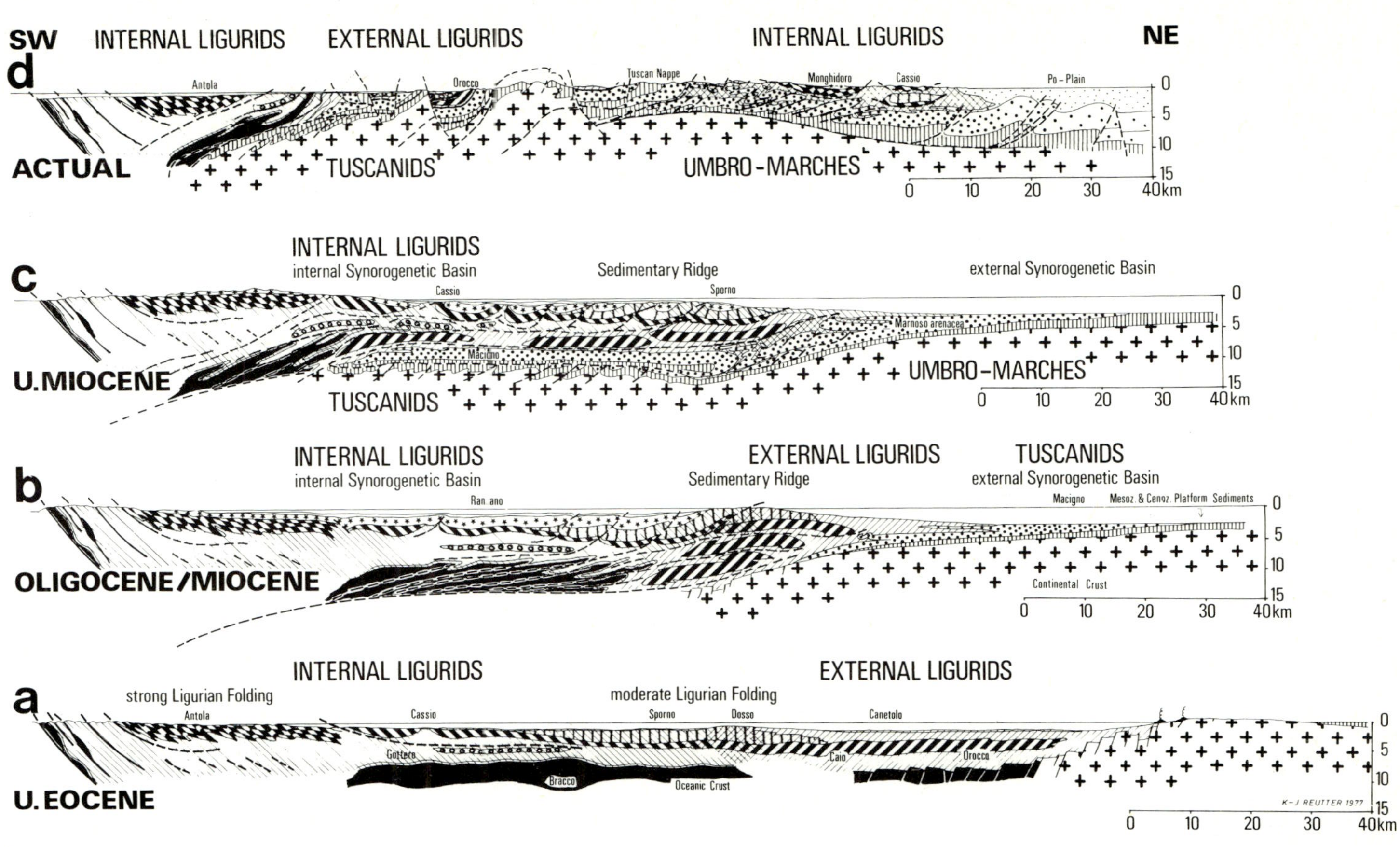

Fig. 3.
Tectonic evolution of the Northern Apennines from the late Eocene to present. It is supposed that the nappes were emplaced as a consequence of the consumption of the oceanic crust and lithosphere of the Ligurian eugeosyncline (a, left & center) and the southwestward marginal subduction of the sialic Adria microplate (a, right s.).

nappe (Decandia & Elter 1972). Where the thrust planes reached the surface, a bulge of eugeosynclinal sediments was piled up, probably within the external Ligurids. Slabs of Orocco-Caio flysch and their basal complexes were pushed over one another and upon the Canetolo sediments. Thus a sedimentary ridge was generated, which separated a shallow internal synorogenic basin (depositional area of the Oligocene/Miocene molasse sediments) from a deep external basin (depositional area of Macigno-type graywackes (fig. 3, section b)).

In the late Oligocene this bulge of Ligurian sediments invaded the deeply submerged border of the continental Tuscan crust, where it acted as the mobile inner wall of the Macigno sedimentary basin. While the downwarping of the continental crust proceeded towards the northeast, the Ligurid nappes advanced in the same direction pushing forward at their front the recently deposited graywackes of the internal margin of the basin. They were deposited upon and between olistostromes of Ligurian material, and they were synsedimentarily deformed and displaced over their more external equivalents to such a degree that they now represent the rather independent Modino-Cervarola unit.

Synorogenic sedimentation also continued in the internal basin. The Upper Eocene and Oligocene sediments of the Antola-Cassio nappe pass upward, locally without any break, into the Miocene deposits. Here the tectonic activity is indicated by local unconformities (Ibbeken, this volume), slumps and olistostromes (Papani 1964).

As a consequence of continental collision, the internal margin of the Tuscan crust was compressed and began to rise (fig. 3, section C). At least in the uppermost tectonic level tectonics became now controlled by gravity. The internal Tuscan sedimentary sequence was decoupled along the Upper Triassic evaporites and slid northeastward together with the independently moving Ligurid nappes on its back.

During the Tortonian, the paroxysmal stage of the Apenninic orogeny, the nappe movement occurred mainly along two thrust planes located at the base of the Tuscan nappe and at the base of the Antola-Cassio unit. At its frontal overturned anticline the Tuscan nappe compressed the Modino-Cervarola unit, which in turn was pushed onto the internal margin of the Marnoso-arenacea.

The Antola-Cassio unit overtook the lower Ligurid units, the Tuscan nappe and the Modino-Cervarola unit and reached the external parts of the Apennines bordering the Po Plain (fig. 1). Later uplift of the internal parts of the Apennines led to the puzzling situation that the lower Ligurid units (Canetolo, Orocco-Caio, Bracco and Gottero) are restricted to the southwestern side of the Apennines, whereas the Cassio, Sporno and Monghidoro flysch subunits of the uppermost nappe are limited to the northeastern side (fig. 3, section D). By this far-reaching and gravity controlled nappe movement parts of the underlying units were detached and dragged along with the Antola-Cassio nappe to the northeast. Thus patches of the Canetolo unit, mostly together with large flakes of Orocco-Caio flysch of several tens of square km (e.g. M. Prampa and M. Rocca in the Reggian and Modenese Apennines) are present in the external parts of the Apennines. Probably the flake of the M. Caio itself can be considered to have been pulled far to the NE at the base of the Cassio flysch. Fragments of the Bracco unit are found near the tectonic base of the Monghidoro flysch (e.g. Sasso di Castro, M. Beni), and remnants of the Gottero unit are present in the same position (e.g. "Frassinoro-Sandstein", Reutter 1968; Rentz 1970, fig. 2).

The postparoxysmal period in the Northern Apennines is characterized by uplift movements and associated tensional faulting of the internal parts, especially of the Tuscan area. During the Pliocene/Pleistocene this phase of tensional tectonics was accompanied by granitic and granodioritic intrusions (e.g. Elba island) and extrusions of alkaline volcanics and pyroclastics, which show a trend of progressively younger age towards the NE (Bortolotti & Passerini 1970). At the external margin, however, compressional tectonics continued leading to large scale folding and upwarding. As contemporaneously the Po Plain basement northeast of the upwarped zone subsided, this configuration gave rise to last movements of allochthonous slabs (e.g. San Marino), and enormous olistostromes in the Po Plain sediments adjacent to the Apennines were built up during the Pliocene (Zanzucchi this vol.).

Gravity Transport, Shifting of Troughs and the Rotation of Northern Italy

The sedimentary troughs of the Oligocene and Miocene graywacke flysches migrated toward the external side of the Northern Apennines. Downbuckling was followed by strong lateral compression. This process was set in motion once the Ligurid units, after subduction of their oceanic basement, had been obducted onto the Tuscan continental margin. By the internal compressional tectonics and consequent upwarping nappe movement was induced and it was facilitated by gravity acting on the slope toward the adjacent flysch trough.

In the northwestern narrow part of the Northern Apennines the allochthon is much more complete than in the broad southeastern part (fig. 2), where it is present only in isolated flakes with tectonically stretched sequences (e.g. "serie ridotta" of the Tuscan nappe). This feature probably is due to the greater velocity of displacement of the above mentioned pair of subsequent downbuckling and upwarping zones in the Southeast. There it occurred at such high rates that the allochthon lost its coherence and only parts of it in special conditions were able to follow the shifting depression toward the Po Plain (e.g. San Marino, fig. 2). The divergent configuration of the Northern Apennines may indicate that the stresses, which were responsible for the downbuckling of the graywacke basins, had the same angular velocity but different radial velocities arount some point in the northwestern Po Plain. This fact may be related with the paleomagnetically observed anticlockwise rotation of Italy.

References

Beck, R.H. & Lehner, P., 1974: Amer. Assoc. Petrol. Geol. Bull. - 58:376-395.

Braga, G., Casnedi, R., Galbiati, B. & Marchetti, G., 1972: Mem. Soc. geol. ital. - 11:547-560.

Bortolotti, V., Passerini, P., Sagri, M. & Sestini, G., 1970: in: Sestini (ed.) Development of the Northern Apennines geosyncline. Sediment. Geol. - 4:341-444.

Bortolotti, V. & Passerini, P., 1970: in: Sestini (ed.) Development of the Northern Apennines geosyncline. Sediment.Geol. - 4:599-624.

Dallan Nardi, L. & Nardi, R., 1975: Cons.Naz.Ricerche,Quaderni, 90:203-255, Roma.
Decandia, F.A. & Elter, P., 1972: Mem. Soc. geol. ital. 11:503-530.
Elter, P., 1973: Acc. Naz. Lincei, 183:97-109, Roma.
Elter, P., Sturani, C.& Weidmann, M., 1966: Lab. geol. mineral. geophys. musée geol. Univ. Lausann, Bull. 167:279-377.
Elter, P. & Raggi, 1965: Boll.Soc.geol.ital. - 84,5:1-14.
Galli, M., Bezzi, A., Piccardo, G.B., Cortesogno, L. & Pedemonte, G.M., 1972: Mem. Soc. geol. ital. - 11:467-502.
Groscurth, J., 1971: Diss. FU Berlin, 215 p.
Iaccarino, S., Papani, G. & Zanzucchi, G., 1974: L'Ateneo Parmese, Acta naturalia, 10,3:381-408.
Merla, G., 1951: Boll. Soc. geol. ital. - 70,1:95-382.
Pagani, G., Papani, G., Rio, D., Torelli, L., Zanzucchi, G. & Zebri, M., 1972: Mem. Soc. geol. ital. - 11:531-546.
Papani, G., 1964: Boll. Soc. geol. ital. 82,3:1-8.
Plesi, G., 1975: Boll. Soc. geol. ital. 94:157-176.
Rentz, K., 1971: Diss. FU Berlin, 333 p.
Reutter, K.-J., 1969: Ateneo Parmense, Acta nat., n.Ser.5(2):1-88.
- , 1968: Ecl. geol. Helv. - 61:183-224.
Sames, C.W., 1970: Beih. geol.Jahrb. - 103, 115 p.
Scholle, P.A., 1971: Geol. Soc. Amer. Bull. - 82:629-658.
Seely, D.R., Vail, P.R. & Walton, G.G., 1974: in: Burk & Drake (ed.) The geology of continental margins:249-260, Springer, Berlin.
Sestini. G. (ed.), 1970): Sedimentary Geology, 4, 647 p.
Van den Berg, J. & Wonders, A.A.H., 1976: Tectonophysics, 33:301-320.

Petrology of the Northern Apennine Ophiolites and their Significance in the Western Mediterranean Area*

L. Beccaluva and G.B. Piccardo

Geological Remarks

The ophiolitic sequences of the Western Alps and the Apennine developed during the Jurassic in the Ligure-Piemontese oceanic basin, separating the paleo-European and Insubrian (Southern Alpine-Apennine) continental blocks.

The late Cretaceous-Eocene (Hunziker 1974) compressive events brought together oceanic sections and segments of both European and Southern Alpine plate margins and thrusted them westwards on to the European foreland along the Alpine arc.

* Financial support by C.N.R. - Italy.

Metamorphic ophiolites of the Voltri Group, which actually represent the connection between the Alpine and Apennine arcuate belts, acquired their tectonic and metamorphic evolution during the Alpine continental collision (Chiesa et al. 1975).

In the Northern Apennines the ophiolitic terrains represent the substratum of the Upper Jurassic - Paleocene sedimentary sequences of the so-called Internal and Transitional Ligurides, whereas they constitute huge olistoliths and olistostromes within the Createceous-Eocene flysch sequences of the External Ligurides.

Fig. 1. Distribution of the ophiolitic sequences (in black) in the Alpine-Apennine system.

Acidic rocks and mafic granulites of continental basement provenance are associated with the ophiolitic materials of the External Ligurides (Eberhardt et al. 1961; Pagani et al. 1972; Beccaluva et al. 1975; Braga et al. 1975): in places, arkosic sandstones are interbedded within the chert section at the top of the volcanites.

The Internal Ligurides are generally considered true oceanic basins, whereas the External Ligurides are interpreted as oceanic to continental basins, contained between oceanic (Bracco ridge) and continental (Insubrian ridge) structural highs, from which they received olistoliths and olistostromes of both oceanic and continental materials, from Cretaceous time. Their sedimentary sequences were subsequently folded and overthrust eastward by the oceanic sequences of the Internal Ligurides.

Petrology

Northern Apennine ophiolitic materials show a lithological homogeneity: they are mainly characterized by tectonite ultramafics, gabbroic intrusives and basaltic volcanites.

Mantle Ultramafics: Tectonic ultramafites in the Internal as well as the External Ligurides show, on the basis of the mineral assemblages, a lherzolitic character.

Mantle ultramafics, where less strongly altered in lizardite (+ chrisotile) - Fe-ores - chlorites assemblages, are characterized by prevailing forsteritic olivine, various amounts of ortho- and clinopyroxenes, and few percents of Al-Cr-spinel and Ca-plagioclase.

They show, on a large scale, compositional homogeneity. Frequently, however, pockets and small bodies of more depleted lithotypes are present. Compositionally banded structures are marked by harzburgitic-dunitic horizons together with less frequent pyroxenetic and gabbroic concentrations.

Lherzolites are characterized by protogranular and porphyroclastic-granoblastic textures, according to the terminology of Mercier & Nicolas (1975), and by a complete metamorphic equilibration in spinel-peridotite facies, according to the classification of O'Hara (1967), commonly followed by an incipient later equilibration at lower pressure conditions (plagioclase-peridotite facies).

Mineral chemistry investigations (Piccardo 1976a; Ernst et al. 1977, in press) indicate that the better preserved spinel-bearing assemblages equilibrated at about 1000°C in the range 10 - 20 Kb, whereas the plagioclase growth suggests a retrograde evolutionary path toward shallower levels ($P<9$ Kb) before the oldest event of serpentinization.

A general lherzolitic character is also indicated by available chemical data. At least the less altered lithotypes (tab. 1), compared with major element estimated compositions for the upper mantle, exhibit a moderately residual character, only considering a pyrolite model for the upper mantle (Green & Ringwood 1967).

Although they cannot be properly interpreted as "primitive" mantle, they do not represent the strongly fractionated harzburgite-dunite material which could be directly linked to the primary tholeiitic melts of the extrusive sequences by means of a simple refractory residua-extracted liquids relationship.

Western Mediterranean alpine peridotites are generally lherzolitic in contrast to the depleted harzburgite-dunite sections of Eastern Mediterranean ophiolites (see for discussion and references: Nicolas & Jackson 1972; Rocci et al. 1975; Menzies 1976).

In the Voltri Massif lherzolites are the most common rock type of the uppermost structural element, the Erro-Tobbio Unit (Messiga & Piccardo 1974; Chiesa et al. 1975), probably derived from the Insubrian pericontinental mantle. Lherzolitic protoliths have been inferred also for the antigoritic serpentinites of the underlying metamorphic ophiolite units.

Intrusives: The intrusives of Northern Apennine ophiolites consist of gabbroic rocks, with rare ultramafic and acidic lithotypes.

The main rock families are represented by cumulus peridotites (dominantly plagioclase-bearing dunites and wehrlites), ol- and cpx-bearing gabbros (Mg-gabbros), Fe-gabbros (and Fe-diorites), dioritic and plagiogranitic (trondhjemitic) rocks (for general revision of their petrographic characteristics, see: Beccaluva et al. 1976b; Piccardo 1976b).

The main mineral phases show the following crystallization order: olivine + chrome spinel - plagioclase - clinopyroxene (Class D, with D_1-D_2-D_3 cyclic sequence, according to the classification of Riccio 1977).

Field relations suggest magmatic connections between cumulus ultramafic and Mg-gabbroic types, and between Fe-gabbroic and more acidic rocks, whereas no primary relations have been so far described between the two major rock groups.

The chemical data (Tab. 1, fig. 2) show that gabbroic rocks can be grouped, according to the petrographic classification, into Mg- and Fe-rich types (see recent revision of chemical data in Beccaluva et al. 1976b).

Table 1. Average chemical composition of ophiolitic rocks from Northern Apennines (oxides as weight %, elements as ppm).

	1	2	3	4	5	6
SiO_2	.48.74 (51)	. 48.89	. 44.75	. 70.77	. 39.07 (36).	42.32
TiO_2	1.62 (63)	0.29	3.88	0.26	-	0.23
Al_2O_3	16.11 (60)	19.60	12.29	14.46	4.35 (36)	3.53
FeO tot.	8.50 (55)	4.60	15.58	1.73	7.23 (88)	7.23
MnO	0.17 (63)	0.07	0.25 (8)	0.01	0.13 (88)	0.12
MgO	7.21 (51)	8.37 (25)	7.57 (15)	0.98 (5)	34.83 (88)	37.52 (2)
CaO	8.19 (51)	9.61	7.17	1.81	1.90 (88)	2.29
Na_2O	4.48 (51)	3.55	2.95	8.61	-	0.17
K_2O	0.39 (51)	0.84	0.41	0.24	-	-
P_2O_5	0.23 (63)	0.02	0.83	0.08	-	-
L.I.	4.01 (51)	-	-	-	11.45 (36)	5.05
Ni	110 (63)	-	-	-	1844 (54)	-
Co	43 (24)	-	-	-	99 (52)	-
Cr	258 (63)	-	-	-	2645 (54)	-
Cu	68 (24)	-	-	-	-	-
Zn	92 (6)	-	-	-	-	-
V	233 (55)	-	-	-	-	-
Sr	218 (24)	-	-	-	-	-
Li	16 (24)	-	-	-	-	-
Rb	8 (24)	-	-	-	-	-
Nb	4 (63)	-	-	-	-	-
Zr	155 (46)	-	-	-	-	-
Y	46 (63)	-	-	-	-	-

1 = Basalts (data from Beccaluva et al. 1976a and Ferrara et al. 1976)
2 = Gabbros (after Beccaluva et al. 1976b; with references)
3 = Fe-gabbros (after Beccaluva et al. 1976b; with references)
4 = Plagiogranites (after Piccardo 1976b, in press)
5 = Ultramafic tectonites (selectes data from Pellizzer 1961; Galli 1964; Giammetti 1968; Beccaluva et al. 1973; Ernst & Piccardo 1977, in press; Ni, Co and Cr data from Beccaluva et al. 1973; Ernst & Piccardo 1977, in press).
6 = Average composition of 2 samples of less serpentinized ultramafic tectonites from M. Aiona and M.Nero (Ernst & Piccardo 1977,in press).
In brackets number of data.

Many gabbros are cumulate products, with cumulus phases and variable amount of trapped liquid, of oceanic tholeiitic melts, at various fractional crystallization stages, in a high diffusion and slow cooling environment under low oxygen fugacity.

The earlier fractionation stages were characterized by separation of olivine $\pm$ chromite, Ca-rich plagioclase and diopsidic pyroxene (cumulus ultramafics, anothosites, troctolites and Mg-rich gabbros), the later stages by Ca-poor plagioclases, augitic pyroxene, Fe-Ti ores, apatite and zircon (Fe-gabbros and Fe-diorites). Diorites and plagiogranites represent the final products of fractionation.

FeO/MgO weight ratio for primary basaltic melts, derived from the composition of the first cumulus phase of olivine in Pl-peridotites, is about 0.67 (Piccardo 1976b) and agrees well with the lowest values of the associated tholeiitic basalts, suggesting analoguous parental liquids for both intrusives and volcanites.

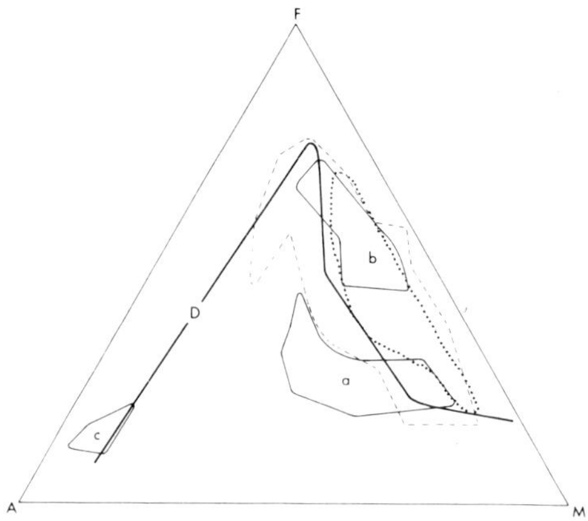

Fig. 2. A (Na_2O+K_2O)-F(FeO tot.)-M(MgO) diagram showing the distribution fields of Northern Apennine gabbros (a), Fe-gabbros (b) and plagiogranites (c); the fields of meta Mg-gabbros and meta Fe-gabbros from the Voltri Group ophiolites (dashed line; after Piccardo 1976b, with references) and of oceanic gabbroic rocks from mid-Atlantic ridge (dotted line; after Thompson 1973) are reported for comparison. The trend (D) of cumulitic sequence from Corsican ophiolites is also shown (after Ohnenstetter et al. 1975).

Cumulate gabbros in Corsican ophiolites are interpreted as products from low-depth fractionation of olivine-tholeiitic magmas in an oceanic environment (Ohnenstetter et al. 1975; Beccaluva et al 1977, in press).

Both Mg- and Fe-rich gabbroic lithotypes have been recognized as protoliths for the exlogitic rocks of the Voltri Group metamorphic ophiolites (Bocchio & Mottana 1974; Messiga & Piccardo 1974; Piccardo 1976b; Cortesogno et al. 1976), and for the high-pressure meta-gabbros of the M. Viso ophiolites (Compagnoni et al. 1976, in press).

Volcanics: The basaltic rocks consist of pillow-lavas, sometimes doleritic lavas and dikes, with a primary paragenesis mainly represented by plagioclase, augitic clinopyroxenes, Fe-Ti ores, ± olivine and ± brown hornblende. Porphyritic textures with olivine and/or plagioclase phenocrysts characterize the dikes, where primary hornblende is generally more abundant.

Although single basaltic dikes and parallel dike associations are widespread and intrude both intrusive and extrusive, no sheeted-dike complexes have been so far found in Northern Apennine ophiolites.

Geochemical evidence, based on elements relatively insensitive to alteration and sub-sea metamorphism (Pearce & Cann 1973; Pearce 1975), show a clear ocean-floor affinity (fig. 3) and a tholeiitic serial char-

acter for Northern Apennine basalts (Beccaluva et al. 1975, 1976a; Ferrara et al. 1976; Bortolotti et al. 1976).

Similar features are also shown by the volcanic sections and their metamorphic equivalents of other ophiolitic associations from the Western Mediterranean area, such as for Corsica (Beccaluva et al. 1977; Gruppo di lavoro sulle ofioliti mediterranee 1977, in press), the Eastern Alps (Bickle & Pearce 1975), Calabria (Pearce 1973; Beccaluva unpubl. data) and the Voltri Group (Mazzucotelli et al. 1976; Piccardo et al work in progress).

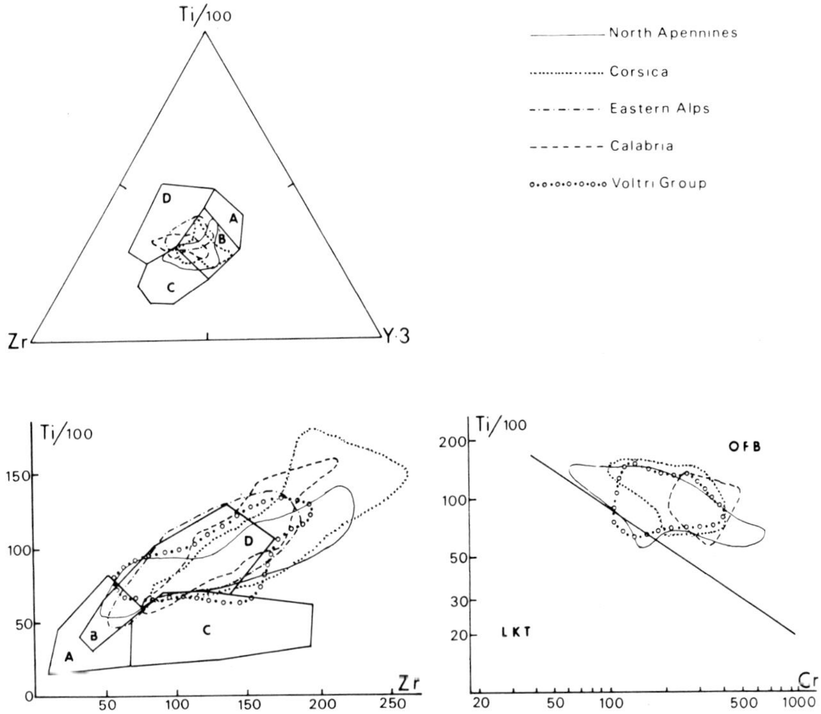

Fig. 3. Ti-Zr-Y discrimination diagram for basic volcanic rocks (Pearce & Cann 1973). Within plate basalts plot in field D, ocean floor basalts in field B, low-potassium tholeiites in fields A and B, calc-alkali basalts in fields C and B. Ti-Zr discrimination diagram (Pearce & Cann 1973). Ocean floor basalts plot in fields D and B, low-potassium tholeiites in fields A and B, calc-alkali basalts in fields C and B. Ti-Cr discrimination diagram (Pearce 1975). OFB: ocean floor basalts. LKT: island-arc tholeiites. - In the diagram the distribution fields of ophiolitic basaltic and metabasaltic rocks from Northern Apennine (Beccaluva et al. 1975, 1976a; Ferrara et al. 1976; Bortolotti et al. 1976), Corsica (Beccaluva et al. 1977; Gruppo di lavoro sulle ofioliti mediterranee 1977, in press), Eastern Alps (Bickle & Pearce 1975), Calabria (Pearce 1973; Beccaluva, unpubl. data) and Voltri Group (Mazzucotelli et al. 1976) are reported.

Exceptions are only represented by some metamorphic mafic rocks from the Eastern Alps (Fusch locality: Bickle & Pearce 1975) and Calabria (e. g. Fuscaldo locality: Pearce 1973) ophiolites, which show an affinity to within plate basalts and have been tentatively interpreted as ocean-island or ridge-crest volcanics.

Northern Apennine basalts show tholeiitic differentiation trends towards ferrobasaltic compositions, produced by fractional crystallization processes mainly involving plagioclase, olivine and minor clinopyroxene, under low P_{O_2} conditions.

The chemical evolutionary paths are characterized by Fe, Ti, V, Zr, P, Nb, Zn increase and Mg, Ni, Cr, Al decrease, according to the abyssal fractionation (fig. 3 and 4).

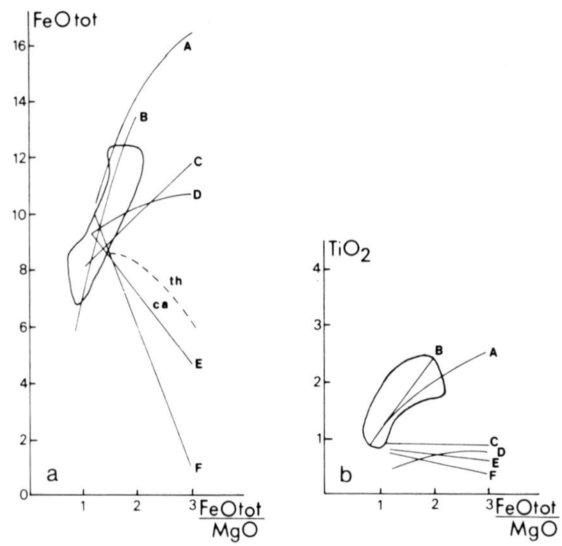

Fig. 4. FeO tot. and TiO_2 versus FeO tot/MgO ratio diagrams. The distribution fields (heavy line) of Northern Apennine basaltic rocks are reported (Beccaluva et al. 1975, 1976a; Ferrara et al. 1976; Bortolotti et al. 1976). For comparison the evolutionary trends of Skaergaard (A), abyssal tholeiites (B), Tofua island (C), Asama volcano (D), Amagi volcano (E) are reported; dashed line is the boundary between tholeiitic (th) and calc-alkalic (ca) series (after Miyashiro 1973).

The less differentiated volcanics have chemistry similar to high Al_2O_3 olivine-tholeiites which, according to geochemical and geophysical interpretations (Green & Ringwood 1967; Kay et al. 1970; Scheidegger 1973; Sleep 1974) could be produced or re-equilibrated at shallow depth in the oceanic mantle (15 - 30 km).

The distribution of some relatively "immobile" and residual elements (fig. 5) shows that the less differentiated basaltic rocks of the Apennine and Western Mediterranean ophiolites, even as members of the "oceanic floor basalts" clan (OFB), cover a range of slightly different degrees of mantle partial melting (PM, as defined by Treuil & Joron 1975).

The lower degree of partial melting, corresponding to Nb, Zr, Nb/Zr

and Zr/Y initial values higher than those of the OFB, appears in the within plate basalts (WPB) of the Eastern Alps, which also have a transitional/alkalic character (Bickle & Pearce 1975).

The diagrams of fig. 5 show the strong increase of the residual elements Nb and Zr during shallow depth fractional crystallization (FC), the Nb/Y and Zr/Y ratios remaining virtually unchanged during this process.

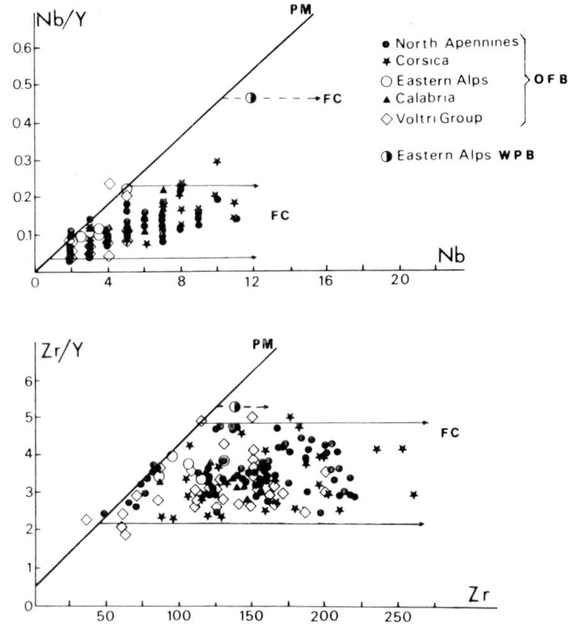

Fig. 5. Nb/Y-Nb and Zr/Y-Zr diagrams of the ophiolitic basaltic and metabasaltic rocks from Northern Apennine (Beccaluva et al. 1976; Ferrara et al. 1976; Bortolotti et al. 1976), Corsica (Beccaluva et al. 1977; Gruppo di Lavoro sulle ofioliti Mediterranee 1977, in press), Eastern Alps (Bickle & Pearce 1975), Calabria (Beccaluva unpublished data), Voltri Group (Mazzucotelli et al. 1976). OFB: ocean floor basalts. WPB: within plate basalts, probable ocean island or ridge crest volcanics (Bickle & Pearce 1975). The heavy oblique line represents the mantle partial melting trend (PM), the horizontal light lines the fractional crystallization trends (FC) according to the model proposed by Treuil & Joron (1975) for the oceanic ridges.

Summary and Discussion

The Northern Apennine ophiolites are characterized by an "atypical" lithological association of lherzolitic mantle material and ocean floor theoleiitic magmatism. In this respect they are different from the oceanic lithosphere which at present forms along active mid-oceanic ridges of well developed basins.

As an environment of these ophiolites, a narrow oceanic area or a pericontinental position could be indicated by the close association with continental crust materials within the External Ligurides basins.

The alternative hypothesis of an intra-oceanic back-arc marginal basin of the ancient Tethys is contradicted by the paleogeographical and geological setting of the Western Mediterranean area, as it is known today (Dewey et al. 1973), which suggests a general continental tensional pattern and a lack of large scale compressional events, such as subduction metamorphism and island arc volcanism.

In accordance with the hypothesis of small oceanic basins between continental plates for the Western Mediterranean ophiolites (Nicolas & Jackson 1972), Northern Apennine ophiolites could be regarded as oceanic sequences produced during an early stage of opening of the basin ("marginal" ophiolites, as proposed for the Othris sequences, Greece, by Menzies & Allen 1974), which probably never reached a quite developed oceanic stage, between rather close continental margins (Piccardo & Riccio 1975; Beccaluva et al. 1975, 1976a; Piccardo 1976a and b) and was strongly affected by rifting and transcurrent mechanisms.

The alkaline volcanism, which suggests active astenospheric upwelling under the continental lithosphere, seems to be confined to restricted areas along the margins of the Western Mediterranean oceanic basins and it has been found only rarely within the Northern Apennine domain (Ricci & Serri 1975).

The separation of Africa and North America brought about the eastward motion of the African plate relative to the Euro-Asiatic plate, producing compressional effects and oceanic crust consumption in the Eastern Mediterranean area, and large scale tensional and transcurrent effects within the epicontinental Western domains (Dewey et al. 1973).

These dominant tensional tectonics, probably associated with active astenospheric upwelling effects in restricted areas, produced lithosperic thinning and rifting along narrow zones (Bortolotti et al. 1976; Piccardo 1976b), which, later on, were integrated into the Western Mediterranean plate system, when lithospheric spreading began to separate the new continental margins.

Along these zones, strongly affected by divergent and transcurrent movements, the lherzolitic uppermost mantle, so widespread within the Apennine ophiolitic terrains, reached shallow levels and characterized the earlier oceanic sequences of the basin.

An oceanic pattern, probably formed by short ridge segments and important fracture and transform zones, should therefore have characterized the evolution of the Ligurian basin as well as the whole Western Mediterranean Jurassic basins.

Acknowledgments

The authors are grateful to M. Galli, G. Zanzucchi, M. Menzies and K.-J. Reutter for the critical reading of the manuscript, to V. Bortolotti, R. Coleman, G.V. Dal Piaz, W.G. Ernst, D. Ohnenstetter and M. Ohnenstetter for many stimulating discussions.

Many thanks are also due to J.A. Pearce, who kindly put at our disposal his Ph.D. thesis containing geochemical data on Calabria ophiolites (samples provided by E.W.F. De Roever).

References

Abbate, E., Bortolotti, V. & Passerini, P., 1970: Sedim. Geol.,4:521-557.
- , 1972: Boll.Soc.Geol. Ital. - 91:239-282.
Barbieri, F., Papani, G. & Zanzucchi, G., 1968: Ateneo Parm.Acta Nat., 4.
Beccaluva, L., Emiliani, F., Venturelli, G. & Zerbi, M., 1973: Ateneo Parmense, Acta Naturalia, 9:69-98.
Beccaluva, L., Macciotta, G. & Venturelli, G., 1976a: Ofioliti I.
Beccaluva, L., Macciotta, G., Piccardo, G.B. & Venturelli, G., 1976b: Ofioliti, 1:107-126.
Beccaluva, L., Ohnenstetter, D., Ohnenstetter, M.& Venturelli, G., 1977: Contr. Mineral. Petrol., 64:11-31.
Beccaluva, L., Venturelli, G. & Zanzucchi, G., 1975: Ateneo Parmense, Acta Nat. - 11:789-802.
Bezzi, A. & Piccardo, G.B., 1971: Mem.Soc.Geol.Ital. - 10:53-63.
Bickle, M.J. & Pearce, J.A., 1975: Contr. Mineral. Petrol.49:177-189.
Bocchio, R. & Mottana, A., 1974: Soc. It. Min. Petr. - 30:855-891.
Bortolotti, V., Cortesogno, L., Gianelli, G., Piccardo, G. & Serri, G., 1976: Ofioliti, 3:331-364.
Bortolotti, V. & Gianelli, G., 1976: Ofioliti, 1:99-105.
Braga, G., Casnedi, R., Giammetti, F., Marchetti, G. & Zerbi, M., 1975: Atti Ist. Geol. Univ. Pavia, 25:89-105.
Braga, G., Casnedi, R., Galbiati, B. & Marchetti, G., 1972: Mem. Soc. Geol. Ital., II.
Chiesa, S., Cortesogno, L., Forcella, F., Galli, M., Messiga, B., Pasquare', G., Pedemonte, G.M., Piccardo, G.B. & Rossi, P.M., 1975: Boll. Soc. Geol. Ital. - 94:555-581.
Cortesogno, L., Ernst, W.G., Galli, M., Messiga, B., Pedemonte, G.M. & Piccardo, G.B., 1976: J. Geol. (in press).
Dal Piaz, G.V., 1974: Boll. Soc. Geol. Ital., 93.
Decandia, F.A. & Elter, P., 1972: Mem. Soc. Geol. It. - 11:503-530.
Dewey, J.F., Pitman, III. W.C., Ryan, W.B.F. & Bonnin, J., 1973: Geol. Soc. Am. Bull. - 84:3137-3180.
Eberhardt, T., Ferrara, G. & Tongiorgi, E., 1961: Bull.Soc.Geol.France 4.
Elter, G., Elter, P., Sturani, G. & Weidmann, M., 1966: Bull. Lab. Géol. Géophys. et du Musée Géol. Univ. Lausann, 167.
Ferrara, G., Innocenti, F., Ricci, C.A. & Serri, G., 1976: Chem. Geol. - 17:101-111.
Galli, M., 1964: Ann.Mus.Civ.St.Nat. Genova. - 74:233-343.
Galli, M., Bezzi, A., Piccardo, G., Cortesogno, L. & Pedemonte, G.M., 1972: 66 Congr. S.G.I., Genova.
Giammetti, F., 1968: Ateneo Parmense, Acta Nat., 4.
Green, D.H. & Ringwood, A.E., 1967: Contr. Mineral. Petrol.,16:103-190.
Hunziker, J.C., 1974: Mem. Ist. Geol.Mineral. Univ. Padova, 31:54p.
Kay, R., Hubbard, N.J. & Gast, P.W., 1970: J. Geophys. Res.,75:1585-1614.
Mazzucotelli, A., Messiga, B. & Piccardo, G.B., 1976: Ofioliti,1:255-278.
Menzies, M., 1976: Geochim. and Cosmochim. - 40:645-656.
Menzies, M. & Allen, C., 1974: Contr. Mineral. Petrol., 45:197-213.
Mercier, J.C. & Nicolas, A., 1975: Geochim.Cosmochim. Acta,39:947-958.
Messiga, B. & Piccardo, G.B., 1974: Mem.Soc.Geol. It. (in press).
Miyashiro, A., 1973: Earth and Planet. Sci. Letters. - 19:218-224.
Nicolas, A., Jackson, E.D., 1972:Schw.Min.und Petr.Mitt.,52:479-495.
O'Hara, M.J., 1967: In: Wyllie, P.J., Ultrabasic and related rocks, Wiley & Sons, New York.

Ohnenstetter, D., Ohnenstetter, M. & Rocci, G., 1975: Pétrologie, 1: 291-317.
Pagani, G., Papani,G., Rio, D., Torelli, L., Zanzucchi, G. & Zerbi, M., 1972: Mem. Soc. Geol. It. - 11:531-546.
Pearce, J.A., 1973: Unpubl. Ph.D. thesis, University of East Anglia.
- , 1975: Tectonophysics, - 25:41-67.
Pearce, J.A. & Cann, J.R., 1973: Earth and Planet. Sci. Letters, 19: 290-300.
Pellizzer, R., 1961: Estr.Atti Acc.Sc.Ist.Bologna,Cl.Sc.Fis.Mem.,ser.I,8.
Piccardo, G.B., 1976a: Ofioliti, 1:279-317.
- , 1976b: SIMP (in press).
Piccardo, G.B., Riccio, L., 1975: Boll. Soc. Geol. It. - 94:693-724.
Reutter, K.J., 1968: Eclogae Geol. Helv. - 61:183-224.
Ricci, C.A. & Serri, G., 1975: Boll. Soc. Geol. It. - 94:1187-1198.
Riccio, L., 1977: Ofioliti, 2:169-177.
Rocci, G., Ohnenstetter, D. & Ohnenstetter, M., 1975: Pétrologie, 1: 172-174.
Scheidegger, K.F., 1973: J.Geol.phys. Res. - 78:3340-3355.
Sleep, N.H., 1974: Geol. Soc. Am. Bull. - 85:1225-1232.
Thompson, G., 1973: Chem. Geol. - 12:99-111.
Treuil, M. & Joron, J.L., 1975: SIMP, 31:125-174.

Metamorphic Evolution of the Ophiolite Sequences and Associated Sediments in the Northern Apennines-Voltri Group, Italy*

L. Cortesogno and G. Venturelli

Introduction

On the Metamorphic Map of Europe (1973), the metamorphic domain of the Western Alps appears sharply separated from the "folded but unmetamorphosed rocks" of the Northern Apennines. The boundary area, in about a N-S direction north of Genova, constitutes the Sestri-Voltaggio Zone. However, at present this traditional view must be considered incorrect. Not only in the Voltri Group/Sestri Voltaggio Zone but also in the Northern Apennines the ophiolite assemblages (ultramafics, euphotide gabbros, basalts and Fe-gabbros and plagiogranites in minor amounts) with associated "eugeosynclinal" sediments underwent metamorphic events. The Voltri Group is mostly characterized by a high pressure/low- to medium-temperature, medium-pressure/low- to medium-temperature and greenschist assemblages whereas the Northern Apennines, east of th Sestri-Voltaggio Zone, mainly by prehnite-pumpellyite and subordinate prehnite-zeolites (Cortesogno et al. 1973; 1976) or their equivalent in sedimentary sequences (Cortesogno & Lucchetti 1976; Venturelli & Frey 1977). In the Sestri-Voltaggio Zone the metamorphic characteristics are sometimes intermediate between the "alpine" (with shear-strain) and "apennine" types. On the basis of the present knowledges, in fig. 1 a very approximate map of the metamorphism has been attempted.

* Financial support by C.N.R., Rome, Italy.

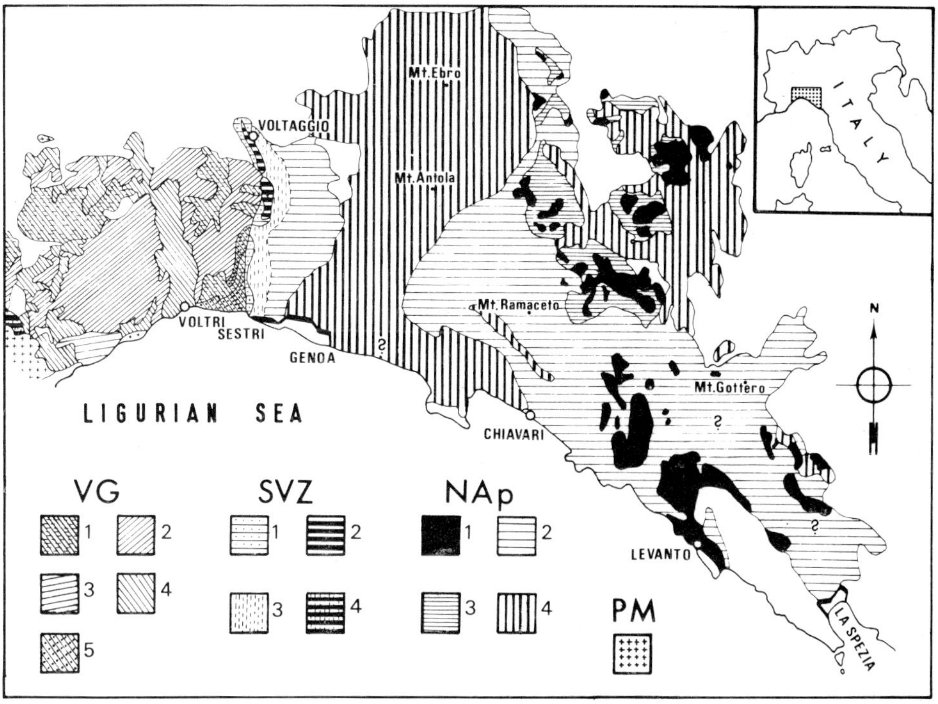

Fig. 1. Attempted sketch-map of the metamorphism in the Voltri Group (VG) Sestri-Voltaggio Zone/Montenotte nappe (SVZ) and in some areas of the Northern Apennines (NAp). PM, pre-Mesozoic polimetamorphic rocks (Arenzano, Savona, Valosio). The white areas are disregarded.

Voltri Group

1.Serpentinized lherzolites (lizardite±antigorite) with scattered gabbro dikes showing mild "alpine" metamorphism. 2. Serpentine-schists (antigorite) and metabasites in minor amounts showing poliphasic assemblages: from high-pressure (garnet+Na-clinopyroxene) to greenschists. 3. Serpentinites (antigorite±lizardite) and associated metagabbros showing only greenschist assemblages. 4. "Schistes lustrés" and associated metabasites with poliphasic metamorphism; from high-pressure (garnet± Na-clinopyroxene in metabasites) to greenschists. 5. "Schistes lustrés" and associated metabasites with high pressure metamorphis (from lawsonite+Na-amphibole±Na-clinopyroxene to Na-amphibole+epidote+albite) and local greenschist overprint.

Sestri-Voltaggio Zone and Montenotte nappe (SVZ)

1. Shelf-sedimentary sequences of Triassic-Jurassic age (Sestri-Isoverde, Voltaggio). 2. Oceanic assemblages (metagabbros, metabasalts, ophiolite breccias, metacherts, detrital limestones and pelitic sequences) under high-pressure conditions. 3. Pillow lavas, hyaloclastites, basaltic dikes subordinate serpentinites (lizardite) showing medium- to high-pressure metamorphism (pumpellyite± lawsonite±prehnite±actinolite) and associated metasediments (cherts,detrital limestones and pelitic rocks). 4. Montenotte nappe constituted by rock types 2 and 3.

Fig. 1. (cont.)
Northern Apennines (NAp)

1. Serpentinites and serpentinized lherzolites (lizardite), gabbros, basalts, ophiolite breccias and plagiogranites and Fe-gabbros in minor amounts showing sub-greenschist metamorphism (mainly prehnite-pumpellyite). 2. Cherts, limestones (*Calpionella limestone*), carbonaceous-pelitic sediments (*Palombini shales*), slates and shales (*Lavagna shales* and *Giariette shales*), sandstones (*Gottero sandstones*) showing sub-greenschist metamorphism (mainly anchizone). 3. Slates and shales (e.g. *Sopralacroce shales*), sandstones, olistostromes belonging to the chaotic *Casanova complex* and showing sub-greenschist metamorphism (anchizone and anchizone/deep diagenesis). 4. Calcareous-sandy turbidites with interbedded shales (*Antola formation* including the Mt. Caio flysch) mainly unmetamorphosed.

Oceanic Metamorphism in Ophiolite Sequences

The ophiolite complexes were subjected to metamorphic recrystallization under sub-sea environment already at an early stage. To understand these processes and their distribution, some environmental conditions have to be taken into account:

- hot centers sited at shallow depth generating steep thermal gradients (Cortesogno & Oliveri 1974; Cortesogno et al. 1975);
- quick cooling of the most central portions of the supposed ridge during the spreading of the sea-floor;
- strong tectonic activity (Bortolotti et al. 1976) which abated only at the time of the deposition of the covering sediments;
- repeated basalt intrusions at different levels;
- marine water circulation (Spooner & Fyfe 1973) facilitated by strong fracturing.

The oldest metamorphic events, leading to a high-temperature assemblage (e.g. brown hornblende) and always connected with strong deformations (blastomylonitic textures), mainly affected gabbros, exceptionally some intruded basaltic dikes or some ophiolite breccias (e.g. Mt. Rossola breccias; Cortesogno & Oliveri 1974). Afterwards new transformations under medium- to low-temperature widely affected gabbros and ophiolitic breccias e.g. blue-green amphibole, actinolite, Fe-epidote, green biotite, stilpnomelane, such as in the Mr. Rossola breccias and the Mt. Capra breccias. Comparable assemblages are not found in the overlying basaltic lavas. At the most shallow levels containing basalt flows, some ophiolite breccias, e.g. Mt. Zenone breccias, and ophiocalcites, the lowest temperature metamorphic transformations seem to vanish to weathering conditions.

During the oceanic metamorphism, the interactions between rocks and circulating waters under hydrothermal conditions (Galli & Cortesogno 1970; Spooner & Fyfe 1973) produced spilitization and intensive mobilization mainly involving alkaline and earth-alkaline elements but also iron, nickel, chromium, copper and others. At the most shallow levels occurred albitization of plagioclase and the formation of haematite and smectites from volcanic glass and mafic minerals. The smectites were transformed into chlorite and illite during the orogenic metamorphism. At the level of the gabbro intrusions, the low-temperature sub-sea meta-

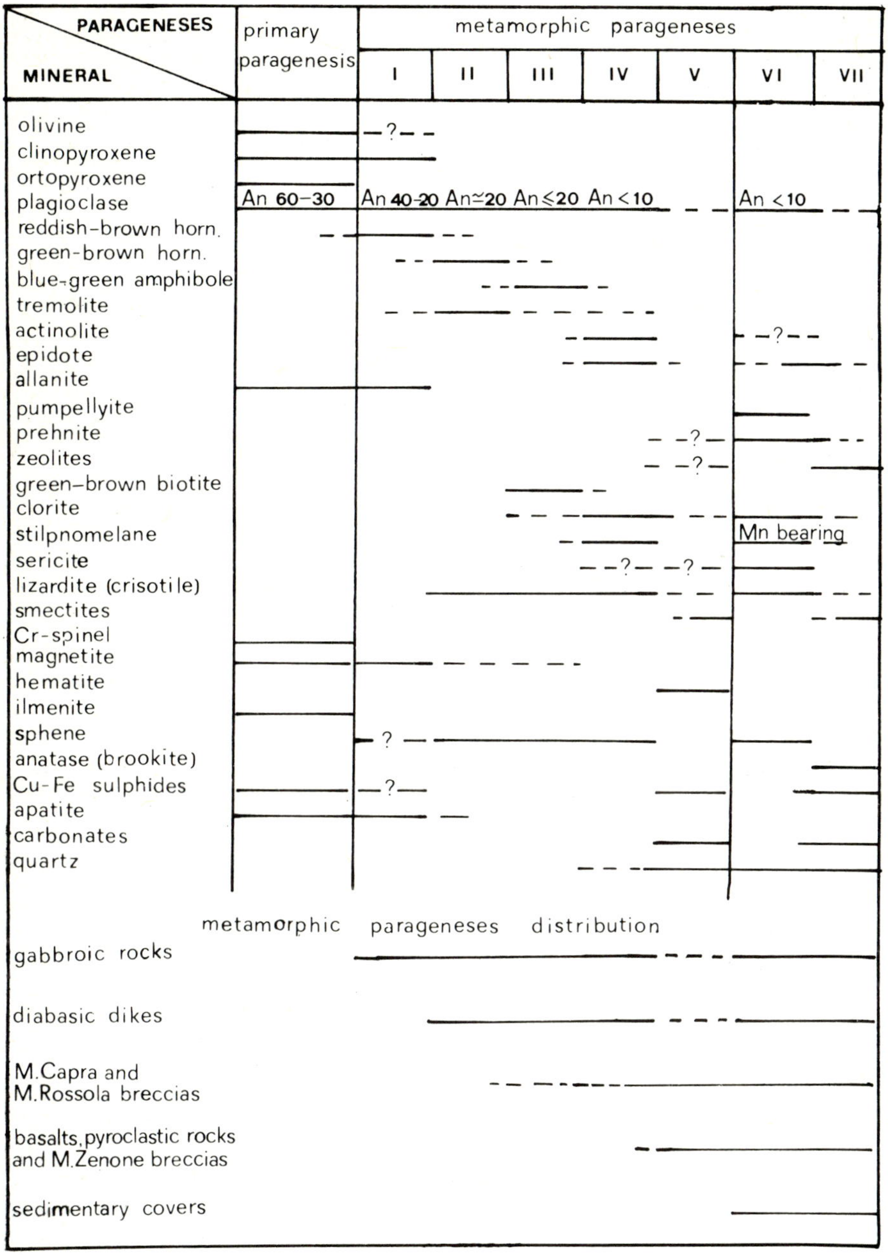

Fig. 2. Mineral development during the oceanic and orogenic metamorphism in the ophiolite sequences of the Northern Apennines. I - V, *oceanic* metamorphism phases; VI - VII, *orogenic* metamorphism phases.

morphism probably was almost inaffective. Actually, the prehnite sometimes found in gabbros can partially be attributed to submarine metamorphism (see discussion in Cortesogno & Oliveri 1974, and in Cortesogno et al. 1975). Except for a few cases (Cortesogno & Oliveri 1974), the sediments associated with the basalts (e.g. cherts) did not undergo sub-sea metamorphism. Bonatti et al. (1976) attribute to this stage the genesis of the Mn-deposits associated with cherts and of the Fe-Cu-Zn-sulfide deposits common in basalts.

Orogenic Metamorphism

The mineral assemblages which have been referred to oceanic metamorphism in the previous section, generally have been obliterated by a further metamorphic overprint. Except for a few cases, the orogenic metamorphism affected all ophiolite complexes and many sedimentary sequences also in the Northern Apennine (fig. 1). A scheme of the metamorphic evolution is given in fig. 2.

In the Voltri Group (Chiesa et al. 1975, 1976; Ernst 1976) most ophiolite sequences underwent a metamorphic evolution from high-pressure to medium-pressure conditions. The oldest assemblage shows high-pressure (10 - 11 kbs) and medium-temperature ($400 < T < 500°C$) (eclogite) conditions; as the pressure decreased, the metamorphism, through Na-amphibole and barroisitic hornblende stability conditions, reached greenschist facies. In particular, in the Palmaro-Caffarella unit the pressure was also high (5-8 kbs) but the temperature lower ($\leqslant 400°C$) and successively the temperature increased while the pressure decreased until greenschist conditions.

In the Sestri-Voltaggio Zone and Montenotte nappe (the last one sited at the eastern side of the Voltri Group) the ophiolite units show both "alpine" metamorphism of HP-LT ($P > 3$ kbs, $250 < T < 400°C$) and metamorphic conditions transitional towards the prehnite-pumpellyite assemblage which characterize the Northern Apennines east of the Sestri-Voltaggio Zone. The shear strain decreases from the Voltri group to the Northern Apennines.

In the Northern Apennines the prehnite-pumpellyite assemblage (Galli & Cortesogno 1970) or its equivalent in the sedimentary sequences (Venturelli & Frey 1977) is dominant. Studies on the mineral parageneses and structural/textural characteristics both in mafic rocks (Cortesogno et al. 1973) and in Mn-ores bearing cherts (Cortesogno & Lucchetti 1976) suggest a main metamorphic phase (prehnite-pumpellyite) and at least a retrograde one (prehnite and zeolites) (fig. 3) probably in relation to the two main tectonic events recently identified by structural investigations in the Lavagna shales (Pertusati & Horrenberger 1975). During the first stage, tensional fractures allowed a strong mobilization at a reduced scale of a few meters (Galli & Cortesogno 1970). During the further events more extended fracturing caused strong mobilization and sometimes led to ore deposits, mainly of Cu-Fe-sulfides but also of zeolites and carbonates.

As already underlined, many sedimentary formations underwent subgreenschist metamorphism (Schamel 1974; Cortesogno & Lucchetti 1976; Venturelli & Frey 1977). Some cherts strictly associated at the time of deposition and at present to ophiolite rocks, mainly to basalts, contain metamorphic minerals; moreover, their Mn-deposits (Bonatti et al 1976),

during the folding accounting for the prehnite-pumpellyite facies, were strongly mobilized. The newly formed minerals are mainly located at the fold nuclei (De Negri & Rivalenti 1971). At this stage many minerals developed: quartz, braunite, rhodochrosite, rhodonite, carpholite, piemontite, sursassite, parsettensite, albite, barite, haematite and others.

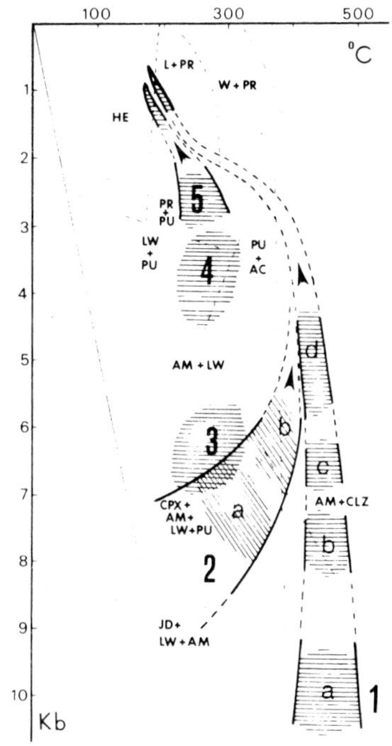

Fig. 3. Pressure/temperature conditions and metamorphic evolution of the ophiolite sequences in the Northern Apeninnes-Voltri Group during the alpine orogenesis.

Voltri Group

1. Metamorphic stage of the eclogitic rocks (modified after Ernst 1976; Beigua, St. Luca-Colma and Ponzema units) a: pre- and post-cinematic eclogitic stage; b: Na-amphibole; c: barroisitic hornblende; d: greenschist stage.
2. Evolution of metabasites associated to the "schistes lustrés" (Palmaro-Caffarella unit): from lawsonite+Na-amphibole±Na-clinopyroxene (a) to Na-amphibole+epidote+albite (b) and local greenschist overprint.

Sestri-Voltaggio and Montenotte nappe

3. High-pressure units: Na-pyroxene+lawsonite+Na-amphibole+pumpellyite.
4. Medium-pressure units: pumpellyite±lawsonite±prehnite+epidote±actinolite.

Northern Apennines

5. From prehnite-pumpellyite to zeolites-prehnite.
Other symbols: HE = heulandite; L = laumontite; PR = prehnite; W = wairachite; PU = pumpellyite; LW = lawsonite; AC = actinolite; AM = Na-amphibole; CPX = Na-clinopyroxene; JD = jadeite; CLZ = clinozoisite.

The clay fraction ($< 2\mu$) of eugeosynclinal sedimentary formations (Abbate et al. 1970) show the assemblage illite-chlorite and illite crystallinity values (fig. 4) which are typical of sub-greenschist metamorphic environment (Venturelli & Fry 1977, and unpubl. data). Although present data are not abundant and although only a few areas have been investigated, the results in fig. 1 are *tentatively* extended to all of the outcrops of the same or similar formations in the Northern Apennines. Obviously the reliability of this extrapolation has to be verified on the basis of new data. The best illite crystallinity was found in the Lavagna shales which in some areas reach the boundary between anchizone and epizone, i.e. approximatively the boundary between sub-greenschist and greenschist facies. The metamorphic conditions described agree well with the prehnite-pumpellyite assemblage characterizing the associated mafic rocks

of the ophiolite sequence. According to the present scarce data, the Antola formation (including the Mt. Caio flysch) belonging to the Helminthoid flysch sequence, is generally unmetamorphosed containing mixed-layer illite/montmorillonite (Schamel 1974; Venturelli & Frey 1977).

In some areas low-grade metamorphosed formations overlie the unmetamorphosed Mt. Caio flysch (see Venturelli & Frey 1977). This proves that metamorphism predated this overthrust.

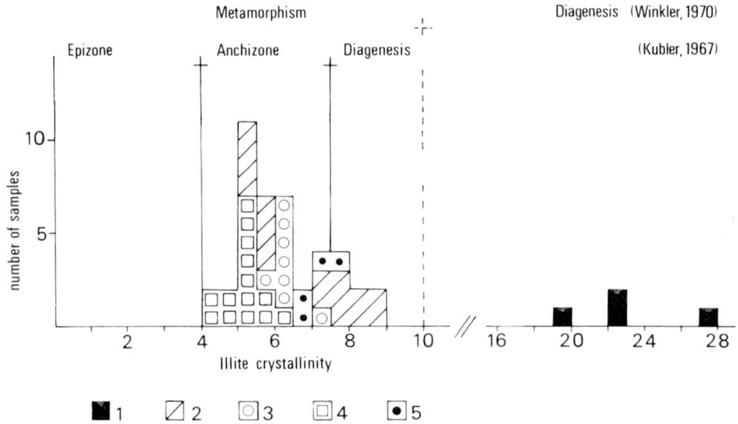

Fig. 4. Illite crystallinity (after Kubler) of the clay fraction (<2μ) from some metamorphosed and unmetamorphosed sedimentary formations of the Northern Apennines (after Venturelli & Frey 1977, and unpubl. data). 1 = Mt. Caio; 2 = Casanova complex; 3 = Gottero sandstones and Giariette shales; 4 = Lavagna shales; 5 = Palombini shales and Calpionella limestone.

Crustal Metamorphism of the Ultramafics

With the exception of the cumulate plagioclase peridotites, which, in some respect, are similar to the gabbroic rocks, it is very difficult to recognize the crustal metamorphic evolution of the ultramafics in the Northern Apennines. Lizardite, sometimes associated to chrisotyle $\pm$ chlorite, is stable during the entire metamorphic evolution starting from the sub-sea serpentinization until the crystallization of serpentine and brucite in fractures during the latest orogenic events. As far as the beginning of serpentinization is concerned, at present it is noteworthy that gabbroic dikes intruded lherzolites before rodingitization, while the basaltic dikes, generally never rodingitized, intruded already serpentinized ultramafics (Bortolotti et al. 1976). Moreover, some minor rodingitization phases were probably also connected to the orogenic environment.

Also in the Sestri-Voltaggio Zone under medium-pressure/low-temperature conditions, lizardite was stable during the orogenic metamorphism. In the Voltri Group, however, during the development of the "alpine" schistosity in the less metamorphosed units, antigorite substitutes lizardite. Under high-pressure conditions only antigorite appears, it is rarely associated with tremolite.

Concluding Remarks

The evolution of the western Tethys starting from the Jurassic time still is a debated problem. The old Jurassic "ocean" was probably different in some way from the present day oceans. Lherzolites are largely predominant over harzburgites and dunites, sometimes (Northern Apennines) some fragments of continental granites are associated with basalts, some basalts from Corsica (Balagna) show little chemical differences to the typical "normal ridge segments" basalts (see data in Gruppo di Lavoro sulle Ofioliti Mediterranee 1976). Moreover, the occurrence of ophiolite breccias in the Casanova complex (whose lower limit is attributed to the Upper Cretaceous) and of ophicalcites may suggest fracturing and early uplift of the ocean floor.

During the closure of the basin, at least in some areas, the subduction produced high-pressure assemblages like those outcropping in the Voltri Group, the Western Alps and the Calabrian Apennines.

The low-temperature metamorphism typical of the Northern Apennines was at least in part connected with the transfer of oceanic crust onto the Insubric margin and successive folding and imbrication. However, not all the eugeosyncline sequences underwent metamorphic transformations (e.g. the Monte Caio Flysch).

References

Abbate, E., Bortolotti,V., Passerini, P. & Sagri, M., 1970: In: Development of the Northern Apennines geosyncline. Sedim. Geol.,4:207-249.

Bonatti, E., Zerbi, M., Kay, R. & Rydell, H., 1976: Geol. Soc. Amer. Bull. 87:83-94.

Bortolotti, V., Cortesogno, L., Gianelli, G., Piccardo, G.B. & Serri, G., 1976: Ofioliti, 1:331-364.

Chiesa, S., Cortesogno, L. & Forcella, F., 1976: Soc. ital. Min. Petr. 33, in press.

Chiesa, S., Cortesogno, L., Forcella, F., Galli, M., Messiga, B., Pasquare', G., Pedemonte, G.M., Piccardo, G.B. & Rossi, P.M., 1975: Boll. Soc. geol. ital. - 94:555-581.

Cortesogno, L., Gianelli, G. & Piccardo, G.B., 1975: Boll. Soc. geol. ital. - 94:291-327.

Cortesogno, L. & Lucchetti, G., 1976: Ofioliti, 1:373-382.

Cortesogno, L., Lucchetti, G. & Penco, A.M., 1973: Ann.Mus. Civ. Stor. Nat., Genova, 79:291-321.

- , 1976: Ofioliti, 1:383-389.

Cortesogno, L. & Oliveri, F., 1974:Rend.Soc.Ital.Min.Petr.,30:931-959.

De Negri, G. & Rivalenti, G., 1971:Atti Soc.Tosc.Sc.nat.Men.A,78:420-437.

Ernst, W.G., 1976: Schweiz. min. petrogr. Mitt. - 56:293-345.

Galli, M. & Cortesogno, L., 1970:Rend.Soc.ital.Min.Petr.,26:597-647.

Gruppo di Lavoro sulle Ofioliti Mediterranee, Sottogruppo Metamorfismo: Meeting on "High-pressure/low-temperature Metamorphism in the Western Alps", Genova.

Kubler, B., 1968:Bull.Centre Rech. Pau-S.N.P.A., 2:385-397.

Metamorphic map of Europe, 1973 (E.Niggli coordinator).UNESCO,Paris.

Pertusati, P.C. & Horrenberger, J.C., 1975:Boll.Soc.geol.ital.94:1375-1436

Schamel, S., 1974: Unpubl. Thesis, Yale University, USA.

Spooner, E.T.C. & Fyfe, W.S., 1973:Contr.Miner.Petrol., 42:287-304.

Venturelli, G. & Frey, M., 1977: Rend. Soc. Ital. Min. Pet.,33:109-123.

Winkler, H.G.F., 1970: N. Jb. Geol. Paläont. Abh., 141:222.257.

Coalification Studies in the Northern Apennines and Palaeogeothermal Implications

K.-J.Reutter, M. Teichmüller, R. Teichmüller and G. Zanzucchi

Introduction

In the Northern Apennines a stack of nappes of predominantly sedimentary rocks of maximum thickness of about 6 km is exposed. Frequent interactions between tectonic and sedimentary phenomena indicate that structures belong to a near-surface tectonic level. Only the lowermost tectonic unit, the Tuscan "autochthon", is epimetamorphic, while the overlying nappes show anchimetamorphic and diagenetic alteration (for more details of nappe tectonics see Groscurth & Reutter, this volume). In order to recognize vertical and lateral variations of diagenesis, the degree of coalification of plant fragments was determined by measuring their optical reflectivity (vitrinite reflectance in oil, method see M. Teichmüller 1975).

Plant fragments are frequently found in the autochthonous and allochthonous flysch and molasse deposits of the Northern Apennines. They are usually present in graywacke turbidites, concentrated in the upper transitional zone of turbiditic to background sedimentation (Bouma 1965: interval d). Comparatively large pieces of woody material of more than 20 cm length were found within coarse graywacke layers (grain size > 2mm). Occasionally the carbonaceous material forms layers more than 10 cm thick.

The aim of this study was to see if the coalification pattern shows inversions due to thrusting or if it corresponds to the present day heat flow distribution.

The Coalification Pattern

In the Tuscan "autochthon" two samples from the Apuan Alps show that in the Palaeozoic (graphite-sericite-marble, Passo del Vestito) as well as in the Cenozoic (Pseudomacigno, Isola Santa) the woody material is metamorphosed to graphite with values of maximum reflectance (Rmax) of 16.9% and 13.3% respectively.

In the overlying nappes coalification does not exceed an anchimetamorphic rank. The Oligocene *Macigno* of the lowermost (Tuscan) nappe yields the highest degree of coalification. In the internal (southwestern) parts the mean reflectance (Rm) reaches 5.1% near La Spezia, 2.5% south of Livorno and 4.9% near Piombino (fig. 1). Heating is markedly less in the external parts of the Macigno, which correspond to the normal flank of the huge recumbent frontal anticline of the Tuscan nappe in the central zone of the Northern Apennines. Here Rm values of only 1.0% (Mommio) and 1.2% (south of Florence) were obtained. Macigno outcrops situated between internal and external parts reveal intermediate reflectance values. The data from Pontremoli clearly indicate that coalification depends on the location of the sample, either internal or external, within the nappe. Thus at the base of the formation, about 4 km west of Pontremoli, Rm values of 1.2 % are obtained, whereas at its top, about 9.5km westsouth-

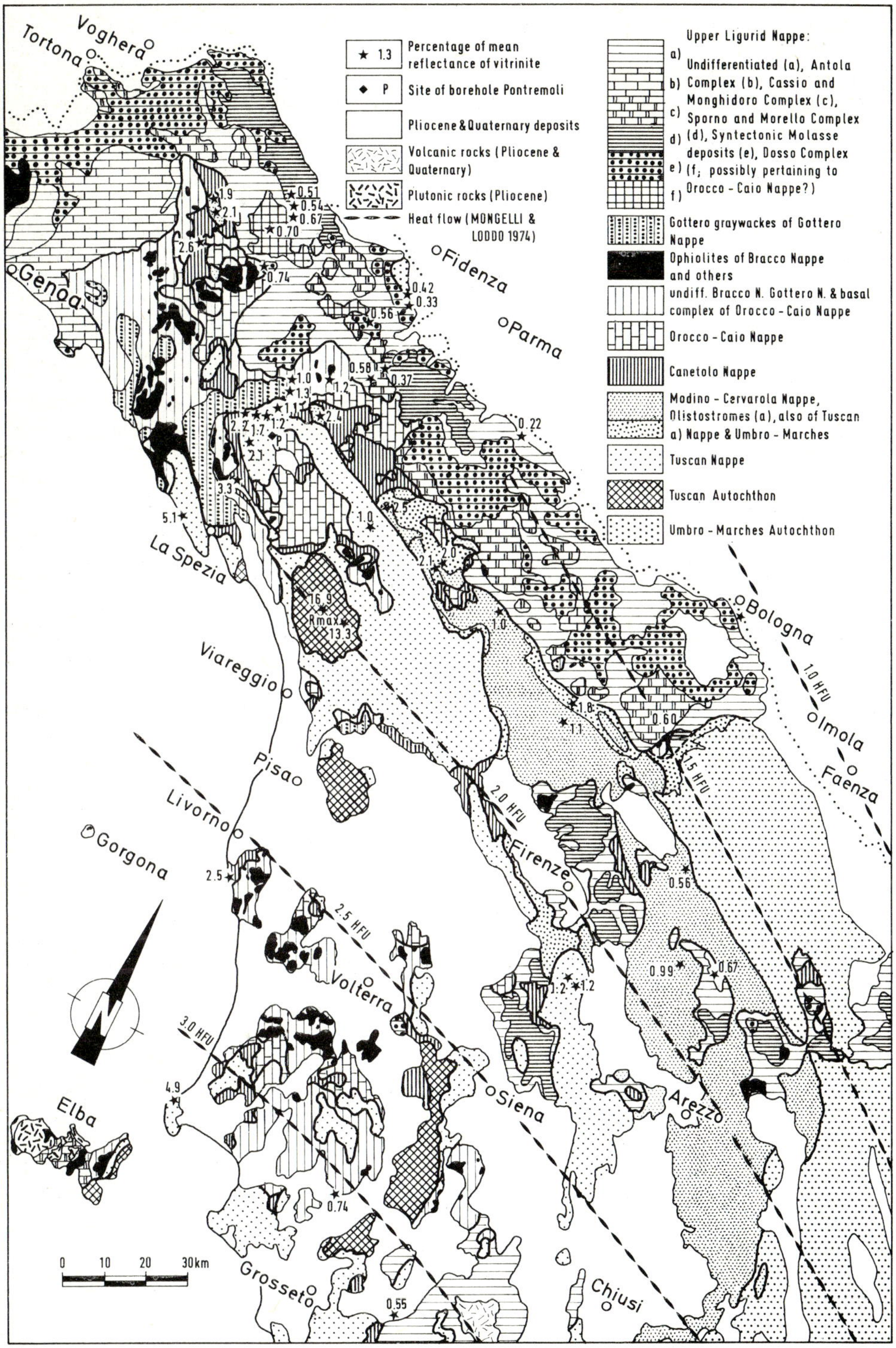
Percentage of mean reflectance of vitrinite
Site of borehole Pontremoli
Pliocene & Quaternary deposits
Volcanic rocks (Pliocene & Quaternary)
Plutonic rocks (Pliocene)
Heat flow (MONGELLI & LODDO 1974)
Upper Ligurid Nappe: Undifferentiated (a), Antola Complex (b), Cassio and Monghidoro Complex (c), Sporno and Morello Complex (d), Syntectonic Molasse deposits (e), Dosso Complex (f; possibly pertaining to Orocco - Caio Nappe?)
Gottero graywackes of Gottero Nappe
Ophiolites of Bracco Nappe and others
undiff. Bracco N. Gottero N. & basal complex of Orocco - Caio Nappe
Orocco - Caio Nappe
Canetolo Nappe
Modino - Cervarola Nappe, Olistostromes (a), also of Tuscan a) Nappe & Umbro - Marches
Tuscan Nappe
Tuscan Autochthon
Umbro - Marches Autochthon
Voghera
Tortona
Genoa
Fidenza
Parma
La Spezia
Bologna
Viareggio
Pisa
Livorno
Gorgona
Imola
Faenza
Firenze
Volterra
Siena
Arezzo
Elba
Grosseto
Chiusi
1.0 HFU
1.5 HFU
2.0 HFU
2.5 HFU
3.0 HFU
Rmax
0 10 20 30km

west of Pontremoli, the values are 2.2%. At Calice, half-way between Pontremoli and La Spezia, the mean reflectance is already 3.3%. All data show a trend of decreasing coalification transverse to the strike and towards the foreland.

Immediately to the NE of the recumbant frontal anticline, coalification of the *Cervarola graywackes*, which belong to the more external Modino-Cervarola unit (Oligocene to M. Miocene), is much more intense than in the Macigno: Rm values of 2.4% are obtained from Pracchiola, 2.5% from Cerreto (in a sequence of somewhat uncertain position), and 1.9% - 3.1% from Bobbio (Piacentine Apennine) and from Civago in the Modenese Apennines. The sudden increase in coalification from SW to NE may be a consequence of an originally higher structural position of the Macigno in the frontal anticline under which the Cervarola graywackes were depressed.

Within the belt of the Modino-Cervarola unit, which becomes broader from NW to SE, the same trend of decreasing coalification towards the external parts can be observed. At its external (northeastern) border, in the Mugello to the NE of Florence, where the Cervarola graywackes are thrust upon the Marnoso-arenacea formation, a mean reflectance of 0.58% is obtained, whereas in a slightly more internal position, near Bibbiena, a value of 1.0% is reached. Furthermore near Montecreto (Modenese Apennine), 8 km to the NE of the frontal fold of the Tuscan nappe a mean reflectance value of 1.0% is recorded.

The same phenomenon of a sudden increase in coalification is observed in an overturned frontal anticline of the Cervarola graywackes near Porretta Terme (Bolognese Apennine). In the normal limb the mean reflectance is found to be 1.1% in contrast to 1.8% 4 km to the North in the slightly overturned Suviana sandstone, which overlies the Cervarola graywackes. The Suviana sandstone can be considered to be the most internal part of the autochthonous *Marnoso-arenacea formation* (Langhian-Tortonian). In a tectonic half-window at Salsomaggiore, where a facies transitional to the molasse deposits of the Po Plain is exposed, reflectance values of only 0.37% are recorded. Thus, the trend of diminishing coalification towards the NE is present in the Marnoso-arenacea as well as in the Macigno and the Cervarola formations.

The Macigno, Cervarola and Marnoso-arenacea formations, which themselves were subjected to considerable dislocation, are overlain by allochthonous deposits of Upper Jurassic to Tertiary age. These deposits are derived from the Ligurian Eugeosyncline, which was originally situated to the southwest of the Tuscan region. It is held that the uppermost unit in the stack of nappes came from the most distant (internal) parts of the Ligurian area whereas the lowermost unit had its origin next to the Tuscan miogeosynclinal facies (Groscurth & Reutter, this volume).

In the Ligured nappes it can be observed once more that coalification increases downwards and also from the external towards the internal parts of the Apennines. The *Aveto sandstone* (Oligocene - Miocene?), which is part of the lowermost Ligurian nappe in a comparatively internal position, is marked by a coalification of 2.6% reflectance, but in an extreme ex-

Fig. 1. Coalification pattern of the Northern Apennines on the basis of vitrinite-reflectance values. Tectonic units after Groscurth & Reutter, this volume.

ternal site, 15 km north ob Bibbiena (Prov. Arezzo), the reflectance level is only 0.67%. The *Ostia sandstone* (U-Cretaceous) of the Orocco-Caio unit yielded a reflectance of 1.2% in a central position (Berceto), but more externally a value of only 0.70% is obtained from the Dosso Complex (Farini d'Olmo flysch, Paleocene). This latter complex, however, is of uncertain tectonic position, as it seems to have been incorporated in the Upper Ligurid nappe. The Gottero graywackes (U-Cretaceous - ?Paleocene) belong to a nappe higher than the Orocco-Caio nappe, but the coalification of 1.3% reflectance can be explained by the more internal position of the sample (some km to the SW of Berceto).

The uppermost unit is made up by several flysch formations of Cretaceous and Palaeogene age, which were involved in the nappe movement together with unconformably overlying molasse type deposits of U-Eocene, Oligocene and Miocene age. Samples from the Ranzano sandstone (Oligocene) yielded coalification values corresponding to 0.74% reflectance at Monteregio (Upper Nure Valley, Piacentine Apennine) in an almost central zone, and in external position, 35 km to the North, 0.51% (Rio Spettine, Lower Nure Valley) and at 55 km to the East (Rio Pessola, Parmense Apennine), 0,56% reflectance.

In the Rio Spettine area, a sample from the underlying Luretta formation (Eocene), about 250-400 m lower in the stratigraphic column but also a little more internal, shows a reflectance of 0.67%. From formations which unconformably underlie the Ranzano sandstones and its equivalents, two other coalification values were obtained. A sample from the Salti del Diavolo conglomerates (U-Cretaceous) in the Baganza valley (Parmense Apennine) gave 0.58% Rm and the other sample from the Monghidoro flysch (Palaeocene) near Passo della Raticosa (Bolognese Apennine) gave 0.60% mean reflectance.

The comparatively high degree of coalification of the Ranzano sandstone, especially at Monteregio, indicates that once it was overlain by a thick insulating cover. Since higher nappes are not known a younger stratigraphic sequence of Miocene age (Bismantova sequence of the Reggian Apennine?) must have been present, but it is not likely that it was more than 2000 m thick. All reflectance values show that the degree of coalification in the upper Ligurid nappe is essentially independent of the stratigraphic position, and that it is controlled by the relative position (internal or external) in the mountain chain.

The Age of Coalification and Orogeny

The coalification pattern suggests that for the most part heating reached its maximum when the stacking of nappes was completed. As the low grade metamorphism of the Tuscan "autochthon" was essentially syntectonic (Günther & Wallbrecher this vol.), the beginning of heating may be considered contemporaneous with the first nappe movements. Therefore, heating should have started earlier in the internal zones (22-20 m.y. ago) than in the external zones (about 11 m.y. ago). In the Upper Ligurid nappe some heating may already have occurred before Miocene time. Strong subsidence preceded the arrival of the first Ligurid nappe on top of the Macigno 22 m. y. ago, but subsidence must have persisted, since the higher nappes continued to slide over the lower ones. Thus a thick pile of allochthonous nappes accumulated upon the Macigno. By analogy with the Carboniferous of the Ruhr, the degree of coalification of the Macigno near Pontremoli

(1.7 - 2.1% reflectance) indicates a thickness of the overlying sediments of 2 to 3 km or more if the continuous movement of the higher nappes towards the external zones of the Apennines is considered. The arrival of the Ligurid nappes in the area of the Macigno, Cervarola and Marnoso-arenacea formations can be dated by means of precursory olistostromes intercalated between normal sediments. Thus nappe movement continued from 22 m.y. to 6.6 m.y. (end of the Tortonian), with the exception of the Po Plain border where still younger mass movements occurred. The nappe movement was preceded by shifting of the axis of the depositional through from the Macigno to the Marnoso-arenacea (Bortolotti et al. 1970) and it was followed by folding and upwarping of the internal zones in such a manner that the slope for the gravity transport was maintained. During these processes the Tuscan nappe sheared off its basement and was overthrust on top of the strongly folded and contemporaneously metamorphosed Tuscan "autochthon".

The inversion of the vertical movement during orogeny was probably connected with introduction of additional heat. In the lowermost tectonic levels rising magmas may have made a contribution. Besides the conductive component, heat transport through the nappes may have occurred by rising gases and liquids. The sharp contrast of metamorphism between Tuscan "autochthon" (low grade) and Tuscan nappe (very low grade) seems to be somewhat contradictory to a posttectonic heating, but it can be explained by secondary processes (see Groscurth, Günther, Reutter & Wallbrecher this vol.).

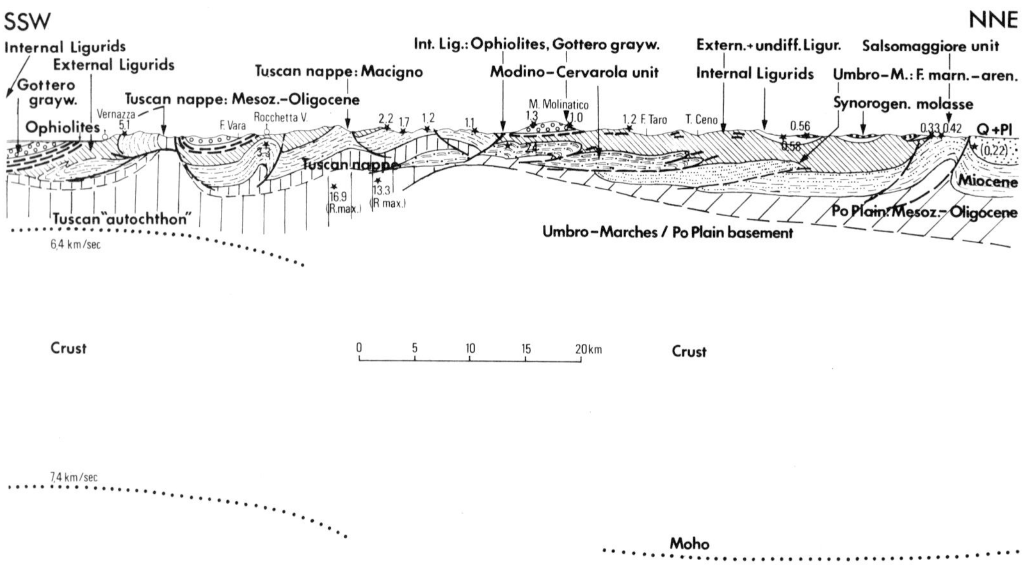

Fig. 2. Schematic cross section through the Northern Apennines with vitrinite-reflectance values (partly projected into the section plane).

The heat flow of the internal zones must have increased continuously during orogeny from minimum values during subsidence to greater values during the inversion giving rise to the present day high values of coalification (fig. 1 and 2). Disregarding the preorogenic coalification which

probably was not relevant in the single nappes, the metamorphism of the lowermost tectonic levels started syntectonically. But, as generally, the heating outlasted the tangential compression and reached its maximum relatively late.

No compelling conclusions about the heat flow during the coalification process in the nappe pile can be drawn so far as no vertical profiles could be sampled. Evidently it was higher in the southwest than in the northeast thus corresponding to the present heat flow pattern (Mongelli & Loddo 1974, fig. 1). Estimates of coalification gradient and of probable thickness of the overlying sediments and nappes, however, lead to higher fossil heat flow values than the present ones. This is emphasized by the temperatures in the borehole Pontremoli, where 3520 m below the present surface only 108°C were measured, which corresponds to a quite normal gradient of 30°C/km or approximately 1.5 HFU. Therefore it might be supposed that in the course of the progressing orogeny the pile of Ligurid and Tuscan nappes was displaced as a whole along a thrust plane which in the central and internal parts of the Northern Apennines was relatively deep seated (fig. 2).

Interpretation of the Regional Differences in Coalification

The decrease of coalification from the SW to the NE is substantially connected with two factors:

1. the reduced thickness of the nappes in the external parts and
2. the decrease of the heat flow towards the NE.

Final uplifting of the internal parts of the Northern Apennines coincided with tensional tectonics which started in late Miocene time. At that time the degree of coalification in the nappes appears to have reached a high level. The coalification proceeded, however, in the marine and lacustrine sediments of Messinian and Pliocene age, which were deposited in subsiding intermontane basins between the rising chains. The peats of the Miocene/Pliocene boundary are already altered to sub-bituminous coals (0.55 % Rm) or, in the coal mine of Ribolla near Grosetto, to high volatile bituminous coals with a mean reflectance of 0.74% Rm although heating did not last longer than 5 m.y. (M. Teichmüller 1962). This extraordinary fact is due to local plutonic intrusions. This young magmatic activity of southwest Tuscany (e.g. granite of Elba, volcanics, young ore deposits, fumaroles) is responsible for the high actual heat flow of more than 3 HFU (Mongelli & Loddo 1974).

The distribution of heat flow intensities in the Northern Apennines from the Miocene to the present is intimately related to the geodynamic evolution which consisted in the elimination of the Ligurian Eugeosyncline by subduction and the final collision of the Corso-Sardian continental fragment with the Adria Microplate (Giese et al. this vol.). Thus intense crustal deformation and melting, together with yet unknown processes in the upper mantle, are the ultimate causes of the heating effects at the surface.

References

Bortolotti, V., Passerini, P., Sagri, M. & Sestini, G., 1970: Sed.Geol.4, (3-4),Development of the Northern Apennines Geosyncline (ed. G.Sestini)

341-444.
Bouma, A.H., 1962: Sedimentology of some Flysch Deposits:168 p., Elsevier, Amsterdam.
Mongelli, F. & Loddo, M., 1974: Paper presented at KAPG-Meeting, Sopron.
Teichmüller, M., 1962: Geol. Jg. 80:69-110.
- , 1975: in: Stach, E., Mackowsky, M. Th., Teichmüller, M., Taylor, G. H., Chandra, D. & Teichmüller, R.: Stach's Textbook of Coal Petrology. Second completely rev. ed.: 303-309, Gebr.Borntraeger, Stuttgart.

Fabric Patterns in the Tectonic Units of the Apuan Alps (Northern Apennines)

K. Günther and E. Wallbrecher

The Apuan Alps consist of a polyphase imbrication of several tectonic units which show different degrees of metamorphism and different strain patterns. The following tectonic complexes can be distinguished (see fig. 1): The autochthonous core of the Apuan Alps, the parautochthonous Massa Zone, the parautochthonous schuppen zones of Stazzema, Isola Santa and Campo Cecina, the parautochthonous Pania Unit, and the Tuscan Nappe.

Older papers mostly deal with macrostructures (Ippolito 1950; Wunderlich 1960, 1963; Hartung 1966; Wachsmuth 1966; Günther 1966, 1973; Nardi 1967; Haccard, Lorenz & Grandjacquet 1972; Dallan-Nardi & Nardi 1974). K/Ar dating by Giglia & Radicati di Brozolo (1970) yielded an age of metamorphism of 14 mio.y. for the Massa Zone and of 11 mio.y. for the autochthonous. Since the timing and kinematic sequence of the orogenic events and the correlation between the tectonic units is controversial, we carried out a structural analysis using the methods of quantitative fabric analysis and microtectonics.

Structural Analysis

In ca. 40 outcrops of the autochthonous core, the Massa Zone, the schuppen of Isola Santa and Stazzema, and the allochthonous frame the following fabric elements were measured: Fold axes in mesoscopic scale, δ -lineations, bedding planes, and foliation planes.

At each outcrop at least 50 small fold axes and lineations were measured, and sphere diagrams were plotted for each type of fabric element using the Fortran IV programme GELI (Adler et al. 1968). The distribution of the resulting maxima of the linear elements is shown in fig. 1 (Arrows represent the position of the maxima).

Linear Elements: Minor folds (mostly drag folds on the flanks of macrostructures) show differences in the distribution of statistical maxima and submaxima (fig. 1). The directions of the main maxima correspond generally with the macrotectonic axes: NW-SE directions as the predominant

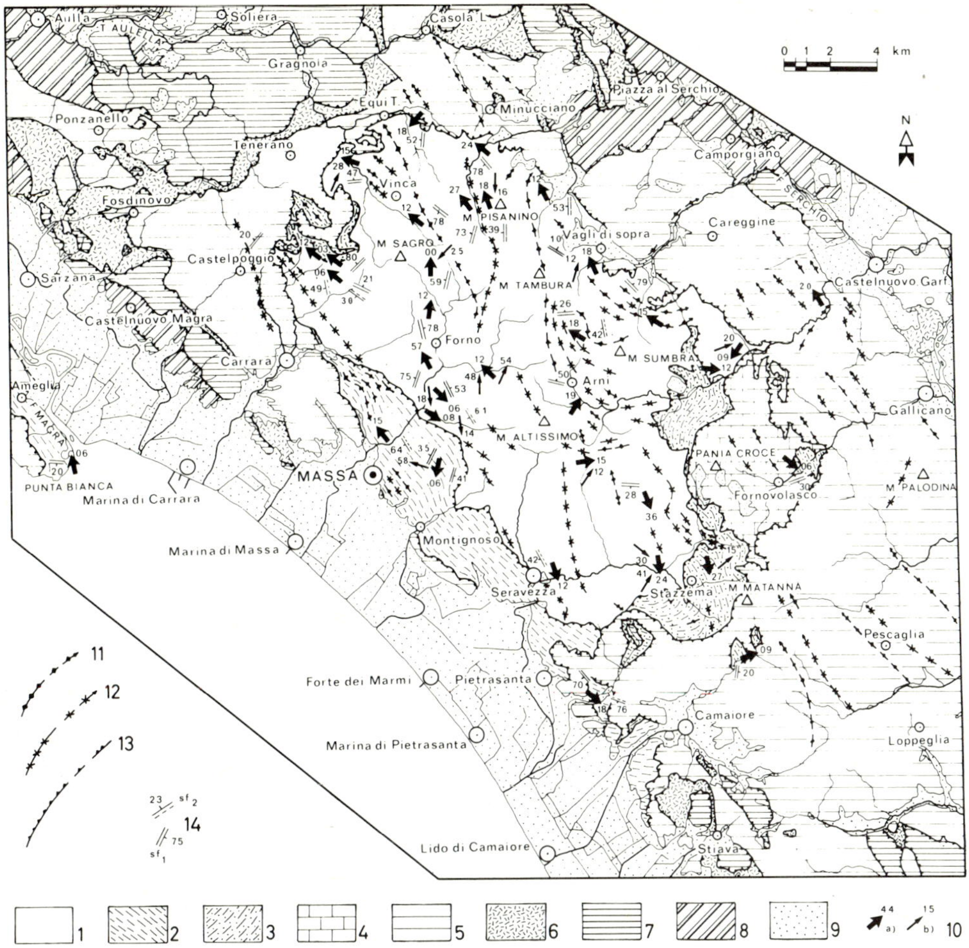

Fig. 1. Tectonic Map of the Apuan Alps. 1 = autochthonous core of the Apuan Alps (Tuscanides I); 2 = parautochthonous Massa Zone, Punta Bianca complex; 3 = parautochthonous schuppen zones of Stazzema, Isola Santa and Campo Cecina, tectonic windows of Fornovolasco, Casoli and Santa Anna; 4 = parautochthonous Pania-unit; 5 = Tuscan nappe (Tuscanides II); 6 = Canetolo-unit (Ligurides I); 7 = Orocco-Caio-unit (Ligurides II); 8 = Ligurian ophiolitic unit (Ligurides III); 9 = neoautochthonous and quarternary deposits; 10 = strike and dip of maxima of linear elements derivated from sphere diagrams (over 50 measurements per diagram) a) main maxima, b) submaxima; 11 = main anticlinal axes; 12 = main synclinal axes; 13 = overthrust, axial plane thrust; 14 = schistosity.

pattern in the autochthonous, N-S directions in the central part of the autochthonous, varying directions in the deepest layers of the core (Calcria del Cipollaio), E-W directions between Vagli di Sotto and Monte Sumbra which twist out of the general Apennine direction.

A synoptic diagram (fig. 2a) of fold axes from the entire autochtho-

nous complex shows an accumulation of data points in the NW direction (Apennine direction); a second accumulation of linears appears in the NE direction. The submaximum in SW direction results from measurements in the Turrite Secca Profile of the schuppen zone of Isola Santa.

In the Massa Zone we also found a main maximum of linear elements in the Apennine direction and a second accumulation in the SSW direction. However, in contrast to the autochthonous, a third system of linear elements is obvious, forming a girdle with another maximum at 336°/56° (fig. 2b). Since these directions are not found in the other diagrams they must be explained as an older system. The girdle distribution and the steep plunge could originate from a rotation of this system around the Apennine direction during the upthrusting of the Massa Zone on the autochthonous core.

Deviating orientations of the main maxima were observed only in the parautochthonous schuppen of Isola Santa and in the window of Casoli (NE-

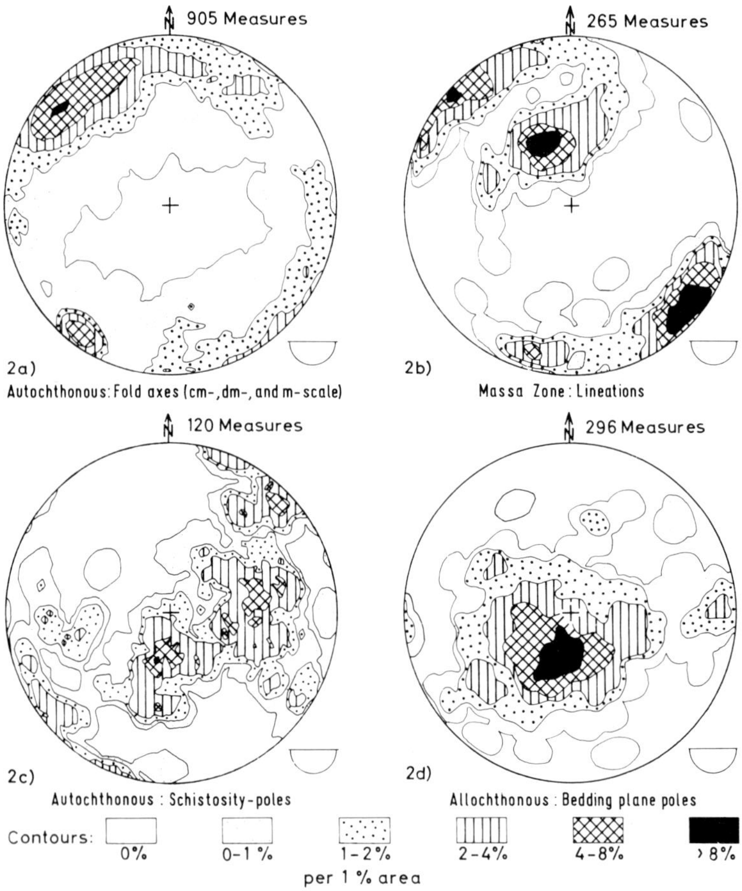

Fig. 2. Synoptic sphere diagrams from the autochthonous core, the Massa Zone and the Tuscan nappe.

SW). The submaxima in the lower tectonic levels of the Apuan Alps show different angles of divergence from the main maxima and an elongation in a girdle towards the main maximum (fig. 2a and 2b). Therefore NE-SW, and other directions in the autochthonous part suggest an older stress system (submaxima) superimposed by a younger shear folding (main maxima). Since the older directions were always observed in the area of the locations where Vai (1972) proved a Silurian age of parts of the phyllite series, and in similar phyllites of the Massa Zone, these directions must be pre-Alpidic structures (Galleria del Cipollaio, Forno, Vinca).

Planar Elements: There are two foliations intersecting at between 30° and 70° almost everywhere in the autochthonous and parautochthonous parts. An older system of narrow schistosity planes is cut and bent sigmoidally (kink bands) by a younger fracture cleavage. In each outcrop the intersection lines of the two systems have the same orientation as the main maximum of the fold axes (fig. 3).

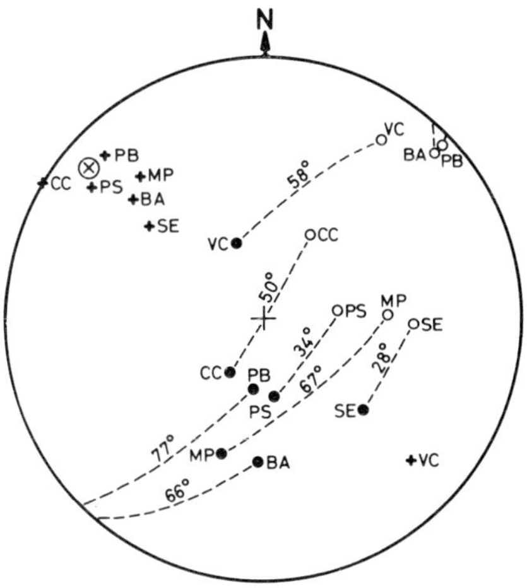

Fig. 3. Relation between the two foliation systems of the autochthonous and parautochthonous parts in single outcrops (full circles = sf_1; open circles = sf_2; + = intersection of sf_1/sf_2; x = maximum of lineations; BA = Poggio Baldozzana; CC = Campo Cecina; MP = Monte Pisanino; PB = Punta Bianca; PS = Passo Sella; SE = San Eustachio; VC = Valdicastello).

In the Tuscan Nappe, framing the Apuan Alps, only one cleavage exists, present only in the incompetent layers. This fracture cleavage shows the same orientation as the younger system of the autochthonous part.

The schistosity in the autochthonous shows a π-circle around the Apennine direction of fold axes (fig. 2a and 2c). The bedding poles in the Tuscan Nappe show the same orientation and indicate a similar stress pattern in the nappe (fig. 2a and 2d).

Vergence: Northeast or east vergence is predominant and can be correlated with the younger cleavage, appearing as an axial plane foliation. Westerly vergence belongs to the older schistosity pattern and is found only in the central part of the autochthonous core.

Conclusions

The Apuan Alps show a mesoscopic axial pattern conformable with the macroscopic NW-SE striking Apenninic direction. In the autochthonous core this direction superposes remnants of an older linear system with a NE-SW

direction of pre-Alpidic age.

During the Apennine orogeny the direction of the main pressure and of the tectonic transport was SW-NE. The Massa Zone overrode the autochtonous in a NE direction, establishing the Apennine direction of fold axes. The pre-Alpidic system of linears was rotated around the Apennine direction and linears with a steeper plunge and the girdle distribution of fig. 2b originated. The emplacement of the parautochthonous schuppen zones of Isola Santa and Stazzema took place in the same time; the Massa Zone might even be the origin of these schuppen.

The older schistosity (sf_1-system) in the autochthonous part originated parallel B with a primary SW vergence and rotated during orogeny into steeper positions still parallel B (fig. 2c and 3).

Probably together with the movement of the Tuscan Nappe towards the NE, foliation systems (sf_2) tautozonal to B with NE vergence began to predominate. The weak folding and the slaty cleavage system in the Tuscan Nappe reproduces the same pattern of stress distribution in a higher tectonic level.

References

Adler, R.E., Krückeberg, W., Pfisterer, W., Pilger, A. & Schmidt, M., 1968: Clausthaler Tekton. H., 8:157 p.

Dallan-Nardi, L. & Nardi, R., 1974: Mem. Accad. Lunigianese Sc. "G.Capellini", 42, (1972), 212 p.

Giglia, G. & Radicati di Brozolo, F., 1970: Bull. Soc. Ital.89:485-497.

Günther, K., 1966: Diss. Freie Univ. Berlin, 179 p.

- , 1973: N. Jb. Geol. Paläont. Abh., 142, 2 (part 1):191-264, 142, 3 (part 2):265-296.

Haccard, D., Lorenz, C. & Grandjacquet, C., 1972: Mem. Soc. Geol. Ital., 11:309-341.

Hartung, W.-D., 1966: Diss. Freie Univ. Berlin, 118 p.

Ippolito, F., 1950: Mem. e Note dell'Ist. Geol. App.Univ.Napoli, 3:49-95.

Nardi, R., 1967: Mem. Soc. Geol. Ital., 6:421-437.

Vai, G.B., 1972: G. Geol., 38, 2:349-372.

Wachsmuth, W., 1966: Diss. Freie Univ. Berlin, 156 p.

Wunderlich, H.G., 1960:Nachr.Ak.Wiss.Göttingen, 2.math.-phys.Kl.6:117-158.

- , 1963: Geol. Rundsch., 53:57-73.

The Significance of the Polymict Rauhwackes (Cargneules) at the Base of the Tuscan Nappe (Northern Apennines)

J. Groscurth, K. Günther, K.-J. Reutter and E. Wallbrecher

In the Northern Apennines greenschist metamorphism of the lower Tuscan unit ("Tuscan autochthon") sharply contrasts with the anchimetamorphism of the overlying Tuscan Nappe. The thrust plane at the base of the Tuscan Nappe is characterized by a thick layer of polymict rauhwackes, which are composed of components of the metamorphic lower unit and the basal parts of the nappe, embedded in an unmetamorphic calcitic cement. The rauhwackes oroginated from an evaporitic polymict breccia.

The rauhwacke layer contains - especially near the contact to the metamorphic lower unit - finer grained flat bodies of some meters of lateral extension which have definite depositional fabrics and approximately the same petrographic composition. Therefore Dallan-Nardi & Nardi (1973), Petacca et al. (1973), Federici & Raggi (1974) maintained that the breccias were sedimentated unconformingly upon the "Tuscan autochthon" after its folding and metamorphism, but before the arrival of the Tuscan nappe. Such an interpretation implies that the metamorphism of the lower Tuscan unit and a following erosional phase preceded the paroxysm of nappe movement in the Northern Apennines. Further Federici & Raggi (1974) postulated that the Tuscan nappe moved across a litoral wave-cut platform.

Groscurth & Reutter (1977) could, however, show that the irregularly shaped, but sharply bounded, deposits are posttectonic young fillings ("sedimentary rauhwackes") of cavities within the surrounding polymict rauhwacke. The primary polymict breccia was formed by friction along the thrust plane and probably consisted predominently of Upper Triassic evaporitic material. It was transformed into the present polymict rauhwacke by subterranous solution of its primary content of gypsum or anhydrite (or even halite). In the course of the solution process cavities were opened and refilled with fine particles, which were washed out of the surrounding collapsing breccia by the circulating groundwaters, and with precipitated carbonates. During this reorganization of the breccia level a NE-vergent cleavage system, which is found in the nappe as well as in the lower Tuscan unit in the form of a late fabric element, was completely obliterated in the breccia (Günther & Wallbrecher this vol.).

The presence of both, metamorphic and nonmetamorphic fragments, in the polymict rauhwacke indicates that the metamorphism was earlier than the last nappe movement. The K/Ar cooling ages of 14 to 11 m.y. (Giglia & Radicati 1970) coincide well with the main period of nappe movements in the Northern Apennines. Furthermore a postmetamorphic emplacement of the Tuscan nappe is very unlikely since, besides the nappe itself, no other cover exists under which the metamorphism of the "Tuscan autochthon" could have taken place. It has been argued (Baldacci et al. 1967) that the unmetamorphic "Modino-Cervarola-Unit" which at present is situated in an allochthonous position external or northeast of the Tuscan nappe, might represent the former top of the "Tuscan autochton" now gravitationally slid off. But its thickness of less than 3000 m does not seem to be sufficient to allow a greenschist metamorphism at its base like that of the "Tuscan autochthon", which requires minimum p/t condictions of 300°C at 2 kbar (Winkler 1976). The only suitable tectonic cover com-

plex, therefore, is the Tuscan nappe. Together with the piggyback stack of already emplaced Ligurian nappes, it would have reached the required thickness of 6000 m. Thus tectonics and metamorphism in the Tuscan units may have developed as follows:

The Tuscan nappe, together with the overburden of at least parts of the Ligurian nappes, sheared off along an Upper Triassic evaporitic layer from its basement rocks as these were affected by compressional tectonics (Late Miocene). It moved, probably controlled by gravity, towards the external parts, overthrusting a similar sequence of rocks which could not shear-off for the lack of an evaporitic layer (Trevisan 1955). During the nappe movement and advancing compressional tectonic, this lower Tuscan unit became strongly folded and underwent a dynamothermal metamorphism, which, however, did not affect the capping nappe. This was possible because the gravitationally gliding nappe could not be stressed in the same way as the lower unit and its relatively cold material did not remain in contact with the hot areas of active metamorphism long enough to become metamorphosed to the same degree as the lower Tuscan unit. Thus the metamorphic components of the polymict rauhwackes must have been derived from those parts of the lower unit which were meatmorphosed beneath frontal parts of the advancing nappe. They were then ripped off and incorporated into the breccia by the continuing movement. This "rabotage" also enhanced the contrast of metamorphism between the two Tuscan units.

The various tautozonal systems of two schistosity planes in the lower Tuscan unit (Günther & Wallbrecher this vol.) also fits this model. They may have been caused by rotation of the main stresses during the migration of the belt of compressional tectonics towards the external (NE) parts. The strong deformation of this unit may indicate that with respect to its crystalline basement, it was also displaced in a final stage of orogeny, together with the overlying nappes. This would best explain why the posttectonic heat-flow was not able to metamorphose the Tuscan nappe extensively. Only a small transitional zone of metamorphism would then have been formed which might have been destroyed by the reduction of thickness of the primary breccia when the evaporites were extracted and the present rauhwackes were formed.

References

Baldacci, F., Elter, P.,Giannini, E., Giglia, G., Lazzarotto, A., Nardi, R. & Tongiorgi, M., 1967: Mem.Soc.geol.ital.,6:213-244, Pisa.

Dallan-Nardi, L. & Nardi, R., 1973: Boll.Sco.geol. ital., 92:435-452,Roma.

Federici, P.R. & Raggi,G., 1974:Boll.Soc.geol.ital.,93:709-722, Roma.

Giglia, G. & Radicati di Brozolo, F., 1970: Boll.Sco.geol.ital., 89: 485-498, Roma.

Groscurth, J. & Reutter, K.J., 1977: N.Jb.Geol.Paläont.Mh.,8:449-465.

Günther, K. & Wallbrecher, E., 1977: Ateneo Parmense, acta nat., 13(2): 175-205, Parma.

Patacca, E., Rau, A. & Tongiorgi, M., 1973: Atti Sco. tosc. Sci.nat. Mem. Ser. A., 80:126-161, Pisa.

Trevisan, L., 1955: Atti Soc.tosc.Sci.nat.,Mem.Ser.A,62(1):1-30, Pisa.

Winkler, H.G.F., 1976: 4th Ed. 334 p, Springer, Berlin.

The Nivione Structure: Contemporaneous Tectonics and Sedimentation between the Ligurian and Tuscan Orogenies

H. Ibbeken

In the course of the Ligurian phase (Alpine orogeny, L.-M. Eocene) the Ligurides were largely joined onto the Paleoeuropean plate (Corsica, Schistes Lustrés, Voltri Group, Antola Group). In the Tuscan phase (Apennine orogeny, Tortonian) subduction took place beneath this intensive sedimentation (Ranzano-Formation) as well as tectonic foreshortening (U.-Oligocene) occurred. During these processes especially the tectonic contacts between the Ligurian sub-units were reactivated and the Ranzano-Formation was synsedimentarily folded.

In the area of the Nivione structure (fig. 1) the Ligurian sub-units Antola and Cassio were shortened by at least 1.5 km. The overlying approx. 800 m thick Ranzano Formation was folded during the sedimentation of the Nivione marls, forming an 11 km long syncline. The anticlinal element adjoining to the north was sheared off. Thus a 7 km long, about 60 m high slope was formed on the sea floor of the Nivione marls, with a slope gradient of approx. 10°. The tectonic shaping of this submarine relief was finished by the beginning of the Nannoplankton Zone NP25 (24.3. m.y.). During the Nannoplankton Zone NP25 the relief was filled up by the turbidity currents of the Castagnola Formation. The turbidites wedge out at angles between 6 and 7° towards the slope. The interbeds are mainly deposited by turbidity currents and not by 'background'-sedimentation.

Post-Ligurian and pre-Tuscan movements can only be recognized when post-Ligurian sediments are overlain unconformably by pre-Tuscan sediments. At the end of the Nannoplankton Zone NP25 (22.5 m.y.) the submarine slope was filled up and the sedimentation continued on into the Miocene.

Post-Ligurian and pre-Tuscan movements can only be recognized by the deformation of Eocene and Oligo-Miocene rocks. If the occurrence of such sediments were not so incomplete on the Ligurides, movements like the ones shown here could probably be shown more often. Thus this probably represents only a minimum activity. A closer continuity of tectonic movements within the orogenic interval cannot be ruled out.

References

Boccaletti, M., Elter, P. & Guazzone, G., 1971: Mem. Soc. Geol. Ital. - 10:371-378.
Ibbeken, H., 1970: Beih. Geol. Jb. - 93:140 p.
Reutter, K.-J., 1968: Eclogae geol. Helv. - 61:183-224.

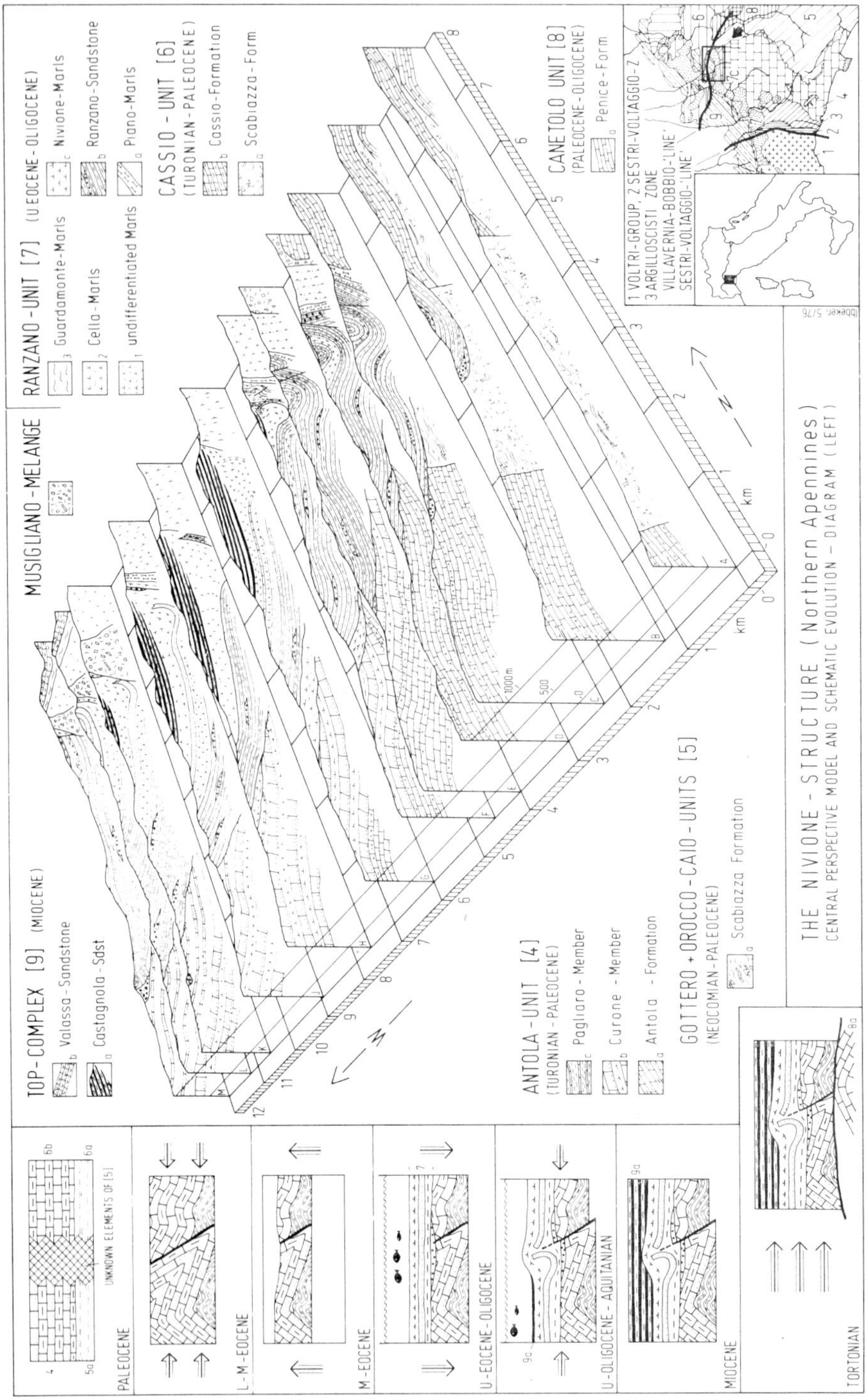
TOP-COMPLEX [9] (MIOCENE)
b Valassa-Sandstone
a Castagnola-Sdst
MUSIGLIANO-MELANGE
RANZANO-UNIT [7] (U.EOCENE-OLIGOCENE)
3 Guardamonte-Marls
2 Cella-Marls
1 undifferentiated Marls
c Nivione-Marls
b Ranzano-Sandstone
a Piano-Marls
CASSIO-UNIT [6]
(TURONIAN-PALEOCENE)
b Cassio-Formation
a Scabiazza-Form.
CANETOLO UNIT [8]
(PALEOCENE-OLIGOCENE)
a Penice-Form.
1 VOLTRI-GROUP, 2 SESTRI-VOLTAGGIO-Z.
3 ARGILLOSCISTI ZONE
VILLAVERNIA-BOBBIO-'LINE'
SESTRI-VOLTAGGIO-'LINE'
ANTOLA-UNIT [4]
(TURONIAN-PALEOCENE)
c Pagliaro-Member
b Curone-Member
a Antola-Formation
GOTTERO + OROCCO-CAIO-UNITS [5]
(NEOCOMIAN-PALEOCENE)
a Scabiazza Formation
THE NIVIONE-STRUCTURE (Northern Apennines)
CENTRAL PERSPECTIVE MODEL AND SCHEMATIC EVOLUTION-DIAGRAM (LEFT)
PALEOCENE
UNKNOWN ELEMENTS OF [5]
L-M-EOCENE
M-EOCENE
U-EOCENE-OLIGOCENE
U-OLIGOCENE-AQUITANIAN
MIOCENE
TORTONIAN
km
1000 m
500
0
N
W

Fig. 1

Tectonics of the Parma Province Apennines

G. Zanzucchi

The tectonic map of the folder, fig. 1, shows the main tectonic units of the Northern Apennines, between the valleys of the Stirone and the Secchia rivers.

The southwestern part of the area contains, from bottom to top, the Pracchiola (Modino-Cervarola) unit (11 on fig. 1) consisting of Eocene to Lower Miocene shales and graywackes with olistostromes, the Tuscan (Orsaro) unit (10) with Middle Cretaceous to Lower Oligocene marls and Middle Oligocene to Lower Miocene graywackes (Macigno), the Argille e calcari (Canetolo) unit (9) of possible Upper Cretaceous to Lower Miocene age, containing flysches and tuffs, the Caio unit (8) with shales, turbidite sandstone, olistostromes and calcareous flysch of Lower Cretaceous to Paleocene age which are unconformingly overlain by marls and sandstones of Upper Eocene to Middle Miocene age, the Bracco and the Gottero units (7) with ophiolites, Jurassic and Lower Cretaceous pelagic sediments and graywackes and shales up to the Paleocene.

The northeastern part of the area contains, also from bottom to top, the Salsomaggiore (Umbro-Marche) unit (5 on fig. 1) with Oligocene to Middle Miocene sands and shales, the Sporno (Luretta) unit (4) with flysches from the Middle Cretaceous to the Middle Eocene and marls and sandstones from the Upper Eocene to the Middle Miocene, the Dosso (Farini d'Olmo) unit (3) with flysches of Middle Cretaceous to Lower Eocene age, the Cassio unit (2) with a Jurassic to Upper Miocene sequence which is partly similar to the Caio unit but shows conglomerates of Middle Cretaceous age, the Solignano (Monghidoro) unit (6) of Upper Cretaceous to Oligocene age which also is similar to the Cassio and Caio units. At the border of the Po Plain this stack of thrust units is covered by a neo-autochthonous unit (1 on fig. 1) of Late Miocene to Quarternary conglomerates and clays.

Units 1-5 only occur in the external Emilian Province. Units 2-4 and 6-9 are at least partially ophiolitic, unit 9 is not ophiolitic at least in the investigated area, while units 1, 5, 10 and 11 are not ophiolitic. In the paleogeographical schemes which have been used by the authors, these units are distinguished in eugeosynclinal and miogeosynclinal sequences (Abbate et al. 1970; Groscurth & Reutter, this vol.).

In the geologic cross sections of fig. 1 there are two different tectonic interpretations which concern the problem posed by the position of the external Ligurid units (Emilianide units of some authors).

In the first interpretation of section A-A', which may also be applied to section B-B', the Caio flysch is taken as the Cretaceous stratigraphic base of the Tertiary "Marne rosate", the Dosso flysch and the Sporno flysch (Braga 1965). Therefore, the Cretaceous Cassio flysch would belong to an independent tectonic unit (Abbate & Sagri 1967), which had been situated originally either internal or external or lateral to the Caio

Fig. 1 (foldout): Apennines of the Parma Province: Tectonic map and geologic cross section (two versions).

flysch. This hypothesis also admits the possibility that the first Ligurid unit (Canetolo complex) may have been partially connected to Caio flysch basin and perhaps also to the outer portion of Bracco basin (Groscurth & Reutter, this vol.; Braga et al. 1973).

The second interpretation, concerning section B-B', but which can also be applied to the section A-A', takes into account the tectonical reassemblage by Anelli (1935) and Monteforti (1972) in which the Cassio flysch is considered as a paleogeographically external variation of the Caio flysch. In this hypothesis the Dosso and Sporno Tertiary flysches, with their Cretaceous ophiolitiferous stratigraphic bases, would have been in even more external positions, perhaps separated by an ophiolitic or crystalline ridge from the Cassio and Caio basin.

The third interpretation, less probable but not impossible, suggests that the mentioned Tertiary sequences occupied a position internal to Caio basin and external to Bracco and Gottero basins; consequently, the Sporno flysch would have been the lateral variation of the coeval and more external Canetolo (Argille e calcari) complex. The latter might have been situated either close to a fossil island arc which separated the Caio flysch basin from Sporno flysch basin, or might have been deposited partially on top of the moving Caio unit. According to these hypotheses it would be difficult to understand the kinematics of the Canetolo unit which in this case, in order to reach the area of the Tuscan Macigno in the Lower Miocene, had to pass over the more external Ligurid units without displacing or disturbing them. By late overthrust movements during the Middle Miocene either the Caio and Cassio units or all external Ligurids would have overridden the Canetolo complex in its allochthonous position.

Another problem is the paleogeographic location of the Solignano-Monghidoro unit. This unit, with its Upper Cretaceous flysch, might be located either in the innermost part of the Ligurid basin or on the southern extension of Bracco zone.

Tectonic Evolution and Comparison between Tuscan and Salsomaggiore Units (Umbro-Marches)

The Canetolo unit overlies the anticlinal structures of M. Orsaro and Pracchiola, while the Salsomaggiore anticline is covered by the Sporno unit (or by variegated clays which tectonically underly the M. Sporno unit). Although the exact age of the stratigraphic tops of the sequences beneath the allochthonous cover has not yet been determined, overriding may have occurred at different times: early Miocene on top of the Macigno, early to Middle Miocene on the Pracchiola unit and Middle to Late Miocene on the Salsomaggiore area. This situation is not confined to the Parma province, but it is common to the whole of the Emilian Apennines. Different possible interpretations can be drawn from the previous assumptions:

a) The Sporno unit, deposited in a more internal basin than the Canetolo unit, overrode the latter unit which was already overlying the Tuscan basin, before overthrusting the Salsomaggiore zone in Middle to Late Miocene time.

b) The Sporno unit, originally in a more external position than the Canetolo basin, overrode the Tuscan basin and the Pracchiola (Modino-Cervarola) basin before the Lower Miocene, without leaving any traces on them,

and then moved outwards to the Salsomaggiore area in the Middle to Late Miocene.
c) The M. Sporno unit, deposited in a more external position than the Tuscan basin, presumably overthrust the Salsomaggiore area, after the Canetolo unit had been emplaces upon the Tuscan and Pracchiola (Modino-Cervarola) units, but before the other Ligurian units had overridden the Canetolo unit and could reach the border to the Po Plain.

Conclusions

In a model of continually evolving tectonics, one might recognize the following movements:
- Lower Cretaceous to middle Upper Cretaceous: The sedimentation of the basal complexes in the Ligurid basins and in part also in the Tuscan basin, is disturbed and affected by gravity phenomena like folds, olistostromes and breccias.
- Upper Cretaceous to Paleocene: There are some unconformities within the Gottero basin and probably also within the Caio, Dosso and Sporno basins, as well as in some parts of the Tuscan basin.
- Lower Eocene to Upper Eocene (Ligurian phase): strong folding occurs in all Ligurian basins, with transport towards the Alps and individualization of the first tectonic units; possible tectonical covering of the Dosso unit.
- Upper Eocene to Lower Miocene: olistostromes and slumping in the Ligurian and Tuscan basins as well as in the Modino-Cervarola basin; earlier overthrust upon the Canetolo complex.
- Lower Miocene to Middle Miocene: arrival of the Canetolo complex upon the Tuscan Macigno and other formations of the Modino-Cervarola unit; further advancement of the Ligurian units over the allochthonous Canetolo complex.
- Middle Miocene to Upper Miocene: the Ligurid units override the Tuscan Macigno and Modino-Cervarola unit, which were already overlain by the Canetolo unit; arrival of the M. Sporno and the M. Cassio units, both connected by a sedimentary cover of Tortonian sediments, upon the Salsomaggiore area after its probable uplift and partly erosion.
- Upper Miocene to Lower Pliocene: strong overthrusts of all Ligurid units towards the Po basin, with formation of olistostromes intercalated in the Tortonian deposits, in the Messinian either at the base of the gypsum or in its top and under the overlying clastic brackish-water sediments, and even at the base of the clastic Lower Pliocene.
- Lower Pliocene to Quaternary: wide spread marine transgression over the Emilian Subapennines, olistostromes in the internal part of Pliocene sedimentary basin, uplift with local erosion up to Middle Pliocene then slow oscillatory upward movements in all of the Northern Apennines.

References

Abbate, E., Bortolotti, V., Passerini, P. & Sagri, M., 1970: Sediment. Geol. - 4:207-249.
Abbate, E. & Sagri, M., 1967: Mem. Soc. Geol. Ital. - 6(1):23-65.
Anelli, M., 1935: Giorn. Geol.,Ann.Museo Geol. Bologna. - 10:1-27.
Braga, G., 1965: Atti Ist. Geol. Univ. Pavia. - 17:3-49.

to p. 278

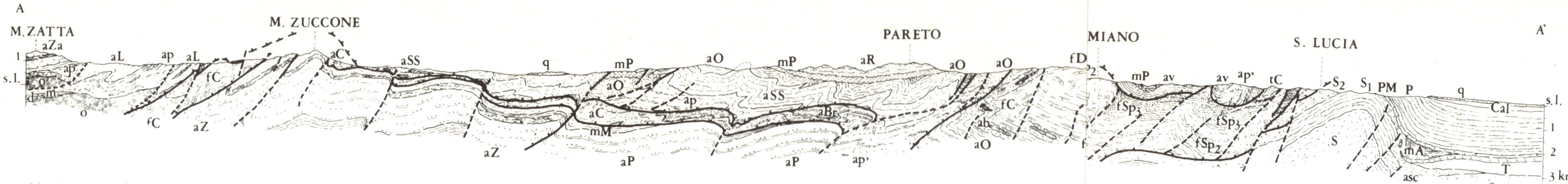

mM - Marra marls (low. Mioc?); aP - Pracchiola sandstones (Olig. - low. Mioc?); mO - M. Orsaro sandstones (Macigno) (Olig. middle- upp.); aC-Canetolo complex (Paleoc. - midd. Eoc.); aBr - Bratica sandstones (Olig.?); aPe - Petrignacola sandstones (Olig.? - low. Mioc?); aSS - S. Siro shales (Alb. - Turon.); aO - Ostia sandstones (Cenom. - Turon.); ab - Caio « complesso base » (Cenom. - Turon); fC - M. Caio flysch (upp. Cret.); fD_{1-2} - M. Dosso flysch and « marne rosate » (Paleoc. - low. Eoc.?); fSP_{1-2-3} - M. Sporno flysch (Paleoc. - midd. Eoc.); o - ophiolites; d - « diaspri » (Malm); m - calpionella limestones (Titon. - Neocom.); ap, ap' - palombini shales (Neocom.); aL - Lavagna shales (upp. Cret.); aZa - Gottero sandstones (upp. Cret. - Paleoc.); av - varicoloured shales (upp. Cret. - Paleoc.?); fCa - M. Cassio flysch (upp. Cret.); mP - M. Piano marls (upp. Eoc. - low. Olig.?); aR, aR' - Ranzano sandstones (Olig.); tC - Contignaco tripoli (low. Mioc.?); mT - Termina marls (Torton.); S - Salso low. sequence (Lang.? - Olig.?); - S_1 - Salso marls (low. Mioc.); S_2 - Salso sandstones (Serraval. - Torton?); T - (Tortonian); asc - « argille scagliose » (oliostostromes); g - (Messin.); PM - (upp. Mioc. - low. Plioc.); P (Plioc.); Cal. (Calabr.); q (Quatern. cont. depos.)

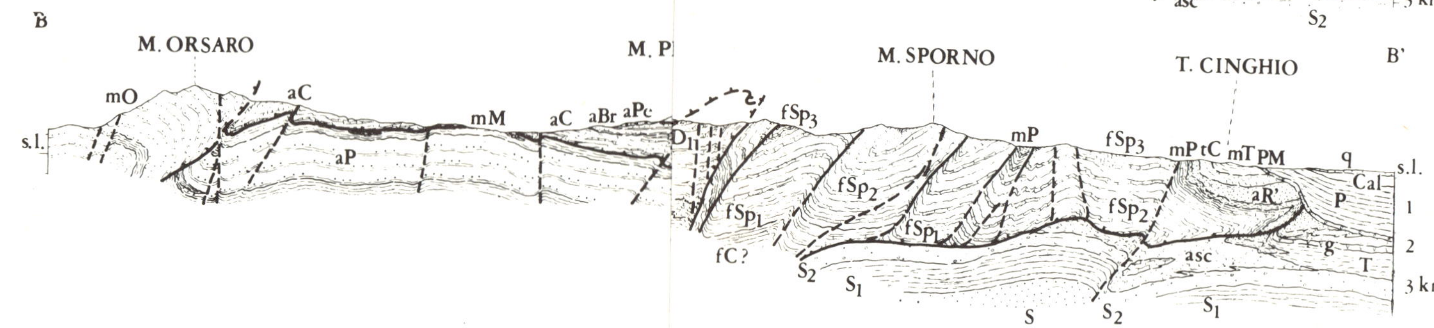

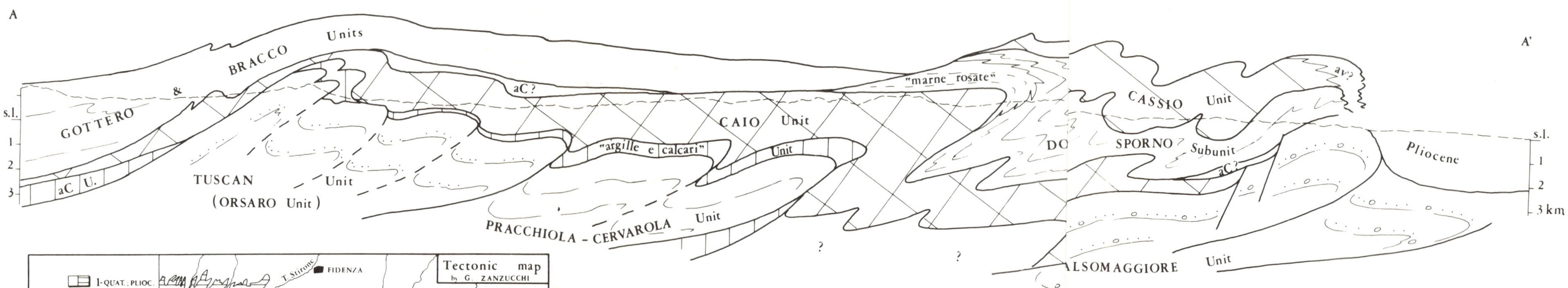

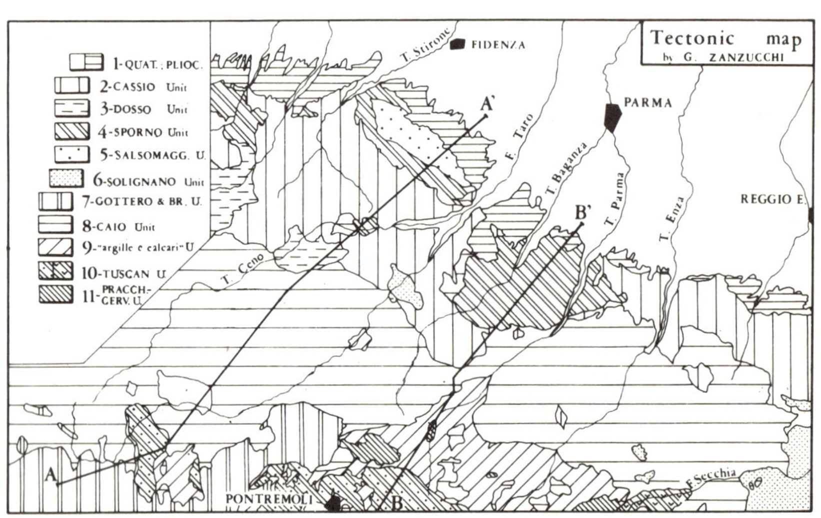

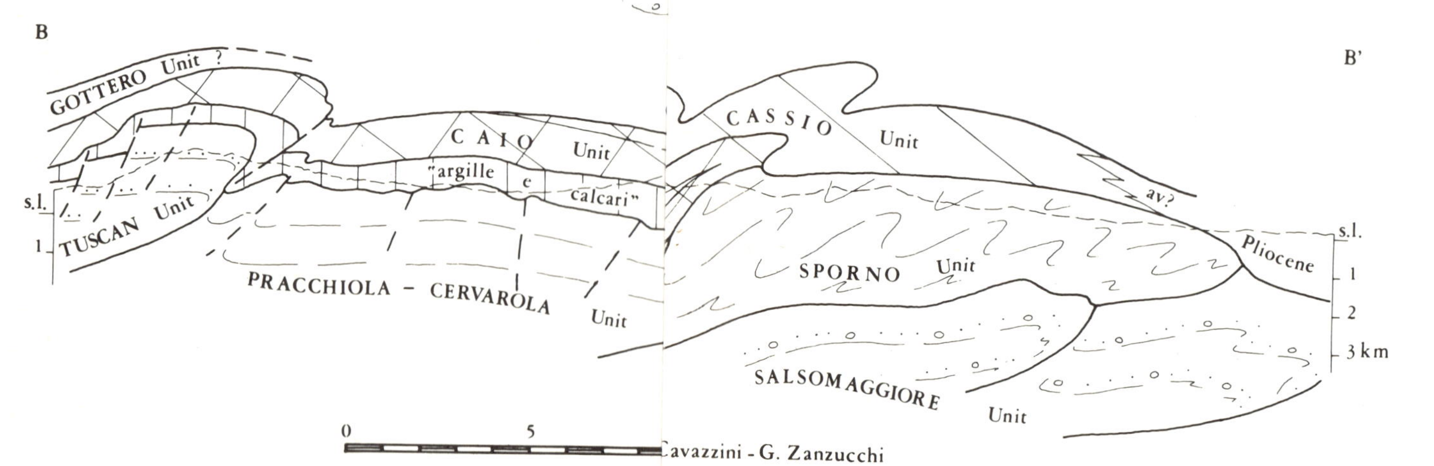

Fig. 1. Apennines of the Parma Province: Tectonic map and geologic cross section (two versions).

Braga, G., Casnedi, R., Galbiati, G. & Marchetti, G., 1973: Atti Ist. Geol. Univ. Pavia. - 23:Tav.22.
Elter, G., Elter, P., Sturani, C. & Weidmann, M., 1966: Arch. Soc. Phys. Nat., Genève, 19(3):279-377.
Iaccarino, S., Papani, G., Rio, D. & Zanzucchi, G., 1974: L'Ateneo Parmense, Acta Naturalia, 10:381-408.
Istituto di Geologia Università di Parma, 1965: Carta Geologica della Provincia di Parma, Litografia Artistica Cartografica, Firenze.
Monteforti, B., 1972: Mem. Soc. Geol. Ital. - 11:47-64.
Mutti, E., Parea, G.C., Ricci Lucchi, F., Sagri, M., Zanzucchi, G., Ghibaudo, G. & Iaccarino, S., 1975: IX Intern. Congr. Sediment., Nice, Field Trip A 11:1-120.
Pagani, G., Papani, G., Rio, D., Torelli, L., Zanzucchi, G. & Zerbi, M., 1972: Mem. Soc. Geol. Ital. - 11:531-546.
Pagani, G. & Zanzucchi, G., 1970: L'Ateneo Parmense, Acta Naturalia, 6(1):3-42.
Reutter,K.-J., 1968: Eclogae Geol. Helv. - 61(1):183-224.
Zanzucchi, G., 1972: Sezioni Interpretative 1:100.000 della Carta Geologica della Provincia di Parma, Litografia Artistica Cartografica, Firenze.

Stratimetric Comparison of Graywacke-Sequences in the Northern Apennines

H. Burger, J. Groscurth and K. Günther

In the Northern Apennines several graywacke-flysch-formations of the Oligo-Miocene age attain accumulative thickness of about 8 to 10 km. These formations are marked by the regular alternation of turbiditic sedimentated strata (graywacke) and of interstrata (pelites) of different origin. By a stratimetric comparison, we investigated whether the individual flysch formations show specific characteristics and whether there are trends common to all formations which would allow conclusions to directed geodynamical processes.

Twelve profile sections in the five investigated formations (Aveto-, Macigno-, Modino-, Cervarola-, Formazione Marnoso Arenacea-Sandstones) were measured bed for bed with more than 12 000 measurements and logs were made of grain size, bed-internal sequences (acc. to Bouma) and petrographic composition. The data were evaluated with the help of electronic data processing methods in order to obtain rhythmograms and autocorrelation functions of strata and interstrata etc. It was then possible to recognize a pattern of development that fits all formations under consideration. This pattern of a completely developed and undisturbed graywacke-formation displays three sections superimposed upon each other (see fig. 1).

Section A represents the actual onset of the turbiditic stage. The onset of the almost exclusively turbiditic sandstones is marked either by thin sandstone beds which gradually get thicker upsection or by a sudden onset of several thick sandstone beds. The average sandstone bed-thickness

of this section and the average grain size of the sandstones are s than those of the section B. The thickness of successive strata resp. ter strata is randomly distributed (autocorrelation coefficient r is zero; see fig. 1).

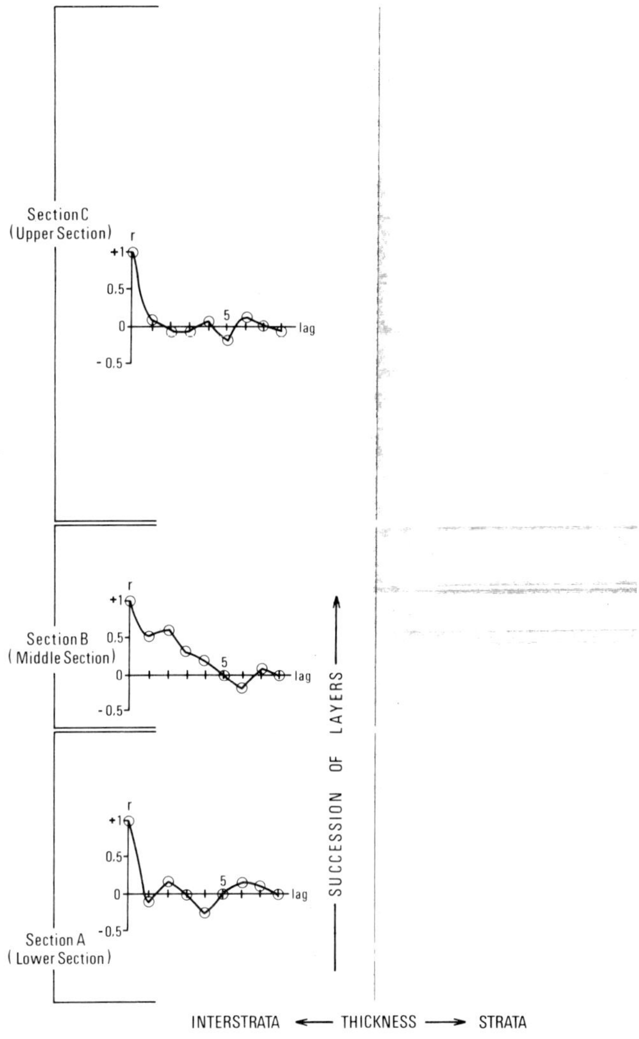

Fig. 1. Rhythmogram of a normal developed graywacke sequence with sliding correlation-coefficient r series for the sandstone-strata. The diagram is compiled of a much longer complete Macigno section, that had been registrated in the Gordana valley. No scale for the thickness is given because the diagram can be taken as representative for the other investigated flysch formations too, even if the absolute thickness of the beds will differ. The seemingly equal thickness of strata in section B is caused by the limited width of the plotter paper.

Inter-Union Commission on Geodynamics. Scientific Report No. 38

Alps, Apennines, Hellenides

Geodynamic Investigations along Geotraverses by an International Group of Geoscientists

Editors: H. CLOSS, Hannover, D. ROEDER, Tennessee, K. SCHMIDT, München

Addendum to page 280, article BURGER, GROSCURTH & GÜNTHER, "Stratimetric Comparison of Graywacke-Sequences in the Northern Apennines"

This figure replaces fig. 1 shown on page 280 as the original drawing has not been printed correctly.

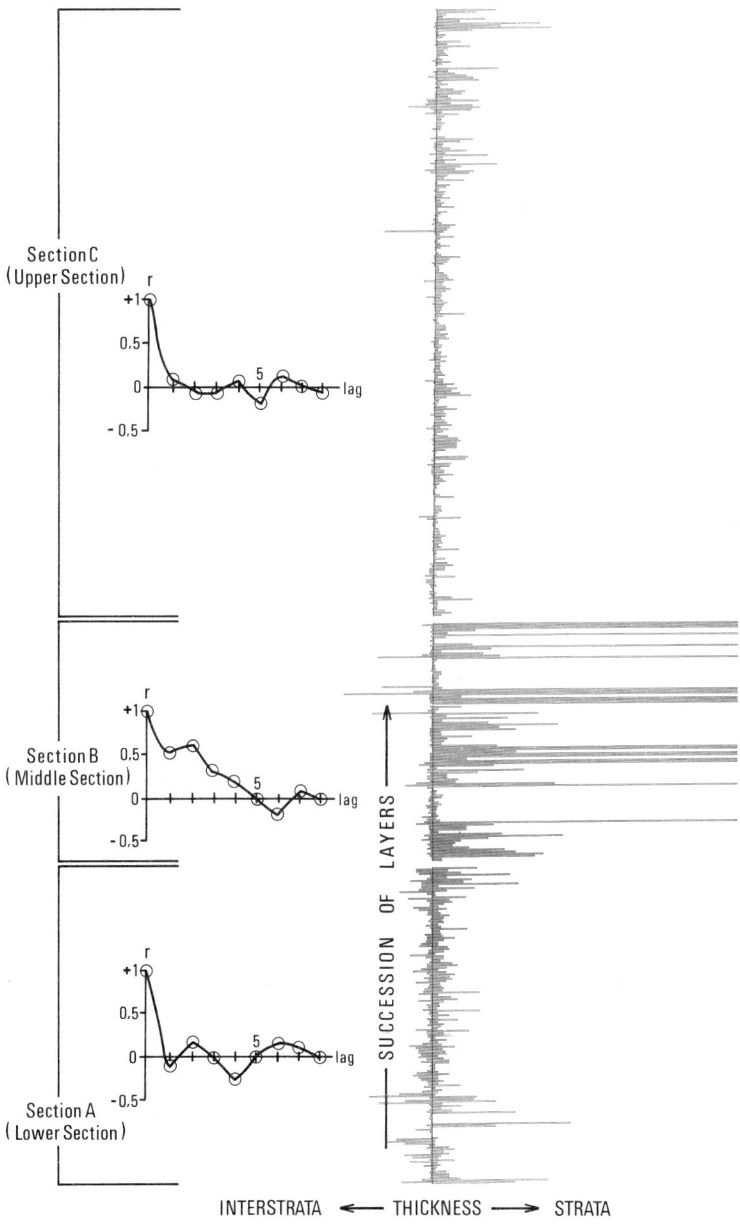

Section B is characterized by the thickest turbidites of the entire series. Furthermore, the sandstones and sometimes the interstrata, too, are grouped clearly in clusters (megarhythms), i.e. they are not randomly distributed. The correlation factor of the thickness of successive strata is generally high. The thickness of the interstrata is often very reduced (amalgamated sandstone beds). The increase of the sandstone thickness is usually accompanied by an increase in grain size up to conglomerates or fluxoturbidites.
Section C is similar to section A. The transition to the almost pure pelitic sedimentation in the upper parts of this section may show sometimes again a distinct increase in thickness of the sandstone beds.

The transition between the sections are not sharp. Even if the absolute thickness of the sandstone layers within the section B are dependent upon the respective trough and the respective position within the trough, the relationship between the three sections are relatively constant.

Even if it could not be fully confirmed because of the lack of sufficient well exposed outcrops, there appear to be lateral heteropic relations between the three sections perpendicular to the striking of the trough-axis. It is assumed that all sections of a flysch-trough under-

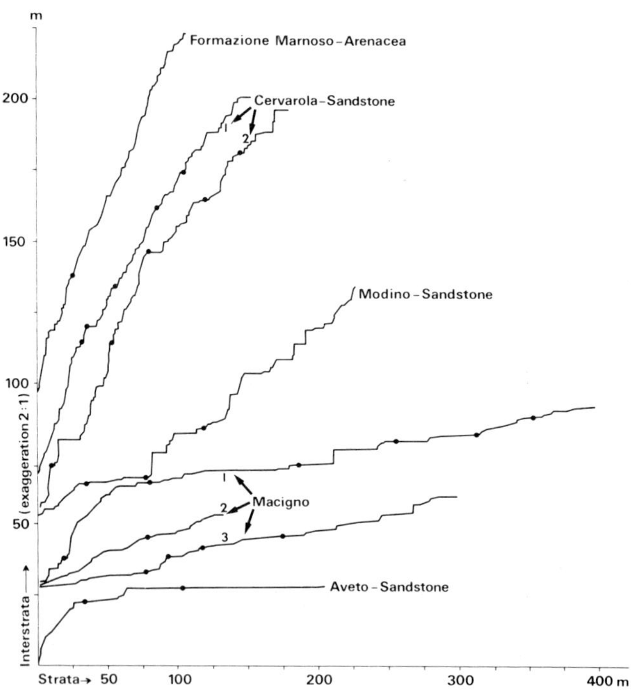

Fig. 2. Stepwise constant functions of the strata-interstrata ratio of five graywacke formations. The dots in the curves are marking short interruptions in the measured sequences. Localities: Form.Marnoso Arenacea at Colli di Casaglia; Cervarola sandstone 1 at Villore (Mugello), 2 at Civago; Modino sandstone at Le Tagliole; Macigno 1 at Val Gordana, 2 at Manarola, 3 at Chianti mountains; Aveto sandstone at Aveto valley.

went all three stages of development if no additional events disturbed the actual flysch-sedimentation, such as the arrival of olisthostromes or gravity nappes.

A comparison of the stepwise constant functions for the individual formations shows another striking characteristic (fig. 2): The strata-interstrata-ratio, represented by the angle of ascent of the functions slides not only from section A through B to C within the individual formations, but also the average angle of ascent of the individual curves is different. The decrease of the strata-interstrata-ratio runs from Aveto- to Macigno- to Modino- to Cervarola-Sandstones up to the Formazione marnoso-arenacea, which means that this trend follows just the probably original palaeogeographic arrangement of the flysch depositional areas from SW to NE.

Even though the origin of all described functions is not the same (not all measurements begin directly above the basal pelites), this trend seems to be too marked to be random.

Conclusions

The fact that the formations show a maximum of turbidite-sedimentation in the middle section allows several conclusions:

1) As the extent of turbidite-sedimentation depends upon the synsedimentary orogenetic events in the source area (backland?), the section B can be interpreted either as a result of the maximal uplift of the source areas of clastic material or as a result of elevation and erosion of source areas that were situated progressively closer to the depositional area than those of section A. In the latter sense the turbidite-sequence of the section A is more distal and of section B more proximal to the source areas. Section C and the normally following turbidite-free pelitic sedimentation in this case, could mark the stage of the progressive involvement (elevation) of the flysch-trough itself into orogeny (shifting of trough-axis to external parts).

2) On the other hand, since it is probable that the main part of a turbidity current is deposited in the deepest parts of the trough, section B can also be interpreted as the result of major subsidence of the individual trough area.

The above mentioned decrease of the strata-interstrata-ratio (fig. 2) from the internal to the external formations may possibly be seen as an effect of redeposition of clastic material from internal to external areas.

In any case the uniform stratimetric patterns within the individual formations and perhaps the change of the strata-interstrata-ratio can be explained by the migration of the flysch deposition in time and space. Since the synorogenetic flysch-series represent, indirectly, an image of the geodynamical processes in the underground, the described patterns are interpreted as the result of the migration of the orogenetic activity from internal to external parts of the orogen.

References

Bouma, A.H., 1962: Sedimentology of some flysch deposits, 168 p., Amsterdam.

Curcio, M., Pranzini, G. & Sestini, G., 1968: Boll. Soc. Geol. Ital. 87(3):623-641.

Kimura, T., 1966: Bull.Earthquake Res. Inst., 44:561-607.
Mutti, E. & Ricci Lucchi, F., 1972: Mem.Soc.Geol.Ital.,11(2):161-199.
Ricci Lucchi, F., 1969: Com. Medit. Neog. Strat. Proc. IV Sess. - Giorn. Geol. ser. 2, H. 35(4):161-198.
Sagri, M., 1973: Boll.Soc.Geol.Ital. - 92,2:233-272.
Sestini, G., 1970: Sedimentary Geology, 4,3/4:559-598.
Walker, R.G., 1967: Journ. Sed. Petr. - 37:25-43.
Wunderlich, H.G., 1966: Hochschultaschenbücher Nr.339,339a,339b:367, Inst. Mannheim.

Evidence of Neotectonics in the North-West Apennines - Po Side

G. Marchetti, G. Papani and M. Sgavetti

Foreword

The study of recent or actual tectonic problems in the northern Apennines has begun some years ago, on the basis of mainly geomorphological and geological researches, for a regional setting of the problem and particularly in a stratigraphical view, in order to date the movements.

The methods applied here are based on photointerpretation at different scales of both satellite images for focusing regional trends, and 1 : 14 000 photos for the solution of local problems; on geological and geomorphological field surveys, and on already known geological data.

At the present state of the research it is possible to draw a morphostructural scheme of the north-west area of the Emilian Apennines and to point out some fracture zones presenting indications of very recent (Quaternary) or active movements.

Morphostructural Scheme of the Emilian North-West Apennines (Papani et al. 1975)

The Po-valley side of the north-west Apennines may be divided into five morphostructural belts (see fig. 1). Their boundaries coincide, mostly, with important dislocations of which only a few seem to be locally active.

The belt of the upper Apennines structures (Belt I) corresponds to the sandstones Unit of the main ridge (Ia) and of the tectonic windows of the Taro, Trebbia and Aveto valleys (Ib) and to the overlying Unit containing the Macigno, Modino-Cervarola and Canetolo Units. It is characterized by a young morphology due to considerable Quaternary uplift, well documented by good examples of incised meanders (Belt Ib).

On the Tyrrhenian side, the Lunigiana and Garfagnana grabens, well known as seismic regions, are linked to Ia. The filling of the Villafranchian lacustrine basins located in these grabens has been caused by the uplift and the following erosion of the Ia belt's rocks; it happened about in the middle Pleistocene and certainly before the last glaciation,

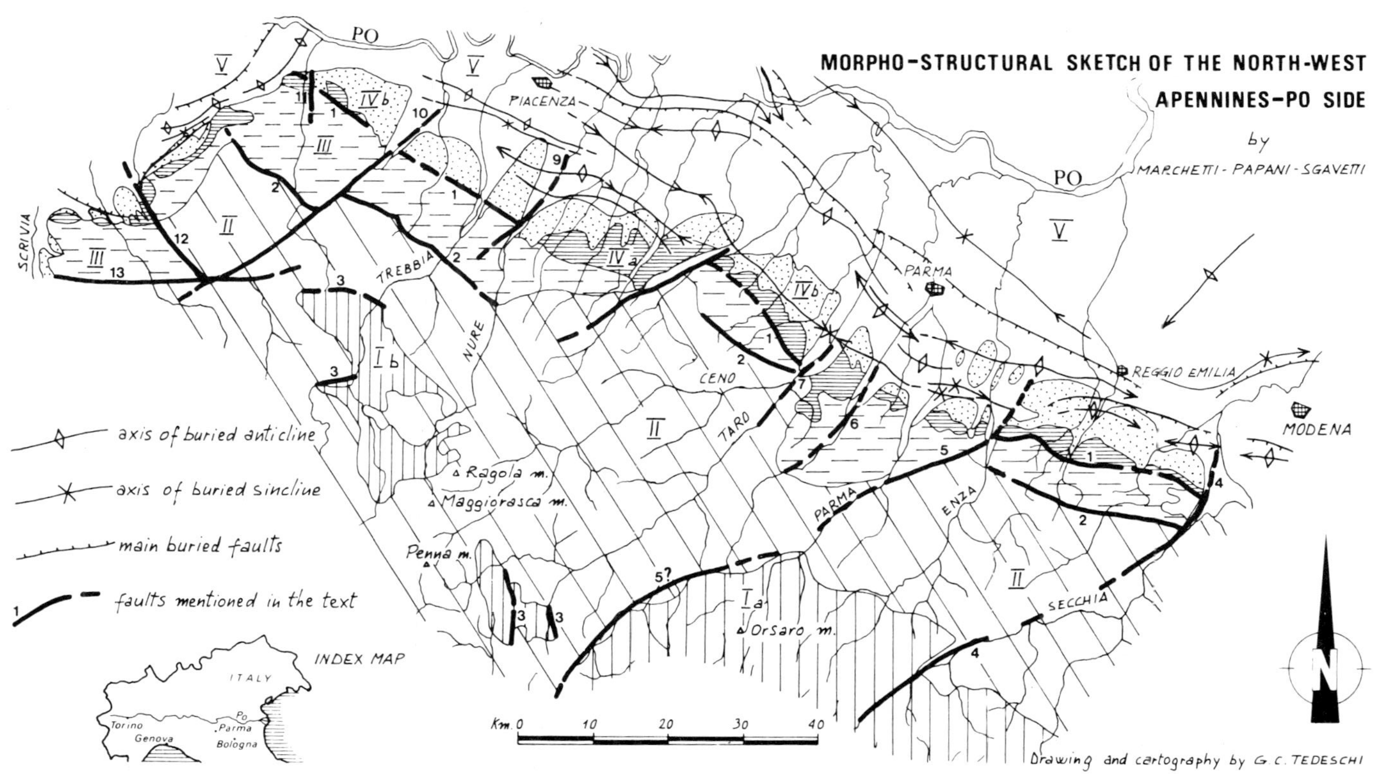

Fig. 1. I = belt of the upper Apennines structures; Ia = main ridge; Ib = tectonic windows of the Taro, Aveto and Trebbia Valleys. II = middle Apennines belt; III = belt of the marginal structures. IV = neo-autochthonous belt; IVa = marine sediments; IVb = fluvial terraces; V = buried pedeapennines folds zone.

but it can be assumed that the belt has continued to rise, reaching an altitude suitable for glacier development only in the Würm glacial period (Losacco 1949).

The middle Apennines belt (Belt II) corresponds to the large outcrop area of the Ligurian nappes and it presents no special morphological character. In this belt several Würmian or pre-Würmian residuals of paleosurfaces are present. They suggest a recent phase of rejuvenation which seems to have been of particular strength in I and III belts, where such paleosurfaces are absent or scarce. Therefore belt II seems structurally lower than the two continuous belts.

The belt of the marginal structures (Belt III) placed at the edge of the Po Plain, is characterized by positive structures in Miocene formations of Padan type and uplifted mainly during the Quaternary. These structures are outcropping such as at Salsomaggiore (Parma), or are located below the Ligurian nappes. It is a rather narrow belt whose upward limit is sometimes uncertain due to the presence of tectonic nappes.

Compared with the adjacent zones of the belt II, it is characterized by a higher mean elevation, by narrower valley cuts and, mainly, by a more intense erosional activity. These elements point toward a Quaternary uplift more marked in the belt III than in the belt II.

Highly rejuvenated sectors of the belt III are present in Stradella zone (Pavia) and next the Nure River, and also in Salsomaggiore and Viano structures.

The neoautochthonous belt (Belt IV) is formed by post-paroxysmal sediments which filled the foothills basin in the Pliocene and in the early to middle Pleistocene. The marine sediments (IVa) are of Tabianian, Piacenzian and Calabrian age; in some places the sequence is continuous (for example: Castell'Arquato, Maiatico, T. Crostolo series) but elsewhere some stratigraphical gaps and unconformities between Tabianian-Piacentian and Piacentian-Calabrian clearly show local differences in tectonic-sedimentary behaviour.

Lacustrine and fluvio-marsh sediments of Villafranchian and/or more recent age outcrop only at river cuts and in fluvial terraces (IVb). Generally they are of Mindel and Riss age. The faults, folds and tilts visible on the terraces (Cremaschi et al. 1975; Marchetti 1974; Marchetti et al. 1975) clearly show that important tectonic movements have occurred in pre-Mindel, Riss-Würm and in post-Würm times.

Pedeapennines (Foothills)

The buried fold zone (Belt V) spreads out transversally along several kilometers. Its surface is formed by Würmian and Holocenic continental sediments of the Emilian plain.

The marine substratum is well known through published results of AGIP exploration (Acc. Naz. Lincei 1959; Reutter 1968). It consists of folded and faulted Tertiary and Quaternary marine sediments with northeastward polarity.

The alluvial cover of belt V shows, in several places, recent and active uplift, coincident with buried anticlines, as is pointed out by isolated terraces which are generally cut by antecedent rivers and by divergent drainage. The subsident zones, coincident with buried synclines, are indicated by convergent rivers.

Moreover, the stratigraphic reconstruction of the upper Pleistocene

and Holocene alluvial deposits (Gasperi et al. 1968; Petrucci et al.1977) allowed to point out plastic and rigid displacements in the alluvial cover. Several river diversions during historical age may be an indirect proof of the still continuing tectonic activity of this belt.

Longitudinal Faults

Longitudinal tectonic faults are present at the belt boundaries as well as inside the single belts, but at present we mention only those which are at the boundary between the belts.

The boundaries between the belts V and IV and between IV and III are marked out by a system of reverse faults and flexures which forms the *Pedeapennines tectonic line*. At the NW margin of the Apennines this line seems to be cut by the equivalent *Voghera fault* (Acc. Naz. Lincei 1959), in NE-SW direction. The importance of these lines, mostly buried under alluvia, may be explained by the fact that locally they show a vertical throw at the base of the Pliocene of about 2 000 m across a width of about 2 km.

The Pedeapennines (Foothills) line marks the southern limit of buried pedeapennines folds zone (V) extending almost to the Po River, where another buried reverse faults system causes its upfaulting over the buried pedealpine structure; within the belt V, at a width of 25 to 30 km, we have a remarkable throw, which in the alignments of Salsomaggiore - Colorno (Parma) and Sassuolo - Modena reaches 5 to 6 000 m and more.

The northern elements of the Pedeapennines tectonic line are covered by fluvial terraces of the belt IVb and affected by the above mentioned dislocations. The inner element, dividing belt III from belt IV, crops out and shows some indications of recent activity, as for example at the Viano structure (Reggio E.), between the Secchia and Enza Rivers; at Salsomaggiore structure (Parma), between the Taro and the Stirone Rivers; in the zone between the Nure and the Tidone Rivers (Piacenza); and in the Pavia Apennines (fault n° 1 of fig. 1).

For the faults separating belt III from belt II, we have no sign of any recent activity (n° 2), although they are well visible geologically. Some short segments of fault - probably active - may be present at the boundary between belt Ib and belt II (M. Zuccone, T. Gotra, Trebbia Valley: n° 3).

Transverse Faults

The sedimentary sequences and the structures of the described five belts present no continuity in longitudinal direction. Sudden changes in along strike suggest the presence of transverse faults.

Some of these faults affect all belts and reach the Tyrrhenian side; others appear to be restricted to the belts III and IV where they can be clearly identified and where, on some of them, an important strike slip component can be seen.

One of the best known lines is the Secchia River fault (n° 4), which in belt I cuts transversally the structures of Macigno and Modino - Cervarola Units, while in belt II it coincides with the triassic questionable outcrop of Secchia Valley. More northward it forms the eastern boundary of belt III, and in belt V it displaces the buried structures of the

plains.

Another fault line can be seen in the lower Enza Valley (n° 5): in belt III it marks the SE margin of Sporno M. structure, in belt IV it divides two areas of different tectonic-sedimentary evolution. In particular, the eastern Messinian sediments are of evaporitic facies, while the western are of continental one. The Plio-Pleistocene serie appears conformable to the east (Barbieri et al. 1967), while to the west it shows considerable gaps and unconformities at the lower Pliocene top and between the upper Pliocene and the Calabrian (Ist. Geol. Univ. Parma 1965, 1966). Furthermore, in belt V the fault cuts off the pre-Würmian terraces as well as some buried structures of the Plains (Acc. Naz. Lincei 1959; Serv. Geol. d'Ital.). Finally, in this zone morphological effects of the known holocenic and historical river diversions are evident.

More westward, still in the Parma Apennines, we are studying at least three transversal faults, for which in belts III and IV meaningful elements have been individualized (fault n° 6, of Baganza Valley; n° 7, of the lower Taro Valley; n° 8 of Stirone-Ongina Valleys).

Also in the Pavia-Piacenza Apennines the actual data have allowed to localize at least five transversal faults. The first of them - fault N° 9 of the lower Nure Valley - shows sharp morphological evidences in belt III, where it displaces continental Quaternary; in the Plains it causes a left lateral offset in the S.Giorgio-Podenzano buried anticlines and cuts off marine Quaternary sediments.

The Tidone Valley fault (n° 10), still under preliminary study, appears to have controlled the terrace evolution in the Sarmato zone and to have caused the displacement of the boundary between belts II and III. Above all, within belt II it causes drainage anomalies, such as deformations and captures which suggest a left-lateral offset.

The lack of neoautochthonous marine sediments, between faults n° 9 and n° 10, shows how much these faults controlled the tectonic-sedimentary evolution in this sector.

The Versa Valley fault (n° 11), remarkable in straightness, concerns belts IV and III giving a sharp displacement in the outcrops of both valley sides.

Finally, the Staffora Valley fault (n° 12) and Villalvernia-Varzi fault (n° 13) may be regarded as transverse in relation to the local structures (of NE-SW direction). They are well known faults in the geological record (Bellinzona et al. 1971; Boni 1963; Gandolfi 1948; Gelati et al. 1974). From the neotectonic point of view we emphasize that Staffora Valley fault causes a planimetric shift in Pliocene sediments (belt IVa) and a marked asymmetry in the fluvial terraces on both sides of the fault itself.

Villalvernia-Varzi fault, besides the well known geological evidence (Gelati et al. 1974), suggesting relatively recent movements, shows notable morphological signs ("fresh" fault scraps, triangular facets, etc.) in the middle Staffora Valley.

Conclusions

Although the northern Apennines sector is not wide and still under consideration it seems that the obtained data agree with the results of other authors (Elter et al. 1975) about the persistence of a compressional style.

During the Quaternary the horizontal component of the slip of the movements have had - and even now have - a remarkable importance. There is direct evidence that many of the mentioned faults (fig. 1) have the character of reverse faults (longitudinal faults) or of faults with doubtless horizontal movement (transverse faults) and that in many cases they are marked by morphological features, slightly modified by the erosion and locally testifying their activity.

Financial support was provided by C.N.R. - Contract No.750106705.

References

Acc. Naz. Lincei - E.N.I., 1959: Atti del Convegno di Milano del 30/9-5/10 1957, Roma.

Barbieri, F. & Petrucci, F., 1967: Mem. Soc. Ital. Sc. Nat. 15(3):181-188

Bellinzona, G., Boni, A., Braga, G. & Marchetti, G., 1971: Serv. Geol. d'Ital., Roma.

Boni, A., 1963: Geol. Rundschau, 57:84-99.

Cremaschi, M. & Papani, G., 1975: Ateneo Parmense, Acta Nat., 11(2): 335-371.

Elter, P., Giglia, G., Tongiorgi, M. & Trevisan, L., 1975: Boll. Geof. Teor. ed. Appl. - 17(65).

Gandolfi, R., 1948: Rend. Acc. Naz. Lincei, Cl. Sc. Fis. Mat et Nat, 8, 5(1-2):67-70, Roma.

Gasperi, G. & Pellegrini, M., 1968: Atti Soc. Nat et Mat. - 99, Modena.

Gelati, R., Bruzzi, D., Catasta, G. & Cattaneo, P.C., 1974: Riv. Ital. Paleont. e Strat. - 80(3):479-524.

Ist. Geol. Univ. Parma, 1965: Guida alle escursioni 63ª Adunanza Estiva S.G.I., Parma-Garda, Parma.

- , 1966: Carta Geol. 1 : 100.000 d. Prov. Parma e zone limitrofe, L.A.C., Firenze.

Losacco, U., 1949: Riv. Geograf. Ital. - 56(2-3), Firenze.

Marchetti, G., 1974: Studi Trentini Sc.Nat., N.S., 51(2A):193-212.

Marchetti, G., Papani, G. & Petrucci, F., 1975: Ateneo Parmense, Acta Nat. 11(4):721-238.

Papani, G. & Sgavetti, M., 1975: Ateneo Parmense, Acta Nat., 11(2): 315-334.

Petrucci, F. & Cavazzini, R., 1977: Gruppo di Studio del Quaternario Padano, Quaderno n° 3, Massaza & Sinchetto, Torino.

Reutter, K.-J., 1968: Ecl. Geol. Helv., 61(1):183-224.

Servizio Geologico d'Italia: Nuova Carta Geol. d'Italia 1:100.000 Fogli 59, 60, 71, 72, 73, 74, 84, 85, 86.

Geotectonic Interpretation of Geophysical Models of the Monferrato Area (Southwestern Po Plain)

K. Günther and K.-J. Reutter

Introduction

Two crustal models have been established of a cross section through the Western Po Plain, the Moferrato hills, and the Piedmontese Tertiary basin by means of magnetotelluric measurements (Losecke et al. this vol.) and deep seismic recordings (Stein et al. this vol.). These models are of great importance as they increase understanding about the poorly known area where the Northern Apennines meet the arc of the Western Alps.

Both mountain chains form parts of the same orogenic system. The eugeosynclinal rocks of Alps and Apennines can be traced back to the same oceanic trough (Penninic-Ligurian eugeosyncline), which was originally situated between the European continent and the Adria microplate. Thus Apenninic Ligurid units are also encountered in the Western Alps (Flysch of Ubaye-Embrunais, Simme-Flysch, Canavese zone; cf. Elter et al. 1966; Sturani 1977). The essential orogenetic phase of the Western Alps is of Upper Eocene age ("Ligurian phase") and, hence, older than that of the Northern Apennines, which had its paroxysm in the Upper Miocene ("Tuscan phase"). The northwesterly and westerly transport directions of the Western Alps are opposed to the northerly or northeasterly directions of the Apennines.

The transitional zones between the two mountain chains were affected by both orogenic phases. There is clear evidence of superimposition of different tectonic movements in the Apennines (Upper Liguride nappe, Groscurth & Reutter this vol.) and in the Western Alps (the famous "retrocharriage"). The area situated between the Western and Maritime Alps and the Monferrato, which constitutes a prolongation of the Northern Apennines, is such a transitional zone where the Apenninic structures should pass into the Alpine ones. The narrow spiral like bending of the Alpine and Apenninic structures in this region seems to be due to the fact that the pivot of the anticlockwise rotational movement of the Adria microplate and the sphenochasmatic opening of the Ligurian Sea during the Miocene was situated here. Unfortunately, direct observation of this geodynamically so important part of the mediterranean orogenic belt is largely hampered by the Neogene sediments of the deep internal depressions of the Piedmonte basin. Geological interpretation mainly depends, therefore, on geophysical data.

Interpretation of the Models

The magnetotelluric resistivity section (Losecke et al. this vol.) extends from the Po Plain south of Milan through the Monferrato hills into the Pliocene Asti depression. The seismic section (Stein et al. this vol.) is longer than the magnetotelluric one, as it reaches the northern border of the Po Plain near Lecco, passes the Langhe region and the Upper Po basin and enters the Western Alps near Cuneo.

The seismic model shows, at the margin of the Southern Alps, Oligocene

to Pleistocene clastic sediments of the Po Plain dipping southwards. The inclination of Mesozoic and Tertiary strata is, however, more homogeneous than indicated. Near Milano the top of the calcareous Mesozoic and Early Tertiary sequence was drilled in a depth of about 5 km.

At the magnetotelluric station Zeme, north of the river Po, four layers of different resistivities were established. Over the lowermost layer with more than 3000 Ohm·m, which corresponds to a crystalline basement, a layer follows of 22 Ohm·m between depths of 16 and 10 km. It is overlain by a low resistivity layer of only 4.7 Ohm·m, which nearly reaches the surface. The latter may correspond perfectly to the Oligocene to Pliocene Po Plain sequence of clays, marls and sandstones, whereas under it the calcareous Mesozoic sequence seems to be present. The layer of 0.9 Ohm·m near the surface obviously corresponds to the youngest sediments of Late Pliocene and Quaternary age. The indicated depths of all mentioned layers are also reasonable from a structural standpoint.

At the next magnetotelluric station south of the river Po, the basement appears at a depth of only 8 km. Indeed, near the river Po, a normal fault is known (AGIP 1959), by which the base of the Pliocene is lowered about 3 km on the northern side. According to the resistivities to the south of the fault, no intermediate layer between the basement and the Po Plain Tertiary sequence is present. If methodical errors are excluded, this indicates that the Mesozoic sequence was eroded before the deposition of the clastic series as a consequence of an early uplifting of the southern block. The most important finding of the stations Frassineto and Casale is a higher resistive layer with values between 830 and 10 Ohm·m between 0 and 2 km (Frassineto) and 1 and 6 km (Casale). Since in the seismic section between the river Po and Asti a low velocity channel beneath a surface layer of 3 km thickness and velocities of about 3.5 km/sec also had to be introduced because of the strong retardation of arrivals in this zone, Stein et al. and Losecke et al. conclude that this anomaly indicates a thrust sheet of Northern Apenninic material, which overlies the young sediments of the Po Plain.

Such a situation would correspond perfectly to that of the external Northern Apennines in the Emilia region, where in boreholes, tectonic windows, and structural setting the allochthonous position of the external Cretaceous and early Tertiary Liguride flysches is demonstrated. The same Liguride material of the external Apenninic zones is found in outcrops of the Monferrato, and just as southeast of Voghera and south of Piacenza it is unconformingly overlain by synorogenetic allochthonous molasse-type sediments of Upper Eocene, Oligocene, and Miocene age (fig.1).

In the resistivity section the top of the basement slopes down to the southwest from 8 km at station Frassineto to 17 km at station Fontane Salera. A similar behaviour of the basement may exist beneath the Emilian Apennines, although it is not necessary to assume it to be more than 10 km deep. In the central zone of the Northern Apennines, where at the surface miogeosynclinal allochthonous units (Tuscan units) appear in tectonic windows beneath the Ligurian eugeosynclinal nappes, a basement high comes into existence at a depth of about 3 or 4 km (borehole Pontremoli).

A corresponding structural basement high seems to exist beneath the magnetotelluric stations Montemagno and Asti, where crystalline rocks with more than 3000 Ohm·m exist at a depth of about only 2 km. The seismic profile also records material with velocities of 6 km/sec at a depth of 3 km south of Asti. Nevertheless, here a different situation than that of the Northern Apennines is indicated by the steep gradient of the crys-

talline basement between the magnetotelluric stations Fontane Salera and Montemagno.

The resistivity model suggests that around the station Fontane Salera the layer of low resistivity, which is referred to as the clastic Tertiary Po Plain sequence, comes near to the surface at the rear of the Apenninic thrust sheet in a position similar to a tectonic window. Geologically such a situation is impossible and the resistivity distribution

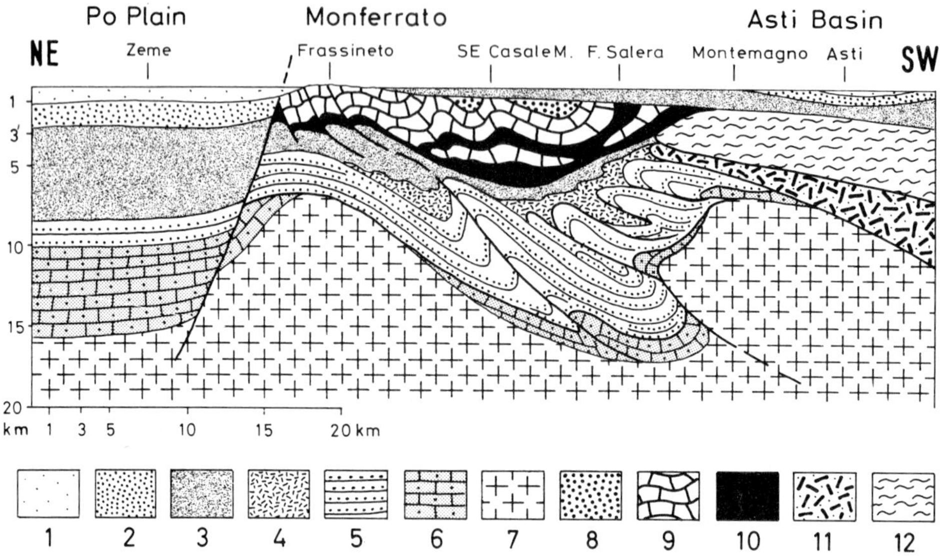

Fig. 1. Geological cross section through the Po Plain, the Monferrato hills and the Asti basin. 1 = Late Pliocene and Quaternary of the Po Plain and the Asti basin; 2 = Lower Pliocene of the Po Plain and the Asti basin; 3 = Oligocene and Miocene of the Po Plain, Molasse series of the Langhe, Ligurian Tongriano; 4 = Tertiary olisthostromes of the Northern Apennines; 5 = Upper Cretaceous to Palaeogene flysch series of the Po Plain, Oligocene to Miocene graywacke series (Salsomaggiore, Bobbio, Cervarola, Macigno) of the Northern Apennines; 6 = Permian to Mesozoic sediments of the Po Plain and the Northern Apennines in Southern Alpine, Umbrian or Tuscan facies; 7 = Crystalline basement of the Po Plain; 8 = Oligocene to Miocene Molasse series of the Monferrato hills, partly folded (Ranzano sandstone, Antognola marls etc.); 9 = Upper Cretaceous to Eocene flysch series of the Monferrato hills (Monteu da Po flysch, Casale cement-stone); 10 = Basal complex of the Monferrato hills ("Argille scagliose" of the Lauriano complex; Albian to Senonian); 11 = Metamorphic basement of the Tuscan units ("Tuscan crystalline nappe"?); 12 = Metamorphic successions of the Western Alpine Piedmont zone ("schistes lustrês" and crystalline nuclei) and of the Liguride complex of the Northern Apennines.

must be otherwise explained. The superficial low resistivity layer of about 500 m thickness corresponds to the Oligocene syntectonic and hence allochthonous Molasse sediments. It must be underlain by Ligurid material which, however, can very well be of argillaceous (low resistivity!) consistence ("Argille scagliose" of the "Lauriano complex", Sturani 1976).

Its thickness should not exceed some km. The remaining 10 or more km of low resistivity material should be attributed to tectonically stacked Tertiary sediments which correspond to a more internal facies than the Po Plain sediments. Shales and sandstones of the Salsomaggiore facies can be expected as well as graywackes similar to those of the Modino-Cervarola unit at Bobbio or even of the Tuscan Macigno, if this unit reaches this far to the northwest. According to the geology of the Northern Apennines these graywacke formations should be mixed with marls and many olisthostromes of Ligurid material. If this interpretation is correct, the tendency of the Northern Apennines to become narrower towards the northwest would continue into the Monferrato area.

The situation near Asti and in the Langhe region is quite different from that of the internal part of the Northern Apennines. In the Langhe region arenaceous and marly sediments of Oligocene and Miocene age, several thousand meters thick (Tertiary Piedmont basin) are found. Towards the south they unconformingly overlie metamorphic rocks of the internal parts of the Western Alps such as the Briançonnais, the Penninic Voltri group, the Sestri-Voltaggio zone and the Antola group, which were deformed and partly metamorphosed during the Ligurian phase. From the Antola group similar sediments extend to the NE over the Ligurian allochthonous units at the external side (near Piacenza) of the Northern Apennines with only a few interruptions. Analogously, the conclusion must be drawn that these sediments of the Langhe are directly connected beneath the Pliocene of the Asti basin with the Oligocene and Miocene deposits of the Monferrato hills.

Therefore, the Langhe region also has to be considered allochthonous, in the same way as the internal parts of the Upper Liguride unit (Groscurth & Reutter this vol.). Since along the strike of the Apennines towards the northwest this unit increases its coherency, it probably forms a unique thrust sheet without any tectonic windows from the Genovese and Piacentine Apennines to the Monferrato and Langhe region.

From this it follows that in the magnetotelluric profile the uppermost low resistivity layer is made up mainly by Oligocene and Miocene Langhe sediments and some Pliocene and Pleistocene above it. The highly resistive, probably crystalline, rocks immediately benath have to be interpreted as allochthonous metamorphic rocks of the most internal parts of the Upper Liguride unit or of the Western Alps. It may consist of schists and ophiolites of the group of Voltri or of the Sestri-Voltaggio zone, or other still unknown units. The thickness of such a thrust sheet is difficult to estimate, but it surely increases towards the south. Under it there might be found allochthonous slices from marginal parts of the Adria microplate (Tuscan units!) which probably have suffered an Apenninic (Miocene) metamorphism. Under them again in a depth which may be supposed between 4 and 6 km an upwarping or even upthrusting of the Po Plain basement in relation to that of the central Northern Apennines can be expected. Thus the whole section of the resistivity model would show a geology similar to that of the Emilian-Ligurian Apennines but more compressed and more accentuated.

Following the seismic model, the top of the crystalline basement remains in a depth of 3 km through the Langhe region. The gently folded Miocene sediments disappear under the Pliocene and Pleistocene cover of the Upper Po basin, northeast of Cuneo, but they are missing where Penninic schistes lustrés and metamorphic sediments of the Briançonnais zone emerge from these young deposits in the Western Alps southeast of Cuneo.

Somewhere in this region the suture of continental collision between the Adria microplate and the European plate or the continental fragments of the Briançonnais must exist. It is hidden, however, under the young sediments or, more probably, under the border of the Western Alps when they suffered the retrocharriage as a hinterland of the Apenninic orogeny. The prolongation of the positive gravity anomaly of Ivrea towards Cuneo delineates its course. A detailed discussion of the complex structures of the deeper crust in this area is dealt with in Giese & Reutter (this vol.).

Conclusion

In both geophysical sections the existence of Apenninic northeastward and northward directed nappe tectonics in the Moferrato area is clearly evidenced. As the syntectonic Oligocene and Miocene molasse deposits of the Langhe region connect the Monferrato hills directly with the Maritime Alps, the main thrust of the Monferrato must continue this far southwards and cut through the metamorphic or crystalline rocks of the basement high, which was detected near Asti by the magnetotelluric and seismic measurements. This implies that just as in the Northern Apennines near Genoa, internal parts of the Western Alps, assuming the role of a hinterland, became involved in the post-Alpidic thrust movements of the Apennines.

References

AGIP-Mineraria, 1959: Atti Conv. Milano, 1957, 2, Accad.naz.Lincei,Roma.

Elter, G., Elter, P., Sturani, C. & Weidmann, M., 1966: Bull. Lab. Géol. Minéral. Géophys. Mus. géol.Univ. Lausanne, 167:279-377.

Giese, P., Günther, K. & Reutter, K.-J., 1970: Z. dtsch. geol. Ges.,120: 151-195.

Sturani, C., 1975: In: Structural Model of Italy, Ogniben, L., Parotto, M. & Praturlon, A. (Ed.):149-174, Quaderni Ric.Sc. 90, Cons.Naz. Ric., Roma.

Tectonic Pattern of the Northern Apennines

J. Bodechtel and A. Smolka

Introduction

Ground-controlled satellite imagery has been used to check and impove a preliminary tectonic model of the Apennines of Tuscany and Umbria (Bodechtel & Nithack 1974). The program included the following steps:

Structural mapping along two traverses through the Northern Apennines. The position of the two sections is indicated on fig. 1.

Detailed evaluation and lineament analysis of SKYLAB-images covering the Northern Apennines.

Results and Discussion

Field Work

About 15 000 joints, most of them were measured along the two sections by use of terrestrial-photogrammetric techniques. They provided an optimum base for digital data processing. Joint-pole configurations on the equal-area-net led to the following results:

The *existence of two (or more) joint fabric systems* has been confirmed for all sites except two. Nearly at all the localities the strike of the b-axes of the two systems is significantly different (fig. 1).

The *differences of the axial trends* vary uniformly between +10° and +25° in the southern section. The amount does not depend on the age of

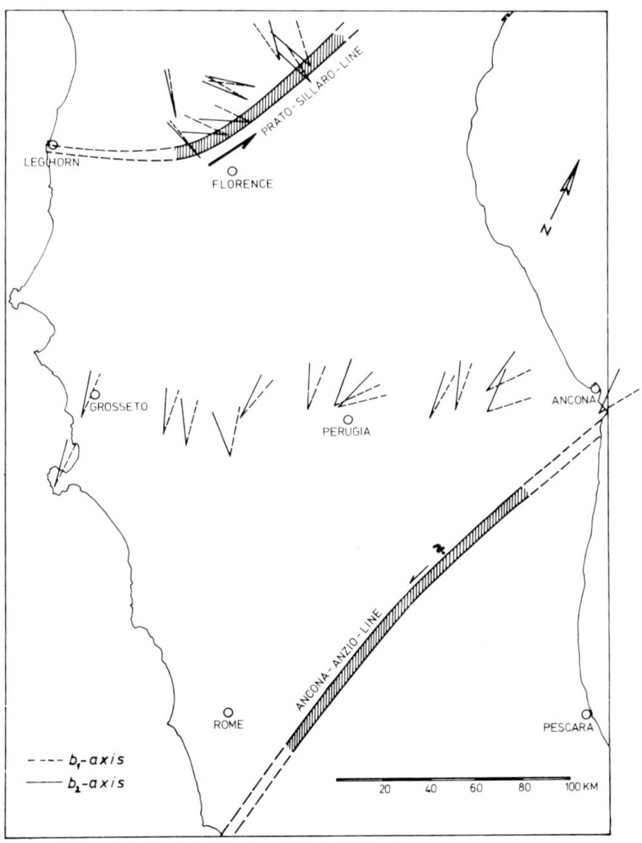

Fig. 1. Axis directions and main fault zones in the Northern Apennines (A. Smolka).

the measured rock-formations. Only on the limbs of overturned or upright folds (Mte. Cetona, Umbrian Apennines) +40°, and even +60° of trend deviation can be observed. This phenomenon can be explained by two mechanisms:

on the outer arc of a fold formed by buckling, the original angle between rotated linears (b_1-axis) and fold axis (b_2-axis) is increased by tangential longitudinal strain (Ramsay 1966);

an additional strain component in the forezone of thrust faults (Cetona thrust, Cervarola thrust) may act in the same way as described above.

In the northern section the pattern exhibited by the axial divergence is not uniform. The pattern is complicated by movements on the Prato-Sillaro-Line (PSL) and by allochthony of some sampled units. Therefore an interpretation is not possible in the same manner as given for the southern section.

The following angle differences could be derived:
for the Marnoso-Arenacea-formation (south of the PSL) + 25°;
for the Cervarola-Sandstone (north of the PSL) -15° - ±0°;
for the Morello-Alberese (within the PSL) +25° - +35°;
for the Macigno of the Mte.Albano (southern prolongation of the PSL) +0° (fig. 1).

Only at the Passo Giogo the situation is masked by the main overthrust of the Cervarola-Sandstone.

At the *localities situated near large faults or fault zones* the b_2-axis (corresponding to the fold axes) differs from the strike direction of the main fold structures. These deviations can be considered as drag effects indicating the sense of movement on the fault. For example, near the fault marked by the Esino valley (Gola della Rossa) and its extension to the W the axial divergence suggests a dextral movement.

In the same way, axial divergence in the Cervarola-Sandstone and the Morello-Alberese nar the PSL indicate a sinistral sense of movement. The same movement was postulated by Bortolotti (1968) based on rather intuitive arguments. Also at the southern end of the Mte. Albano a similar axial divergence can be observed.

Evaluation of SKYLAB-imagery

A lineation map at a scale of 1:500.000, showing only the lineations longer than 5 km, has furnished the following results:

There are few *lineaments crossing several tectonic units*, e.g. the PSL, which touches Macigno, Morello-Alberese, Cervarola-Sandstone, Cantiere-Monghidoro-Unit, and the Pliocene. Another lineament begins N of La Spezia and extends to the ENE along the Magra valley, over the Passo di Lagastrello to the Enza Valley. This fault crosses the Tuscan formation and the overlying Ligurian units. Nearly all the other lineaments, even very long ones as the normal fault accompanying the Cetona-Orsaro-Structure, are confined to one tectonic unit or separate tectonic domains (e.g. Ancona-Anzio-Line). These lineations must be interpreted as basement structures penetrating the sedimentary cover.

There is *no geological evidence for a large fault zone in the Arno valley* and its extension to the E, a photolineament, which was postulated by Bodechtel & Nithack (1974).

The following *main lineament trends* can be derived from fig. 2.

A perpendicular set of lineaments (65°/155°) occupies the entire map area. This set corresponds to anti-Apenninic and Apenninic directions S of the Arno River.

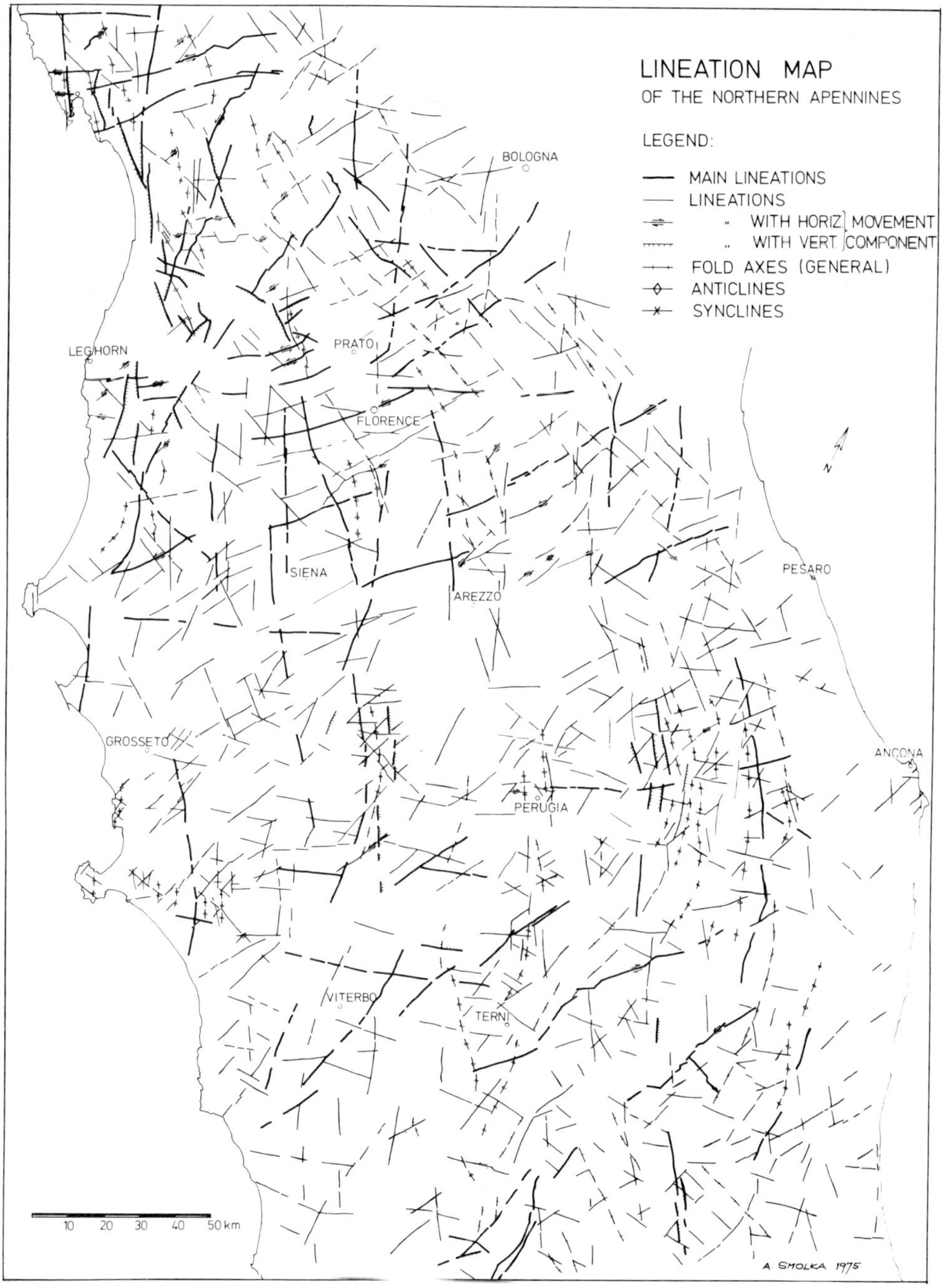

Fig. 2. Lineations longer than 5 km obtained from SKYLAB-imagery.

Another perpendicular set of lineaments shows a preferred 35°-45°-trend and a subordinated 125°-135°-trend. This set is well developed N of the Arno River and corresponds to the anti-Apenninic and Apenninic directions in this region. The 35°-45°-trend is still prominent in the southern Tuscany and the Nera valley (Umbria), whereas the 130°-direction disappears almost totally S of the Arno River.

Some curved lineaments strike between 350° and 25° between the PSL and the Ancona-Anzio-Line. The long lineation begins in the Campiglia mountains and extends to the eastern border of the Mesozoic nucleus of the Lima valley. Also two large lineaments in the young volcanites belong into this group. These lineations are not strongly coordinated with the orogenic structures. Therefore they may be reactivated old structures of the pre-Apenninic basement. In the Umbrian arc the lineament pattern is generally more complicated because of the bending of the fold axes from a NW/SE-direction to a NS-direction.

On the satellite imagery the main element of the *Prato-Sillaro-Line* appears as a well recognizable lineament, which follows the Idice valley and limits the Cantiere-Monghidoro-Unit to E. The lineament extends southwestward cutting the Morello-Alberese mountains in the direction of Calenzano.

In the Albano mountains the anti-Apenninic direction follows the PSL. There is also a well developed trend in prolongation to the PSL-trend in the N (20°).

In the region S of the Cervarola thrust, the PSL-direction becomes distinct and much more prominent than the anti-Apenninic direction.

North of the Cervarola thrust, the PSL-direction changes to a strike direction of 5 to 10°.

In all three zones subordinate maxima are developed at 75°, 95° and 160°. The first two directions have to be considered as strike slip faults obligue to the 40°- and 65°- (anti-Apenninic) directions. Mainly left-lateral strike slip faults are developed. The 160°-trend is common in the entire Northern Apennines.

Conclusions

The clockwise rotation of the Tuscan-Umbrian region (Bodechtel & Nithack 1974) is well established by tectonic field data and satellite-image interpretation. However, new results show that the north boundary of the rotated block is not the *Arno fault zone* but the *Prato-Sillaro-Line*. This changes the shape of the rotated block from a triangle to a parallelogram. Consequently, the movement causing the rotation of the b_1-axis likely is a torsion rather than a rotation (fig. 1). The horizontal component of movement on the Ancona-Anzio-Line is considered rather small. Concerning the continuation of the PSL to the SW, the sinistral faults and the axial divergences in the southern Albano mountains suggest an extension of the fault zone to the Tyrrhenian coast near Livorno with a slightly curved trend. The existence of two lineation systems visible on the satellite images (40/130° and 65/155° resp.) agrees well with the conception of sinistral torsion of the block S of the PSL. Considering the extension of these two systems to N of the PSL, Bodechtel & Nithack (1974) inferred a similar movement on this northern block.

The conception proposed here does not contradict the idea of the anticlockwise rotation of Italy (Alvarez 1974, 1976; Soffel 1974, 1975). This

rotation ceased in the Eocene according to Soffel. The above described torsion, however, took place between the two main orogenic phases in the Northern Apennines, that is between Aquitanian and Tortonian (Baldacci et al. 1966).

References

Abbate, E.,Bortolotti,V.,Passerini,P.& Sagri,M.,1970:Sed.Geol.4:207-249.
Adler, R.E., 1973: Ber.Ld.Nordrh.-Westf.,2385:135 p.
Alvarez, W. & Lowrie, W., 1974: Boll.Soc.Geol.Ital.,93:883-891.
Anderson, E.M., 1951: Dynamics of faulting and dyke formation with application to Britain, Oliver & Boyd:206 p.
Bock, H., 1972: Geol. Rdsch., 61:824-849.
Bodechtel, J., 1968:Bayer.Akad.Wiss.Math.Nat.Kl.Abh.N.F.,136:1-36.
Bodechtel, J. & Nithack, J., 1974: Geoforum, 20:11-24.
Bortolotti, V., 1966: Boll. Soc.Geol. Ital, 85:529-540.
Channell, J.E.T. & Tarling, D.H., 1975:Earth Planet.Sci.Lett.,25:177-188.
Dallan,Nardi, L., Elter, P. & Nardi, R., 1971:Boll.Sco.Geol.Ital.90: 203-211.
Fuchs, A., 1944: N.Jb. Min. Abh., Abt. B., 88:337-373.
Gerlini, R., 1965: Boll. Soc. Geol. Ital.,84:169-212.
Groscurth, J. & Hemmer, K., 1973:N.Jb.Geol.Paläont.Abh.,144(2):181-205.
Hafner, W., 1951: Stress distribution and faulting. Geol.Soc.Amer.Bull., 62:373-398.
Hemmer, C., 1971: Diss. FU Berlin:234 p.
Hiller, K., 1973: Unpubl.Dipl. Arb. Univ. München: 79 p.
Illies, H., 1974: Oberrh. geol.Abh., 23:1-24, Karlsruhe.
Nithack, J., 1974: Diss. Univ. München: 102 p.
Ogniben, L., 1969: Mem. Soc. Geol. Ital. 8(4):453-764.
Ogniben, L., Parotto, M. & Praturlon, A., 1975: Structural model of Italy Maps and explanatory notes. Ric. Sc. 90:502 p. Roma.
Ramsay, G., 1967: Folding and fracturing of rocks: 568 p., McGraw-Hill, New York.
Reutter, K.J., 1968:Ecl. Geol. Helv., 61(1):183-224.
Sander, B., 1948: Einführung in die Gefügekunde der geologischen Körper, Teil I: 215 p., Springer Wien, Innsbruck.
Sestini, G. (ed.), 1970: Sed. Geol. 4 (3,4):

Part 3

Synthesis

An Approach to the Geodynamics of the Corsica - Northern Apennines Double Orogene

K.-J. Reutter, K. Günther and J. Groscurth

Introduction

If the Adriatic microplate is considered a part of the African continent, the Mediterranean mountain chains can be classified either as "Europe-vergent" or "Africa-vergent" with regard to their characteristic asymmetric structure and polar development. However, the segment comprising the Western Alps, Corsica and the Northern Apennines forms a double orogene in which both structural trends are realized.

In the Western Alps and on Corsica structures and thrusts are directed essentially towards the European foreland and its Corso-Sardinian fragment to the West, whereas the structures and thrusts of the Apennines, which were formed later, trend to the Adriatic foreland to the Northeast. The chains cannot be considered to be independent of each other, as they must have originated from the same geosynclinal disposition, which, during the Jurassic, consisted of the pre-Alpidic European continent, the continental Adriatic microplate,and an intermediate oceanic basin, the so-called Penninic-Ligurian eugeosyncline. The essential effect of orogeny was the elimination of this oceanic basin and the consequent collision of the two continental plates. Sediments and oceanic crustal fragments (ophiolites) from the Penninic-Ligurian eugeosyncline are found in the Western Alps and Corsica as well as in the Northern Apennines. Some units, which were already dislocated during the main Alpine orogenetic event ("Ligurian phase", or, Trümpy 1973, "Meso-Alpine event") in the Upper Eocene, also participated in the Apenninic thrust movements in the Upper Miocene ("Tuscan phase").

The opening of the Ligurian and Balearic Seas as well as the anticlockwise rotational movements of the Corso-Sardinian continental fragment and the Adriatic microplate, which have been proved by many palaeomagnetic studies, belong to the geotectonic framework of mountain building in the West Alpine and Apenninic segment of the Mediterranean belt. Boccaletti, Elter & Guazzone (1971) and Boccaletti & Guazzone (1972, 1975) thoroughly discussed the kinematics of these complex and intercalated geotectonic processes with the aid of the plate tectonic concept, which proved very helpful. Now further studies, especially the geophysical investigations reported in this volume, shed more light on some detail of this double orogene and the conditions of its formation.

Northwestern Limit of the Apennines

The development of the Apennines out of the edifice of the Western Alps is so gradual that no line can be determined as the northwestern limit of the Apennines. Moreover, the placement of a boundary depends on the definition of what is called Apenninic.

The presence of certain eugeosynclinal rocks can be used just as little as a directional trend towards the Adriatic microplate because based on this paramter an essential part of the Western Alps would have to be considered as Apennines. It seems better to adopt a combination of several parameters: the presence of Oligocene/Miocene graywacke flysches with northeast trending development and structures, as well as far-reaching overthrusting of older partly geosynclinal rocks in the same direction and during the same time span (fig. 1).

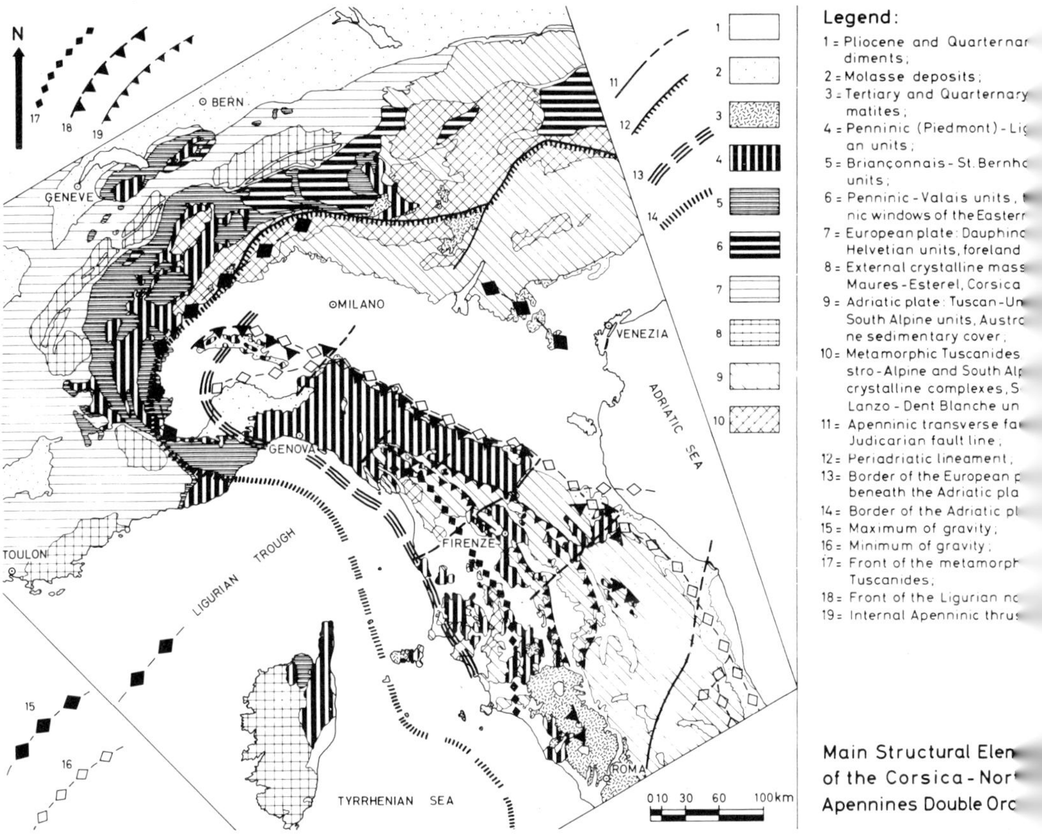

Fig. 1. Main structural elements of the Corsica - Northern Apennines double orogene. The trace of the border of the Adriatic microplate is based on refraction seismic data (Giese, Morelli & Nicolic this vol.).

With respect to the latter argument, at least, the Monferrato and Colli Torinesi region represents a true extension of the Northern Apennines. By means of refraction seismic and magnetotelluric measurements their allochthonous position over Miocene sediments of the Po Plain was proved (Stein et al., Losecke et al. this vol.) and it was recognized as

a part of a huge northward carried nappe, which includes rocks belonging to the edifice of the Western Alps at its tail (Günther & Reutter this vol.).

In the vicinity of Turin the front of this Apenninic nappe seems to bend backwards into a southwestern strike before it disappears beneath the Neogene sediments of the Western Po Plain. North of Cuneo, on the southern tip of the Dora Maira massif, northeast striking structures turn with a very narrow loop in the West Alpine direction. However, as these rocks belong to a much deeper structural level than those of the Colli Torinesi, it is not clear if the connection between Alps and Apennines is realized here. The well-known retrocharriage of the Penninic nappes in the Western Alps and the fan structure of the Grand St. Bernhard nappe, which both are of post-Meso-Alpine, i.e. Oligocene age, can be considered as a prolongation of the Apenninic trend into the Alps.

"Europe-vergent" Thrusting

The westward direction of transport of nappes in the Western Alps and Corsica is explained by most geologists with a plate tectonic model which postulates the subduction of the Penninic-Ligurian oceanic basin beneath the Adriatic microplate. The thrust planes are thus considered as synthetic branches of the subduction plane, which even survived the collision with the pre-Alpine European continent. The moment of continental collision is not yet quite clear. Dal Piaz et al. (1972) and Compagnoni et al. (1975) suppose, based on radiometric data, that it occurred already during the Mid-Cretaceous (Trümpy 1973: "Paleo-Alpine event"), others believe that it occurred only during the Upper Eocene. The presence of early Alpine HP/LP metamorphics shows at least that subductory processes were active already during the Cretaceous (Laubscher 1975).

In the Sestri-Voltaggio zone, ophiolite bearing sediments, which are tectonically stacked along east-dipping thrusts, show an upwards and eastwards decreasing HP/LT metamorphism (Cortesogno & Venturelli this vol.). The geological situation of northeastern Corsica is quite similar in that at least two nappes of ophiolites with eugeosynclinal sediments have been recognized (Seidl & Zacher this vol.; Nardi 1968). According to Ohnestetter et al. (1976) the lower unit shows a strong HP/LT metamorphism, ascribed to the early Alpine event, and a younger low grade metamorphism of possible Lower Tertiary age. The upper unit is not or very low grade metamorphic.

In Corsica as well as in the Sestri-Voltaggio zone, the metamorphism may have been caused by oceanic subduction. In both areas there was no continental collision because no crystalline basement nappes overlie the ophiolitic units. The presence of non-metamorphic ophiolites in the Northern Apennines indicates that an oceanic area persisted to the east of the metamorphic ophiolite zone from the Upper Cretaceous to the early Oligocene (Elter & Pertusati 1973; Boccaletti et al. 1971). The metamorphism therefore may have been caused by subduction and accretion to an oceanic arc, which collided with the European continent during the Eocene (fig.3).

At the beginning of the Oligocene the "Europe-vergent" orogene fringed the European plate from the Central Alps to the Betic Cordillera along the not yet disrupted and rotated Corso-Sardinian block. Somewhere in the southern part of the Western Alps the geotectonic character of the Meso-Alpine cordillera must have changed from a continent/continent collision

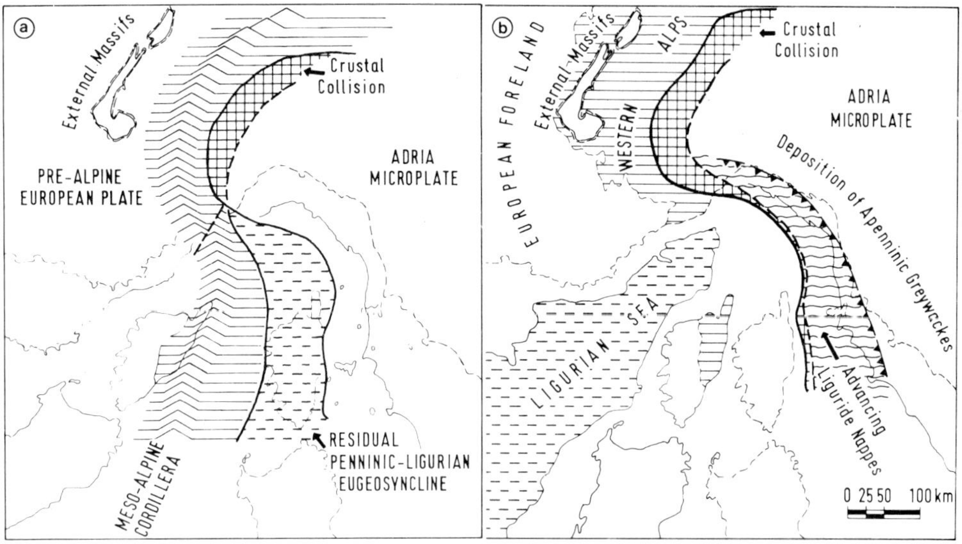

Fig. 2. Sketch maps representing the paleogeographic situation at the end end of the Eocene with the Meso-Alpine Cordillera and (b) in the Lower Miocene after the closure of the Penninic-Ligurian oceanic basin.

zone to an arc/continent collision zone. To the south of this transition the residual Penninic-Ligurian ocean was situated. Possibly this limit was marked by the angular contact existing between two mountain arcs, which might account for the first irregularities of strike directions within the loop of the western Alps and Corsica (fig. 2). To the south-east of the Meso-Alpine cordillera, the Adria microplate as well as the residual oceanic basin, remained submerged.

"Africa-vergent" Thrusting

In the Northern Apennines pelagic and flysch deposits which belong to the sedimentary cover of the Adriatic microplate are overlain by Penninic-Ligurian eugeosynclinal sediments and ophiolites. Sediments of the uppermost nappe were obviously involved in Alpine movements of the "Ligurian phase" before they became displaced during the Apenninic movements. To the north of Genoa and in the Asti region (Günther & Reutter this vol.) this uppermost nappe is still affixed to the Western Alps, which can be considered the hinterland of the Northern Apennines. Overthrusting was directed northeastward (present direction after rotation!), onto the Adriatic foreland. It took place essentially during the Miocene with a paroxysm in the Upper Miocene ("Tuscan phase").

Following the argument of synthetic thrust and subduction planes, the residual Penninic-Ligurian oceanic lithosphere must have been subduced along a southwest (rotation of plates not considered!) dipping plane. This means that at the end of the Eocene the subduction system, which had originated the West Alpine and Corsican orogeny, flipped in the opposite direction. Thus the newly formed Alpine Cordillera became the

leading edge of the Corso-Sardinian continental fragment which drifted apart from Europe. It acted as an upper plate or as an island arc, if the opening Ligurian-Balearic Sea is considered as a marginal oceanic basin in this process (Boccaletti et al. 1971; Boccaletti & Guazzone 1972, 1975). The inverted movement not only affected the remaining oceanic area, but also extended with fading intensity into those parts of the Western Alps, where collision had already occurred, causing the Penninic retrocharriage and other "Africa-vergent" structures (fig. 2,3).

In the Northern Apennines, the pattern of structures and distribution of allochthonous materials and of synorogenetic sediments deposited at the front of the advancing nappes and upon their top can be compared with recent subduction zones in areas of great sedimentary thickness (Seely et al. 1974; Beck & Lehner 1974), e.g. with the processes at the front of the Calabrian arc (Görler this vol.). The hypothetical southwest dipping subduction zone is supposed to have passed upwards into almost flat lying faults which caused the formation of an accretionary wedge consisting of the peeled-off uppermost oceanic crust (Bracco unit, Decandia & Elter 1972) and of pretectonic and syntectonic sediments. During the early Miocene this bulge of sediments and ophiolites was pushed upon the deeply submerged border of the Adriatic microplate which then became involved into the subduction process.

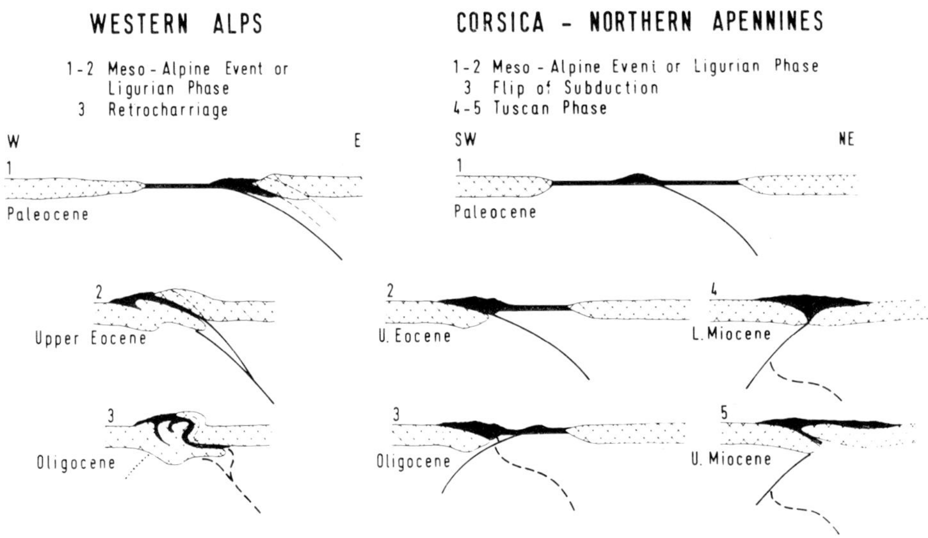

Fig. 3. Schematic comparison of the geotectonic evolution of the Western Alps beyond the termination of the Northern Apennines and of the Corsica-Northern Apennines double orogene.

The allochthonous sediments of the accretionary wedge spread out from the border of the Adriatic microplate towards the external side of the Apennine by means of gravity tectonics. Throughout the Apennines an increasingly younger age of the Neogene graywacke flysches can be observed which corresponds to a migration of the depositional trough from the internal to the external side. At the border of the Po Plain the flysch se-

diments pass laterally into the molasse of the Po basin. At the internal side of the flysch troughs folding and upwarping took place, thus providing the slope for gravitational gliding of the allochthonous units (fig. 3).

There are strong differences in the tectonic setting of the Western Alps and the Northern Apennines which remind us that the Apenninic orogeny was not only a mirror-inverted repetition of the Alpine orogeny: The Liguride nappes are not overlain by crystalline thrust sheets; and there is no HP/LT metamorphic mineral assemblages in the ophiolitic nappes of the Northern Apennines, except for some very small occurrences (Argentario) of uncertain affiliation. Carbonification studies indicate that the pile of nappes in the internal Apenninic zone never exceeded 6 or 8 km (Reutter, Teichmüller et al. this vol.). Moreover, in most parts of Apennines near surface tectonic structures are present. In Tuscany the thin Liguride allochthon is distributed only in discontinuous patches, and fensters of the Tuscan units, which belong to the cover of the Adriatic microplate, appear everywhere. Even at the southwesternmost point of the Northern Apennines on the island of Elba, the Tuscan autochthon crops out. To the north, where the Northern Apennines are not as broad as in Tuscany, the Liguride allochthon is thicker and more continuous. Even there, on the shelf of the Ligurian Sea, the top of the Apenninic basement was detected by means of refraction seismic investigations at a depth of only 5 km (Letz et al. this vol.). Hence, the internal border of the Adriatic microplate was uplifted after loading with the sediments of the accretionary wedge of the Upper plate and after suffering some deformation and low grade metamorphism.

This structural pattern of the Northern Apennines shows a sharp contrast to that of the Western Alps. There the geological and geophysical data prove that the European foreland dips beneath and increasingly thick cover of allochthonous units which originated from the margin of the European plate, from the Penninic realm and from the overriding Adriatic hinterland (Giese, Günther & Reutter 1970).

Crustal Structure

Typical cross sections through the Western Alps based on refraction seismic data show a clearly asymmetric crustal structure. The crust/mantle boundary dips from 35 to more than 50 km from the foreland towards the hinterland, where it is superseded by another shallow crust/mantle boundary in a depth of about 25 km. Here a crustal doubling can be recognized which indicates that the European foreland plate was widely overridden by the Adriatic hinterland plate, whereby the intervening oceanic area was completely eliminated. Hence, this structure represents the geometry of a subduction model which reflects synthetic thrusting and subduction.

Based on this argument, it could be expected that towards the south but still within the West Alpine-Apenninic double orogene as a result of the "Africa-vergent" Apenninic orogeny, the Apenninic hinterland consisting of the European plate with the Alps at its margin would overly the Adriatic foreland plate (Boccaletti et al. 1971). It was therefore surprising that refraction seismic studies revealed quite another crustal setting.

According to Giese, Morelli & Nicolic and Letz et al. (this vol.) the Moho-discontinuity of Corsica, which is found at a normal continental

depth of 30 km, can be traced eastwards under the island of Elba and western Tuscany into a depth of about 60 km. Underneath the Corsican basin, situated between Corsica nad Elba, this eastward dipping Moho must be updomed to a certain degree in order to gravitationally compensate a basin with Neogene sediments of more than 5 km thickness. Beneath Elba and Western Tuscany at a depth between 15 and 25 km another crust/ mantle boundary was detected which gradually increases in thickness towards the central parts of the Apennines. It evidently constitutes the crust of the Adriatic microplate. Hence, the western border of this continental plate which forms the geotectonic base and foreland of the Apennines, obviously overlies the European or, more precisely, the Corso-Sardinian continental fragment, instead of underlying it (Giese, Morelli & Nicolic this vol.). In this respect the structure is quite similar to that of the Western Alps near Ivrea and apparently does not reveal any effects of the Apenninic orogeny (fig. 4).

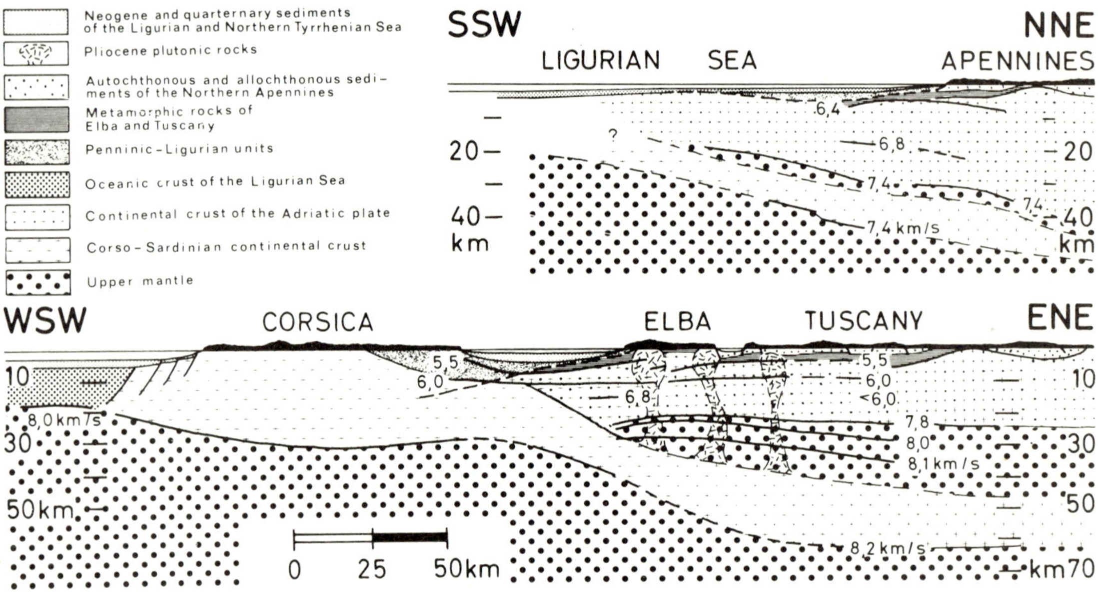

Fig. 4. Geological interpretation of the refraction seismic profiles Corsica-La Spezia and Corsica-Elba-Siena (after Letz et al. this vol.).

This crustal structure is not merely a local effect since also along two other profiles, which from Corsica pass into the Apennines near Viareggio and to the north of La Spezia (Letz et al. this vol.) an identical crustal overlapping was observed. Updipping of the crust/mantle boundary from the Apennines towards the shelf edge of the Ligurian Sea is also confirmed by a N-S directed profile near Chiavari. Therefore, it is evident that the western border of the Adriatic microplate extends at least as far as the Ligurian shelf with a rather updipping tendency (see basement contourmap, fig. 1 cf. Giese, Morelli & Nicolic this vol.).

The problems posed by the Apenninic orogeny and posed by seemingly contradictory geological and geophysical observations, are best put in evidence by a refraction seismic profile from Lac Nègre, in the autoch-

thonous West Alpine Argentera Massif, to Genoa (Röwer et al. 1978). Here, a crustal doubling is encountered fom the Briançonnais unit SE of Cuneo, at least up to the Penninic crystalline rocks near Savona. According to surface geology the whole segment of crustal overlapping belongs to the structural edifice of the Western Alps. The lower crust/mantle boundary, about 50 km deep, is that of the European plate, whereas the upper one, about 23 km deep corresponds to that of the Adriatic microplate. Hence, this latter one appears to be inserted tectonically between the uppermost and the lower tectonic level of the Western Alps. This situation is also reasonable from a geological standpoint since towards the east as well as to the north, in the Apennines and in the Monferrato hills, rocks which were deformed with the Western Alps and which still are connected with them, are allochthonous over Neogene sediments of the Adriatic plate (fig. 5).

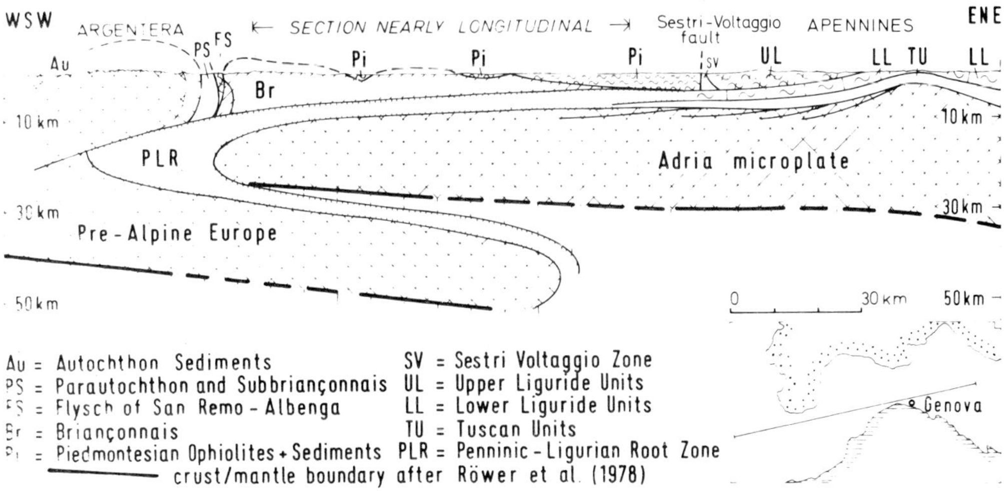

Fig. 5. Hypothetical setting of the major structural elements in a WSW/ ENE section through the Maritime Alps from the Argentera to the Northern Apennines.

The crustal overlapping of the Apenninic foreland over the crust of the hinterland seems to correspond perfectly with the "flake tectonics" model of Oxburgh (1972), except for the level of crustal splitting which is much higher than proposed for the model.

Continental Collision

According to the refraction seismic data from the Western Alps and the Northern Apennines, continental collision occurred in the entire double orogene. The lithosphere of the Penninic-Ligurian ocean was completely consumed, and a crustal overlapping of the bordering European and Adriatic plates was even achieved (fig. 3 of Giese, Morelli & Nicolic this vol.).

We can only conjecture the reason why in the moment of collision the crust of the Corso-Sardinian fragment was underthrust under the border of the Adriatic microplate antithetically to the supposed direction of

subduction of the oceanic lithosphere and the direction of obduction of the accretionary wedge of oceanic sediments.

An eastward or southeastward inclined position of the European plate margin inherited from the preceding collision with the oceanic arc (Corsica) may have been one of the causes. Another cause was possibly the preceding crustal overlapping in the Northern part of the Western Alps which prescribed the mode of continental collision after the consumption of the residual oceanic area analogously to the blades of a pair of scissors.

Dating of the moment of continental collision is not very clear. Following the model, it should have occurred after the emplacement of the accretionary wedge of sediments upon the margin of the Adriatic microplate in the Aquitanian. Indeed, the Lower Miocene of the Apennines represents a tectonically active period in the Northern Apennines. First deformation of the Apuan Alps (Kligfield et al. 1977), the Montenotte thrust to the north of Savona (Pasquaré 1968), and unconformities (Ibbeken this vol.) correspond to this phase and may be consequences of the collision. The filling of the Corsican basin must be intimately connected with the collision but it is not exactly known if it started as early as the Lower Miocene. The retrocharriage of the Western Alps may have been produced somewhat earlier but it certainly had the same cause. The collision also caused the steepening of the West Alpine root zones and the backward overturning and disappearance of the Ivrea body under more external units towards the south.

The first underthrusting of continental crust under the margin of the Adriatic microplate must have caused its upbending. Thus, it was prevented that crystalline nappes from the upper plate were thrust upon the Liguride units. The further far-reaching underthrusting for several tens of km which is revealed in the refraction seismic profiles, must be connected with the Miocene orogenic development of the Apennines. It certainly also influenced the later magmatic events in Tuscany (fig. 3).

It is curious, however, that only very few structures can be related to the process of underthrusting. Among these the huge west-facing anticline of La Spezia, local east-dipping schistosity planes in the metamorphic Tuscan sequence, and perhaps the young gravity tectonics of Elba have to be mentioned.

Rotational Movements

As already mentioned, the paleodirections during the evolution of the Corsica-Apennines orogene changed considerably in the course of rotational movements of the Adriatic microplate as a whole, of parts of it with respect to the main plate, and of the Corso-Sardinian continental fragment. Many paleomagnetic determinations of this area have been made (Pucher et al. and Soffel this vol.; Van den Berg 1977; Westphal et al. 1976) which prove these mostly anticlockwise movements. These data seem to be in good accordance with the geological findings.

The belt of crustal doubling between Corsica and the Apennines provides further confirmation of the movement of the Corso-Sardinian continental fragment towards the Italian peninsula. This migration is connected causally with the splitting-off from the European continent and the sphenochasmic opening of the Ligurian and Balearic Seas. If in the course of this movement final closing of the Penninic-Ligurian oceanic

basin was achieved in the Lower Miocene, the Corso-Sardinian microcontinent must have drifted apart from the southern coast of France at least in the Oligocene (fig. 2). Indeed there are several facts which favour this timing (e.g. formation of the Rhone graben, and of apport of clastics for the Grès d'Annot, etc. Auzende et al. 1973).

The Adriatic microplate was also subjected to rotation, but it could be shown palaeomagnetically that it did not behave as a really rigid plate. Parts of it rotated more or less than others to such a degree that there must have been a certain internal mobility along wrench faults and thrusts (Görler & Giese this vol.; Catalano et al. 1976). Thus, in the northern part of Italy at least two units of different amounts of anticlockwise rotation exist, namely the Southern Alps with 15° during the Upper Cretaceous and 15° between early Oligocene and Lower Miocene, and Umbria which rotated 30° in the Upper Cretaceous and 30° from the early Oligocene to the Pliocene (Van den Berg 1976; Van den Berg & Wonders 1976).

The rotation of Corsica amounts to 30°; that of Sardinia to 60°. According to the model here proposed, the absolute velocity of Corsica during rotation and translation must have been greater that that of the Adriatic microplate (Umbria) because collision seems to have already occurred in the Aquitanian. This means that from this moment both plates were probably coupled and together carried out the residual rotational movement until the early Pliocene.

Some structural features within the Adriatic microplate are evidently related to the rotation of plates after continental collision. The tectonic structures of the Northern Apennines diverge from NW to SE to such a degree that the Apennines in central Tuscany are three times as broad as, for example, in a cross section from Chiavari to Parma. The evident migration of the Neogene flysch troughs and the following compressional zone from the internal to the external side, therefore, covered a greater distance in the southeast. In terms of a rotational movement this shifting configuration had an overall equal angular velocity and, according to the structural divergence, an increasing radial velocity towards the SE. Since in this direction nappes become thinner and less and less complete it may be supposed that the motion of gravity nappes was hampered by very high velocities.

Several transverse faults crossing the Northern Apennines (the best known is the Leghorn-Prato-Sillaro line) seem to be tangential elements of this rotational migration of the tectonic stress field (fig. 1). They supported the trend of faster movement in the southeastern part of the Northern Apennines. Bodechtel & Smolka (this vol.) described relative clockwise rotations of single fault-bounded blocks within the superordinated anticlockwise rotating Adriatic plate, as revealed by different lineation systems and lateral slips.

The sphenochasmic opening of the Ligurian-Balearic Seas, the rotation of Corso-Sardinia, and the rotational migration of structures in the Northern Apennines all point to a common pivot which must have been situated somewhere to the southeast of Turin. The vortex of structures which is formed by the Monferrato hills, the Western Alps, and the Maritime Alps, must, therefore, have been generated in the course of these rotational movements by the same geodynamic cause and during the same Oligocene-Miocene time span (Boccaletti & Guazzone 1970). Thus it may be considered an Apenninic structure.

Present State

In the course of Cretaceous and early Tertiary movements between the European plate and the Adriatic plate, at least two different subduction and island arc systems existed which terminated with the Paleo-Alpine phase in the Mid-Cretaceous and the Meso-Alpine or "Ligurian phase" in the Upper Eocene. After the flip of the subduction plane, the Corso-Sardinian continental fragment moved towards the Adriatic plate, giving rise to a geotectonic situation which again can be compared with a trench/arc/back-arc model (Boccaletti & Guazzone 1972).

The post-collisional period in the Northern Apennines was characterized by the above mentioned migration of the Neogene flysch troughs and a following belt of compressional tectonics through the Apenninic area of the Adriatic plate from the internal to the external side. This new orogenic system persisted from the Lower Miocene to the early Pliocene and, with reduced velocity and intensity, seems to be still active in the border zone of the Po Plain and the Apennines (Elter et al. 1975).

Obviously, after the passage of the compressional belt Tuscany and its Tyrrhenian shelf became affected by tensional tectonics with strong vertical movements and contemporaneous magmatic activity of crustal origin.

Seismic and gravity data show that from the Ligurian and Tuscan shelf the thickness of the crust of the Adriatic microplate increases from 20 to 40 km (Giese, Morelli & Nicolic this vol.). This contrast between internal and external parts of the Northern Apennines is emphasized structurally by tensional tectonics where the crust is thin, and compressional tectonics where it is thick (Elter et al. 1975). Boccaletti & Guazzone (1972) have interpreted this pattern as another still active trench/arc/back-arc model, in which the depositional area of the Po Plain and the Adriatic Sea represent the trench, the belt of compressional tectonics the arc, and the Tuscan area of tensional tectonics, thin crust, high heat flow, and magmatism the back-arc or hinterland region.

Although this comparison is rather compelling, the fact must be stressed, that no important thrust has been found which could have separated the Po Plain basement from the external Apennines. This geotectonic setting exists within one single plate which is the foreland plate of the Northern Apenninic orogeny. It seems to represent a transitional feature to the geotectonic situation in the Southern Apennines (Görler & Giese this vol.).

Conclusions

The complicated structures of the Corsica-Northern Apennines double orogene are a consequence of the repeated movements with changing directions of the Adriatic microplate, or, basically, the African plate with respect to the European plate (Smith 1971; Dewey et al. 1973). As the continental plates approached each other, the main effects were the consumption of the intervening oceanic crust and continental collision. These effects were evidently achieved by a series of separate spreading and subduction systems, which successively developed between the two plates. The tectonic and crustal structures they generated, were superimposed on one another. The structures of the Corsica-Northern Apennines double orogene are more complex than those of many other parts of the Mediterranean mountain belt

system because movement patterns with opposed directions became superimposed and the geotectonic functions of foreland and hinterland were inverted.

The crustal structure at the suture between Corso-Sardinian and the Adriatic microplate is clearly characterized by a crustal overlapping. If all assumptions based on geological data are correct, the asymmetry of this crustal overthrust does not correspond to the asymmetry of the preceding subduction system. The antithetic disposition of subduction as well as obduction and structural trend of the Apennines on the one hand, and crustal thrusting on the other, should not be taken for a second flip of the subductional system in the moment of collision: Rather, it seems to be a collisional effect which may have been determined by secondary circumstances. In the case of the Corso-Sardinian fragment, it may have been a certain nose-heaviness of the crust of the upper plate due to a sedimentary load at its front and rifting at its rear. The comparison with the northern part of the Western Alps shows that continental collision can produce different crustal structures.

The outlined relationship between the Western Alps and the Northern Apennines demonstrate that it is not possible to draw a boundary line without defining the structural level under consideration. For surface geology the presence of pre-Oligocene "Europe-vergent" structures can be used as a diagnostic feature, although in the transitional zone these Alpine structures were possibly displaced by the Apenninic movements. A surface boundary might be traced near Genoa, but, considering the crustal structure, the contact between Alps and Apennines would be situated perhaps to the south of Cuneo in the center of the Western Alps.

The evolution of basins of the back-arc type in Tuscany and the Tyrrhenian sea is also superimposed over preceding crustal structures of Alpine and Apenninic age and trend. These adjacent and coeval basins were formed behind the compressional belts of the Northern and the Southern Apennines. The boundary between these chains which is marked by the prominent Anzio-Ancona fault zone, is comparable to the re-entrant angle between two arc systems.

References

Auzende, J.M., Bonnin, J. & Olivet, J.L., 1973: Jh. geol. Soc. London, 129:602-620.

Beck, R.H. & Lehner, H., 1974: Amer.Assoc.Petrol.Geologists,Bull.58:376-395.

Boccaletti, M. & Guazzone, G., 1975: in: C.H. Squyres (ed.) Geology of Italy. Earth Sci.Soc. Libyan arab.Rep.:143-163, Tripoli.

- , 1972: Mem. Soc. geol. Ital., 11:201-216.

- , 1970: Mem. Soc. geol. Ital., 9:177-195.

Boccaletti, M., Elter, P. & Guazzone, G., 1971: Nature Phys. Sci., 234 (49): 108-111.

Catalano, R., Channel, J.E.T., D'Argenio, B. & Napoleone, G., 1977: Mem. Sco. geol. Ital., 15:95-118.

Compagnoni, R., Dal Piaz, G.V., Hunziker, J.C., Gosso, G., Lombardo, B. & Williams, P.F., 1975: Ital.-USA coop. Project on high Pressure-low Temp. Metamorphism Western Alps, internal Report, 2:1-54, Bologna.

Dal Piaz, G.V., Hunziker, J.C. & Martinotti, G., 1972: Mem. Soc. geol. Ital., 11:433-466.

Decandia, F.A. & Elter, P., 1972: Mem. Soc. geol. Ital., 11:503-530.
Dewey, J.F., Pitman, W.C., Ryan, W.B.F. & Bonnin, J., 1973: Geol. Soc. Amer. Bull, 84:3137-3180.
Elter, P., Giglia, G., Tongiorgi, M. & Trevisan, L., 1975: Boll. Geofis. teor. appl., 17(65):3-18.
Elter, P.& Pertusati, P., 1973: Mem.Soc.geol. Ital., 12:359-375.
Giese, P., Günther, K. & Reutter, K.-J., 1970: Z.dtsch.geol.Ges.:120.
Kligfield, R., Hunziker, J. & Schamel, S., 1978: Manuscript in review, Earth Planet. Sci. Lett.
Laubscher, H.P., 1975: Amer. Jh. Sci., 275:865-876.
- , 1971: Amer. Jh. Sic., 271:193-226.
Nardi, R., 1968: Mem. Soc. geol. Ital., 7:323-344.
Ohnestetter, D., Ohnestetter, M. & Rocci, G., 1976: Bull. Soc. géol. France, 18:115-134.
Oxburgh, E.R., 1972: Nature, 239:202-204.
Pasquaré, G., 1968: Riv. ital. Paleont., 74:1257-1273.
Röwer, P., Prodehl, C. & Giese, P., 1978: Boll.Geofis.teor.appl., in press.
Seely, D.R., Vail, P.R. & Walton, G.G., 1974: in: Burk & Drake (ed.) The geology of continental margins: 249-260, Berlin.
Smith, A.G., 1971: Geol. Soc. Amer., Bull, 82:2039-2070.
Trümpy, R., 1973: in: De Jong & Scholten (ed.) Gravity and Tectonics: 229-251.
Van den Berg, J., 1977: Mem. Soc. geol. Ital., 15:83-89.
Van den Berg, J. & Wonders, A.A.H., 1976: Tectonophysics, 33:301-320.
Westphal, M., Orsini, J. & Vellutini, P., 1976: Tectonophysics, 30:141-157.

Chapter 4

The Geotraverse Sicily - Southern Italy - Ionian Sea (IIIa+b)

Introduction

P. Giese and K. Görler

Calabria maintains a special position within the Apennine-Maghrebide orogen, as shown by the following features:

1. The Calabrian-Peloritanian Massif is the largest surface exposure of deep continental crust within this mountain range. The bulk of rocks in the other sections of the Apennine-Maghrebide system are Mesozoic and Tertiary series which have almost everywhere been proved to be allochthonous in their present position.
2. Whereas, in the other sections, the Apennine-Maghrebide orogen borders externally on continental crust, the Calabrian section lies between two areas having an oceanic and intermediate crust respectively. The highest Bouguer-gravity values in the entire Apennine-Maghrebide orogen are observed in Southern Calabria.
3. Calabria shows an extraordinarily sharp bend from the NW - SE trend of the Apennine to the E - W trend of Sicily and the mountains farther west.
4. Regarding recent crustal movements, Calabria proves to be a special region. Particularly in Southern Calabria, abnormally high rates of uplift occur together with a corresponding seismicity and erosion rate.
5. The occurrence of deep-focus earthquakes under the Tyrrhenian Sea, the high heat flow, and the volcanism of the Eolian Islands indicate active subduction in the region of the Calabrian Arc.

These special features make Calabria appear as an anomalous region within the Apennine-Maghrebide orogen. On the other hand, similarities of facies and tectonic structure exist in the Southern Apennine and in Northern Sicily. At any rate, the present position of Calabria as a narrow continental fragment, mostly of Hercynian age, must be fundamentally different from that of former times.

Therefore, investigations within a geodynamic program in this central region of the Mediterranean should have the following objectives:

- an analysis of the seismotectonics and neotectonics in Calabria,
- an attempt to trace back to the past the sequences of movements detected for the present,
- an attempt to reconstruct the paleogeographic development of Calabria from the end of the Hercynian orogenesis up to the present,
- a translation of the knowledge obtained into hypotheses regarding the geodynamic development of Calabria.

This chapter deals especially with conditions in Southern Calabria. By interdisciplinary research and the application of new investigation methods, new data and ideas regarding the geodynamic development of the Calabrian Arc were to be obtained.

The tectonic activity of Calabria suggested to devote special efforts to the investigation of these recent events. Thus, by means of various

procedures, attempts were made to obtain information about the history of uplift and erosion in Southern Calabria and to find indications of the mass redispositions. Studies of faultplane solutions complement the data of recent movements. Another group of studies dealt with the structure and composition of the Calabrian crust by means of petrographical and geophysical investigations.

Studies of the Mesozoic, transgressing the crystalline, and of the older sediments, redeposited during the Younger Tertiary in olistostromes, helped to reveal the paleogeographic development of Calabria.

A synthesis of the results obtained with the numerous data already available contributes to an understanding of the very complex geodynamic development in the central Mediterranean.

Part 1

Geophysics

Crustal Structure of Southern Italy

K.-G. Schütte

Introduction - Net of Profiles

Seismic refraction measurements in Southern Italy cover a complete orogenic system. The Tyrrhenian Sea forms the hinterland of the Apennine-Sicily orogene composed of Mesozoic and Tertiary sediments in which the Calabrian Massif with its highly metamorphic rocks and its quick rise, continuing at present (Brogan et al. 1975), occupies a key position. The foredeep of the S-Apennines is formed by the Bradanian graben and the Gulf of Taranto, filled since the Miocene with thick sediments becoming younger towards the SE and partly folded up to their Pleistocene parts. The foreland is formed by the Puglian platform with its undeformed mesozoic sediments, over 6 km thick.

In this heterogeneous region a total of 10 seismic refraction profiles and 2 fans were measured from 1965 until 1973 (fig. 1) in close cooperation with French, Italian, and German geophysical institutions. In this paper the results of these measurements are summarized in a contour map of the crust/mantle boundary, but only selected profiles are discussed.

Puglia

In Puglia, the foreland of the Apennines, seismic refraction measurements were made as early as in 1965 and 1966 (Finetti et al. 1966; Bellemo et al. 1967) and reinterpreted by Eilert (1976) in connection with an investigation of Profile 3/71. The data from Profile 10/72 in S-Puglia have been evaluated by Morelli et al. (1975).

Although Profile 3/71 runs about parallel to, and Profile 10/72 almost normal to the strike, both yield, for S-Puglia, very similar seismograms in which, after the mentioned authors, besides P_g, P^M, and P_n, a reverse phase can be correlated which is interpreted as intercrustal reflection P^I (fig. 2a). In respect to P_g, it occurs with a delay. The P^M phase, reflected at the crust/mantle boundary, also shows a delay in time, relating to P^I.

The correlated travel-time curves suggest an upper crust with a total thickness of about 17 km (figs. 3 and 5), divided by a low-velocity layer. The transition to the lower crust is marked by the beginning of a second low-velocity layer which reaches down to a depth of approx. 25 km and has a minimum velocity of v_{min} = 5,5 km/s. Total crustal thickness amounts to about 30 km.

Considering recent geological ideas it seems possible that P^I is a re-

flection from the upper edge of the crystalline. Consequently the wave, described above as P_g, would have passed sediments only and should then be described as P_s phase.

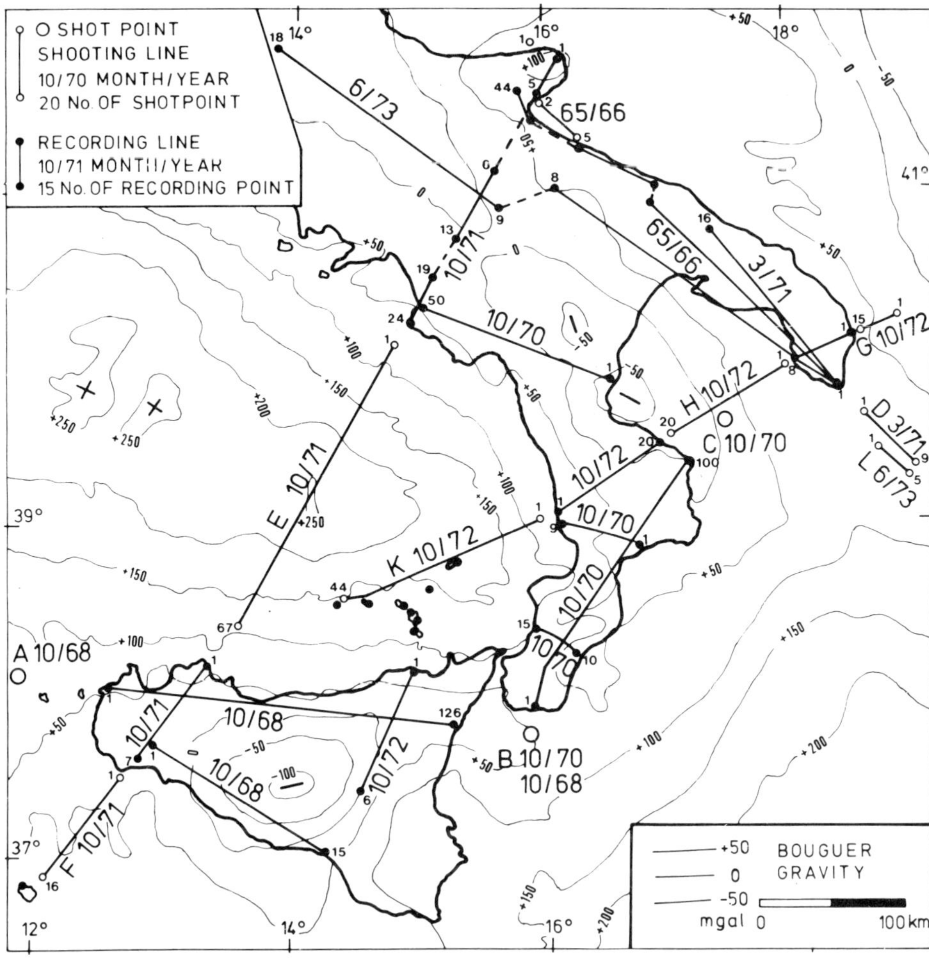

Fig. 1. Position map of seismic refraction measurements in Southern Italy.

Measurements in N-Puglia in 1965/66 result in a similar crustal structure (fig. 3), but the intercrustal reflector was not observed. The transition from upper to lower crust is marked by a low-velocity layer (v_{min}= 5,6 km/s). Crustal thickness is about 32 km. The occurrence of zones with marked low-velocity inversions in the foreland of the Apennines may be an effect of the presumed recent orogenesis in the Ionian Sea.

The structure of the upper mantle, also shown in fig. 3, has been derived from the measurements in June 1973 (Cassinis et al. 1976).

Transition Zone Campania/Tyrrhenian Sea

On the profile 10/71, besides P_g (observed up to Δ = 65 km) and P_1^M ($60 \leq \Delta \leq 120$ km), an intercrustal reflection P^I ($50 \leq \Delta \leq 80$ km) can be observed at the stations 21 to 19. At the stations 21 to 9, in a distance range from 130 to 180 km, very late onsets are recorded ($t - \frac{\Delta}{6}$ up to 8 sec) with large amplitudes which are being interpreted as reflections P_2^M (fig. 2b). The P_n phase can be clearly recognized only in the record sections of the stations situated on the Gargano.

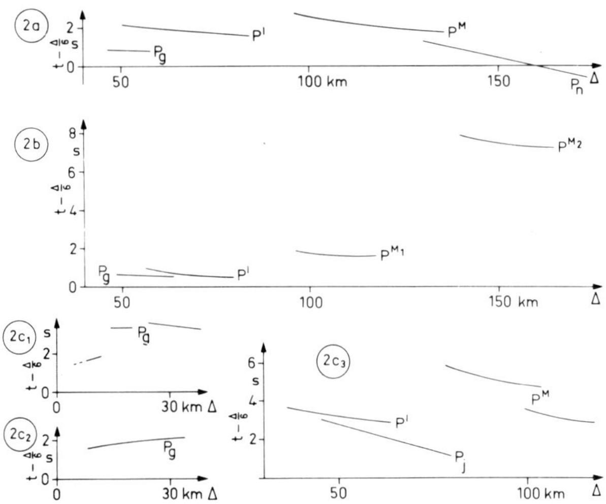

Fig. 2. Examples of travel-time curves. Further explanations see text.

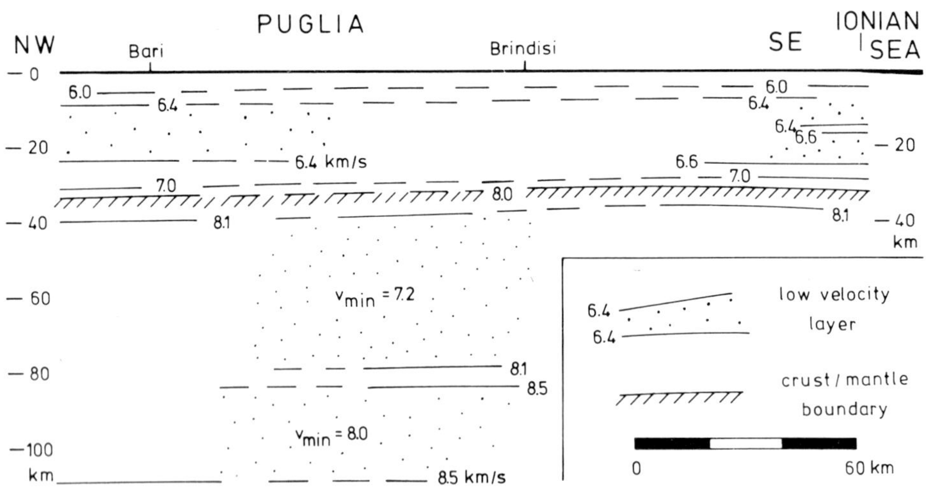

Fig. 3. Structure of the crust and upper mantle in Puglia, the foreland of the Apennines.

The evaluation of the observations of the two most southwestern stations (nrs. 24 and 23) reveals a crustal thickness of 22 km about 50 km offshore Campania. A weak velocity inversion between 7 and 11 km depth marks the transition from upper to lower crust, both of which show a simple structure (fig. 4).

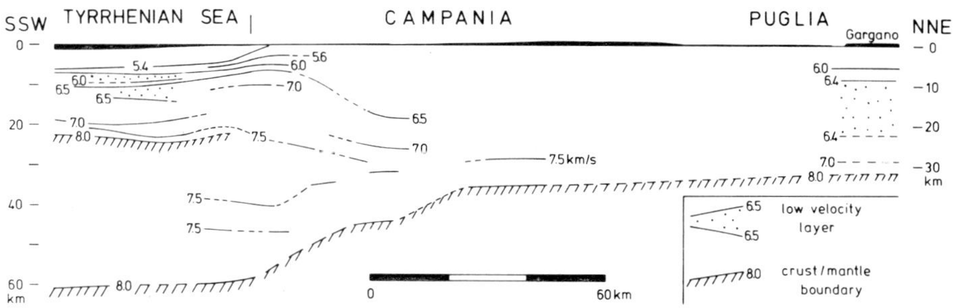

Fig. 4. Crustal structure from the Campania to the Tyrrhenian shelf region.

The comparison of record sections of stations situated further inland is more difficult although in each one of them the mentioned wave groups are clearly recognizable. With a fixed observation distance, travel times of each of the phases P_g, P^I, and P_1^M vary from station to station. Evaluation hence results in a complex crustal structure with changing stratum slopes in the shelf region near the coast and under the south-western Campania. The absence of the P_n phase in the record sections of the stations 24 to 6 suggests a velocity inversion in the shelf and coastal region, immediately under the crust/mantle boundary. The evaluation of the P_2^M phase with the x^2, t^2 method leads to a reflector at a depth of 35 km under the Apennines crest, dipping towards SW and reaching a depth of about 60 km, 45 km offshore. Including the results of the measurements in 1965/66 in N-Puglia, this reflector must be interpreted as crust/mantle boundary of the Adriatic plate which is overlain by crust and uppermost mantle of the Tyrrhenian shelf region.

Profile Tyrrhenian Sea - Calabria - Gulf of Taranto

Shots in the Tyrrhenian Sea were recorded on the Eolian Islands and in N-Calabria. All record sections based on these observations show a simple pattern. On Stromboli, Panarea, and Salina, only the phases P_g and P^M were observed because shotpoint distances were too small. With the measurements on the two southwestern islands Filicudi and Alicudi, the P_n phase, too, could be recorded. In these two record sections, further later onsets occur in a distance range up to 60 km; they can, however, be correlated only over short distance intervals (Morelli et al. 1975). The record sections of all stations in N-Calabria show a clear and simple arrangement of the wave groups P_g, P^M, and P_n. Contrary to the observations in the Campania, the P_n phase is clearly visible in a distance range from 80 to 200 km even with 50 kg shots.

The interpretation of these data results in a crustal thickness of 16 to 18 km in the region of the Eolian Islands. About 30 km offshore Calabria, it amounts to 20 km, then thickens quickly up to 29 km under the SW border of the Sila. In this whole profile section, the crust/mantle transition is characterized by a strong velocity increase while the lower crust shows a very small positive velocity gradient or even a weak velocity inversion (fig. 5).

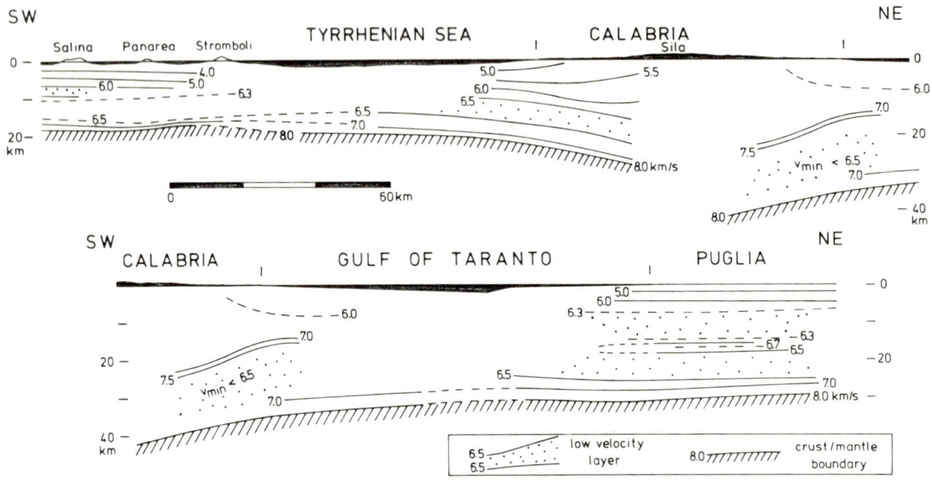

Fig. 5. Crustal structure in the region of the Eolian Islands, the Tyrrhenian shelf, N-Calabria, the Gulf of Taranto and S-Puglia.

The upper crust, too, increases in thickness from SW to NE, and the velocity lines, in the coastal region of Calabria, point upwards to the region of low temperature - high pressure metamorphites as well as granulites and kinzigites respectively.

The observations of the shots in the Gulf of Taranto, carried out in Calabria, result in extremely complex record sections (fig. 2c). When the data of one station are compiled to a record section, the P_g travel-time curve falls into short curved pieces staggered as to time (fig. $2c_1$). Compiling, however, all records of one shot to a record section (fig. $2c_2$), a continuous P_g phase does appear in each section but the travel-time curves from the different shots show a relative staggering in time. This observation can only be interpreted as a consequence of a high variability in thickness of sediments in the Gulf of Taranto, this thickness generally increasing over the first 20 km offshore and decreasing again further seaward.

In all record sections, the P^M group is dominant and shows a delay in respect to P_g. A phase correlation shows a strong splitting up of the P^M group (fig. $2c_3$). P_n was observed only at the remotest stations near the Tyrrhenian coast. At the stations 4 to 14 a first phase P_j is recorded in the distance range from 50 to 100 km, having an apparent velocity of more than 8 km/s. Also a corresponding reversed branch P^I is clearly observed (fig. $2c_3$). This high apparent velocity seems to be caused by the mentioned variation of thickness of sediments in the Gulf of Taranto. In re-

versed record sections comparable phases P^I, P_j are clearly visible, too, but P_j shows a velocity of about 7 km/s. In respect to P_j the P^M phase shows a delay of more than 2 sec.

The evaluation of data reveals that the crust/mantle boundary under the centre of the Gulf of Taranto lies at a depth of 32 km and dips in the direction of Calabria where crustal thickness amounts to 40 to 45 km. Under the Ionian coast, at a depth of about 15 km, exists a zone of high velocity which dips towards SW. It is located above a well developed low-velocity layer. Although the parameters of this channel were determined only with some uncertainty, its existence is clearly proved by the long time delays of the P^M groups (fig. 5).

Main Features of the Crust/Mantle Boundary in S-Italy

The morphology of the crust/mantle boundary in the Southern Italian area including the adjacent sea regions appears to be influenced less by the mountain chain of the Apennines than by the transition from the region of the Tyrrhenian Sea to that of the Adriatic plate (fig. 6). The crust/mantle boundary continuously dips from the foreland towards the central

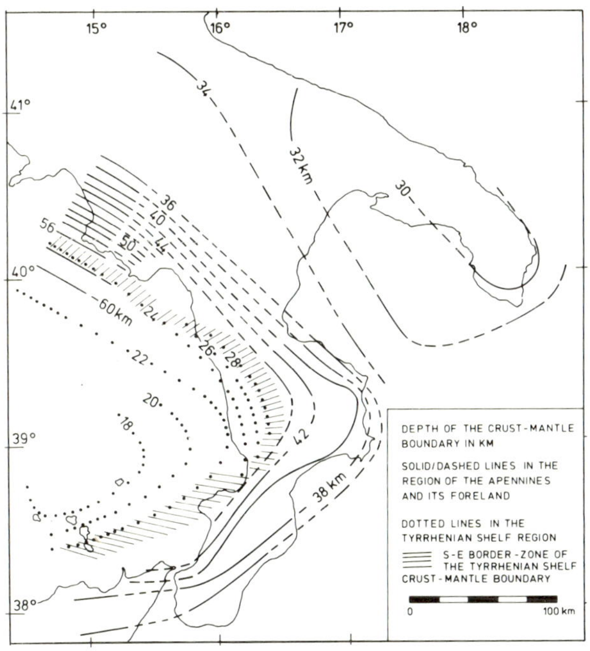

Fig. 6. Contour map of the crust/mantle boundary in Southern Italy.

zone of the Apennines and continues to dip in the direction of the hinterland to an ultimate depth of 45 to 60 km. This results in a clear asymmetry of crustal thickness in respect to the Apennine chain.

Also in the Tyrrhenian shelf region, the hinterland of the Apennines, the crust/mantle boundary dips toward the Apennines but extends to a depth of only 24 to 29 km. The transition between these two parts of the crust is characterized by an abrupt change of thickness. The transition of the

internal crustal structure is quite complex. The thin crust of the Tyrrhenian shelf region verges directly on the thick crust of the Apènnine area or directly forms its upper part.

References

Bellemo, S., Finetti, I., Mechler, P., Morelli, C., Rocard, Y., & de Visintini, G., 1967: Boll. Geofis. Teor. Appl., 9:108-119.

Brogan, G.E., Cluff, L.S. & Taylor, C.L., 1975: Tectonophysics,29:323-330.

Cassinis, R., Giese, P., Morelli, C. & Röwer, P., 1976: Report at the 15th General Assembly of the E.S.C. (in press).

Eilert, H., 1976: Dipl.-Arbeit, Freie Universität Berlin.

Finetti, I., Bellemo, S. & de Visintini, G., 1966: Boll. Geofis. Teor. Appl., 8:21-39.

Morelli, C., Giese, P., Cassinis, R., Colombi, B., Guerra, I., Luongo, G., Scarascia, S. & Schütte, K.-G., 1975: Boll. Geofis. Teor. Appl., 17:183-210.

Magnetic Measurements in Northern Calabria

G. Schwarz

While the greater part of the Apennines is built up by sediments, in Calabria metamorphic and crystalline rocks crop out. A detailed study of the structure of the crystalline massif reveals further peculiarities. At the western border of the Sila, the central massif of northern Calabria, rocks of the lower crust are outcropping, grading towards the E into metamorphites and magmatites of the upper crust. This complex is of Hercynian age. On the other hand, towards the W, high-pressure, low-temperature rocks of Alpine age are in contact with the deep crustal rocks of the Sila massif.

Results of seismic refraction measurements also revealed an anomalous structure for northern Calabria (Schütte 1974; Morelli et al. 1975). Coming from Puglia, crustal thickness increases under the Sila to more than 40 km while decreasing abruptly near Cosenza to 25 - 30 km. The Sila itself forms a crustal nappe with a thickness of 20 - 25 km, as shown by seismic data (Schütte this vol.). In order to obtain further information about this anomalous structure, magnetic measurements have been carried out in northern Calabria in summer 1975. Widely spaced magnetic surveys have already revealed the existence of a weak anomaly in the region of the Sila (Gantar et al. 1961; Morelli 1973).

The total magnetic intensity survey was carried out with an ASKANIA proton magnetometer. Four profiles were measured, each running from the Tyrrhenian to the Ionian coast with a point distance of 0.5 to 2 km. The values obtained have been corrected for the daily variations of the

earth's magnetic field by means of registrations of the Geomagnetic Observatory of L'Aquila. Finally, the normal magnetic field (fig. 1), derived from the regional magnetic field of Southern Italy (Morelli 1973), has been eliminated. As the altitudes of the observation points vary rather strongly (maximum altitude differences about 1600 m), the decrease of the intensity of the normal magnetic field with increasing height was taken into account ($\Delta T/\Delta z = -21\ \gamma$/km). All measuring stations were reduced to 800 m above sea level. The measured values were reproducible within $\pm 3\ \gamma$.

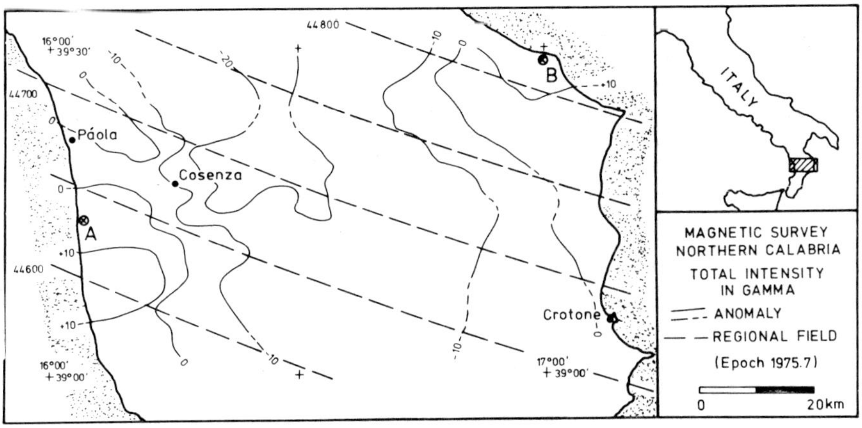

Fig. 1. Total magnetic intensity anomaly and regional field (in γ). Reduction level (anomaly) 800 m a.s.l. AB: Profile for model computations.

Except for a local anomaly near a diabase in western Calabria, no other local magnetic anomalies could be detected, not even in the area of exposed rocks of the lower crust.

The total intensity anomaly in the region of the Sila is nearly striking SSE - NNW, with a half-value width of about 35 km (fig. 1). The course of the lines of equal intensity justifies the assumption of a NS striking body with an inverse remanent magnetization. The magnetic source layers will have their largest thickness under the central Sila of northern Calabria and wedge out towards N as well as towards S.

For the two-dimensional model computations a smoothed profile (figs. 1 and 2) was chosen, running WSW - ENE. In principle, the Curie temperature limits the depth of the magnetically effective body to about 40 km (El-Ali 1977). However, as a possible bottom limitation, the interface at a depth of 20 - 25 km, derived from seismic data, was accepted. The magnetization of model bodies was assumed to be homogeneous. The measured anomaly can be approximated by disturbing bodies with a magnetization between $50 \cdot 10^{-5}\ \Gamma$ and $250 \cdot 10^{-5}\ \Gamma$. Numerous models were computed from which two extreme ones are shown (fig. 2) to demonstrate the limits of "reasonable solutions". The anomaly can be explained by a long, slab-shaped body, but also by a dome-like, wedge-shaped body. The body 'a' in fig. 2 - with a magnetization of $J = 50 \cdot 10^{-5}\ \Gamma$, having a declination of $\delta = 180°$ and in inclination of $i = -30°$ - requires a connection between the lower crustal rocks at the surface and the layer of higher velo-

city at a depth of about 20 km, while the body 'b' (J = 130 · 10^{-5} Γ, δ = 240°, i = -30°) coincides with this layer.

The approximation of the measured anomaly by these bodies can be improved by assuming another local disturbing body (Schwarz 1977).

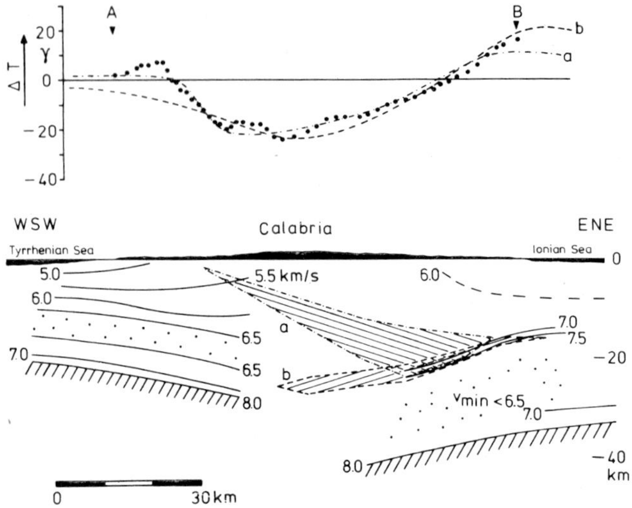

Fig. 2. Two-dimensional model computations for profile AB. Crustal structure - lines of equal velocity - after Schütte (1977). Model 'a': Magnetization J = 50 · 10^{-5} Γ, declination δ = 180°, inclination i = -30°. Model 'b': Magnetization J = 130 · 10^{-5} Γ,, declination δ = 240°, inclination i = -30°. Dots represent measured, smoothed anomaly.

References

El-Ali, H., Gantar, C., Morelli, C., Pisani, M., Segre, A.G. & Zampieri, L., 1961: Boll. Geof. Teor. Appl., III:338-366.

Morelli, C., 1973: In: Structural Model of Italy, Scale 1 : 1 000 000, Grafica Editoriale Cartografica S.p.A., Roma.

Morelli, C., Giese, P., Cassinis, R., Colombi, B., Guerra, I., Luongo, G., Scarascia, S. & Schütte, K.-G., 1975: Boll. Geof. Teor. Appl., XVIII:183-210.

Schütte, K.-G., 1974: Report at the ESC Meeting, Trieste.

Schwarz, G., 1977: Dipl.-Arbeit, Freie Universität Berlin.

A Geothermal Profile Between the Adriatic and the Tyrrhenian Seas

H. El-Ali and P. Giese

A number of heat flow data are available from Southern Italy and the adjacent seas (fig. 1). In order to compute the temperature distribution, a profile along a seismic refraction line between the Adriatic Sea and the Tyrrhenian Sea has been chosen (Schütte this vol.; Finetti & Morelli 1973). From the clustered heat flow data, corresponding mean values are projected on to that line. In general, a small decrease of heat flow from Puglia to Calabria is revealed (fig. 3).

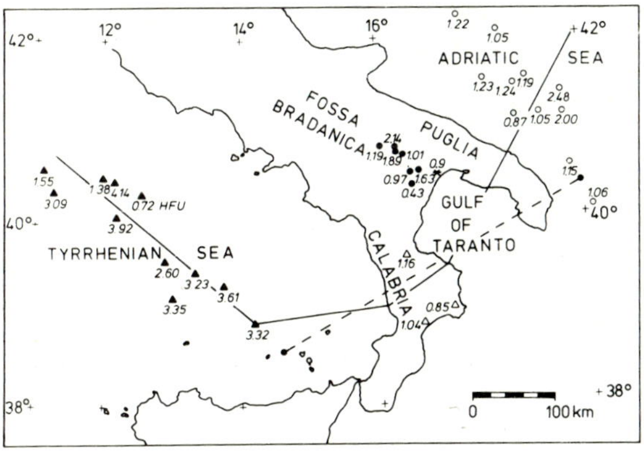

Fig. 1. Map showing the position of heat flow measurements.
——— geothermic profile
- -'- seismic refraction line (Giese & Morelli 1975)
o Lavenia 1967
● Mongelli & Ricchetti 1970
x, △ Lodo & Mongelli 1973,75
▲ Erickson 1974

A one-dimensional approach under steady-state conditions has been used at first approximation for temperature computation. Two types of models have been treated, an exponential model (Lachenbruch 1968) and a layered model.

As heat production, 3, 5 and 7 HGU (heat generation units) have been used for the upper crust in both models. If 7 HGU are used, a negative temperature gradient results for the layered model which is, therefore, regarded as unrealistic.

Assuming a heat production between 3 and 5 HGU under the southern Adriatic Sea and Puglia, a temperature between 400°C and 500°C can be expected at the crust/mantle boundary at a depth of 30 km. Correspondingly, a heat flow between 0.25 and 0.74 HFU resp. passes the crust/mantle boundary (fig. 2A).

In the Gulf of Taranto as well as in Northern Calabria a heat production of 5 HGU already generates a negative temperature gradient in the lower crust. A temperature of 300°C and a heat flow of 0.22 HFU at the crust/mantle boundary prevails in the Gulf of Taranto. For Calabria, a nearly constant temperature of 300°C is found for the depth range between 35 and 50 km. The heat flow through the crust/mantle boundary becomes nearly zero (fig. 2B).

In the Tyrrhenian Sea, the extremely high heat flow has a mean value of 2.81 HFU. At the crust/mantle boundary at 11.5 km depth, an anomalous high heat flow of 2.75 HFU passes the crust/mantle boundary where a temperature of 480°C is reached (fig. 2C). This high heat flow indicates a

convective heat transfer within the upper mantle.

The great lateral temperature gradient between the Tyrrhenian Sea and the Calabrian Massif requires a correction of the one-dimensional model (fig. 3A). By a simple approximation method, two-dimensional temperature profiles are determined.

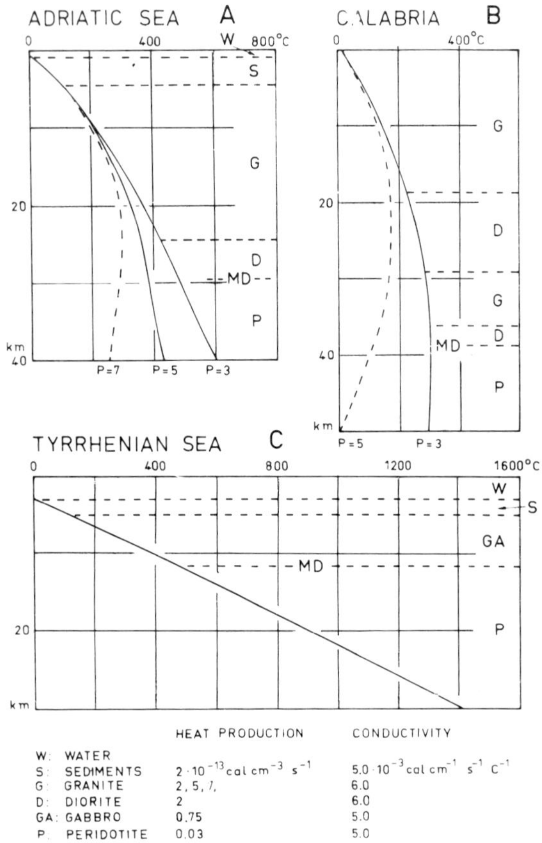

Fig. 2. Calculated temperature depth functions
P = heat production (HGU)
A = Adriatic Sea
B = Calabria
C = Tyrrhenian Sea

By this improvement of the exponential model (fig. 3C) as well as of the layered model (fig. 3B) the temperature distribution is nearly the same for temperatures of up to 300°C. Only the 400°C isotherm is different in both models. Independent of the method applied, the crust under Calabria is of lower temperature than at corresponding depth ranges under the foreland. This is mainly due to the low heat flow value in Calabria and the Gulf of Taranto. In Calabria, the heat flow has been measured in very young sediments. Thus it cannot be excluded that the steadystate in these sediments has not yet been reached.

This low temperature under Calabria contradicts the seismic results which show a distinct velocity inversion in the depth range between 30 and 40 km, implying high temperatures (Schütte this vol.). As the tectonic development of Calabria took place in the late Tertiary, nonstationary processes must be considered also. In order to form an idea whether a

distinct positive heat flow anomaly can be expected within only a few million years, the following extreme model is considered.

The melting temperature of granitic rocks (about 700°C) can be regarded as a reasonable maximum temperature for the following estimation. A plate with a thickness of 40 km (homogeneous crust) is rather quickly

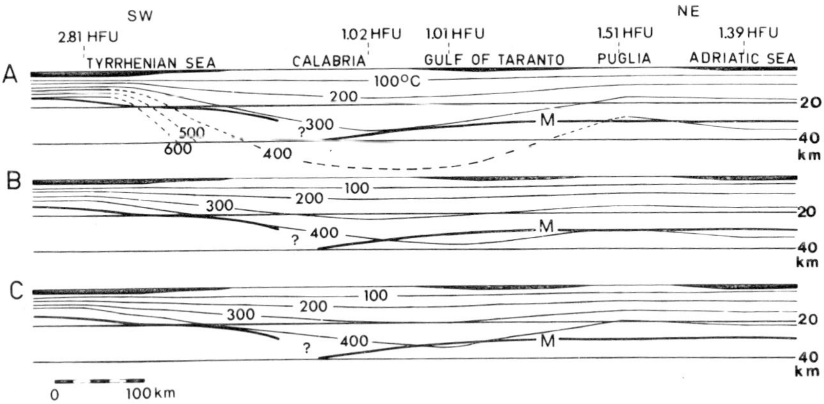

Fig. 3. Profiles with isotherms. A = composed of one-dimensional distributions based on a layered model; B = layered model corrected for the two-dimensional case; C = exponential model, corrected for the two-dimensional case: MD = Mohorovičić Discontinuity.

heated up at its bottom from 300°C to 700°C. The following values show the percentage by which the surface heat flow increases with time:

Time after start of heating	Percentage of increase of surface heat flow
$10 \cdot 10^6$ a	37%
$15 \cdot 10^6$ a	52%
$20 \cdot 10^6$ a	58%

The present state and position of Calabria were formed in the late Miocene, 10 to 20 m.y. ago. A distinct high anomalous surface heat flow cannot yet be expected from the temperature increase due to this young orogenetic process. A completely different situation is bound to exist under the Tyrrhenian Sea. Although the heat flow fo 2.81 HFU is extremely large, hardly any remarkable mass of heat-producing granitic crust exists. The computation of steady-state temperatures would lead to an unreasonably high temperature of 1400°C at a depth of only 30 km, that is, the melting temperature of basic and ultrabasic rocks. Therefore, the heat transfer must also include a convective component in the upper mantle.

Assuming that 2/3 of the total heat flow, i.e. about 1.8 HFU, must be related to convection, a simple balance equation allows to estimate the minimum velocity v (cm/s) of the upwelling material

$$v = \frac{q_m}{\rho \cdot c \cdot \quad T}$$

q_m = heat flow through the crust/mantle boundary, 1.8 HFU; ρ = density of upwelling material, 3.3 g/cm³; c = specific heat 0.3 cal g^{-1} C^{-1}; ΔT = temperature of the upwelling material against the cooling surface, 600°C.

The velocity reaches a value of 0.15 cm/y, which is within a reasonable order or magnitude.

We assume a connection between the upwelling upper mantle material and the descending slab of the Ionian plate which is postulated by seismological and geochemical arguments. Andrews & Sleep (1974) have explained the high heat flow in marginal basins and island arc volcanism by numerical modelling of flow induced by a downgoing slab. Fig. 4 shows the position of the isotherms 5 and 10 m.y. after the starting of the subduction process. In a first approximation the isotherms can be identified with effective surfaces of ascending material. For the period of 5 to 10 m.y., a mean velocity of 1 cm/y results for the upwelling material. Bearing in mind that the velocity derived from the balance equation is too small, both estimations, resulting from quite different considerations, yield comparable results.

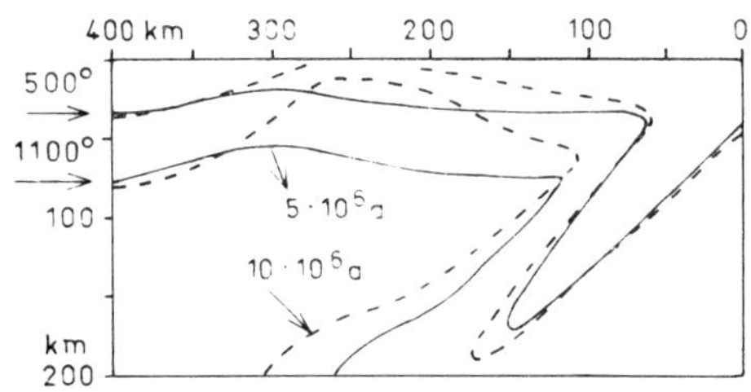

Fig. 4. Section showing the isotherms (in 500°C and 1100°C) after 5 and 10 m.y. after beginning of the subduction process.

Finally, the following main results of this geothermal study are concluded. There is a weak decrease of heat flow from the foreland to the central zone. Under steady-state conditions there is no significant temperature difference recognizable between the outer and inner zones. Considering an unsteady heat transfer, no significant anomalously high heat flow is detectable at the surface if the temperature has increased from 300°C to 700°C 10 or 20 m.y. ago. Under the Tyrrhenian sea in the Upper mantle a convective transfer must be postulated, having an upwelling velocity of more than 15 cm/y.

References

Andrews, D.J. & Sleep, N.H., 1974: J.R.astr. Soc., 38:237-251.
Erickson, A.J., 1970: Ph.D. Thesis, Massachusetts, Inst.Technology, Cambridge.
Finetti, I. & Morelli, C., 1973: Boll. Geof. teor. appl, XV:263-341.
Giese, P. & Morelli, C., 1975: Boll. Geof. teor. appl. XVII:184-191.
Lachenbruch, A.H., 1968: J. Geophys.Res., 73:6977-6989.
Lavenia, A., 1967: Boll. Geof. teor. appl., IX:323-332.
Loddo, M. & Mongelli, F., 1974: Report EGS-Meeting, Triest.
- , 1973: Nature:244.

A Tectonic Model of the Northern Ionian Sea from Refraction Seismic Data

W. Weigel

During the Meteor-cruises 17 (1969) and 22 (1971), the Ionian Sea between the Malta Escarpment and the Peloponnesus (fig. 1) has been investigated by deep seismic sounding on three WSW/ENE profiles (Weigel & Hinz 1970; Hinz 1974; Weigel 1974). In addition to the seismic refraction measurements by anchored sonobuoy stations at sea (Kebe 1971), the shots in 1971 with charges between 50 kg and 1 t, fired by FS "Meteor", have been recorded by 25 land stations on the Peloponnesus. The entire program was carried out in cooperation with Italian and Greek institutions.

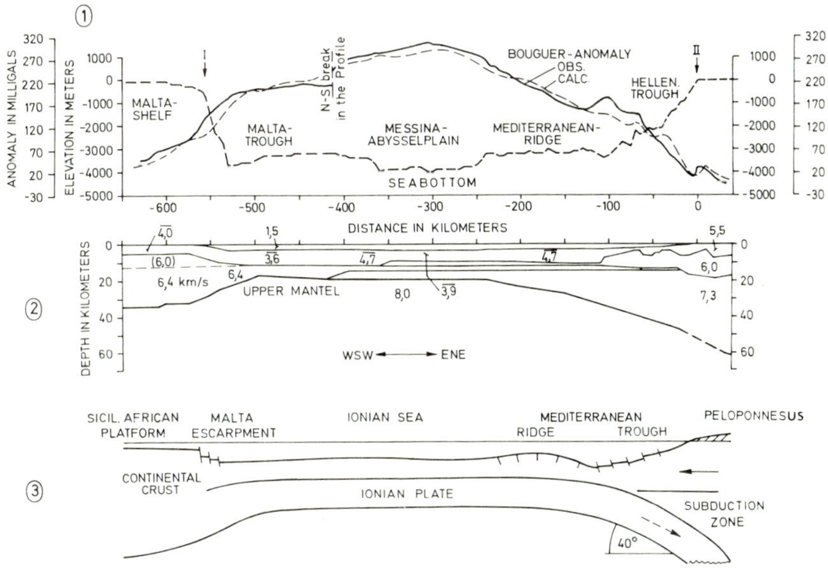

Fig. 1. (1) Topography of the sea bottom and Bouguer Anomaly on a section between Malta Shelf and Peloponnesus. I: λ = 15° 29,4'; ϕ = 35°54,7'. II: λ = 21° 36,1'; ϕ = 37° 12,45'. (2) Crustal section crossing the Ionian Sea. Mean compression velocities for the sediments (Detailed section: Weigel 1974). (3) Schematic plate tectonic model.

The topography of the Ionian Sea bottom reveals several geological units: the very steep Malta Escarpment, the Messina Abyssal Plain, the Mediterranean Ridge, and the Hellenic Trough (fig. 1, 1). The Bouguer anomaly indicates an updoming of mantle material with a maximum in the area of the Messina Abyssal Plain, verified by the seismic results presented in fig. 1, 2. The two dimensional calculated Bouguer anomalies from this model give a good agreement with the observed gravity values. The undulations in the area of the Hellenic Trough are mainly caused by

strongly fractured basement with a p-velocity of 6.0 km/s. These fractures result from compressional stress as evident in fault-plane studies (Galanopoulos 1973; Gertig 1973; Makris 1975). Intermediate material with a p-velocity near 7,3 km/s is typical for areas with tectonic mobility and may be related to the subduction of the Ionian Plate under the Peloponnesus (fig. 1, 3).

The Mediterranean Ridge appears to be updomed by ENE-WSW directed post Tertiary compressional stress, which caused strong faulting in mainly evaporitic material (Finetti, Morelli 1974; Sigl et al. 1973; Hieke 1972). The thickness of sediments (6-7 km) is nearly the same as in the Messina Abyssal Plain, a formerly shallow sea, indicated by a sedimentary low velocity channel (Hinz 1974) underlaying the evaporites. - The structure of the basement surface in the area of the Malta Escarpment points to fault-bounded subsidence of the Ionian Sea bottom of more than 5000 m.

As in other Mediterranean basins, the crustal type under the central Ionian Sea is neither an oceanic nor a continental one; it is intermediate, and different in the West and East. In the West (Sicilian African Platform) we find a typical continental crust and in the East (Peloponnesus) a continental crust modified by tectonic mobility.

The schematic plate tectonic model (fig. 1, 3) is an arc-continent collision type (after Dewey & Bird 1970), with the subduction beneath the island arc accompanied by crustal movement due to rising mantle material (Makris 1975) in the Aegean area.

References

Closs, H., 1972: "Meteor"-Forsch.Ergeb. R.A., 10:31-50.
Comninakis, P.E. & Papazachos, B.C., 1972: Geol.So.Am.Bull.82:1083-1101.
Dewey, J.F.& Bird, J.M., 1970: J.Geoph., R., 75,14:2625-2647.
Finetti, I. & Morelli, C., 1973:Boll.d.Geof.Teor.ed Appl.,15, 60.
Galanopoulos, A.G., 1973: Anal. di Geof., 26,1:85-105.
Gertig, H.: Unpubl. Dipl. Thesis, Inst.f.Geoph. Univer. Hamburg.
Hieke, W., 1972: Zeitschr.d.deutsch.Geol.Ges., 123:567-570.
Hinz, K., 1974: Geol. Jahrb., R.Geoph., 2:33-65.
Jacobshagen, V. & Makris, J., 1974: Nachr.Dtsch. Geol. Ges., Sonderh. Geowiss., Tag. 1973, Frankfurt a.M.
Kebe, H.-W., 1971: "Meteor"-Forsch.Ergeb. R.C. 6:14-20.
Makris, J., 1973: Bull. of the Geol.Soc. of Greece, 10:206-213.
Makris, J., Mavridis, L.N., Menzel, H., Stavrou,A. & Veis, G., 1973: Zeitschr. f. Geoph. 39:929-936.
Makris, J., 1975: J. Geophys., 41:441-443.
Papazachos, B.C., 1973: Geoph. Jour. Roy. Astr. Soc., 33:421-430.
Ritsema, A.R., 1970: XII Ass. Gen. Comm.Seismol. Europ., Luxembourg.
Sigl, W., Hinz, K. & Gardé,S., 1973: "Meteor"-Forsch.Ergeb. R.C., 14:51-54.
Weigel, W., 1974: Journ. Geophys. 40:137-140.
Weigel, W. & Hinz, K., 1970: XII Ass.Gen.Comm.Seismol.Eur. Luxembourg.
Weigel, W., 1974: Hamb. Geophys. Einzelschr., 26:140 p.

Isostatic Behaviour of the Crust and Upper Mantle in Southern Italy, the Tyrrhenian Sea and Sicily

M. Snoek

Isostatic considerations are presented, based on a crust-mantle model derived mainly from refraction seismic data obtained on land and at sea. The area under investigation is marked by vertical and horizontal displacements (Damiani 1970; Caputo et al. 1972, 1974) and by active volcanism (Barberi et al. 1974). The analysis of seismicity data suggests the existence of an inclined zone of seismicity, interpreted as a Benioff zone (Ritsema 1972; Papazachos 1973; see also Schick this vol.).

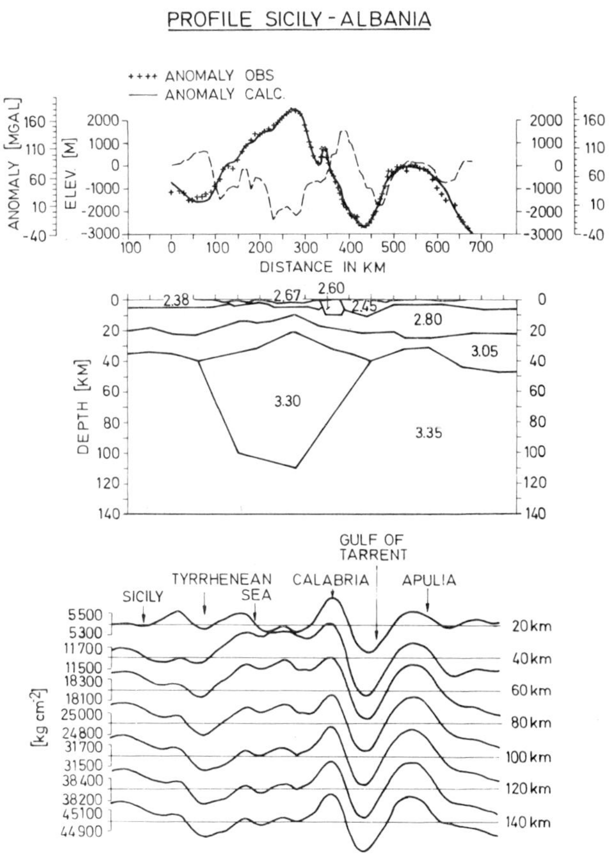

Fig. 1. Cross section Sicily-Albania. Upper part: Observed and calculated Bouguer anomaly. The intersecting line represents the topography. Middle: The 2-D density model (values unit is gcm^{-3}) representing seismic layers. The 3.05 gcm^{-3} to 3.30/3.35 gcm^{-3} interface is regarded as the crust-mantle boundary. Lower part: Graph of the summation of mass as function of distance and depth.

Seismic Measurements and a Two-dimensional (2-D) Crust-Mantle Model

Explosives in charges between 25 and 450 kg were fired to generate seismic energy. The maximum distance for seismic energy recorded with buoys at sea was 70 kg, revealing a P wave velocity of 7.4 km^{-1} for the deepest refractor in the Tyrrhenian Sea. For technical reasons no reversed profiles were recorded. The intercept-time method was applied to calculate thickness and depth of seismic layers for the mainly short profiles. More sophisticated methods were not used due to the fact that some of the seismic results were poor (Snoek 1975).

Results of Morelli et al. (1975) were integrated for land areas. From the obtained seismic data a 2-D crust-mantle model was compiled. To close the information gap on the deeper structures results of geological, seismological and geothermal studies were used. The configuration of the Calabrian block is mainly based on the geological studies performed by Giese & Görler (1974) and Damiani (1970). The tilted crustal body was calculated assuming a mean density of 2.6 gcm^{-3}. The high heat flow values in the SE-Tyrrhenian Sea (Erikson 1970) suggest the existence of a shallow-depth heat source. It coincides with a body attenuating seismic waves as derived from earthquake data by Caputo et al. (1972) for this region. The upper boundary for this body is expected to be identical with the deepest detected refractor. For the calculations therefore a reduced density of 3.30 gcm^{-3} was assumed representing intermediate material, whereas the normal Upper Mantle density is considered to be 3.35 gcm^{-3}. For the sediment coverage mean densities of 2.38 gcm^{-3} in the W and 2.45 gcm^{-3} in the E were used. These values were obtained by converting the P wave velocities into density values applying the Nafe & Drake density-velocity relation (Grant & West 1965). The mean values for the sediments include results calculated by Finetti & Morelli (1973) from seismic reflexion measurements.

Study of Isostasy

Gravimetric data were taken from published Bouguer anomaly maps (Finetti & Morelli 1973). For the density model theoretical Bouguer anomalies were calculated (Talwani Method) and fitted to the observed anomalies, (see fig. 1 upper part). In order to obtain the regional trend of isostatic behaviour the method of mass summation (Janle 1973) was applied for the following reasons:

- The basic isostatic 3-D models of Pratt & Airy cannot be used as the area under investigation is too inhomogeneous.
- The summation of mass takes density alterations of the crust into consideration. In contrast to the isostatic reduction there is no need to calculate with average densities.

The distribution of mass of the adjusted model was summed for a column of one square kilometer and a depth increasing to 140 km in steps of 20 km. In order to eliminate local effects and to omit small scale structues, the data were smoothed by computing a simple running average using a window of 10 km. The results were plotted as function of mass, distance and depth (see fig. 1 lower part).

Results and Interpretation

The plot of the calculated summation of mass shows the structural differences of the crust-mantle system. It gives evidence of the state of isostasy as a function of depths and shows lateral alterations. The surplus or deficit of mass can be taken as a direct indication of a tendency for compensational isostatic motion. The main results of the seismic and isostatic considerations are:

- The SE-Tyrrhenian Sea has an intermediate crust of approximately 23 km thickness.
- The Upper Mantle in the Tyrrhenian Sea is laterally inhomogeneous, partly consisting of low velocity and low density material. Heat flow values as well as the attenuation of seismic waves support the 2-D density model.
- A significant feature of the plot is the high deficit of mass between Apulia and Calabria indicating a strong disequilibrium of the crust-mantle system. The great accumulation of sediments deduced from seismic data and the gravity minimum (-30 mgal) at the SW part of the Gulf of Tarrent and the E-coast of Calabria mark a downbuckling in the crust, possibly due to the subduction of the Ionian lithosphere below Calabria.
- The distribution of mass below Apulia, Calabria and the Tyrrhenian Sea shows relative mass-surpluses for the entire range of depth with the exception of the Tyrrhenian Sea, where an equilibrium seems to exist below 80 km. To obtain a balanced crust-mantle system downgoing movements must be assumed beneath Apulia and Calabria, if the system is considered to be free of additional geodynamic stress.
- With increasing depth the mass summation shows a surplus under Sicily and a mass deficit under Albania.

Vertical movements on this profile are only reported in W-Calabria, where the uplift of the coastal region can be shown by recent marine terraces (Damiani 1970; Görler this vol.) and in the Tyrrhenian Sea, where the subsidence is deduced from the sediment analysis (Selli et al.1973). This is partly in contradiction to the postulated isostatic compensational movements of these regions.

Studies of horizontal displacements have only started recently. Measurements with several Laser-geodimeters across the Strait of Messina undertaken by Caputo et al. (1974) indicate a relative S-N displacement of Sicily with respect to Calabria.

Based on these facts one can assume strong lateral forces causing horizontal displacements and preventing vertical compensational isostatic movements in this area.

Acknowledgement

This program was supported by the Italian and German Research Associations CNR and DFG. I would like to thank Dr.K. Hinz of the BGR Hannover for the cooperation as well as master and crew of NO "F.S.Marsili" and the working group for the operations at sea of the Geophys. Institute of University Hamburg.

References

Caputo, M., Panza, G. & Postpischl, D., 1972: Tectonophysics,15:219-231.
Caputo, M., Folloni, G., Pieri, L. & Unguendoli, M., 1974: Geophys. J.R. astr. Soc., 38:1-8.
Damiani, A., 1970: Boll. Soc. Geol. Ital., 89:145-158.
Erikson, A.J., 1970: PhD Thesis, MIT, Cambridge, Mass. USA.
Finetti, I. & Morelli, C., 1973: Boll. Geof.Teor.ed Appl., XV:60-79.
Giese, P. & Görler, K., 1974: Nachricht.Dtsch.Geol.Ges., 9:67-77.
Grant, F.S. & West, G.F., 1965: Interpretation theory in applied Geophys. Inter. Series in the Earth Science McGraw-Hill.
Janle, P., 1973: Hamb. Geophys. Einzelschr., 19:95 p.
Morelli, C., Gantar, G. & Pisan, M., 1975: Boll.Geof.Teor. ed Appl., XVII:39-58.
Papazachos, B.C., 1973: Geophys. J.R. astr. Soc., 33:421-430.
Ritsema, A.R., 1972: Geol. Minjb., 51:540-573.
Snoek, M., 1975: Hamb. Geophys. Einzelschr., 28:77 p.

Palaeomagnetism of Mesozoic Volcanic Rocks in Sicily

A. Schult

Palaeomagnetic research in Sicily was carried out in the southeastern tip at Capo Passero on Upper Cretaceous rocks and in West Sicily on Upper Cretaceous and Middle Jurassic volcanic rocks. Details of sampling, measurements and results were given in previous papers (Schult 1973, 1976). The final results are given in tab. 1.

Results for the Upper Cretaceous at Capo Passero were confirmed by Barberi et al. (1974) and Gregor et al. (1975). The best estimation of magnetization direction and pole position is probably a combination of all three collections. This result is also given in tab. 1. The palaeomagnetic data for West Sicily are less reliable than those for South East Sicily due to a restricted number of suitable volcanic rocks and a relative unstable magnetization of the rocks (Schult 1976), but the data are reliable enough to give some useful information.

The most reliable K - Ar age of the Capo-Passero basalts (Barberi et al. 1974) of 71 my is consistent with stratigraphic evidence indicating Maestrichtian age. The ages of the volcanic rocks in West Sicily are from stratigraphic data only.

In fig. 1 Cretaceous and Jurassic pole positions for West Sicily, South East Sicily, Africa and "stable" Europe are compiled. The coincidence of the Upper Cretaceous pole positions for South East Sicily and Africa is consistent with the geological evidence that the Ragusa platform of South East Sicily is an extension of the Sahara platform (e.g. Caire 1973; Barberi et al. 1974).

Table 1. Mean palaeomagnetic results after AF-cleaning of Mesozoic volcanic rocks in Sicily

Location	Age	N	D	I	95	k	°N	°E	A_{95}
South East Sicily (36.7°N,15.1°E)	Ku (71 my)	19[2)]	347	22.0	7.1	22.8	62	223	5.8
Mean of all collections [1)]	Ku (71 my)	41[2)]	345	25.3	4.6	24.6	63	229	3.8
West Sicily (38.06°N,12.66°E)	Ku	4[3)]	75	28.7	17.4	29	21	100	14.9
West Sicily (37.9°N, 13.5°E)	Jm	6[4)]	73	66	21.6	10.5	38	67	31.5

N total number of sites, D declination, I inclination, k precision parameter, 95 A_{95} radius of the 95% circle of confidence, °N °E pole position. [1)] Mean of all collections in South East Sicily by Schult (1973), Barberi et al. (1974), and Gregor et al. (1975) (giving unit weight to each site direction). [2)] All sites have reversed direction of magnetization (here projected to normal direction). [3)] All sites are normal. [4)] One site is reversed, the other normal.

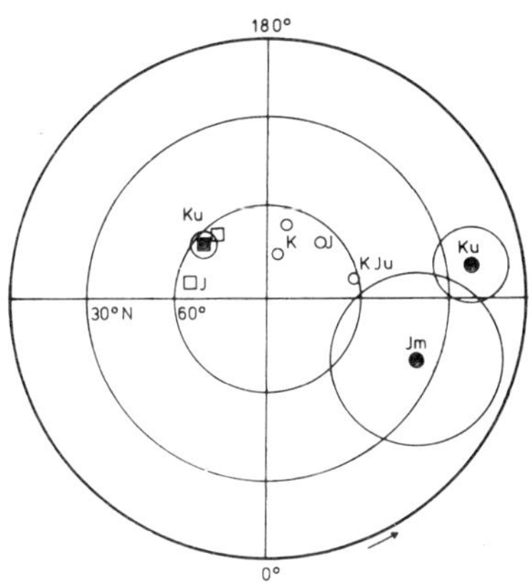

Fig. 1. Cretaceous and Jurassic Palaeomagnetic pole positions with their respective circles of confidence (when given) after Schult (1976).
K = Cretaceous, J = Jurassic, u = Upper, m = Middle, ● West Sicily, ■ South East Sicily, o "stable" Europe, □ Africa.

For West Sicily and South East Sicily the Upper Cretaceous pole positions are quite different, in consistence with the suggestion that old plate boundaries cross Sicily (e.g. Caire 1973). The directions of magnetization of Cretaceous rocks in West Sicily and South East Sicily (see tab. 1) imply a clockwise rotation of about 90° of West Sicily relative to South East Sicily since the Upper Cretaceous. The same angle of clockwise rotation of West Sicily since the Upper Cretaceous is also found relative to Africa. - A similar clockwise rotation of 70° of West Sicily with respect to Africa and South East Sicily since the Mesozoic was proposed by Caire (1973) in a hypothetical reconstruction of the central Mediterranean region.

With respect to *stable* Europe the palaeomagnetic data imply also a clockwise rotation of West Sicily in the order of 30 - 60° (fig.1). This figure is only tentative because the Cretaceous pole positions for Europe are not very reliable. - Similar amounts of clockwise rotation of West Sicily relative to Africa and therefore also with respect to South East Sicily and also relative to *stable* Europe are indicated by the Jurassic pole positions (fig. 1).

The sense of rotation of West Sicily is opposite to any known rotation of crustal blocks in the Central Mediterranean according to palaeomagnetic data (Soffel this vol.). Recent palaeomagnetic data from sediments in West Sicily also indicate a clockwise rotation since the Mesozoic (Catalano et al. 1975).

Acknowledgement

I thank Prof. G. Vianelli, Palermo for his generous help and advice and Prof. G. Angenheister and Prof. H. Soffel for many stimulating discussions.

References

Barberi, F., Civetta, L., Gasparini, P., Innocenti, F., Scandone, R., & Villari, L., 1974: Earth Planet. Sci. Lett., 22:123-132.

Caire, A., 1973: The Calabro-Sicilian arc. In: Gravity and Tectonics (K.A.de Jong & R.Scholten, Eds.), J. Wiley & Sons, New York.

Catalano, R., Channel, J., D'Argenio, B. & Napoleone, G., 1975: Soc. Geol. Ital. Seduta 12,12 (Abstract only).

Gregor, C.B., Nairn, A.E.M. & Negendank, J.F.W., 1976: Geol. Rundschau, 64:948-958.

Schult, A., 1973: Earth Planet. Sci. Letters, 19:97-100.

- , 1976: Earth Planet. Sci. Letters , 31:454-457.

Seismotectonic Survey of the Central Mediterranean

R. Schick

The Central Mediterranean and Southern Italy have been of seismotectonic interest since long. Suess (1885) suggested a correlation between the distribution of the earthquake foci and the arc-like structure of the cristalline massifs of Calabria and Northern Sicily. Peterschmitt (1956) proposed that the distribution of deep earthquakes in the Tyrrhenian Sea may be represented geometrically by the mantle of a cone, of which the apex is located at about 40°N, 12°E and at a depth of 700 km. Following Peterschmitt, the existence of a Benioff-zone was widely accepted or rejected (Van Bemmelen 1972). Ritsema (1972) assumed that there is neither a clear connection of the Tyrrhenian deep shocks with the coast-line arc of S. Italy nor is it evident that the ocean-side lithosphere of the Ionian Sea

is sub-thrusting the Calabrian arc and its hinterland of the Tyrrhenian basin. Riuscetti & Schick (1975) re-examined all available seismograms from earthquakes in S. Italy since 1905. The authors evaluated source parameters of deep and shallow earthquakes, and presented a critical re-calculation of focal depths and fault-plane solutions. A part of the seismotectonic picture which is given in the following sections is based on the data presented by Riuscetti & Schick (1975). Fig. 1 shows the studied area.

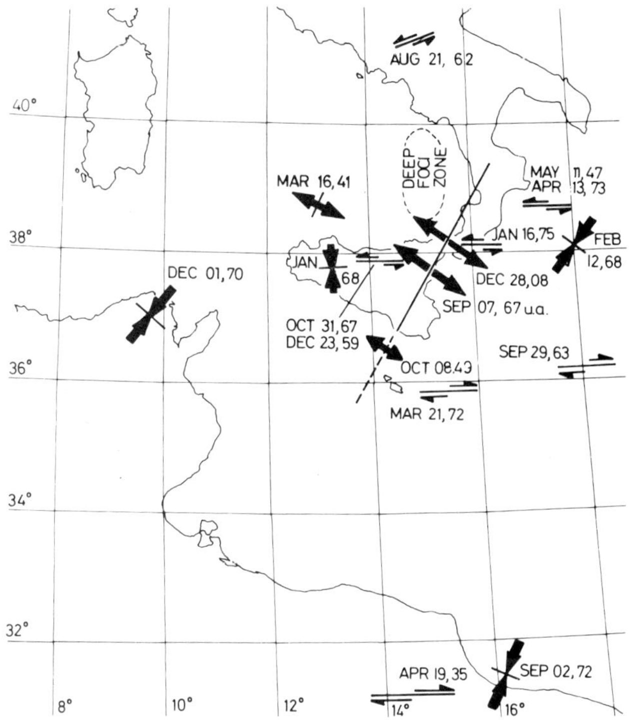

Fig. 1. Schematic illustration of fault-plane solutions for crustal earthquakes. The dates represent the origin times of the earthquakes. In case of strike-slip motion the preferred fault-planes and dislocations are given, otherwise the horizontal force components of the stress system corresponding to the fault-plane solution are presented.

Crustal Seismicity

The distribution of epicenters in the Central Mediterranean is rather random, and it is not possible to trace fault lines by connecting epicenters. The region with the most dense population of earthquakes is Southern Italy. Here the regions of low seismicity and high seismicity with magnitudes clearly exceeding Ms = 6 and reaching $M_B \approx M_S \approx 7$ are well defined. The Caltanisetta basin and the Apulia platform show an extremely low seismicity, while the earthquakes with maximum magnitudes are

aligned along the northern branch of the Comiso-Messina-S.Eufemia fault-line, with a concentration between the Straits of Messina and Cosenza. An unusual type of seismic activity is observed in Northern Lybia (fig. 1), where on APR 19, 1935, four earthquakes occurred within a few hours, ranging in magnitude between $M_S \approx 6.5$ and $M_S \approx 7$, in a region with very low average seismicity. It is suggestive that the shape of the magnitude-frequency relation (b-factor) is much smaller than usual. Possibly it is an indication that the earthquake fractures propagate into a rather homogeneous medium, and that the slippages do not occur along old, pre-fractured fault-planes but are generating new ruptures.

Fault-plane Solutions (F.P.S.) and Stress Patterns

The horizontal stress directions derived from F.P.S. of crustal earthquakes and the preferred fault-planes in case of strike-slip motions are illustrated in fig. 1. The F.P.S. are divisible into three separate groups:

a) Earthquakes with strike-slip motion preferably occur along E-W striking fault-planes. The sense of motion is not uniform, however. The dominant strike-slip motion in Northern Sicily, Calabria and the adjoining sea regions shows left-lateral dislocation. South of Sicily, the sense of dislocation motion is preferably right-lateral, in agreement with the motion of earthquakes observed in the Tellian Atlas of Algeria.

b) Earthquakes aligned along the Comiso-S.Eufemia fault-line, such as the Messina shock of Dec. 28, 1908 (Schick 1977), show predominant normal faulting. The mechanism of the Messina shock may be represented by normal faulting along a 30 km long fault-plane roughly parallel to the Straits of Messina. The strike of the fault-plane is given as N 15°E, with a dip of 70°W. Other earthquakes along the Comiso line show a similar mechanism. There is evidence that the basaltic volcanism of Etna is linked with the line of normal-fault shocks in Eastern Sicily and Calabria, and that furthermore this fault is accompanied by an offset of the Moho in Calabria (Schütte 1974). The F.P.S. of the earthquake near the Island of Ustica on Mar 16, 1941, supports the same mechanism as the earthquakes along the Comiso line. The volcanism of Ustica also shows basaltic activity. It is suggestive that the tectonic setting of the Ustica and the Messina earthquakes is embedded in large tectonic structures covering the Tyrrhenian Sea. Following Heezen et al. (1971), the topography of the Tyrrhenian Sea floor is marked by a linear system of ridges and troughs oriented approximately N 15°E.

c) The third category of earthquakes in the Central Mediterranean shows compressional axes which are roughly oriented NE-SW. Riuscetti & Schick (1975) suggest a global compressional stress-field in the Central Mediterranean which is acting in about this direction. This compressional stress-field is acting longitudinal to major fault-lines in Eastern Sicily and the Tyrrhenian, thus causing a widening perpendicular to the Comiso fault and leading to the observed normal faulting mentioned in (b). The model is acceptable for Sicily, Calabria and the adjoining sea region, but it does not take into account the right-lateral strike-slip observed east of Malta and in Lybia. The primary driving forces may thus be E-W oriented shear motions associated with a stress-field that causes normal faulting when already existing faults as zones of weakness are longitudinal to the compression axis, and thrust faulting when the faults are transversal to the compressional force components.

Shallow-focus Seismicity of Volcanic Terrane

A typical volcanic-terrane seismicity is observed along the slopes of Etna. The shocks are very shallow, with focal depths seldom exceeding 1...2 km. The macroseismic intensity of these shocks can be quite high (Riuscetti & Distefano 1972), although the corresponding body-wave magnitude might be low. Furthermore, the ratio between surface-wave magnitude and body-wave magnitude is significantly higher than "normal"; typical values are observed with the earthquake of Apr 21, 1971: Body-wave magnitude 3.0, surface-wave magnitude 4.0, maximum macroseismic intensity in the Mercalli-scale 7-8. As the spectral energy of these shocks is shifted to longer periods, these events are sometimes called red earthquakes. It has been demonstrated by Riuscetti & Schick (1975) that the red spectrum of these earthquakes is caused by an anomalously low rupture velocity.
(Note: By error the rupture velocity of the shock of Apr 21, 1971, was given as 0.7 km/s in the paper of Riuscetti & Schick (1975), the value should be 0.07 km/s).

Tyrrhenian Deep Earthquakes

Based on a study of selected quakes with high quality records, Riuscetti & Schick (1975) believe that the main energy release of these deep earthquakes occurs in a relatively small zone stretching about N-S, geographically near a line connecting Etna and Vesuvius. Also, the depth range of significant energy release is limited to 280-300 km of depth.

Few of the deep earthquakes in the Tyrrhenian yield reliable F.P.S. One of the more reliable solutions is the one of the earthquake of Apr 13, 1938, which had a magnitude $M_B = 7.0$. The F.P.S. shows a nearly horizontal mean stress axis stretching N-S. The preferable fault-plane is vertical with a dislocation vector acting downward on the Eastern side of the fault-plane and upward on the Western side. Other deep earthquakes in the Tyrrhenian show a comparable dip-slip mechanism, but the horizontal mean stress axis can deviate considerably from the N-S direction.

References

Heezen, B.C., Cray, C., Segre, A.G. & Zarudski, E.F.K., 1971: Nature, 229:327-329.

Peterschmitt, E., 1956: Ann. Geof. 9:305-335.

Ritsema, A.R., 1972: Geologie en Mijnbouw, 51:541-545.

Riuscetti, M. & Distefano, R., 1972: Boll.Geof. Teor.Appl., 14:150-164.

Riuscetti, M. & Schick, R., 1975: Boll. Geof. Teor. Appl., 17:59-78.

Schick, R., 1977: Geol. Jb., E11, in press.

Schütte, K.G., 1975: Rapports et Proces-Verbaux des Réunions C.I.E.S.M., 23, 4a:51-52, Monaco.

Suess, E., 1885: Das Antlitz der Erde, vol.1, F. Tempsky (Prag) and G. Freytag (Leipzig), 310 p.

Van Bemmelen, R.W., 1972: Developments in Geotectonics 2, Elsevier Publ. Comp., Amsterdam-London-New York, 257 p.

Satellite Lineaments of the Central Mediterranean Region (Sicily/Calabria)

J. Bodechtel and U. Münzer

The mapping of lineaments on aerial photographs and on satellite imagery is an important approach in photogeological studies. The lineament pattern commonly reveals tectonic phenomena at a small scale and can be used to investigate the aerial distribution of joint- and fracture systems on a statistical basis.

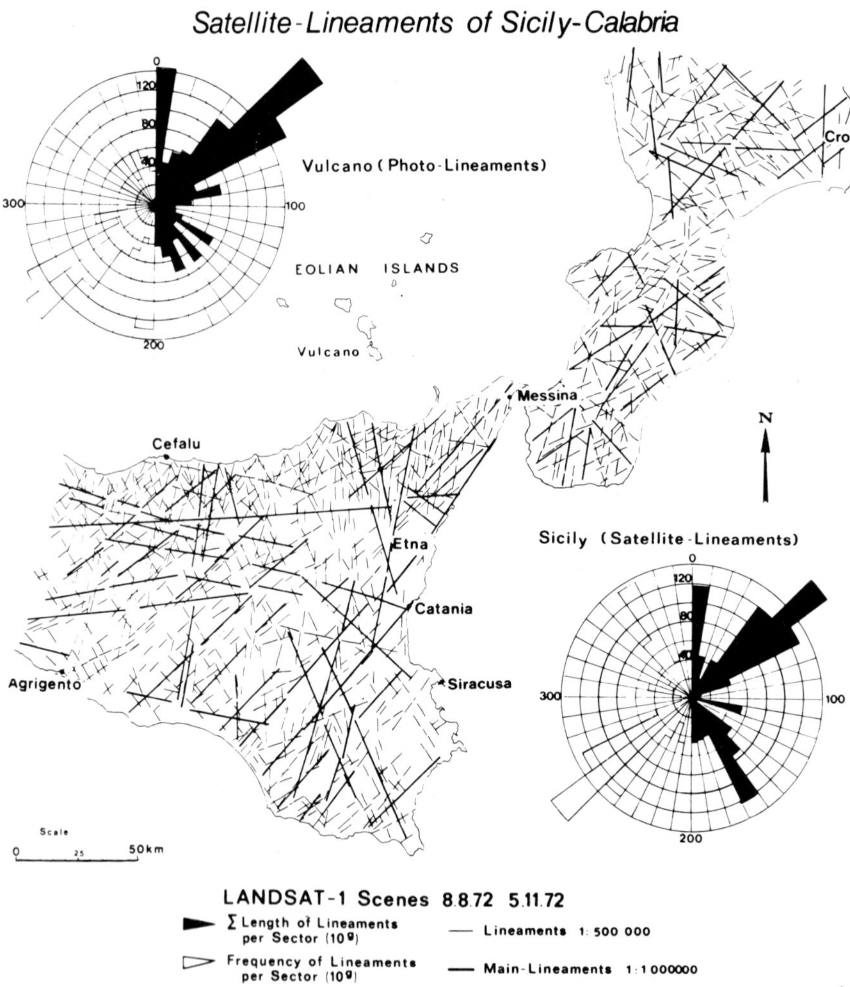

Fig. 1.

Several investigations have already proved significant correlations in the statistical orientation of lineaments. Nevertheless, a careful interpretation of photo-lineaments is required. Illumination effects may cause a directional artificial enhancement of lineaments. Therefore the examples demonstrated have been mapped on satellite imagery and aerial photographs taken during various seasons and thus showing different illumination conditions.

The interpretation on the basis of Landsat-1 scenes (fig. 1) covers an area of approximately 4900 km². The lineaments were mapped on the basis of the MSS bands 5 (0,6 - 0,7 μm) and 7 (0,8 - 1,1 μm) displayed at a scale of 1:500.000 from overpasses dated 8/8/72 and 5/11/72. In total 1254 lineaments were counted with an overall length of 6410 km. Due to the morphology and grey tone contrasts, other major lineaments were traced and mapped at a scale of 1:1.000.000.

It was possible to trace some of the major lineaments over the entire island of Sicily. Magnification to a scale of 1:250.000 demonstrates that these lineaments are composed of parallel minor lineations. This proves the existence of large fracture zones (Cassinis et al. 1975). A concentration of major lineaments striking NE-SW is also evident.

Evaluation of satellite images has shown that the major lineations are commonly accompanied by parallel striking elements, either in the direct vicinity of the main structure or as parallel traces at greater distances. A histogram evaluation of the lineaments clearly reveals 3 maxima to the N, NE and SE (fig.1).

The lineations derived from the satellite images indicate a simplified structural pattern which is of special significance in defining the young fault tectonics of the southern Tyrrhenian block.

The main directions of photolineaments coincide with seismotectonic results in the central Mediterranean (Schick this vol.) and confirm earlier results (Medwenitsch 1967; Nickel 1964; Pichler 1968; Adler 1966).

The interpretative satellite lineations were checked with an intensive photogeological study within the area of the Eolian islands and especially on the island of Volcano. The results of this project included the tracing of 1243 lineations on 27 stereo photographs on a scale of 1:10.000. It clearly demonstrates the correlation of this lineament system with those derived on small-scale satellite images (fig. 1).

References

Adler, R., 1966: N.Jb. Geol. Paläont., 11:641-665, Stuttgart.

Carolis, C.de, Giudice, E.Lo & Tonelli, A.M., 1975: Bull.Volcanologique, 39:1-14, Rome.

Cassinis, R., Lecchi, G.M. & Tonelli, A.M., 1975: Proc. of the NASA Earth Resources Survey Symposium: 851-867, Houston-Texas.

- , 1974: Proc. Symp. at Frascati, C.N.R. Publ., 95:185-195.

Keller, J., 1970: Ber.Naturfreunde Ges. Bd. 57:22 p., Freiburg.

Medwenitsch, W., 1967:Mitt.Geol.Ges.Wien, Bd.59,1:120 p.

Morelli, C., Giese, P., Cassinis, R., Colombi, B., Guerra, J., Luongo,G., Scarascia,S.& Schütte,K.-G.,1975:Boll.di Geofis.ed Appl.17,67:183-210.

Münzer,U.,1977: Geol.Aufnahme der Insel Vulcano mit Unterstützung geophotogrammetrischer Methoden. Unpubl.Dipl.-Arbiet, Univ. München,109 p.

Nickel, E., 1964: Verein d.Freunde d.Mineralogie u.Geologie e.V., 118 p. Heidelberg.

Pichler,H., 1968: Geol. Mitt. Bd.7:229-332, Aachen.

Part 2

Petrography

Granulite Facies Metamorphism in the Northern Serre, Calabria, Southern Italy

V. Schenk and W. Schreyer

The crystalline massif of Calabria contains, in its northern and northwestern parts, high-pressure/low-temperature metamorphic rocks formed synkinematically during thrusting of alpine age (De Roever 1972). These rocks are overlain by the highly metamorphic gneisses of Sila (see fig.1 inset) which towards the east are overlain by granites, micaschists and phyllites, that is by rocks formed at shallower depths than the gneisses.

Giese and Görler (1974 and in this volume) observed in 20-25 km depth below the crystalline basement of Sila a seismic velocity inversion which they explain by subduction of material of the upper crust from the west under material of the lower crust along a deep, steeply dipping thrust zone. This subduction and subsequent uplift may be responsible for the present surface exposure of deep crustal rocks in Sila.

Thus both geologically and geophysically, the Calabrian Massif shows similarities to the Ivrea Zone of the Alps (Berckhemer 1968; Reinsch 1973). Therefore, a small, relatively well exposed study area was selected in the northwestern Serre (see inset of fig. 1), which is geologically similar to Sila. The highly metamorphosed basement containing granulite-facies rocks dips gently (30°) towards SSE under the Serra San Bruno-granite of Upper Carboniferous age (Ferrara & Longinelli 1961) which is in turn overlain by Palaeozoic phyllites affected by contact metamorphism. The northern end of the Serre basement is marked by a series in greenschist facies in which lenses of relict granulite-facies rocks suggest retrograde metamorphism. Petrographic work in the Serre has also been performed by Piccarreta et al. (1973), Borsi et al. (1976), and Lorenzoni et al. (1976).

Geology

Fig. 1 is a geologic map of the small area investigated in detail. Two main units were recognized: The Lower Unit exposed in the north consists predominantly of metabasites, ultramafics, and light granulites. It is overlain in the south by the Upper Unit made up of a series of metapelites with silicate marbles and numerous layers of metabasic rocks.
Between the two units there is a structural concordance but an apparently sedimentary discordance, as the metapelites of the Upper Unit overlie the granulites as well as the metabasites of the Lower Unit. In the east strong post-crystalline mylonitization affects the basal metapelites of the Upper Unit. The boundary between the two units consisting of rocks

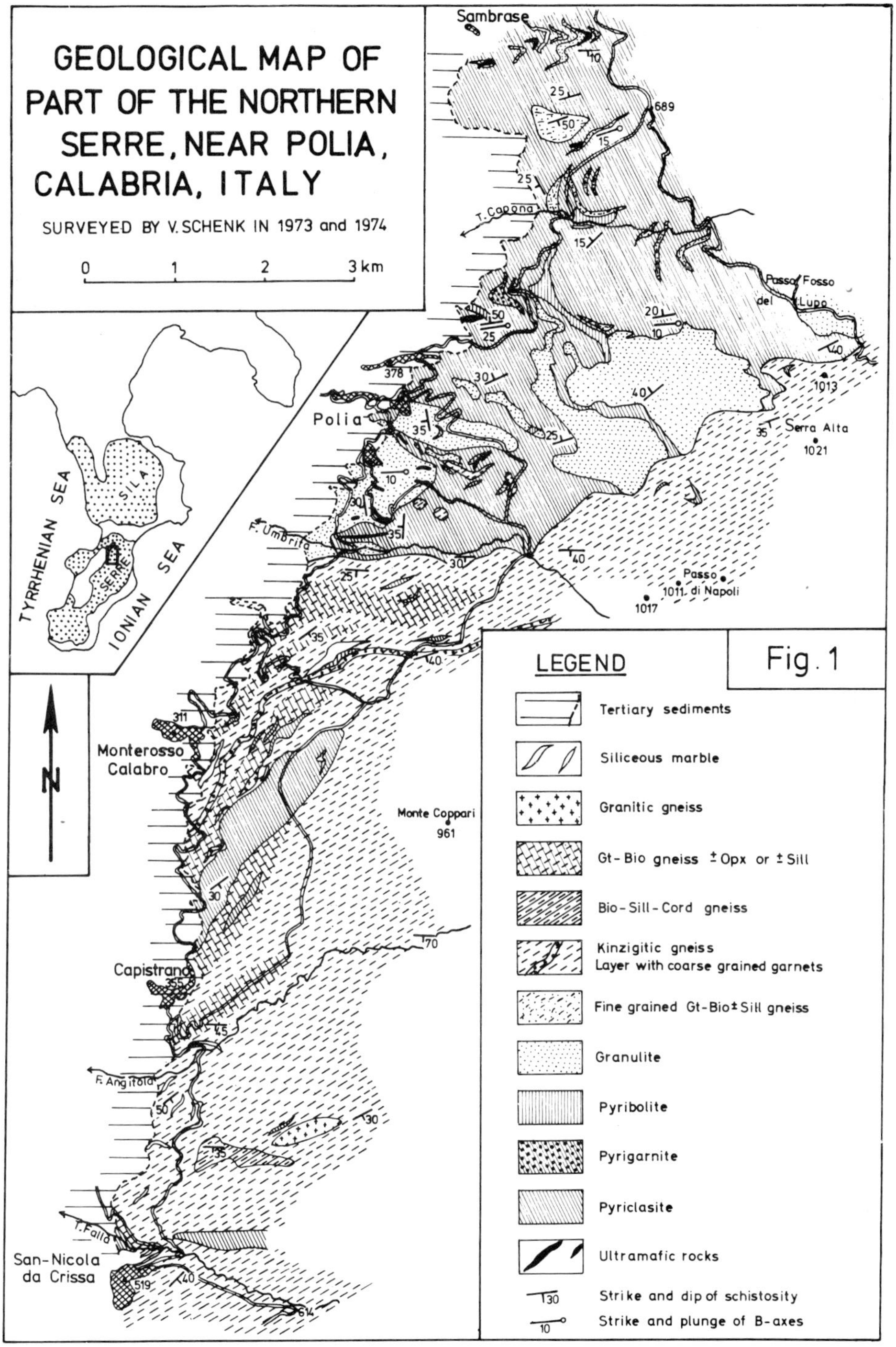
GEOLOGICAL MAP OF PART OF THE NORTHERN SERRE, NEAR POLIA, CALABRIA, ITALY
SURVEYED BY V. SCHENK IN 1973 and 1974
0 1 2 3 km
TYRRHENIAN SEA
SILA
SERRE
IONIAN SEA
N
Sambrase
Polia
Monterosso Calabro
Capistrano
San-Nicola da Crissa
Monte Coppari 961
Serra Alta 1021
Passo di Napoli
Passo Fosso del Lupo
T. Capona
F. Umbrita
F. Angitola
T. Falla
LEGEND
Fig. 1
Tertiary sediments
Siliceous marble
Granitic gneiss
Gt-Bio gneiss ± Opx or ± Sill
Bio-Sill-Cord gneiss
Kinzigitic gneiss
Layer with coarse grained garnets
Fine grained Gt-Bio ± Sill gneiss
Granulite
Pyribolite
Pyrigarnite
Pyriclasite
Ultramafic rocks
Strike and dip of schistosity
Strike and plunge of B-axes

of different rigidity may have been used for fault movements only later, during the young uplift of Calabria.

In the Lower Unit the rocks are folded with axes plunging WSW (250°/5-10); the most common direction of lineation due to mineral orientation is similar (235°/15). In the Upper Unit the fold axes in highly migmatized rocks show greater scatter with general plunge towards the south but without a pronounced maximum. The mean attitude of the schistosity is 140°/35.

The most common metabasites of the Lower Unit are *banded pyriclasites* consisting of pyroxene and plagioclase with or without biotite, garnet, hornblende, and quartz. One layer particularly rich in garnet may be called *pyrigarnite* after Mehnert (1972). One body of a coarse grained, generally massive rock containing hornblende is named *pyribolite*. In addition, there are small lenses of *scapolite rocks*. Practically concordant with the metabasites are the large bodies of light *granulites* and smaller layers and lenses of *metapelitic* rocks within the Lower Unit. *Ultramafic* rocks are confined to this unit. They occur also as lenses of rarely more than 100 m extension and are mainly made up of banded *peridotites*; however the largest body, some 400 m long, occurring S of Polia is mostly *pyroxenite*.

In contrast to this rather varied series of rock types in the north, the metapelites of the Upper Unit exhibit macroscopically a much more monotonous character. The most common rock is a coarse grained garnet- and biotite-rich *kinzigite* with cordierite and visible prisms of sillimanite exhibiting stromatitic to phlebitic textures. One rock type characterized by very large garnets could be mapped as a continuous layer of 4-5 km extension thus proving a rather well preserved stratigraphy also in the Upper Unit. Coarse *silicate marbles* containing diopside, garnet, tremolite, clinozoisite, forsterite, scapolite, and graphite occur in lenses up to 30 meters thick which occasionally seem to follow the stratigraphic layering. *Metabasites* form also concordant layers. They are most prominently represented by the large, about 200 meters thick body of dioritic pyriclasite SE of Monterosso, but there are numerous layers with thicknesses of less than a meter. Common in, and confined to the pyriclasites of the Upper Unit is secondary growth of amphibole, mostly cummingtonite, after pyroxene.

Age

No contribution can be made at this stage to the knowledge of the geological age of the metamorphism. It is clear, however, from indirect reasoning that the granulite facies metamorphism to be discussed must be at least of Palaeozoic age, which is again similar to the geological relations of the Ivrea Zone. Isotopic data on biotites of high grade gneisses of the Serre which gave ages of 127-147 m.y. (Borsi et al. 1976) may be caused by thermal events that are younger than the granulite facies metamorphism.

Granulites

The granulites of the Serre are light rocks (mafic minerals less than 30 vol.%) with grain diameters below 3 mm, metamorphosed in granulite fa-

cies, i.e. exhibiting mainly water-free mafic minerals (de Waard 1973). They all contain plagioclase (Plg), K feldspar (KF), and quartz (Q). Commonly the three mafic minerals garnet (Gt), orthopyroxene (Opx), and biotite (Bio) appear, although Opx and/or Bio may be missing. Rarely Gt and Bio are accompanied by sillimanite. The main opaque phases are pyrrhotite and ilmenite. Occasional graphite suggests a sedimentary origin. The

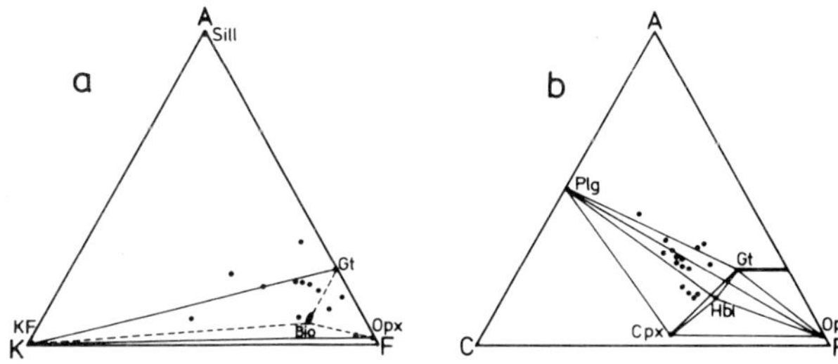

Fig. 2. a) AKF plot showing chemical compositions (small dots) and mineralogical phase relations of some Calabrian granulites. b) ACF plot showing chemical compositions and alternative phase relations in some Calabrian metabasites.

bulk chemistry of some granulites and their resulting mineralogy may be represented in the AKF-diagram of fig. 2a using the requisite mineral analyses. Both Bio and Gt may appear in two and three generations, respectively, depending on the availability of water during the metamorphic history. Bio-Q-symplectites form late and at the partial expense of Opx. Biotites are Mg-rich (phlogopites), garnets have compositions $Alm_{59-66}Pyr_{26-35}$, and Opx are hypersthenes (En_{70-58}) with some 3 weight per cent Al_2O_3. The plagioclase has An_{20-37}.

Metabasites

On the basis of their mineralogy three main types of metabasites may be distinguished (Mehnert 1972): *Pyriclasites* consisting mainly of plagioclase and orthopyroxene with or without additional clinopyroxene (Cpx), hornblende (Hbl), biotite, quartz, and garnet; *pyrigarnites* consisting of plagioclase, garnet, and orthopyroxene (± Q, Bio); and *pyribolites* containing plagioclase, amphibole, and orthopyroxene (± Cpx, Bio, Gt). Further subdivisions are based on the very variable fabric of these rocks, which are most commonly well banded due to changing ratios of plagioclase to mafic minerals. Plagioclase amounts to 40-60 vol. %; its changing basicity (An_{54-97}) cannot be related systematically to any feature of the geological map. Opx is a hypersthene with generally En_{68-62}; Cpx is a ferrous diopside. The amount of amphibole may vary from 5 to 85 vol.%. It is a brown-green hornblende (with ilmenite lamellae) in the north, whereas in the south appear also cummingtonite and/or a yellowish green hornblende as alteration products of Opx. Opaque phases are pyrrhotite and/or ilmenite, but also magnetite; additional accessories are most frequently apatite (up to 3 vol %) and some zircon, sphene, and carbonate.

The bulk chemistry of the metabasites may be represented in an ACF-

diagram (fig. 2b) which also exhibits possible phase relations. The formation of Gt $\pm$ Cpx at the partial expense of Plg, Opx, and Q was noted in some rocks. In the metabasites of the northern area Bio-Q-symplectites similar to those of the granulites were observed. However, in a later stage these biotites were replaced by garnets, and some hornblendes broke down into symplectitic intergrowths of Opx and Plg. The dehydration reaction of the hornblende is missing in the southern Upper Unit, where the partial breakdown of Opx into cummingtonites marks the last stage of metamorphism.

Ultramafic Rocks

These rocks may be subdivided into peridotites, pyroxenites, and hornblendites.

Following the IUGS (1973) nomenclature proposal, the *peridotites* lie in the fields of dunite and harzburgite. They exhibit a banding due their changing contents of olivine (Fo_{90}), Opx (En_{90}), and clinoamphibole (pargasitic hornblende). Some layers are rich in phlogopite. Strong enrichment of Opx and amphibole in layers within the mm- and cm-range may lead to an actual interlaying of dunite with pyroxenite and hornblendite. Opx may form porphyroblasts of about 1 cm diameter in the generally fine to medium grained peridotites. Due to strong serpentinization fresh peridotites are rare. A notable accessory phase is brownish to opaque spinel (picotite).

The *pyroxenites* are, in the IUGS (1973) nomenclature, websterites, bronzites, and olivine-clinopyroxenites. In addition to Opx and Cpx the websterites contain always brown-yellow clinoamphibole and phlogopite and occansionally rutile. Except for magnetite there is no spinel. Olivine is missing. The olivine-clinopyroxenites sometimes contain picotite in addition to magnetite.

The *hornblendites* consist of 35-70 vol. % of a pargasitic hornblende in addition to olivine (Fo_{80-90}), bronzite (En_{80-90}), diopside, and a light green aluminous spinel (ceylonite) containing some Cr. The distribution of olivine and pyroxene is not homogeneous, but lens-shaped portions rich in these phases are indications of some (relic?) banding.

Metapelites

Despite their monotonous character in the field, the mineralogy of the metapelites is rather variable, and as many as 15 different mineral assemblages were recognized within these essentially kinzigitic rocks. Most commonly they are medium to coarse grained, folded, migmatic garnet-biotite-sillimanite gneisses with or without cordierite. The leucosomes consist of quartz and two feldspars, but these phases may also appear in the mafic melanosomes, although rarely quartz-free kinzigites were also observed. In the lowermost (northern) parts of the Upper Unit biotite is rare or absent. These highly restitic, degranitized rocks are particularly rich in Al_2O_3 (20-30 weight %), whereas SiO_2 is generally below 50 w.%.

Although the large garnets (up to 5 cm diameter) show zoning with regard to the nature and grain size of their inclusions, they are virtually homogeneous chemically. Garnet compositions in different specimens were found to lie in the range Alm_{58-79} Pyr_{13-36} $Gross_{3-5}Spess_1$. Cordierites

are rather rich in Mg (Mg-$Cord_{75-90}$), and the same is true for most of the biotites which are still phlogopites despite their reddish brown color in thin section. Plagioclase is an andesine with An_{32-45}. K-feldspar appears as microcline or orthoclase (Ab_{14}). Ore minerals are pyrrhotite, chalcopyrite, pentlandite, and ilmenite; further accessories are zircon, rutile, apatite, monazite, and graphite. Graphite-rich layers have been mined locally.

Conclusions

A thick series, with predominantly basic rocks at the bottom grading into mainly pelitic rocks at the top was metamorphosed under granulite-facies conditions. Because of the occasional reaction Opx + Plg → Gt ± Cpx + Q, conditions near the border of the medium-pressure to high-pressure granulite range were attained (Green & Ringwood 1967), and a mineralogy of the hornblende granulite facies (De Ward 1965) developed. Temperature estimates on the basis of Al-contents in orthopyroxenes (Anastasiou & Seifert 1972; Obata 1976), the two pyroxene thermometer (Wood & Banno 1973; Boyd 1973; Herzberg & Chapman 1976), and the garnet-cordierite thermometer (Currie 1971) yield some 800-850°C. The application of the barometer after Wood (1974) and Ghent (1976) gave pressures of about 7,5 - 10,5 kb.

Present work is directed towards the problem of correlating the observed sequence of re-hydration (Opx replaced by Bio + Q) and re-dehydration (Bio replaced by Gt) reactions of the granulite-facies rocks with changes of pressure, temperature and water fugacity.

Ackmowledgements: Sincere thanks are due to Prof. Claudio D'Amico and his former coworkers at Messina for the excellent introduction into the problems of Calabrian petrology.

References

Anastasiou, P. & Seifert, F., 1972: Contr. Mineral. and Petrol. - 34:272-287.
Berckhemer, H., 1968: SMPM. - 48:235-246.
Borsi, S., Hieke-Merlin, O., Lorenzoni, S., Paglionico, A. & Zanettin-Lorenzoni, E., 1976: Boll. Soc. Geol. It., - 95:26p.
Boyd, F.R., 1973: Geochem. Cosmochim. Acta. - 37:2533-2546.
Currie,K.L., 1971: Contr. Mineral. and Petrol. - 33:215-226.
De Roever, E.W.F., 1972: GUA Papers of Geology, Ser. 1:3.
De Waard, D., 1965: Amer. J. Sci. - 263:455-461.
- , 1973: N. Jb. Mineral. Mh.: 381-392.
Ferrara, G. & Longinelli, A., 1961: Boll. Soc. Geol. Ital. - 80:25.
Ghent, E.D., 1976: Am. Mineral. - 61:710-714.
Giese, P. & Görler, K., 1974: Nachr. Deutsch.Geol.Ges. - 9:67-77.
Green, D.H. & Ringwood, A.E., 1967: Geoch.et Cosmoch.Acta. - 31:767-833.
Herzberg, C.T., Chapman, N.A., 1976: Am. Mineral. - 61:626-637.
IUGS Subcommission on the systematics of Igneous Rocks, 1973: N. Jb. Miner. Mh.:149-164.
Lorenzoni, S., Paglionico, A., Zanettin-Lorenzoni, E., 1976: Boll. Soc.

Geol. It. - 95:31 p.
Mehnert, K.R., 1972: N. Jb. Mineral. Mh.:139-150.
Obata, M., 1976: Am. Mineral. - 61:804-816.
Piccarreta, G., Morelli, L.A. & Paglionico, A., 1973: Boll. Soc. Geol. It. - 92:861-889.
Reinsch, D., 1973: N. Jb. Miner. Abh. - 119, 3:266-284.
Wood, J.B., 1974: Contr. Mineral. Petrol. - 46:1-15.
Wood, J.B.& Banno, S., 1973: Contrib. Mineral. Petrol. - 42:109-124.

Part 3

Stratigraphy and Tectonics

New Stratigraphical and Petrological Data on the Paleozoic Sediments of the Stilo-Pazzano-Area (Southern Italy)

G. Knoche

Along the south-eastern side of the Serre massif - especially in the area between Pazzano and Guardavalle (fig. 1) - remnants of Paleozoic metamorphic sediments are preserved. They are discordantly overlain by Mesozoic limestones and Tertiary sandstones and conglomerates. These metamorphites are mainly composed of NE-SW-striking fine-grained terrigeneous clastics, extremely poor in fossils but occasionally interbedded by fossiliferous limestones.

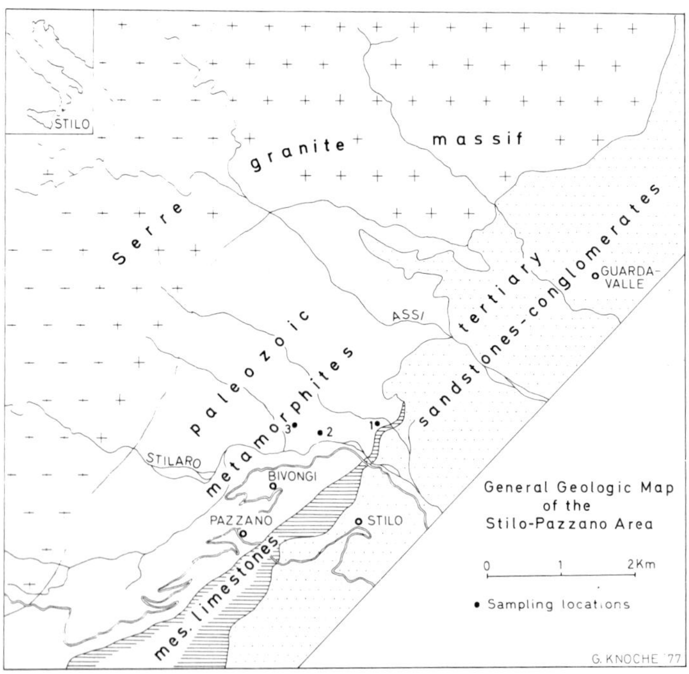

Fig. 1.

This hitherto undivided Devonian sediment (Afchain 1970; De Capoa-Bonardi 1970; Görler & Ibbeken 1970) underwent an intensive tectonic and metamorphic evolution during the Upper Paleozoic (Variscan orogeny).

It is thus rather difficult to subdivide the sequence stratigraphically and to determine its thickness.

The larger western part of the sediment belt is located adjacent to the Serre massif and of quite higher metamorphism than the eastern one. Therefore its stratigraphy has remained undivided. The small eastern part, hardly or not at all metamorphosed, can be divided into the following units (Knoche 1977):

graywackes; cherts, siliceous schists; (covered interval) graywackes; cherts, siliceous schists; flaser limestones; graywackes; crinoid-limestones.

This sequence reaches a thickness of up to 300 - 500 metres. On the basis of bryozoa, conodonts and tentaculites, these sediments can be determined as ranging from the lower Devonian (Emsian) to the Upper (Nehdenian) at least and possibly younger (Knoche 1977). A Silurian age of some parts cannot be excluded.

This sequence represents deep-water deposits accumulated in a trough-like offshore marine basin far below effective wave base.

The association of cherts, siliceous schists, pelagic limestones (flaser limestones) and terrigeneous turbidity deposits (graywackes) of flysch-like facies can be considered as an indication of geosynclinal sedimentation, but the absence of ophiolites, tuffs, or tuffaceous breccias - usually associated with radiolarian cherts and related sediments in many geosynclines - prevents a more detailed classification of the former geosyncline.

One can distinguish an older regional metamorphism and a subsequent contact metamorphism caused by the intrusion of the adjacent Serre S. Bruno granite (Colonna et al. 1973). Both metamorphic events overlap in a large area. In the area clearly without contact metamorphism the grade of metamorphism is described with the help of illite crystallinity of pelitic rocks (Knoche 1977).

In one profile from W to E the grade of metamorphism increases by Kubler-indices, diminishing from 7,0 - 6,2 (sampling location 1) and 4,3 - 3,5 (sampling location 2) to values of 2,5 - 2,0 (sampling location 3) see fig. 1.

Metamorphic grades in Calabrian metasediments of the Stilo-Pazzano-Guardavalle-area (Colonna et al. 1973; Di Pierro et al. 1973) range from diagenesis to low-grade metamorphism sensu Winkler (1974), provided that the values of illite crystallinity are not altered by weathering or tectonic deformation.

References

Afchain, C., 1970: Somm. Soc. Géol. France 4:150-151.

Colonna, V., Lorenzoni, S., Lorenzoni, E.Z., :Bull.Soc.Geol.Ital. 92:803-830.

De Capoa Bonardi, P., 1970:Rdc.Acc.Sc.Fis.Mat.Soc.Naz.Sc.Lett.Arti,Napoli, 37:127-128.

Di Pierro,M.,Lorenzoni,S.& Lorenzoni,E.Z.,1973:N.Jb.Miner.Abh.119:57-64.

Görler,K.,Ibbeken,H.,1970:N.Jb.Geol.Paläont.Mh.:424-426.

Knoche,G.,1977: Berliner geowiss.Abh.,Reihe A.,Bd.1, 67 p., Berlin.

Winkler,H.G.F.,1974:Petrogenesis of metamorphic rocks:320 p.Berlin-Heidelberg-New York.

Critical Review of Postulated Nappe Structures in Southern Calabria

K. Görler

Ever since Lugeon & Argand (1906) and Limanowsky (1913) all geological syntheses of Calabria have postulated a more or less complex nappe structure. The various authors differ considerably, however, on the limits of the tectonic units, on their vergences and on the date of their transport. All, however, tend to depart from conditions in Sicily and the Peloritani Massif on the one hand, the Calabrian-Lucanian Apennines, the Sila and the northwestern Calabrian Coastal chain on the other, and apply the complicated tectonic structures of these regions and their respective interpretations to the Aspromonte and the Serra S. Bruno. Recent syntheses of this sort are given by Caire (1973), Ogniben (1973, 1975), Vezzani (1973) and Amodio-Morelli (in press).

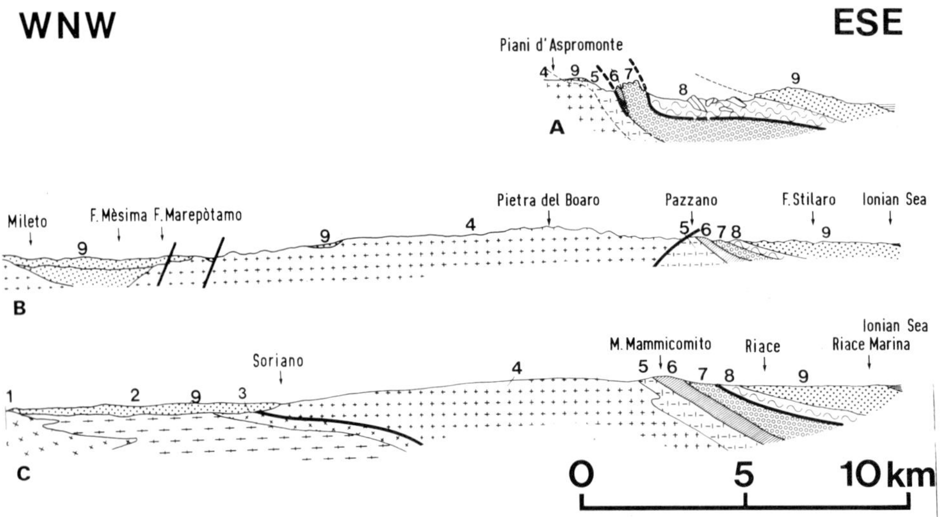

Fig. 1. Schematic sections of Southern Calabria. A according to Caire (1973): (4) Aspromonte-Serra San Bruno basement; (5) phyllites (parautochthonous); (6) slices of the limestone chaine; (7) discordant Oligo-Miocene molasse; (8) chaotic flysch of internal origin (multicolored argille scagliose) with internal Numidian; (9) postorogenic middle to upper Miocene, Pliocene and Quaternary. B according to Ogniben (1973): (4) Aspromonte Nappe; (5) Galati Nappe, low grade phyllites; (6) Galati Nappe, Mesozoic limestone; (7) late orogenic Calabride flysch; (8) Antisicilide Complex; (9) Postorogenic Complex. C according to Amodio-Morelli et al. (in press): (1), (2), (3) Polia-Copanello Unit; (4) Stilo Unit, granodiorites and granites; (5) Stilo Unit, phyllites, metagraywackes and metalimestones; (6) Stilo Unit, neritic limestones; (7) Stilo Unit, detrital sediments; (8) Stilo Unit, argille varicolori; (9) Transgressive sediments, attacked by orogenic transport.

These hypotheses of a complex nappe structure in Southern Calabria were not confirmed by field investigations. The postulated nappe boundaries are dealt with in order of occurrence, from the base up (fig. 1 A, B, C).

Postulated overthrust of the Polia-Copanello Unit over the Bagni-Fondachelli Unit and the Castagna Unit in the northern Serra (Paglionico & Piccareta 1976; Borsi et al. 1976; Amodio-Morelli et al. in press).

According to Schreyer (pers. comm.) the metamorphites occurring between Girifalco and Maida are retrograde metamorphic rocks, which are formed secondarily from the high-grade metamorphites belonging to the Polia-Copanello Unit. There is no evidence supporting the nappe overthrust postulated in this area by the Italian colleagues listed above.

Postulated overthrust of the Stilo Unit over the Polia-Copanello Unit (Borsi et al. 1976; Amodio-Morelli et al. in press).

According to these authors, the Stilo Unit with its magmatic and metamorphic rocks was overthrust over the magmatic and metamorphic rocks of the Polia-Copanello Unit (mainly of the Castagna Subunit) during the Neogene. The following factors contradict this hypothesis:
1. On both sides of the postulated overthrust contact there outcrop very similar and probably coeval rocks.
2. The contact between the two units only outcrops in a very few places and "the inclination is generally quite strong (even 70°)" (Borsi et al. 1976) for an overthrust plane.
3. There is no direct indication that the few tectonic contacts are of Tertiary age. On the contrary, Upper Miocene and Pliocene transgress beyond both units, and on top of the supposedly overthrust Stilo Unit there is a strata series stretching from the Jurassic to the Pleistocene and unaffected by compression tectonics. The inclination of these sediments in the eastern Serra S. Bruno, which becomes steeper with increasing age, only documents a regular tilting movement in this part of Calabria, beginning in the Oligocene and probably continuing up to the present (fig. 2).

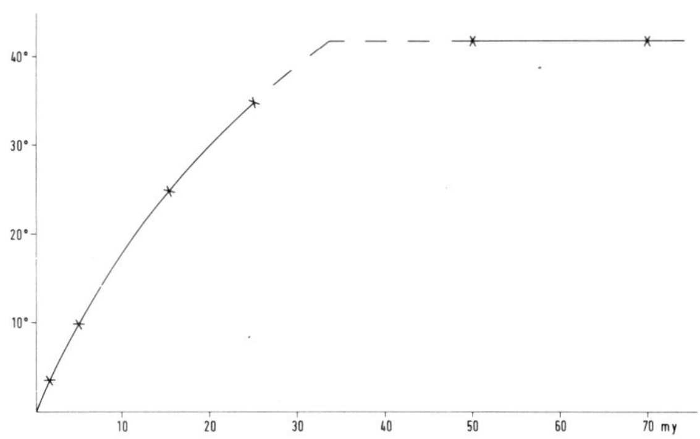

Fig. 2. Tilting of Southern Calabria. Inclination of different strata plotted against their age. The dotted line marks the critical time interval not documented by sediments. The point of intersection of the two curve segments indicates the probable start of tilting and perhaps also of subduction. Dip angles based on the profile published on sheet "STILO" of the geological map of Calabria.

4. According to Borsi et al. 1976, the Rb/Sr ages of biotites from the rocks of the Stilo Unit lie between 193 and 279 m.y. (Lower Permian - Lower Jurassic); those of the dioritic-kinzigitic Unit (= Polia-Copanello Unit) lie between 127 and 147 m.y. (Upper Jurassic - Lower Cretaceous). Since the analysed rock samples are all Hercynian or even older formations, at most the highest values give an approximate indication of the date of consolidation of the acidic magma in the Stilo Unit. In the west of the Serra S. Bruno, however, there are rocks outcropping which were formed in deep parts of the crust (Lorenzoni et al. 1976; Schenk & Schreyer 1977), whilst on the eastern edge of the Serra S. Bruno the Devonian sediments are low-grade or anchimetamorphic (Afchain 1970; De Capoa Bonardi 1970; Görler & Ibbeken 1970; Knoche this vol.). This superficial pattern of rocks with varying metamorphic grades can only be due to differences in the speed of uplift and thus in the rate of erosion.

Thus the pt-conditions for the start of the Rb/Sr clock should have been reached considerably earlier in the east of the Serra, with its slight and therefore slower uplift, than in the west. There are therefore no grounds for assuming a nappe structure because of this difference in Rb/Sr ages between the rocks of the eastern and western Serra S. Bruno.

Postulated overthrust of the Aspromonte Nappe over the Galati Nappe (Ogniben 1973, 1975; Vezzani 1973)

According to these authors, the deepest tectonic unit outcropping in Southern Calabria is the Galati Nappe, which is characterized by paleozoic lowgrade phyllites and transgressive reefoid Upper Jurassic and Cretaceous limestones, whereas the Aspromonte Nappe is mainly built up of high-grade metasediments with large intruded plutons.

The following points speak against the view expressed by Ogniben (1973, 1975) and Vezzani (1973) that these two rock complexes moved against each other as nappes during the Alpidic period:

1. In all the outcrops known to the authors we don't find Ogniben's postulated geometric situations that phyllites and transgressive Mesozoic are overlain by highly metamorphic series. The opposite is the case - only rarely do both rock complexes adjoin locally, with more or less vertical dislocation planes.
2. Both complexes are thermal metamorphised by the intrusion of the Serra S. Bruno granite (Colonna et al. 1973), which absolute age-datings (260 m. y.) have proved beyond doubt to be Variscan (Ferrara & Longinelli 1961).
3. Investigations by Knoche (this vol.), in the Stilo Pazzano area suggest a probable gradation between the two metamorphic complexes, also shown in numerous places on the geological map 1:25.000.

Measurements of the illite crystallinity showed that over a distance of only 1.5 km the Kubler-index changes continuously from E to W from 6-7 (anchimetamorphism) to 2 (epimetamorphism).

These results support the assumption that the approx. 40° tilting since the Oligocene proved in the sediments of the Calabrian southeast slope has also affected the lower parts of the crust and is also responsible for the unusually close proximity of metamorphic zones in the present land surface caused by deep-reaching erosion. The results of Schreyer & Schenk (this vol.) also fit into this picture. The conditions of origin which they established for metamorphic rocks on the western edge of the Serra S. Bruno (700 - 800°, 6-8 Kb) show that today rocks outcrop here on the surface that were formed in the subcrustal area. Such rocks are in fact to be expected if we assume that the entire crust of the Serra S. Bruno til-

ted by about 40°. All in all, the succession of rocks with a rapidly increasing grade of metamorphism from E to W seems to be adequately explained without postulating big nappe thrusts.

4. At the base of the Mesozoic, which, according to Ogniben (1973), is characteristic for the Galati Nappe, Lower to Middle Jurassic bright brown coal has been found in the Canolo area (Görler, Grebe & Teichmüller in prep.). Increased tectonic stress or superposition by a pile of nappes may therefore be excluded.

Postulated overthrust of the Castagna Nappe over the Aspromonte Nappe (Ogniben 1973, 1975; Vezzani 1973)

It proved practically impossible to check the occurrence of the Castagna Nappe under the Aspromonte Nappe, as postulated on a bibliographical basis by Ogniben (1973), since no particulars are given of its occurrence and the details on the coloured general map, in a general sketch (fig. 97, Ogniben 1973) and in the text differ widely. Field investigations in the areas where, according to Ogniben, this nappe occurs, brought no indication of its existence.

Postulated tectonic contact between the crystalline basement and "slices of the limestone chain" (Caire 1973)

The contacts between the crystalline basement and the predominantly limestone Mesozoic in Southern Calabria are mostly only badly outcropped, and are characterized by sulphidic mineralizations. In two good outcrops between Canolo and Canolo Nuovo (locality Torrente Pacchina and Vallone Roncatoni) the contacts are clearly stratigraphical. In the Vallone Roncatini outcrop, fluviatile quartz conglomerates, filling in the scours in the phyllite substratum, form the base. In the Torrente Pacchina the lowest horizon above the phyllite consists of a heavily pyritized rootled bed with carbon remains. In these two sections at least the contact between paleozoic phyllites and mesozoic sediments is not tectonically overprinted.

Postulated overthrust of "chaotic flysch of internal origin" with internal Numidian over Oligo-Miocene molasse (Caire 1973)

Postulated overthrust of Antisicilide Complex over late orogenic flysch (Capo d'Orlando Flysch) (Ogniben 1973, 1975; Vezzani 1973)

In Southern Calabria a coarsely clastic Upper Oligocene to Lower Miocene, becoming flysch-like towards the top, transgresses directly the crystalline or the karstified Mesozoic. This "Capo d'Orlando flysch" (Ogniben 1960) is overlain by argille scagliose with blocks of Numidian flysch. The Miocene sedimentation continues over the argille scagliose-intercalation, which is between 0 and 1000 m thick. Caire (1973) interprets this unconsolidated mélange of Cretaceous-Oligocene rocks as a nappe of internal origin, i.e. coming from the West. Ogniben (1973, 1975) and Vezzani (1973), on the other hand, consider this "Antisicilide Complex" (Ogniben 1969) to be a nappe transported from E or SE against the general vergency.

In the author's opinion, however, it is a typical olistostrome. This interpretation is supported by:

1. The totally unconsolidated state of the argille scagliose.
2. The low degree of coalification of the Jurassic and Oligocene coal in the underlying parts of this *nappe*.
3. The obviously sedimentary contacts, against sole and top, e.g. multiple interbidding of normal sedimentation and argille scagliose in the range of dm to m.
4. The occurrence of nanno- and microfossils belonging to the missing in-

terval of normal sedimentation within the argille scagliose and the occurrence of fossils of the argille scagliose in the normal sedimentation directly beneath the argille varicolori.
5. The non-simultaneous breaking off and re-starting of normal sedimentation before and after the argille scagliose deposition. The simplest explanation of this effect is a slow forward building up and replacement of the mudflow front from E or SE against the Calabrian Massif.
6. The imbrication of the competent rock fragments (olistoliths) is only explainable by mudflow transport. It also indicates an E-S movement of the argille scagliose material (Görler this vol.).

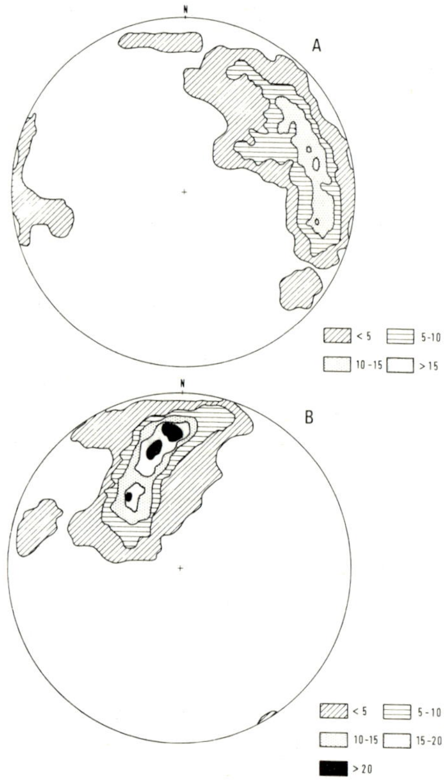

Fig. 3. Face of poles distribution of joint planes in a Jurassic limestone near Canolo (Reggio Calabria), according to Yonan (1977). (A) Oligocene; 110 values of joint cracks refilled with detrital sediments during the Oligocene transgression. (B) Actual (postoligocene) joint system; 120 values.

Conclusions

According to present information, an Alpidic (Eocene) nappe structure in Southern Calabria is not only unproved but also unlikely. Even the tilting which is typical for the eastern edge of the continental crust only began in the Oligocene and seems still undiminished (fig. 2). The only indication of pre-Oligocene tectonics is a different joint system from today, which was followed by the pre-Upper Oligocene karstification of the Portlandian limestones in the Canolo area (fig. 3).

The few indisputable overthrusts also affecting Tertiary sediments can at most be traced over a few metres, for example near Sideroni and 3 km

west of Moticella. In these cases, phyllites or augen-gneisses locally overlie Oligo- or Miocene Capo d'Orlando Flysch, so that these localized upthrusts can only be Apenninic (Miocene) in age. In Southern Calabria at least, there seems to be no nappe structure belonging to this period. But suggested by Görler & Giese (this vol.) underthrust structures seem probable in the Calabrian Ridge off the east coast of Calabria.

In comparison with the areas of the Apennine-Maghrebide-Orogene which adjoin to the north and west, Southern Calabria itself seems like an oasis of rest. This applies above all to the Sila and the neighbouring Apennines, where in the west the sudden side-by-side of highpressure-low temperature metamorphites and highpressure-high temperature metamorphites and refraction-seismically determined duplication of continental crust (Giese & Görler 1974, this vol.) clearly show that, as in Sicily, extensive over- or underthrust must be present. Thus, in the very area where there is a subduction zone still active today (Görler & Giese this vol. fig. 1), the tectonic structure is relatively simple. In neighbouring areas, however, where the collision of continental crust (Adriatic plate and African plate) with the Apennine-Calabria-Atlas Arc caused subduction to come to a halt, much more complex structures occur. The cause of the complex orogenic strucutre would appear to be not just the subduction of oceanic crust but above all the collision of sections of continental crust.

References

Afchain, C., 1970: C.R. somm.Séances Soc. géol. de France, 5:150p, Paris.

Amodio-Morelli,L.,Bonardi,G.,Colonna,V.,Dietrich,D.,Giunta,G.,Ippolito, F.,Liguori,V.,Lorenzoni,S.,Paglionico,A.,Perrone,V., Piccarreta,G., Russo,M., Scandone,P.,Zanettin-Lorenzoni,E.& Zuppetta,A.: Mem. Soc. gol. Ital., in press.

Borsi,S.,Hieke Merlin,O., Lorenzoni,S.,Paglionico,A.& Zanettin-Lorenzoni, E., 1976:Boll. Soc. geol. Ital.,95:1-26.

Caire, A., 1973: In: De Jong,K.A. & Scholten,R.(Ed.): Gravity and Tectonics:157-173, Wiley & Sons, London, New York, Sydney, Toronto.

Colonna,V.,Lorenzoni,S. & Lorenzoni,E., 1973:Boll.Soc.geol.Ital.92:801-830.

De Capoa Bonardi,P., 1969:Mem. Soc. Naturalisti Napoli, 78:1-130.

Ferrara, G. & Longinelli, A., 1961: Boll.Soc.geol.Ital.80(2):25-36.

Giese, P. & Görler, K., 1974:Nachr.dtsch.geol.Ges.,9:67-77.

Görler, K.,Giese,H. & Teichmüller,M.: in preparation.

Görler, K. & Ibbeken, H., 1970:N.Jb.Geol.Paläont.Mh., 7:424-426.

Limanowsky, M., 1913: Bull.Int.Acad.Sc.Cracovie,Cl.Sc.Math.Nat.Serie A, (6A):370-385.

Lorenzoni,S.,Paglionico,A. & Zanettin-Lorenzoni,E., 1976: Boll.Soc.geol. Ital.,95:1-31.

Lugeon, M. & Argand,E., 1906:C.R.Acad.Sci.Paris,142:1107-1109.

Ogniben,L.,1975: La Ricerca Scientifica, 90:365-408.

– , 1973: Geologica rom. 12:243-585.

Paglionico,A. & Piccarreta,G., 1976:Boll. Soc. geol. Ital.,95:1-11.

Vezzani, L., 1973: Lincei, Probl. attuali Sc. Cult.,Quad.183:15-48.

Yonan, S., 1977:Dipl.-Arbiet, Freie Universität Berlin (unpubl.).

Neogene Olistostromes in Southern Italy as an Indicator of Contemporaneous Plate-Tectonics

K. Görler

From the Molise in Mid-Italy along the east coast of Calabria, in Sicily and in the North African Maghrebides as far as the Atlantic coast near Tanger we find *argille varicolori* (variegated shales), a bouldery mudstone.

The argille varicolori (a.v.) and their related boulders and blocks contain Cretaceous and Lower Tertiary fauna, however they are usually under- and overlain by normal Neogene sediments. For this reason they are usually interpreted as nappes (cf. Ogniben 1973; Vezzani 1973: *Sicilide*, *Antisicilide* or *Crotonide*; Amodio-Morelli 1976: *Sicilide*; Caire 1973: *flysch interne chaotique* and Burton 1971: *argille scagliose*).

Although a tectonic deformation of these sediments is clearly visible, an interpretation as olistostromes fits in better with the following observations:

The base-top relationships are time transgressive or conformable.

The contacts between the a.v. and normal sedimentation are often transitional without any tectonisation, but with a mixing of material and faunas.

A tectonic transport of the unconsolidated argillaceous material without cleavage seems impossible.

The lithology and fauna of the a.v. and related sediments indicate - as far as the Cretaceous and Paleocene-Eocene parts are concerned - a primary deposition in an off-shore deep-sea area below the carbonate compensation depth.

The directions of mudflow-movements in Southern Italy were reconstructed from the time transgressive base or from the sense of imbrication of olistoliths (fig. 1). Since mudflows are found in and against the direction of tectonic vergency, axes of origin may be reconstructed for the a.v. and the related series in the area from the Molise to the Gulf of Taranto and through Sicily. A hypothetical prolongation of these axes of origin is given in fig. 2. According to this hypothesis, the series classified as *Sicilide* in the *Structural Model of Italy* (Ogniben 1975), is not material transported in a large nappe from the W over the Campania-Lucania limestone platform to its present position, but more or less parautochthonous sediments which were deposited in an oceanic basin between the limestone platforms of the Apennines and Apulia.

Even a rough estimate of the Cretaceous and Lower Tertiary sedimentary masses found today in the shape of olistostromes intercalated in Neogene sediments leads, when their pelagic facies are taken into account, to the conclusion that at the turn of the Eocene/Oligocene a sea with a breadth of several hundred kilometres must have existed between the two limestone platforms in order to provide room for these deep-sea sediments to accumulate. Today there is no trace left of the former oceanic crust between the Molise and the Gulf of Taranto. However the corresponding sediments were apparently squeezed together during the Neogene, forming a submarine sedimentary ridge with mudflows on each side. It is, of course, only natural to attribute this compression of sediment into a sedimentary ridge and the simultaneous disappearance of ocean crust to the effect of a sub-

duction zone. And indeed examples of a sedimentary ridge being formed instead of a trench over a zone of subduction have been found in just such areas of thick sedimentary cover of an oceanic crust (Jacobs, Russel &

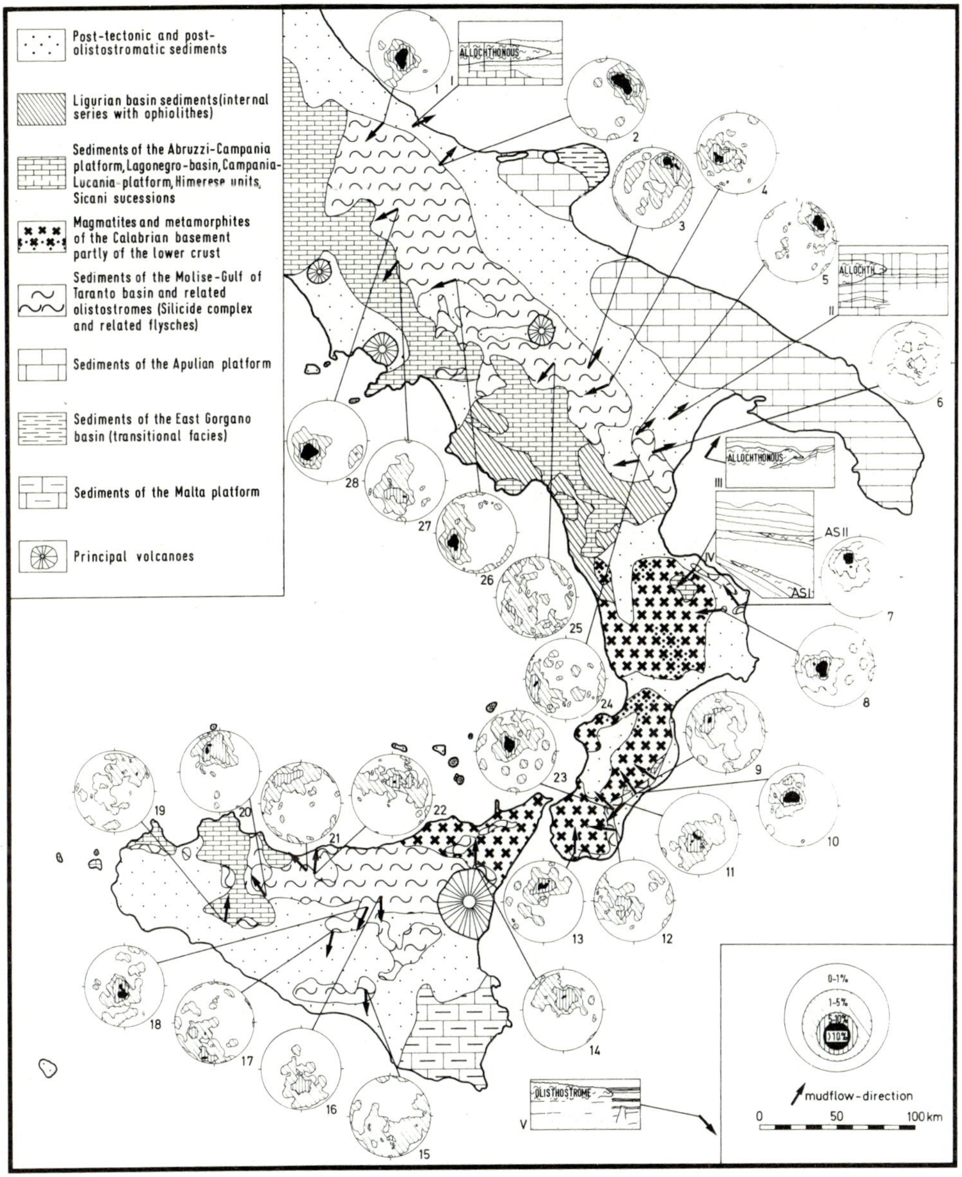

Fig. 1. Directions of movements of mudflows in southern Italy. Measured in olistostromes using their stratigraphic position (sketches with roman figures) and the imbrication of olistolithes. Sketch I and II according to Carissimo et al. 1963, sketch III according to Finetti & Morelli 1973, sketch IV according to Ogniben 1962, sketch V according to Mulder et al. 1975.

Wilson 1974; Chase & Bunce 1969; Westbrook 1975). Accordingly, the fossil Neogene subduction zone would coincide more or less with the axis of origin of the olistostromes, i.e. it would come from the northern Molise and join the Ionian Sea in the Gulf of Taranto. The same would be true of Sicily, except that here the Hyblean or Malta platform south of this axis of origin belongs to the African plate. Here too, one of the extensions of the fossil subduction zone along the strike should be looked for in the Ionian Sea. Both axes of origin are characterized by marked negative gravitational anomalies which point not only to the accumulation of unconsolidated sediments, also proved by drillings, but, according to Morelli (1975), to a "deepening of the sial and of the sima" as well.

Outside Calabria, paleomagnetic measurements in the limestone platforms point to different movements on both sides of the axes of origin. Significantly different values are available for the direction and extent of rotation for the Limestone Apennine as compared to the Apulian platform, and for the Malta- as compared to the Panormides-platform in NW Sicily (Schult 1973, 1976; Barberi et al. 1974). The extensions into the Ionian Sea of the probably Neogene subduction zones in the Molise Gulf of Taranto zone and through Central Sicily are, however, linked together by the still active subduction zone parallel to the Calabrian arc (Peterschmitt 1956; Ritsema 1970, 1975; Barberi et al. 1973, 1974). There is no trench formation in this recent subduction zone either. Instead there is the sedimentary Calabrian Ridge (Belderson et al. 1974) where there are still mudflows and the formation of olistostromes (Mulder 1973; Mulder et al. 1975). It thus seems very likely that the still active subduction zones and the Calabrian Arc are only the sparse remains of an originally much wider island arc-system (fig. 2).

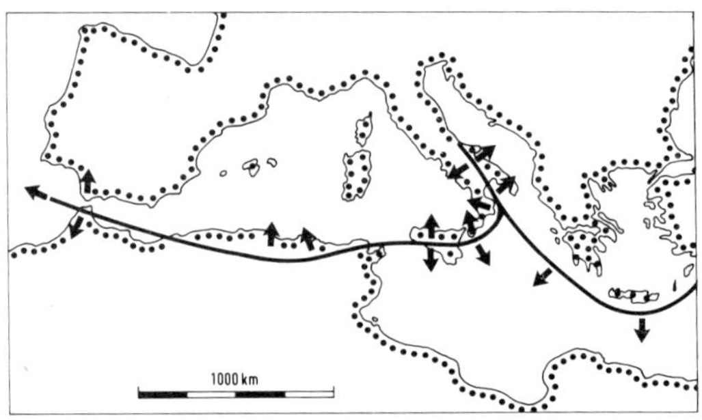

Fig. 2. Hypothetical axes of mudflow origin in the Mediterranean during the Neogene. These axes may indicate also the approximate recent plate boundaries between African and Eurasian plates. - Based on Mulder et al. (1975); Boulin & Raoult (1971); Magné & Raymond (1971); Lajat et al. (1975) and the data of fig. 1.

The available seismic results for the Ionian Sea indicate a total crustal thickness of 19 km (Hinz 1974), of which 14 km are water and sediments (Finetti & Morelli 1973). The high gravity values of this region also support an interpretation as oceanic crust with a very thick sedimentary cover.

Because of the bipolarity of the mudflows and gravity tectonics it may be supposed that, on the outer part of the sedimentary ridge, there were,

and are, externally directed mudflow movements near to the surface and, right at the bottom, a counteracting underthrusting in the opposite direction provoked by subduction. Therefore, it is probable that some stacks of sediments have been subjected several times to the effects of subduction and resedimentation. The result will be a mélange which could equally well indicate tectonic action or resedimentation, like the argille varicolori of southern Italy. For this reason olistostromes on the descending plate can only be preserved from the latest period of subduction, whereas on the hanging plate mudflow sediments can be preserved from the initial phase of subduction. Applying this criterion, and for other reasons, it is supposed that subduction against Europe and the related formation of an island arc-system began during Oligocene time. During the Neogene, subduction must have been halted here wherever there was a collision of this island arc with continental crust, such as the Apulian platform (Adriatic plate) or the Malta platform (African plate). Subduction continued only where oceanic crust is still available, as is the case in Calabria with the adjacent Ionian Sea (Giese & Görler 1977). The thick sedimentary cover is squeezed to form the Calabrian Ridge (Belderson et al. 1974) and still active mudflows and the recent formation of olistostromes can be observed (Mulder 1973; Mulder et al. 1975).

References

Amodio-Morelli, L., Bonardi, G., Colonna, V., Dietrich, D., Giunta, G., Ippolito, F., Liguori, V., Lorenzoni, S., Paglionico, A., Perrone, V., Piccarreta, G., Russo, M., Scandone, P., Zanettin-Lorenzoni, E. & Zuppetta, A.: Mem. Soc. Geol. Ital,, in press.

Barberi, F., Gasparini, P., Innocenti, F. & Villari, L., 1973: J. geophys. Res. 78,23:5221-5232.

Barberi, F., Innocenti, F., Ferrara, G., Keller, J. & Villari, L., 1974: Earth and planet. Sci. Lett., 21:269-276.

Belderson, R.H., Kenyon, N.H. & Stride, A.H., 1974: Nature,247:453-454.

Boullin, J.-P. & Raoult, J.-F., 1971:Bull.Soc.géol.France (7),13:338-362.

Burton, A.N., 1971: Quart.J.geol.Soc. London, 126:369-381.

Caire, A., 1973: in: De Jong, K.A. & Scholten, R. (Ed.): Gravity and Tectonics:157-173, Wiley & Sons, London,New York,Sydney,Toronto.

Chase, R.L. & Bunce, E.T., 1969: J.geophys.Res. 74:1413-1420.

Carissiomo, L., D'Agostino, O., Loddo, C. & Pieri, M., 1963: 6. Wld. Petroleum Congr.Frankfurt,Sect. 1, Paper 27:1-26.

Finetti, I. & Morelli, C., 1973: Boll.geofisc.teor.appl.15,60:263-341.

Görler, K., 1975: IXth int.Congr.Sedimentol.Nice 1975,theme 4 Tectonique et Sédimentation, 1:162-175.

Görler, K. & Reutter, K.-J., 1968: Geol.Rdsch.,57:484-514.

Hinz, K., 1974: Geol. Jb. E 2:33-65.

Jacobs, J.A., Russel, R.D. & Wilson, J.T., 1974: Physics and Geology, 622 p., McGraw-Hill Book Comp., New York.

Lajat, D., Biju-Duval, B., Gonnard, R., Letouzey, J. & Winnock, E., 1975: Bull. Soc. géol. France (7),17:481-485.

Magné, J. & Raymond, D., 1971: Bull. Soc. géol. France (7),13:363-370.

Morelli, C., 1975: Consiglio naz. Ric.Quad. "la Ricerca sci." 90 (Structural Model of Italy): 427-447.

Morelli, D., Cornea, I. & Lazarescu, V., 1975:Boll.Geofis.teor. ed appl., 18:169-178.

Mulder, C.J., 1973: In: Drooger (Ed.): Messinian Events in the Mediterranean:44-59, Konink1.Nederl.Akad.Wetensch., Amsterdam.
Mulder, C.J., Lehner, P. & Allen, D.C.K., 1975: Geol.en Mijnb.,54,3-4; 208-221.
Ogniben, L., 1962: Geol. Rom., 1:255-282.
- , 1969: Mem.Soc.geol.Ital.8,4:453-763.
- , 1973: Geol.Rom., 12:243-585.
- , 1975: Quad. La Ricerca Scient., 90:365-408.
Peterschmitt, E., 1956: Ann.Geofis. 9,3:303-334.
Ritsema, A.R., 1970: XII.Assem.gén.Comm.seismol.europ.,:1-5, Luxembourg.
- , 1975: Progr.in Geodynamics, 13:1-12.
Schult, A., 1973: Earth and planet.Sci.Litt., 19:97-100.
- , 1976: Earth and planet. Sci. Lett., 31:454-457.
Vezzani, L., 1973: Accad.naz.Lincei,Probl.attuali Sc.Cult.,Quad.183 (Moderne vedute sulla geologia dell'Appennino):15-48.

Calabrian Arc and Ionian Sea: Vertical Movements, Erosional and Sedimentary Balance

P. Ergenzinger, K. Görler, H. Ibbeken, P. Obenauf and J. Rumohr

Contents

1) Introduction

The rate of uplift and the rate of erosion in a particular area must be approximately equal, at least for periods of millions of years. In a marine environment, subsidence is documented by sediment thickness if the water depth remains constant. If vertical movements are not balanced by sedimentation, facies changes or regressions or transgressions are bound to result. Here, the rates of erosion and sedimentation have been determined and integrated over time intervals of varying length in order to determine vertical movements in Southern Calabria and thus to set up an account of recent erosion and sedimentation in the Ionian Sea. The resulting data on erosional activity in Southern Calabria are summed up in fig. 1.

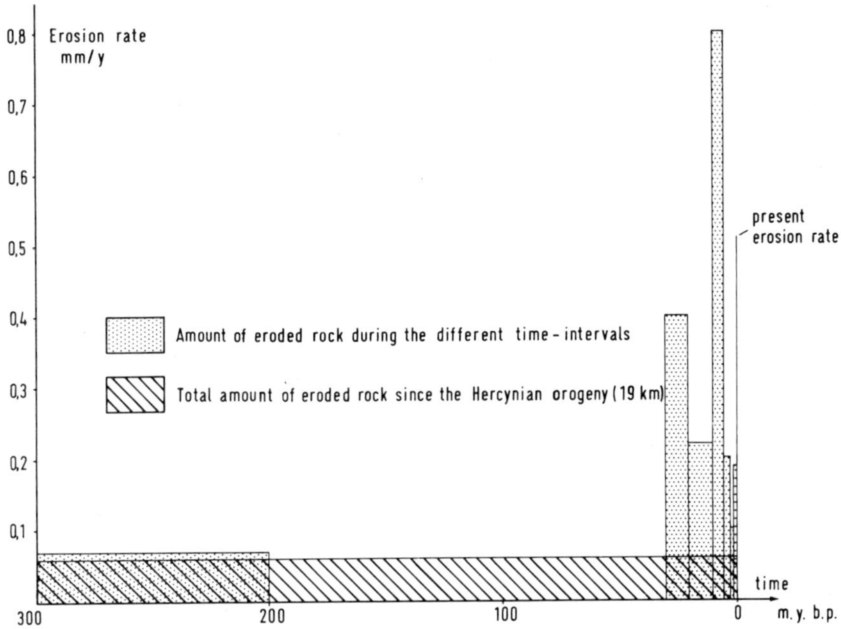

Fig. 1. Average rates of erosion in Southern Calabria during the last 300 m.y. according to the results of the different methods of investigation. The diagonally hatched area represents the total eroded rock masses since the Hercynian orogeny. The dotted area, the same size, shows the probable distribution over different periods.

2) The last 300 m.y. (Görler)

Method: Evaluation of the recent superficial distribution of magmatic and metamorphic rocks originating in different levels of the crust.
Results: Highest local rate for uplift and erosion: 0.1 mm/y; average rate: 0.07 mm/y; both values refer to the total period of 300 m.y.

Only large vertical movements and a corresponding erosion can explain the fact that in Southern Calabria today rocks outcrop which have been proved to originate at depths of over 20 or 30 km (Borsi et al. 1976; Lorenzoni et al. 1976; Schreyer & Schenk this vol.). According to geological evidence, the rocks are probably Hercynian, although an older formation cannot entirely be ruled out.

In the eastern part of Southern Calabria, Upper Devonian sedimentary rocks (approx. 350 m.y. b.p.) are affected by regional metamorphism and locally retrograded in contact with the intrusion of the Serra S. Bruno granite. Since the granite has a radiometric age of at least 279 m.y. (Lower Permian) (Borsi et al. 1976), an age of approximately 300 m.y. can be assumed for the Hercynian regional metamorphism.

In the western part of Southern Calabria, where outcropping rock was metamorphized at a depth of about 30 km, a mean uplift and erosion rate of

approx. 0.1 mm/y over the last 300 m.y. may be assumed. Since, however, rocks of low or very low-grade metamorphism outcrop in the eastern part of Southern Calabria, an average value of only approx. 18 km erosion or 0.07 mm/y must be estimated for the whole of Southern Calabria.

This erosion, integrated over 300 m.y., did not take place at a constant rate. A first phase of relatively rapid upthrust and erosion could have occurred at any time during the Carboniferous-Triassic period, but sediments from this period are not preserved in Southern Calabria. Sedimentation resumed in the Jurassic, but over low-grade metamorphites. The non-metamorphic sedimentary cover of these metamorphites must therefore have been eroded in the meantime. If we estimate its thickness at approx. 7 km, an average erosion rate of 0.1 mm/y results for the period between 300 m.y. and 200 m.y.

The Jurassic basal conglomerate is fluviatile in origin. Extremely mature sediments indicate a wide hinterland with little relief and without uplift tendencies. The latter is probably true up to the Middle Oligocene (30 m.y.), since an episodic sedimentation of shallow-water limestones without angular unconformities or inserts of clastic sediments is found for this period in the eastern part of Southern Calabria. According to these findings, the rate of uplift and erosion must have been very small or zero between 200 m.y. and 30 m.y. ago.

3) The last 30 m.y. (Görler)

Method: Petrographic analysis of coarse sediments of different ages.
Results: Highest local rate for uplift and erosion: 0.65 mm/y. Average value, 0.35 mm/y; both values refer to the last 30 m.y.

Thick terrigenous sediments of the Oligocene-Pleistocene along the east coast of Southern Calabria are the main evidence for a rapid uplift and erosion in the central and western parts of Southern Calabria. Following the estimates above, an average of 11 km and a maximum of 20 km of erosion remain for the period from the Middle Oligocene to the present. By order of magnitude these eroded rock masses correspond to the large piles of sediment shown to exist on both sides of Calabria in the Crotone-Spartivento Basin and the Eastern Calabrian coastal region, as well as in the Gioia Basin.

In order to obtain more exact information on the development of erosion during this period, a total of more than 20 000 separate pebbles in 56 outcrops in the Mesozoic, Tertiary and Quaternary conglomerates of Southern Calabria were measured in tangential sections and determined macroscopically. Tab. 1 shows approximately coeval conglomerates from different regions of Southern Calabria. Here those rocks were grouped together which are roughly typical of the different crustal stockwerks of Southern Calabria. The selection of the deposits to be investigated was also limited by the relatively small outcrop surfaces of the respective conglomerates and by the conditions of outcrops. Differences in the transport resistance had to be ignored. The values obtained show the tendency towards an increase in the proportion of rocks from deeper crustal levels with decreasing conglomerate age.

The repeated facies changes of the Tertiary sediments in Eastern Calabria, the synsedimentary displacements demonstrated by different dip angles, the common rapid changes in the composition of the conglomerates and, above all, the proportion of vein-quartz content indicate a very

	Pretertiary sedimentary cover, volcanics, low and very low grade metamorphites (%)	Acid plutonites, their contact metamorphic cover and medium grade metamorphites (%)	highgrade metamorphites (%)	Vein quartz (%)
Pleistocene - Upper Pliocene	10,2	53,9	30,1	5,8
Upper Miocene	8,8	80,8	8,1	2,3
Oligocene	42,4	48,7	4,2	4,7
Jurassic	7,0	0	0	93,0

Tab. 1. Composition of approximately coeval conglomerates from Jurassic to Pleistocene. Increasing of rocks from deeper crustal levels with decreasing conglomerate age.

time b.P. (m.y.)	uplift (km) Max.	Average	Erosion (km) Max.	Average	rate of uplift (mm/y) Max.	Average	rate of erosion (mm/y) Max.	Average
1-0	1,8	0,9	0,4	0,2	1,8	0,9	0,38	0,19
2-1	0	0	0,2	0,1	0	0	0,2	0,1
5-2	0,2	0,1	0,9	0,5	0,07	0,03	0,3	0,2
10-5	7,5	4	8	4	1,5	0,8	1,6	0,8
20-10	2,5	2	3	2,2	0,25	0,2	0,3	0,22
30-20	9	5,0	7,5	4	0,9	0,5	0,75	0,4
200-30	0	0	0	0	0	0	0	0
300-200	10	7	10	7	0,1	0,07	0,1	0,07
300-0	31	19	30	18	0,1	0,06	0,1	0,06

Tab. 2. Uplift and erosion in Southern Calabria.

intermittent sedimentation. Tab. 2 and fig. 1 represent the probable course of uplift and erosion in Southern Calabria, based on all these factors.

4) The last 1 m.y. (Görler)

Method: Volumetric estimate of the drainage system cut into the Pleistocene abrasion plain.
Results: Uplift rate varying regionally between 0 and 1.8 mm/y. Average erosion rate, 0.2 mm/y. Valid for areas of at least approx. 10 km² varying between 0 and 0.38 mm/y.

During the Pleistocene there must have been a relatively long period of tectonic inactivity in Southern Calabria during which all older structures were cut off by an abrasion plain or a peneplain. The remains of

this plain can be traced from the coast up to an altitude of almost 2000 m (see geological map of Calabria 1 : 25 000). After uplift and tilting movements had begun again about 1 m.y. ago, a pronounced vertical erosion led to the present very steep relief in Southern Calabria.

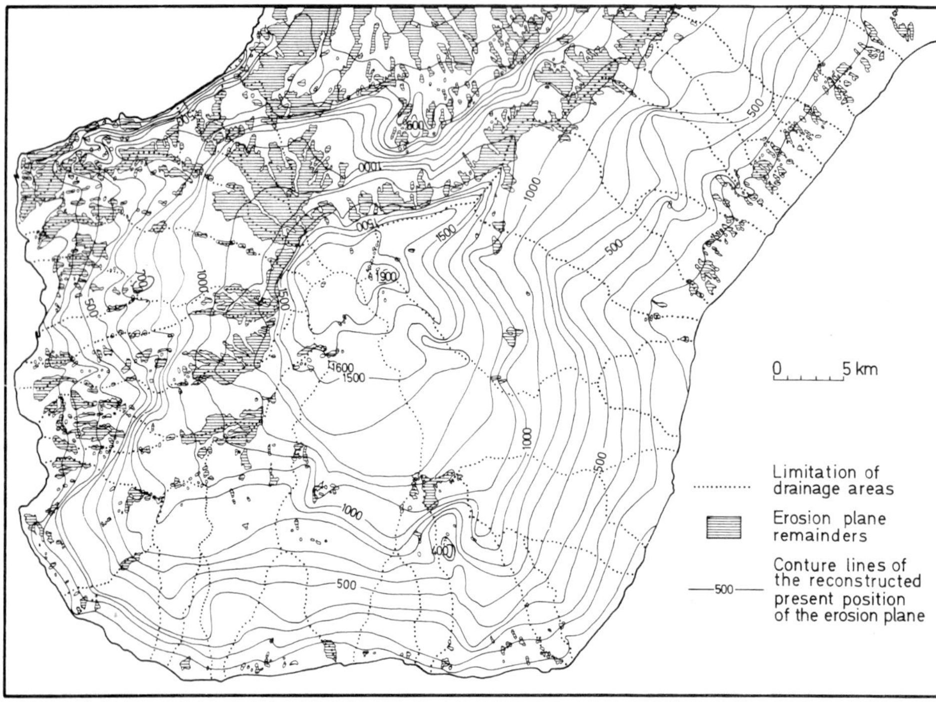

Fig. 2. Reconstruction of the hypothetical present position of the Pleistocene abrasion plain, disregarding interim erosion. Assuming that the plain originally lay practically horizontal near sea level, the map also gives an impression of neotectonic processes in Southern Calabria.

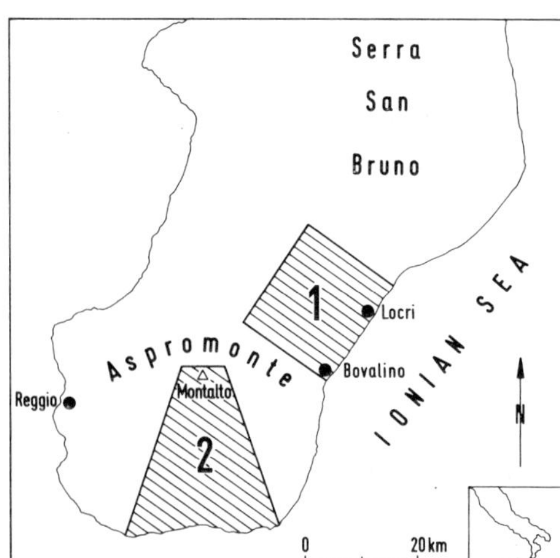

Fig. 3. Position map of the two areas of erosion shown in fig. 4 and 5.

From the remaining parts of the plain and from stratigraphic data the hypothetical present-day position of this abrasion plain was reconstructed (fig. 2). A comparison with the actual relief should provide information on erosion activity since the formation of the plain. For this purpose series of profiles approximately parallel to the coast were drawn and the area between the present-day relief and the reconstructed abrasion plain was measured by planimeter (fig. 3, 4 and 5). This method furnished data showing the dependence of the erosion rate on the distance to the coast and the watershed.

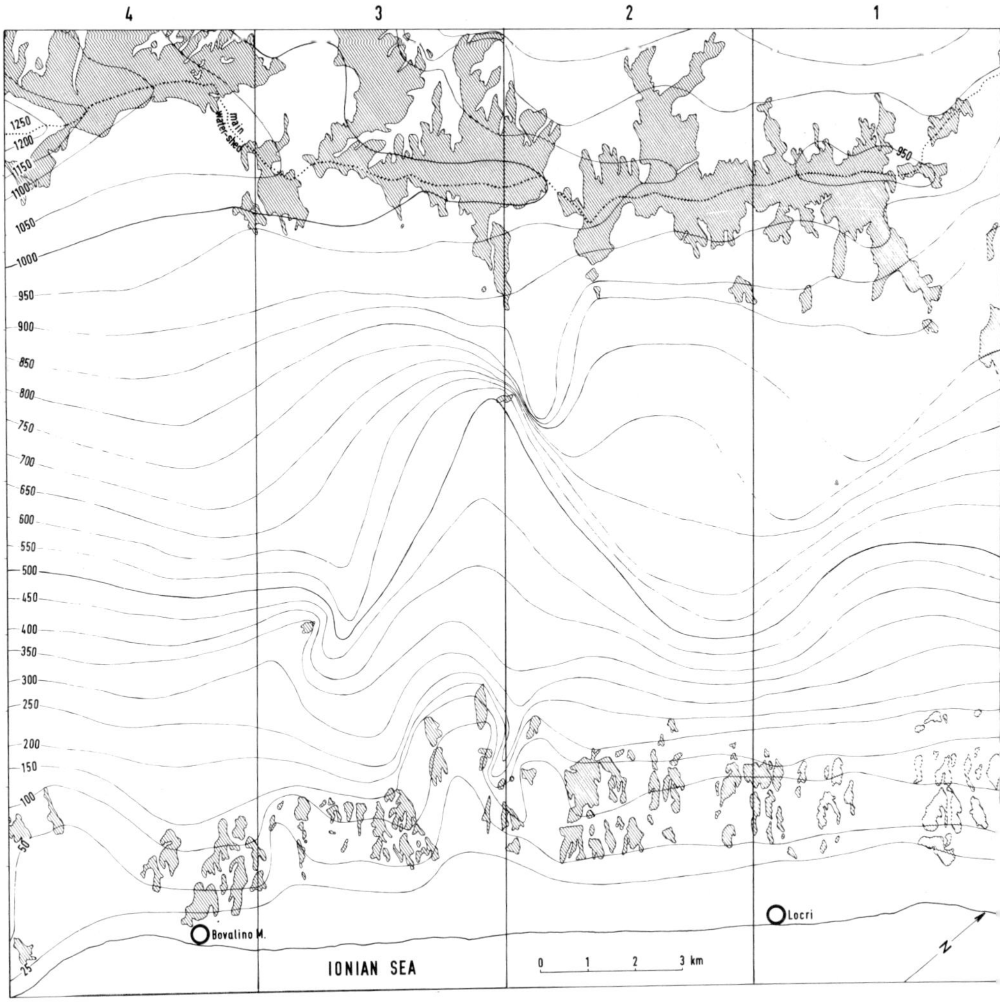

Fig. 4. Preserved remains of plain and reconstructed abrasion plain of area 1 in fig. 3.

Fig. 5. Profile series of roughly longshore profiles through area 1. The upper limit of the profiles is the reconstructed abrasion plain, the lower limit the present-day relief. The dotted areas represent the area eroded since the formation and re-lifting of the abrasion plain, measured planimetrically. For position of profiles see fig. 4.

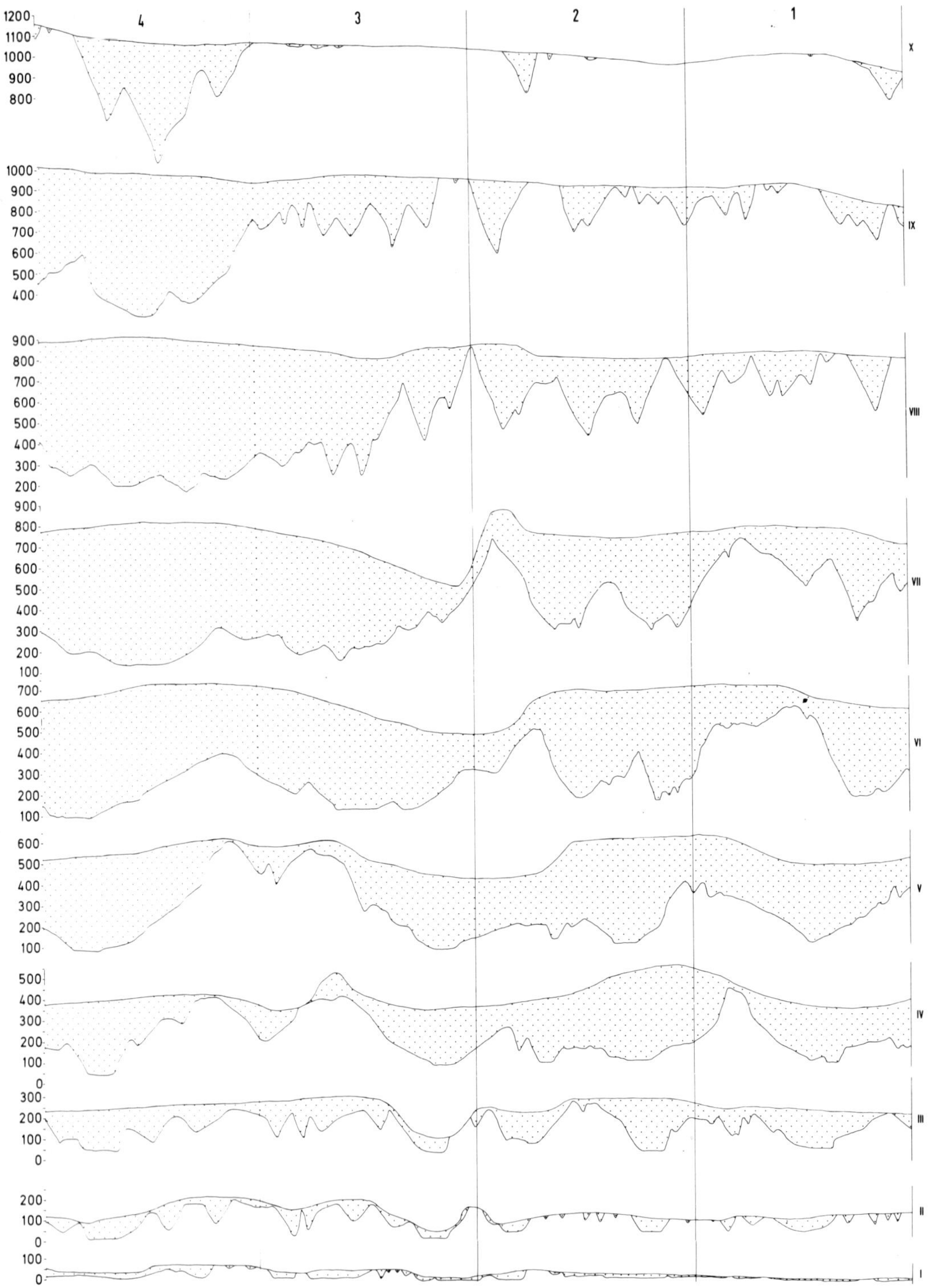

4
3
2
1
1200
1100
1000
900
800
X
1000
900
800
700
600
500
400
IX
900
800
700
600
500
400
300
200
VIII
900
800
700
600
500
400
300
200
100
VII
700
600
500
400
300
200
100
VI
600
500
400
300
200
100
V
500
400
300
200
100
0
IV
300
200
100
0
III
200
100
0
II
100
0
I

Corresponding investigations were carried out in all of Southern Calabria south of the southern Serra S. Bruno. The most important results are: The average rate of erosion for this area over the last 1 m.y. is 0.2 mm/y. The area with the highest erosion values now lies east of the watershed at about half to a third of the distance between watershed and coast (fig. 6). During the last 1 m.y. cumulative uplift took place considerably quicker than erosion, but the mean rate for this period is probably composed of low values at the start of the renewed uplift and of much higher values for the most recent past. A sharp increase in erosion rates is to be reckoned with in the future, since in the long term the rates of uplift and erosion must be approximately equal.

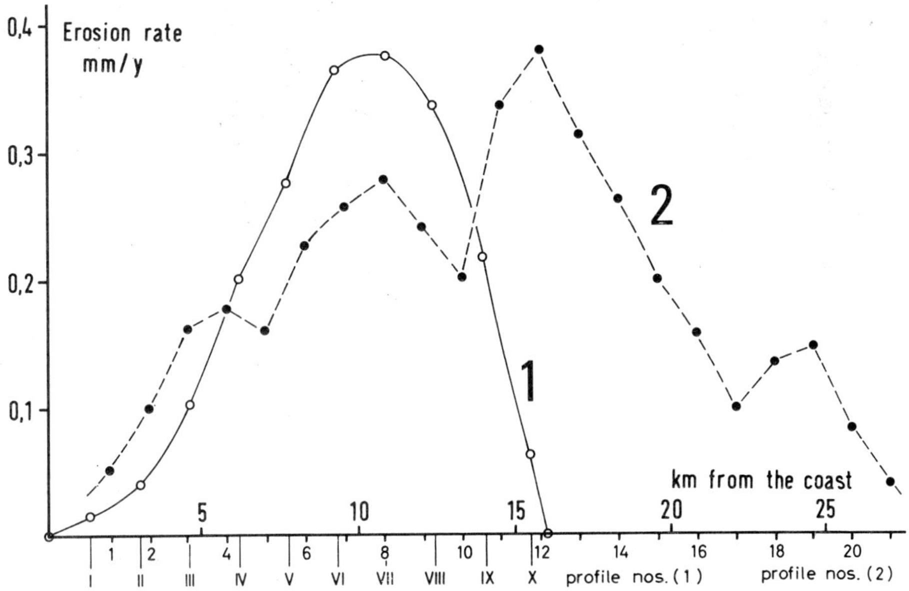

Fig. 6. Average erosion rates of individual profiles plotted against distance from the coast or watershed. Curve 1 represents an area along the SE coast where neotectonic activity is apparently mostly limited to a SE-tilting. Curve 2 represents an area with the marked young fracture tectonics typical of the S and W coast of Southern Calabria. (see fig. 3).

5) The last 100 years (Rumohr, Ibbeken)

Method: Balancing of the sediment supply on the basis of changes in shelf morphology.
Results: Average erosion rate of 10 mm/y for the period 1876-1976 in the La Verde drainage basin.

The area of investigation between Capo Bruzzano and Siderno (25 km) and 6-12 km offshore is supplied by 3 small and 3 large rivers (Buonamico, Condoianni, Laverde), and has the normal stock of Calabrian submarine morphology, with a Pleistocene shelf, 120 m of shelf edge, an active canyon and a narrow shelf.

480 km echosounding profiles were compiled to a bathymetric chart (Rumohr & Ibbeken 1975). More than 100 sediment samples and cores gave a

general picture of the sediment distribution.

A comparison of 470 soundings made in 1876 with the recent depth recording reveals the amount of sedimentation and erosion in the past 100 years (fig.7 foldout).

Results: Bedloads from the Buonamico and Careri rivers are carried directly into the canyon head during floods. The longshore transport carries the Laverde bedload into a canyon branch. Running through the 7-9° inclined canyon floor to depths of 400-500 m, the material undercuts the 10-25° steep canyon flanks. Creeping masses several meters thick and 50-500 m in diameter continue the sediment flux on a 2-3° inclined canyon floor.

During storm periods fresh water suspension jets move several kilometers offshore. Most of the suspended load settles on the outer shelf in water depths greater than 50 m where wave action is low and clay content in sediments high. On the slope itself, suspensions accelerate downhill with increasing erosion efficiency and form deep valleys (Rumohr & Ibbeken 1975).

In the last 100 years the coastline and the submarine morphology have been changed by a combination of these processes: The bedload from the Laverde has formed a delta 300 m seaward and has deepened the canyon branch. The canyon branch itself has moved 150 m landward and the coastline changed respectively.

The floor of the canyon head has deepened to 100 m in places; the coastline has retreated. In the canyon head is still an undigested bulk of Buonamico material 40-50 m thick (= 0.07 km^3). A similar smaller mass lies in front of the Careri river mouth. Parts of the southern canyon flank slumped and formed an accumulation below. In the canyon head a budget of sedimentation and erosion produces a net figure of 17 m/100 y over an area of 10.6 km^2 erosion.

The submarine morphology of the smaller shelf in the north suggests accumulation, but on the whole this area is undergoing destruction. It seems here that sedimentation alternates with turbidity current formation. A new canyon is being formed in front of the Condoianni river mouth. The average erosion on the small shelf was 14 m in the last 100 years. In consequence the coastline moved 50-120 m landward. The average sedimentation in the north was 11 m/100 y and in this case the coastline has moved seaward. Sedimentation on the outer Laverde shelf reaches a thickness of 10-40 m in 100 years; under the effect of this load 0.7 km^3 of sediment have slid down the continental slope.

If we regard the outer Laverde shelf as a natural sediment trap for suspended load, we can calculate the minimum order of magnitude of erosion in the total Laverde and part of the Buonamico river system in the last 100 years:

Fig.7 (foldout). Centralperspective model of the sea floor off SE-Calabria and the hinterland. Bathymetric chart of the Buonamico Canyon area.

Sed. rate (m)	area ($10^6 \times m^2$)	x sediment (m)	= Volume ($10^6 \times m^3$)
20 - 50	2.76	25	69.0
10 - 20	6.09	15	91.4
5 - 10	7.85	7.5	58.9
2 - 5	9.37	3.5	32.8
0 - 2	7.49	1	7.5
		Total	259.6
		- 35% water content	90.9
		soft rock equivalent	168.7 $\times 10^6$ m^3

This volume may be composed of 2/3 Laverde (= 112.5 x $10^6 m^3$ and 1/3 Buonamico-Careri freight (= 56.2 x $10^6 m^3$). Measurements in the rivers showed the relation: suspended load/bed load = 2/1 to be a reasonable estimate. To obtain the total Laverde river discharge in our case, the amount of suspended load must be increased to include the bed load portion which has been carried into the canyon.

Laverde suspended load	112.5 x $10^6 m^3$
+ 50 % bed load	56.2 "
Total Laverde discharge/100 y	168.7 x $10^6 m^3$
- 30% reduction to hard rock	50.6 "
hard rock equivalent	118.1 x $10^6 m^3$

If we correlate this to the Laverde drainage area (117 x $10^6 m^2$) we arrive at an erosion rate of 10 mm/y during the last 100 years.

This cannot be assumed to be a normal 100 years rate and can hardly be explained merely by the effects of damming up the rivers and other human activities. In addition, however, the main position of sediments derives from the upper course of the river void of extensive agricultural influence.

The above mentioned average erosion rate since the Early Pleistocene is 0.2 mm/y. Thus, 100 cm of erosion represent the result of 5.000 years and this is in the order of magnitude of a recurrence interval of the greatest earthquake in historic times: the 1908 Messina earthquake (Brogan et al. 1975).

The breakdown of slope stability is a common secondary phenomenon of earthquakes. Extreme precipitation might release such instabilities and can deliver the mobilized masses into the Ionian Sea as in January 1973.

Erosion and sedimentation in the Calabrian arc - Ionian Sea system is highly episodic. Processes and rates over the last 100 years may give us an impression of the geological events of several thousand years.

6) The last 4 years (Ergenzinger, Obenauf)

Method: Measurement and estimation of total fluvial load.
Results: Erosion rate of 0.5 mm/y for the drainage basin of the Fiumara Buonamico, assuming 4 extremely heavy rainfalls per century.

The Buonamico was chosen as test area for determining the order of magnitude of erosion in the Aspromonte.

Investigations

1. Survey of present slope erosion. A comparison of aerial photographs

taken in 1941 and 1953 with mapping by Sijmons (1975) established the amount of slope material mobilized at different periods.
2. Estimate of correlative sedimentation in the Ionian Sea Mobilized slope material is brought to the valley floors mainly through gully erosion and into the Ionian Sea by fluvial transport. The order of magnitude of imported material in the valley floors may be calculated by measuring the rate of sedimentation in the natural lake Lago di Costantino. The amount of sediment deposited on the valley floors of the Buonamico is determined by annual geodetic surveys of river bed cross-sections. (fig. 9). The difference between the material import and the amount of fluvial accumulation corresponds to the volume transported into the Ionian Sea.

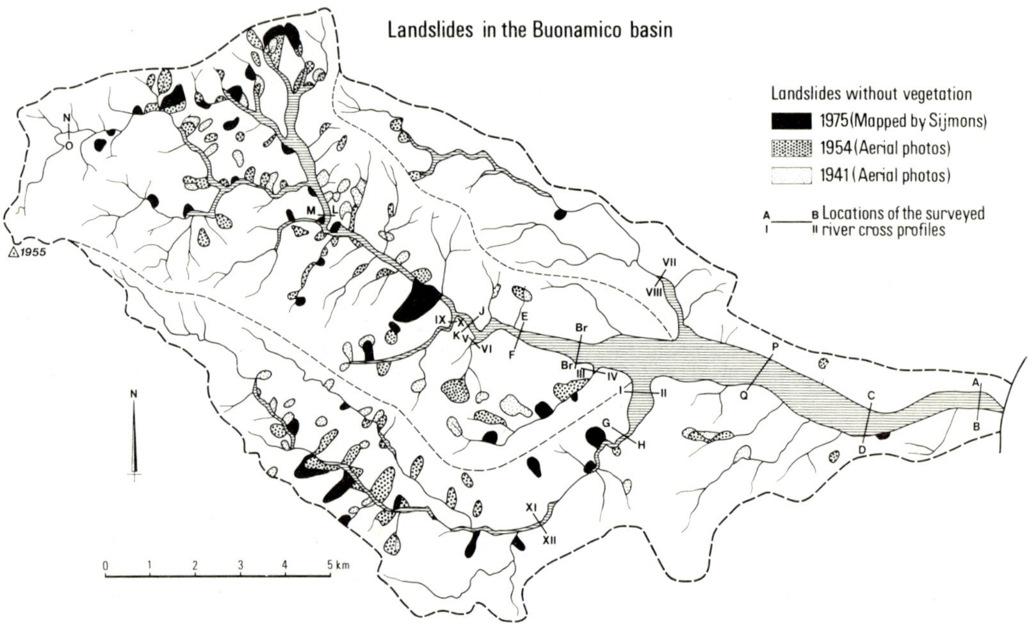

Fig. 8. Landslides in the Buonamico basin.

Results
1. During the period 1954-1975, land slumps etc. affected an area of over 3.15 km² in the crystalline area of the Buonamico drainage basin. At least 16 x 10^6m³ slope material must have been mobilized and about 5 x 10^6m³ of this material was transported into the valley floors.
2. Since 4.1.1973 the erosion volume of the basin (50 km²) above the Lago di Costantino has been approximately 2 x 10^6m³ solid material. If this volume of erosion is extrapolated for the total mountain area of the Buonamico basin (120 km²), there was a volume of erosion of more than 4 x 10^6m³ during 1973-1976.
3. During the period 1973-1976 the volume of erosion was distributed as follows:

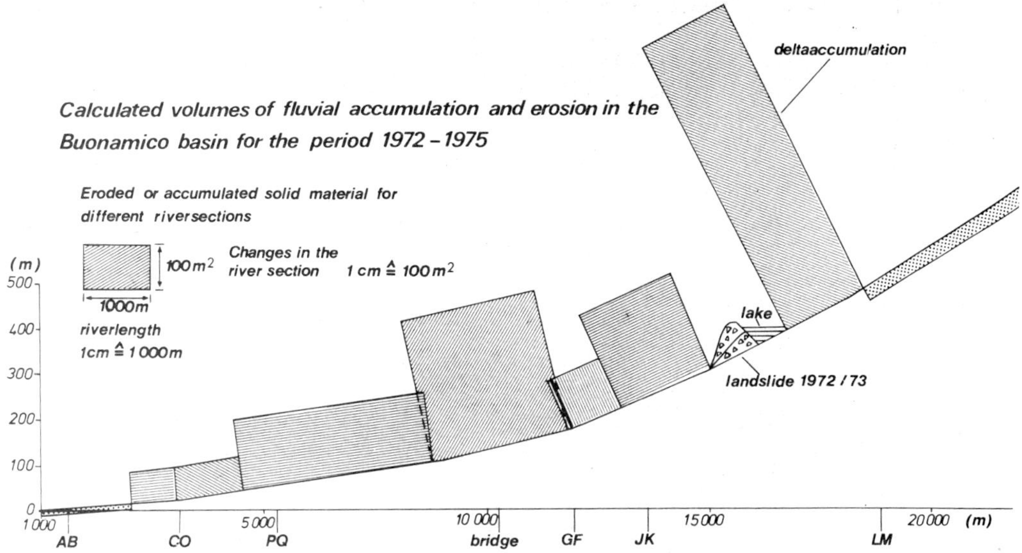

Fig. 9. Calculated volumes of fluvial accumulation and erosion in the Buonamico basin for the period 1972-1975.

Volume of erosion	+ 4.8 x 10^6 m³
Delta accumulation (Lago di Costantino)	- 1.2 x 10^6 m³
Basin accumulation (Lago di Costantino)	- 0.3 x 10^6 m³
Deposition in valley floors	- 2.4 x 10^6 m³
Transport into the Ionian Sea	0.9 x 10^6 m³

Because the Lago di Costantino was formed halfway through the thunderstorm period of 1972-1973 (Guerrichio & Melidoro 1973), thus trapping only half the water, this calculation gives only a minimum of the real amount. A more realistic estimate of the transport into the Ionian Sea is 2 x 10^6 m³.

4. Precipitation data for the Buonamico drainage basin suggest that in 1972 and 1951 at Polsi (1903 mm, 1534 mm) and San Luca (856 mm, 875 mm) the threshold values for slope erosion and fluvial transport were exceeded, that is, twice in 50 years.

5. Under present conditions we may therefore assume an erosion rate in the mountain area of more than 20 x 10^6 m³ solid material. Following the currently prevailing accumulation trend, more than 8 x 10^6m³ of this will be transported into the Ionian Sea. For the entire Buonamico drainage basin of 146 km² we thus arrive at an erosion rate of approximately 0.5 mm/y.

7) The quantitative composition of the recent erosional debris of the Calabrian Massif (Ibbeken, Rumohr)

The sediments of the river mouths and of several longitudinal profiles of 19 rivers in Southeastern Calabria are extremely immature. During the maximally 25 km long fluvial transport no important petrographical dif-

ferentiation takes place; there is no change in sphericity but a distinct rounding. The mean quantitative petrographic composition of the river mouth sediments correlate with the mean quantitative geological composition of the drainage areas (fig. 10).

Results

1. Particle size distribution: The sediments are predominantly distinctly bimodal. The primary mode is found in the 63 mm range (maximum particle size 800 mm), the secondary mode in the 0.63 and the valley between in the 2 mm range. The mean composition of the river mouth sediment is: 75% gravel, 23% sand, 2% silt/clay. For total fluvial load, effective discharge, the proportions are reversed.
2. Petrographic distribution: > 16 mm: the gravels can be assigned macroscopically to the separate rock series of the supply areas (fig. 10).
16-2 mm: this range is characterized by rock fragments, the first occurrence of feldspar particles and increasing proportion of quartz.
2-0.5 mm: considerable increase in quartz and feldspar.
0.5-0.063 mm: increase in quartz, heavy minerals up to 10%, very instable, in metamorphite drainage areas up to 50% hornblende.
< 0.063 mm: quartz, mica and feldspar decreasing, chlorite increasing.
3. Particle shape distribution: the shape of all fractions > 16 mm is clearly independent of particle size and petrography. The entire material is on the average bladed to compact bladed (Sneed & Folk 1958).
4. Roundness: the roundness increases regularly in the longitudinal river profile towards the river mouth, 0.55 (pegmatite) and 0.8 (marble) (Krumbein 1941).

In Southeastern Calabria there are 3 sorts of marine transport: direct transport from river to canyon head, longshore transport of sand and gravel and suspended transport on the narrow shelf, followed by sediment movement by slumpings and turbidity currents.

No change in the petrographic composition is to be expected from the pebbly mudflow transport on the canyon floor. The quantitative petrographic composition of the pebbles > 16 mm in the submarine fans at the foot of the slope ought thus to reflect the quantitative geological composition of the supply areas, referred to pebble-producing units. Since no significant increase in rounding occurs during the longshore transport in the short coastal sections between deltas and canyons, it is not possible to use pebble roundness in submarine fans to establish whether the material comes directly from river mouths or from longshore transport.

8) Conclusions (Görler)

Uplift and erosion in Southern Calabria are extremely episodic. As a result of the Hercynian orogeny erosion averages about 7 km of rock. This is distributed evenly over 100 m.y. in fig. 1 (0.07 mm/y) only because of the lack of data. The climax of orogenic development in the Alps during the Eocene was a period of complete rest in Southern Calabria, but the intercalation of thick olistostromes in the Miocene normal sediment and the high uplift and erosion values (0.8 mm/y) are records of the Apennine orogeny nearby.

Pliocene and Quaternary movements were varied. The last million years show an average rate of erosion of only 0.2 mm/y but an uplift rate of 0.9 mm/y. Since high rates of uplift are necessarily followed by corres-

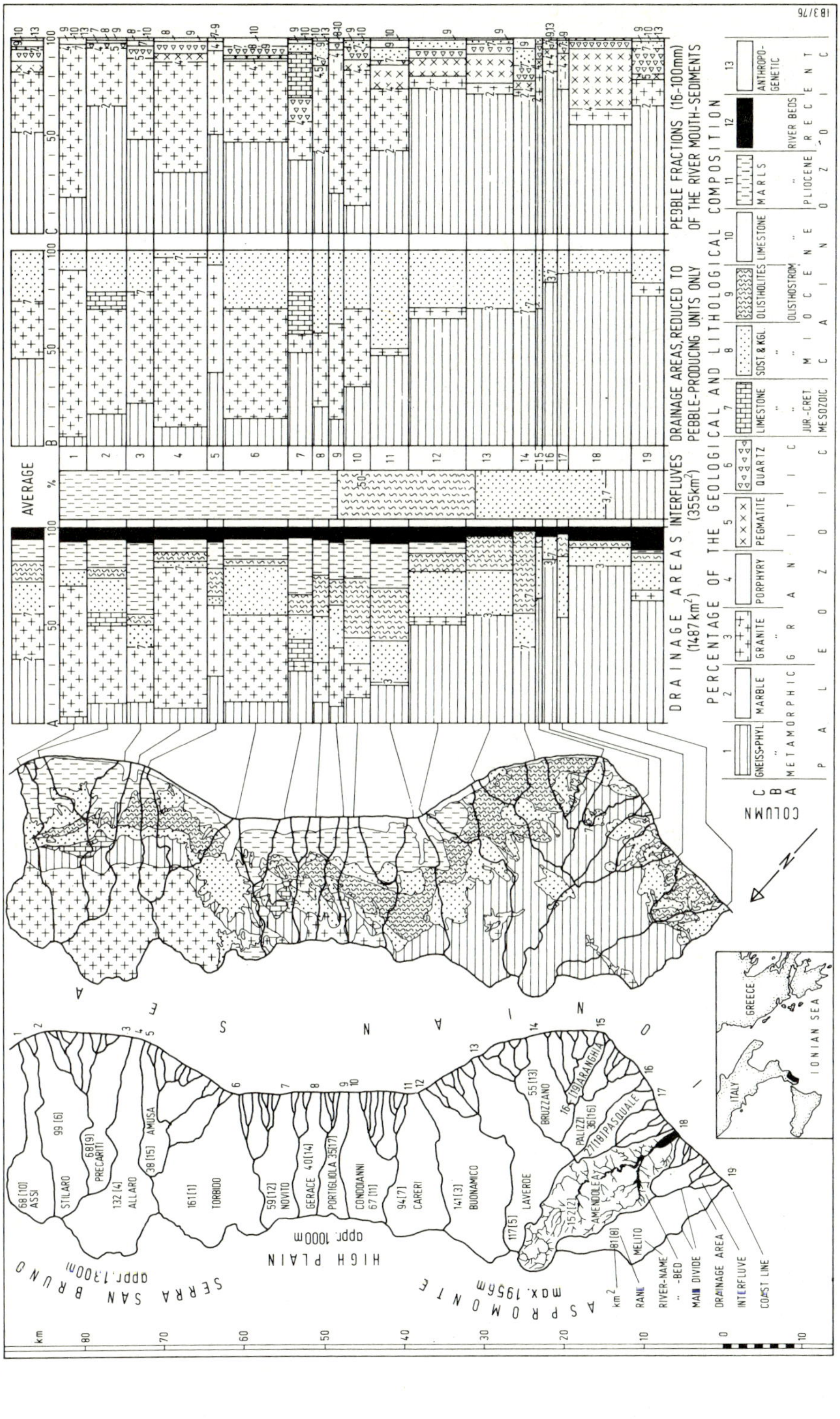
183/76
DRAINAGE AREAS
(1487 km²)
INTERFLUVES
(355km²)
DRAINAGE AREAS, REDUCED TO PEBBLE-PRODUCING UNITS ONLY
PEBBLE FRACTIONS (16-100mm) OF THE RIVER MOUTH-SEDIMENTS
PERCENTAGE OF THE GEOLOGICAL AND LITHOLOGICAL COMPOSITION
AVERAGE
%
COLUMN
A
B
C
GNEISS+PHYL
MARBLE
GRANITE
PORPHYRY
PEGMATITE
QUARTZ
LIMESTONE
SOST & KGL.
OLISTHOLITES
LIMESTONE
MARLS
RIVER BEDS
ANTHROPOGENETIC
METAMORPHIC
GRANITIC
JUR.-CRET
MESOZOIC
OLISTHOSTROM
MIOCENE
PLIOCENE
RECENT
PALEOZOIC
CAENOZOIC
ITALY
GREECE
IONIAN SEA
IONIAN SEA
SERRA SAN BRUNO
appr. 1300m
HIGH PLAIN
appr. 1000m
ASPROMONTE
max. 1956m
68 [10] ASSI
99 [6] STILARO
68 [9] PRECARITI
132 [4] ALLARO
38 [15] AMUSA
161 [1] TORBIDO
59 [12] NOVITO
40 [14] GERACE
35 [17] PORTIGLIOLA
67 [11] CONDOIANNI
94 [7] CARERI
141 [3] BUONAMICO
117 [5] LAVERDE
55 [13] BRUZZANO
[19] ARANGHIA
36 [16] PALIZZI
27 [18] PASQUALE
152 [2] AMENDOLEA
81 [8] MELITO
km²
RANK
RIVER-NAME
-BED
MAIN DIVIDE
DRAINAGE AREA
INTERFLUVE
COAST LINE
km
N

ponding rates of erosion, the extremely high erosion rates obtained for the present (0.5 and 10.0 mm/y) no longer seem so improbable, but special factors such as anthropogenic influence and the great Messina Earthquake of 1908 may play an important part.

References

Borsi, S., Hieke Merlin, O., Lorenzoni, S., Paglionico, A. & Zanettin-Lorenzoni, E., 1976: Boll. Soc. geol.Ital., 95:1-26.

Brogan, G.E., Cluff, L.S. & Taylor, C.L., 1975: Tectonophysics,29:323-330.

Guerricchio,A. & Melidoro,G.,1973: Geologica appl.e idrogeol.,8:315-337.

Krumbein, W.C., 1941: J. Sed. Petrol. 11:64-72.

Lorenzoni, S., Paglionico, A. & Zanettin-Lorenzoni, E., 1976: Boll. Soc. geol. Ital., 95:1-31.

Rumohr, J. & Ibbeken, H., 1975: Rapp.Comm.int. Mer Médit.,23,4a:245-246.

Sijmons, K., 1977: Unpubl. manuscript.

Sneed, E.D. & Folk, R.L., 1958: J.Geol., 66:114-150.

Fig. 10. Correlation of the mean quantitative petrographic composition of the river mouth sediments with the geological composition of the drainage areas.

Part 4

Synthesis

Aspects of the Evolution of the Calabrian Arc

K. Görler and P. Giese

Introduction

The geodynamic development of the Calabrian Arc, like the entire Mediterranean region, is governed by relative movements between the European and African plates, which, in turn, are caused by the different spreading rates and directions of the North and South Atlantic (Pitman & Talwani 1972; Francheteau 1973; Le Pichon et al. 1977; Kristoffersen 1977).

As these processes are still going on, it is reasonable to make the present the starting point for further considerations. Present and recent past also offer the best possibilities to study the course and interaction of the different exogene and endogene processes with their transportation of large geological masses. The first part shall, therefore, be devoted to the present orogenic state (Wunderlich 1965) and to the movements that can presently be observed.

The second part deals with the interpretation of fossil geological structures and the development of the Calabrian Arc from the Cretaceous to the present. Just how speculative all statements on the development of an orogene are is demonstrated by the example of the Calabrian Arc and the adjacent areas. The Ionian Sea, for instance, is seen as an oceanic area on one hand and as a quasi-continental promontory of the African plate, on the other.

The Present Geotectonic State of the Calabrian Arc

The following points are important for critical examination of the present geotectonic state of the Calabrian Arc:

Morphology

a) The strike of the Apennine-Maghrebide Orogene, in Calabria, bends sharply from a NW-SE to E-W course.

b) The young sedimentary basins of the plain of Sibari, the Isthmus of Catanzaro and the Strait of Messina interrupt the Apennine-Maghrebide Orogene perpendicular to the strike.

c) South Calabria, a continental zone with a width of no more than 50 km, is situated between the Tyrrhenian and the Ionian Seas, both having a water depth greater than 3.000 m. Only east of North Calabria, in the Gulf of Taranto, a transition zone from the Ionian Sea to the Bradanic trough and the Molise Basin exists.

d) Off the eastern Calabrian coast lies a *Calabrian Ridge*, with a very irregular relief (Belderson et al. 1974).

e) The Eolian Islands form a volcanic arc.
f) The central Tyrrhenian Sea is an abyssal plane with submarine volcanoes.

Geological Structure

In contrast to the most of the Apennine-Maghrebide, Calabria is characterized by Hercynian or even older magmatites and metamorphites. Generally, when going from east to west, within the Hercynian crystalline there is a continuous transition of rocks from the upper to the lower parts of the crust.

Along the Ionian coast this crystalline is covered by transgressive Mesozoic and, above all, Upper Tertiary sediments.

Calabria is bordered on both sides by submarine troughs with thick young sediments (Selli & Fabri 1971; Finetti & Morelli 1972; Selli & Rossi 1975). While very thick sediments were also revealed in the central Ionian Sea (Finetti & Morelli 1972; Sancho et al. 1973) only a thin sediment cover was detected in the central Tyrrhenian Sea (Selli & Fabri 1971).

In Calabria, as well as in the Bradanic trough and on Sicily, the Upper Tertiary sediments show intercalated olistostromes. Such olistostromes still flow down from the Calabrian and the Mediterranean ridges (Mulder 1973; Mulder et al. 1975).

Although a great number of common features exists in the Calabrian Arc, a detailed study reveals that Southern Calabria differs appreciably from Northern Calabria and Sicily. For example, Southern Calabria shows no intensive deformation during Mesozoic and Tertiary, nor do Liguride Nappes exist south of the Isthmus of Catanzaro.

Gravity Distribution and Crustal Structure

Calabria is situated between the gravity highs of the Tyrrhenian and the Ionian Seas, therefore, appears as a relative minimum (Morelli 1975) which became weaker towards the South. Southern Calabria shows gravity values that are, on an average, 10 mgal higher than those of Northern Calabria, however, due to the proximity of the gravity high of the Ionian Sea. On the whole, the isolines of the Bouguer gravity follow the strike of the Calabrian Arc. A number of local minima are to be found on or immediately before the eastern coast of Calabria, indicating thick young sediments and a thick crust. A number of local anomalies point to lateral inhomogeneities in the crust along the Calabrian Arc. The Strait of Catanzaro is indicated by a minimum across the general strike direction.

Details on the crustal structure of Calabria and the adjacent areas can be gathered from the results of the seismic measurements in these regions (Schütte this vol.). The Calabrian Crystalline shows a crustal thickness of 40-45 km. In the adjacent Southern Apennines, to the north, as well as on Sicily to the west, the same crustal thicknesses have been measured.

From Northern Calabria through the Gulf of Tarent towards Puglia, crustal thickness decreases continuously from 45 to 30 km. It is still unknown how the transition from Southern Calabria to the Ionian Sea occurs, because no seismic crustal data are available. Since a recent active subduction zone beneath the Calabrian Ridge is presumed, the change from continental crust in Southern Calabria to oceanic crust in

the Ionian Sea may be expected in this area.

The shelf region of the Tyrrhenian Sea is characterized by a crust of no more than 20-25 km thickness which continues 20 km inland into Northern Calabria, i.e. the neighbourhood of Cosenza. The change from thin to thick crust is abrupt. In Southern Calabria, although the exact location of this abrupt change cannot be determined because of the lack of data, this transition has to be expected on or immediately before the western coast.

An unusual feature is present in the Sila. Here, a weakly indicated crust/mantle boundary, which disappears in the E, can be proved at a depth of about 20 km. However, its relation to the shallow crust/mantle boundary of the Tyrrhenian shelf is unknown.

Seismicity and Neo-Tectonics

Calabria is characterized not only by catastrophic shallow earthquakes but is also one of the few European areas with deep earthquakes. Most authors (Peterschmitt 1956; Ritsema 1970, 1975; Morelli et al. 1975) interpret the deep earthquakes west of Calabria as belonging to a Benioff plane dipping NWN with 55°. But according to Ritsema (1970) "there is a clear gap in which up till now no earthquake foci have been detected from 220 km to roughly 100 km depth. One earthquake at a depth of about 450 km is likewise separated from the main body of data by a depth interval of about 100 km". This discontinuity of the earthquake foci is also emphasized by Meszaros & Hedervari (1974) and Schick (this vol.).

According to Wilson (1973), a passive, merely gravitational subduction seems probable in such a case, however, after the termination of a subduction, for instance after a collision, such tearing of the downgoing plate must also be expected. The shallow earthquakes show different fault plane mechanisms (Schick this vol.).

For South Calabria morphological and stratigraphic data indicate intensive neo-tectonism (Brogan et al. 1975; Ergenzinger et al. this vol.). Throughout the entire Tertiary sediment sequence along the eastern Calabrian coast it can be seen that the eastward inclination increases with age. Thus an eastward tilting of Calabria since Oligocene/Miocene time is inferred (Görler this vol., fig.2). It obviously is still taking place in connection with a rapid uplift of the central parts of Calabria and intensive fracture tectonics to the west and south.

Volcanism

According to Barberi et al. (1973), the K-content and the Sr-isotope proportion of the Eolian Islands (volcanic rocks) are caused by subduction and melting of oceanic crust. Because of the large cation and the lead-isotope distribution, it is even possible that they originated from mantle material (Barberi et al. 1974).

The composition of the basaltic rocks of Etna suggests as origin upper mantle material. There is strong evidence that basaltic volcanism is linked with the line of normal faulting in Eastern Sicily and Calabria (Schick this vol.).

According to Selli & Fabri (1971) several of the NS elongated central Tyrrhenian Seamounts are fissural volcanoes or typical oceanic tholeitic lavas (Del Monte 1972; Selli 1974).

Conclusions Concerning the Present Geotectonic State of the Calabrian Arc

The distribution of earthquakes and recent andesitic volcanism of the Eolian Islands implies a still active subduction zone starting off the eastern coast of Southern Calabria dipping at about 55° against the Tyrrhenian Sea. The Bouguer gravity (Morelli 1975) as well as the heat-flow (Loddo & Mongelli 1975; Loddo et al. 1976) are in agreement with such an active subduction zone. This subduction zone seems to be exactly limited to the region where Calabria borders directly on the Ionian Sea with its oceanic crust (Görler this vol.).

In conclusion this narrow strip of continental crust of Calabria can be seen as a non-volcanic arc, the Eolian Islands as a volcanic arc and the Tyrrhenian Sea as a back arc basin (Boccaletti & Guazzone 1972; Barberi et al. 1973; Boccaletti et al. 1976).

In comparison with other active subduction zones, for instance the Pacific, the Southern Calabrian subduction shows some peculiarities. With a length of only about 100 km, this zone is relatively small when one considers that the greatest known foci depth there is 450 km. Probably here, only a rest of a once much more extensive subduction zone exists. In the same way, the Calabrian Arc must be regarded as the rest of a formerly larger island arc.

Usually a subduction zone is connected with a trench but here off the SE-coast of Calabria, a well-defined Calabrian Ridge could be detected (Belderson et al. 1974). Similar to the Barbados Ridge (Chase & Bunce 1969; Westbrook 1975) the downgoing plate is covered by thick sedimentary ridge built up by the stacking of wedge-shaped sediment slices (Accretionary wedge).

A Model of the Geodynamic Evolution of Southern Italy

The attempt to reconstruct the geodynamic history of Calabria gives, at first, rise to the question of when the subduction, still active in the Calabrian Arc, began.

A first clue is given by the depth of the deepest earthquake foci at about 450 km. With a dip angle of 55°, a length of 550 km for the downgoing slab can be calculated. Assuming a subduction velocity of 2 cm/yr this subduction occurred over a period of 25 m.y. (Upper Oligocene/Lower Miocene). The first occurrence of olistostromes, intercalated in the normal sediments of the same age in Eastern Calabria and Sicily, supports the assumption that the subduction process began at that time.

The tilting of Southern Calabria began at about the same time (Görler this vol.) and was probably also caused by the beginning subduction.

The occurrence of volcanic components within the Calabrian Arc sediments indicates a new period of subduction activity. Until now such volcanics of possibly Oligocene or Miocene origin were known only from psammites in Sicily. The study of Upper Oligocene and younger conglomerates of Southern Calabria (Ergenzinger et al. this vol.) revealed the existence of andesitic and dacitic rocks of unknown age in almost all psephites. The source area for these volcanics must be assumed to have been situated outside the present Calabrian Arc, because no feeding channels for the volcanics are known. In conclusion Wezel (1977) stated that "the composition and petrochemistry indicate that source terrains of the andesitic detritus were located on the concave side of an arc within the

Tyrrhenian (or the Alboran) sialic blocks. At present the only Oligo-Miocene volcanoes are found in western Sardinia. Present distance of the Sardinian eruptive centers varies from 300 km (central Apennines) to about 500 km (Sicily). However, these distances resulted from a later crustal lengthening caused by the continental distension of the Tyrrhenian block (Selli & Fabri 1971)".

A further sign that Calabria must have once belonged to a larger continental mass is given by the occurrence of extremely mature limnetic Jurassic conglomerates at the base of the Mesozoic sediments in Southern Calabria.

Hardly any palaeomagnetic data from the Calabrian Arc are available and palaeomagnetic data from the Calabrian Massif itself are non-existent. Such data from the neighbouring limestone platforms, however, offer some interesting results: Since Senonian time the Abruzzi-Campania Platform rotated 32°-36° (Channel & Tarling 1975; Catalano et al. 1977) and the Campania-Lucania Platform about 73° (Manzoni 1975; Catalano et al. 1977) both counter-clockwise relative to stable Europe. In contrast thrust sheets in northwestern Sicily show Post-Cretaceous clockwise rotations of 30°-60° (Schult 1973, 1976, this vol.; Barberi et al. 1974; Catalano et al. 1977).

The palaeomagnetic data from southeastern Sicily prove the connection of this area with the African plate (Schult 1973, 1976, this vol.; Barberi et al. 1974; Channel & Tarling 1975; Gregor et al. 1975).

The following fig. 1 shows a possible palaeogeographic reconstruction of the geodynamic evolution of the area under study.

This reconstruction is based on the following:

1. The geotectonic development of the Mediterranean region since the Mid-Cretaceous is governed primarily by the approach towards one another of the African and the European plates.

2. This approach causes the development of subduction zones.

3. If the velocity of these two advancing plates does not reach the minimum required, for instance, for an unhindered subduction from gravity, friction, and thermal convection, an external shifting of the subduction zone occurs. This leads to breaking-off of continental fragments from the European or African block and thus to the formation of island arcs.

Between the shifting fragments on one side and the original block on the other side, a back arc basin develops.

4. The collision of such an island arc with a continent or another island arc causes extensive nappe thrusting and a simultaneous dying out of the subduction. Further extensive subduction of continental crust does not take place.

5. If one subduction system had died but there is still oceanic crust available, a new system can develop.

In our model these rules for plate tectonics are combined with geological and geophysical data:

1) The movement of Africa relative to Europe since the Cretaceous is determined by analysis of the sea floor spreading history of the Atlantic Ocean (Pitman & Talwani 1972; Francheteau 1973; Le Pichon et al. 1977; Kristoffersen 1977).

2) The palaeomagnetic data of Central and Southern Italy were compiled by Soffel (this vol.), Catalano et al. (1977) and Van den Berg (1977). Only data from platforms are used because the intercalated basins must, in most cases, just be considered mobile zones.

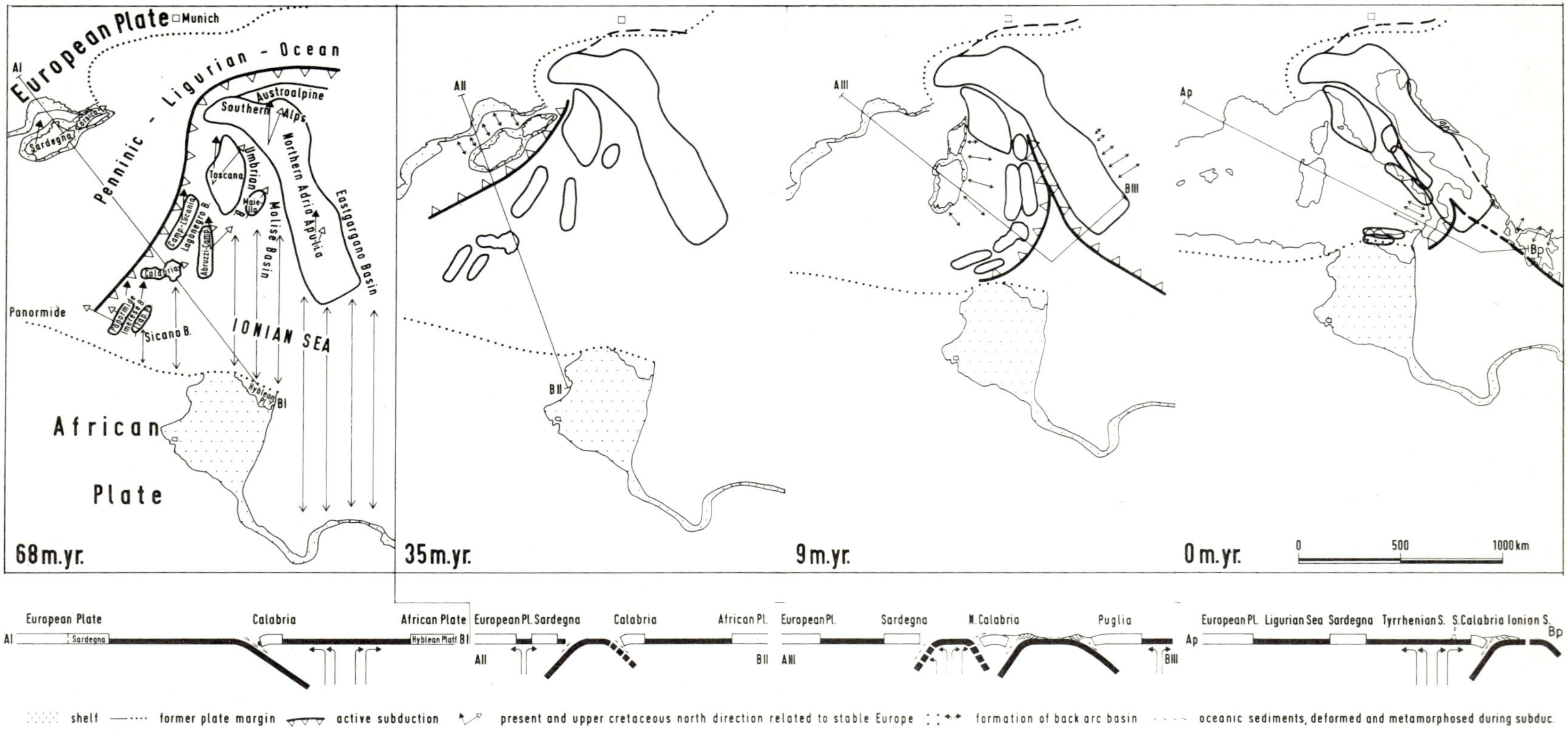

Fig. 1. Tentative reconstruction of the geodynamic evolution of the central Mediterranean region from Upper Cretaceous to present. Based on published models for the Africa displacement with respect to Europe (Pitman & Talwani 1972; Kristoffersen 1977; Biju Duval et al. 1977) and the available paleomagnetic data (s. text).

3) The reconstruction uses common paleogeographic ideas concerning the relative position of the several platforms and basins (e.g. Catalano et al. 1977). The only greater disagreement occurs with regard to the palinspastic position of the Calabrian platform of which no palaeomagnetic data are available.

Smith (1971), Boccaletti & Guazzone (1972, 1975), Dewey et al. (1973), Biju-Duval et al. (1977) started with partly similar assumptions but obtained different results. The closing of the Thetys in the central Mediterranean area took place in three main phases each of which was characterized by different subduction zones, island arcs, and marginal seas.

Phase 0 or Rifting Phase

Almost all authors agree that, after an initial phase of distension tectonics during the Triassic, the opening of the Penninic-Ligurian Ocean started in the Lower Jurassic. This opening phase continued up to the Middle Cretaceous, followed by the approach of the African and European plates.

Phase I or Alpine Phase (Middle Cretaceous - Upper Eocene)

Because of the different spreading rates and directions in the Atlantic Ocean, the two plates began to move towards one another, which in turn created a subduction dipping under the African Plate. This led to the splitting off of parts of the African Shelf and the formation of new small oceanic basins.

Just before the splitting off of the continental fragments, the African shelf had shown fracturing in basins and platforms, proved by different facies. A number of the split-off continental fragments were of remarkable size. Different rotations and drifts can be partially reconstructed by means of palaeomagnetic and geological data. Present conditions in the Pacific area indicate that these split-off elements may have been rearranged to form an island arc. A core (Austroalpine, Po-Plain, Puglia platform, East Gargano Basin) of this Island Arc I followed a rather restricted course and, therefore, can be considered a microcontinent, the "Adriatic microplate" (Giese & Reutter this vol.).

Rapid drifting towards the North during the Upper Cretaceous is proved by palaeomagnetic studies on rocks of the Umbrian basin and the Southern Alps (Van den Berg 1977).

Differential movements among various fragments of this Island Arc I are documented by palaeomagnetic data. Especially during the Upper Cretaceous the Umbrian Basin and the Southern Alps rotated in the same direction but at different rates (Van den Berg 1977).

The smallest rotations related to the African Plate are documented for the Puglian Platform, the Eastern Gargano Basin and the Southern Alps (Channel & Tarling 1975; Van den Berg 1977; Catalano et al. 1977). The strong change of inclination of the Adriatic microplate during Upper Cretaceous implies a large movement towards the European block. As a consequence of these northwards directed movements at the rear of this micro-continent a new back arc basin (Marginal Sea I), the Ionian Sea, was opened. This idea is supported by the observation that the oldest rocks in the olistostromes, coming from the Ionian Sea, are Cretaceous in age.

Towards the end of phase I the most northern part of Island Arc I (Southalpine, Austroalpine) collided with the European Plate. This collision provoked the Alpine orogeny and stopped the African-dipping subduction I as no further oceanic crust was available.

Phase II or Northern Apennines Phase (Upper Eocene to Lower Miocene)
The state in the beginning of phase II corresponds approximately to fig. 1 b. Only in the southwestern continuation of the Alps, a remainder of the Penninic-Ligurian geosyncline which becomes broader in the SW still exists (Reutter et al. this vol.). For this reason, eugeosynclinal series metamorphosed during the Mid-Cretaceous to Upper Eocene subduction can be expected, as well as still subductible oceanic crust. The series compiled in the "Structural Model of Italy" as "Liguride Complex" are interpreted as originated from this geosyncline remainder. Phase II was characterized by a European-dipping subduction II within this geosynclinal remainder. Consequences were the splitting off of Corsica and Sardinia (Island Arc II) from the European Plate, the formation of the Ligurian Sea (Back Arc Basin II) and, finally, the collision of the Corso-Sardinian Block with the western part of Island Arc I (Toscana, Campania-Lucania, Calabria). This collision was responsible primarily for the nappe structure in the Northern Apennines (Reutter et al. this vol.) but also the Lower Miocene overthrusts in the Southern Apennines and Northwest Calabria. According to the available data (Amodio-Morelli et al. 1976; Carrara & Zuffa 1976) rocks metamorphosed during Subduction I were involved.

According to this interpretation, it is only logical that rocks of the Ligurian geosyncline are to be found in the S only as far as the Isthmus of Catanzaro. Beyond this region no collision took place, because south of Sardinia Island Arc II ended abruptly, and, therefore, no possibility for overthrusting of Liguride material existed.

Phase III or Southern Apennines Phase
(Upper Oligocene/Lower Miocene to present)
After the closing of the Ligurian-Penninic synclinal remainder during phase II, the subduction jumped to the SE-side of the western part of Island Arc I still dipping in the same direction. The subduction currently active in the Calabrian Arc has to be regarded as a remainder of Subduction IIIa. At present, a second active subduction zone (IIIb) exists in the Hellenic Arc. These two active subduction zones are morphologically characterized by the Calabrian Ridge and the fragmentary Calabrian Trench on one side and the Eastern Mediterranean Ridge and the Hellenic Trough on the other side. These ridges merge south of the Gulf of Taranto (Belderson et al. 1974) and continue northward as the Molise Basin.

The Hellenic Trough continues as the Taranto Trough and further northward as the Bradanic Trough.

This interpretation suggests that formerly subduction of the Hellenic Arc continued further N at the SW side of the eastern part of Island Arc I (fig. 1). Because of the shape of Island Arc I, which tapered towards the N, the possibilities of outward migration near the European Plate for Island Arcs IIIa and IIIb before arc-arc-collision were very limited. In this model the southern Adria has to be seen as a marginal sea (IIIb) which did not have the chance to develop beyond the earliest stage, and the Apulian Platform as part of an Island Arc IIIb. The strange morphology and the high gravity and heat flow values (Morelli 1975; Loddo & Mongelli 1975) of the Southern Adria and Northeastern Apulia support this hypothesis. From N to S the hypothetical line of collision between Island Arcs IIIa and IIIb, would then be located in the region of Umbria (?), the Molise Basin and the Gulf of Taranto. Finally, Subduction IIIa seems to have predominated Subduction IIIb, as the existence of the Tyrrhenian Back Arc Basin demonstrates. Also in accordance with the crustal seismic

profiles of Campania and Northern Calabria (fig. 3), crust of the Apulian platform can be found today under Island Arc IIIa.

Geophysical and geological data indicate that Subduction IIIa probably started as early as the Upper Oligocene before Subduction II was entirely extinguished. Subduction IIIa caused the separation of the Western part of the former Island Arc I (Toscana, Campania-Lucania and Calabria) from Corsica and Sardinia. With this separation the development of the Tyrrhenian Sea began (Marginal Sea IIIa). Meanwhile in the Molise Basin, the Ionian Sea and the Sicano Basin, subduction, compression; and stacking of a sedimentary ridge were taking place. In the Ionian Sea these processes are in part going on today.

Olistostromes and gravity nappes slid down the above mentioned ridge. In the "Structural Model of Italy", they are interpreted as "Sicilide", "Antisicilide" and "late-orogenic Apennine flysch". By determining the direction of flow in olistostromes it was shown that they originated in the Molise basin, Ionian Sea, and Sicano Basin themselves.

During the period from Upper Miocene to Lower Pliocene, Island Arc IIIa collided with Island Arc IIIb in the northern part of Subduction Zone III. These two Island Arcs consisted of remobilized elements of Island Arc I. This collision caused the main orogenic phase in the Southern Apennines to start in the north and then wander southward. This can be seen in the crustal profiles of Northern Calabria and Campania determined from explosion seismics (Giese & Reutter this vol.; Schütte this vol.).

According to Wezel (1975), the collision in Sicily between the "Calabro-Peloritanian Block" and the "North African Block" took place during the "Tectonic Phase of the Langhian", as movements of flysch series and varicoloured clays either towards the foreland or partly towards the Peloritani Massif can be observed during this period. These features are considered the result of active subduction before the collision. The collision itself is believed to have taken place during the compressive phase of the Upper Miocene.

With these collisions the subduction in the region of the Southern Apennines and Sicily came to a standstill. Only Southern Calabria had not yet collided with a continental foreland, and for this reason, subduction is still active here. Some peculiarities in the geological structure of Southern Calabria can be explained by this fact. This applies, above all, in contrast to the activity in Sicily and Northern Calabria - to the abnormally quiet conditions during the Mesozoic and Tertiary as documented by the facies and lack of deformation of the sediments (Görler this vol.).

Events of Phases I and II in Southern Calabria are not known. Just as in the Southern Apennines, Northern Calabria, and Sicily a trace of phase III exists in the form of olistostrome intercalations in the normal Miocene sediments, whereas the collision of the Southern Apennines and and Sicily with their respective forelands resulted in a strong compression with large overthrusts. This has yet to occur in Southern Calabria.

The crustal structure of Northern Calabria presents a special problem. According to the results of refraction seismics (Schütte this vol.), a local crustal doubling in the Sila exists. Indeed, most authors assume a large crystalline nappe here, the "Calabrides" (Quitzow 1935).

The latter generally interpreted as being an extremely internal element which coming from the far west, overthrust the Apennine sequences. The crustal profile in connection with other geological data makes the

explanation of the Sila crustal doubling, shown in the sequence of movements in fig. 2 ,plausible. It is supposed that, during the final stage of subduction IIIa, caused by the collision with the Apulian Platform, the Calabrian Island Arc fragment broke into two pieces, the edge of the western segment being overthrust by the eastern one. Increasingly deeper parts of the crust are revealed in outcrops along an E-W line as far as the tectonic contact with the slightly metamorphosed Ligurides. According to this interpretation the Calabride Complex is not seen as an extremely internal alpine element, rather as having occupied a palaeogeographic position which corresponds roughly to the Campania-Lucania platform and the Panormide platform. A neritic limestone sequence, related to the aforementioned elements, is known to exist in the NW under the Ligurides, which were thrust up during collision II, as well as in Eastern Calabria for instance near Stilo and Gerace as remainders of a transgression over the Hercynian Crystalline.

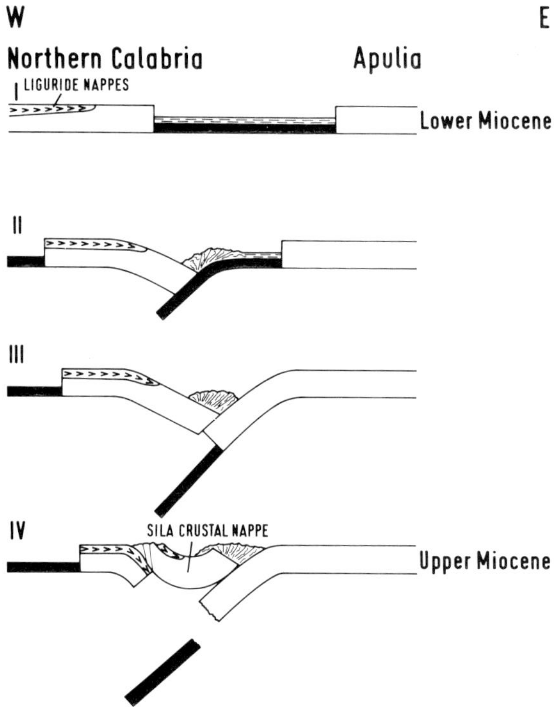

Fig. 2. Sequence of profiles illustrating the proposed model of development of the crustal structure in northern Calabria. Based on all available data (s. text).

The interpretation of the Hercynian crystalline in the Calabrian Arc as a large nappe originating in the far west seems very unlikely since, according to current seismic investigations, this crustal nappe under the Sila shows a thickness of 20 km. Considering the rapid erosion in Calabria this crustal nappe must have been even thicker at the time of overthrust. The degree of metamorphism of the internal units of NW Calabria is much too weak to be explained by the overthrusting of a crystalline nappe of such an enormous thickness.

This interpretation does not clarify why, in Calabria and North-eastern Sicily, such a rapid rising and erosion, at some points down into the

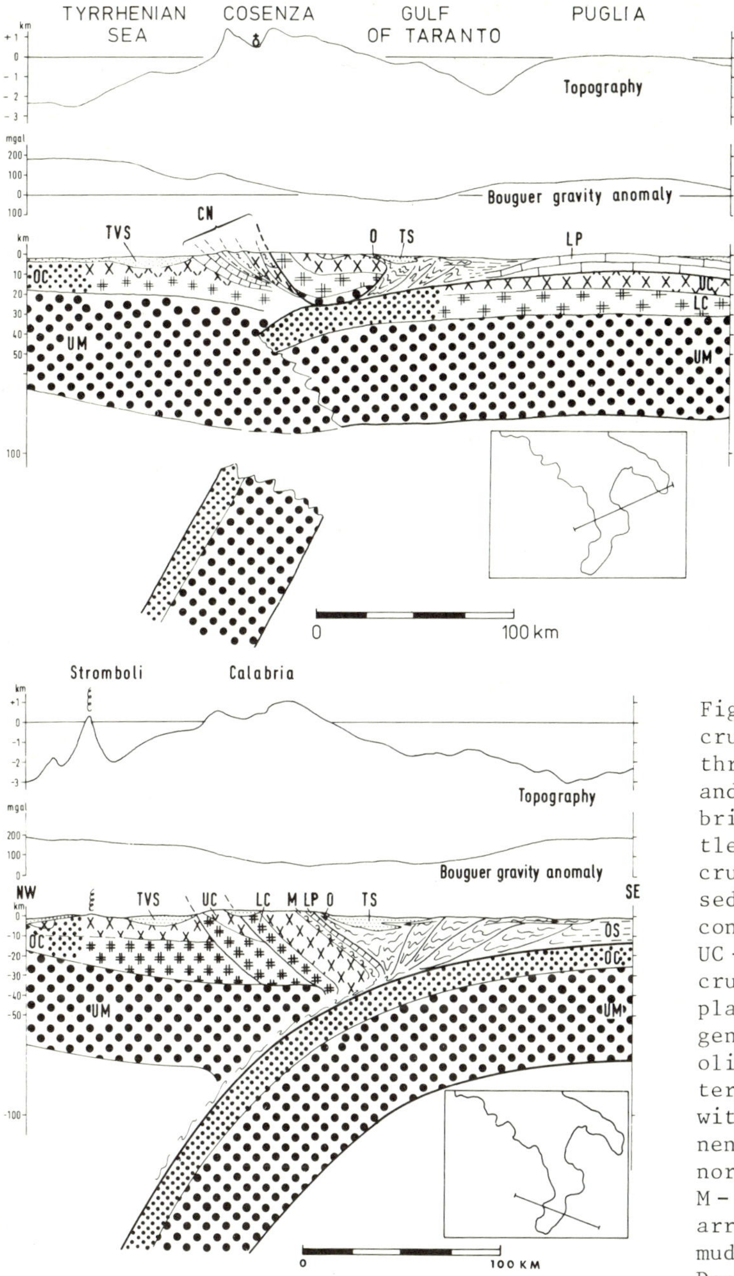

Fig. 3. Schematic crustal sections through Northern and Southern Calabria. UM - upper mantle, OC - oceanic crust, OS - oceanic sediments, LC - lower continental crust, UC - upper continental crust, LP - limestone platform, TS - terrigenous sediments, O - olistostromes, TVS - terrigenous sediments with volcanic components, CN - nappes of northwestern Calabria, M - metamorphites, arrows indicate the mudflow directions. Depth of Moho according to Giese & Morelli (1975); Ahmed (1976); Schütte (this vol.). Thickness of sediments based on Finetti & Morelli (1972, 1973); Selli & Fabri (1971); Selli & Rossi (1975); Sancho et al. (1973). Inclination of Benioff plane according to Ritsema (1970); Barberi et al. (1973). Gravimetry according to Morelli (1975).

lower crust, has occurred since Oligocene-Miocene. The appearance of deep crustal rocks in the zone of maximum bending is known also from other arc systems. The relatively young opening of the sphenochasms of the Strait of Messina and the Gulf of Squillace is evidence of the unusual mechanical stress in this region.

Conclusions and Probable Future Evolution of the Central Mediterranean Region

According to the proposed speculative model of plate tectonic evolution in the central Mediterranean during the last 70 m.y., there do not exist any remainders of primary Tethys oceanic crust. The existing oceanic basins are, created by repeated island-arc activity, brought about by the advance of the African and European plates towards one another from Middle Cretaceous time up to the present. During this process old oceanic crust was subducted several times and new marginal seas were created. Therefore, the area of sedimentation between Africa and Europe decreased continuously.

The smaller the area of sedimentation becomes, the more obvious it is that the events in the Mediterranean differ from those in the Pacific. With the advance of Africa toward Europe and the decrease of the area of sedimentation the rate of sedimentation of clastic sediments increases. The process is even accelerated by the increase of relief intensity, due to the tectonic activity caused by the movement of the African and European plates. Indeed, there seems to be a tendency of the sedimentation rates to increase with time. Reflection seismic investigations, too, revealed enormous thicknesses of sediments (e.g. Finetti & Morelli 1973; Sancho et al. 1973). It is probably due to this particular feature of the Mediterranean that here, over subduction zones, instead of a trench, a sedimentary ridge can be built up by stacking of such thick sedimentary masses. Consequently, large submarine mudstreams occur flowing down both sides of these sedimentary ridges (Görler this vol.). In accordance with this model many olistostromes are intercalated especially in Upper Tertiary sediments.

Even the rare occurrence of vulcanites could be caused by the fact that the products of volcanic activity are hidden by the intensive sedimentation. If the present development continues, even the behaviour of oceanic crust, in spite of being covered by extremely thick sediments, approaches that of a continental one. Recent examples are the Ionian Seas and the Nile Cone. The southern Adria could probably be included here, as well.

Looking into the near geological future, it seems likely that the region of arc-arc collision will continuously advance into the central Ionian Sea through the outward migration of the Calabrian and Hellenic Arcs, until finally, the Hellenic Arc will collide with the African plate. Ritsema (1975) estimates that this will happen within 3-5 m.y. At that time the oceanic crust of Marginal Sea I will again be totally subducted.

Then, at the absolute latest, a total reorganization of the subductory system in the central Mediterranean would have to be exptected, always under the assumption that Africa and Europe continue to approach each other. If the processes continue in their usual way there may be a new Africa-dipping subduction within the Marginal Seas II and III, forming new island arcs, new marginal seas, and new collisions. Theoretically,

	Phase 0 Rifting Phase Triassic - M. Cretaceous	Phase I Alpine Phase M.Cretaceous - U.Eocene	Phase II Northern Apennines Phase U.Eocene - L.Miocene	Phase III Southern Apennines Phase U.Oligocene - present
Rifting or subduction	Opening of the Penninic-Ligurian Ocean	African dipping subduction I within the Penninic-Ligurian Ocean	European dipping subduction II within the residual Penninic Ligurian Eugeosyncline	European dipping subductions within the Marginal Sea I beneath the western (IIIa) and eastern branch of Island Arc I
Elements of the developing Island Arc	---	Island Arc I: East Gargano Basin, Apulian Platf.,Austroalpine, Southalpine Umbrian Bas., Toscana,Molise Bas., Maiella Platf.,Abruzzi-Campania Platf.,Lagonegro Bas.,Campania-Lucania Platf.,Calabrian Platf.,Sicano Bas.,Trapanese Platf., Imerese Bas.,Panormide Platf.	Island Arc II: Corsica, Sardinia	Island Arc IIIa: Toscana (?), Campania-Abruzzi Platf., Maiella Platf.,Lucania-Campania Platf., Calabrian Platf., Panormide Platf.,Trapanese Platf. Island Arc IIIb: Apulian Platform,Eastgargano Bas.,Crete
Creation of new oceanic regions	Penninic-Ligurian Ocean	Marginal Sea I: Ionian Sea and parts of Molise Basin, Lagonegro Basin, Imerese Basin, Sicano Basin	Marginal Sea II: Ligurian Sea	Marginal Sea IIIa:Tyrrhenian Tyrrhenian Sea Marginal Sea IIIb: Southern Adriatic Sea, Aegean Sea
Assumed movements of the Calabrian Platform rel. to stable Europe	As part of African Plate	NW-drift, counterclockwise rotation	stationary	SE-drift, counterclockwise rotation
Orogenic events in Calabria		greenschist metamorphism of Liguride units found in Northwest Calabria	Europe-vergent overthrusting of Liguride units,metamorphosed during phase I above NW.Calabrian Platf.Metamorphism of the overridden limestone sequences in NW Calabria	Main compressive phase in the S.Apennines,N Calabria and Sicily with Africa-vergent overthrusts.Mudflow transport of varicol.clays from Molise Bas. Ionian Sea and Sicano Bas. towards the bordering Platf.

this to and fro migration of island arcs could continue until Europe and Africa would finally be united as one plate and, perhaps, only small remainders of oceanic crust covered by thick sediments and lying between complicated orogenes will mark the former suture.

References

Ahmed, H., 1976: unpubl. Frankfurt.

Amodio-Morelli, L., Bonardi, G., Colonna, V., Dietrich, D., Giunta, G., Ippolito, F., Liguori, V., Lorenzoni, S., Paglionico, A., Perrone, V., Piccarreta, G., Russo, M., Scandone, P., Zenettin-Lorenzoni, E. & Zuppetta, A., 1976: Mem. Soc. geol. Ital., in press.

Barberi, F., Civetta, L., Gasparini, P., Innocenti, F., Scandone, R. & Villari, L., 1974: Earth and planet. Sci. Letters, 22:123-132.

Barberi, F., Gasparini, P., Innocenti, F. & Villari, L., 1973: J. geophys. Res., 78(23):5221-5232.

Barberi, F., Innocenti, F., Ferrara, G., Keller, J. & Villari, L., 1974: Earth and planet. Sci. Lett., 21:269-276.

Belderson, R.H., Kenyon, N.H. & Stride, A.H. 1974: Nature, 247 (5441): 453-454.

Biju-Duval, B., Dercourt, J. & Le Pichon, X., 1977: Int.Symp. struct. Hist. mediterr.Basins (Biju-Duval, B. & Montadert, L., eds.), Editions TECHNIP, Paris.

Boccaletti, M. & Guazzone, G., 1972: Mem. Soc. geol. Ital., 11:201-216.

- , 1975: in: Squuyers, C.H. (Ed.): Geology of Italy. Earth Sci. Libyan arab. Rep.:143-163, Tripoli.

Brogan, G.E., Cluff, L.S. & Taylor, C.L., 1975: Tectonophysics,29:323-330.

Carrara, A. & Zuffa, G.G., 1976: Bull. geol. Soc. Amer.,87:1229-1246.

Catalano, R., Channel, J.E.T., D'Argenio, B. & Napoleone, G., 1977: Mem. Soc. geol. Ital., 15:95-118.

Channel, J.E.T. & Tarling, D.H., 1975: Earth planet. Sci. Lett.,25:177-188.

Chase, R.L. & Bunce, E.T., 1969: J. geophys. Res., 74:1413-1420.

Del Monte, 1972: Giorn. Geol. 38, 1:231-252.

Finetti, I. & Morelli, C., 1972: Boll. geofis. teor. appl. 14(56):291-342.

- , 1973: Boll. Geofis. teor. appl. 15(60):263-341.

Francheteau, J., 1973: In: Implication of Continental Drift to the Earth Sciences (ed. Tarling, T.H. & Runcorn, S.K.), 1, Academic Press: 197-202, London.

Gregor, C.B., Nairn, A.E.M. & Negendank, J.F.W., 1975: Geol. Rdsch., 64: 948-958.

Le Pichon, X., Sibuet, J.C. & Francheteau, J., 1977: Tectonophysics, in press.

Kristoffersen, Y., 1977: Earth planet. Sci. Lett., in press.

Loddo, M. & Mongelli, F., 1975: Boll. Geofis. teor. ed appl., 18:115-122.

Loddo, M., Mongelli, F. & Roda, C., 1976: Nature Physical Science, 244: 91-92.

Manzoni, M., 1975: Geophys. Res. Lett., 2:427-429.

Mészáros, S. & Hédervári, P., 1974: Ann.Geofis., 22:367-380.

Morelli, C., 1975: Sonsiglio naz. Ric.Quad. "la Ricerca sci." 90 (Structural Model of Italy):427-447.

Morelli, C., Cornea, I. & Lazarescu, V., 1975: Boll. Geofis. teor. ed appl. 18:169-178.

Mulder, C.J., 1973: In: Drooger (Ed.): Messinian Events in the Medi-

terranean:44-59, Koninklijke Nederlandse Akademie von Wetenschappen, Amsterdam.
Mulder, C.J., Lehner, P. & Allen, D.C.K., 1975: Geol. en Mijnb., 54 (3-4): 208-221.
Quitzow, H.W., 1935: Abh. Ges. Wiss. Göttingen phys. Kl. (3), H.13: 63-179.
Peterschmitt, E., 1956: Ann. Geofis., 9(3): 303-334.
Pitman, W.C. & Talwani, M., 1972: Bull. geol. Soc. Amer., 83:619-646.
Ritsema, A.R., 1970: XII. Assem. gén. Comm. seismol. europ: 1-5, Luxembourg.
- , 1975: Progr. in Geodynamics, 13:1-12.
Sancho, J., Letouzey, J., Biju-Duval, B., Courrier, P., Montadert, L. & Winnock, E., 1973: Earth and planet. Sci. Lett., 18:189-204.
Schult, A., 1973: Earth and planet. Sci. Lett., 19:97-100.
- , 1976: Earth and planet. Sci. Lett., 31:454-457.
Selli, R. & Fabri, A., 1971: Rend. Cl. Sci. fis., math. e nat. Accad. naz. Lincei (VIII), 50:104-116.
Selli, R. & Rossi, S., 1975: Rapp. Comm. Int. Mer Médit. 23,4a:115-116.
Smith, A.G., 1971: Bull. geol. Soc. Amer., 82:2039-2070, Boulder, Col.
Van den Berg, J., 1977: Mem. Soc. geol. Ital., 15:83-90.
Westbrook, G.H., 1975: Geophys.J. r.astron. Soc., 43:201-242.
Wezel, F.C., 1977: Int. Symp. Structural History Mediterranean Basins. Split 25-29 Oct.1976. Biju-Duval, B. & Montadert, L. (Eds.), Editions TECHNIP, Paris:287-302.
Wilson, J.Z., 1973: Tectonophysics, 19:149-164.
Wunderlich, H.G., 1965: Tectonophysics, 1:495-516.

Chapter 5

Aegean Region (Geotraverses IIIc and IV)

Investigations in the Aegean Region: an Introduction

V. Jacobshagen, K.-O. Kopp and J. Markis

One of the major aims of the German geodynamics program in the Aegean area was to establish a sound knowledge of the physical state of this zone and combine it with the geological information to an evolution model.

The geophysical data available during our planning stage were fairly scarce and consisted mainly of earthquake studies. From early publications it was clear that several similarities exist between the Hellenic arc and the Pacific island arcs:

- the *hypocentral distribution* of the earthquakes in the Hellenic arc pointed to the existence of a Benioff zone dipping towards the interior of the arc (Caputo et al. 1970);
- the *Cretan island arc* is internally accompanied by an active volcanic arc;
- the *gravity data*, located in marine areas only, were too thinly distributed to permit any reliable conclusions; they were, however, interpreted in such a way that it more or less fitted the above given picture of an island arc (Wunderlich 1968, 1969);
- *fault plain solutions* of some Greek earthquakes were interpreted as indicating a southwesterly drift of the Aegean sea (Ritsema 1969).

This situation made it obvious that a fundamental geophysical study including gravity and magnetostatic surveys, deep seismic soundings, high precision seismicity observations as well as high precision gravity nets were of utmost importance. In order to achieve such an extensive operating plan, close cooperation among many German and Greek institutions had to be established. Under the coordination of the University of Hamburg more than 6 German geophysical institutes worked jointly with the Geological Survey of Greece, the Greek Navy and Army and the Universities and Technical Universities of Athens and Thessaloniki in surveying the entire country gravimetrically and magnetically and in performing 4 deep seismic sections. The Seismological Service of Greece and the University of Frankfurt completed a high precision seismicity study of the Peloponnesos. The Technical University of Hannover, together with the Technical University of Thessaloniki, established a high precision gravity net.

The evaluation of all these studies helped in developing crustal models of the Hellenides and the Aegean sea, in establishing facts about the physical state of the matter below the Aegean crust and in reliably locating the Benioff zones. Residual travel-time anomalies were also studied.

Regarding Aegean geology, a palaeogeographic and tectonic conception of the Hellenides was already drawn out by C. Renz (1940, 1955) and later on developed mainly by French authors (e.g. Brunn 1956; Aubouin 1959; Celet 1962; Dercourt 1964; Mercier 1966; Godfriaux 1970). New geotectonic

outlines were presented by Aubouin et al (1963, 1970). Aubouin (1965) had also deduced from them a general scheme of a geosynclinal development.

According to Philippson (1959) the term "Aegean region" ("Ägäis") includes the whole Greece, the Aegean Sea, and Western Anatolia. The Hellenides may be defined to be the Alpidic mountain ranges between the Apulian/pre-Apulian platform (foreland) and the crystalline complexes of the Serbomacedonian Massif and the Rhodope Mountains respectively (hinterland). To the Dinarides in the northwest, a geotectonic limitation may be drawn with the transverse Zone of Skutari-Peć according to Kober (1929) whereas to the Taurides of Anatolia no geological boundary exists.

A better understanding of the Hellenide evolution depends upon a more precise knowledge of the

- palaeogeography of the depositional areas
- structural pattern of the Hellenide belt
- ages of tectonogeneses and of flysch deposition
- age and mode of emplacement of ophiolites
- age, character and distribution of metamorphism and related plutonism
- chemical properties, distribution pattern and age of subsequent volcanism
- late- and post-orogenic sedimentation and tectonics.

These aspects were predominantly considered by our geological studies and based on the experience of several German research teams that had been working for several years in

- eastern Macedonia and Thrace (e.g. F. Kockel and co-workers since 1965; Meyer 1968; Birk et al. 1970; Kopp et al. 1969).
- Aegean islands (e.g. Chios: Besenecker et al. 1968; Cyclades: Dürr since 1966; Santorini: Pichler et al. 1972).
- Crete (e.g. Creutzburg & Papastamatiou 1968, 1969; Kuss 1963, 1970; Seidel 1968).
- Peloponnesos (e.g. Bender et al. 1960; Bannert & Bender 1968).
- Corinth (e.g. Freyberg 1952, 1973; Schröder 1970).

A review of the results of these studies is presented in the following papers. Our results, as well as voluminous research work carried out by colleagues mainly from Greece, France, Great Britain and the Netherlands, enabled the compilation of a brief synthetic essay (Jacobshagen et al., this volume).

All participants of the German research program feel deeply indebted to the Greek authorities and to their colleagues of the Institute of Geological and Mining Research, Athens, and the Greek universities, who supported the project. Their contribution was not only organisatory but essential in interpreting and discussing the geological problem with us. Without their active cooperation and kind hospitality our investigations would have been impossible.

References

Aubouin, J., 1959: Ann. géol. Pays hell. - 10:1-484, Athens.

- , 1965: Geosynclines. 350 p., Amsterdam (Elsevier).

Aubouin, J., Brunn, J.H., Celet, P., Dercourt, J., Godfriaux, I. & Mercier, J., 1963: Livre Mém. Paul Fallot 2:583-610, Paris.

Aubouin, J., Bonneau, M., Celet, P., Charvet, J. and 13 al., 1970: Ann. Soc. géol. Nord 90:277-306, Lille.

Bannert, D. & Bender, H., 1968: Geol.et Palaeont. 2:151-162, Marburg.

Bender, H., with contributions of Leuteritz, K., Mänz, H. & Suesskoch, H., 1960: Ann. géol. Pays hell. - 11:201-213, Athens.
Besenecker, H., Dürr, S., Herget, G., Jacobshagen, V., Kauffmann, G., Lüdtke, G., Roth, W. & Tietze, K.-W., 1968: Geol. et Palaeont.- 2:121-150, Marburg.
Birk, F., Boer, H.U.de, Kronberg, P., Meyer, W., Pilger, A. & Schenck, P.-F., 1970: Beih. Geol. Jb. - 88:3-179, Hannover.
Brunn, J.H., 1956: Ann. géol. Pays hell. - 7:1-358, Athens.
Caputo, M., Panza, G.F. & Postpischl, D., 1970: J. geophys. Res. - 75:4919-4923.
Celet, P., 1962: Ann. géol. Pays hell. - 13:1-446, Athens.
Creutzburg, N., 1958: Freiburger Universitätsreden,N.F., 26:5-47.
- , 1966: Erdkunde 20:20-30, Bonn.
Creutzburg, N.& Papastamatiou, J., 1966: Geol. geophys. Res. - 11:173-185, Athens.
Dercourt, J., 1964: Ann. géol. Pays hell. - 15:1-418, Athens.
Epting, M., Kudrass, H.R., Leppig, U. & Schäfer, A., 1972: N. Jb. Geol. Pal., Abh. 141:259-285.
Freyberg, B.v., 1952: Ann. géol. Pays hell. - 4:157-188, Athens.
- , 1973: Erlanger geol. Abh. - 95:1-183.
Godfriaux, I., 1970: Ann. géol. Pays hell. - 19:1-280.
Ivanov, R. & Kopp, K.-O., 1969: Geol. et Palaeont. - 3:123-151, Marburg.
Kauffmann, G., Kockel, F. & Mollat, H., 1976: Bull. Soc. géol. France, VII, 18:227-232.
Kober, L., 1929: Centr.-Bl. Miner. Geol. Paläont., 1929:425-437.
Kockel, F. & Walther, H.W., 1968: Bulgar. Acad. Sci., Bull. geol. Inst. - 18:133-142, Sofia.
Kockel, F., Mollat, H. & Walther, H. W., 1971: Geol. Jb., 89: 529-551.
Kopp, K.-O., Pavoni, N. & Schindler, C., 1969: Beih. Geol. Jb. - 76:1-136.
Kuss, S., 1963: Prakt. Akad. Athinon. - 38:431-436, Athens.
- , 1970: Ber. naturf. Ges. Freiburg i.Br. - 60:35-83.
Lebküchner, H., 1972: Geological Map of Greece 1 : 50 000, Samothraki Sheet, Athens.
Mercier, J., 1966: Thèse Fac. Sci. Univ. Paris, 573 p.
Meyer, W., 1968: Geol. et Palaeont. - 2:173-192, Marburg.
Philippson, A., 1959: Die griechischen Landschaften IV. Das ägäische Meer und seine Inseln. - 412 p., Frankfurt.
Renz, C., 1940: Pragm. Akad. Athinon, 8:1-171, Athens.
- , 1955: Die vorneogene Stratigraphie der normalsedimentären Formationen Griechenlands. - 637 p., Athens.
Ritsema, A.R., 1969: Geol. Rundschau. - 59:36-56.
Schröder, B., 1970: N. Jb. Geol. Pal., Mh. 1970:27-33.
Seidel, E., 1971: N. Jb. Geol. Pal., Abh. 137:443-460.
Wunderlich, H.G., 1968: Repts. 23rd internat. geol. Congr. Prague. - 1:57-68.
- , 1969: Z. dtsch. geol. Ges. - 118:266-284.

Part 1

Aegean Crust and Upper Mantle

A Geophysical Study of Greece Based on: Deep Seismic Soundings, Gravity, and Magnetics

J. Makris

In the following paper, the geographical term "Aegean Area" is used to describe the region of Greece, the Aegean Sea, and the eastern Aegean islands.

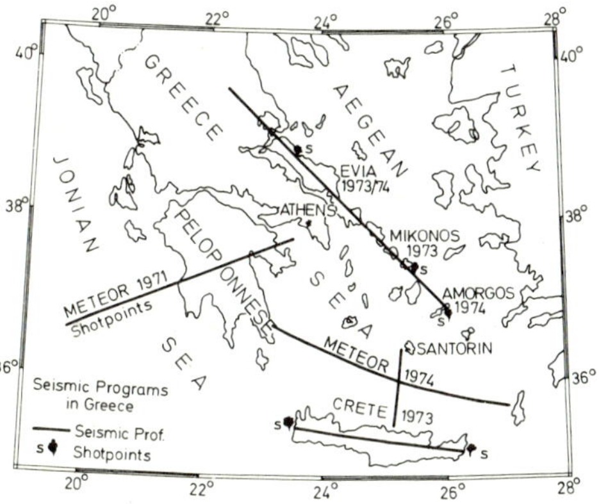

Fig. 1. Locations of seismic lines in Greece, observed in the period 1971-1974.

Deep-seismic Soundings

Deep seismic surveys have been conducted on 5 profiles (fig. 1). The main results may be summarized as follows:

The velocity-depth distribution along all seismic sections reveal a pure continental crust of varying thickness and state of deformation, underlain by Upper Mantle material of lower p-wave velocity than normal. The west Hellenic Mountains (Ionian-, Gavrovo- and Tripoliza-Zones) have a thick crust of the order of 46 km below the western Peloponnesos (fig. 2). Also the sedimentary cover in these zones is fairly thick and has maximum values of the order of 8 km at the area of Tripolis, central Peloponnesos. The crustal deformation along the west Hellenic chains is of very great intensity and the crust nearly doubles in size within only 10 km horizontal distance from the western coast of Peloponnesos. It decreases gradually to the east to 32 km below Aigina in the Saronic bay and 30 km at South Evia in the Pelagonic zone.

The other sections show a normal continental crust approximately 30 - 35 km in thickness below the Pelagonic zone with a strongly attenuated part in the Cretan Sea. The minimum crustal thickness coincides with the deep Cretan Sea trough and is only 18 - 20 km thick. The crust below Crete increases again to nearly 32 km (fig. 3), (Makris et al. 1977).

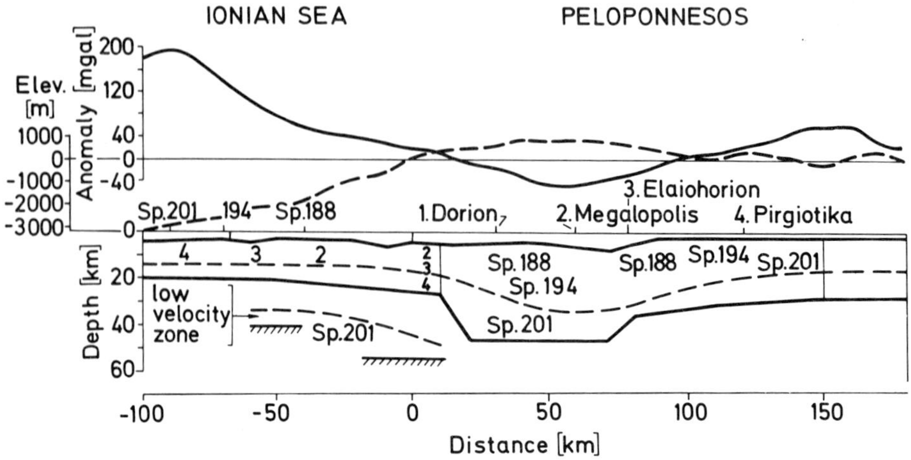

Fig. 2. Seismic section: Ionian Sea-Peloponnesos. The crustal model has been developed by evaluating 2.500 seismograms, compiled in 16 montages.

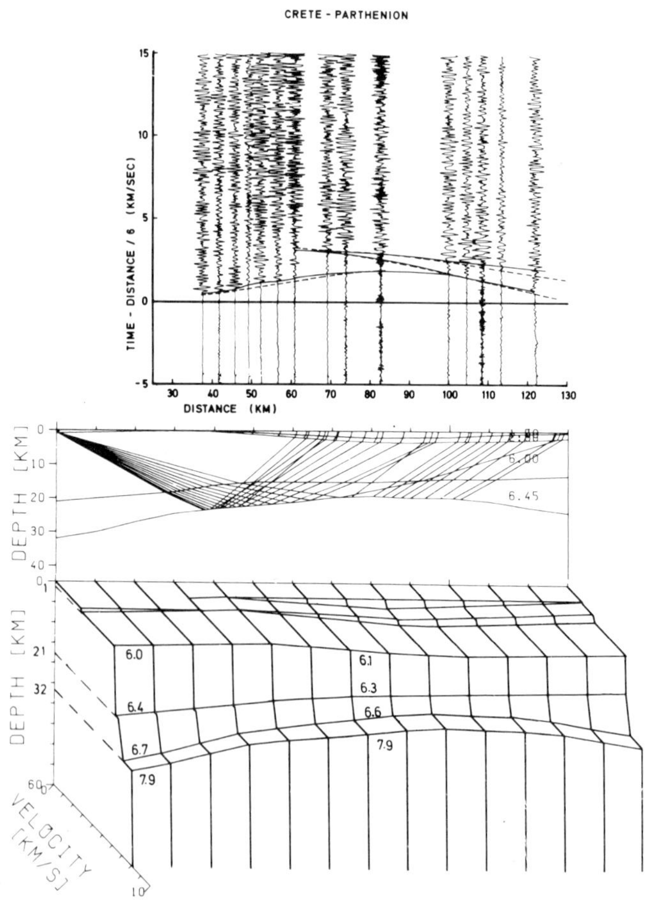

Fig. 3. Seismic section across the Cretan Sea.

The observed p-wave velocities are 6 km for the crystalline basement and 7.5 - 7.7 km for the Upper Mantle. A Conrad discontinuity has not been observed as a first order boundary within the crust. Most sections, however, require a gradual velocity increase from 6.4 - 6.8 km at its lower part. Strong P_MP-reflections have been observed at nearly all seismic sections and indicate a good seismic contrast between the lower crust and the Upper Mantle. Deeper reflections originating at discontinuities within the Upper Mantle have also been observed. Their implications will be discussed in section 3.

The Bouguer Anomalies

On a gravity map by Makris (1976), (fig. 4) the Greek Mainland and the Peloponnesos show negative Bouguer anomalies with maximum negative values of -140 mgal confined to the Ionian and Gavrovo zones of the area of the western Hellenides. The gravity field rapidly changes to positive values at the Pelagonian zone, which covers the western Macedonian, the Thessalian and the Attic Cycladic massives, and also parts of the Tripolitza zone on the SE Peloponnesos, with maximum values at South Attica (+100 mgal) and the Parnon Mountains, SE Peloponnesos (+80 mgal). The trend of the gravity anomalies is NNW - SSE, parallel to the main geological and morphological structures. The Aegean Sea and all of the Greek islands including Crete have also positive Bouguer anomalies with maximum values in the Cretan Sea (+175 mgal).

The gravity map shows a great number of smaller anomalies which strongly disturb the regular regional gravity trends. In spite of this, one can still recognize that in the Aegean Sea the gravity trends change from the NNW - SSE direction of the Greek Mainland to ENE - WSW at the Cretan Sea and the Saros graben of the North Aegean Sea.

At the Greek Macedonia, particularly the Macedonian massif, the Vardar zone, the Servo-Macedonian massif and the Serais graben the Bouguer anomalies strike parallel to the NNW - SSE trend of the Greek Mainland. The E-W trends of the Rhodopi massif seem to dominate only in Thrace, where the field gradually bends in ENE direction to join the Black Sea strike.

Crustal and Upper Mantle Structure from Gravity and Seismic Data

Two 2-dimensional (2-D) crustal and Upper Mantle models were developed from the seismic results and were computed with 2-D techniques (fig. 5 and 6).

Section 1 (fig. 5) extends from Turkey to the Ionian Sea and the Malta Shelf over the Aegean Sea and the Peloponnesos. It crosses the seismic line Mikonos - Evia and coincides with the seismic section Aigina - Kiparissa - Ionian Sea and the seismic profiles shot by "FS Meteor" during the cruises Nrs. 17 and 22 (Hinz 1974; Weigel 1974). The seismic control along this section is very reliable and the interpretation of the observed gravity anomalies is very well constrained. In computing the gravity effect from the known crustal structures only, it is not possible to fit the computed to the observed anomalies.

A gravity difference of approximately 50 mgal out of total of 360 mgal between the Aegean region and the Ionian Sea cannot be explained by the

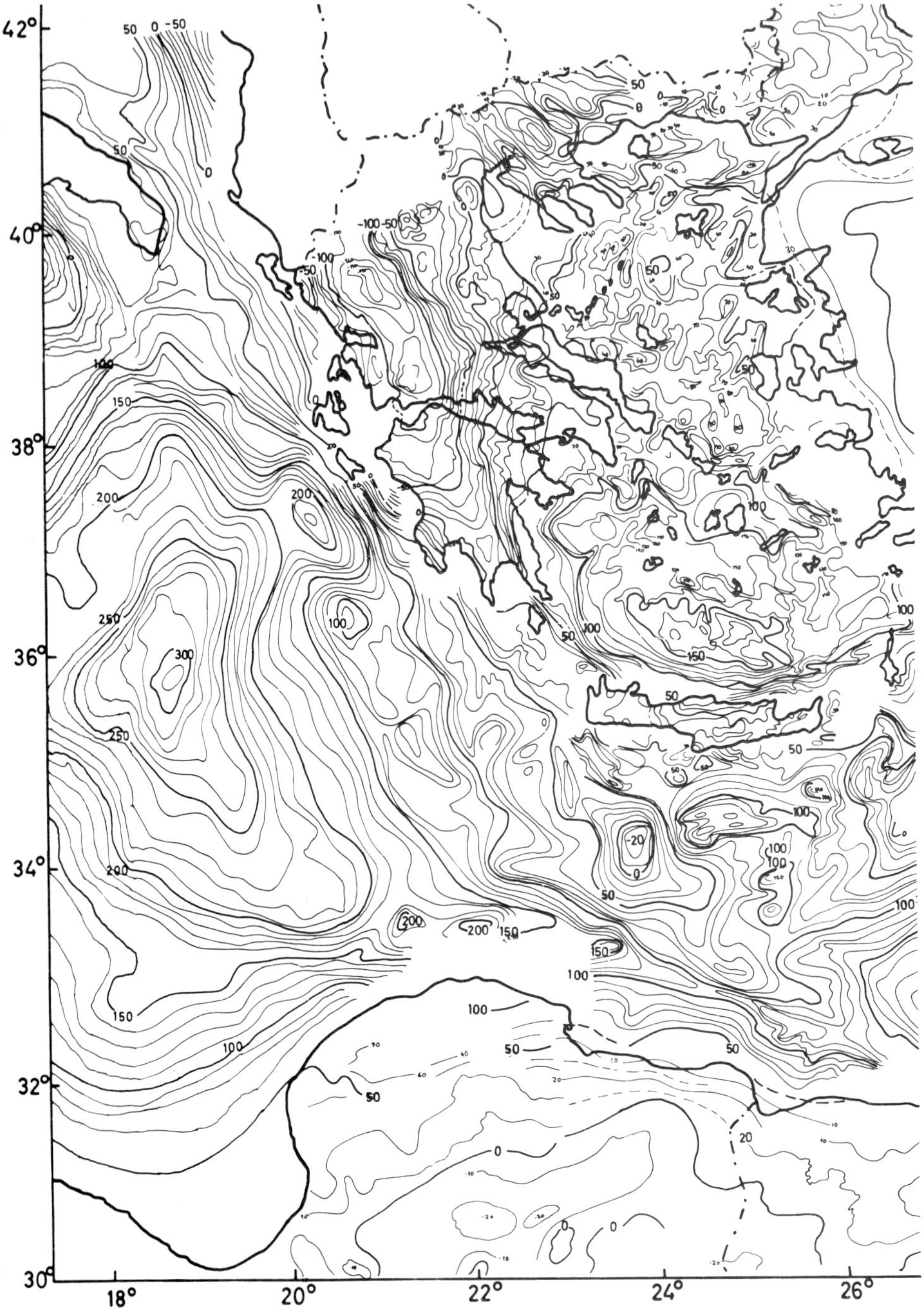

Fig. 4. Bouguer gravity map of Greece and the central Mediterranean Sea compiled after Öcelci (1974), Makris (1974, 1977), Finetti et al. (1974), and publications of the "Bureau Gravimétrique International", Paris.

seismically adjusted crust only. This was achieved by introducing a low density body below the crust of the Aegean area, extending to the Mediterranean Ridge in the Ionian Sea. The low density and as well as low velocity zone has penetrated into the lithosphere from the asthenosphere and seems to consist of at least two zones. This is seismically evident by deep reflections observed along the Peloponnesos-Ionian Sea line. The 20 1-ton charges fired by "FS Meteor" in the Ionian Sea produced sufficient energy to permit penetration below the Moho. Reflected arrivals from a sub-crustal event can be clearly identified in most of the seismic records along this section (Makris 1977).

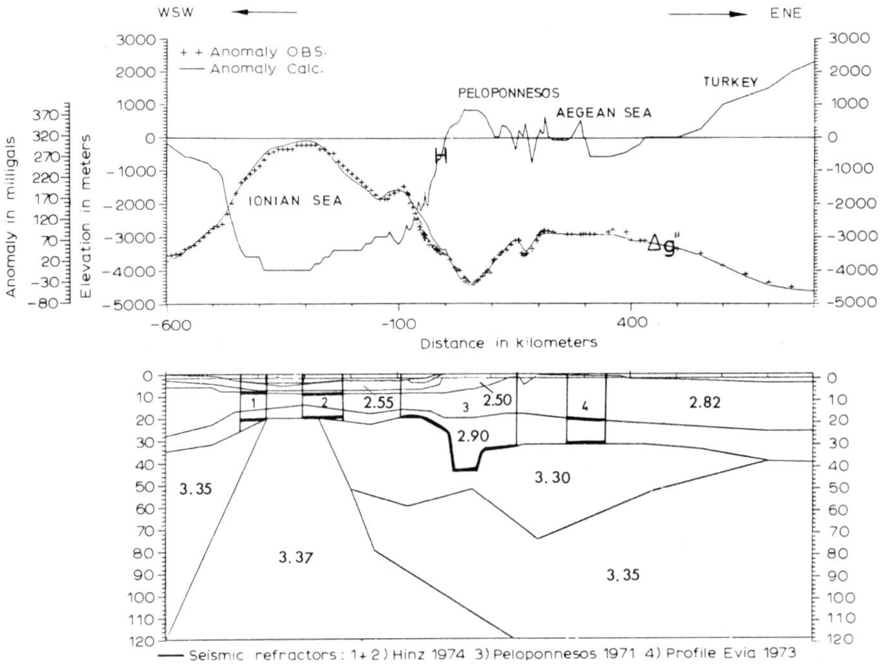

Fig. 5. 2-dimensional crustal model Turkey-Malta shelf obtained by combination of gravity data and seismic observations. Curve H gives the topography along the model. Density values are in g/cm³.

Profile 2, fig. 6, crosses from Bulgaria, the Aegean Sea and the Mediterranean Sea to Libya. It crosses three independent seismic lines in the Aegean area and the seismic profile R_1 of Lort et al. (1974) in the Mediterranean Sea south of Crete. To the North the model is adjusted to the Moho depth map of Bulgaria (UNESCO 1973).

The results obtained are quite similar to those of profile 1, fig. 5. The crust is also continental with maximum attenuation at the Cretan Sea, where its thickness is only 18 - 20 km. Crete has also a fairly thick crust of approximately 31 - 32 km. Towards the north Aegean Sea and the Rhodopi massif , the crustal thickness increases to nearly 44 km. The Mediterranean Ridge south of Crete has also a fairly thick crust of the or-

der of 32 km and thick sediments of the order of 10 km (Lort et al.1974). The crustal thickness below Libya increases also to more than 34 km. The main difference in the crustal structure and deformation along section 1 compared to that of section 2 is that the crust of section 2 resembles a continental type of crust at all its length and that the deformation of the foreland between Crete and Libya is of very strong shortening. The isostatic minima are therefore very narrowly spaced.

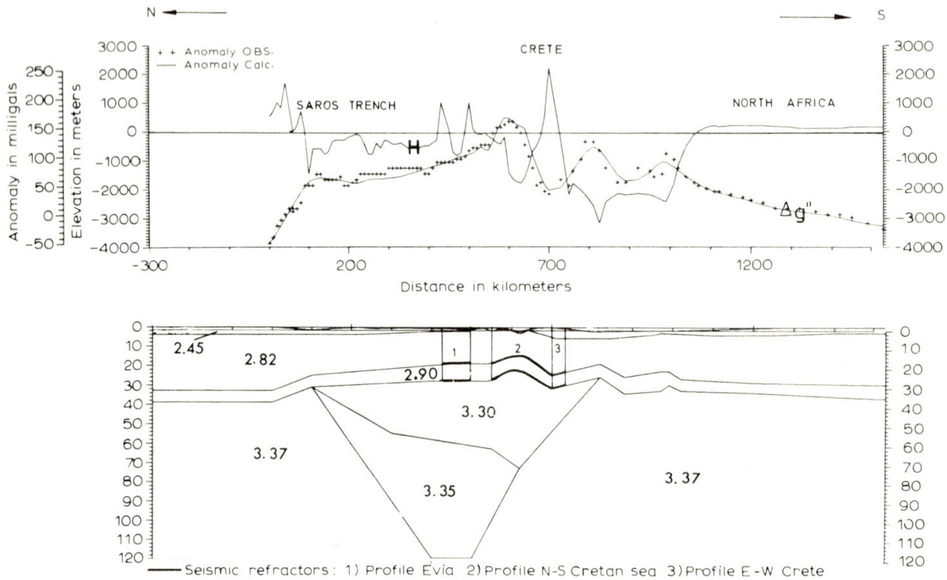

Fig. 6. 2-D crustal model between Bulgaria and Cyrenaica, North Africa. Explanations as in figure 5.

The density distribution below the Moho discontinuity is similar to that obtained at section 1. The Upper Mantle below the Aegean area has a low density structure, extending to the basis of the lithosphere. The seismic velocities as well as the gravity gradients require the subdivision of the low velocity zone into two parts, an upper part with ρ = 3.30 g/cm³ corresponding to V_{pn} = 7.7 km and a lower part with ρ = 3.35 g/cm³. The low velocity zone extends in this case also beyond the Hellenic Arc to the Mediterranean Ridge.

Discussion of Magnetic Geothermal and Seismicity Data

The seas around the Aegean area have no linear magnetic anomalies. In particular, the Ionian and Levantine Seas show hardly any magnetic patterns, whereas the Aegean Sea and the eastern Greece (Pelagonic zone) are convered by a great number of local anomalies that can be correlated to geological and mineralization structures. The ophiolites, the volcanic arc and also known basement highs are some of the structures that show up magnetically. The Saros graben of the northern Aegean Sea shows up very

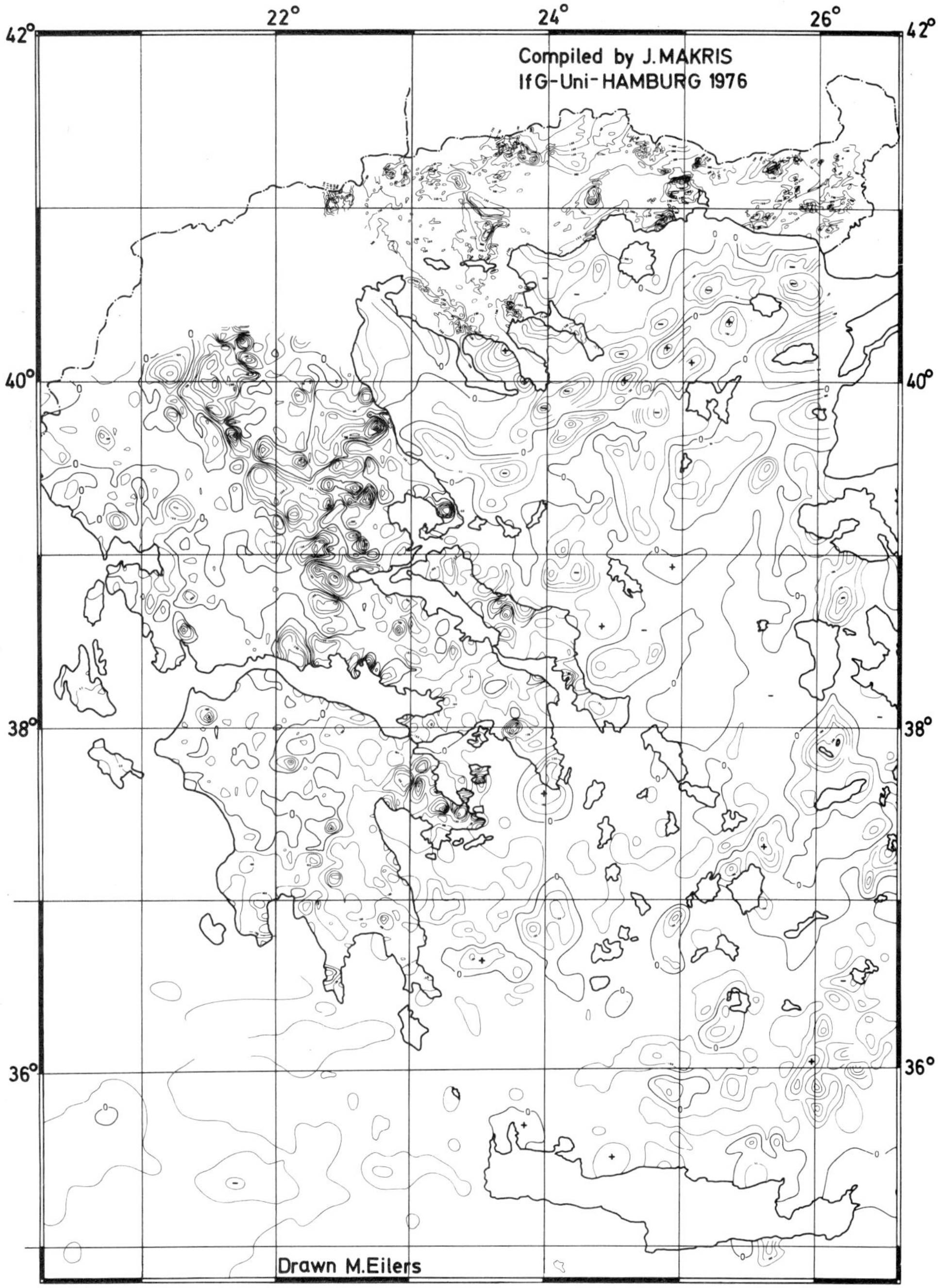

Fig. 7. Magnetic map of Greece, rest field anomalies. For detailed information, see text.

strongly magnetically*.

The mean heat flow shows values of about 0.88 HFU for the entire eastern Mediterranean region except for the Aegean Arc (Jongsma 1974; Erickson 1970; Ryan et al. 1970). The various provinces do not show up geothermically, except for small deviations in areas to tectonic activity. The data obtained in the Aegean area by Jongsma (1974) show much larger values with mean heat flow of about 2.08 HFU. Comparing these results to the world mean of 1.45 HFU, one could conclude that the eastern Mediterranean outside the Aegean Arc is at a cooler state than behind the arc, where the mean heat flow values are higher than the world average. A possible explanation for this is that the Aegean area has a heated upper mantle due to convective movements of hot asthenospheric material penetrating the Aegean lithosphere. This process might have resulted due to subduction of cooler plate boundaries below the Aegean Arc, thus disturbing the thermal equilibirum and initiating the rising of the asthenosphere. The thermal regime, however, could also be explained by the vertical movement of thermal instabilities as described by Gass (1972) or Schuiling (1973), and as has been observed in intercratonic tectonic activity, where no plate-boundary processes are involved. The differences of the thermal regime observed inside and outside the Aegean Arc from surface observations can be explained by the mechanical uplift of the isotherms due to the morphological elevation of the area behind the arc and the strong erosion involved. Outside the arc the crust is sinking and the isotherms remain at a lower level.

A Lithothermal Model of the Aegean Region

The model proposed for the present tectonic activity in the Aegean area is that of a hot plume expanding gradually from the inner zones outwards. It extends presently to the Mediterranean Ridge.

The existence of such an Upper Mantle structure is supported by the following geophysical observations:

Lower density than normal as required by 2-D crustal and Upper Mantle models below the Aegean region (fig. 5 and 6).

P-wave velocities in the Upper Mantle lower than normal. Sub-crustal reflections at 45 - 65 km depth support the gravity model.

A travel-time delay of +2 sec for distant earthquakes recorded at the Hellenic seismological net (Ekonomides 1972; Agarwal et al. 1976; Jacoby et al. this vol.). Gregersen's p-wave travel-time residual study (1977) for earthquakes occurring in the Aegean region clearly shows the inhomo-

* The magnetic map, fig. 7, consists of three sets of data:

a) Anomalies at Greek Mainland and the Peloponnesos are ΔZ-residuals (Makris, Tödt, in preparation).
b) Anomalies at Macedonia and Thrace are ΔT-residuals of an aeromagnetic investigation flown at 270 m above sea-level by Atlas Copco Abem Ab, Stockholm on behalf of Institute of Geological and Mining Research, Athens. (IGME - personal communication).
c) Data at sea are ΔT-residuals from ship borne investigations, compiled by the Mobil-Oil, 1967.

Space of isolines: 50 γ .

genity of the structure within and at the borders of the Hellenic arc. He concludes, in agreement with Agrawal et al. op. cit.; Richter et al. this vol., that a fast slab is sinking below the arc at the Rhodos-Karpathos section of it. However, the situation below Crete is unclear, because of a diffuse seismicity also south of the arc, in the Mediterranean Ridge (Galanopoulos 1977).

The low velocity/density structure transporting abnormal heat into the crust is responsible for the following processes:

Upward movement of the Aegean crust, denudation of the sediments by erosion and gravity gliding. The extensional tectonics and the present sinking of the Aegean Sea is due to isostatic readjustment.

The *crustal thickening* along the Hellenides is due to crustal downbuckling and thickening at a compressional front. The collision is responsible for the high seismicity and the destructive processes occurring at the outer zones of the Hellenides. The Aegean crust is forced to override part of the Ionian - East Mediterranean Lithosphere, creating the Hellenic Trench and causing subduction.

The great *accumulation of light material* causes isostatic undercompensation as shown at the western Peloponnesos and the Mediterranean Ridge.

The *process of material accumulation* at the orogenic front brakes down as soon as the buoyancy of the continental crust becomes too large to permit any further sinking. The downbuckling process thus is converted into an uprising phase and a new orogenic compressional front developes at a new position of the foreland. (This is the present situation of the Mediterranean Ridge).

The *deep seismicity* is caused by crust and Upper Mantle fragments subducted into the low Q Asthenosphere. Horizontal compression is not excluded. In this way more complicated Benioff-Zones of the Pacific could be explained due to the interaction of continental - continental or continental - subcontinental blocks.

The *present morphology of the deep Ionian Sea* is caused by foundering of the heavy Ionian crust and Upper Mantle below the Aegean area.

The model discussed above provides a possible explanation for the various geophysical observations made so far in the Aegean area. It is very probable that the present alpine tectonics are due to thermal processes which might have been initiated by the relative approach between Europe and Africa.

References

Agarwal, N.K., Jacoby, W.R. & Berckhemer, H., 1976: Tectonophysics, 31:33-57.

Ekonomidis, A., 1970: Diss. Athens.

Erickson, A.J., 1970: PhD thesis, Mass. Inst. Techn.

Dürr, S., Geol. Jahrbuch, Hannover, in press.

Finetti,I.,& Morelli,C.,1973:Boll.di Geofis.Teor.ed Appl.XV,60:263-285.

Galanopoulos, A., 1960: Greece: A catalogue of shocks with I_0 VI or M 5 for the years 1801-1958, Athens.

- , 1963: Ann. Geofis., 16,1:37-100.

- , 1973: Ann. di Geofis., 20,1:109-119.

Gass, J.G., 1968: Nature, 220:39-42.

Hinz, K., 1973: Bull. geol. Soc. Greece, 10/1:203.

Jongsma, D., Wissmann, G., Hinz, L. & Garde, S., 1977: "Meteor"-Forsch.

Ergeb., Reihe C-No.27:3-31.
Karnik, V., 1974: Proceedings of the Seminar on the seismotectonic map of the Balcan Region. UNESCO, Skopje.
Leydecker, G., 1975: Diss. Univ. Frankfurt, 75 p.
Lort, J.M., Limond, W.Q. & Gray, F., 1974: Earth and Pl. Sci. Letters, 21:355-366.
Makris, J., 1977: Geophysical Investigations of the Hellenides. Hamburger Geophysikalische Einzelschriften, Nr.33:128 p.
- , 1977: Deep Seismic Soundings at the Eastern Jonian Sea, in prep.
Makris, J. & Vees, R., 1977: J. Geophys., 42:329-341.
Makris, J. , Weigel, W. & Koschyk, K., 1977: "Meteor"-Forsch.Ergeb., Reihe C Nr.27:31-44.
Neprochnov, Yu.P., Kosminskaya, I.P. & Malovitsky, Ya.P., 1970: Tectonophysics, Vol. 10, 5/6:517-539.
NIGMR, 1971: National Institute of Geological and Mining Research, Pers. Comm.
Özelçi, H.F., 1973: Bull. Miner. Res. and Expl. Inst. Turkey, No. 90.
Papazachos, B.C., 1973: Bull. geol. Soc. Greece, 10/1:165-168.
Papazachos, B.C. & Comninakis, P.E., 1973: J. Geophys. Res. - 76: 8517-8533.
Philippson, A., 1959: Die griechischen Landschaften. 4: Das Aegaeische Meer und seine Inseln. - 412 p.
Ronov, A.B. & Yaroshevsky, A.A., 1969: Geophysical Monograph 13, American Geophysical Union:37-57.
Ryan, W.B.F., Stanley, D.J., Hersey, J.B., Fahlquist, D.A. & Allen, T.D., 1970: In: The sea, Vol. 4, Maxwell, A.E., ed., New York, Wiley-Interscience:387-492.
Said, R., 1962: The Geology of Egypt. Elsevier Publishing Co., Amsterdam, New York.
Schuiling, R.D., 1973: In: Jong, K.A. de & Scholten, R. (ed.): Gravity and Tectonics:35-47, Wiley, New York.
Bulgarian Academy of Sciences ATLAS, 1973: Maps of the Territory of Bulgaria, Base for the Seismotectonic Map 1 : 1.000.000 (Balkan Seismicity Project - UNESCO), Sofia.
Vogt, P.R. & Higgs, R.H., 1969: Earth and Planet. Sci., Letters,5:439-448.
Weigel, W., 1974: J. Geophys. 40:137-140.
Woodside, J. & Bowin, C., 1970: Bull. Geol. Soc. Am. - 81:1107.

Crustal and Upper Mantle Structure of the Aegean Arc from Travel Time Residuals and Gravity

W.R. Jacoby, N.K. Agarwal and H. Berckhemer

Based on morphological, geological, and geophysical data, the Aegean region has been interpreted to be a trench - island arc - marginal sea system. To test the hypothesis, we have attempted to investigate the Up-

per Mantle velocity structure by the use of teleseismic P wave travel time residuals. In particular we have searched for a high-velocity slab presumably dipping below the Aegean Sea as suggested by intermediate depth seismicity (Papazachos & Comninakis 1971; Richter & Strobach this vol.). Negative travel time residuals for rays passing through the slab have been found consistently in typical island arcs (e.g. Utsu & Okada 1968; Davies & McKenzie 1969; Jacob 1972). Another typical feature for which we have looked is the low-velocity Upper Mantle above the dipping slab.

Method, Data, Results

The theoretical background and computational procedures have been described by Agarwal et al. (1976). We have taken the U.S.C.G.S. preliminary determinations of epicenters (PDE) as focal parameters for our study. We have accpeted most of the P arrival times as listed in the preliminary bulletins for 1966 to 1970, issued by the Seismological Institute of the National Observatory at Athens, Greece.

A representative set of 500 of the 5000 original records were reread. We used the stations Athens (ATH), Archangelos (ARG) on Rhodes, Vamos (VAM) on Crete, Valsamata (VLS) on Kephalenia, Joanina (JAN) and Polygyros (PLG) on the mainland, and Paraskevi (PRK) on Lesbos. As our reference travel time function $t(\Delta, h)$ of epicentral distance Δ and focal depth h we have chosen the P68 tables by Herrin et al. (1968).

The *absolute* travel time residuals are defined as

$$\Theta_{ik} = T_{ik} - t_{ik} \tag{1}$$

where T_{ik} is the observed travel time of the P phase from the earthquake source i to the station k, and $t_{ik} = t(\Delta_{ik}, h_i)$ is the theoretical P68 travel time across the distance Δ_{ik} for focal depth h_i. There are problems both in travel time standard as the one chosen and in interpretation of the residuals because of lateral variations of P velocity along the whole ray path (see e.g. Lilwall & Douglas 1970; Jacob 1972; Julian & Sengupta 1973). To minimize the effects of variations in the source regions and deep mantle and thus enhance the effects of structure in the Aegean sub-receiver region, we have also computed *relative* travel time residuals τ_{ikA} with respect to the central station ATH:

$$\tau_{ikA} = (T_{ik} - T_{iA}) - (t_{ik} - t_{iA}) \approx \Delta G - \frac{\delta t}{\delta \Delta}\Big|_{\Delta_{iA}} \cdot (\Delta_{ik} - \Delta_{iA}) \tag{2}$$

where the index A pertains to ATH and ΔG is the measured travel time difference between stations k and ATH. To avoid the influence of regional Upper Mantle anomalies and to avoid identification problems of PKP phases, we have limited our epicentral distance range to $30° \leqq \Delta \leqq 100°$.

The scatter of the P residuals both *absolute* and *relative* is typically at least ± 5 s. This requires some form of statistical treatment. We have found simple averaging to be the most adequate method. Beside the overall mean station residuals we have taken averages of 30° sectors of azimuth of approach to study the azimuthal variation of the residuals (figs. 1 and 2).

One of the most obvious results is the average *absolute* residual of about +2 s at ATH and more at the other stations (fig. 1). In azimuthal variation we note the relatively smaller residuals from the NE quadrant

at all stations except ARG where the smallest residuals are from the NW.

The *relative* P residuals (fig. 2) are generally positive, i.e. the arrivals are at all stations delayed more than at ATH. Along the sedimentary arc at JAN, VLS, VAM, and ARG the azimuthal variation is represented by polar curves elongated parallel to the arc, i.e. rays arriving in a direction parallel (perpendicular) to the arc are delayed more (less) than average.

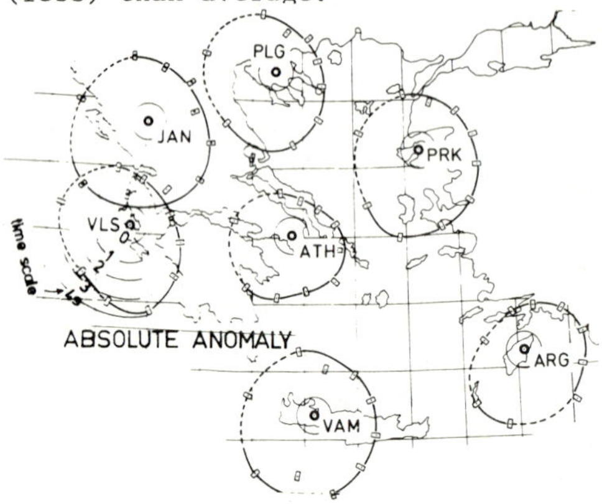

Fig. 1. Smoothed 30° sector means of the *absolute* travel time residuals at the Greek stations.

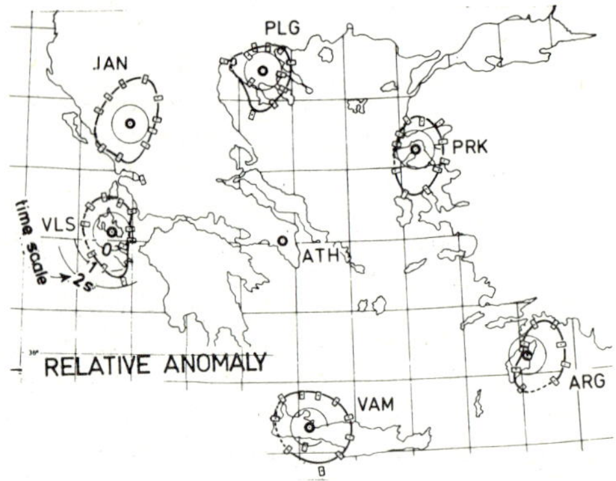

Fig. 2. Smoothed 30° sector means of the *relative* travel time residuals at the Greek stations.

Interpretation

We believe that the *absolute* travel time delay of about 2 s is caused by a generally low-velocity Upper Mantle in the Aegean region, while the *absolute* azimuthal variations are largely a source-region and/or lower-

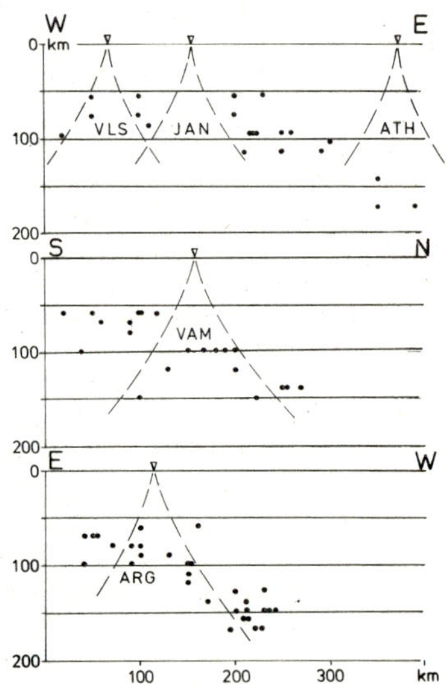

Fig. 3. Projection of intermediate-depth seismicity onto vertical planes crossing the Aegean arc, according to Papazachos & Comninakis (1971). Also projected are the nearest Greek stations with sketches of rays appropriate to the epicentral distance range of this study.

mantle effect; and that the *relative* residuals, both average and in azimuthal variation, are caused by crustal and Upper-Mantle island arc velocity structure.

Very simple models suffice. If we construct a three-layer model of crust, high-velocity Upper Mantle lid, and low-velocity zone (lvz) to an arbitrary depth of 300 km, a 2 s P delay can be explained e.g. by a 220 km lvz of 7.55 km/s instead of 8.13 km/s. This is a remarkably low velocity which does not change significantly if the reference model is changed. The average *relative* residuals at the various stations can be explained by thicker crust and/or lvz than below ATH. Crustal thickness variations of the right direction and order have been found by refraction seismology for the line VLS-ATH and by gravity for JAN-ATH (Makris 1973 and pers. comm. 1975).

The azimuthal variations of the *relative* residuals shown by the elongated curves of fig. 2 at JAN, VLS, VAM, and ARG suggest two possible explanations: (1) P velocity anisotropy of a few percent with the maximum velocity perpendicular to the arc, and (2) crustal thickening of 5 to 10 km (30 to 60 km laterally across the arc below the stations).

The azimuthal variation of the *relative* residuals at ARG, Rhodes, is particularly important, since it clearly displays the effect of a high-velocity slab that coincides with the earthquake distribution (fig.3). The slab which dips below the Aegean should be characterized by a P velocity 5 to 15 % higher than outside velocity. Fig. 3 further shows that JAN and VLS are not located strategically to see the effect of the slab, while at VAM, Crete, the situation is complicated by the intermediate earthquakes south of the arc.

If Birch's velocity-density relation

$$\rho = a^* + b^* \cdot v_p \qquad (3)$$

(Birch 1960, 1961) holds, we should expect that gravity may constrain our simple models further. However, we found that reconciling Bouguer anomalies and P residuals with corresponding models is difficult. There is no problem to fit a high density to the high-velocity dipping slab (fig. 4), but there appears to be an inconsistency between positive P residuals, indicating low-velocity Upper Mantle, and positive gravity anomalies on the concave side of the sedimentary arc. Computed model gravity profiles A and C (fig. 4), based on the model geometry shown, but with low-velocity, low-density Upper Mantle below the Aegean Sea consistent with the P delays, are far too low. For curves A and C, Birch coefficients b* as low as 0.25 and 0.15 gcm^{-3}/kms^{-1}, respectively, have been assumed. As shown in fig. 4, additional mass had to be assumed in the form of high-density lower crust and high-density Upper Mantle lid below the Aegean Sea to obtain a better fit (curves C_1, C_2, and C_3). The detailes of the various models are listed in the figure caption.

If our P residuals are correctly determined and if they do arise in the sub-Aegean Upper Mantle, we have to conclude either that Birch's law (3) does not hold in this case or that, beside low velocities, the length of the ray path (extended by at least 3 km over its norm by uplift of the surface) contributes to the P delays. Both arguments suggest thermal expansion by a lateral temperature anomaly of about +250 K and probably partial melting in the Upper Mantle below the Aegean.

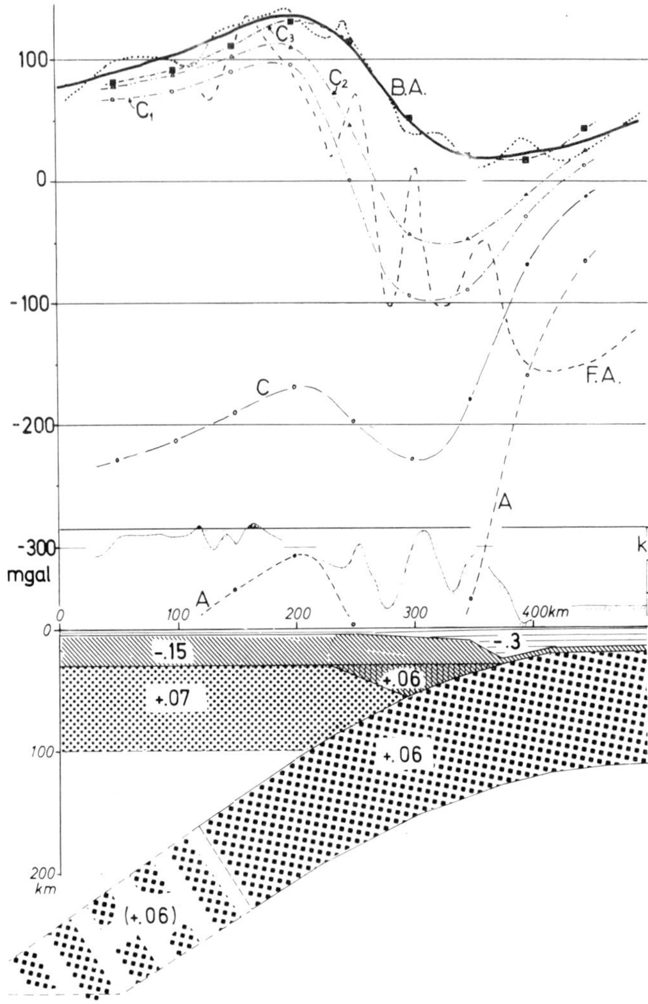

Fig. 4. Upper Mantle density section crossing the Aegean arc near Karpathos. Densities are consistent with observed Bouguer anomalies (B.A.) and free air anomalies (F.A.) according to Allen & Morelli (1971), but not consistent with travel time-based velocity model. Models A and C are consistent with velocities with Birch coefficients $b^* = 0.25$ and $0.15\ gcm^{-3}/kms^{-1}$, respectively. Mass added below Aegean to model C leads to C1, C2, and C3. The density contrasts assumed are listed below in the order:

listed below in the order:	A	C	C1	C2	C3
upper crust	-0.5	-0.3	-0.3	-0.3	-0.3
lower crust	-0.3	-0.2	-0.2	-0.15	-0.15
crustal wedge near trench	-0.3	-0.2	-0.2	-0.15	-0.15
Aegean mantle lid	0	0	+0.1	+0.08	+0.07
Mediterr. lithosphere	+0.09	+0.06	+0.06	+0.06	+0.06
extension to 300 km depth	0	0	0	0	+0.06

Acknowledgments

The help of Prof. Galanopoulos and his colleagues at Athens and of colleagues at Frankfurt, especially J. Drissler, is gratefully acknowledged.

References

Agarwal, N.K.,Jacoby,W.R.& Berckhemer,H.,1976:Tectonophysics,31:33-57.
Allen, T.D. & Morelli, C., 1971: Boll. Geofis. Teor. Appl. - 13:99-142.
Birch, F., 1960: J. Geophys. Res. - 65:1083-1102.
 - , 1961: J. Geophys. Res. - 66:2199-2224.
Davies, D. & McKenzie, D.P., 1969: Geophys. J. Roy.Astr. Soc.- 18:51-63.
Jacob, K.H., 1972: J. Geophys. Res. - 77:2556-2573.
Julian, B.R. & Sengupta, M.K., 1973: Nature, 242:443-447.
Lilwall, R.C. & Douglas, A., 1970: Geophys. J. Roy. Astr.Soc.19:165-181.
Makris, J., 1973: Bull. Geol. Soc. Greece. - 10:206-213.
Papazachos, B.C. & Comninakis, P.E., 1971: J. Geophys. Res. - 76:8517-8533.
Utsu, T. & Okada, H., 1968: J. Fac. Sci. Hokkaido Univ., Ser. VII, 3:65-84.

A Study of Seismicity in the Peloponnesus Region by Precise Hypocenter Determinations

G. Leydecker, H. Berckhemer and N. Delibasis

The Hellenic arc system contains the strongest seismo-tectonic activity in Europe. This region resembles Pacific island-arc structures by showing high earthquake activity, a trench on the convex side, an inner volcanic belt active in historic time, large free air anomalies and a marginal sea, the Aegean Sea. But the diffuse distribution of shallow and intermediate earthquakes (Papazachos & Comninakis 1971) is not consistent with what would be expected of an island-arc Benioff zone.

Data Collection

The lack of precise focal depth determinations was the main reason for the installation of three short period seismograph stations on the Peloponnesus in the villages Kastri (KAS), Ithomi (ITH) and Tropeia (TRP) (fig. 1). It is a tripartite array with 60 km sides. Each of the stations consists of a vertical Willmore seismometer Mark II with a free period of 1.4 s, a crystal clock, a time signal receiver and an ink recorder. The magnification of the whole system has been chosen to be 50 000 at 1.5 Hz.

The nearest stations of the Greek seismograph network are Valsamata (VLS), Athens (ATH), Penteli (PTL), Vamos (VAM) on Crete and Janina (JAN) in N-Greece. The records of these and other Greek stations were placed at

our disposal by the National Observatory of Athens, Seismological Institute. Our three stations had been in operation for two years, from April 1972 until March 1974, recording a few hundred Greek shocks.

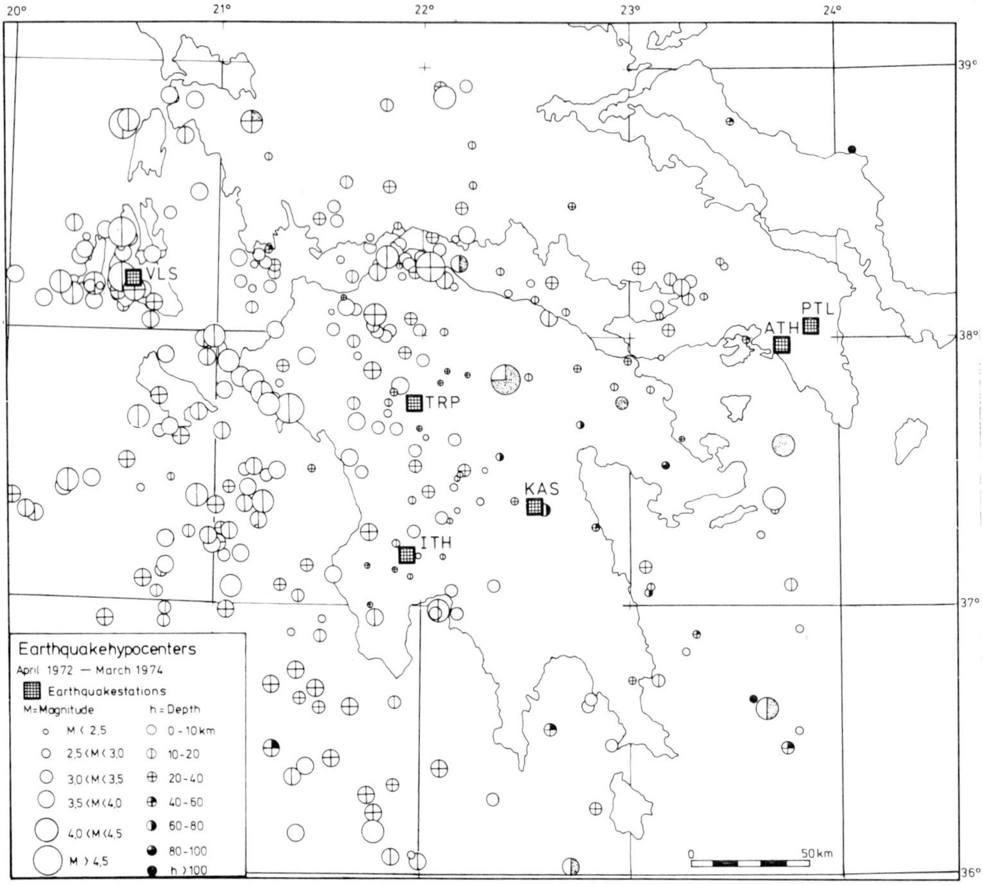

Fig. 1. Epicenters in the Peloponnesus region.

Computational Procedure

For hypocenter computation, we used a modified version of the program HYPOLAYR of Eaton (1969). It assumes a horizontally stratified velocity-depth model with downward increasing velocity. For the crust, the model of Makris (1975), derived from German seismic refraction measurements, was used. The iteration procedure of the program minimizes the travel time residuals by improving the four focal paramters ϕ_o, λ_o, h_o, T_o. Small earthquakes inside the triangle were computed by using P- and S-arrival times. Earthquakes of larger magnitude could be analysed by also using the nearby stations of the Greek network. A formula was developed for estimating unknown magnitudes on the basis of the duration time of the signals (Tsumura 1967; Lee et al. 1972). This equation was found by comparing the signal duration time of earthquakes with

the magnitudes determined from recordings by a Wood-Anderson seismograph at Athens.

A special investigation (Leydecker 1975) was made to determine the accuracy of the computed hypocenters. The standard deviation of the arrival-time residuals and the standard errors of the four focal parameters are a poor measure of the quality of the hypocenter. By application of statistics on each hypocenter, the expectation values of the standard errors can be estimated. These are better criteria for hypocenter accuracy than merely the actual standard errors. The accuracy was generally found to be ± 5 km for the epicenter and ± 5 - 10 km for the depth in the area of the Peloponnesus. While some of the depths reported in this paper are similar to those calculated by the USCGS, others were found to differ by more than 100 km. This is the improvement gained by stations close to the hypocenters.

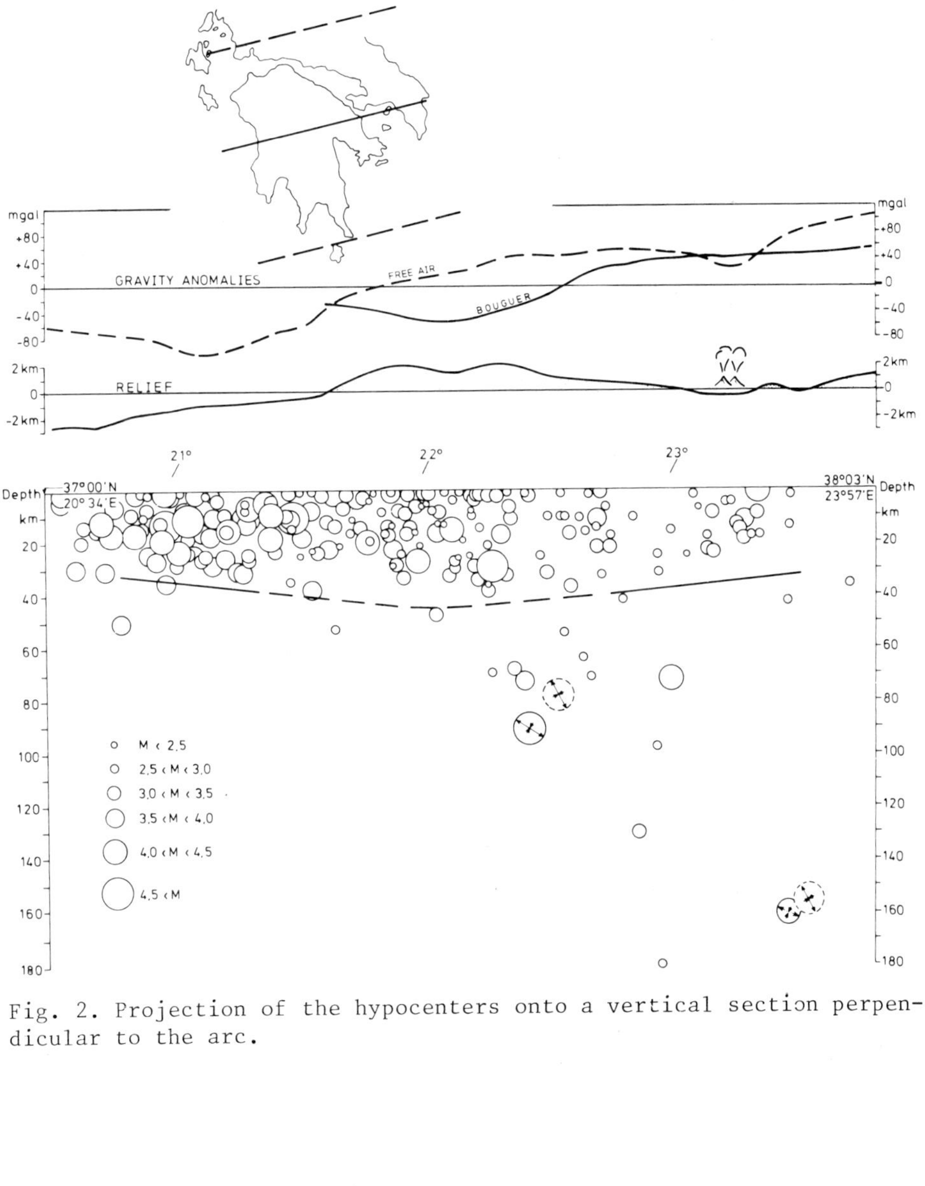

Fig. 2. Projection of the hypocenters onto a vertical section perpendicular to the arc.

Results

During the observation time 321 earthquakes in the Peloponnesus region were analysed (fig. 1). Surprisingly, there are no intermediate earthquakes in the Ionian sea and west of 22° E in contrast to previous publications (Galanopoulos 1972; Papazachos & Comninakis 1971). Of course, the observation time is too short to exclude definitely the occurrence of intermediate shocks west of 22° E. East of 22° E we find intermediate deep events with increasing depth towards the east. However, in this region there are fewer shallow events than in the western part.

The most important results can be seen in a hypocenter projection onto a plane perpendicular to the island-arc (fig. 2). The gravity anomalies and the trend of the Moho are taken from Makris (1973, 1975); two of the earthquakes (dotted lines) with fault plane solution are taken from Isacks & Molnar (1971).

More than 90% of the events are inside the crust, 0 - 40 km deep. The almost uniform distribution throughout the whole crust is a clear indication of deep fracturing. The intermediate earthquakes to 180 km depth are located on a plane with an average dip angle of 45°. We interpret it as a Benioff plane which, in terms of plate tectonics, indicates downgoing lithosphere. This interpretation is supported by the fault plane solutions: all axes of maximum tension are in the dip of this plane. The axes of maximum pressure, however, do not show any consistency. This can be understood as a principal force system due to gravitational pull on the slab.

Discussion

A comparison (Leydecker 1975) with typical and thoroughly studied arcs reveals that the similarities are more than superficial. In the Kurile arc we find the same general picture (Fedotov 1968): the bulk of the earthquakes is almost uniformly distributed throughout the crust; there is a small number of shallow events in the inner part of the island-arc; the intermediate shocks begin at 130 km distance from the trench, just as in the Peloponnesus region. Also the stress distribution is comparable with some other Benioff zones (Isacks & Molnar 1971).

Precise focal depth determinations have greatly improved the picture of a descending lithosphere. Similar results should be expected at other parts of the Hellenic island-arc.

Acknowledgment

The authors thank Prof. Galanopoulos and coworkers, Seismological Institute of the National Observatory of Athens for their help during the installation of the stations and by preparing of the records.

References

Eaton, J.P., 1969: U.S. Geological Survey, Open files report.

Fedotov, S.A., 1968: In: Knopoff,L., Drake, C.L. & Hart,P.J. (Editors), The crust and Upper Mantle of the Pacific Area. Am. Geoph. Union,

Washington D.C.:131-139.
Galanopoulos, A.G., 1972: Bull. Geol. Soc. Greece. - 9:266-285.
Isacks, B.L. & Molnar, P., 1971: Rev. Geoph. - 9:103-174.
Lee, W.H.K., Benett, R.E. & Meagher, K.L., 1972: U.S. Geological Survey, Open file report.
Leydecker, G., 1975: Veröff. Inst.f.Meteorologie-Geophysik Frenkfurt/Main no 29.
Makris, J., Mavridis, L.N., Menzel, H., Stavrou, A. & Veis, G., 1973: Ztschr. f. Geophys. - 39:929-936.
Markis, J., 1975: J. Geoph. 41:441-443.
Papazachos, B.C. & Comninakis, P.E., 1971: J. Geoph. Res.76:8517-8533.
Tsumura, K., 1967: Bull.Earthqu.Res. Inst. - 45:7-18.

Benioff Zones of the Aegean Arc

I. Richter and K. Strobach

Introduction

In the western and southern earthquake belts of the Aegean sea, several authors (Caputo et al. 1970; Galanopoulos 1973; Leydecker 1975; Morelli et al. 1975; Papazachos et al. 1971, 1973) found indications for a dipping zone of the earthquake foci, but, with the exception of the results of Leydecker (1975), the presented cross-sections include data which are summed up along broad horizontal zones demonstrating the existence of a dipping zone, but hiding the finer details of the three-dimensional earthquake distribution and the corresponding tectonic strucutres and processes.

The Model

In order to get a better imagination of the earthquake distribution, we constructed a three-dimensional model of the foci. This was completed by wooden pearls with different colours for the magnitudes fixed on vertical steel needles. Both the horizontal and the vertical scale is 1:1 000 000. The covered area is the same as in fig. 1.

The data used for representation in the model are listed in the following table. The total number of foci is 837. Only foci of depths $>$40 km were used to avoid an unnecessary masking of the essential parts of the model.

Table of used data: 1901-1955: Catalogue of Kárnik (1969); 1953-1965: Catalogue of Rothé (1969); 1964-1973: Regional Bulletin of the ISC.

The coordinates, depths, and magnitudes are not homogeneous in time. The older data appear to give depths and magnitudes which are both too great. But this supposition does not seriously influence the essential results which are mostly based on the overwhelming number of the newer data.

Preliminary Results

The model gives a clear insight into the three-dimensional distribution of the foci. The dashed line EB in fig. 1 marks the outer border of the earthquake belt; the other dashed line BDE is the border of the deep foci; it encircles an area which is totally free from deep shocks. The most remarkable feature is a deep cluster of foci which is situated west of the island Kos (*Kos cluster*). Southeast of it is another cluster of shallower sub-crustal earthquakes. Both clusters are marked by shaded areas in fig. 1, also the third noteworthy assemblage of deep seated foci with the shape of a slender tongue and which is situated beneath Athens. The other deep foci are distributed more disintegrated, but with a clear tendency of deepening towards the inner border of deep shocks (line BDE in fig. 1). After a careful observation of the model five cross-sections were drawn (fig. 2 - 6). Their positions are shown in fig. 1 by the demarcated rectangels, which include all foci presented in the cross-sections. The symbols for the magnitudes are shown in fig. 3, the orientation of the sections is marked by the letters A - B, C - D, etc.

Cross-section A - B (fig. 2) summarizes foci of a 85 km wide zone, including the Kos cluster. A clear Benioff zone appears with a dip angle of 32 degrees. Fig. 3 shows the same area, but a much smaller zone of 30 km was cut out to demonstrate the remarkable concentration and isolated position of the cluster and the gap between this and the sub-crustal cluster above. Cross-section E - F (fig. 4) illustrates the area more northeast of the foregoing sections. The Kos cluster appears again, but the other foci show the continuation of the Benioff zone following the arcuate system to the northeast.

A broad section G - H (fig. 5) covering most of the southern part of the Aegean arc and including Crete gives no significant indication for a Benioff zone, although one cannot exclude its existence. The foci with depths of about 150 km belong to the older determinations and are not very reliable concerning both depths and magnitudes, as stated above. Here the underthrusting plate may be split up, as will be discussed below.

The fifth cross-section J - K (fig. 6) covers a part of the Peloponnesus and Attica. Here again a clear Benioff zone exists and reaches a depth of 175 km forming a slender tongue. The dip angle is 35 degrees. The foci additionally entered by crosses are special determinations by Leydecker (1975 and this vol.). These very certain data fit exactly in the dipping foci zone of our model.

The existence of an underthrusting plate reaching maximum depths of 175 km is clearly indicated by the sections exhibiting Benioff zones, but these zones are restricted to certain regions and are interrupted by diffuse earthquake distributions.

Tectonics

If the African plate underthrusts the Aegean arc, strong horizontal tensional stresses should develop in the plate in the strike direction of the arcuated features. This is a geometrical consequence of the strong curvature of the bending line along the system of the Hellenic, Pliny and Strabo trenches (fig. 1). In order to form a conical downgoing slab it is inevitable that the border of the slab, which reaches a depth of about 175 km, undergoes a process of intensive stretching in its strike direction by the order of 25 per cent. This value follows from the equation (2) in Strobach (1973)

$$\varepsilon = \frac{1}{R_o} \cdot \frac{\cos\beta - \cos(\alpha - \beta)}{\sin\beta - (1/R_o)\cos\beta},$$

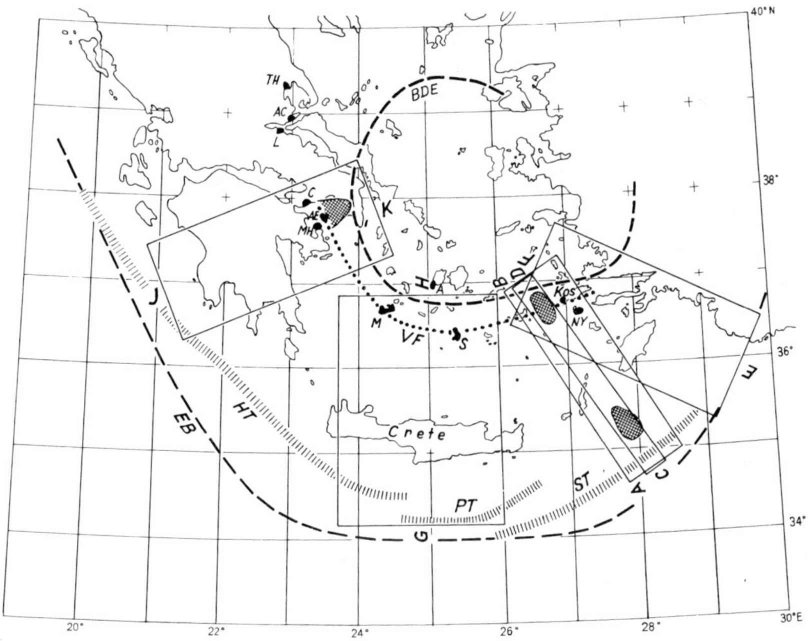

Fig. 1. Map of the investigated area. The positions of the cross sections correspond to the rectangles A-B, C-D etc. EB = outer border of earthquake belt; BDE = border of deep earthquakes; HT = Hellenic trench; PT = Pliny trench; ST = Strabo trench; VF = volcanic front (after Fytikas et al. 1976): TH = Thebe; AC = Achilleon; L = Lesvos; C = Crommyonia, AE = Aegina; MH = Methana; M = Milos; A = Antiparos; S = Santorin, Kos; NY=Nysiros.

Fig. 7. Focal mechanism solutions for the deep Kos-cluster (above) and the adjacent sub-crustal cluster. Equal area projections of the lower hemisphere of the radiation field. Quadrants of tensional stresses are black. T = tensional axis; P = compressional axis. The arrows indicate the mean horizontal direction of tensional stress for the Kos cluster.

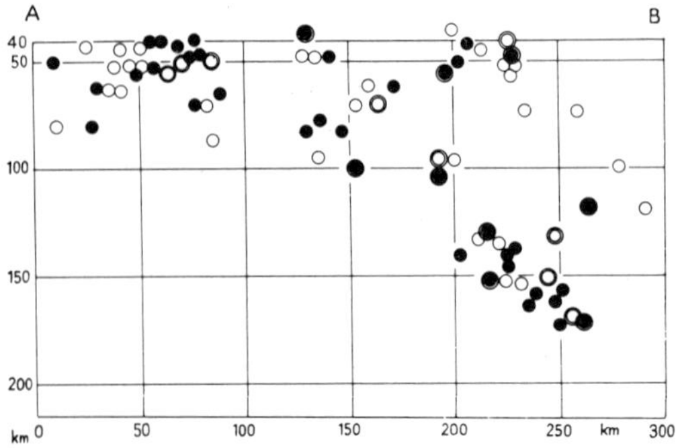

Fig. 2. Cross-section A-B, showing the projection of all foci of the area A-B in fig. 1 on a vertical plane. For magnitude symbols see fig. 3.

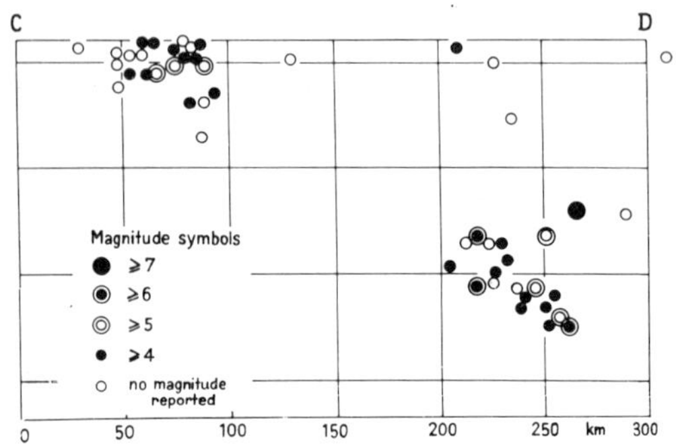

Fig. 3. Cross-section C-D. See text of fig. 2.

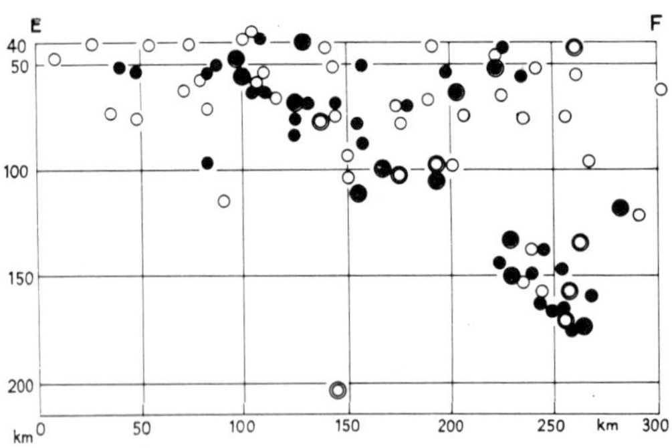

Fig. 4. Cross-section E-F. See text of fig. 2.

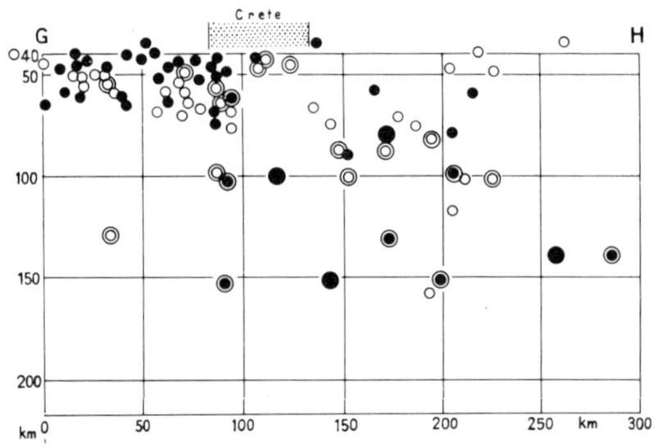

Fig. 5. Cross-section G-H. See text of fig. 2.

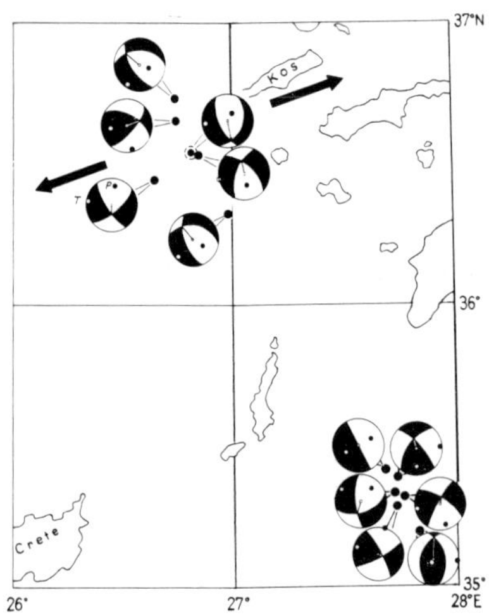

Fig.7.

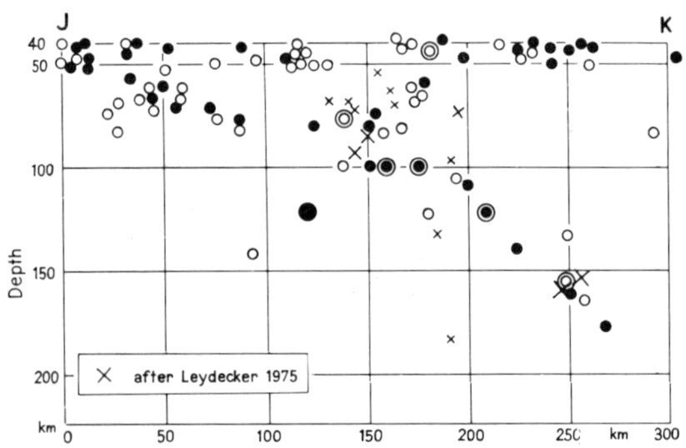

Fig. 6. Cross-section J-K. The crossses are determinations by Leydecker (1975). See text of fig. 2.

where ε = relative extension, l = length of the slab (290 km), R_o = earth's radius, β = radius of the island arc (3°8 or 423 km), α = dip angle of the slab (32°).

It is clear that no lithospheric plate can resist tensional deformations of 25 per cent without breaking up in pieces or extending by shear fractures.

Therefore we used all suitable informations of the ISC to get fault-plane solutions for the deep-focus shocks of the Kos cluster. Fig. 7 shows the results. In spite of the uncertainties of the reported directions of the first motions, there is a clear tendency in the solutions of predominating tensional stresses with a mean horizontal direction of N 70°E, which nearly coincides with the strike direction of the slab. The motions represent normal faulting, i.e. extension in the sense discussed above; only one of the six solutions indicates thrust faulting.

The existence of only one pronounced cluster (fig. 3) and of one other sufficiently documented tongue (fig. 6) among the distribution features of the deep shocks, and the irregular grouping beneath Crete (fig. 5), seem to be indications for the splitting up and disintegration of the underthrusting African plate.

The disappearance of the slab, insofar it is marked by shock positions, at a depth level of 180 km, is consistent with the data of Benioff zones of the Pacific area. If a slab sinks into the mantle with a velocity of 10 cm/yr, it vanishes after 10 million years due to temperature equalization, and during that time its leading edge has passed a distance of 1000 km. The slab of the African plate reaches a length of about 300 km. If it also vanishes after 10 million years of subduction, its velocity should be of the order of 3 cm/yr, a value which is consistent with the calculations of Le Pichon (1963) who got a relative velocity between the African and the Eurasian plate of 2,6 cm/yr for the Aegean region.

Furthermore, the results of this investigation are in good agreement with the position of the volcanic front in island arcs. Many authors have shown, that the volcanic front corresponds to the line where the dipping plate reaches a depth of 150 km. In fig. 1 the volcanic front (VF) is marked by the dotted line. It includes the volcanic island of Santorin and passes all cross-sections exactly at that point where the Benioff zone has reached a depth of 150 km.

References

Caputo, M., Panza, G.F. & Postpischl, D., 1970: J. Geophys.Res. 75: 4919-4923.

Fytikas, M., Giuliani, O., Innocenti, F., Marinelli, G. & Mazzuoli, R., 1976: Tectonophysics, 31:T29-T34.

Galanopoulos, A.G., 1973:Annali di Geofisica, 26:85-105.

Kárník, V., 1969: Seismicity of the European area, part 1, Dordrecht.

Leydecker, G., 1975: Diss. Frankfurt.

Morelli, C., Pisani, M. & Gantar, C., 1975: Boll. di Geofisca teor. ed appl. - 18:127-168.

Papazachos, B.C.& Comninakis, P.E.,1971:J-Geophys.Res.,76:8517-8533.

Papazachos, B.C., 1973: Geophys.J.R. astr.Soc., 33:421-430.

Le Pichon, X., 1968: J. Geophys. Res. - 73:3661-3697.

Rothé, J.P., 1969: The seismicity of the earth 1953-1965, UNESCO.

Strobach, K., 1973: Z. f. Geophys. - 39:819-831.

Part 2

Geology of the Hellenides

Alpidic Development and Structure of the Peloponnesus

V. Jacobshagen, D. Richter and J. Makris, with Contributions of G.H. Bachmann, P. Giese and H. Risch

Introduction

On the Peloponnesus and the adjacent islands the well-known isopic zones of the Hellenides (Renz 1955; Aubouin et al. 1963) are related to at least four tectonic levels according to Jacobshagen et al. (1976b). The lowermost Plattenkalk (= Talea Ori) series is believed to be the eastern continuation of the pre-Apulian zone and therefore should be autochthonous. The second tectonic level (West Hellenic nappe system), consists of two units: the lower is represented by the Phyllite series whereas the upper comprises the sediments of the Ionian and the Tripolitza zones. The Tripolitza sediments are overlain by yet another tectonic level henceforth called the Central Hellenic nappe system. On the Peloponnesus it comprises the Pindos nappe and the Mesozoic of the Argolis Peninsula (Argolicum); in continental Greece, the Parnassos unit as well as the Subpelagonic/Pelagonic nappes belong to the Central Hellenic systems, too.

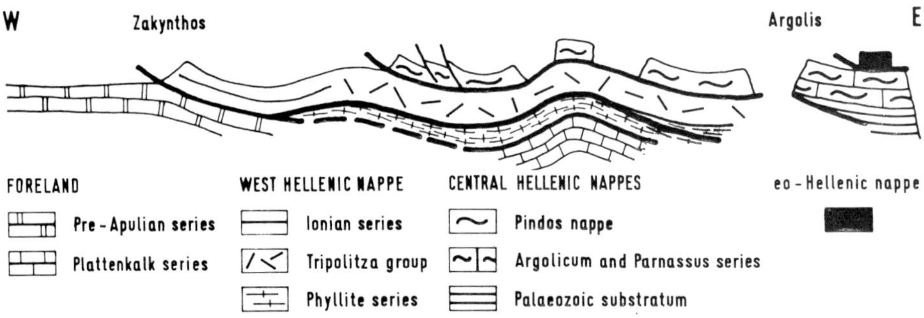

Fig. 1. Schematic cross-section through the nappe sequence of the Peloponnesus.

The Argolicum is overthrust by an oceanic nappe of ophiolites and pelagic sediments during the Late Jurassic - Early Cretaceous Eo-hellenic phase (Jacobshagen et al. 1976a) being the uppermost tectonic unit. Both the Argolicum and the Eo-hellenic nappe relics have subsequently been covered by Cretaceous limestones and the Palaeogene flysch. The described nappe sequence is shown schematically in fig. 1.

Our model is based on arguments concerning stratigraphy, facies and metamorphism which are recorded in the description of the different units (see below). Furthermore, it is supported by

the *ubiquity of tectonic contacts* between the pre-Apulian and the Ionian zone: wherever contacts are exposed on the Ionian islands overthrusts have been described (e.g. Bornovas 1964; I.F.P./I.G.S.R. 1966; B.P. Co. 1971);

the fact that *the Ionian zone* and *the Gavrovo-Triplitza zone are covered by a uniform flysch* (Richter 1976) which shows no signs of a nappe boundary in its interior (compare e.g. the detailed maps in B.P.Co. 1971). Therefore, both zones belong to only one nappe contrary to the proposal of Dercourt et al. (1976).

For most of the previous authors, the essential criterion for grouping the isopic zones into *external* and *internal* units was the age of the main alpidic tectogenesis within each zone. With this criterion, the Pindos zone had been attributed to the external zones. We propose to restrict the term *external zones* to the pre-Apulian zone, the Plattenkalk Series, and the West Hellenic nappes, based on their tectonic position as well as to the beginning of their main flysch sedimentation.

In the following, the major tectonic units are briefly reviewed.

Autochthonous (pre-Apulian Zone and Plattenkalk Series)

The *pre-Apulian zone* is only exposed on the island of Zakynthos where flat lying neritic limestones of Cretaceous to Eocene age occur, overlain by Oligocene marls and limestones. The youngest beds are Neogene sandstones and marls (B.P.Co.1971). This zone is believed to be the eastern margin of the Apulian platform.

Recent publications have shown that the Plattenkalk series exposed in the Taygetos and Parnon Mountains and on the Mani Peninsula of the southern Peloponnesus must be separated from the Phyllite series. The total thickness of the Plattenkalk series is unknown; at least several hundred meters are exposed consisting of marbles which are rich in chert lenses or bands in the higher part. The uppermost Plattenkalk is built up by varicoloured chloritic marbles with calcschists which have yielded an Oligocene microfauna (Bizon & Thiébault 1974). On the top of the Plattenkalk series flyschoid sediments were cited from some places (Bizon et al. 1976).

Based on the lithological character and the tectonic position, the Plattenkalk series on the Peloponnesus resp. the Talea Ori series on Crete have been attributed to the Ionian zone by Kuss & Thorbecke (1974).

In opposition to this interpretation, we consider the Plattenkalk series to have been located topologically between the pre-Apulian zone and the Ionian zone, for the following reasons:

- the *top of the Plattenkalk* carbonates is of Oligocene age whereas in the Ionian zone this stage is completely represented by flysch ;
- the *pelitic and sandy sediments* above the varicoloured Plattenkalk described by Bizon et al. (1976) cannot be identified with the very thick sequence of typical flysch in the Ionian zone but may have an intermediate position between the Ionian flysch and the clastics covering the pre-Apulian carbonate pile.
- the *Plattenkalk series is overthrust* by the Phyllite series (Lower West Hellenic nappe) which is covered by the Tripolitza limestones. As

the latter passes over into the Ionian sequence laterally (e.g. B.P.Co. 1971), the tectonic position of the Ionian zone is clearly higher than that of the Plattenkalk series;
- a *lateral transition* between the pre-Apulian and the Ionian zone as has been assumed by all authors in the past *cannot* be observed anywhere.

Lower West Hellenic Nappe Unit: Phyllite Series

The Phyllite series (Tripolitza phyllites auct.) is situated between the Plattenkalk series below and the Tripolitza limestones. It comprises preponderantly phyllites, chlorite schists, quartzites and meta-volcanites. On the top of the Phyllite series gypsum was found at some places which may belong to the Lower Triassic. High pressure metamorphism is indicated by glaucophane in several places (Paraskevopoulos 1964; Richter & Thorbecke unpubl.). The deposition of the metasediments is dated by Carboniferous and Permian foraminifera (Paraskevopoulos 1964; Fytrolakis 1971; Lys & Thiébault 1971). Therefore, the metamorphism is of Alpidic age.

The above mentioned facts prove the importance of the tectonic contact between the Phyllite series and the Plattenkalk sequence without any doubt. The upper contact to the Tripolitza limestones, however, is discussed intensely at present. Generally, it is of undoubted tectonic origin; nevertheless, sedimentary contacts have been found in some places suggesting local areas of pre-Middle Triassic metamorphism of the phyllites and urging to conclude that the Phyllite series is not uniform but comprises both Alpidic and pre-Alpidic metamorphosed parts. If this is true, the bulk of the Phyllite series must be a separate tectonic unit as it is supposed in Crete (Thorbecke 1974; Creutzburg & Seidel 1975). This view is supported by the high pressure metamorphism which cannot be explained by the present cover of sedimentary nappes. A metamorphic discontinuity exists between the Phyllite series and the underlying Plattenkalk series since the latter is only weakly metamorphosed in comparison (Richter & Thorbecke, unpubl.).

The palinspastic position of the Phyllite series is enigmatic. The existence of gypsum on the top of the Phyllite series as well as at the base of the Ionian sequence (I.F.P./I.G.S.R. 1966) and of the westernmost Tripolitza series of the Peloponnesus (B.P.Co. 1971) may suggest that at least parts of the Phyllite series could have been the basement of the Ionian and the Tripolitza series. This assumption implies that the Phyllite series must have been separated from its Mesozoic carbonate cover by Alpidic events and metamorphosed during a subduction.

Upper West Hellenic Nappe Unit: Ionian and Tripolitza Sediments

Sediments of the Ionian zone are exposed in a few small areas west at Cape Pappas and near Kyllini (Northwest Peloponnesus, fig. 2) with Upper Cretaceous and Palaeogene limestones. On Zakynthos, a few outcrops of Ionian sediments show the same stratigraphy and facies as in continental Greece (vide Renz 1955; I.F.P./I.G.S.R. 1966; B.P.Co. 1971). Elsewhere in the Western Peloponnesus the Ionian zone is covered by Neogene sediments.

The same is true for the Tripolitza zone which rarely crops out on the Western Peloponnesus. The sedimentary sequence of this zone, here called

Tripolitza series, consists of the Tripolitza Limestone (probably Middle Triassic to Upper Eocene) with a thickness of 1500 m and the flysch cover (Upper Eocene to Aquitanian). The flysch lies largely unconformably on the carstified Tripolitza limestone (Richter this vol.)and reaches a thickness of 500 m in the eastern and central Peloponnesus and of more than 1200 m in the western areas (south of Patras and north of Pylos). The main tectogenesis of the Tripolitza series occurred probably during the Lower Miocene (Kowalczyk et al. 1977).

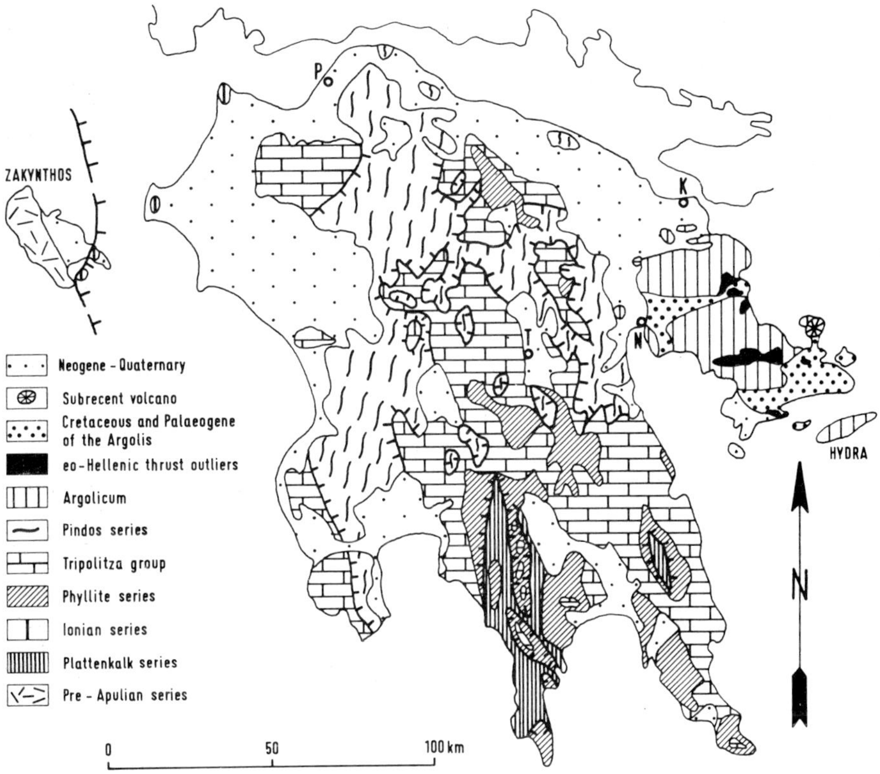

Fig. 2. Tectonic sketch-map of the Peloponnesus.

A conspicuous recrystallization in the lower parts of the Tripolitza limestone of central Laconia (Richter 1975) suggests an alpidic metamorphic event, but it is not clear whether it occurred together with the recrystallization of the Plattenkalk series i.e. after the overthrust of the Tripolitza series or before.

The tectonic structure of the Upper West Hellenic unit is characterized by broad anticlines and synclines and by fractures.

Pindos Zone

The Pindos nappe moved more than 80 km from the east to the west over the Tripolitza series. The pelagic Pindos series (Renz 1955; Dercourt 1964; Aubouin et al. 1970; Lalechos 1974) shows an arenaceous intercalation with extremely variable thickness, the First Pindic Flysch (Barremian to Lower Turonian), and, at the top, a final flysch (Palaeocene to Eocene). The Pindos rocks form coherent mountain chains along the western coast of the Argolian Gulf and in the western part of the Peloponnesus (Olonos Range). In the central Peloponnesus there are only some thrust outliers resting on the Tripolitza series.

The Pindos series commonly shows a stronger internal deformation than the underlying Tripolitza series. In the eastern mountain chain recumbent west-facing folds are the characterizing structural elements which document westward movement of the Pindos thrust sheet (Doert, this vol.). Upright to oblique, partly overturned multifacing small-scale folds are superimposed on the main folding (Doert, Richter & Mariolakos 1977). The western chain is made up of numerous thrust slices gently dipping to the east, each of which is itself strongly folded. Contrary to the eastern mountain chain a first generation of folds shows axial planes which are vertical or dip steeply to the east (Richter 1977). It is evident that folding preceded thrusting. The structural cast of the Pindos nappe indicates that is represents a superficial tectonic level.

Argolian Mesozoic and Eo-Hellenic Ophiolites

According to Dercourt (1964) the Argolis Peninsula belongs to two different zones of the Hellenides: The Trapezona Massif in the north was supposed to be a part of the Parnassus Zone and the southern mountains of the Argolis were regarded as parts of the sub-Pelagonian Zone with lateral transition between both areas.

The stratigraphic sequence of the southern region (Bannert & Bender 1968; Kalkreuth et al. 1978; Bachmann & Risch 1976) consists of neritic limestones with local transitions to cherty limestones and red nodular limestones (Hallstatt facies) passing upward into Ammonitico rosso (Upper Lias to Lower Dogger) and a series of radiolarites, cherty limestones and shales with intercalations of tuffites and breccias. They are believed to be of Upper Jurassic and perhaps lowermost Cretaceous age.

The sequence of the Trapezona Massif is similar to that of the southern region but shows more and wider stratigraphic gaps; the *Diabase - Radiolarite - Tuffite Series*, at least, has been found in various places. The most important hiatus includes nearly the complete Cretaceous. Consequently, the Trapezona Massif may be interpreted as an area of temporary uplift and denudation. Facies connections to the Parnassus zone, however, exist only in the westernmost Trapezona Massif (Bachmann & Risch 1976).

In this paper, the Triassic-Jurassic sequences of the southern region and of the Trapezona Massif are united with the term *Argolicum* since the former "sub-Pelagonian zone" cannot be defined clearly (Jacobshagen et al. 1976a).

The Argolicum has been overthrust by obducted oceanic crust about the Jurassic/Cretaceous boundary (Jacobshagen et al. 1976a). During this Eohellenic tectogenesis the area was exposed to erosion, and thus only small thrust outliers of serpentinites, diabases, and radiolarites are

preserved. Thus, the *Diabase-Radiolarite-Tuffite Series* has been divided by a thrust plane into two parts. During the upper part of the Lower Cretaceous the entire Argolis region was resubmerged. Neritic limestones (Barremian to Cenomanian) pass up into pelagic limestones and marls (Turonian to early Palaeocene). Flysch deposits (Late Palaeocene to Eocene) overlie rocks of varying age, commonly with a hiatus.

Palaeogeographical Connection between the Pindos Zone and the Argolian Mesozoic

Throughout Mesozoic time, the south Argolian depocenter and the Pindos trough were closely adjacent, as can be inferred from facies relationships. The Pantokrator Limestone grades laterally into cherty limestones, and the Upper Jurassic of the Argolis differs from the normal Pindos sediments only by the presence of volcanic intercalations. Particularly the Upper Cretaceous sediments are nearly alike in both areas.

Furthermore, an *Orbitolina*-bearing horizon occurs, with the same age and lithofacies, both at the top of the Eo-hellenic flysch in the Western Argolis and at the top of the First Pindic Flysch (Bachmann & Risch 1976). East of Navplion, the Eo-hellenic flysch overlaps the Pantokrator limestone whereas west of Argos it may underlie the Upper Cretaceous Pindos Limestones.

Tectonic Structure of the Argolicum

Apart from the Eo-hellenic thrust outliers there are conspicuous folds and minor overthrusts with varying vergence within the Argolicum. The prevailing direction of fold axes varying about W-E is locally crossed by younger N-S running folds (e.g. Kalkreuth et al. 1978). Both deformations are younger than the Palaeogene flysch and probably originated by the Late Eocene tectogenesis (Jacobshagen & Skala 1977; Jacobshagen et al. this vol.).

The predominant E-W direction of the Argolian tectonic structures fits perfectly into the large scale pattern of the late Eocene Central Aegean arc (Jacobshagen & Skala 1977). It is not necessary to assume a block rotation of the Argolis Peninsula (Pucher et al. 1973, 1974) especially since there is no fracture evidence of such an event either in the field or on aerial and space photographs.

Also in the eastern Pindos mountain chain (s. above) near Astros folds with E-W running axes occur (Doert this vol.) as a further argument for a primary connection between both of these sites of sedimentation areas.

Crustal Structure of the Peloponnesus

The crustal structure of the Peloponnesus has been investigated with gravimetric, seismic, and magnetic measurements.

From a gravimetric point of view, Greece can be divided into two regions: the western Hellenides, characterized by a strong negative Bouguer anomaly, and the Aegean area s.str. showing a positive Bouguer anomaly. The gravity high of the Ionian Sea reaches nearly the western margin of Greece (Makris 1976 and this vol.; Makris et al. 1973).

The southern part of the negative anomaly continues onto the Peloponnesus but its amplitude decreases from 120 mgal in the north to about zero in the south.

Details of the crustal structure have been revealed by a seismic refraction profile crossing the Peloponnesus from southwest to northeast (Hinz 1974; Weigel 1974; Makris 1976). In the pre-Apulian area below the Ionian Sea the crust is 20-22 km thick. This thin foreland crust continues inland 30-40 km under the Ionian zone. A gentle easterly rise of the crustal base is possible.

30 to 40 km east of the coastline the crustal structure changes profoundly. The crust/mantle boundary becomes unrecognizable from the coastal regions, and another separate crust/mantle interface was found by reflected waves at 45-50 km depth. It cannot be excluded that the upper boundary overlaps the deeper one. The relatively low average crustal velocity derived from the deep crust/mantle boundary reflections indicates a considerable amount of sialic material. Thus, the large crustal thickness beneath the Peloponnesus may be seen as composed of two sialic layers of reduced thickness. Toward the east, the crustal thickness decreases in accordance with the gravimetric data. Because no seismic observations are available from the eastern end of the profile details of this up-dip of the crust/mantle boundary are still unknown. A strong gradient in the Parnon Mountains on the eastern Peloponnesus points to complications of the crustal structure beneath that area. Below the Aegean Sea the crust is 25-30 km thick.

The magnetic field of the Peloponnesus is rather regular with small anomalies concentrated in the southern regions and in the Argolis (Makris 1976).

Orogenic Development of the Peloponnesus

Nearly all early Mesozoic sediments of the Peloponnesus were deposited on a shelf furrowed only by the Pindos through which widened successively during the Jurassic (Jacobshagen 1972). During the Tithonian, ophiolite sheets were transported from the northeast to the southwest over the Pelagonian platform (Kauffmann 1976; Bachmann & Risch 1976) into the Argolian basin. Due to this early tectonic event, deposition of the Eo-hellenic flysch and of the First Pindic Flysch took place in front of the moving thrust sheet. Thus, the rising cordillera may be regarded as the source of the flysch detritus. This orogenic event was followed in the Argolis by a phase of extension and subsidence: The Upper Cretaceous sea flooded the Argolian area, and limestones were deposited. In the Ionian, Tripolitza and Pindos zones the carbonate sedimentation continued without interruption during the Upper Cretaceous. Palaeogene tectonic events are testified by the main flysch of the Argolis and of the Pindos zone.

In the Eocene the movement of the Central Hellenic nappe system started. This cannot be proved on the Peloponnesus but must be inferred from a comparison with the structure of Northern Greece (vide synthesis Jacobshagen et al. this vol.) and of the Cyclades (Dürr et al. this vol.). The weak metamorphism of the Hydra Triassic (Richter this vol.) may also have happened during this paroxysm.

At the same time, a major regression of the Eocene sea in the Tripolitza zone caused by tectonic uplift and accompanied by faulting terminated the Tripolitza limestone deposition. Subsequently, still during the

late Eocene, the Tripolitza zone subsided and a flysch sea flooded transgressivly the carstified and faulted surface of the Tripolitza limestone (Richter this vol.).

The late Eocene tectogenesis continued probably into the Oligocene. Thick olisthostromes in the flysch of the Tripolitza zone indicate, at least that parts of the Pindos zone had already risen to a cordillera at that time (Richter 1976a, b; Kowalczyk et al. 1977).

The beginning of a Miocene tectogenesis is indicated by the termination of flysch deposition in the Tripolitza zone during the Aquitanian.It ended before or at the beginning of the Tortonian (Kowalczyk et al.1977). In the course of these movements the Tripolitza sediments were folded and then partly denudated (Richter 1974). Subsequently, the Central Hellenic nappe system advanced further towards the west shown by the Pindos overthrust. It is uncertain whether the West Hellenic nappe moved over the Plattenkalk series at that time. On the island of Levkas, where Ionian rocks are overthrust on pre-Apulian Upper Miocene (I.F.P./I.G.S.R. 1966), this nappe movement might have taken place during the late Neogene. It must be taken into account, however, that the Levkas overthrust was the result of a later compressional deformation of minor intensity which was locally enhanced by halokinetic effects (thick Triassic evaporites!).

On the whole, the tectonic structure of the Peloponnesus is characterized by more or less flat lying nappe sheets and long distances of transport. This scheme is in accordance with the quoted geophysical results. The recent Benioff plane (Leydecker 1975; Leydecker & Berckhemer this vol.) is believed to be of very young origin. Up to now, it cannot be related to structural features of the described Alpidic orogenesis.

References

Aubouin, J., Bonneau, M., Celet, P., Charvet, J. et al., 1970: Ann. Soc. géol.Nord, 90:277-306, Lille.

Aubouin, J., Brunn, J.H., Celet, P., Dercourt, J., Godfriaux, J. & Mercier, J., 1963: Extr. Livre Mém. Prof. Paul Fallot, Mém.Soc.géol. France, 2:583-610.

Bachmann, G.H. & Risch, H., 1976: N.Jb.Geol.Paläont.Abh.,152:137-160.

Bannert, D. & Bender, H., 1968: Geologica et Palaeontologica,2:151-162.

Bizon, G. & Thiébault, F., 1974: C.R.Acad.Sci.Paris, (D), 278:9-12.

Bizon, G., Bonneau, M. et al., 1976: Bull.Soc.géol.France, (VII), 18:337-345.

Bornovas, J., 1964: Inst. Geol.Subsurface Res., 10:1-142, Athens.

B.P.Co., 1971: Inst.Geol.Subsurface Res., 10:1-73, Athens.

Creutzburg, N. & Seidel, E., 1975: N.Jb.Geol.Paläont.,Abh., 149:363-383.

Dercourt, J., 1964: Ann.géol.pays hell., 15:1-418.

Dercourt, J., Wever, P.de & Fleury, J.-J., 1976: Bull.Soc.géol.France, VII, 18:317-326.

Doert, U., Richter, D. & Mariolakos, I., Ms. 1977.

Hinz, K., 1974: Geol. Jb., E2:33-65.

I.F.P./I.G.S.R., 1966: Étude géologique de l'Épire (Grèce nord-occidentale). 306 p.

Jacobshagen, V., 1972: Z.dt.geol.Ges., 123:445-454.

Jacobshagen, V., Risch, H. & Roeder, D., 1976a: Z.dt.geol.Ges.,127:133-145.

Jacobshagen, V., Makris, J., Richter, D. et al.,1976b: Z.dt. geol. Ges., 127:337-363.

Jacobshagen, V. & Skala, W.: Ann. géol. Pays hellén., in press.
Kalkreuth, W., Wallner, P., Jacobshagen, V. & Risch, H.: Bull. geol. Soc. Greece, in press.
Kauffmann, G., 1976: Bull. Soc. géol. France VII, 18:308.
Kowalczyk, G., Richter, D., Risch, H. & Winter, K.P., 1977: N. Jb. Geol. Paläont., Mh.:549-564.
Kuss, S.E. & Thorbecke, G., 1974: Ber. Naturforsch.Ges. Freiburg i.Br., 64:39-75.
Lalechos, N.S., 1974: Geol. Reconn., 53,94 p., Athens.
Leydecker, G., 1975: Diss. Univ. Frankfurt, 75 p.
Lys, M. & Thiébault, F., 1971: C.R. Acad. Sci.Paris, (D), 272:1-2.
Makris, J., 1973: Bull. geol. Soc. Greece, 10(1):206-213.
- : Hamburger geophys. Einzelschr., in press.
Makris, J., Mavridis, L.N., Menzel, H., Stavrou, A. & Veis, G., 1973: Z. Geophys., 39:929-936.
Paraskevopoulos, G.M., 1964: N. Jb. Miner., Abh., 101:195-209.
Pucher, R., Bannert, D. & Fromm, K., 1973: Bull. geol. Soc. Greece, 10,1:169-171.
- , 1974: Tectonophysics, 22:31-39.
Renz, C., 1955: Inst.Geol.Subsurface Res., Athens, 637 p.
Richter, D., 1974: N. Jb. Geol. Paläont., Abh., 145:96-128.
- , 1976: Z.dt.geol. Ges., 127:467-483.
- : Rapp. Comm. internat. Mer. Medit., in press.
Thiébault, F., 1968: Ann. géol. Soc. Nord, 88:209-214.
Thorbecke, G., 1974: N. Jb. Geol. Paläont., Mh.:246-256.
Weigel, W., 1974: Hamburger geophys. Einzelschr., 28, 140 p.

Paleomagnetism in Greece: Indications for Relative Block Movement

R. Pucher, D. Bannert and K. Fromm

There is a difference of 120° between the strike of the Pindos mountain chain and the structural grain of the Argolis peninsula (Aubouin 1959; Bannert & Bender 1968). Both consist of Mesozoic to Paleogene age. The difference in strike may have been caused by block rotation, and consequently it was of interest if paleomagnetic data would suggest a similar amount of rotation between both areas.

The data derived from the paleomagnetic samples are summarized in table 1. Only igneous rocks of Jurassic age were selected. From the declination of the remanent magnetization with dip correction, a clockwise rotational block movement with an angle of about 110° could be assumed. On the other hand, the corresponding virtual pole position of the Pindos samples agrees fairly well with that of the African plate. This fact is of great interest, since Turkey showed the same position with respect to Africa for a long time (Van der Voo 1968).

Table 1: Data of the magnetic samples

	Pindos area	Argolis peninsula
Type of rock	gabbro	diabase
Sampling sites	2	4
Lava strata	4	not specified
Specimen	7	9
Samples	23	24
Remanence direction	+ and -	+ and -
Natural remanent magnetization (NRM)	1.22 A/m	2.22 A/m
Susceptibility	$1.51 \cdot 10^{-3}$	$1.49 \cdot 10^{-3}$
Remanent magnetization H∿=32 kA/m	$2.11 \cdot 10^{-1}$ A/m	$1.82 \cdot 10^{-1}$ A/m
Mean-site, dipcorrected:		
Declination D	334°	82°
Inclination I	22°	19°
Fisher's statistics α_{95}°	9°	17°
Virtual pole position	112°W 53°N	110°W 12°N

References

Aubouin, J., 1959: Ann. géol. Pays hellén., 10:1-525.
Bannert, D. & Bender, H., 1968: Geologica et Palaeontologica, 2:151-162.
Pucher, R., Bannert, D. & Fromm, K., 1974: Tectonophysics, 22:31-39.
Van der Voo, R., 1968: Geophys. J. roy. astron.Soc., 16:543-547.

Late Mesozoic and Paleogene Development of the Argolis Peninsula (Peloponnesos)

G.H. Bachmann and H. Risch

Introduction

The Argolis peninsula is characterized by a complicated pattern of different facies (fig. 1). Therefore any correlation with the scheme of isopic zones of the Hellenides was up to now of doubtful value. The development of facies and paleogeography, however, shows the peninsula to be situated at the western rim of the Pelagonian Platform (Bachmann & Risch 1976; Subpelagonian Zone of various authors), with strong influence of the adjoining Pindos Trough.

Norian - Middle Jurassic

During Norian to Middle Liassic times the Argolis was situated at the western rim of an extensive carbonate platform (cf. Jacobshagen 1972), with deposition of 500 - 1000 m of the Pantokrator Limestone. Influence of the Pindos Trough is indicated by small basins with sedimentation of pelagic carbonates.

From Upper Liassic the platform was subject to more rapid subsidence. Locally, red limestones of Ammonitico rosso type were deposited.

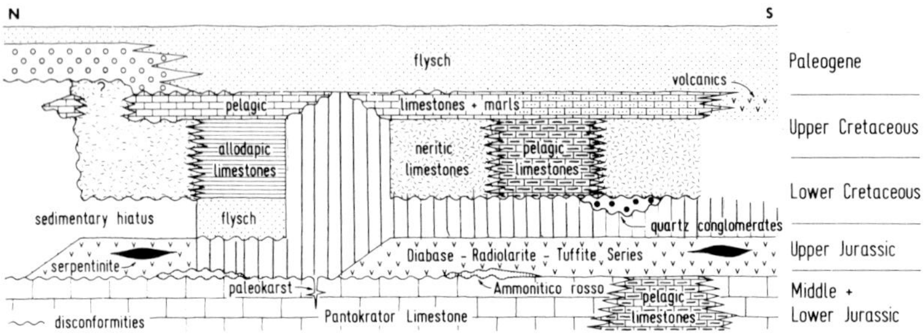

Fig. 1. Stratigraphic cross section of Argolis area, schematic, see text.

Upper Jurassic - Lower Cretaceous (Barremian)*

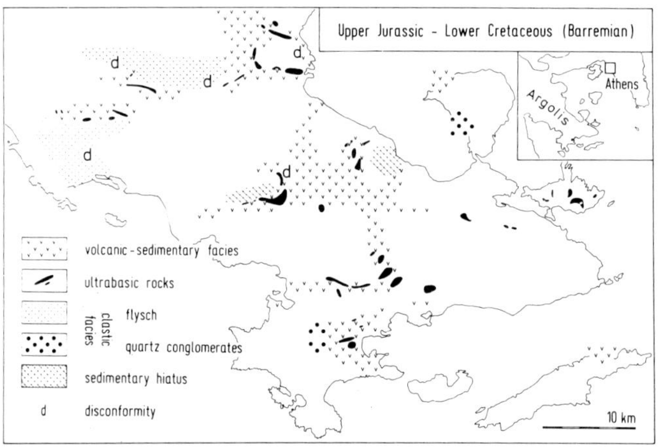

Fig. 2. From Upper Jurassic to Lower Cretaceous, the Diabase-Chert-Tuffite-Series were deposited. They are up to 1000 m thick and consist of diabases, often with pillow structure, associated with tuffites and tuffitic breccias, and red, commonly silicified pelites. Radiolarites occur especially in the lowermost parts and suggest persistent deepening of the sea floor. Allochthonous fragments of older rocks indicate local relief and tectonic activity.

Ultrabasic rocks (peridotites, serpentinites) occur mainly in the upper part of the Series. They have been explained as parts of a large

* The figs. are not palinspastic.

nappe which originated by obduction of oceanic crust from the Vardar Zone during the Eo-hellenic phase (Jacobshagen et al. 1976).

At the front of the obduction, a flysch trough developed in the northwest Argolis (Bachmann & Risch 1976). This flysch closely resembles flysch deposits of approximately the same age in continental Greece and Jugoslavia (Boetian and Bosnian Flysch resp.). Therefore, deposition of these flysches in a connected trough is assumed.

Lower Cretaceous (Barremian) - Cenomanian

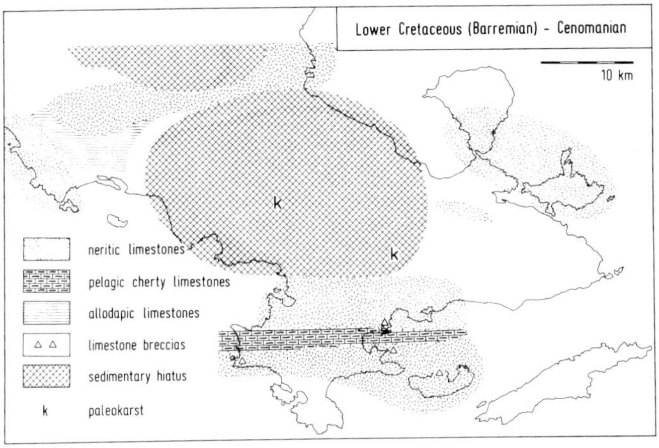

Fig. 3. During the Lower Cretaceous to Cenomanian, large parts of the central and northern Argolis peninsula were subject to erosion at local uplifts. Locally, the Pantokrator Limestone was reexposed and karstified (Bannert & Risch 1977). At the margins of these areas several 100 m of neritic carbonates were deposited. In a trough in southern Argolis, elongate east-west, pelagic carbonates were deposited simultaneously. This suggests an influence by the Pindos Zone.

In a small area in northwestern Argolis sedimentation of allodapic limestones, quartz conglomerates and limestone breccias took place. These allodapic limestones closely resemble the well-known Orbitolina Horizon (Renz 1955) of the Pindos Zone (pers. comm. Doert 1975).

Turonian - Maestrichtian

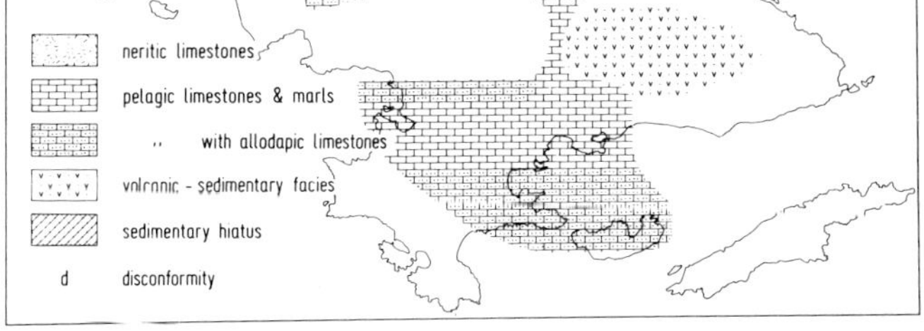

Fig. 4. In the course of the Upper Cretaceous, the Argolis peninsula underwent more rapid subsidence again. The character sediments are red or whitish cherty limestones, and red marls with pelagic foraminifers, up to several 10 m thick (cf. Kalkreuth et al. 1976). They form the transition to the Paleogene flysch and may thus be

compared with the *couches rouges* facies of the Alps. In some areas they contain abundant beds of allodapic limestones. This kind of sedimentation took place on former uplifts as well. Only small uplifts remained; they were subject to erosion before the deposition of the flysch. In southeastern Argolis, the limestones and marls interfinger with a sequence of andesites, tuffites and flyschoid sandstones. In the north, neritic carbonates were deposited, a facies which persisted locally up to the Lower Paleocene.

Paleogene - Eocene

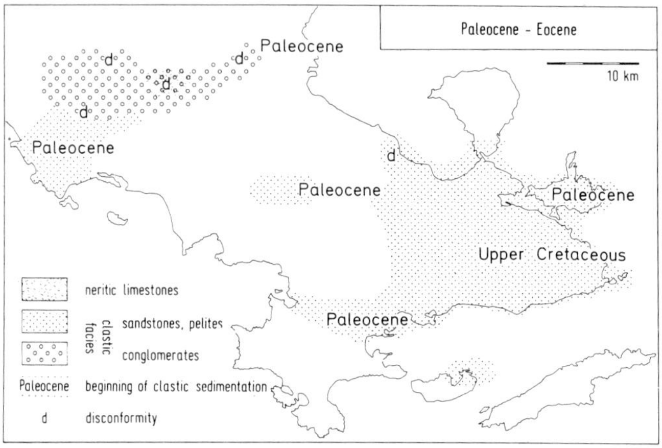

Fig. 5. Flyschoid sediments were deposited in southeastern Argolis as early as Late Cretaceous. In most parts of the peninsula, however, flysch deposits did not begin before the Paleocene. It came to a halt in the course of the Eocene. In the northern Argolis, olistostrome-like masses of conglomerates and breccias consisting of reworked material, interfinger with the flysch.

References

Bachmann, G.H. & Risch, H., 1976: N. Jb. Geol. Paläont. Abh.,152:137-160.
 - , in press.
Bannert, D. & Risch, H.: Z.dt. Geol. Ges., in press.
Dercourt, J., 1974: Ann. Geol. Pays Hellén., 15:1-413.
Ferriere, J., 1974: C.R. somm. Soc. Geol. France, 3:53-59.
Jacobshagen, V., 1972: Z. dt. Geol. Ges., 123:445-454.
Jacobshagen, V., Risch, H. & Roeder, D., 1976: Z.dt.Geol.Ges.,127:133-145.
Kalkreuth, W., Risch, H. & Wallner, P., 1976: N.Jb. Geol.Paläont. Mh.: 350-360.
Kalkreuth, W., Wallner, P., Jacobshagen, V. & Risch, H., 1976: Bull. Soc. Geol. Greece, in press.
Renz, C., 1955: Inst. Geol. Subs. Res., Athens, 637 p.

Mesoscopic Fabric Studies in the Pindos Zone of the Central Peloponnesus

U. Doert

On the Peloponnesus, pelagic rocks of the Pindos facies crop out mainly within two separate ranges (Jacobshagen et al. this vol.). Since a long time it has been discussed whether

both areas are parts of a uniform Pindos nappe (Blumenthal 1933; Renz 1940; Dercourt 1964) or

these ranges must be derived from two similar pelagic troughs separated by a central Tripolitsa shelf region (Kiskyras 1963; Fytrolakis 1972).

To answer this question the structures and the tectonic position of the Pindos rocks in some critical areas of the Central Peloponnesus have been investigated with mesoscopic fabric analysis.

In all areas the Pindos rocks overlie the Tripolitsa group (Jacobshagen et al. this vol.) with mostly flat-lying tectonic contacts.

In the eastern margin of the *western Pindos chain* and the bordering parts of the Tripolitsa zone, mapped in detail, the predominant mesoscopic structures are overturned to recumbent minor folds (fig. 1), both in the overlying Pindos series and in the underlying Tripolitsa flysch. These folds are vergent to the S, W or in some places to the NW. This is confirmed in the sedimentary structures.

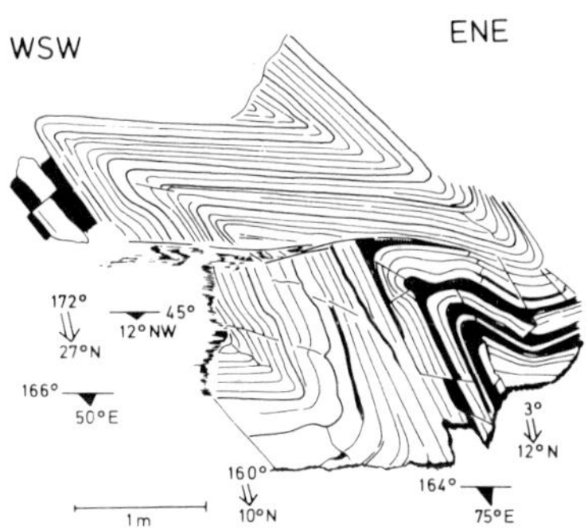

Fig. 1. Typical structure found in the Pindos rocks on the road Megalopolis - Kalamata S of Paradishia. Thin bedded platy limestones and reddish marls (marked in grey colour) of Upper Cretaceous age are deformed into overturned to recumbent minor folds. Vergencies are going to the W.

Folds and thrusts are intersected and dislocated by younger, steeply dipping normal faults.

The western Pindos mountain ranges are underlain by numerous thrust sheets dipping in general steeply to the E, each of them being intensely folded and disturbed. Minor folds with steeply dipping axial planes are dominating.

The cross sections west and north of Megalopolis demonstrate a gradual change from flat stratification with horizontal thrust planes and recumbent minor folds, predominating in the area of outcropping Tripolitsa facies, to the steeper structures of the Pindos sheets of the western chain.

Within the Tripolitsa zone some *isolated complexes of Pindos rocks* appear, between the western and eastern Pindos chains, in a flat-lying allochthonous tectonic position. Mesoscopic fabric observations lead to the

conclusion that these tectonic outliers have arrived at their present position from eastern or northeastern regions.

West of Argos the Tripolitsa series is exposed in a dome surrounded by Pindos rocks "le bombement interne de la zone de Gavrovo-Tripolitsa" (Dercourt 1964), studied with emphasis on mesoscopic fabric analysis by Kunz (1975). Here again evidence was found that the Pindos series was thrust towards the W or SW at the eastern border of the outcropping Tripolitsa group as well as at the western one. Thus the Tripolitsa rocks should be a tectonic window.

The *eastern Pindos chains* have been studied in several cross sections. Recumbent folds and thrusts to the W predominate between the Gulf of Argos and the eastern border of the Tripolitsa zone. These main structures are superposed by upright to inclined, partly overturned multi-vergent minor folds.

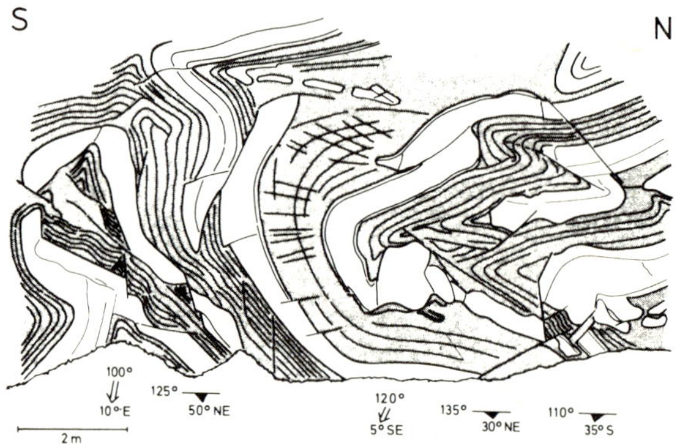

Fig.2. Typical structure within the eastern Pindos chain N of Astros on the road Astros - Argos. Upper Cretaceous light limestones and reddish marls (marked in grey colour) are deformed into disharmonic recumbent minor folds. Vergencies are going to the S.

West and north of Astros, fold axes scatter predominantly about E-W (fig. 2) similar to the deformation of the Argolian Mesozoic (Jacobshagen et al. this vol.).

These tectonic results confirm the opinion of Blumenthal, Renz, Dercourt and others that all Pindos areas of the Peloponnesus belong to one uniform thrust sheet. This interpretation is based on the following facts

the apparent *flat-lying allochthonous* tectonic position of the Pindos rocks in pelagic facies upon the neritic Tripolitsa rocks in extensive areas;

tectonic contacts between the Pindos rocks and the Tripolitsa series without any exception;

general vergence of folds and thrusts towards the W/SW;

connection between Pindos outliers and the western Pindos chain by tectonic structures and continuous facies transitions.

Consequently, the Pindos nappe must be derived from the areas E or NE. of the Tripolitsa zone and bordering the Argolian Mesozoic area. This question is discussed in more detail by Jacobshagen et al. (this vol.).

The internal tectonic structure of the Pindos rocks suggests that the confining pressure during tectogenesis was not very high, and that the Pindos nappe is a structural unit transported at a superficial tectonic level and which received its internal structure by an interaction of folding and gravity sliding as proposed by Temple (1968). The present distribution of the Pindos rocks on the Central Peloponnesus within two se-

parate ranges and isolated klippes may have resulted not entirely from erosion of a once uniform thrust plate, but possibly from an early dissection by gravity sliding.

References

Blumenthal, M.M., 1933: N. Jb. Min. etc., Abt. B, Beilagebd., 70:449-514.
Dercourt, J., 1964: Ann. Géol. Pays Hellén., 15:1-418.
Fytrolakis, N., 1972: Bull. Geol. Soc. Greece, 9:122-132.
Kiskyras, D., 1963: Bull. Geol. Soc. Greece, 5:1-21.
Kunz, E., 1975: unpubl. Diplomarbeit Erlangen.
Renz, C., 1940: Verh. Akad. Athen, 8:1-171.
Temple, P.G., 1968: Geol. Soc. Amer. Bull., 79:687-700.

Structural Tectonics of the Parnassus-Ghiona Mountains in the Central Hellenides

W. Schwan

Among the 7 main zones of the Hellenic orogen the Parnassus zone is of particular geotectonic interest because of its central location. As everywhere in the Hellenides, the main structural problem is the definition of the style and the relative importance of thrusting and folding. Therefore, the Parnassus zone has been studied in its widespread and best exposed central region.

Structures

(Fig. 1 and 2). Upper Triassic to Eocene sediments are orogenically deformed. The calcareous and low neritic facies indicates ridge character; the Eocene is developed as flysch. Depending on lithology, competency, and bed thickness, large, open to juratype folds or small tight folds exist. Rigid massifs, partly with reef character, show flat bedding, such as in the Central Parnassus. Folds of thick-bedded Jurassic-Cretaceous limestone have wave lengths of 2 to 3 kilometers. They are flat and upright or oblique.

Major synclines with less competent flysch are narrower. Minor folds are well developed. Fold structures, in general, have NW-SE axes and verge to the SW. In the northern Ghiona, however, they run N-S and are overturned to the west, and in the southern Parnassus E-W striking axes and southern vergencies predominate. Antivergent backfolds occur north of Karouti and elsewhere. In the eastern part of the flysch syncline of Arachova, minor crossfolds may be caused locally by mechanical adaptation to the west dipping axis of a large anticline of Upper Cretaceous limestone (Keupp 1974).

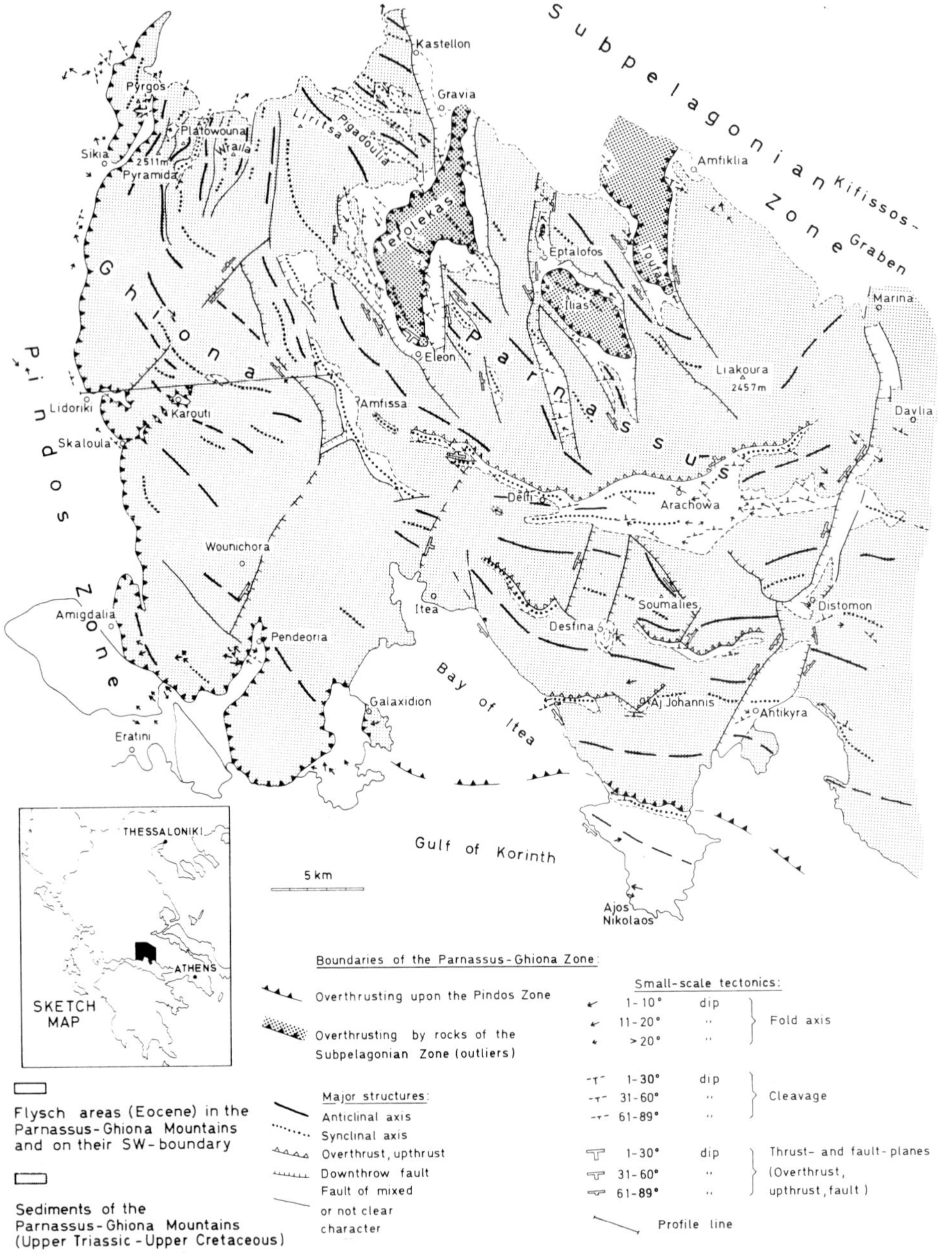

Fig. 1. Main structure elements of the Parnassus-Ghiona Mountains (Schwan 1971-1973).

Slaty cleavage (s_1), well developed in the shale of the flysch, shows NE-dip. At some places a second cleavage (s_2) with knick folds is evident.

Imbricate structure is developed in the southern Parnassus and in the northern Ghiona where Mesozoic calcareous masses are thrust over the flysch synclines.

Thrusts with SW-vergence exist partly at the southwestern borders of the Parnassus-Ghiona unit towards the Pindos zone. In the southwestern area of the Ghiona, near Karouti and Galaxidon, Mesozoic inliers and half-inliers of flysch indicate thrust transport of 5 to 10 kilometers. In the northeastern marginal area of the Parnassus, Subpelagonian outliers of Jurassic-Cretaceous rocks (Celet 1962) lie with thrust contact and with transport distances of up to 15 kilometers on Parnassus flysch.

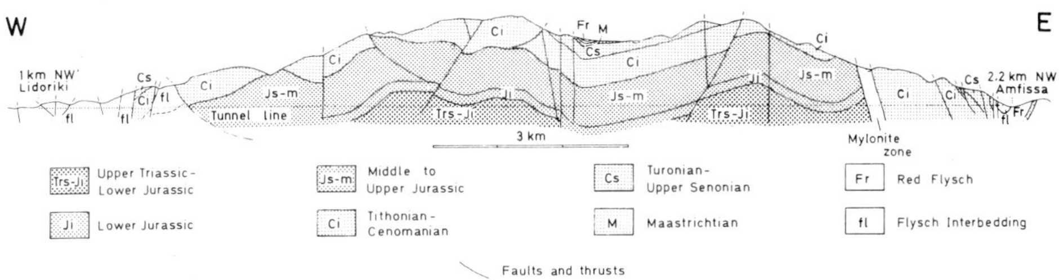

Fig. 2. Profile of the Ghiona (Galeos 1969; Schwan 1973).

Typically associated with the thrust structures are horizontal cleavage and recumbent folds in the flysch as well as strong mylonizations of dolomites underlaying the outliers. These characteristic features point to gravitational gliding.

The Subpelagonian outliers above the Parnassus flysch have been protected from erosion by their position within graben-like structures. Mixed forms of nappe and graben structures are called *nappe-graben structures* (Deckengräben). In the southern Parnassus near Desfina a limestone wedge thrust onto the flysch has also sunk graben-like into the surrounding limestone (Nöth 1975).

The rigid calcareous masses of the Parnassus-Ghiona mountain range are strongly fractured by normal faults, forming a block mosaic which runs crosswise or diagonal to the B-axis. Major tension faults run NNE and have developed macrotectonic forms with a fault offset of 1 to 3 kilometers and with thick mylonites, e.g. in the flysch-graben of Distomon and in the tilted Triassic block of Wounichora and north of Desfina.

Kinematics

During the orthogeosynclinal stage, the epirogenic trend of the Parnassus-Ghiona region as a ridge is evident in its calcareous lithology, the biofacies during Upper Triassic to Cretaceous time, and in the limestone conglomerate within the flysch, as well as indirectly by the absence of basic magmatites.

The orogenic kinematics of the Parnassus zone can be inferred from the macrotectonic and the related microtectonic forms. The shortening vectors oscillate around the NE-SW line. Compression in the flat and juratype folds of the carbonate beds was moderate, but was slightly stronger in

the flysch. The marginal zones of the Parnassus-Ghiona complex show strong shortening. The tensional vectors of the normal faults oscillate around the NW-SE line. The effects of NE-SW shortening and NW-SE extension were combined during the main orogeny, but the extension continued posthumously. The kinematic sequence includes three stages. 1. Early, mainly gravitational gliding of the Subpelagonian nappe onto the northeastern Parnassus-Ghiona complex, and of the latter toward the SW onto the Pindos zone. 2. Folding, cleavage, and imbrication caused by NE-SW compression. 3. Normal faulting by NW-SE tension during and after the main orogeny. The main orogeny took place after the flysch sedimentation and may correspond to the late Eocene Pyrenean phase. Most block movements occurred in Late Tertiary and Quaternary time.

Tectonic Position

The Parnassus-Ghiona mountains represent a large rigid thrust mass limited at front and back by northeastward dipping thrusts, in the NE against the Subpelagonian zone and in the SW against the Pindos zone. The Parnassus zone may be partly allochthonous. The distance of transport of its carbonate rocks cannot be determined. Strong horst-like upward movements along young normal faults have caused alpine morphology with elevations of up to 2500 meters.

References

Aubouin, J., 1958: Bull. Soc. Géol. France 6,8:731-750.
Aubouin, J. et al., 1970: Ann. Soc. Géol. Nord, 40,4:277-306.
Bender, H., Hirschberg, K., Leuteritz, K. & Mänz, H., 1960: Ann. Géol. Pays Hellén., 11:201-213.
Bizon, G. & Thiébault, F., 1974: C.R. Séanc. Acad.Sc., 278 sér.D:9-12.
Blumenthal, M.M., 1931: Eclog. Geol. Helv., 24:347-372.
B.P. Co. Ltd., 1971: Geol. Greece, 10, 73 p.
Celet, P., 1957: Bull. Soc. Géol. France 6,7:571-583.
- , 1958: Bull. Soc. Géol. France 6,8:471-486.
- , 1960: Bull. Soc. Géol. France 7,2:427-434.
- , 1962: Ann. Géol. Pays Hellén., 13:1-446.
Fleury, J.J. & Godfriaux, I., 1974: Ann. Soc. Géol. Nord, 94,4:149-156.
Galeos, A., 1969:Profile du tunnel de Ghiona.
Godfriaux, I., 1962: C.R. Acad. Sc., 255:1761-1763.
- , 1970: Ann. Géol. Pays Hellén., 19:1-282.
Keupp, H., 1974: Diplomarbeit:149 p.
Nöth, H.-O., 1975: DFG-Abschlußbericht II:11 p.
Papastamatiou, J., 1960: Bull. Soc. Géol. France 7,2:398-409.
Papastamatiou, J. & Tataris, A., 1962: Bull. Soc. Géol. Greece,5:83-88.
Papastamatiou, J., Vetoulis, D.& Tataris, A., 1963:Bull.Soc.Greece, 5: 83-88.
- , 1963: Bull. Soc. Géol. Greece, 5:43-51.
Paraskevaidis, I., 1959: Ecolg.Geol.Helv., 52:881-896.
Renz, C., 1940: Verh. Akad. Athen, 8:1-171.
Schwan, W., 1973/74:DFG-Zwischenbericht.

The Main Flysch Stages of the Hellenides

D. Richter
with Contributions of I. Mariolakos and H. Risch

Flysch and Molasse of the Ionian Zone

The flysch deposition of the Ionian zone commences in the Upper Eocene with a pelitic-carbonate transitional series which overlies a pelagic limestone succession of Palaeogene age (Richter 1973a, 1974a). The thickness of the transitional series in continental Greece diminishes from 50 to 60 m in the external Ionian zone to less than 10 m at the east margin of the Ionian zone. Its widespread distribution shows that deep water (pre-flysch) conditions existed in the Ionian furrow before the beginning of the true flysch facies.

The flysch consists of mudstones, siltstones and (graded) sandstones (turbidites), in thicknesses ranging from 1000 m in the west to 4000 m in the east of the Ionian zone. At the beginning of the Oligocene, the first tectonic movements formed three parallel minor troughs (subzones) in the Epirus region, each with its own facies (Richter 1976a, b). This disposition of the sea-floor controlled the flow direction of turbidity currents as evident in flute casts and other sole marks. In continental Greece the transport of psammitic material was mainly towards the south, but on the Peloponnesus it was towards the north. Flysch deposition ended in the Aquitanian with the beginning of the main tectogenesis. On Levkas Island the flysch succession was intensely folded and partly eroded during this time, with an Upper Aquitanian molasse (Levkas Series) unconformably resting on the folded flysch. In the Epirus region the flysch sea was restricted to a basin of 50 km by 100 km covering the axial Ionian subzone. A Burdigalien flyschoid series, deposited here, shows some features of a molasse facies. Along the margins of this special basin the Burdigalian strata lie in part unconformably on the flysch and locally on Liassic limestones, suggesting strong local movements, uplift and erosional activity during the Aquitanian (I.F.P./I.G.S.R. 1966). Marine deposition in the Epirus ceased at the end of the Burdigalian. The main tectogenesis in the Ionian zone occurred between the end of flysch deposition and early Middle Miocene, since on Corfu Island and in the Parga area a Middle/Upper Miocene molasse lies unconformably on strongly folded flysch. These older molasse strata show signs of a weak folding. Since they are unconformably overlain by Lower Pliocene beds (younger molasse) a further weak tectogenesis took place in late Miocene time.

Flysch of the Gavrovo-Tripolitsa Zone

The flysch of the internal Ionian subzone passes without any facies boundary into that of the Gavrovo-Tripolitsa zone. The area of the anti-

Fig. 1. Diagram illustrating the pre-flysch development and the flysch periods of isopic zones in continental Greece.

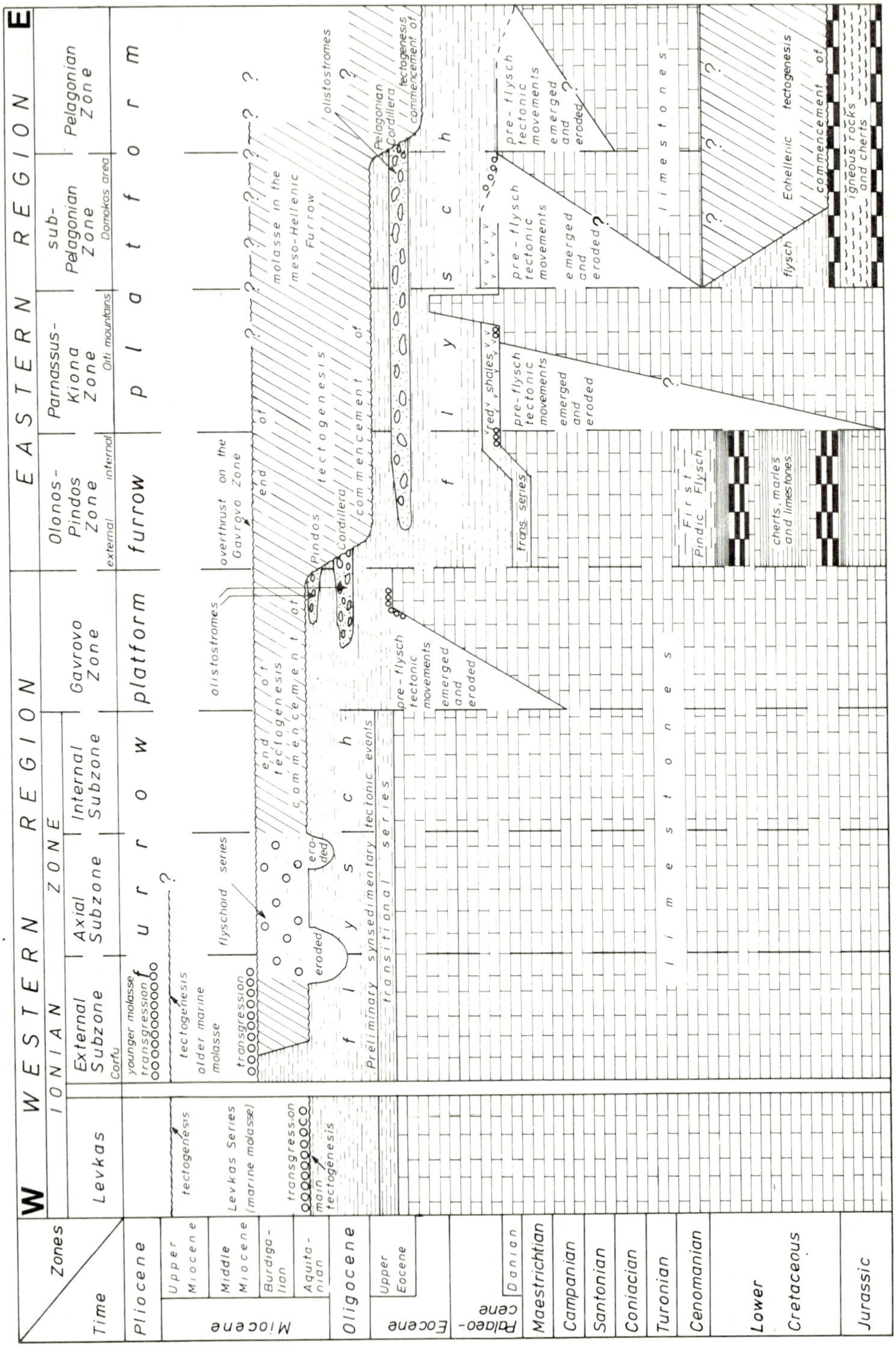
W
WESTERN REGION
EASTERN REGION
E
Zones
Time
Levkas
IONIAN ZONE
External Subzone
Corfu
Axial Subzone
Internal Subzone
Gavrovo Zone
Olonos-Pindos Zone
external
internal
Parnassus-Kiona Zone
Oiti mountains
sub-Pelagonian Zone
Domokos area
Pelagonian Zone
Pliocene
Miocene
Upper Miocene
Middle Miocene
Burdigalian
Aquitanian
Oligocene
Eocene
Upper Eocene
Palaeocene
Danian
Maestrichtian
Campanian
Santonian
Coniacian
Turonian
Cenomanian
Lower Cretaceous
Jurassic
furrow platform furrow platform
flysch
limestones
tectogenesis
Levkas Series (marine molasse)
transgression
main tectogenesis
younger molasse
older marine molasse
flyschoid series
olistostromes
eroded
Preliminary synsedimentary tectonic events
transitional series
pre-flysch tectonic movements
emerged and eroded
overthrust on the Gavrovo Zone
Pindos Cordillera
commencement of tectogenesis
end of tectogenesis
molasse in the meso-Hellenic Furrow
Pelagonian Cordillera
red shales
trans. series
First Pindic Flysch
cherts, marles and limestones
Eohellenic tectogenesis
commencement of
igneous rocks and cherts

Fig. 1

clines of Gavrovo and Klokova and probably other parts of the Gavrovo-Tripolitsa zone in continental Greece, presently under the flysch cover, emerged at the beginning of the Upper Eocene prior to flysch deposition, and resulted in extensive erosion and karstification of the pre-flysch limestones. In the central and western parts of the Gavrovo massif the flysch lies transgressively on the rugged surface of the Tertiary or even Upper Cretaceous limestones. At the beginning of the flysch deposition in the Upper Eocene the western part of the Gavrovo Massif rose strongly. Numerous faults resulted and huge blocks, more than one kilometre in length, broke off and slid downslope into the flysch sea where they came to rest near the foot of the massif and became buried under flysch. Richter & Mariolakos (1973a) suggest the term *olisthothrymma* (= slide fragment) for this phenomenon of combined tectonic breaking, gravitational sliding and resedimentation.

Similar, but more widespread events took place in the Gavrovo-Tripolitsa zone on the Peloponnesus. In the early Upper Eocene, strong uplift, faulting, and the formation of horsts and grabens accompanied a major regression and the termination of Tripolitsa carbonate deposition (Richter & Mariolakos 1972, 1973a, b, c; Richter 1974b, 1976d). Extensive erosion created a peneplain with karst phenomena developed in many areas. In late Upper Eocene time the Gavrovo-Tripolitsa platform subsided, and the flysch sea advancing from both the Ionian zone and the Olonos-Pindos zone flooded transgressively the karstified and faulted surface of the Tripolitsa Limestone. From the horst areas still above sea-level, huge blocks broke off and slid downward to be slowly covered with flysch deposits (Richter 1976c). The flysch of the Gavrovo-Tripolitsa zone consists of rocks similar to that of the Ionian zone, attaining a thickness of more than 5000 m in continental Greece, but of only 500 - 1000 m on the Peloponnesus.

The Oligocene and Aquitanian parts of the flysch of the Gavrovo-Tripolitsa zone contain thick olisthostromes. Rounded olistholiths of Olonos-Pindos material suggest a fully developed Cordilleran topography in part or all of the source area.

The youngest flysch beds of the Gavrovo-Tripolitsa zone in continental Greece and on the Peloponnesus contain an Aquitanian microfauna (Richter 1976a, b). Thus it seems likely that the main tectogenesis of the Gavrovo-Tripolitsa zone occurred during early Miocene, i.e. probably sychronous with the main folding of the Ionian zone.

Flysch of the Olonos-Pindos Zone

The flysch begins with a transitional series of several ten meters thickness conformably overlying Maestrichtian limestones (Brunn 1956; Aubouin 1959, 1964; Fleury 1968; Koch & Nicolaus 1969). In the lowest flysch mudstones over the transitional series, a Palaeocene assemblage of microfossils has been obtained. The flysch series consists of a thick sequence of sandstones, siltstones, grey and occasional red mudstones. There are many olisthostromes with pebbles originating from the Pelagonian zone s. l. (sub-Pelagonian and Pelagonian zone). The flysch sedimentation continued until the Upper Eocene. The deepest part of the flysch trough was situated in the Epirus region where the thickness exceeds 4000 m.

The tectogenesis started probably in the late Upper Eocene as suggested by the olisthostromes in Oligocene flysch strata of the Gavrovo-Tripolitsa

zone. It came to an end with the thrust transport of the Olonos-Pindos nappe over the Gavrovo-Tripolitsa zone. This final movement must have occurred after the folding of the Gavrovo-Tripolitsa zone since the Olonos-Pindos nappe lies on a denudation relief of the latter (Richter 1974b).

Flysch of the Parnassus-Kiona Zone

The flysch of the Olonos-Pindos zone passes without facies boundary into that of the Parnassus-Kiona zone (Richter & Mariolakos 1974a). The sedimentation of the latter begins in Palaeocene time with characteristic red calcareous shales of 45 - 100 m thickness (Papastamatiou 1960; Celet 1962). Significant uplift and faulting took place before the sedimentation of the basal red beds. In several areas erosion reached down to the Jurassic rocks. In less emerged areas an interruption of the sedimentation caused the formation of a hardground. Other parts remained as sea floor and sedimentation continued without interruption, leading to transitional beds of several metres thickness between the Upper Cretaceous/ Palaeocene limestones and the red shales (Richter & Mariolakos 1974b, 1975a).

Later in the Palaeocene the whole Parnassus-Kiona zone was buried under a 800 - 1100 m thick flysch succession of sandstones, siltstones and mudstones. The flysch sedimentation came to an end during the Upper Eocene. Several olisthostromes in Upper Eocene deposits contain rounded pebbles consisting of sediments and crystalline rocks of the Pelagonian zone s. l. They resemble those in the Olonos-Pindos flysch and indicate that mudstreams moved from the rising Pelagonian/sub-Pelagonian cordillera far towards the west into the deepening flysch trough over the Parnassus-Kiona (and Olonos-Pindos) zonal substratum.

Flysch of the Pelagonian Zone s. l. (sub-Pelagonian and Pelagonian Zones)

The flysch deposits of the Parnassus-Kiona zone and of the Olonos-Pindos zone pass without facies boundary into the flysch of the Pelagonian zone s. l. In the Pelagonian zone s. str. flysch sedimentation began in latest Maestrichtian time (Bignot, Fleury & Guernet 1971), but in the sub-Pelagonian zone not until the Palaeocene (Richter 1976d; Richter & Mariolakos 1977). In the western parts of the sub-Pelagonian zone red shales form the basal beds of the flysch succession. A widespread pre-flysch emergence of the Upper Cretaceous limestones was accompanied by faulting, weathering and karstification (Richter & Mariolakos 1977). Later the flysch sea advanced over the karst-shaped surface. The flysch deposition ceased in the east (Pelagonian zone s.str.) within the Middle Eocene, while in the west (sub-Pelagonian zone) it continued until Upper Eocene time.

Conclusion

The main (Tertiary) flysch of the Hellenides can be divided into two regions. In the Eastern Region the flysch deposition began in the latest Cretaceous or in the Palaeocene and ended in the Upper Eocene (or in the eastern part of the Eastern Region as early as the Middle Eocene). In the Western Region the flysch sedimentation began in the Upper Eocene and

continued into the Lower Miocene. For a very short time a uniform flysch sea covered almost the entire Hellenides. The flysch development shows that the main tectogenesis moved in steps from east to west. Except for the Ionian basin and the Olonos-Pindos trough, uplift, block faulting and erosion preceded the flysch stage in most areas.

Since the lithology of the Greek flysch was not influenced by its pre-flysch substratum, it can be regarded as a sedimentary sequence independent of the Hellenidic facies zones.

References

Aubouin, J., 1959: Ann. géol. pays hellén., 10, 483 p.
Brunn, J.M., 1956: Ann. géol. pays hellén., 7, 358 p.
Bignot, G. Fleury, J.-J. & Guernet, G., 1971: Bull. Soc. Géol. France, 7,13, 446 p.
Fleury, J.-J., 1968: C.-r. Séance. Soc. Géol. France, 9:334-335.
I.F.P./I.G.S.R., 1966: Inst. Franc. Petr., 360 p., Paris.
Koch, K.E. & Nicolaus, H.J., 1969: Inst. Geol. Subsurf.Res., 9, 140 p. Athens.
Papastamatiou, J., 1960: Bull. Soc. Géol. France, 7:398-408.
Richter, D., 1973a: N. Jb. Geol. Paläont. Mh.:80-98.
- , 1973b: N. Jb. Geol. Paläont. Abh., 143:304-344.
- , 1974a: N. Jb. Geol. Paläont. Mh.:149-174.
- , 1974b: N. Jb. Geol. Paläont. Abh., 145:96-128.
- , 1976a: N. Jb. Geol. Paläont. Abh., 151:73-100.
- , 1976b: N. Jb. Geol. Paläont. Abh., 151:224-252.
- , 1976c: N. Jb. Geol. Paläont. Mh.:237-249.
- , 1976d: Z. dt. Geol. Ges., 127:467-448.
Richter, D. & Mariolakos, I., 1972: Bull. Geol. Soc. Greece, 9:206-228.
- , 1973a: N. Jb. Geol. Paläont. Abh., 142:165-190.
- , 1973b: Ann. géol. pays hellén., 25:1-12.
- , 1973c: Prakt. Akad. Athinon, 48:29-44.
- , 1974a: Bull. Geol. Soc. Greece, 10:413-424.
- , 1974b: Prakt. Akad. Athinon, 48:407-426.
- , 1975a: Ann. géol. pays hellén., 26:417-434.
- , 1975b: Prakt. Akad. Athinon, 50:377-390.
- , 1976: N. Jb. Geol. Paläont. Mh.: 591-602.
- , 1977: Z. geol. Wiss. Berlin, 5:363-373.

Part 3

Crete

Stratigraphic and Tectonic Sequence of Crete

K.-O. Kopp

The island of Crete consists of a stack of six major thrust units representing sediments from three or four paleogeographic provinces, with the highest thrust unit representing obducted oceanic crust and/or its derivatives. The thrust sheets attain an estimated cumulative thickness of 6.5 km and represent an unknown palinspastic width amounting to at least several 100 km, imbricated during several phases since the late Jurassic, and presently forming the upper plate of a subduction zone with extensional faulting and predominantly vertical movements. The age of some of the mappable units is unknown, and some of the contacts are in dispute and interpreted either as unconformities or as thrust faults. Several units contain greenschist-grade or high-pressure type metamorphism or granodioritic plutonism. There is no confirmed occurrence of Hercynian-age or pre-Permian basement.

Other reviews on the Cretan nappe pile and its rocks were given by Creutzburg & Seidel (1975) and Bonneau (1976). Both reviews contain a geological map of Crete, but these maps differ remarkably from one another. From this it may be seen how incoherent the actual state of knowledge still is.

The nappe sequence is partly covered by Neogene and Quaternary (Neo-Autochthonous), consisting of red beds (Creutzburg 1963), marine sandstones, marls, limestones, and gypsum (Meulenkamp 1971). They are in places tilted up to vertical or uplifted up to 700 m above sea level.

The nappe sequence from the top down is as follows:

Ophiolite nappe. What has been called Ophiolite nappe before, must be called otherwise now on Crete, because the crustal age of this unit is 160 - 120 m.y. in the Pelagonian type area but 70 m.y. in Crete and Lykia (Seidel et al. 1976). The new names introduced in Crete are:Asterousia nappe (Bonneau 1972), Serpentinite-Amphibolite association (Creutzburg & Seidel 1975).

This unit consists of diabases, serpentinites, ultrabasites, amphibolite, gneiss, marble, quartzite, greenschist, and phyllite. Granodiorites occur at Melambes, Kali Limenes (Asterousia mountains), and Kalo Chorio (Baumann et al. this vol.; Baranyi et al. 1975); for the last two radiometric ages of 70 m.y. have been measured.

These complexes often form isolated masses, and there are two additional complexes: the Miamou unit (Bonneau et al. 1974) consisting of late Jurassic limestone-flysch with abundant microfauna and ophiolite pebbles, and the Arvi unit (Bonneau 1973a), consisting of Triassic to late Cretaceous sediments of intern Pindic character with major masses of spilite (Bonneau 1976). This unit resembles the "Diabas nappe" of Western Lykia (de Graciansky 1968, 1972).

The *Olonos-Pindos Series*, separated from the above complexes by a major thrust, consist of:

several 100 m Pindos Flysch (Paleocene to Eocene),

200-300 m Upper Cretaceous limestone with cherts,

up to 50 m First Pindos Flysch (Cenomanian to Turonian),

pelagic limestones with cherts and radiolarites (Triassic to Lower Cretaceous), and

pink limestones and marls with *Halobia* (late Triassic).

For details see Seidel (1971).

The Pindos series is thrust into numerous sheets, the lowest of which can sometimes be individualized as "Ethiá Series" (Renz 1947), a transitional facies between the pelagic Pindos series and the now underlying neritic Tripolitza series from the external slope of the Pindos trough.

The *Tripolitza Series* consists of the following members:

up to 1000m preserved Tripolitza Flysch of Eocene to Oligocene age. The flysch lies with an erosional unconformity on

about 1000 m neritic, mostly massive, often bituminous Tripolitza Limestone of Middle Triassic (Kopp & Ott 1977) to Middle Eocene age (Leppig, this vol.). A sedimentary transition leads to

several 100 m Rabdoúcha beds (Kopp & Ott 1977) of Triassic age, consisting of black shales, light sandstones, and violet marls with *Myphoria*, *Equisetites* a.o. This member was formerly attributed to the metamorphic Phyllite Series by all authors except Sannemann & Seidel (1976).

A nappe thrust plane, which is doubted by the French authors and by Kuss & Thorbecke (1974), separates this series from the underlying

Phyllite Series, consisting of carbonates, schists, metadiabases, and quartzite phyllites metamorphosed under high pressure/low temperature conditions (Schubert & Seidel 1972; Seidel, Okrusch & Schubert 1975). Some authors try to subdivide it, while others take this for premature. The present writer believes to have some evidence for the following succession in westernmost Crete. It is based on Cayeux (1902), under discussion now and not generally accepted.

Upper Group (Quartzite Phyllite formation) = Facies of Kakópetros of Aubouin & Dercourt (1965):

6. Upper Quartzite Phyllite, 300 - 400 m,
5. Cipollines and conglomerates, up to 400 - 500 m,
4. Lower Quartzite Phyllite, 600 m.

Lower Group = Facies of Sklavopoúla Aubouin & Dercourt:

3. Dark, platy, fossiliferous Triassic limestones alternating with lawsonite-bearing schists (Seidel & Okrusch 1976), until 500 m,
2. Dolomite and rauwacke, 400 - 500 m,
1. Gypsum, up to 200 m near Stomion.

This succession seems to be, at least partly, rather a facies succession as suggested by Aubouin & Dercourt, than a lithostratigraphic column.

In this volume, two maps of mineral facies are given by Seidel & Okrusch. Strong shearing and deformation of the Phyllite series to a sort of mélange in eastern Crete is stated by Baumann et al.

At the eastern and northern flanks of the Levka Ori a nappe thrust plane separates the Phyllite series from the underlying

Trypali Limestone, a neritic, mostly massive carbonate series. 700 - 800 m are to be seen (see cross sections in Creutzburg & Seidel 1975). Some algae found at the type locality at Asigoniá (Ott 1965; Kopp & Ott 1977) indicate Liassic age there, other biostratigraphic evidence is

lacking until now. West of the Levka Ori it seems possible that the gypsum and dolomite of the Phyllite series are an equivalent of the Trypali series, as suggested by Creutzburg & Seidel (1975) and Kopp & Ott (1977).

Under a nappe thrust plane, which is doubted by Xavier (1976), follows the

Talea Ori Series as defined by Epting et al. (1972). Younger synonyms ("Plattenkalk series", "Ida series")should be avoided. The series is regarded to be, but must not be, autochthonous. Most French authors and Kuss & Thorbecke (1974) connect it with the Ionian zone of western Greece; this is doubted by Jacobshagen (this vol.).

The succession of the Talea Ori series is as follows:

20-30 m preserved Kalavros beds (Bonneau 1973b), consisting of phyllitic, marly limestones with Eocene and/or Oligocene forams (Fytrolakis 1972; Bonneau l.c., Bizon & Thiebault 1974). They are exposed e.g. at Kalavros and Cape Sideros (both Eastern Crete), near Idaion Andron (Psiloritis), and at the Volika refuge hut (Levka Ori).

At least 1500 m Plattenkalk (Platy Limestone), a pelagic limestone with layers or nodules of chert. The well banked limestones is recrystallized and has not furnished any fossils. Creutzburg (pers. comm.) found brownish quartzitic schists near the base in the Levka Ori.

About 1200 m Stromatolithic Dolomite furnishing a Norian microfauna near the base (Epting et al. 1972). An angular unconformity, with a weathered karst surface below, separates it from

about 600 m Sisses beds, consisting of light, barren, often oölithic dolomites, and phyllites.

About 2000 m Fodele beds consist of dark, Permian limestones with Fusulinids, Crinoids, corals, Bellerophon, etc., and phyllites ("Galinos Phyllite").

The single almost complete type section of the Talea Ori series is exposed in the Talea Ori, near the northern coast of central Crete. The whole series, more than 5000 m thick, form the strongly overturned southern limb of an anticline 30 km in length, the northern limb of which is cut off by a nappe thrust. Concerning the amount of overturn movement, the series is rather not, in a strong sense, in an autochthonous position here.

References

Aubouin, J. & Dercourt, J., 1965: Bull. Soc.géol. France, 7:787-821.

Baranyi, I., Lippolt, H.J. & Todt, W., 1975:N.Jb.Geol.Pal.Mh.,257-262.

Bizon, G. & Thiebault, M.F., 1974:C.r. Acad.Sci., 278:9-12.

Bonneau, M., 1972: C.r. Acad. Sci., 275:2303-2306.

- , 1973a: C.r. Acad. Sci., 276:1249-1252.

- , 1973b: C.r. Acad. Sci., 277:2453-2456.

- , 1976: Bull. Soc. géol. France, 18:351-353.

Bonneau, M., Beauvais, L. & Middlemiss, F.A., 1974: Ann. Soc. géol. Nord, 94:71-85, Lille.

Cayeux, L., 1902: C.r. Acad. Sci., 134:1116-1119.

Creutzburg, N., 1963: Ann. géol. Pays hell., 14:357-404.

Creutzburg, N. & Seidel, E., 1975: N.Jb.Geol.Pal. Abh., 149:363-383.

Epting, M., Kudrass, H.R., Leppig, U. & Schäfer, A., 1972: N. Jb. Geol. Pal. Abh., 141:259-285.

Fytrolakis, N., 1972: Bull. geol. Soc. Greece, 9:81-100.

Graciansky, P. de, 1968: Bull. MTA, 71:42-62.
- , 1972: Thèse Orsay.
Kopp, K.O. & Ott, E., 1977: N.Jb. Geol. Pal. Mh.:217-238.
Kuss, S. & Thorbecke, G., 1974: Ber.naturf. Ges. Freiburg, 64:39-75.
Meulenkamp, J., 1971: in: Strid, A., ed., Opera Botanica,30:5-12, Lund.
Ott, E., 1965: N.Jb. Geol. Pal. Mh.:683-693.
Renz, C., 1947: Ecl.geol. Helv., 40:379-384.
Sannemann, W. & Seidel, E., 1976:N.Jb.Geol. Pal. Mh.:221-228.
Schubert, W. & Seidel, E., 1972: Z.dtsch. geol. Ges., 123:371-384.
Seidel, E., 1971: N. Jb. Geol. Pal. Abh., 137:443-460.
Seidel, E., Okrusch, M. & Schubert, W., 1975: Contr. Miner. Petrol., 49:105-115, Berlin-Heidelberg-New York.
Seidel, E., Okrusch, M., Harre, W., Kreuzer, P. & Raschka, H., 1976: Contr. Miner. Petrol., 57:259-275.
Seidel, E. & Okrusch, M., 1976: Bull. Soc. géol. France (7) 18:347-350.
Xavier, J.P., 1976: Thèse de Docteur de 3me Cycle, 101 p., Paris.

Geological Cross Sections through Western Crete

K.-O. Kopp

Continuing the work of Creutzburg & Seidel (1975) and Xavier (1976), two other cross sections through Western Crete are compiled and discussed here. *Western Crete* includes the Levka Ori and the region north and west of them. For the general stratigraphic succession see the foregoing article.

The sections run parallel to the western coast of Crete at distances of 9 and 38 km.

Section I

cuts the Levka Ori and is representative for their entire length of 40 km. The lowermost unit is the *Talea Ori series*, exposed mainly at the steep escarpments of the southern coast and in the Samaria gorge where it forms a SW - NE trending anticline. It was drawn after an unpublished map of Thorbecke (1974). A fault striking from Chora Sfakion to the east forms the steep southern escarpment of the Levka Ori and separates them from a débris-covered coastal plain underlain by Phyllite series.

The eastern, northern, and western flanks of the Levka Ori as well as great parts of their summits above 2000 m are underlain by the next higher unit, the *Trypali Limestone*. Since this unit is known in Western Crete only, the authors prefer to attach it to another, better known unit: Xavier (1976) to the Taleo Ori series (as an uppermost member, notwithstanding its Liassic fossils at Asigoniá, Ott 1965), Creutzburg & Seidel (1975) to the Phyllite series (as a basal member,the Phyllite being

partly younger than Liassic, then), Kuss & Thorbecke in Leppig (this vol.) to the Tripolitza Limestone as identical, and Kopp & Ott (1977) to the same, as an analogous and synchronous platform sediment from beyond a Penninic Phyllite trough. Tectonic syntheses in the South Aegean ought to take into account this actual state of disagreement.

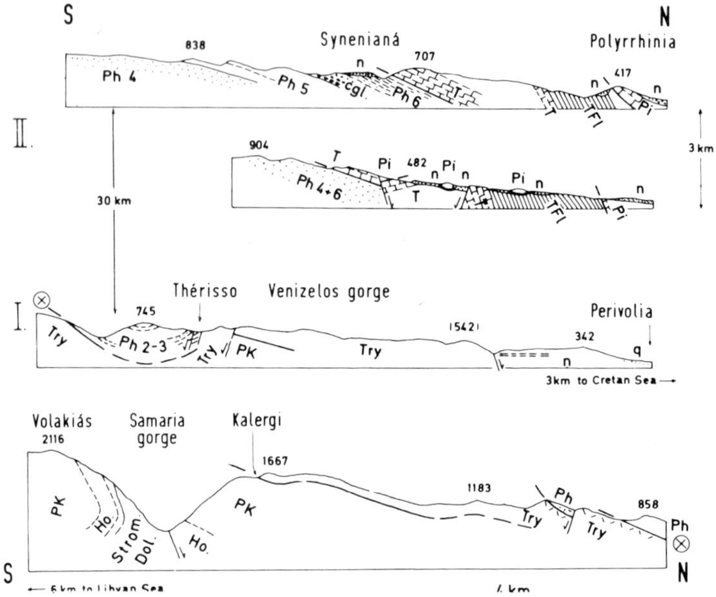

Fig. 1. Geological Cross sections through Western Crete.

q = Quaternary,
n = Neogene,
Pi = Pindos Series,
TFl = Tripolitzsa Flysch,
T = Tripolitzsa Limestone,
Ph = Phyllite Series, subdivision (numbers) according to Cayeux (1902) and Kopp (this vol.),
Try = Trypali Limestone
PK = Plattenkalk,
Ho. = Quartzitic schist near the base of PK; Strom. Dol. = Stromatolithic Dolomite; PK + Ho + Strom.Dol. = Talea Ori Series.

Within the Levka Ori an axis of highest tectonic elevation runs from the Kryoneritis range to the Volakiás - Gigilos mountains. It extends farther west to the gulf of Stomion where the deepest members of the Phyllite series (or Trypali series?) crop out. The southern flank of this 70 km long anticline is steeper than the northern flank which extends more than 15 km wide with gentle north and northeast dipping in the central part of the Western Crete section.

At the eastern, northern, and western morphological border of the Levka Ori the Trypali limestone dips and plunges below the next higher unit, the *Phyllite Series*. In our section, the Phyllite forms part of the Highland of Chaniá (Creutzburg 1963) which extends between the Levka Ori and the northern coastal plains around Chaniá.

The thrust zone between Trypali series and Phyllite series is characterized by mylonitized Trypali limestone, the "breccia of Omalos" of Xavier (1976). The phyllite consists of platy marble with brachiopods, dark pelitic schists, and chloritic schists (metadiabase, probably simultaneously extruded). It forms a westerly plunging syncline containing successively the Upper or Quartzite Phyllite Group, the Tripolitza Series, and Pindos Series.

At the northern flank of the syncline the Trypali Limestone forms the Malaxa mountain chain, the hills south of Chaniá, and the chain of hills between Alikianoû and Gerani, as part of a 50 km long anticline between Vamos and the southern part of the Kisamos gulf. Similar to the southern anticline, it has a steep south limb and a gently dipping north limb. This asymmetry in the Western Crete section suggests antithetic rotation of large blocks.

Section II

cuts the southern flank of the syncline 25 km farther west, with the core of the syncline covered by Neogene.

The section begins with the *Upper Group* of the *Phyllite Series*, called Quartzite Phyllite formation by Wurm (1950), with massive quartzite beds alternating with quartzitic phyllites. Their age is unknown, due to lack of fossils. Carbonate beds underlying them have furnished Triassic fossils (Cayeux 1902; Seidel 1968). Furthermore, a facies transition between the fossiliferous carbonates of the Lower Group and the basal parts of the quartzitic-phyllitic Upper Group (Aubouin & Dercourt 1965) may suggest a Triassic and possibly younger age for the Upper Group of the Phyllite Series. The age of metamorphism is Oligocene/Miocene (Altherr & Seidel, this vol.).

In our section the Quartzite Phyllite is divided into two parts by an intercalation of a white and light brownish dolomite marble. Wurm (1950) called it Mana dolomite. It is overlain by a coarse conglomerate with quartzite and carbonate pebbles. Where the dolomite thins out the conglomerate also becomes finer and thinner.

Tripolitza Series is the next higher unit. Its basal parts are probably older than large parts of the Phyllite Series (Kopp & Ott 1977). There are about 1000 m Tripolitza Limestone and 1000 m Tripolitza Flysch preserved south of Polyrrhinia. Between this series and the overlying Neogene many small and scattered remants of

Pindos Limestones are preserved, pebbles of which are abundant in Neogene conglomerates.

References

Aubouin, J. & Dercourt, J., 1965: Bull.Soc.Géol.France (7)7:787-821.
Cayeux, L., 1902: C.r. Acad. Sci., 134:1116-1119.
Creutzburg, N., 1963: Ann.géol.Pays hell., 14:357-404, Athens.
Creutzburg, N. & Seidel, E., 1975: N.Jb.Geol.Pal.Abh., 149:363-383.
Kopp, K.-O. & Ott, E., 1977: N.Jb.Geol. Pal. Mh.:217-238.
Ott, E., 1965: N. Jb. Geol. Pal. Mh.:683-693.
Seidel, E., 1968: Diss. Univ. Würzburg, 102 p.
Wurm, A., 1950: N.Jb.Geol. Pal. Mh.:206-239.
Xavier, J.-P., 1976: Thèse de Docteur de 3me Cycle, 101 p., Univ. Paris.

The Eastern Crete Nappe Pile - a Result of Gravity Tectonics

A. Baumann, G. Best, W. Gwosdz and H. Wachendorf

The phenomenon of *Stockwerk* tectonics characterizes the Sitia Ori of eastern Crete. Four allochthonous units with different style of deformation are piled up above the autochthonous Cherty Limestone (*Plattenkalk*). The following tectonic units are distinguished:

	Stockwerk	Style of deformation
Neoautochthonous	Molasse: Neogene	Block-faulting
Allochthonous	4. Volcano-sedimentary complex: Mesozoic - Paleogene	Mélange
	3. Pelagic sediments and flysch (Pindos Series): Triassic-Eocene	Imbrication
	2. Neritic limestones and flysch (Tripolitza Series): Triassic-Oligocene (?)	Rigid plate
	1. Phyllitic Series: Permian-Triassic	Mélange
Autochthonous	Cherty Limestone: Permian-Oligocene (?)	Concentric folding

Autochthonous

From the Talea Ori Series of central Crete (Epting et al. 1972) only the uppermost member, the well-bedded, pelagic, crystalline Cherty Limestone (*Plattenkalk*) is exposed in broad anticlinal culminations with a thickness of about 1 500 m. The top of this series is restricted to the flanks of anticlines and consists of greenish to reddish calcareous phyllites and flyschoid sandstones with an average thickness of 200 m. In these low-grade metamorphic *Kalavros-Beds* (Bonneau 1973) globigerinids of probably Oligocene age and chondritic trace fossils were found. The whole sequence is deformed into folds with southward vergence and with wavelengths between meters and kilometers.

Phyllitic Series

The lowermost allochthonous unit is represented by the Phyllitic Series, which is about 600 m in thickness. The Permian age of some parts of this unit is proven by one fossil occurrence in phyllites (Papastamatiou & Reichel 1956). Stratigraphically younger horizons with a Triassic/Jurassic (?) fauna, known from western Crete (Cayeux 1902), are presently not documented in eastern Crete.

The Phyllitic Series is a tectonic mixture bound by shear surfaces and is therefore interpreted as a mélange in the sense of Greenly (1919). The relatively flat lying mélange includes components of fairly different origin and is composed of fragments of competent blocks embedded in a

pervasively sheared phyllonitic or pelitic matrix of incompetent behaviour. The lenticularly deformed blocks and phacoids ranging in diameter from centimeters to several hundreds of meters, consist of marbles, red limestones, quartzites, gypsum, andesites with their tuffs, and amphibolites enveloped by micaschists. Parts of the Phyllitic Series are completely metamorphosed, others are unmetamorphosed or contain metamorphics as exotic blocks only.

A high temperatur/low pressure metamorphism of the amphibolite phacoids is older than the nappe transport. A younger high pressure/low temperature metamorphism is probably due to the nappe transport. The phacoids are mostly east-westerly elongated and are orientated normal to the tectonic transport. They represent tectonically separated and isolated blocks which originated by shearing and rotation. The deformation of the Phyllitic Series is caused by translatation on densely spaced shear planes which tend to be parallel to the edges of the phacoids and subparallel to the bedding planes.

Tripolitza Series

The Tripolitza Series comprises massive unbedded to thick bedded, partly dolomitized, in some sites laminated reefal limestones of Jurassic (?) to Eocene age with an average thickness of 300 m. This unit rests with an unconformity on the Phyllitic Series and locally on the Cherty Limestone. The base of the Tripolitza Series is mostly brecciated to a mylonite up to 25 m in thickness. In other parts (loc. Arnikou) a horizon of black shales and siltstones with Myophoria whateleyae are identified as the sedimentary base of the Tripolitza Series. This horizon may be equivalent to the "Triassic sequence of Ravdoucha/NW-Crete" of Sannemann & Seidel (1976).

A sedimentary transition into late Eocene ot Oligocene (?) lithic flysch sediments of minor thickness on top of the Tripolitza-limestone is exposed only in graben-like positions.

Pindos Series

Equivalents of the Pindos Series have been preserved as relics of minor extension along step faults. Several hundreds of meters of well bedded limestones with cherts, allodapic limestones and flysch sandstones of predominantly Eocene age were deposited in a trough. Bonneau & Zambetakis (1975) identified fossiliferous horizons of Triassic and Jurassic age as parts of the Mangassa Series, a subfacies of the Pindos Series, which is distributed in Eastern Crete only. Upper Jurassic volcanics and the first flysch, which are typically developed in Central and Western Crete are missing in Eastern Crete. The Pindos nappe with a fairly flat lying base is internally imbricated and shows listric surfaces which form wedges dipping generally to the north.

Volcano-Sedimentary Complex

Remnants of the uppermost unit are confined to down-faulted graben depressions, which intersect the Eastern Cretan nappe pile. The Volcano-

Sedimentary Complex is equivalent to the "Subpelagonian Zone" (Aubouin 1965) of central Greece and consists of discontinuous strata of laterally non persistent beds. The heterogeneous assemblage contains serpentinized ultramafics, pillow lavas (spilites) with red limestones, radiolarites, sandstones and black shales in a chaotic mixture involving a classical Steinmann trinity, similar to a coloured mélange in the sense of de Graciansky (1968). The diorite complex of Kalo Chorio with marbles in thermal contacts forms a massif of more than 10 km in diameter. It is believed that this complex belongs to the same tectonic unit due to radiometric datings of the hornblende in the diorite which yields an Upper Cretaceous age (Baranyi et al. 1975; Wachendorf et al. 1975), as do low to medium pressure/temperature metamorphic rocks in the Asterousia mountains (Seidel et al. 1976).

Approach to a Geodynamic Interpretation

The dynamics of the Eastern Cretan nappe pile is controlled by flat lying thrust sheets moved from north to south. A palinspastic reconstruction for the allochthonous units results in a horizontal displacement of more than 100 km. The movements started in the internal parts presumably in the Upper Jurassic (Aubouin 1973) and resulted in a successive emplacement of the nappe pile towards the external parts. The orogenic wave reached Crete after deposition of the autochthonous Oligocene (?) Kalavros beds. The driving force for the nappe transport is interpreted as a progressively expanding diapiric mantle uplift spreading outwards from the Aegean Sea. The regional uplift induced supratectonics consisting of detachment and gravitational sliding of the crustal cover. Sliding of coherent masses is dominant in the massive neritic Tripolitza Limestone, which was moved as a rigid plate; whereas the well bedded pelagic Pindos Series were imbricated to wedge shaped thrust slices. Intercalated argillaceous and flyschoid series with more "plastic" properties acted as lubricants. Furthermore, reduced basal friction and very low cohesion in the ductile series due to fluid overpressure under an overburden favoured the large-scale nappe transport.

References

Aubouin, J., 1965: Geosynclines. In: Developments in Geotectonics 1, 335 p., Amsterdam.

- , 1974: Bull.Soc.géol.France 7,15:426-460.

Baranyi, I., Lippolt, H.J. & Todt, W., 1975: N.Jb.Geol.Pal. Mh.:257-262.

Baumann, A., Best, G., Gwosdz, W. & Wachendorf, H., 1976: Tectonophysics, 30:33-40.

Baumann, A., Best, G. & Wachendorf, H., 1977: Geol. Rundschau 66:492-521.

Bonneau, M., 1973: C.R.Acad.Sci.,Série D.277:2453-2456.

Bonneau, M. & Zambetakis, A., 1975:C.R. Acad.Sci., Série D, 281:17-19.

Cayeux, L., 1902: C.R. Acad. Sci., 134:1116-1119.

Epting, M., Kudrass, H.R., Leppig, U. & Schäfer, A., 1972: N. Jb. Geol. Paläont., Abh., 141:259-283.

Graciansky, P. de, 1968: M.T.A. Enst.Bull., 71:42-62, Ankara.

Greenly, E., 1919: Mem. Geol. Surv., 980 p., London.

Papastamatiou, J. & Reichel, M., 1956: Eclogae geol.Helv., 49_147-149.

Sannemann, W. & Seidel, E., 1976: N. Jb. Geol. Paläont.,Mh.:221-228.
Seidel, E., Okrusch, M., Kreuzer, H., Raschka, H. & Harre, W., 1976: Contrib., Mineral. Petrol. 57:259-275.
Wachendorf, H. Baumann, A., Gwosdz, W. & Schneider, W., 1974: Z. dtsch. Geol. Ges., 125:237-251.
Wachendorf, H., Best, G. & Gwosdz, W., 1975: Geol. Rundschau, 64:728-750.

Regional Distribution of Critical Metamorphic Minerals in Crete

E. Seidel and M. Okrusch

For the geodynamic interpretation of the eastern Mediterranean, the Alpine metamorphism, its age relative to the main tectonic movements and the development of metamorphic grade in space and time, are of special interest. A main prerequisite for the solution of these problems is a detailed knowledge of the regional distribution of critical metamorphic assemblages (fig. 1). This paper gives a review on the critical metamorphic mineral assemblages on the islands of Crete and Gavdos, mainly based on own observations. Additional data were collected from the literature and from personal communications.

Crete

In the island of Crete, we distinguish six pre-Neogene geological units including one probably autochthonous, lowermost element and a pile of at least four nappes (Creutzburg & Seidel 1975), the serpentinite-amphibolite association, the Pindos series, the Tripolitza series, the phyllite-quartzite series, and the (?) autochthonous plattenkalk series. So far, the Tripali unit which is intercalated between the plattenkalk series and the phyllite-quartzite series in western Crete and in the western part of central Crete is assumed to be combined with the phyllite-quartzite series to form one overthrust unit.

The rocks of the plattenkalk series, the Tripali unit, the phyllite-quartzite series, and the serpentinite-amphibolite association are metamorphosed. This metamorphism is older than the overthrust tectonics. So far, no detailed information is available on the metamorphic grade of the plattenkalk series and the Tripali unit. This is partly due to the fact that relatively pure carbonate rocks, unsuitable to indicate P-T conditions, are predominant in these series.

The *phyllite-quartzite series* was metamorphosed under high-pressure conditions and under temperatures which increased from very low to low grade during the temporal evolution of the metamorphism in western Crete. The first low-temperature, high-pressure assemblage in the meta-basalts of western Crete is

glaucophane + lawsonite + pumpellyite + chlorite + albite + aragonite.

Judging from pertinent experimental data, this assemblage was formed at about 300°C and 7 kilobars water vapour pressure (Schubert & Seidel 1972).

The subsequent increase of metamorphic grade is documented by the assemblages

glaucophane + epidote ± pumpellyite + chlorite + albite + calcite

in the meta-basalts and

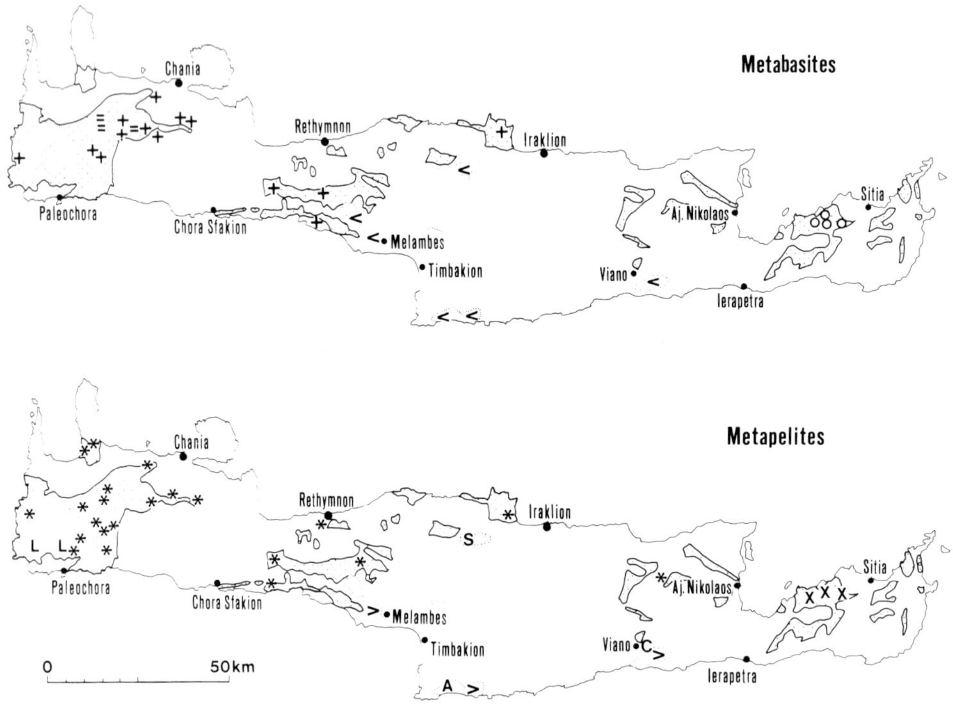

Fig. 1. Metamorphic mineral assemblages within the phyllite-quartzite series (solid lines) and the serpentinite-amphibolite association (dashed lines) of Crete. Outline of the tectonic units after Creutzburg.
Metabasites: = glaucophane + lawsonite + pumpellyite + chlorite + albite ± aragonite ± quartz; + glaucophane + epidote ± pumpellyite + chlorite + albite ± calcite ± quartz (for central Crete see text); o common green hornblende + plagioclase + epidote (crossite, albite, chlorite); < common green or brown hornblende + plagioclase ± diopside.
Metapelites: * chloritoid + chlorite + phengitic muscovite ± paragonite + quartz; C ferrocarpholite + chloritoid + chlorite + phengitic muscovite + quartz; L lawsonite + paragonite + phengitic muscovite + chlorite + quartz ± calcite ± dolomite; X garnet ± staurolite ± kyanite + biotite + muscovite + plagioclase + quartz (albite, chlorite); A andalusite + garnet + biotite + muscovite + plagioclase + quartz; > andalusite + sillimanite + cordierite ± garnet + biotite + K-feldspar (+ muscovite) + plagioclase + quartz; S sillimanite + cordierite ± garnet + biotite + K-feldspar (+ muscovite) + plagioclase + quartz.

chloritoid + chlorite + phengitic muscovite ± paragonite + quartz

in the adjacent metapelites. This second prograde stage was accompanied by strong deformation (Seidel et al. 1975).

Recently, epimetamorphic marls with the assemblage

lawsonite + paragonite + phengitic muscovite + chlorite + quartz ± calcite ± dolomite

were detected in the phyllite-quartzite series of SW-Crete (Seidel 1976). These lawsonite schists are intercalated with fossiliferous marbles of Triassic age indicating that the temperatures did not exceed about 400°C.

In the western part of central Crete, chloritoid schists of the phyllite-quartzite series are associated with meta-basalts characterized by the greenschist facies assemblages

actinolitic hornblende + epidote + chlorite + albite + quartz.

Glaucophane/crossite is present only in subordinate amounts while lawsonite and pumpellyite are absent. Thus a regional increase of metamorphic grade in an eastward direction, though probably under the same confining pressure, is indicated.

In eastern Crete, the meta-basalts of the phyllite-quartzite series are characterized by the assemblage

common green hornblende + plagioclase (An 30 - 40) + epidote.

In a second, retrogressive stage, albite, crossite, and parts of the epidote were formed. Associated with the amphibolites are phyllonitic micaschists composed of

garnet ± staurolite ± kyanite + biotite + muscovite + plagioclase + quartz.

This assemblage indicates temperatures up to 650°C and water vapour pressures of 7 - 8 kilobars. The problem is still open to discussion whether these rocks are the high-grade equivalents of the glaucophane-bearing meta-basalts and the related chloritoid schists in western Crete.

It should be noted, that the tectonic unit of the phyllite-quartzite series in eastern Crete not only contains these relatively high grade metamorphic rocks, but also sequences of considerably lower, probably anchi-metamorphic grade. From a petrofabric study, Wachendorf et al. (1974) recognized this unit as a tectonic mélange. The anchi-metamorphic part is characterized by frequent quartz veins with ferrocarpholite.

In the village of Viános, south-eastern part of central Crete, ferrocarpholite is present in veins as well as in metapelitic schist with the assemblage

ferrocarpholite + chloritoid + chlorite + phengitic white mica + quartz.

These ferrocarpholite-chloritoid schist could well be the lower-grade equivalents of the chloritoid-bearing metapelites in western Crete (Seidel & Okrusch 1977).

The *serpentinite-amphibolite association* consists of two parts, tectonically separated. The upper part is composed of serpentinite bodies of different size, locally intruded by diabase dikes (Thorbecke 1973) while the lower part includes amphibolites, gneisses, micaschists, calc-silicate rocks, and marbles formed under high-temperature, and relatively low-pressure conditions. According to Bonneau (1970) and Thorbecke (1973) rocks of considerably lower metamorphic grade also occur in this unit.

Typical assemblages are:

common green or brown hornblende + plagioclase ± diopside

in the metabasites and

sillimanite ± andalusite + cordierite ± garnet + biotite ± K-feldspar + muscovite + plagioclase + quartz

in the metapelites. From this assemblage, P-T conditions of about 700°C and 5 kilobars water vapour pressure are estimated. Thus the metamorphic pressure was markedly lower in this unit than in the phyllite-quartzite series.

Judging from biostratigraphic evidence, the high-pressure metamorphism in the phyllite-quartzite series is post-upper Triassic and, consequently, of Alpine age. A question of paramount interest is whether the low-pressure metamorphism of the high-grade rocks in the serpentinite-amphibolite association has the same age. A radiometric dating program to solve this problem gave concordant K/Ar dates on hornblendes and biotites from the high-grade rocks of the upper unit of about 70 m.y.(Seidel et al. 1976a). So far, no reliable data are available for the phyllite-quartzite series.

Gavdos

(Geographical position south of Crete, see fig. 1, Dürr et al. this vol.).

The metamorphic rocks exposed on the island of Gavdos were supposed to be equivalent, in their tectonic position, to the serpentinite-amphibolite association (Vicente 1970). However, marked differences are recognized with respect to the metamorphic assemblages and the bulk rock composition of individual constituents. The metabasites of Gavdos contain the assemblage

blue-green hornblende $\pm$ glaucophane/crossite + epidote $\pm$ chlorite $\pm$ garnet $\pm$ phengite + albite,

while the adjacent metapelites are characterized by the assemblage

garnet + chlorite + phengite + albite + quartz.

These assemblages point to a metamorphism under temperatures of 500-550°C and pressures of 6-8 kb (Schliestedt 1976; Schliestedt & Seidel 1976). The metamorphism must be older than Turon-Senon since unmetamorphosed sediments of this age rest, with transgressive unconformity, upon the crystalline basement (Vicente 1970). Recent K/Ar dates on phengites and amphiboles indicate, indeed, a Late Jurassic age of metamorphis (Seidel et al. 1976b).

References

Bonneau, M., 1970: Bull. Soc. géol. France (7) 12:1124-1129.

Creutzburg, N. & Seidel, E., 1975: Neues Jahrbuch Geol. Paläont. Abh., 149:363-383.

Schliestedt, M., 1976: Mineral. Petrogr. Inst. T.U. Braunschweig.

Schliestedt, M. & Seidel, E., 1976: Fortschr. Mineral. 54, Beiheft 1: 79-81.

Schubert, W. & Seidel, E., 1972: Z. Dtsch. Geol. Ges., 123:371-384.

Seidel, E., 1976: Fortschr. Mineral. 54, Beiheft 1:92-93.

Seidel, E., Okrusch, M. & Schubert, W., 1975: Contr. Mineral. and Petrol. 49:105-115.

Seidel, E., Okrusch, M., Kreuzer, H., Raschka, H. & Harre, W., 1976a: Contr. Mineral. and Petrol., 57:259-275.

Seidel, E., Okrusch, M., Schliestedt, M., Kreuzer, H., Harre, W. & Raschka, H., 1976b: XXVth Congr. and Plen. Ass. Split (22-30 Oct.) Marine Geol. and Geophys.
Seidel, E. & Okrusch, M., 1977: Contr.Mineral. & Petrol., 60:321-324.
Thorbecke, G., 1973: Ber. Naturfr.Ges. Freiburg i.Br., 63:81-92.
Vicente, J.C., 1970: Bull. Soc. géol. France, 7, XII:481-495.
Wachendorf, H., Baumann, A., Gwosdz, W. & Schneider, W., 1974: Z. dtsch. Geol. Ges., 125:237-251.

Microfauna in the Tripolitza Limestone in Northern Middle Crete

U. Leppig

In northern middle Crete the Tripolitza Limestone at about 1000 m thickness and the conformably overlying Tripolitza Flysch represent the sediments of the Tripolitza Nappe. Carbonate sedimentation persisted from late Triassic to late Middle Eocene. The period from the Bathonian (Upper Dogger) to the Biarritzian (upper Middle Eocene) can be documented with fossils. The strata below the Bathonian are about 100 m thick and consist of stromatolithic dolomites and structureless marbles. They have been encountered at a depth of 237 m in the well FAO that was located 9 km west of Iraklion between the mountains Chéri and Metóchi Koryfí. Because of their thickness they probably include the Lower Dogger as well as parts of the Liassic and perhaps some Triassic. In the lowermost parts of the core the calcite content decreases and the quartz-mica content increases. A similar transition between Tripolitza Limestone and underlying clastics was found by Thorbecke (personal comm.) at the road southwest of Argyroúpolis, Nom. Rethymnis. In fig. 1 these transition beds are called "couches de passage".

The faunistic associations studied and compiled in fig.1 consist of the following elements:

VIII. Biarritzian: *Alveolina div. sp.:* northern slope of Trikotiá mountain, 1080 m above sea level.

VII. Lower and Middle Ilerdian: *Alveolina div. sp.:* western slope of Trikotiá mountain, 1090 m.

VI. Campanian - Maestrichtian: *Orbitoides cf. media:* Gaidarou mountain, 730 m, *Rhapydionina liburnica, Broeckinella arabica, Laffitteina marsicana:* 20 m below Radar station of Goniés, 860 m.

V. Turonian - Campanian: *Murciella cuvillieri:* western slope of Gaidarou moutain, south of road Týlissos - Goniés, 550 m, *Accordiella conica:* southern slope of Guguliéta, 560 m.

IV. Upper Cenomanian - Lower Turonian: *Pseudolituonella reicheli, Nezzazata simplex, Reicaella ramonensis, Coxites zubairensis:* southern and eastern slope of Guguliéta mountain, 580 and 460 m.

III. Barremian - Albian: *Coskinolina sunnilandensis:* above the old road Iraklion - Rethymnon, between km 8 and 9, 160 m; *Pseudotextulariella?*

scarsellai: Niftas mountain, 540 m.

II. Kimmeridgian: *Clypeina jurassica:* northern slope of Chéri mountain, 200 m; *Cladocoropsis mirabilis:* above Gérgeri, southern slope of Psiloritis, 1150 m (Thorbecke legit).

I. Bathonian - Callovian: *Praekurnubia cretica* , *Pseudocyclammina mayncі*, *Parurgonina caelinensis:* FAO bore hole, depth 40 m.

For the Liassic, *Dissocladella cretica* (Ott 1965) from the Trypali limestone near Sigoniá may be added here, if the Trypali Limestone is identical with the Tripolitza Limestone as suggested by Kuss & Thorbecke for the following reasons (pers. comm.):

1. The limestone containing Dissocladella is neritic and very similar to the Tripolitza limestone.
2. It overlaps cherty limestone (Plattenkalk, Talea Ori Series), as does the Tripolitza Limestone.
3. There is no argument against the suggestion that it is Tripolitza Limestone here as well.

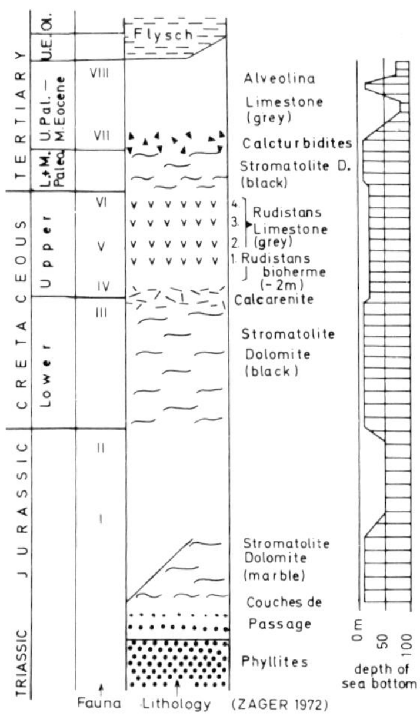

Fig. 1. Stratigraphy of the Tripolitza Limestone of northern middle Crete between the villages Márathos, Krousóna and Goniés, deposited within the central part in the Tripolitza platform. Vertical scale: thickness, about 1000 m. I - VIII: faunistic assiciations, see text.

Crete occupies an important position in paleobiogeographic respect as a connecting link between Europe and the Near East. In the Upper Cretaceous, within the Tethyan area, faunistic differences can be seen. The fauna in the Tripolitza Limestone belongs to the Near Eastern rather than to the European faunal province.

The results of the sedimentological (Zager 1972) and the microfaunistic (Leppig 1974) data show the different environmental conditions which episodically prevailed during the sedimentation of the Tripolitza Limestone. We distinguish: tidal region (a) - calm lagoon (b) - shallow open lagoon (c) - transition (a-c) - slope of rudist biostroms (d) - slope of bioherms (e) - transition (c-d-e) - rudist bioherms (g) - blue-

green algae bioherm (h) - deep open lagoon (i).

Structural analysis of two foraminiferal species (*Praekurnubia cretica n.sp.* and *Coxites zubairensis* SMOUT) was made (Leppig 1976).

References

Leppig, U., 1974: Diss. Freiburg i.Br.
- , 1976: Ecl. geol. Helv., 69/3.
Zager, D., 1972: Diss. Freiburg i.Br.

Part 4

Central Aegean Region

The Median Aegean Crystalline Belt: Stratigraphy, Structure, Metamorphism, Magmatism

St. Dürr, R. Altherr, J. Keller, M. Okrusch and E. Seidel

Contents

1. Introduction

1.1. Generalities

Within the Aegean orogenic arc, four crystalline complexes - the West-Macedonian, the Thessalian, the Attic-Cycladic. and the Menderes Complex - form a distinct belt. This "Pelagonian" (s.l.) belt has a median position between the internal Serbo-Macedonian and Rhodope crystalline areas to the northeast, and the external metamorphic rocks on the Peloponnesos and on Crete Island to the southwest and south. We prefer to say Median Aegean Crystalline Belt; the name "Pelagonian" s.str. should be restricted to a definite geologic unit (and corresponding palinspastic-paleogeographic zone) within this belt, as will be pointed out in the final chapter.

The crystalline complexes of the Median Aegean Belt currently are called "massifs". The character of these massifs - pre-Alpidic versus Alpidic - and, consequently, their role in the geodynamic evolution of the Aegean realm is controversial, because we still lack sufficient know-

ledge on stratigraphic ranges and tectonic structures of the metamorphic series, and on the age and character of metamorphic and magmatic events.

According to current interpretation, summarized in the syntheses of Philippson (1898, 1901, 1918, 1959), Aubouin et al. (1960-1963), Aubouin (1965), Brinkmann (1971), and numerous others, the medium to high grade metamorphic cores of the complexes probably represent the Precambrian, the low grade metamorphic cover series the Lower Paleozoic. Both together were regarded as the rigid pre-Alpidic basement of the surrounding (mostly non-metamorphic) younger Paleozoic-Mesozoic-Paleogene sequences of the Alpidic orogen. Playing thus the role of "Zwischengebirge" (median masses) or Tethyan microcontinents, the Median "massifs" seemed to fit well into the modern microplate concepts (e.g. Dewey & Bird 1970; Dewey et al. 1973).

However, contrasting opinions were stated repeatedly, assuming a Mesozoic age for the crystalline complexes as a whole or for parts of them (e.g. Bittner 1880; Kober 1929; Marinos & Petrascheck 1956; Andronopoulos 1962; Katsikatsos 1970). Our own results tend to confirm the latter view, with all its geodynamic consequences.

Our paper covers the central part of the Median Aegean Belt, which means the Attic Cycladic Complex and the western marginal part of the Menderes Crystalline Complex.

Both complexes consist of migmatites, various gneisses, micaschists, and phyllites, interbedded with marbles, frequently of considerable thickness; smaller granites intruded here and there. Some young depressions are filled with Neogene and early Quaternary, mostly terrestrial sediments and locally volcanic rocks. Lateron, during Quaternary time, wide parts of our segment have been submerged under sea level. A chain of recent volcanoes marks the southern border of the crystalline area. The structure seems to be simple at first sight: a set of mostly NNE-SSW elongated domes, cut by the network of postorogenic distractive faults.

Our special interest is focused on:

stratigraphy and structures along the southern margin of the Cyclades and Menderes Crystalline Complexes (Dürr 1975);

the mineral assemblages indicating high-pressure metamorphism in the Cyclades archipelago (Seidel & Okrusch 1976; Okrusch et al. in press);

the magmatic activity, especially the conditions of emplacement and radiometric ages of granitic intrusions (Altherr et al. 1976a, 1976b; Besang et al. 1976) on islands of the central and eastern Aegean Sea.

For a better understanding of our results, let us consider briefly what is known up to now about some key regions within the Median Belt, beginning in the northwest of continental Greece.

1.2. The Thessalian Crystalline Complex

Up to the early sixties, the great majority of the crystalline rocks in eastern Thessaly had been regarded as a paradigm of an autochthonous sialic basement, covered by the Upper Paleozoic and Lower Mesozoic neritic sediments of the Pelagonian Zone. At the margins of the "massif" these sediments occasionally are slightly metamorphosed (Brunn 1956; Aubouin et al. 1960-1963).

Godfriaux (1962, 1970), however, recognized the Mt. Olympus dome to be a tectonic window. It exposes a vaulted, "non-metamorphic", entirely neritic sequence of carbonate rocks, ranging from Triassic up to Eocene. This neritic sequence most probably belongs to the external Hellenic Gavroro Zone (Fleury & Godfriaux 1975), or - as Godfriaux previously had

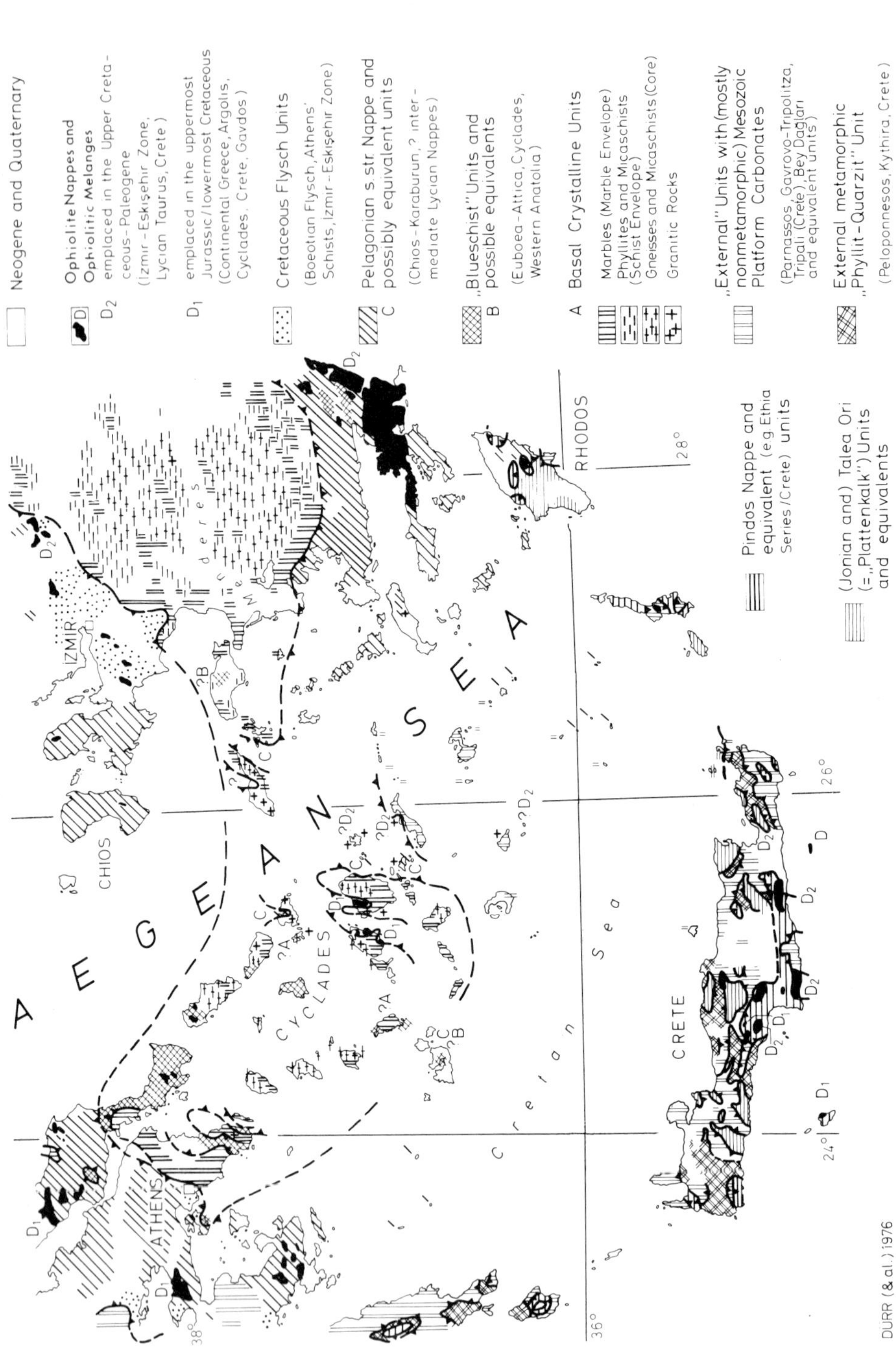

Fig. 1. Geological Units in the Central and Southern Aegean Area.

maintained - to the Parnassos Zone. Hence, the entire surrounding metamorphic terrane represents a huge basement nappe.

Further studies (Derycke & Godfriaux 1976) revealed this basement nappe to be subdivided. A lower unit, with marbles, ophiolitic rocks, and a flyschoid sequence, is characterized by blueschist metamorphism with glaucophane, lawsonite, and jadeite. An upper unit, characterized by (Alpidic) greenschist metamorphism, contains, beneath Mesozoic carbonate rocks, a variety of (pre-Alpidic) metamorphic schists and meta-granites, which seem to be what remains from the pre-Alpidic (Hercynian - Yarwood & Aftalion 1976) basement of the (Subpelagonian-) Pelagonian zone (sensu stricto; cf. chapter 5).

It should be remembered that, since the general acceptance of plate tectonics, the ophiolite masses resting upon the Triassic-Jurassic sediments of the Pelagonian Zone, e.g. in northwestern Thessaly, are considered as remnants of an ophiolite nappe, emplaced in uppermost Jurassic and/or Lower Cretaceous times (Dercourt 1971; Bernoulli & Laubscher 1972; and others).

1.3. Southern Euboea and Attica

South of the Thessalian crystalline area, sediments of the (Subpelagonian-) Pelagonian Zone, in most parts non-metamorphic, covered by remnants of the Ophiolite Nappe, build up the eastern Othrys Mountains, Northern and Middle Euboea, eastern Lokris and Boeotia. Beneath them, metamorphic rocks reappear in Southern Euboea and Attica.

Both regions have been studied repeatedly for more than a century; nevertheless, no agreement on their interpretation has been attained.

Following Marinos & Petrascheck (studies in Laurion 1956; for Southern Euboea see Andronopoulos 1962; Katsikatsos 1970; Guernet 1971, 1972) two groups may be distinguished.

The upper group contains well bedded marbles, quartzites (partly meta-radiolarites), metamorphosed ophiolites, and metamorphic flyschoid sequences. Mostly a strong deformation is visible.

Glaucophane - epidote - chlorite - white mica - albite - quartz (- lawsonite - calcite) is the characteristic mineral assemblage in Southern Euboea (Guernet 1972). The original rocks have been regarded to be of Mesozoic to ?Paleogene age, based on lithostratigraphic comparison (Andronopoulos, Katsikatsos).

The lower group consists of the well-known Attic marbles with intercalations of micaschists. Within this group, badly preserved Mesozoic fossils have been described by various authors (see Marinos 1955), doubted by others. Katsikatsos (1970), Argyriadis (1967), and Papadeas (1970), however, conclusively proved parts of the marbles to be of Upper Triassic and (Upper) Cretaceous age in Northern Attica and the northwestern part of Southern Euboea.

Hence the two groups should represent two different units, both predominantly Mesozoic in age (compare Katsikatsos et al. 1976).

Guernet (1971, 1972), on the contrary, regarded both as one group representing the Paleozoic basement of the Pelagonian Zone and thus reaffirmed the "autochthonist's" point of view (see however Guernet 1975).

For the grandodiorite of Laurion which intruded the lower unit Marakis (1970) reports a K-Ar age on biotite of 10^{+} ? m.y.

2. The Menderes Crystalline Complex

2.1. The Menderes Complex in Anatolia

The Menderes Complex is bordered, to the south and the east, by the Lycian Nappes (Brunn et al. 1971), to the northwest and to the north, by the İzmir-Eskişehir (-Ankara) Trough (Brinkmann 1966). So far, the Menderes Complex is not clearly delimited to the northeast. Some of the Greek islands off the Anatolian coast (Samos, Agathonisi, the Phourni Islands, Ikaria) seem to complete it in southwestern direction.

The core of the complex, exposed over vast areas, consists of micaschists and paragneisses which pass gradually into metablastic augengneisses and, finally, into anatectic granitic gneisses. The critical mineral assemblage in the micaschists and paragneisses is (Schuiling 1962; Izdar 1971; see also de Graciansky 1966):

kyanite - staurolite - almandine - biotite - muscovite - plagioclase - quartz. The parent rocks are generally believed to be of Precambrian and/or lowermost Paleozoic age and to be transformed by an old (Caledonian or older?) metamorphism.

The core is surrounded by an envelope of lower grade micaschists, phyllites, and marbles characterized by numerous emery and diasporite deposits. The schists are placed to the Lower Paleozoic by lithostratigraphic comparison (Brinkmann 1971). A Devonian (Önay 1949; Schuiling 1962), Carboniferous (Brinkmann 1967), or Permo-Triassic (Wippern 1964) age has been conjectured for the marbles.

The metamorphic grade within the envelope increases in a remarkably regular way from top to bottom and is thought to be related to the migmatization in the core (Schuiling 1962; de Graciansky 1966; Başarir 1970). Critical assemblages point to the greenschist facies of various metamorphic grade, e.g.

garnet - biotite - muscovite - albite - quartz,
chloritoid - muscovite - paragonite - quartz.

The diaspore/corundum isograd is crossed within the marbles (Önay 1949).

According to the different opinions on the stratigraphy and the regional geological context, tectogenesis and metamorphism of the envelope are supposed to have happened at the end of the Hercynian (Önay 1949; Schuiling 1962) or, alternatively, at the beginning of the Alpidic era (Wippern 1964; Brinkmann 1967).

Ketin (1966) alone resolutely believed in a continuous pile of sediments, ranging up to the Upper Cretaceous and, hence, placed tectogenesis and metamorphism of the whole complex into late Cretaceous to early Tertiary times.

A mediating interpretation, combining Hercynian with Alpidic events, has been advanced by Izdar (1971, 1975) and substantiated by Dora (e.g. 1976).

Our studies concentrated on the envelope, best exposed in the mountainous southwestern part of the complex.

Simple only in appearance, the structure of the envelope in fact is dominated by isoclinal folds and imbrications; it is thus in harmony with the syntectonic metamorphism. The large open folds described, for instance, by Önay (1949), figured by Ketin (1966), belong to a later generation.

In spite of these complications, a kind of stratigraphy may be established. First, a distinction should be made between the Lower or Schist Envelope, and the Upper or Marble Envelope.

The Lower Envelope comprises various types of low-grade metamorphosed pelites to psammites, with mostly lenticular intercalations of marbles, meta-lydites, and meta-conglomerates.

The Upper Envelope is almost exclusively constituted of carbonate rocks. Their original thickness probably amounted to several thousand meters; the facies can be assumed as predominantly neritic. Dolomitic marbles seem to constitute lower parts of the pile. In the Labada Dağ (İlbira Dağları) area, NW of Milâs, less recrystallized domains yielded algae and ?*Triasina*, indicating an Upper Triassic to ?Liassic age of these dolomites. The main and upper part of the Marble Envelope is built of thick-bedded, sometimes almost massive calcareous marbles. They contain emery and diasporite lenses (metabauxites) as key-horizons; from the stratigraphic point of view, it cannot be decided whether the primary sediment pile contained only a single or several bauxite horizons. Quite frequently, the meta-bauxite deposits allow to distinguish top from bottom. Rudists (amongst them *Hippurites* s.l.) have been found southwest, west, and farther westnorthwest (near Akbük) of Milâs, in marbles which are stratigraphically above a main bauxite horizon. In the same region, the Rudists bearing marbles are followed immediately by the Kizilağaç Formation (s.str., comp.Brinkmann 1967): it consists of 100 - 250 m of thin-bedded, multi-coloured marbles, most probably metamorphic pelagic limestones of the Scaglia type. West of Milâs, the Marble Envelope, finally, is covered by muscovite-chlorite schists which may represent relics of a flysch.

These discoveries have important corollarries:

1. Meta-dolomites and - most decisive - meta-bauxites bearing marbles are of widespread occurrence at the margins of, and locally within, the southern and central parts of the Menderes Crystalline Complex. They testify to the extension of the Triassic to Cretaceous Marble Envelope, thought to form the normal sedimentary cover of the Lower Envelope and the Core. Hence they mark a geologic unit which may be called Menderes Massif s.str. (see below).

2. In age and facies, the Marble Envelope, as developed west of Milâs, can be compared best with the Mesozoic to Paleogene autochthonous-para-autochthonous formations of the Western Taurids (see Brunn et al. 1971). The Kizilağaç marbles probably represent an uppermost Cretaceous to lowermost Tertiary interval. A connection underneath the inner Lycian nappes is not unlikely. In any case, the region of the present Menderes Massif s.str. had been a huge platform with neritic carbonate sedimentation throughout most of the Mesozoic time.

3. The Lycian Nappes should have come from north of the Menderes Massif, like the nappes of the Occidental Taurus, which came from northeast of the Sultan Dağ (Brunn et al. 1971). Internal structures, observed in Lycian units SW of Milâs, illustrate southward movements. The lowermost Lycian "Unit of the Eocene Flysch" (Brunn et al. 1971) is possibly derived, at least in part, from the top of the Menderes Marble Envelope.

4. The Cretaceous Flysch of the İzmir-Eskişehir (-Ankara) Zone (Verdiers 1963, stratigraphy accepted) should repose with tectonic contacts upon the Marble Envelope. Observations south of İzmir do confirm this assumption. Constituents of the İzmir-Eskişehir Zone (ophiolites, glaucophane rocks, Upper Cretaceous flysch) have counterparts in the Lycian Nappes.

5. If the base of the Marble Envelope is of Upper Triassic age, the Lower Envelope may comprise the whole Paleozoic and even the Lower Tri-

assic. This view is supported by lithostratigraphic comparison of some characteristic rock types, e.g. metamorphosed sedimentary iron-ores connected with quartzites (Brinkmann 1967), graphitic phyllites, meta-lydites, and meta-conglomerates of distinct habitus.

6. All investigations agree, that only one metamorphism affected both, the Lower and Upper Envelope of the Menderes Massif, and that this metamorphism caused the migmatization of the core and parts of the envelope. This main tectogenetic and thermal event cannot have started prior to the end of Cretaceous time; more probably, it took place during the Eocene. The Rb-Sr determination on micas carried out by Jäger (Berne) and reported by Izdar (1975) and Brinkmann (1976), muscovite 66 ± 4 m.y., biotite 22,2 ± 1,3 m.y.*, document Alpidic metamorphism, although the muscovite age seems to be somewhat high.

For all that, it must be emphasized:

The new picture drafted here applies only to the Menderes Massif s. str., characterized by the metabauxites-bearing marbles. Where they lack, at present the Massif cannot be delimited. More or less considerable parts of the Menderes Crystalline Complex ("Menderes Massif" s.l.), especially the northeastern ones (beyond the upper Gediz river), but also some portions in the western border regions (e.g. south of Selçuk/Ephesos) may be built of metamorphic rocks belonging to other, structurally higher units. The presence of glaucophane bearing rocks and of ultramafics is a certain indication in this respect. Near Tíre (northern part of the Menderes Crystalline Complex) such a unit has been found by Argyriadis et al. (1976).

2.2. Samos, the Phourni Islands, Ikaria

On the island of *Samos*, which still is poorly known (Philippson 1959), two different units may be distinguished. To the lower one we assign gneisses and micaschists with the Barrow type assemblage

kyanite - garnet - biotite - muscovite - quartz

near Palaeomylos (northern part of central Samos) and Vathy (E-Samos); in the latter area, also chloritoid was recorded (Schneider 1914). Diasporite bearing marbles occur in the eastern and western part of the island. These formations may represent core and envelope of the Menderes Massif s. str. The upper unit, characterized by rocks with

glaucophane - epidote - chlorite - white mica (Schneider 1914)

(Ampelos Oros in the central northern part of Samos), can be compared to the upper group in Southern Euboea (see Guernet 1972). A glaucophane-epidote bearing meta-gabbro is described by Butz (1912) from Kosmadei (NW Samos).

The marbles and crystalline schists of the *Phourni Islands* may resemble those of the basal unit of Samos, according to the descriptions of Köhne (1937). Mineral assemblages (including kyanite and chloritoid) point to metamorphic conditions of the upper greenschist facies. On the western island of Thymaina, these series are overthrusted by a non-metamorphic Triassic to ?Liassic limestone with Megalodontids.

Ikaria was described by Ktenas (Marinos ed. 1969). A lower group with gneisses, micaschists, and marbles containing an emery deposit, underwent amphibolite facies metamorphism as indicated by the assemblage

kyanite - staurolite - garnet - biotite - muscovite - plagioclase - quartz

* Berne uses $\lambda_{^{87}Rb} = 1.47 \times 10^{-11} y^{-1}$.

in metapelites (Altherr unpubl.). An upper group of phyllites and well-bedded marbles is clearly separated by its contrasting, low greenschist metamorphic grade; fossils have not been found.

The western part of the island consists of a deformed granite. K-Ar dating on biotites yielded a model age of about 9 m.y. (Altherr et al. in prep.).

The former presence of a unit with non-metamorphic sedimentary rocks is certified by the components of breccias (the age of which is still

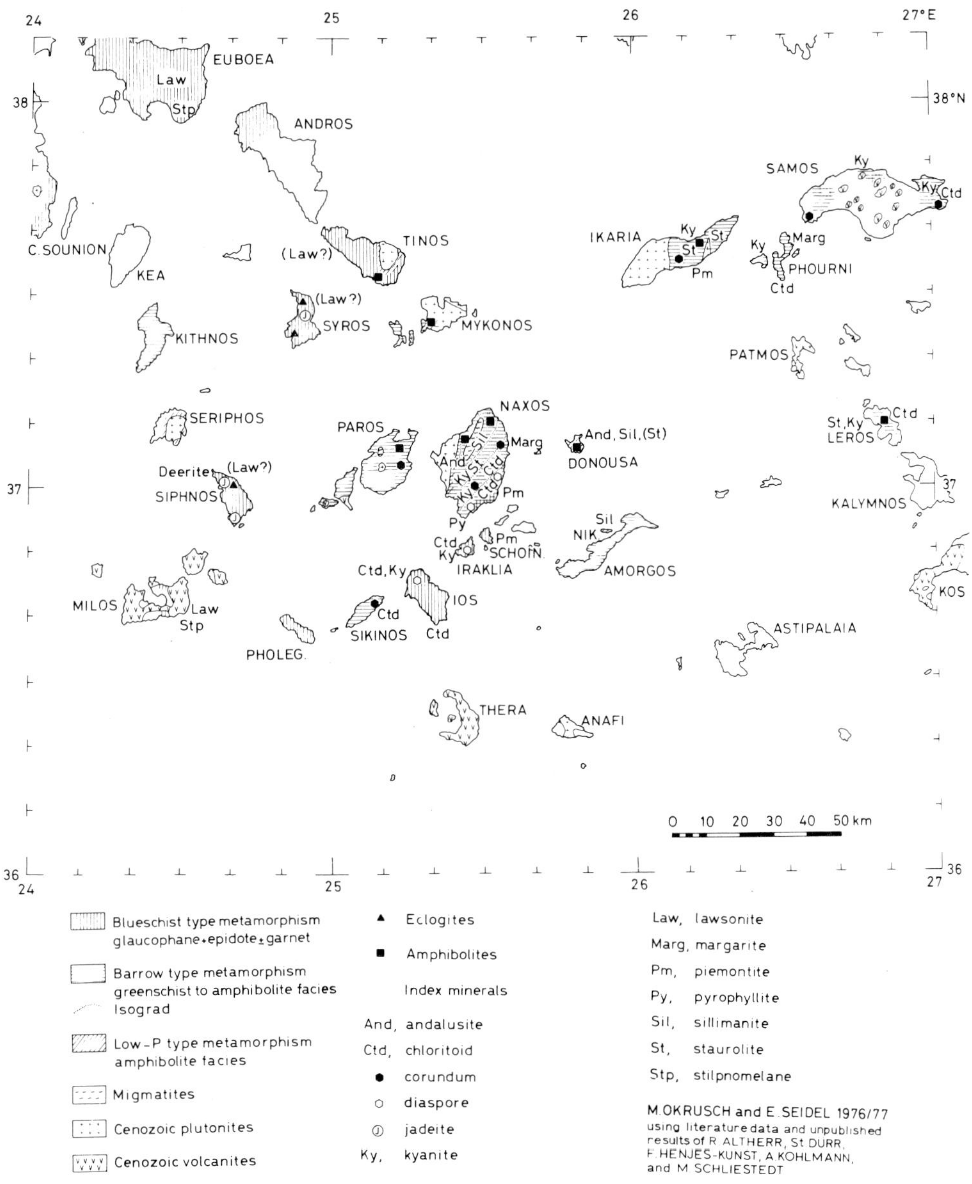

Fig. 2. Types of Metamorphism in the Cyclades Archipelago.

unknown) and overlying Pliocene (Ktenas 1927) conglomerates, on the southeastern coast of Ikaria.

3. The Cyclade Islands

3.1. Generalities

Amidst the Median Aegean Crystalline Belt, the Cyclade Islands take a linking position between the crystalline areas of Southern Euboea - Attica and the Menderes Complex. The scattered distribution of these islands and the great variety of rock types complicate the geological interpretation.

Lithologically, similarities exist to the series of the Menderes Massif s.str. as well as to both units in Southern Euboea - Attica.

Evidently, the rocks were formed under highly contrasting geothermal gradients (e.g. Schuiling 1973). This may be due to either a) a temporal succession of different baric types of metamorphism (sensu Miyashiro 1973) within one unit and/or b) to the fact, that the rocks belong to different tectonic units.

By comparison with Attica, the marble-rich series of the Cyclade Islands were considered as Mesozoic by some authors.

To support this, the emery occurrences on Naxos, Paros, Ios, and Sikinos have been cited, which are supposed to be equivalents of the Mesozoic bauxites in continental Greece and Yougoslavia (e.g. Renz 1955; Petrascheck 1960). Direct evidence, however, is wanting. Only Négris (1914-1919; see PIA 1922, for a discussion) described ?Triassic algae in a Naxotian marble, while Permian fossils were found at one point within the marbles on Antiparos (Anastopoulos 1963) and in marbles of NW Andros (Papanikolaou 1977).

Other facts, however, have led to regard the Cycladic metamorphic rocks as Hercynian or even older basement. The gneisses of Mykonos are locally covered by relics of non-metamorphic Upper Paleozoic (Papastamatiou 1963) and Triassic (Cayeux 1911) carbonates. On the eastern coast of Naxos, the marbles are bordered by Upper Permian limestones, but with serpentinite at the contact (!) (Marks & Schuiling 1965). Remnants of a non-metamorphic ophiolite suite, transgressed by Barremian limestones (Papageorgakis 1969), rest beside (? upon) marbles in northeastern Paros and western Naxos (see Roesler this vol.).

Also, the (mainly N-S to NNE-SSW) orientation of fold-axes and lineations has been interpreted as characterizing an ancient (Attic-) Cycladic nucleus within the alpidic mountain range crossing the Archipelago (Renz 1940; Trikkalinos 1954; Philippson 1959; Schuiling 1962, 1973). A part of this range, limiting the Attic-Cycladic "Massif" to the east and southeast, has been thought to be visible in the non-metamorphic Mesozoic sediments of the Makares Islands east of Naxos and on Amorgos (see Renz 1955).

3.2. The Central Cyclades

Our discussion will start with a reasoning on *Naxos*. With its central migmatite and the surrounding domed schists and marbles, the island can well be called a small counterpart of the Menderes Massif s.str. Indeed, Naxos' emery bearing marbles have been compared repeatedly with those of the Menderes Marble Envelope (cf. Önay 1949). If this comparison is valid in the new belief, sketched in chapter 2.1., the marbles of Naxos should

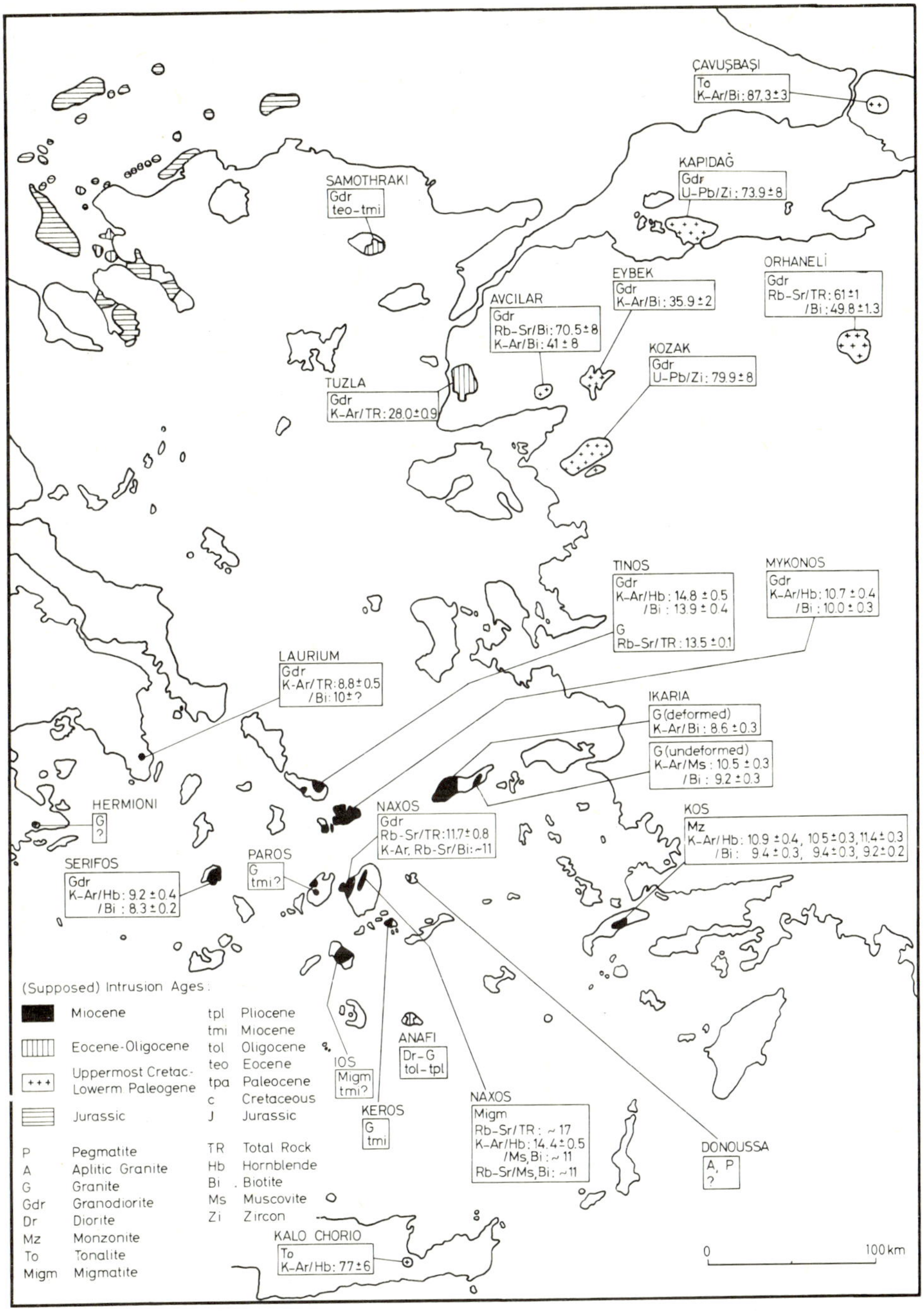

Fig. 3. See caption fig. 4.

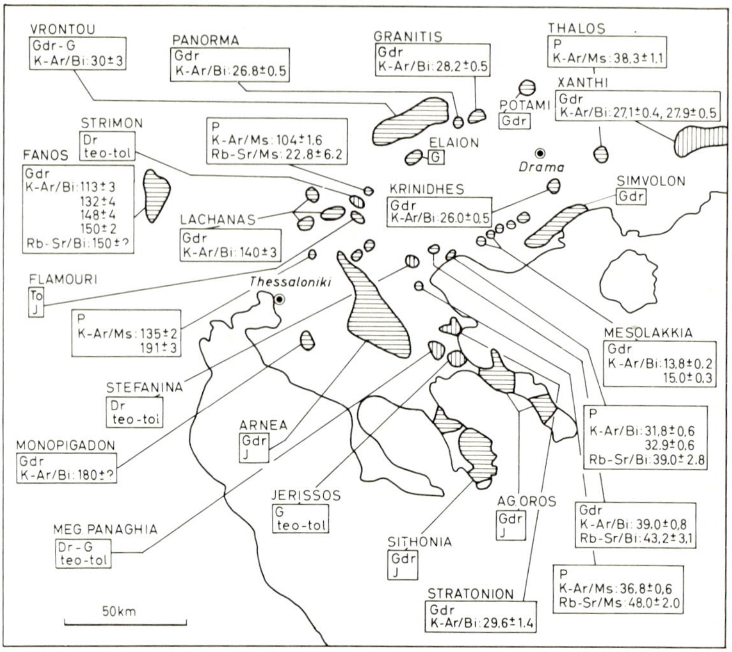

Fig. 4. Granitic rocks in the Aegean area. Compiled by R. Altherr & J. Keller (1976), using literature data and unpublished results of St. Dürr, H. Kreuzer & I. Wendt.

represent a metamorphic sequence of dolomites and neritic limestones, being Triassic to, at least, Upper Cretaceous in age, originally deposited onto an older basement.

On Naxos, no direct proof for this hypothesis seemed available; the ?Triassic algae (Négris 1914-1919) had not been confirmed by other workers. However, these algae have been rediscovered recently (Dürr in prep.). The general lack of fossils, however, is easily understood, considering the strong deformation and the metamorphism.

The mostly open, upright folds with NNE-SSW oriented axes, figured by Papavasiliou (1909) and Jansen et al. (1973), are folds of a later, (?) second generation; they bend the foliation. The foliation in turn certainly denotes isoclinal recumbent folds. Such folds may be divined on Jansen's map, they can be seen in the field, but still remain to be investigated. This (?) first isoclinal folding possibly was connected with a first, synkinematic metamorphism (under low temperature/medium to high pressure conditions), which is documented by the albite-glaucophane parageneses in the outer envelope. According to K-Ar and Rb-Sr determinations on muscovites, this event was terminated about 45 m.y. ago (Andriessen et al. 1976).

Isoclinal folding, as assumed already by Papastamatiou (1951), can explain the repeated alternation of schists and emery bearing marbles. This folding excludes a neat separation of Lower and Upper Envelope as practicable in the Menderes Massif s.str. The central migmatite probably re-

places the main part of the Paleozoic schists supposed below the marbles, which form huge resisters within the migmatite.

The formation of the migmatite was a late-kinematic one. Locally, the migmatite is directly connected with intrusive, "synkinematic" aplitic granites (Jansen et al. 1973), the largest of which was dated, by Rb-Sr total rock analysis, at about 17 m.y. (Andriessen et al. 1976). A K-Ar determination on hornblende of an amphibolite within the migmatic core yielded a Middle Miocene age of 14.5 ± 0.5 m.y. (Altherr et al. in prep.). For micas of the central migmatite, Rb-Sr and K-Ar ages of around 11 m. y. have been obtained with some K-Ar biotite ages ranging down to about 5 m.y. (Andriessen et al. 1976).

The isograds of the thermal aureole, surrounding the central migmatite dome, cut the fold structures of the late (?) second generation obliquely. Temperatures within the aureole ranged from about 700°C or more within the core to less than 420°C (corundum/diaspore isograd) in the outer parts of the envelope (Schuiling & Oosterom 1967; Jansen et al. 1973; Jansen & Schuiling 1976). No pressure gradient can be detected: the formation of kyanite (and/or sillimanite) next to the core, the lasting stability of glaucophane in the outer envelope, possibly indicate an overall state of medium pressure at around 6 kbar.

The late-kinematic anatexis was followed, underneath the actually visible level, by later phases of magma generation, creating a granodiorite which intruded the metamorphic rocks in western Naxos and, lateron, became deformed. Mineral assemblages in the zone of contact metamorphism contain andalusite and thus testify to lower pressure. Rb-Sr total rock analyses on rocks of the granodiorite massif gave an isochrone of 11.7 ± 0.8 m.y. (Andriessen et al. 1976), essentially based on likewise deformed aplitic and pegmatitic dykes, cutting the granodiorite.

We consider all these ages to date a Tertiary epoch of metamorphism, starting in Paleogene times with rapidly increasing pressure but still low temperature. This stage was followed by an increase of temperature which culminated in anatexis and generation of magmas in Miocene times.

After the cooling and erosion of the crystalline complex, during Upper Miocene to Lower Pliocene times, the non-metamorphic rocks described by Roesler (this vol.) were emplaced tectonically on to the denuded crystalline basement. These rocks most probably represent elements of the Pelagonian Unit s.str. (cf. chapter 5) and of the Ophiolite Nappe (cf. 1.2.), both covered by meso-autochthonous Cretaceous, Paleogene, and ?uppermost Oligocene to Middle Miocene molasse.

Allochthonous elements and crystalline basement together have been affected by a moderate deformation during Upper Miocene (and/or Lower Pliocene ?) times. Its maximum is confined to a narrow corridor on western Naxos ("zone of structural weakness" of Schuiling & Oosterom 1967; compare Jansen et al. 1973): Here the molasse shows occasionally isoclinal folds and imbrications. Its eastern border is a high angle thrust, accompanied by a zone of retrograde metamorphism within the metamorphic schists. All structural elements trend N-S to NNE-SSW.

Erosion of the Naxotian basement is recorded in an Upper Pliocene to ?Lower Quaternary fluvial to lacustrine sequence (cf. Roesler this vol). Vaulting (with NNE-SSW axis) and mostly normal faulting movements continued during Quaternary times; they soon led to the ingression of the sea.

The picture of Naxos, sketched so far, corresponds to the hypothesis developed for the Menderes Massif s.str.; it is quite different from the

hitherto accepted opinions. Nevertheless, we believe it to consider coherently all known facts.

Additional evidence for this new picture is gained on neighbouring islands.

Doubtless it is the envelope of Naxos which constitutes the two westernmost islands of the small *Erimonisia* (= lonely islands), a few kilometers south-southeast of Naxos: *Iraklia* with diasporite bearing marbles, and Schoinousa. *Schoinousa* (see Dürr 1975) consists of marbles and schists with the critical assemblages

glaucophane - garnet - epidote - chlorite - white mica - albite - quartz.

Also with respect to the structure, it is much alike southeastern Naxos. On the southern satellite island, Agrilos, within the marbles, a dolomite was detected with *Triploporella* cf. *remesi* (Steinmann), an algue which is known from the Upper Jurassic and mainly the Lower Cretaceous, thus corroborating the Mesozoic age assigned to the Naxotian marbles.

Paros, the western neighbour of Naxos, consists of a sequence of migmatites, gneisses, metamorphic schists, and marbles with an emery deposit, as well as of more or less deformed granitic intrusions (according to Philippson 1901, 1959; Frangatos 1951; the studies of Schuiling et al. still being unpublished).

On the island of *Ios* (Philippson 1959; Marinos 1947) a core of orthogneiss is covered by an envelope of schists and diasporite bearing marbles; parts of this envelope seem to constitute *Sikinos* (with chloritoid bearing emery, Papavasiliou 1913) and *Pholegandros*. Glaucophane-bearing rocks are known from the islands of Ios (Marinos 1947) and Pholegandros (Schuiling 1973). Characteristical assemblages on northern Ios are (Henjes-Kunst in prep.):

glaucophane - phengite - epidote - quartz (- garnet - albite - chlorite), diaspore - chloritoid - kyanite - hematite (- rutile - calcite - white mica).

Mykonos and *Delos*, north of Naxos, are built of a deformed granodiorite which resembles the granodiorite of western Naxos. K-Ar dating on biotite yielded 10.0 ± 0.3 m.y. for Mykonos and 10.8 ± 0.3 for Delos. A hornblende from Mykonos gave a preliminary age of about 11 m.y. (Altherr et al. 1976b).

Besides there occur gneisses with

plagioclase - quartz - biotite (- garnet - muscovite),

amphibolites with

hornblende + plagioclase + epidote,

and marbles. The metamorphic series were overprinted by contact metamorphism leading to assemblages of higher grade. The remnants of non-metamorphic sediments (see 3.1.) rest upon the crystalline rocks with tectonic contacts (!) and must be considered as relics of the Pelagonian (s.str.) Nappe and its meso-autochthonous molasse conglomerates (Dürr in prep.).

Equally, remnants of this nappe certainly are the Upper Paleozoic rocks on the eastern coast of Naxos (cf. 3.1.) and the Upper Triassic - ?Liassic limestones on the *Makares Islands* off this coast. Southeast of Naxos, on the *Erimonisia Keros and Antikeri*, similar sequences are found more or less metamorphosed, due to a first very low to low grade synkinematic, regional metamorphism and a second post-kinematic thermal metamorphism bound to granitic intrusions (Dürr 1975).

Relics of the uppermost tectonic unit in the Aegean area, the Ophio-

lite Nappe s.l., covered by the already mentioned meso-autochthonous molasse, occur also on two of the Makares Islands and on the Erimonisia Kato-Koupho and Keros.

Interestingly, the uppermost unit of Crete (the serpentinite-amphibolite association of Creutzburg & Seidel 1975) which may be confined to the Ophiolite Nappe s.l. (comp. 5.) contains a distinct series of high-grade amphibolites, marbles, calcsilicate rocks, gneisses and micaschists. Mineral assemblages in the metapelites of this series including garnet, cordierite, andalusite, and sillimanite (with K-feldspar and/or muscovite) point to a temperature maximum of about 700°C and a pressure of about 5 kb. According to K-Ar dating on hornblendes and biotites, the metamorphism took place in late Cretaceous time (Seidel et al. 1976). Similar crystalline rocks build, according to Dürr (1975), the islands of *Donoussa* east of Naxos and *Nikouria* (see 3.4) off Amorgos, perhaps also Makra south of Anaphi (Melidonis 1963). On Donoussa, the critical assemblage

andalusite/sillimanite - muscovite - biotite - quartz

was recognized in metapelites. Andalusite may include staurolite.

Radiometric K/Ar dating on one hornblende and one muscovite from Nikouria gave cooling ages similar to those mentioned above from Crete (Dürr et al. in prep.).

3.3. The Western Cyclades

Special problems are involved with the interpretation of the western and northwestern Cyclades Islands. They are built up by a variegated series of metamorphosed basic (and acid) volcanites, gabbros, ultramafics, silicic and carbonaceous sediments.

Glaucophane-bearing rocks are known from the islands of Andros (Marinos 1954), Tinos, Syros (Ktenas 1908; Dixon 1968, 1976; Seidel & Okrusch 1976), Siphnos (Ktenas 1908; Davis 1966; Seidel & Okrusch 1976; Okrusch et al. in press), and Milos. On Milos glaucophane bearing rocks occur as xenolites in volcanic rocks (Keller & Hoffmann in prep.) as well as in the basement of the island; K-Ar determinations carried out on glaucophane and white mica from basement rocks yielded 64.2 ± 6.50 m.y. and 33.2 ± 1.03 m.y. respectively (Fytikas et al. 1976).

Our own investigations concentrated on *Syros* and *Siphnos* where glaucophane is very frequent not only in metabasites but also in metaacidites, metapelites, calsilicate rocks, and even in marbles. The assemblages

jadeite - glaucophane - epidote - garnet - chlorite - white mica - quartz,

glaucophane - epidote - garnet - chlorite - white mica - albite - quartz - calcite,

omphacite - garnet - phengite

are typical for both islands.

The critical mineral assemblages indicate that, during prograde metamorphism, temperatures of 450 - 500°C were reached at a confining pressure of at least 14 kilobars. During the increase of metamorphic grade, the upper stability limit of lawsonite which was recognized in relics by Dixon (1976), was crossed. According to K/Ar and Rb/Sr data on white micas, the high P/T metamorphism took place in Late Eocene times. (Altherr et al. in prep.). A later retrogressive overprint led to the breakdown of jadeite + quartz to form albite. The eclogite lenses intercalated within the glaucophane rocks of Syros, are unusually rich in TiO_2 (up to about 6 %). Starting from the margins of individual bodies and along joints,

the eclogites were gradually transformed into garnet glaucophane rocks. Associated with the eclogites of Syros are metagabbros and serpentinites. Eclogitic rocks are alo present in the northern part of Siphnos.

On Tinos metabasic glaucophane schists intercalated with glaucophane-bearing marbles are widespread.

Glaucophane - garnet - epidote - chlorite - white mica - albite - quartz

is a typical mineral assemblage. Garnet porphyroblasts contain inclusions formed by aggregates of epidote and white mica, pseudomorph, presumably, after lawsonite. Marakis (1973) reported two discordant K/Ar dates on white micas of 33 and 27 $\pm$ 1 m.y. Granitic intrusions which are non-deformed occur on Tinos and Seriphos. They yielded Middle Miocene Rb/Sr and K/Ar whole rock and mineral ages (Altherr et al. 1976b). These intrusions led to a local overprint of the country rocks.

Like the other islands of the Western Cyclades Serifos and Kythnos are built up of variegated sequences, which originally also might have been affected by high P/T metamorphism.

3.4. Amorgos and the Levitha Islands

Amorgos generally is believed to represent the southeastern border of the Cycladic Crystalline Complex (Renz 1940, 1955).

According to Philippson (see 1959) and Renz (1933, 1955, Geol. Map of Greece 1 : 500.000), the island comprises, in basal position, a group of "semi-metamorphic" schists with limestones and conglomerates, their age being unknown. They have been compared with the Paleozoic of Chios and/or with the Athens' schists formation. This "basement" has been thought to be covered, with sedimentary or tectonic contact, by a flat-lying, ostensibly slightly disturbed carbonate sequence, in which "Permian" (Renz & Reichel 1945, see below) and Upper Triassic (Renz 1933) had been substantiated, whereas the existence of Jurassic and Cretaceous have been supposed on account of the presence of metabauxites. A (?) second group of schists covers the carbonate rocks in the northeastern part of the island.

Contrary to traditional assumptions, the structure of Amorgos is highly complicated and characterized by isoclinal recumbent folds of various dimensions, the limbs of which often are drastically thinned out by plastic flow and/or shear. Curiously, these folds are overturned to the north, that means, they are directed to the interior of the orogen. The entire sequence of rocks was affected by very low to low grade metamorphism.

The metamorphic schists can be derived from a Paleogene (middle to upper Eocene - ?Oligocene) flysch as indicated by nummulite breccias. The "Permian" formation yields Megalodontids; it had been mistaken for Permian by an erronous determination of poorly preserved *Triasina* cf. *hantkeni* Majzon (1945 still unknown) as *Nankinella* (written communication of Prof. M. Reichel/Basel).

Hence the stratigraphy of Amorgos, from bottom to top, begins with a conglomerate, rich in quartz and quartzite pebbles, the Kapsala Formation (several tens of meters). It upwards passes gradually into the Richti Formation (? 10 - 30 m), an alternation of some coarser clastic sediments with mainly shales and marls containing lenses and blocks of coral limestones. These limestones have been dated as Upper Carnian - (?)Lower Norian. They are followed by the dolomites and coral-bearing limestones of the Kryoneri Formation (s.l.), the former "Permian", now recognized

as Upper Triassic - (?)Liassic (up to 300 m and more). This neritic formation is overlain by pelagic, well bedded limestones with replacement cherts, the Chozoviotissa Formation (? 80 - 150 m; ? Jurassic - ? Lower Cretaceous). The last and most important (? more than 1000 m) carbonate formation builds up the ridge of the island from the central part up to the eastern cape and, therefore, has been called "Kammkalk" ("crest limestone") Formation. It consists of mostly thick bedded, sometimes massive limestones, which contain several small deposits of diasporites, originally one or (?) two bauxite horizons of probably Cretaceous age. Palimpsests of large nummulites suggest that the uppermost part of the limestone sequence is of Middle Eocene age; the overlying flysch thus begins at about the same time as on the neighbouring island of Astypalaea (Christodoulou 1967; Leontaris 1974). All the three main occurrences of "crystalline schists" on Amorgos belong to the Eocene - (?)Oligocene flysch; in every case, they represent the cores of large recumbent synclines.

The stratigraphic sequence just described resembles, in its upper parts, that of the Hellenic Parnassus Unit by its bauxites. Yet, judging from the late appearance of the flysch, it should have had, like Astypalaea, a more external position, for example that of the Gavrovo-Tripolitza Zone. Compared to both zones/units, Amorgos is distinguished by its Triassic part of the stratigraphic column and by its complicated folding and the metamorphism.

The small island of *Nikouria*, immediately off the northwestern coast of Amorgos, is entirely built of high-metamorphic rocks of the amphibolite facies (see 3.2.). From the structural point of view, it seems that the flysch of Aigiali in northeastern Amorgos dips gently to the west-southwest underneath the crystalline rocks of Nikouria.

The eastern promontory of Amorgos points to a west-eastern alignment of small islands, the *Levitha group*. These islands are constituted of a sequence of anchi-metamorphic, partly dolomitized limestones. Their folding seems to be gentle on Levitha, but on Kinaros it is nearly isoclinal and upright; the axes are oriented roughly north-south.

Fossils of (?)uppermost Jurassic to Lower Cretaceous age were found immediately above a formation of well bedded pelagic limestones with cherts. Within (?)overlying thick bedded limestones Upper Cretaceous is certified by Hippuritids. The formations of the Levitha islands have some features in common with the upper formations of Amorgos and Kalymnos. They cannot be attached unequivocally, either to one of the Lycian or to one of the Hellenide units.

4. Dodecanese Islands

The northern Carian (and Dodecanese) islands, north of Kos, were described by Desio (1931).

Agathonisi, near the mouth of the Büyük Menderes, may belong to the Marble Envelope of the Menderes Massif s.str.

Further south, the larger islands *Leros* and *Kalymnos* are the most interesting ones.

Both islands, according to Desio (1931), Christodoulou (1966, 1970) and Dounas et al. (1972), have a basement of medium grade metamorphic rocks and two overlying, "semi-metamorphic", Upper Paleozoic groups. These are covered, transgressively, by Upper Mesozoic carbonate rocks,

which are overthrusted, according to Dounas et al., by similar carbonate rocks of nearly the same age.

Relying on these descriptions, Leros and Kalymnos as well as the adjacent Lipsi and Arki islands have been considered (e.g. by Brinkmann 1971) as a border region of the Menderes Crystalline Complex.

Own investigations (Dürr 1978) lead to propose a different, tectonic, subdivision.

The crystalline rocks, comprising amphibolites, kyanite-staurolite micaschists (Katagas & Sapountzis 1977a), and some marble intercalations, evidently affected by an Alpidic retrograde metamorphism, form the basal element of a nappe ("Marina-Unit"). They are transgressively overlain by the Val-Camere formation (? Permo-Triassic); this in turn grades upwards into dark dolomites and limestones with (?)Megalodontids (Upper Triassic-Liassic); followed by well bedded limestones with replacement cherts (? Upper Jurassic, Christodoulou 1970, to ? Lower Cretaceous). All these three formations well resemble the Karaova, Gereme, and Çal Dağ formations within the Lycian (Köyceğiz-) units southwest of Milâs (see Brinkmann 1967; Bernoulli et al. 1974). They are covered (with tectonic contact ?, Dounas et al. 1972) by thick-bedded to massive, light coloured limestones of Upper Jurassic to (? Upper) Cretaceous age.

Beneath the nappe, a greenschist metamorphic (Katagas & Sapountzis 1977b), Paleozoic group represents a lower unit ("Teménia Unit"). It consists of at least three formations: schists with intercalations of partly fossiliferous, mostly thin-bedded and impure limestones, proved to be of Upper Paleozoic age by Christodoulou (1966, 1970) and Dounas et al. (1972); a second formation of well bedded marbles; finally a formation of quartzites. This unit, if regarded as being overturned on Leros, resembles the Kara Dağ Unit in the Lycian Taurus (see Brunn et al. 1971; de Graciansky 1972). Katagas and Sapountzis (1977b) report the occurrence of three isolated blocks of blueschists on central Leros.

Thus, Leros and Kalymnos allow to enlarge the extension of the Lycian Nappes within the southeastern Aegean Sea (comp. Bernoulli et al.1974).

Similarly, the small island to the north of Leros, *Arki* and *Lipsi*, seem to be constituted by formations of the Marina Unit (cf. Desio 1931).

On the island of *Kos*, an intrusive, quartz-bearing monzonite is bordered by the following lithologic units:

1. A paleozoic series of phyllites, marls, and limestones, most probably belonging to the "Teménia Unit" of Leros and Kalymnos. Contact metamorphism can be traced over a few kilometers from very low and low grade quartz-muscovite-chlorite-phyllites to high grade hornfelses with coexisting almandine + cordierite + sillimanite besides parageneses with wollastonite + grossularite and with hypersthene + hornblende + plagioclase. From these mineral assemblages maximum temperatures of about 800°C and a pressure range of 4-6 kb can be estimated for the contact metamorphism. Therefore, the depth of intrusion must be accounted as at least 13 km. Before the intrusion the temperature of the country rocks was below 400°C (Altherr et al. 1976a).
2. Mesozoic-Eocene limestones do not show any sign of contact metamorphism even in the immediate vicinity of the monzonite. The contact between both units is a tectonic overthrust plane. Thus, the nappe-like emplacement of the Mesozoic-Cenozoic limestones took place after intrusion, uplift, and surface erosion of the monzonite (Altherr et al.1976a).
3. The monzonite massif and its tectonic cover are overlain by autochthonous lacustrine and fluviatile sediments of Neogene age. Locally,

some ignimbrite intercalations occur in the basal parts of this Neogene. K-Ar dating on biotite of such an ignimbrite gave an age of 10.4 ± 0.3 m.y.

K-Ar dates on hornblendes and biotite of the monzonite were reported and discussed by Besang et al. (1976) and Altherr et al. (1976a). The highest ages are 11.9 ± 0.4 m.y. for hornblendes and 10.5 ± 0.5 m.y. for the corresponding biotite. In the outer contact aureole, metamorphic temperatures did not reach 400°C which is less than the closing temperature of the K-Ar system in amphiboles (500°C). Therefore, it is concluded that the K-Ar dates on hornblende represent magmatic cooling ages and the intrusion occurred some 12 m.y. ago. We are unable to postulate a later metamorphic event with higher temperatures.

A geodynamic model for this part of the Eastern Aegean area has to account for deep-seated intrusive phenomena only 12 m.y. ago. In a very limited time span within the Upper Miocene, uplift and erosion of the intrusive body amounted to at least 13 km. This was followed by the tectonic emplacement of unmetamorphic limestones (which possibly are of "Tripolitza-Gavrovo type", Blondeau et al. 1975), still before the sedimentation of the Neogene sediments. Yet, we have difficulties to explain the identical K-Ar data determined on the two biotites, separated from the monzonite and the ignimbrite.

5. Geodynamic Conclusions

Our results, combined with the data reported in the literature, may be summarized as follows.

At the margins of the Median Aegean Crystalline Belt, some regions, formerly assumed to be border regions, must be excluded because they belong to other, adjacent units: In southwestern Turkey, the norhtern part of the Lycian Nappes has been separated from the Menderes Complex by de Graciansky (Brunn et al. 1971). Within the Aegean Sea, the northern Carian Islands (Arki, Lipsi, Leros, Kalymnos) and probably Kos (see 4.) belong to these Lycian Nappes; Amorgos (3.4.), like Astypalaea and some of the small islands in the eastern Cretan Sea (Bernoulli et al. 1974), are assigned tentatively to the Parnassos and/or Gavrovo-Tripolitza realm.

The median crystalline complexes themselves no longer can be regarded as rigid "massifs" of pre-Alpidic basement. In fact, they are Alpidic complexes, composed of different tectonic units.

By lithologic and lithostratigraphic correlation of the units which can be individualized in the particular complexes, one may distinguish four groups (comp. fig. 1, 2).

1. The group of basal units (A) is best represented by the southern and central parts of the Menderes Complex (Samos and the Phourni Islands included), i.e. the Menderes Massif s.str. with Core, Lower or Schist Envelope, and Upper or Marble Envelope. From the stratigraphic point of view, the Marble Envelope can be well compared with the "Autochthonous" and "Paraautochthonous" of the western Taurids. Perhaps both domains should be linked in palinspastic reconstructions.

In the central Aegean region, crystalline rocks equivalent to those of the Menderes Massif s.str., reappear on Ikaria (2.2.), and on the central Cyclade Islands (3.2.). The basal units of the Cyclade Islands may be correlated with the basal crystalline units in Attica and South-

ern Euboea (1.3.).

The basal units of Attica and Southern Euboea resemble the unit visible within the window of the Thessalian Olympus, with regard to stratigraphy and tectonic position (1.2.). This unit, in turn, is Gavrovo-like in the late Eocene onset of the flysch sedimentation. An earlier beginning (Paleogene) of the flysch may be supposed for the basal crystalline units from southern Euboea to the Menderes massif s.str.; thus a comparison, in Greece, with the Parnassos zone is possible.

The units under consideration show two common features: Neritic facies predominates from Late Triassic to Late Cretaceous times, and they occupy a basal tectonic position in all cases. Hence, we consider the described units to denote a series of windows in median position along the Aegean orogenic arc, exposing Mesozoic carbonate sequences deposited on "external" platforms.

The carbonate sequence of the Thessalian Olympus, according to Godfriaux (1962, 1970), is non metamorphic. The basal crystalline units in southern Euboea and Attica suffered an Alpidic greenschist facies metamorphism. In the Cycladic archipel, however, the basal units underwent a high P/T metamorphism of Late Eocene age which was followed by a development of thermal aureoles, culminating in the formation of migmatites, and of intrusive granitic magmas in the Miocene. Barrovian metamorphism and migmatization in the Menderes Complex seem to have occurred somewhat earlier in the Tertiary (see 2.1.).

2. The units of the next higher, second group (B) are characterized by variegated sequences including ophiolitic rocks affected by high P/T metamorphism. We confine to this group:

The Blueschist Unit around the window of the Thessalian Olympus (1.2.).

The upper crystalline unit in southern Euboea and Attica (1.3.), which seems to be correlated to the Thessalian Blueschist Unit.

The units with variegated sequences in the Cycladic area, originally metamorphosed under high P/T conditions and locally overprinted by the Miocene Barrovian metamorphism and related Plutonic activity (3.3.).

With doubts - the crystalline group of units with glaucophane assemblages on Samos and in the marginal regions of the Menderes Crystalline Complex (2.1.).

We still lack sufficient knowledge of these units, but we may speculate that they comprise Mesozoic rocks of continental margin and oceanic origin.

3. The third group of units (C), overlies (A) and (B). In Thessaly, group (C) is represented by a unit composed of crystalline rocks, at the base, which are characterized by (Alpidic) greenschist metamorphism, and low grade or non-metamorphic Upper Paleozoic to Middle Triassic sediments of varying nature as well as Upper Triassic to Upper Jurassic neritic carbonate rocks, at the top. This unit and its southeastern equivalents, including the corresponding parts of the obsolete Subpelagonian Unit, should be named Pelagonian Nappe or Unit s.str. The restricted use of the term Pelagonian retains the essentials of the general accepted definition of the Pelagonian Zone (see Aubouin et al. 1960-1963): neritic carbonate sedimentation during Upper Triassic-Jurassic times on a pre-Alpidic crystalline basement.

The underlying, lower crystalline units (groups (A) and (B) of this chapter) as well as the overlying Ophiolite Nappe (s.l.) (see below) with its cover of mesoautochthonous Cretaceous and Tertiary sequences (comp. 1.2 and Kauffmann this vol.) must be detached.

The basal crystalline rocks of the Pelagonian Nappe s.str. are generally absent in southeastern continental Greece, Euboea, and the Aegean archipelago. In the eastern part of the Aegean, the units possibly substituting the Pelagonian s.str. (e.g. Chios-Karaburun, intermediate Lycian units ?) do not participate in the constitution of the Median Aegean Crystalline Belt.

4. The fourth and uppermost group of units is formed by remnants of the Ophiolite Nappe s.l. which is subdivided in

D_1, the "Eo-hellenic" (Jacobshagen et al. 1976) Ophiolite Nappe s.str. in the Greek mainland which was obducted in late Jurassic to early Cretaceous times, and

D_2, the nappe of the Serpentinite-Amphibolite Association including the Unit of Arvi in Crete (Creutzburg & Seidel 1975) as well as the Lycian Peridotite and Diabase Nappes (de Graciansky 1968; Brunn et al.1971), the ophiolites of which were obducted not later than in the late Cretaceous.

Frequently, slices of metamorphic rocks are recognized as parts of, or associated with these nappes. Some of the metamorphics are characterized by the presence of glaucophane (e.g. in the Lycian Taurus, van der Kaaden 1966), while others were formed under low to medium pressure/ high temperature conditions (e.g. on Crete, Seidel et al. 1976).

The formation of the blueschists in units A and B is assumed to be related to subduction processes which took place during the Eocene.

Subsequently, a huge pile of nappes was built up. During the Miocene, Barrovian metamorphism took place in the center of the Aegean arc, culminating locally in anatexis and magma generation. These processes were followed by rapid uplift of the crust towards the end of the Miocene, combined with a decompression of the equally rising upper mantle. The still ongoing arching and faulting movements are consequences of these processes. At the outer rim of the orogenic arc, however, the crustal thickening still seems to continue at present.

References

Altherr, R., Keller, J. & Kott, K., 1976a: Bull. Soc. géol. France, (7), 18:403-412.

Altherr, R., Keller, J., Harre, W., Höhndorf, A., Kreuzer, H., Lenz, H., Raschka, H. & Wendt, I., 1976b: 25th Congr. and Plen. Ass.C.I.E.S.M., Split.

Altherr, R., Harre, W., Kreuzer, H., Okrusch, M. & Seidel, E., 1976c: 25th Congr. and Plen. Ass. C.I.E.S.M., Split.

Anastopoulos, J., 1963: Geol. geophys. Research 7:231-375, Inst. Geol. Subsurface Research, Athens.

Andriessen, P.A.M., Boelrijk, N.A.I.M., Hebeda, E.H., Priem, H.N.A., Verdurmen, E.A.Th. & Verschure, R.H., 1976: 4th European Coll.Geochr. Cosmochr. and Isotope Geol., Amsterdam.

Andronopoulos, B., 1962: Geol. geophys. Research 7:102-234, Inst.Geol. Subsurface Research, Athens.

Argyriadis, I., 1967: C.R.Acad.Sci.Paris, (D), 264:438-441.

Argyriadis, I., Mercier, J.L. & Vergely, P.: C.R. Acad.Sci. Paris, in press, June 1976.

Aubouin, J., 1965: Geosynclines. - Developm. Geotectonics, 1:335 p., Elsevier, Amsterdam.

Aubouin, J., Brunn, J.H., Celet, P., Dercourt, J., Godfriaux, J. & Mercier, J., 1960-1963: Livre Mém.Prof.Paul Fallot, 2:583-610, Paris.
Başarır, E., 1970: Ege Üniv.Fen Fak. İlmi Rap./Scient.Rep. Fac. Sci. Ege Univ., 102:42 p., İzmir-Bornova.
Besang, C., Eckhardt, F.-J., Harre, W., Kreuzer, H. & Müller, P., 1976: Geol. Jb. in press.
Bernoulli, D. & Laubscher, H., 1972: Eclogae geol. Helv.,65:107-118.
Bernoulli, D., Graciansky, P.Ch.de & Monod, O., 1974: Eclogae geol. Helv., 67:39-90.
Bittner, A., 1880: Denkschr. kais. Akad. Wiss.Wien, math.-natw. Kl., 40:1-74.
Blondeau, A., Fleury, J. & Guernet, C., 1975: C.R.Acad.Sci. Paris, (D), 280:817-819.
Brinkmann, R., 1966: N. Jb. Geol. Paläont., Mh.:603-618.
- , 1967: Ege Üniv. Fen. Fak. İlmi Rap./Scient. Rep. Fac. Sci. Ege Univ., 43:12 p., İzmir-Bornova.
- , 1971: Geol. Rdsch., 60:886-899.
- , 1976: Geology of Turkey, Enke Stuttgart.
Brunn, J.H., 1956: Ann. géol. Pays hellén.,(1)7:358 p.
Brunn, J.H., Dumont, J.-F., Graciansky, P.Ch.de, Gutnic, M., Juteau, T., Marcoux, J., Monod, O. & Poisson, A., 1971: in: Campbell, A.S. (ed.): Geology and History of Turkey:225-255, Petrol.Explor.Soc.Libya,Tripoli.
Butz, J., 1912: Centralbl. Min.:609-615.
Cayeux, L., 1911: C.R.Acad.Sci. Paris, 152:292-293.
Christodoulou, G., 1966: Bull.geol.Soc.Greece, 6:385-399.
- , 1967: Aristoteleion Panepist. Thessalonikis, epistim.Epetiris fisikomath. Skolis, (Aristotoles-Univ. Thessaloniki, wiss.Jb. phys.-math. Abt.), 10:171-180.
- , 1970: Ann.géol.Pays hellén., 1,21:307-319.
Creutzburg, N. & Seidel, E., 1975: N.Jb.Geol.Paläont.Abh.,149:363-383.
Davis, E.N., 1966: Geol. geophys.Research, 10:161-220, Inst.Geol. Subsurface Research, Athens.
Dercourt, J., 1971: Bull. Soc. géol. France,(7)12:261-317.
Derycke, F. & Godfriaux, I., 1976: Bull.Soc.géol. France,(7)18:252.
Desio, A., 1931: Carta geol. Italia, 24:1-347, Rom.
Dewey, J.F. & Bird, J.M., 1970: J.geophys.Research, 75:2625-2647.
Dewey, J.F., Pitman III, W.C., Ryan, W.B.F. & Bonnin, J., 1973: Bull. geol. Soc. America, 84:3137-3180.
Dixon, J.E., 1968: Comm. Min. Soc. Cambridge, Jan.4th 1968.
- , 1976: Bull. Soc. géol. France,(7)18:280.
Dora, O.Ö., 1976: N.Jb. Miner. Abh., 127:289-310.
Dounas, A.G., Kakavas, N.J. & Tassios, N.P., 1972: Hydrolog. hydrogeol. Invest., 6:153 p., Inst. Geol. Subsurface Research, Athens.
Dürr, St., 1975: Habil.Schr., 105 p., Marburg, Geol.Jb.(B), in press.
- , 1976: Bull.Soc.géol.France,(7)18:429.
- , 1978: Manuskr., Marburg, Z.dtsch. geol. Ges., in press.
Fleury, J.-J. & Godfriaux, I., 1975: Ann.Soc.géol.Nord,94:149-156,Lille.
Frangatos, D., 1951: Geol.Reconnaissance, Rep.,13:21 p.,Minist. Coord. Subsurface Research Depart.(Inst.Geol.Subsurface Research),Athens.
Fytikas, M., Giuliani, O., Innocenti, F., Marinelli, G. & Mazzuoli, R., 1976: Tectonophysics, 31:29-34.
Godfriaux, I., 1962: C.R. Acad.Sci.Paris, 255:1761-1763.
- , 1970: Ann.géol. Pays hellén.,(1) 19,1968A:1-281.
Graciansky, P.Ch.de, 1966:Rev.Géogr.phys.Géol.dynam.,(2)8:289-306, Paris.

- , 1968: Bull. Min. Research Explor.Inst.Turkey, for.Ed.,71:42-62.
- , 1972: Thèse Univ.Paris-Sud (Orsay), 896:762 p., Paris.
Guernet, C., 1971: Thèse, typewritt., 395 p., Paris.
- , 1972: Z.dtsch. geol. Ges., 123:353-364.
- , 1975: C.R. Acad.Sci. Paris, (D),280:1525-1527.
Izdar, E., 1971: in: Campbell, A.S. (ed.): Geology and History of Turkey: 495-500, Petrol. Explor. Soc. Libya, Tripoli.
- , 1975: Ege Üniv. Mühendislik Bilimleri Fak. Yayınları, 8:59 p., İzmir-Bornova.
Jacobshagen, V., Risch, H. & Roeder, D., 1976: Z.dtsch.geol. Ges., 127:133-145.
Jansen, J.B.H. et al., 1973: Nat. Inst.Geol.Min.Researches, Athens.
Jansen, J.B.H. & Schuiling, R.D., 1976: Am.J.Sci.,276:1225-1253.
Kaaden, G.v.d., 1966: Bull. Min. Research Explor.Inst. Turkey, foreign Ed., 67:36-64.
Katagas, C. & Sapountzis, E., 1977a: N.Jb. Miner.Abh.,129:100-112.
Katagas, C., 1977b: Tschermak's min. petrogr. Mitt.,24:39-55, Wien.
Katsikatsos, G., 1970: Praktika Akad. Athens, 44:223-238.
Katsikatsos, G., Mercier, J.L. & Vergely, P.: C.R.Acad.Sci.Paris,in press.
Ketin, I., 1966: Bull. Min.Research Explor.Inst. Turkey, foreign Ed., 66:23-34.
Kober, L., 1929: Sitzber.Akad.Wiss. Wien, math.-naturw.Kl., Abt. 1, 138:299-327.
Köhne, F., 1937: N.Jb. Min. etc., Beil.Bd. 73, Abt.A:1-78.
Ktenas, K.A., 1908: Tschermak's min.petrogr. Mitt., 26:257-320, Wien.
- , 1927: C.R. Acad. Sci. Paris, 184:756-758.
- , 1969: geol.geophys. Research, 13:58-85, Inst.Geol.Subsurface Research, Athens.
Leontaris, S.N., 1974: Ann.géol. Pays hellén.,(1)25:33-104.
Marakis, G.I., 1970: Ann. géol. Pays hellén.,(1)19,1968A.
- , 1972: C.R. Séances, SPHN Genève, 7:52-58.
Marinos, G., 1947: Ann.géol. Pays hellén.,(1)1:60-96.
- , 1954: Geol. geophys. Research, 3:201-226, Inst.Geol.Subsurface Research, Athens.
- , 1955: Bull. geol. Soc. Greece, 2:1-13.
Marinos, G. & Petrascheck, W.E., 1956: Geol.geophys.Research, 4:1-247, Inst. Geol. Subsurface Research, Athens.
Marks, P. & Schuiling, R.D., 1965: Praktika Akad. Athenon, 40:96-99.
Melidonis, N.G., 1963: Geol. geophys. Research, 8:53-308, Athens.
Miyashiro, A., 1973: Metamorphism and metamorphic belts. Allen & Unwin, London.
Négris, P.K., 1914/1919: Roches cristallophylliennes et Tectonique de la Grèce. 310 p., Athens.
Okrusch, M., Seidel, E. & Davis, E.N.: N.Jb. Miner. Abh., in press.
Önay, T.S., 1949: Schweiz. min. petrogr. Mitt.,29:357-491.
Papadeas, G.D., 1970: Praktika Akad. Athenon, 44:10-18, Athens.
Papageorgakis, J., 1969: Praktika Akad. Athenon, 43:163-174, Athens.
Papanikolaou, D.I.: Prakt. Akad. Athens, in press, 1977.
Papastamatiou, J.N., 1951: Geol. geophys. Research, 1:37-69, Minist. Coord., Subsurface Research Dept., (Inst. Geol.Subsurface Research), Athens.
- , 1963: C.R.Acad. Sci. Paris, 256:5167-5169.
Papavasiliou, S.A., 1909: Z.dtsch.geol.Ges.,61:134-201.
- , 1914:Z.dtsch.geol.Ges.,65:1-123.

Petrascheck, W.E., 1960: Abh.dtsch.Akad.Wiss.Berlin,Kl.III,H.1, (E.-Kraus-Festschr.):9-18.
Philippson, A., 1898: La tectonique de l'Égéide. Ann.Géogr.:112-141,Paris.
- , 1901: Petermanns Mitt., 134:172 p., Gotha.
- , 1918: Handbuch der regionalen Geologie,5/2:183 p.,Winter,Heidelberg.
- , 1959: Das Aegaeische Meer und seine Inseln.:412 p., Klostermann, Frankfurt a.M.
Pia, J.v., 1914/1919: Review of Negris,Ph. Roches cristallophylliennes, Athens.
- , 1919: N.Jb. Min. etc.:320-328.
- , 1922: N.Jb. Min. etc.:112-115.
Renz, C., 1933: Eclogae geol. Helv., 26:131-154.
- , 1940: Pragmateiai Akad. Athinon, 8:171 p.
- , 1955: Geol.Greece, 657 p., Inst.Geol.Subsurface Research, Athens.
Renz, C. & Reichel, M., 1945: Eclogae geol. Helv., 38:211-313.
Renz, C., Liatsikas, N. & Paraskevaidis, H., 1954: Inst.Geol.Subsurface Research, Athens.
Schneider, K., 1914: Diss. Münster: 49 p., Greven und Bechthold, Köln a.Rh
Schuiling, R.D., 1962: Bull. Min. Research Explor.Inst.Turkey, foreign Ed., 58:71-84.
- , 1973: in: Jong,K.A.de & Schuiling, R. (ed.): Gravity and Tectonics: 35-47, Wiley, New York.
Schuiling, R.D. & Oosterom, M.G., 1967: Nederl.Akad.Wetensch.Amsterdam, Proc., (B), 70:165-175.
Seidel, E. & Okrusch, M., 1976: Bull.Soc.géol. France,(7)18:347-350.
Seidel, E., Okrusch, M., Kreuzer, H., Raschka, H. & Harre, W., 1976: Contrib. Mineral. Petrol., 57:259-275.
Trikkalinos, J.K., 1954: Pragmateiai Akad.Athinon,18:46 p., Athens.
Verdier, J., 1963: Bull. Min. Research Explor.Inst. Turkey, foreign Ed., 61:38-40.
Wippern, J., 1964: Bull. Min. Research Explor. Inst. Turkey, foreign Ed., 62:74-82.
Yarwood, G.A., 1976: Bull. Soc. géol. France,(7)18:259-264.

Very Low Grade Metamorphism in Middle Triassic Red Limestones from Hydra

D.K. Richter

A metamorphic alteration of the sedimentary rocksequence of Hydra (Upper Carboniferous ? - Jurassic after Römermann 1969) was described by Renz (1925) from the eastern part of the island. Renz mentioned spotted slates in the Middle Triassic red limestone formation near Cape Surva and in 1940 (see Renz & Reichel 1945) attributed this part of Hydra as a metamorphic area to the Cycladic Massif. Römermann (1968, 1969) did not find any metamorphic alteration and therefore refused the interpretation by Renz. Recently, Hydra has been ascribed to the *Subpelagonic Zone* W of the

Cycladic Massif (Jacobshagen & Makris 1974).

As metamorphic minerals are rare (see below), and continuous pelitic layers are not known from Hydra, 38 samples of Illyric red limestones ($<$ 2 % clay content) of similar microfacies (biomicrite and biopelmicrite up to biopelsparite) were investigated to clarify the contradiction mentioned above. The samples were taken at 17 localities distributed over the island (fig. 1) and examined under the following aspects:

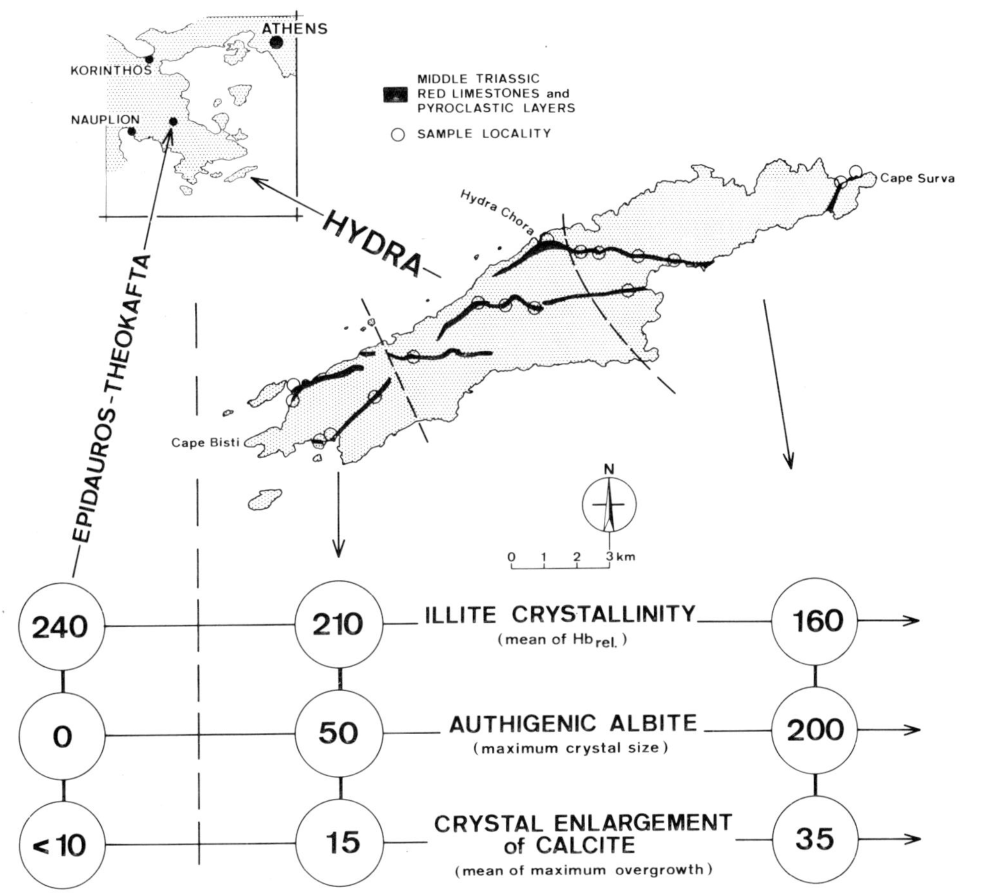

Fig. 1. Upper: geographic situation. - Middle: distribution of the Middle Triassic red limestone formation of Hydra (after Römermann 1968) and sample localities. - Lower: regional change of illite crystallinity (Hb_{rel}), size of authigenic albite (μ), and crystal enlargement of calcite fibres on thin shells (μ).

crystallinity of illite of the non-carbonate fraction 2-6μ
size of *authigenic albite* crystals
crystal enlargement of calcite fibres oriented perpendicular to the surfaces of thin shells ("aggrading crystallization").

The crystallinity of illite increases as shown by decreasing relative width at half height (Hb_{rel} according to Weber 1972a) from W to E across the island. The maximum size of authigenic albite crystals and the maximum crystal enlargements on thin shells, averaged for different areas, increase in the same direction. The stage of diagenesis of the Middle Triassic red limestones from Hydra is more advanced in comparison with the Middle to Upper Triassic red limestones from Epidauros where the microfacies was investigated by Bachmann & Jacobshagen (1974).

Pumpellyite and prehnit in Middle Triassic pyroclastic rocks in the western part of Hydra indicate the very-low-grade metamorphism sensu Winkler (1974). Recently detected antigorite and talcum in a serpentinite from the central part of the island and paragonite found in shallow water limestones underlying the Middle Triassic red limestone formation at Cape Surva point to a metamorphic stage close to low-grade metamorphism.

The results prove a slight metamorphic alteration of the rock sequences of Hydra. An increase in post-Triassic heat flow can be shown approaching the Cycladic Massif. The increasing crystallinity of illite is in good accordance with the results of Weber (1972b) who by this method found a heat dome in the eastern part of the Rhenish Massif (Germany). The more intensive authigenesis of albite due to the increased heat flow in the eastern part of Hydra correlates with the results of Patnaik & Füchtbauer (1975) who observed an increasing growth of albite in Middle Triassic limestones towards the Bramsche Massif (Northern Germany).

References

Bachmann, G.H. & Jacobshagen, V., 1974: Z.dtsch.Geol. Ges., 125:195-223.
Jacobshagen, V. & Makris, J., 1974: Nachr.dt.geol. Ges., 9:78-85.
Patnaik, P. & Füchtbauer, H., 1975: IXth Intern.Congr.Sedimentology, Theme 7:163-168.
Renz, C., 1925: Ecl. geol. Helv., 19:363-372.
 - , 1940: Mém. Acad. Athénes, 8:1-171.
Renz, C. & Reichel, M., 1946: Ecl. geol. Helv., 38:211-313.
Römermann, H., 1968: Geologica et Palaeontologica, 2:163-171.
 - , 1969: Diss., 62 p.
Weber, K., 1972a: N. Jb. Miner. Mh., Jg.1972:267-276.
 - , 1972b: N.Jb. Geol. Paläont. Abh., 141:333-363.
Winkler, H.G.F., 1974: 3.Edition (Springer-Verlag):320 p.

Relics of Non-Metamorphic Sediments on Central Aegean Islands

G. Roesler

On the Central Cyclades, crystalline rocks predominate. Ophiolites and non-metamorphic sediments only occur in small regions up to 150 m above sea level, whereas the metamorphic rocks rise up to 1000 m.

On Paros and Naxos, rocks of the ophiolitic suite including remnants of red and green chert rest with tectonic contact on the crystalline basement. On the Makares Islands however, the ophiolites overly with tectonic contact, slightly metamorphic limestones of Upper Triassic age. On eastern Naxos, non-metamorphic Upper Permian rocks occur as a small allochthonous residue beside ophiolites and the metamorphic basement.

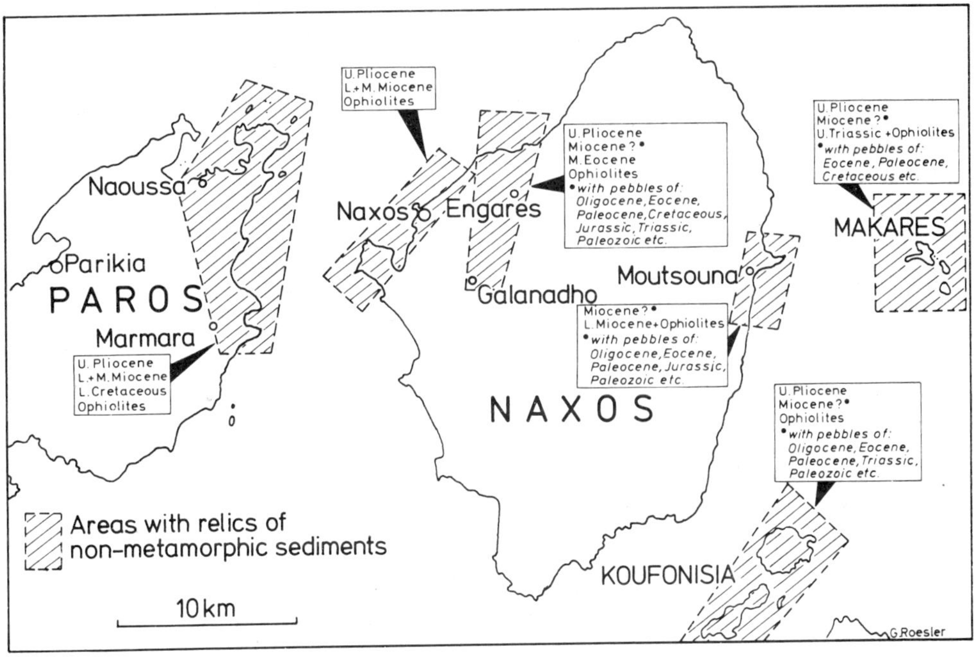

Fig. 1. The study area. For further geographical information see Dürr et al. fig. 3, this vol.

On eastern Paros, the ophiolites are partially covered by transgressive Lower Cretaceous limestones (Papageorgakis 1969). On western Naxos, three remnants of Middle Eocene limestones were observed in tectonic contact with the subjacent ophiolites, but occasionally they contain ophiolite detritus as well.

Marine Miocene deposits are found on Paros and on western and eastern Naxos. These sediments are associated with ophiolites, and locally a tectonic contact with the crystalline basement can be seen. The flyschoid, blue-black to dark grey series is approximately 200 m thick and, where

exposed, starts with a coarse conglomerate upon diabase. Foraminifera and molluscs give an age of early Lower Miocene up to later Middle Miocene. The upper boundary is erosional.

Similar in colour and detritus to the marine Miocene but coarser, molasse-like, paralic and fluvial deposits occur in the western island of Naxos, on eastern Naxos, on the Makares and Koufos Islands (known on the latter as the "Quaternary Pesoulia Conglomerate" of Renz 1928). Due to the absence of datable fossils, the exact age is not yet established. Oligocene pebbles in conglomeratic strata and unconformably overlying Upper Pliocene however bracket the age as between Early Miocene and Early Pliocene. Age differences between the separate deposits cannot be excluded. Of great importance is the composition of the conglomerates. Pebbles of low metamorphic rocks are rare, whereas pebbles of ophiolites and non-metamorphic sediments predominate. Carbonate pebbles of Upper Paleozoic, Triassic, Jurassic, Cretaceous, and, very abundantly, Paleogene age were found. In the Cyclades all these sediments occur at best in small remnants.

Light coloured, up to 200 m thick sequences of conglomerates, sandstones and freshwater limestones occur on Paros, western Naxos, the Makares and the Koufos Islands. These strata are nonconformable and transgressive on the older Neogene and can be dated by means of freshwater gastropods as Upper Pliocene. In contrast to the older Neogene, the pebbles of these conglomerates are derived exclusively from the presently exposed metamorphic Cycladic basement. Locally the freshwater limestones are altered to chert.

Quaternary, light to red coloured fluvial conglomerates alternating with sandstone are found on several coastal areas, as well as even younger, thin marine sandstones.

Due to the tectonic nature and the predominately exotic pebble content, Jansen (1973) postulated an allochthonous origin for the marine Miocene and the similar Pesoulia-type deposits on Naxos.

According to recent radiometric dating (Besang et al. 1976; Altherr et al. 1976; Andriessen et al. 1976; Dürr et al. this vol.) metamorphism and associated plutonism in the Cyclade Crystalline Complex happened about 17-11 m.y. ago, that is, during the late Miocene.

Metamorphic mineral assemblages indicate a depth of at least 15 km. Thus, within a short time during the Upper Miocene and the Lower Pliocene uplift and erosion of at least this depth took place and the denuded crystalline basement was overthrust by ophiolites and non-metamorphic sediments. Due to continued uplift the once more denuded metamorphic basement became a source for the Upper Pliocene conglomerates.

References

Altherr, R., Keller, J. & Kott, K., 1976:Bull.Soc.géol.France,7,18:403-412

Andriessen, P.A.M.,Boelrijk, N.A.I.M., Hebeda, E.H.,Priem, H.N.A., Verdurmen, E.A.Th. & Verschure, R.H., 1976: 4th Europ.Coll.of Geochronology, Cosmochronology, and Isotope Geology, Amsterdam, Abstract.

Besang, C., Eckhardt, F.-J., Harre, W., Kreuzer, H. & Müller, P.: Radiometrische Altersbestimmungen an neogenen Eruptivgesteinen der Türkei. Geol. Jb., in press.

Jansen, J.B.H., 1973: Geological Map of Greece, Island of Naxos, Athens.

Papageorgakis, J. 1969: Prakt. Acad. Athens, 43:368-376.

Renz, C., 1928: Prakt. Acad. Athens, 3:552-557.

The Central Aegean Part of the Pelagonian Realm during Permian and Triassic Time

G. Kauffmann

The initial events of the development of the Hellenic mountain belt happened in latest Paleozoic and earliest Mesozoic time. This paper reviews the paleogeographic variations during this critical period of time in east-central Greece, the Argolis peninsula, the islands of Euboea, Lesbos, Chios, and the peninsula of Karaburun. The area is underlain by two tectonic units: The ophiolite nappe above and the Pelagonian unit below.

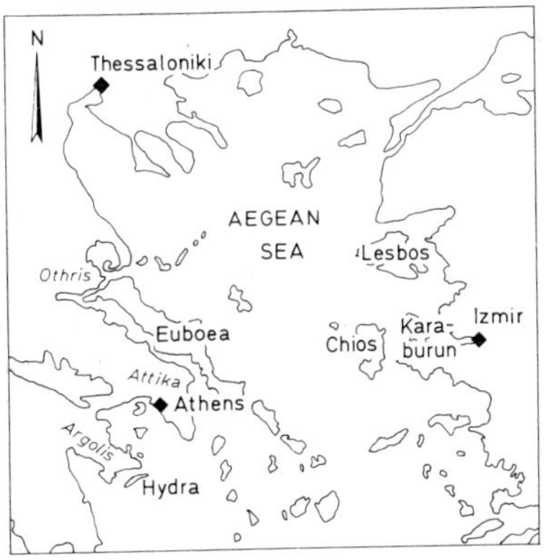

Fig. 1. The study area in the Aegean region.

The latter also includes Paleozoic and Triassic-Jurassic rocks of the obsolete Subpelagonian zone (Jacobshagen et al. 1976; Kauffmann 1976).

Permian: In northern Greece, north of the Othris mountains, Permian time probably was a period of continental sedimentation with clastics and red beds - whereas much of the Pelagonian region to the S and SE was at times covered by a shallow epicontinental sea.

Lower Permian sediments in Attica and on the islands of Euboea, Hydra, and Chios are nearshore deposits of a shallow marine environment consisting of clastics and intercalated fossiliferous limestones with algae, fusulinids, sponges, corals, and brachiopods. The fossils indicate all stages of the Lower Permian and also the Kungurian stage of the Middle Permian. During lower *Middle Permian* time (boundaries sensu Kahler 1974) the sea regressed. A period of strong erosion was accompanied or followed by sedimentation of clastic sequences with pebbles of crystalline, Lower Paleozoic, Carboniferous, and Lower Permian rocks. Red beds, mainly in the upper part are at least partly of continental origin (Dereuder 1973).

The *Middle Permian* Noeschwagerina limestones from the island of Hydra and from the Othris mountains may be marine equivalents of the detrital series.

The late *Middle Permian* and *Upper Permian* marine carbonates document a widespread transgression over the Pelagonian region. They have been found in Attica, in the Othris mountains, on the islands of Hydra, Euboea, Chios, Lesbos, and on the Karaburun peninsula. The deposition startes with dark algal dolomites and limestones (with Gymnocodiaceae and Dasycladaceae) of a lagoonal environment. At some localities (Hydra, Attica, Eubo-

ea) these layers grade into massive grey or reddish reef limestones (with calcisponges, Tubiphytes, and brachiopods). Basin sediments have been observed on Salamis island.

During *latest Permian* the sea again regressed and intense erosion and karstification affected the Pelagonian region. There is no evidence of an oceanic Tethys during Permian time, neither in the Pelagonian nor in the Vardar region.

Upper Skythian - Anisian: During *Upper Skythian* time (Smithian - early Spathian) the sea transgressed over the denudated and karstified Pelagonian region. In the north it advanced to the western border of the Serbomacedonian massif. There its deposits indicate a subsidence of the Vardar region. During *latest Skythian and Anisian* the area was divided into irregularly bounded facies realms. Thick neritic carbonates accumulated on platforms and in adjoining basins (e.g. Chios, Pateras mountains, northern part of Pastra mountain). In interspaced zones with tectonic activity and volcanism submarine lava flows and ashes, red shales, cherts, cherty limestones, and red limestones of the Hallstatt type were deposited (e.g. Euboea, southern part of Pastra mountain, eastern part of Chios). Transitional zones show both facies types mixed and interfingered. The tectonic and volcanic activity in parts of the Pelagonian region may indicate the breakup of the continental plate ("cassure tethysienne" Agyriadis 1975) in Lower and early Middle Triassic time.

Ladinian - Jurassic: During this interval, the Pelagonian region constituted an elongate shelf platform between the deep or oceanic troughs to the SW (Pindos) and to the NE (Almopias). The continuous quiet sedimentation of thick, predominantly neritic carbonates on the platform, interrupted only by few emersions, making major tectonic events on the shelf and in its vicinity during this time rather unlikely.

Triassic sediments in the ophiolite nappe: The ophiolite nappe, thrust over the Pelagonian platform in the eohellenic phase (Jacobshagen et al. 1976), contains sporadically Triassic pelagic sediments, such as radiolarian cherts and cherty limestones, tightly connected with basic volcanics (e.g. the northeastern Pindos mountains: Terry 1971, in the western Othris mountains: Koch & Nicolaus 1969; Ferriere 1974, in the Knimis mountains, near Exarchos/Lokris, and in northern and Central Euboea: Kauffmann 1976). These pelagic sediments of an eugeosynclinal facies type may have been deposited on oceanic crust and may indicate the first opening of a Tethys ocean in the Aegean region.

References

Agyriadis, J., 1975: Bull.Soc.Géol.France, 7,17:56-67.
Aubouin, J. et al., 1970: Ann. Soc. Géol. Nord, 90:277-306.
Bernoulli, D. & Laubscher, H., 1972: Eclogae Geol.Helv., 65:107-118.
Besenecker, H. et al., 1968: Geol. et Palaeont. 2:121-150.
Clement, B., 1968: C.R. somm. Soc. Géol. France, 9:332-333.
Clement, B., Guernet, C. & Lys, M., 1971: Bull. Soc. Géol. France, 7,12:88-91.
Dereuder, J.-C., 1973: DEA Univ. Lille, 108 p.
Ferriere, J., 1974: Bull. Soc. Géol. France, 7,16:543-562.
Guernet, C., 1971: Thèse Univ. Paris, 395 p.
Hecht, J., 1972: Z. dtsch. geol. Ges. 123:423-432.
Jacobshagen, V., Risch, H. & Roeder, D., 1976: Z. dtsch. geol. Ges.,

127:133-145.
Kahler, F., 1974: Rep. Sci. Exp. Sven Hedin, 52:147 p.
Kauffmann, G., 1976: Z. dtsch. geol. Ges., 127:387-398.
- : Das Paläozoikum des östlichen Mittelgriechenland, in press.
Kauffmann, G., Kockel, F. & Mollat, H., 1976: Bull. Soc. Géol. France, 7,18:227-232.
Koch, E. & Nicolaus, H., 1969: Eidikai Meletai Geol. Ellados, 9:190 p.
Kockel, F., Mollat, H. & Walther, H., 1971: Geol. Jb., 89:529-551.
Marinos, G. & Reichel, M., 1958: Geol. of Greece, 8:16 p.
Renz, C. & Reichel, M., 1946: Eclogae Geol. Helv., 38:211-313.
Römermann, H., 1969: Diss. Univ. Marburg: 62 p.
Spiliadis, T., 1961: Pract. Acad. Athinon, 36:254-263.
Tataris, A., 1967: Bull. Geol. Soc. Greece, 7:36-51.
Terry, J., 1971: C.R. somm. Soc. Géol. France, 9:384-385.

Alpine Structure and Development of the Southern Pelion Peninsula and the North Sporades

V. Jacobshagen, W. Skala and E. Wallbrecher

All pre-Neogene rocks of this region are to some extent metamorphized. They belong to two different tectonic units: the deeper one is part of the Pelagonian zone s.str.; and the higher is part of the Eohellenic nappe system (Jacobshagen et al. 1976a). Both systems are covered unconformably by Upper Cretaceous breccias, limestones, and flysch (Upper Maestrichtian-Eocene?).

Stratigraphy

The stratigraphical sequences have been described in some detail by Jacobshagen & Skala (1976) who refer to Greek authors (Papastamatiou & Marinos 1938, 1940; Marinos 1961; Papastamatiou 1963; Ferentinos 1974; Kelepertsis 1974) and by Wallbrecher (1976).

a) Pelagonian Zone

The Pelagonian rocks are exposed mainly on the southern Pelion peninsula (fig. 1) and on the islands of Skiathos, Skopelos and perhaps Skyros. Their stratigraphic column begins with the very thick clastic Skiathos Series: microconglomerates, quartzites, phyllites, and gneisses in quartz-albite-epidote-biotite subfacies. On Skiathos, garnet and plagioclase have been found in this series. The upper part consists of pencil structured phyllites of the quartz-albite-muscovite-chlorite subfacies. The age of the Skiathos Series is probably late Paleozoic.

The Skiathos Series is overlain by up to 200 m of massive calcite or dolomite marbles. Their Middle Triassic to Upper Jurassic age has been

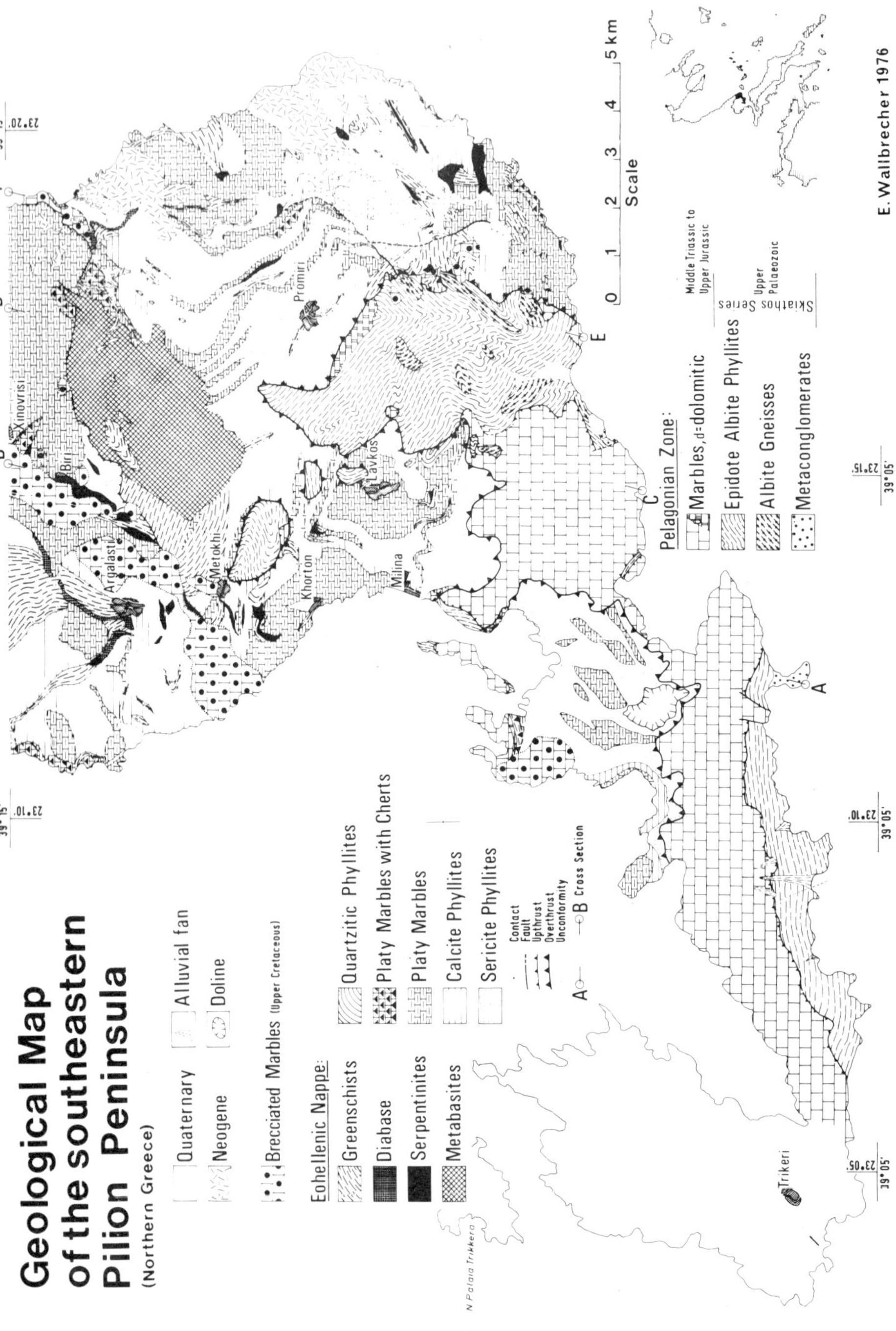

Fig. 1. Geological map of the southeastern Pilion Peninsula.

proven by some fossils on Skopelos (Papastamatiou 1963). Between the Skiathos Series and the marbles a small horizon of coloured schists and volcanic rocks (porphyritic andesites according to Ferentinos 1974) which represents the Werfen Beds (Lower Triassic) has been observed in Skiathos. Small relics of probably Upper Jurassic radiolarite breccias and ophiolites have been preserved in southern Skopelos; the breccias rest with sedimentary contacts on the marbles.

b) Eohellenic Nappe

The upper tectonic unit crops out on the southern Pelion Peninsula and on the islands of Alonnisos, Peristera (Kyra Panagia), Skantsoura and Skyros. It also consists of metamorphic rocks of the greenschists facies: sericite phyllites and calc-phyllites passing laterally into platy marbles, locally with cherty layers.

The typical constituent of this sequence comprises concordant metabasites, metaserpentinites, and ophicalcites. At some places they are clearly lined up along fault planes (thrusts?) (fig. 1). An exactly comparable association with metamorphic rocks has been described from Alonnisos under the term Mortero Series (Kelepertsis 1974). The deeper parts of Kyra Panagia, the greenschists of Skantsoura and the ophiolites of Skyros may also be attributed to this series.

Of special geodynamic importance is the presence of glaucophane within the calcschist - metabasite series both in the serpentinites and in calcschists or platy marbles (Schütz 1975; Schwandt 1976). Usually the glaucophane crystals contain a core of green hornblende. This suggests that this region suffered very high pressures after the greenschist metamorphism. Lawsonite has not, however, been found together with the glaucophane.

On Alonnisos, the Mortero Series is conformably overlain by meatmorphic Upper Jurassic Cladocoropsis limestones. This leads to the assumption that the origin of the Eohellenic nappe may be located within the inner parts of the Vardar Zone.

As a whole the Eohellenic nappe is thought to belong to a sheet of oceanic crust which was obducted from the NE at about the Tithonian/Cretaceous boundary (Bernoulli & Laubscher 1972; Jacobshagen et al. 1976a).

c) Upper Cretaceous-Palaeogene Cover

The nappe tiers described above are covered disconformably by conglomerates of Upper Cretaceous age with varying composition and thickness. These basal conglomerates are followed by rudist limestones (fossiliferous at least on the North Sporades) of Cenomanian to Turonian age. In eastern Skopelos and in Alonnisos they fade laterally into an alternance of platy limestones, psammitic layers, and pelites (Palouki Series). A similar flyschoid facies of the Upper Cretaceous is seen in the Almopias zone of Macedonia (Mercier 1973a).

The Upper Cretaceous is overlain conformably by a thick flysch series which appears to be laterally continuous with the Palouki Series but is deposited on a ferruginous crust on the top of the rudist limestone marking a stratigraphical gap of an unknown amount. The age of the flysch may be latest Cretaceous to Eocene.

Structure and Tectonic Evolution

Probably during the late Eocene orogenesis, the whole sequence was subjected to low-grade metamorphism and was deformed in a very complicated way:

Whereas in the eastern and central Aegean islands (Lesbos, Chios, Psara, Sykros) NE-striking fold axes predominate, in the North Sporades and the southern Pelion Peninsula a cross-folding with both NE and NNW striking elements is characteristic (fig. 2).

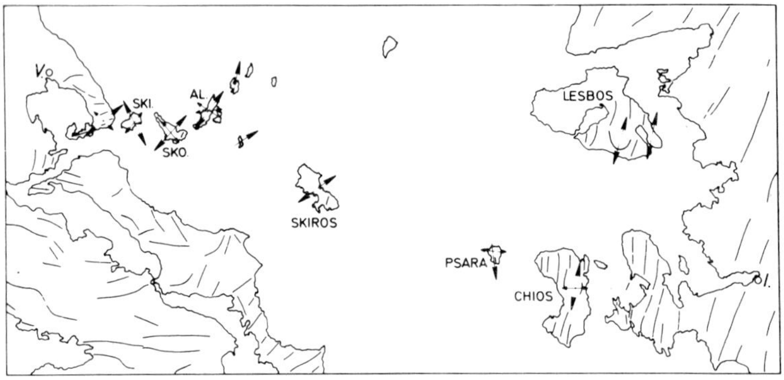

Fig. 2. Directions of alpidic fold axes on Middle Aegean islands. The lengths of the arrows correspond to the importance of the axial systems.

In most cases NE-SW striking B-axes can be identified as the younger directions, by fabric analysis or directly in the field based on the age relationships of the s-surfaces (mostly axial surfaces) of these folds.

Within the incompetent series, such as the flysch or the crystalline Glossa Series of Skopelos, both directions are distinct ; whereas in the more competent beds (Skiathos Series, marbles) only one axial system can be observed.

In general, the kinematic sequence is uniform over the whole area: the NE striking B is younger than the NNW axes.

In a few outcrops, however, a NW-SE direction which is usually older is found to be younger. B_1 and B_2 occur everywhere from the deepest tectonic levels up to the flysch of probable Upper Cretaceous - Lower Tertiary age. The same directions are also found in gently-flanked monoclinal and in soft-sediment folds within the Neogene of Skyros and Alonnisos. In general, the folds of both directions are of the same type. In both systems flexural slip folding predominates. The size of the folds varies from a few millimeters to some meters.

As for the origin of this cross-folding two possibilities have been discussed (Jacobshagen et al. 1976b; Jacobshagen & Skala 1977):

The *deformations* took place in the course of one tectonical phase, but *not strictly synchronous*. A younger B orientated in the sense of the $B' \perp B$ tectonics of Sander (1948).

The *deformations* were *strictly synchronous*. This interpretation is based on experiments of Bhattacharji (1958) and observations of Sutton & Watson (1954) and Ramberg (1959).

Looking at a sketch-map of the main tectonic structures of the Aegean

region (Dürr, Jacobshagen et al. this vol.), one can see that the North Sporades and some other areas with similar fold patterns (Delfi,Marathon, Hymettos, Argolis) are located within the narrow Central Aegean fold arc (Jacobshagen et al. 1975; Jacobshagen & Skala 1977) which runs from Northwest Anatolia across the Aegean Sea to central eastern Greece where it swings to the N/NW in the direction of the Pindos ranges. In the central Aegean Sea, it crosses the Miocene median crystalline belt (Dürr et al. this vol.) nearly perpendicularly. The cross-folding may be due to extreme lateral shortening in the interior of this arc which, based on radiometric ages of metamorphism and plutonism and on stratigraphic superposition, was formed during the Late Eocene orogeny. The early Neogene volcanic rocks of the North Aegean Sea (Borsi et al. 1973) are thought to have been produced subsequent to this orogenic event.

In the course of that tectogenesis an upper nappe system was thrust onto the Palaeogene flysch of the Sporades and the Pelion, as must be inferred from the metamorphism in this flysch. In fact, some outliers of an ophiolite nappe resting on the flysch were found recently in the critical area (Jacobshagen, Märtz & Reinhardt 1977).

Recent strike-slip movement along the North Anatolian fault and the west edge of the Vardar zone suggests that the Rhodope block did not participate in the tight bending of the Central Aegean arc from the middle Tertiary to the present.

References

Bernoulli, D. & Laubscher, H., 1972: Eclogae geol. Helv.,65:107-118.
Bhattacharji, S., 1958: J. Geol., 66:625-667.
Borsi, S., Ferrara, G., Innocenti, F. & Mazzuoli, R., 1973: Bull. volcanol., 36:473-496.
Ferentinos, G.C., 1974: Bull. geol. Soc. Greece, 10:323-358.
Jacobshagen, V., Märtz, J. & Reinhardt, R., 1977: N. Jb. Paläont., Mh. 10:613-620.
Jacobshagen, V., Risch, H. & Roeder, D., 1976a: Z. dtsch. geol. Ges., 127:133-145.
Jacobshagen, V. & Skala, W., 1977: Ann. géol. Pays hellén. in press.
Jacobshagen, V., Skala, W. & Wallbrecher, E., 1975: Rapp. Comm. int. Mer Médit., 23,4a:173.
- , 1976b: Bull. Soc. géol. France (VII),18:281-286.
Kelepertsis, A., 1974: Z. dtsch. geol. Ges., 125:225-236.
Marinos, G., 1961: Bull. geol. Soc. Greece, 4:14-28.
Mercier, J., 1973: Ann. géol. Pays hellén. (I), 20.
Papastamatiou, J. & Marinos, G., 1938: Prakt. Akad. Athens, 13:45-49.
- , 1940: Prakt. Akad. Athens, 15:344-346.
Papastamatiou, J., 1963: Bull. geol. Soc. Greece, 5:52-74.
Ramberg, H., 1959: Norsk geol. tidsskr., 39:99-151.
Sander, B., 1948: 215 p.
Schütz, W., 1975: Unveröff. Diplom-Kartierung F.U. Berlin, 21 p.
Schwandt, K.H., 1976: Unveröff. Dipl.-Arbeit F.U. Berlin, 43 p.
Sutton, J. & Watson, J., 1954: Quart. J. geol. Soc. London, 110:21-48.
Wallbrecher, E., 1976: Z. dtsch. geol. Ges., 127:365-371.

Geochemical Implications for the Santorini Island Group (Aegean Sea, Greece)

H. Puchelt

Major, minor and trace element determinations and correlation patterns have established that the Christiani islands, Santorini, and the submerged volcano Colombos north-east of Santorini belong to a coherent series of volcanics, ranging in composition from high-aluminum basalt to dacite and rhyolite. This is apparent in samples of all areas, if the K_2O values are plotted versus silica as proposed by Taylor (1969) for grouping calc-alkaline rocks (fig. 1). Further support for this relationship

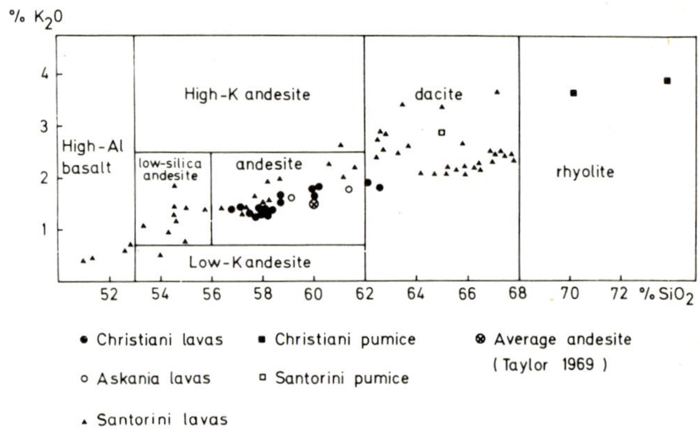

Fig. 1. K_2O/SiO_2 of volcanics on Santorini and Christiani.

is given by the positive correlations of Rb, Ba, Zr, Hf versus K_2O and Sc and Ti versus iron in all sequences and by the presentation in the AFM triangle (fig. 2).

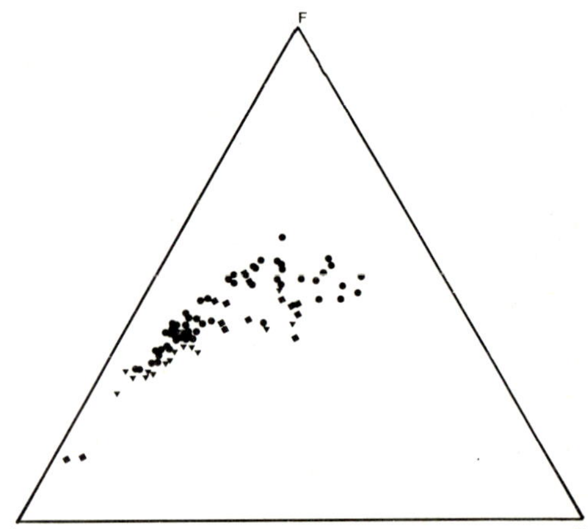

Fig. 2. AFM triangle for Santorini and Christiani volcanic rocks; circles = Santorini lavas; squares = Christiani lavas, triangles = Santorini pumice.

Four cycles of volcanic activity have been observed on Santorini as reported by Nicholls (1971) and Pichler & Kussmaul (1972). The three previous cycles ended up with strong pumice eruptions of dacitic to rhyodacitic composition.

The main part of the materials for the magmas originate from partial melting of oceanic floor basalts which during their subduction underneath the Aegean microplate were metamorphosed to eclogites. Participation of water is probable in this process which should occur at depths below about 90 km (Green & Ringwood 1968). The water content of the downgoing slab could most probably participate in the additional partial melting in the overlying mantle thus providing the high aluminum basalt which cannot easily be explained otherwise. In addition, a two-stage process is possible which starts from an initial partial melt of eclogite and creates an intermediate magma volume within the mantle. This magma is subject to differentiation and segregation (Ringwood 1975).

All Santorini lavas are enriched in light rare earth elements (LREE). The lanthanum-enrichment factor (calculated versus chondrite concentrations) gives values of close to 30 for high aluminum basalts and values of near 100 for all other lavas. Heavy rare earths are enriched by factors of 8 to 20 (fig. 3). No Eu-anomalies have been found in lavas, but all pumice layers both of Santorini and of Christiani (rhyolite in fig. 3) exhibit a more or less pronounced relative Eu depletion. The pumice layers always represent low-pressure differentiation products with plagioclase separation which in turn is responsible for the negative Eu anomaly (cf. distribution-coefficient in fig. 4). All pumice eruptions in this group of islands produced a large amount of this material which was windcarried to large areas SE of Santorini, and hence, pumice was found on the neighbouring islands and in deep sea sediments. Pumice can easily

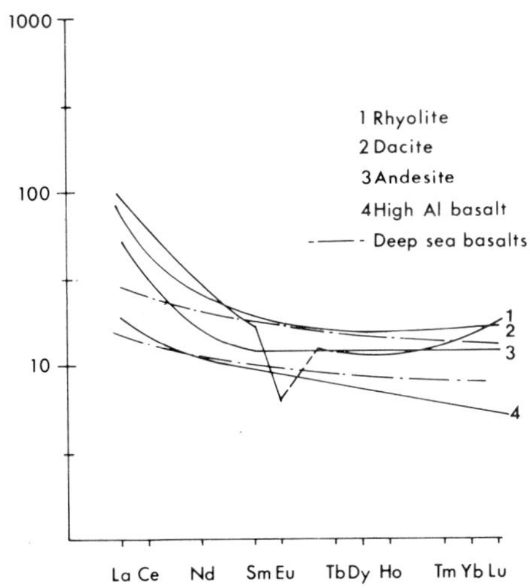

Fig. 3. Normalized REE patterns of volcanic rocks from the Santorini group (Puchelt 1977) and selected deep sea basalts (Puchelt & Emmermann 1977).

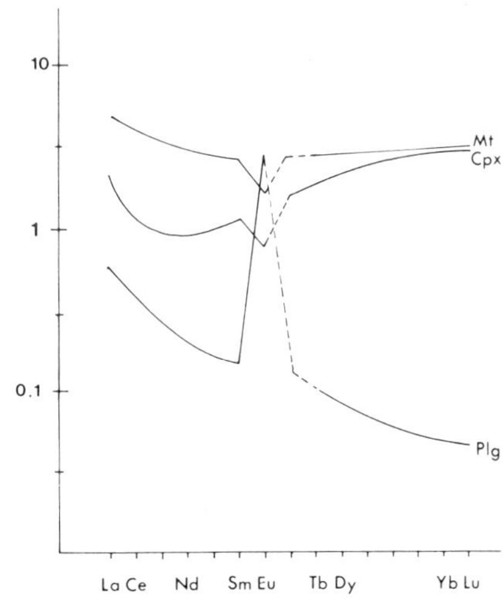

Fig. 4. REE distribution coefficients for magnetite, clinopyroxene and plagioclase rhyolitic pumice glass (Puchelt & Kramar 1978).

be traced to its sources by geochemical characterization and thus be used for the purposes of tephrochronology (Puchelt & Murad 1978).

The REE enrichment increases in the sequence from high aluminum basalt towards andesite and dacite. The rhyolithic pumice of Christiani has a lower REE level than the Santorini pumice layers. This might be due to separation of clinopyroxene phenocrysts or magnetite, both of which were found to have distribution coefficients above unity for all of the REEs safe Eu in clinopyroxene (cpx). Thus crystallization of these minerals would decrease the level of REEs in the remaining melt under the prevailing conditions. This model works also for the removal of several other trace elements: Sc (distribution coefficient [d] between cpx and matrix = 20; [d] magnetite = 12); Co ([d] cpx/melt = 69; [d] mt/melt = 113); and Hf ([d] cpx/melt = 2.6); [d] mt/melt = 20). The Hf did not decrease as much as expected for the magnetite distribution. Therefore, coefficients measured and the low iron content observed in the rhyolitic pumice separation of this phase cannot have been of large influence.

It must be pointed out, that distribution coefficients highly depend on the composition of the parental melt and that small changes largely modify the values (cf. Puchelt & Kramar 1978). This is true especially for REE distribution in plagioclase. From distribution coefficients determined on phenocrysts from Christiani pumice (fig. 4), it is to be concluded that clinopyroxene and plagioclase fractionation was involved in the development of this pyroclastic.

The remnants of Christiani volcano indicate a formerly much larger volcanic edifice. Its age has not been determined yet, although it is assumed to belong to the Quaternary outer arc of volcanics in the Argean (Fytikas et al. 1976). At least the Christiani pumice must have penetrated granitic rocks, phyllites, and limestones or skarns from the submerged land connection between Peloponesos and Asia Minor as witnessed by the xenoliths found in and underneath the pumice layer (Puchelt et al. 1978). The geochemistry of the xenoliths precludes any large contamination of the magmas by these materials.

Conclusions

Santorini, Christiani and the submerged Colombos volcano are genetically related.

In accordance with the Hatherton-Dickinson model (1969) the normalized K_2O/SiO_2 ratios rise with increasing depth of the Benioff zone. The calculated depth of the downgoing slab should be *135 km below Christiani, 145 km below Santorini and, very tentatively 165 km below Colombos*. These depths are in fair agreement with the geophysically found Benioff zone.

Trace-element data of the various lavas imply that most of the volcanic material was mobilized by partial melting obviously of (quartz)-eclogite, the metamorphic equivalent of ocean floor basalt. An additional process may have occurred involving mantle material above the underthrust slab which generated the high-alumina basalts by hydrous partial melting of mantle when the subduction mechanism together with metamorphic processes provided the necessary water (Nicholls 1974). Lavas generated in both ways may have been contaminated in various forms:

direct incorporation of mantle material by the more acidic calc-andesite magmas indicated by occasionally high Cr, Ni, and Mg values which can be perfectly correlated;

small-scale contamination by crustal material (phyllites and gneisses); *occasionally incorporation* of sea floor sediments or the dissolved content of penetrating solutions, which was assumed by Pe and Gledhill (1975) to explain certain unusually high $^{87}Sr/^{86}Sr$ values in the Aegean arc.

This seems to be the only applicable explanation to introduce the necessary amount of radiogenic Sr (cf. Puchelt & Hoefs 1971).

Differentiation within the lavas reached distinct influence in the last phases i.e. in the pumice stages of Santorini and Christiani, where negative Eu-anomalies point to a removal of plagioclase.

Magnetite removal cannot have played an important role otherwise Co, Hf and Th would have been more depleted. On the other hand do Hf and Th increase with decreasing iron. A slight decrease in REE as in Christiani pumice can indicate either small magnetite removal or clinopyroxene fractionation.

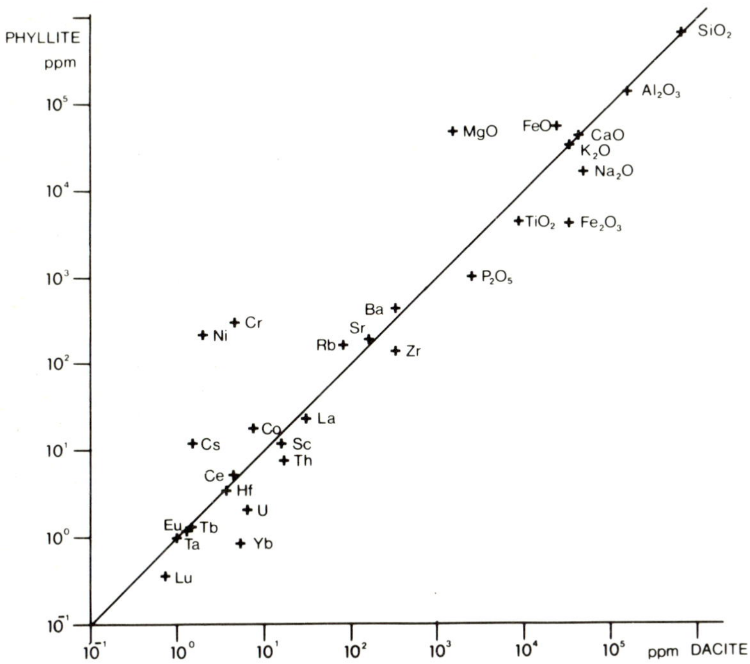

Fig. 5. Concentrations in phyllite versus concentrations in dacite of same main element content.

Xenoliths (granites, skarn and/or limestones, and phyllites in case of Christiani and limestone and phyllites on Santorini) prove that these rocks have been penetrated. Major and minor element geochemistry including REE patterns rule out that phyllites and granites have served as source material for rhyodacites, dacites and latiandesites (fig. 5).

At least locally solid lava and pyroclastics have been leached by hot carbondioxide-bearing or acidic waters, which is usually described

as hydrothermal or fumarolic alteration. In most cases the only reagents provided by the volcano are carbonic or sulfuric acid. All other substances are leaching products of the specific rocks. The generated solutions can precipitate part of their dissolved substances at suitable conditions of Eh and pH. Thus recently ironrich sediments are formed within the Santorini caldera. Favorable places prodive the environment for sedimentation of several meters thickness. A multidisciplinary investigation of these ironrich sediments and the conditions of their formation has been carried out (Puchelt & Schroll 1978).

References

Fytikas, M., Giuliani, O., Innocenti, F., Marinelli, G.& Mazzuoli, R., 1976: Tectonophysics, 31:T29-T34.
Green, T.H. & Ringwood, A.E., 1968: Contr. Mineral. Petrol.,18:105-162.
Hatherton, T. & Dickinson, W.R., 1969: J.Geophys.Res., 74:5301-5310.
Nicholls, I.A., 1971: J. Petrol. 12:67-119.
- , 1974: Contrib.Mineral.Petrol., 45:289-316.
Pe, G.G. & Gledhill, A., 1975: Lithos, 8:209-214.
Pichler, H., Günther, D., Kussmaul, S., 1972: Naturwissenschaften, 5: 188-197.
Pichler, H. & Kussmaul, S., 1972: N.Jb. Miner. Abh., 116:268-307.
Puchelt, H. & Hoefs, J., 1969: Acta first intern.sci.congr.on the volcano of Thera:318-327.
Puchelt, H., Murad, E. & Hubberten, H.W., 1977:N.Jb.Min.Abh., in print.
Puchelt, H. & Kramar, U., 1978: submitted to N.Jb.Min. Mh.
Puchelt, H. & Emmermann, R., 1977: Contrib.Mineral. Petrol,62:43-52.
Puchelt, H., 1978: to be submitted to N. Jb. Min.
Puchelt, H. & Murad, E., 1978: to be submitted to N. Jb. Min.
Puchelt, H. & Schroll, E., 1977: Genesis of marine iron sediments from Santorini, Greece, Springer Verlag, Heidelberg, in print.
Ringwood, A.E., 1975: Composition and Petrology of the Earth's Mantle. McGraw Hill, N.Y., 618 p.
Taylor, S.R., 1969: Proceed. Andesite conf.; Oregon, Dept. Geol. Miner. Ind. Bull. vol. 65:43-63.

Part 5

Late Cenozoic Stratigraphy and Geodynamics

Late Cenozoic Stratigraphy and Paleogeography of Greece - a Review

D. Büttner and G. Kowalczyk

The following compilation presents the stratigraphic and paleogeographic background for studies on the Neogene evolution of the Hellenides. Uncertainties and difficulties in this essay are due to discrepancies between marine and continental Neogene biostratigraphy and to the remarkable progress in Neogene stratigraphy during the last years. Older stratigraphic classifications can be utilized only with reservations. This review is based on the chronostratigraphic scale by Lüttig & Steffens (1976).

Miocene Depositional Areas

In the Ionian Sea and on the islands of Kefallinia, Zakynthos, and Korfu marine sedimentation prevailed throughout the entire Miocene and gradually developed from fully marine to neritic. In the Messinian, a short period of evaporation followed. Younger Pliocene beds unconformably overlie the evaporites.

Levkas Island as well as the coastal regions of Epirus and Akarnania still bear marine Miocene sediments, whereas in most of its central and eastern parts a Miocene cover is lacking, with the exception of the allochthonous West Hellenic Flysch of early Burdigalian age.

In the Mesohellenic Trough, early marine sedimentation continued to an ill defined date in the early Miocene (Helvetian) marked by interfingering continental beds.

Neogene sedimentation in the intramontane basins of the Greek mainland began in an imprecisely dated period of the late Miocene, termed Pontian or Turolian, or middle to late Tortonian, on the basis of Pikermi faunas in Thessalia and Boeotia. These basins contain exclusively limnic and fluviatile sediments with local lignite interbeds.

On the Peloponnesus, sediments have not yet been confirmed as Miocene, except for the allochthonous flysch. However, since in the western part, in Elis, indications of late Miocene strata are reported, this part of the Peloponnesus probably belonged to the Ionian Sea in late Miocene time.

On the Island of Kithira, deposition of limnic and fluviatile beds started in the Late Miocene (Pontian). However, on Crete, fluviatile sediments of late Serravallian age have been found, whereas contemporaneous marine deposits are present on Gavdos Island. In Late Tortonian and Messinian time, marine facies dominated on Crete, which, however, interfingered with continental beds in the direction of the former coast.

The western and central Cretan Sea entered the marine realm in the

Tortonian at the latest. Probably, sedimentation did not start synchronously; thus locally, an earlier beginning is possible, such as in the Karpathos Trench, on the eastern coast of the Peloponnesus. This geophysically determined sedimentary sequence, which was dated in analogy to the events on Crete, has recently been confirmed by means of core analysis (DSDP Leg 42 A).

Marine sedimentation on Cos Island started in early Miocene (Besenecker & Otte this vol.). It seems to continue up to the middle Miocene in the western part (Bignot & Guernet 1976), whereas in the east, fluviatile and limnic sediments were deposited. Pyroclastics in this sequence have been dated radiometrically as approx. 10 m.y. old (Altherr et al. 1976) which confirms a late Middle Miocene (Sarmatian) age. Thus, the environmental change from marine to continental conditions must have taken place between the Serravallian and the Sarmatian. During the Sarmatian and the entire Pliocene, continental sedimentation took place on Cos Island. Pollen-bearing lignite intercalations in limnic strata suggest a Pliocene to early Pleistocene age (Akça assemblage) and vertebrates of Villafranchian age partly confirm this. Finally, in early Pleistocene time, fully marine sediments covered this island.

On Samos Island limnic and terrestrial to fluviatile deposits of the middle Miocene (Sarmatian) to the Pleistocene have been identified. The Pikermi fauna from Samos Island and radiometric data (8-10 m.a.) indicate a late Miocene age for parts of this sequence. A lithostratigraphic correlation of the east Aegean coastal region between Chios, Samos, and Cos, with the subdivision established on the Anatolian mainland (Becker-Platen 1971) has been carried out (Meissner 1976).

The central Aegean region between Euboea, Lesbos, and Chios belonged to a mainland during the Miocene (Besenecker & Büttner this vol.). However, since on Skyros Island marine to brackish sediments of late Miocene (Tortonian) age are present, an approximately N - S running seaway (Strimon Basin, Skyros, Attica (?), Crete) can be postulated at that time interval.

On the Cycladian Islands of Paros and Naxos, marine Early Miocene to late Middle Miocene remains are documented (Roessler this vol.).

In Thessalia, the Neogene has been subdivided in predominantly limnic and fluviatile Lower Thessalian beds and fluviatile to exclusively continental red deposits (Upper Thessalian beds, Schneider 1972). Continental sedimentation appears to start in late Miocene (Pannonian) time (Pikermi fauna) and to extend into the Pliocene and early Pleistocene.

To the North, a comparison of the Late Cenozoic basins (Gramann & Kockel 1969) shows marine Early Miocene (Burdigalian) clastics in the central part of the Salonica Basin and uncertain Sarmatian above. On the contrary, neither the northern nor the southern well in the Thermaic Gulf has given indications of a marine interval during early Miocene to early Pleistocene (Faugères & Robert 1976).

In the Strimon Basin, sedimentation started with limnic, fluviatile, and terrestrial clastics of Late Miocene age which locally contain Hipparion remnants, and with travertine in the Orphanou Basin. Marine and hypersaline sediments of Late Miocene age followed. Up to 400 m thick evaporites are reported from a well east of Thasos which are probably of similar age (West 1973). Contemporaneous and partly older "Sarmatian" marine deposits are found in the southern part of the Thrace Basin.

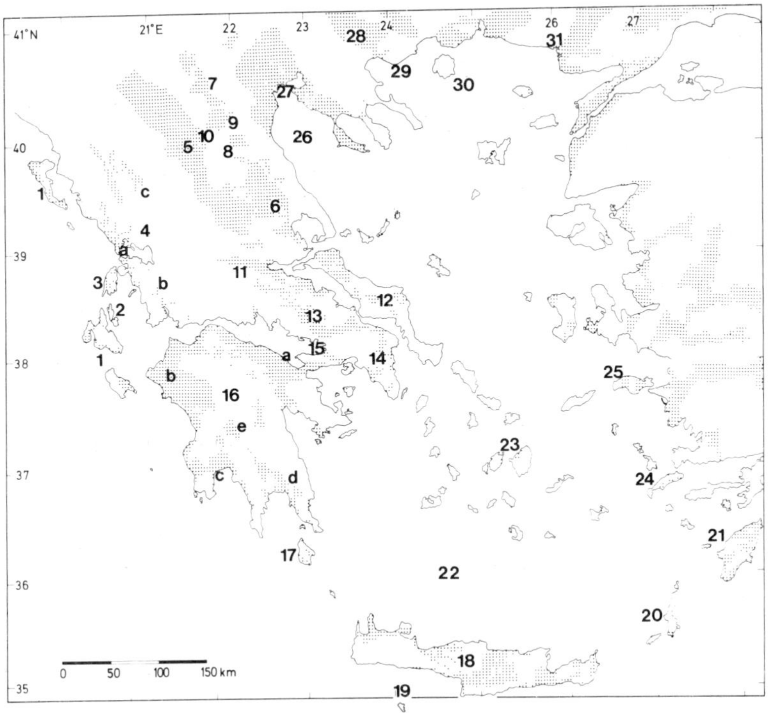

Fig. 1. Regional distribution of Neogene sediments (dotted areas) in Greece (in part with Quaternary and in the Mesohellenic Trough with Oligocene). Numbers show position of sections outlined in fig. 2.

Pliocene Depositional Areas

During the early Pliocene, the sedimentary pattern at the end of the late Miocene persisted in many of the above-mentioned areas. On the Greek mainland, limnic and fluviatile deposition started or continued with local lignite intercalations (Basins of Servia, Kozani, and Ptolemais, Mesohellenic Trough, eastern Thessalia, Boeotia, and Attica). The continental sedimentation in these intramontane basins continued, with sporadic diastems, into the Quaternary.

In the Ionian Sea and on its islands, marine facies continued throughout the Pliocene up to the Pleistocene (Calabrian).

In the coastal regions of the Peloponnesus, marine ingressions took place in the early Pliocene (Elis, Messinia) and in other places not until late Pliocene time (northern Peloponnesus, Corinthian Gulf, eastern Laconia). The marine deposits are underlain by continental beds and are marginally interbedded with limnic and fluviatile clastics. In the northern Peloponnesus marine sediments of late Pliocene age were covered by continental deposits which also occur in the southern part of the Argolis Peninsula. The beginning of the deposition in the intramontane basins of the Peloponnesus is dated as late Pliocene.

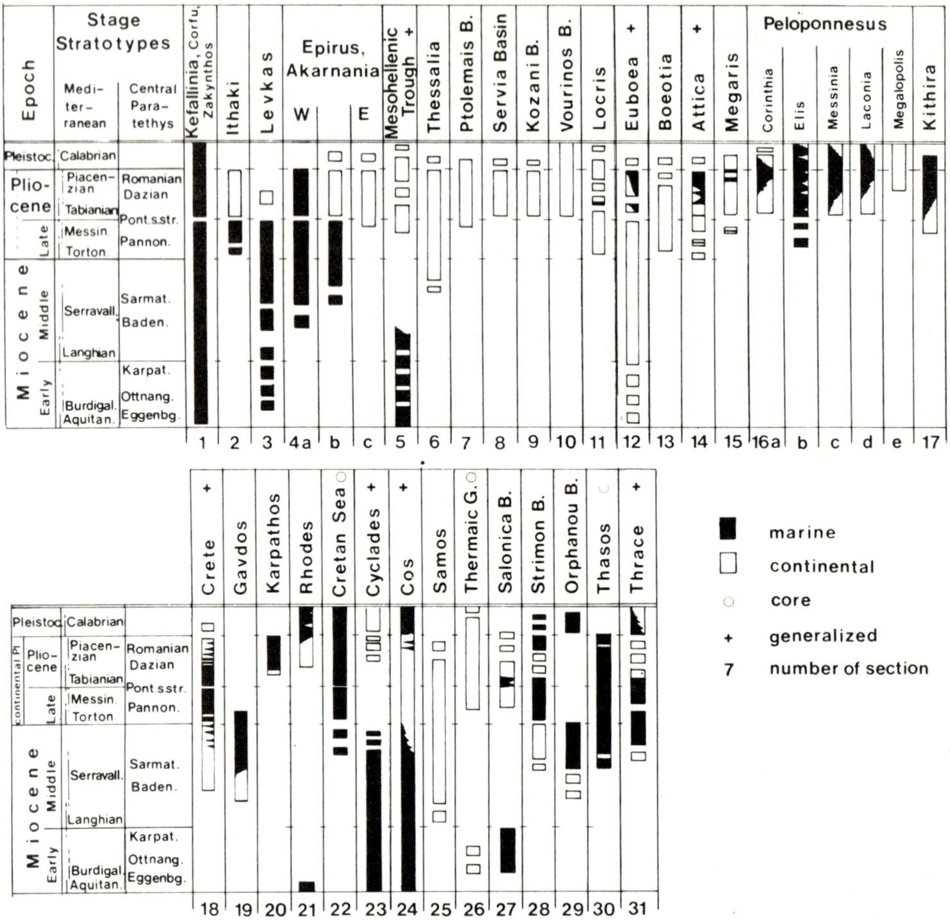

Fig. 2. Tentative correlation of Neogene sections in Greece (positions s. fig. 1). A detailed description of the area between the Magnisia Peninsula and Turkey is given by Besenecker & Büttner this vol.; see also special references.

Continental facies reached the basal layers of early Pliocene age on Kithira Island. Here, and in the southern Peloponnesus, they were followed - also in early Pliocene time - by marine sedimentation which lasted up to the Calabrian.

The early Pliocene of Crete is largely characterized by marine sediments. They are locally overlain by limnic and fluviatile deposits of late Pliocene age. In the eastern and southern part of Crete, synchronous marine sediments have been found. Transitions between marine and continental facies during the entire Neogene have been reported from Crete.

The Neogene of Karpathos Island is restricted to marine sediments of Pliocene age.

In the western and central Cretan Sea, marine sedimentation prevailed during the entire Pliocene. In the late Pliocene, the eastern part of the Cretan Sea was finally integrated into the marine realm.

Consequently, Neogene sedimentation on Rhodes Island began with limnic to fluviatile facies, passing into marine facies at the end of late Pliocene time. On Cos temporary ingressions occur at the end of the Pliocene, followed by fully marine Calabrian deposits.

In the North Sporades and Euboea area, there is but little chronostratigraphic evidence of continental Pliocene facies (e.g. Magnisia Peninsula, Norhtern Euboea), whereas on the islands west of the Anatolian coast entirely continental deposits accumulated (uppermost part of the Yatagan- and Milet Member).

In late Pliocene time, marine neritic sediments were deposited at the edge of eastern central Euboea.

In the Thermaic Gulf, the central and southern Salonica Basin, and the Strimon Basin, Pliocene limnic, fluviatile and terrestrial clastics have been accumulated on the top of the Pontian (Novorossian) brackish to marine ingression. However, in the Orphanou Gulf, and in a graben structure between Thasos and Samothraki Islands, subsidence with alternating marine and terrigenous accumulation persisted throughout Pliocene time, even transgressing the Miocene depositional area (West 1973).

The Pliocene sedimentation in the northern part of the Thrace Basin is continental and passes laterally to the South into a sequence with brackish and partly marine layers. Farther to the South, on Samothraki Island, marine Pliocene has been reported (Davis 1963).

Paleogeographical Development

The paleogeographic setting of this area can be reconstructed only since Tortonian/Messinian time (in the Ionian Sea perhaps only since early Pliocene time) because of nappe and fold tectonics during the Neogene. For example, the main tectonic deformation of the external zones took place during the Lower/Middle Miocene, and even in the internal zones, considerable compressional movements can be inferred (Jacobshagen, Dürr et al. this vol.).

The sedimentation of the West Hellenic Flysch (Richter 1976) continued up to the early Miocene (Burdigalian). The Ionian Sea area, as a part of the authochthonous Preapulian Zone, developed into an archipelago with increasingly flat marine conditions during Miocene time (de Mulder 1975).

Meanwhile, the Mesohellenic Trough remained in the marine realm. Crete and the Cretan Sea were in early and in early middle Miocene part of a mainland and underwent erosion, whereas the central Cyclades area, with Rhodes and Cos Island, belonged to a sea. It is not known if Early Miocene marine deposits in the Salonica Basin were connected to the Mesohellenic Trough or the Cyclades area.

In the middle Miocene the sea generally retreated from the Greek mainland and marine sedimentation ceased in the Mesohellenic Trough. The marine depositional area of the Ionian Sea persisted and marginally marine sediments were deposited on top of the dislocated West Hellenic Flysch.

The Greek mainland and the Peloponnesus were not yet separated from each other and the middle Aegean area between Euboea and the islands from Lesbos to Samos formed a continent. Northeastern Greece, as far as to the Strimon Basin, probably belonged to this mainland. However, the area be-

tween the Orphanou Gulf, Thasos Island and Thrace were part of a sea with a marine to hypersaline environment.

Intensive tensional fault tectonics formed the internal basins on the Greek mainland and led to the present configuration of the Peloponnesus gulfs.

In the late Middle Miocene (Serravallian), tectonic tension caused the break up of the southern Aegean region, and the western and central parts of the Cretan Sea were formed. This ingression was not synchronous over the whole area. Very probably, it reached as far north as Attica and connected this southern marine area with the marine region in northeastern Greece in late Middle to Late Miocene time.

After a general regression in the Late Miocene (Messinian salinity crisis), another ingression took place during early Pliocene time. It culminated in the Late Pliocene and covered the marginal area of the Greek mainland and the southern Aegean islands. Moreover, the coastal regions, which had a continental environment during late Miocene and early Pliocene time, were flooded (e.g. Attica, Euboea). Only in the internal basins continental sedimentation persisted.

Finally, in late Pliocene time, the eastern part of the Cretan Sea was incorporated into the marine area so that the land connection between Rhodes Island and the mainland was cut off and Cos Island marginally submerged.

A new gradual regression at the Pliocene/Pleistocene boundary followed, which mainly affected the Greek mainland and the northern Peloponnesus. Thus, in the area of the Corinthian Gulf, continental sedimentation again prevailed. The following sea-level rise at the beginning of the middle Pleistocene is documented only by marine terraces near the present shoreline , but it created the present distribution of land and sea.

In considering the paleogeographic development since Pliocene time, eustatic and glacio-eustatic factors have been taken into account. Nevertheless, the distribution of land and sea since the late Miocene might have been determined essentially by large vertical movements and tensional faulting, which controlled the local configuration, sedimentation rates in the internal basins and erosional rates in the related regions of denudation.

References

Altherr, R., Keller, J. & Kott, K., 1976: Bull. Soc. géol. France, 18,7: 403-412.

Becker-Platen, J., 1971: Newsl. Stratigr., 1,3:19-22.

Bignot, G. & Guernet, C., 1976: Géologie méditerr., 3:15-26.

Davies, E.N., 1963: Ann. géol. Pays hellén., 14:133-212.

Faugères, L. & Robert, C., 1976: Géologie méditerr., 3,4:209-218.

Gramann, F. & Kockel, F., 1969: Geol. Jb., 87:445-484.

Lüttig, G. & Steffens, P., 1976: Explanatory notes for the paleogeographic atlas of Turkey from the Oligocene to the Pleistocene,Hannover.

Meissner, B., 1976: N. Jb. Geol. Paläont. Abh., 152:161-176.

Mulder, E.F.J. de, 1975: Utrecht Micropaleont. Bull., 13:139 p.

Richter, D., 1976: Z. dtsch. Geol. Ges., 127:467-483.

Schneider, H.E., 1972: N.Jb. Geol. Paläont. Mh.:183-195.

West, J., 1973: The Oil and Gas Journal:128-130.

Special References
for fig. 1 and 2 (numbers indicate localities resp. sections)

(1, 2, 3, 4) I.G.R.S./I.F.P., 1966: Étude géologique de l'Epire.: 306 p., (Technip) Paris.
BP Co Ltd., 1971: Inst. Geol. Subsurf. Res., 10:73 p.
Mulder, E.F.J. de, 1975: s. above.
Braune, K., Fabricius, F. & Heimann, K.O.:this vol.
(5) Brunn, J.H., 1965: Ann. géol. Pays hellén., 7:358 p.
Aubouin, J.J., 1959: Ann. géol. Pays hellén., 10:483 p.
(6) Schneider, H.E., 1972: s. above.
(7, 9) Vetoulis, D., 1957: Ann. géol. Pays hellén., 8:48-79.
Brunn, J.H., 1956: s. above.
(8) Desprairies,A. & Faugères, L., 1971: Bull. Soc. géol. France, 7,13:67-84.
(10) Faugères, L. & Vergely, P., 1974: C.R.Acad.Sci. Paris,D 278:1313-1316.
(11) Mercier, J.-L., Carey, E., Philip, H. & Sorel, D., 1976: Bull. Soc. géol. France, 7,18:355-372.
(12) Besenecker, H. & Büttner, D.: this vol.
(13) Celet, P., 1962: Ann. géol. Pays hellén., 12:446 p.
Mitzopoulos, M.K., 1961: Ann. géol. Pays hellén., 11:301-314.
Rümke, C.G., 1976: Kon.nederl.Akad.Wetenschp., B 79:245-255.
(14) Marinos, G. & Symeonides, N., 1973:Bull.Geol.Soc.Greece,9,2:160-176.
Guernet, C. & Sauvage, J., 1970: Bull.Soc. géol. France, 7,12:241-245.
Schütt, H., 1976: Arch. Moll., 107:35-51.
(15) Theodoropoulos, D., 1968: Erlanger geol. Abh., 73:1-23.
(16a) Freyberg, B.v., 1973: Erlanger geol. Abh., 95:183 p.
(16b) Streif, J.: this vol.
(16e) Vinken, R., 1965: Geol. Jb., 83:97-148.
(17) Theodoropoulos, D., 1973: Thesis: 94 p., Athens.
(18) Drooger, C.W. & Meulenkamp, J.E., 1973: Bull. Geol. Soc. Greece, 10,1:193-200.
Benda, L., Meulenkamp, J.E. & Zachariasse, W.J., 1974: Newsl. Stratigr., 3,3:205-217.
Zachariasse, W.J., 1975: Utrecht Micropaleont. Bull., 11:171 p.
(19) Freudenthal, T., 1969: Utrecht Micropaleont. Bull., 1:208 p.
Vincente, J.-C., 1970: Bull. Soc. géol. France, 7,12:481-495.
(20) Aubouin, J., Bonneau, M. & Davidson, J., 1976: Bull. Soc. géol. France, 7,18:385-401.
(21) Meulenkamp, J.E., Mulder, E.F.J. de & Weerd, A. van de, 1972: Z. dtsch. Geol. Ges., 123:541-553.
Broekman, J.A., 1974: Utrecht Micropaleont. Bull, 8:131 p.
(18, 19, 20, 21) Schmidt, R.R., 1973: Kon. Nederl. Akad. Wetenschp., B 74:287-310.
Sissingh, W., 1976: Kon. Nederl. Akad. Wetenschp., B 79, 4:271-299.
Jongsma, D., Wissman, G., Hinz, K. & Gardé, S., 1977: "Meteor" Forsch.-Ergebnisse, C 27:3-30.
(23) Roessler, G.: this vol.
(24) Besenecker, H. & Otte, O.: this vol.
Bignot, G. & Guernet, C., 1976: s. above.
(25) Besenecker, H. & Büttner, D.: this vol.
(26) Faugères, L. & Robert,C.,1976:s. above.
(27) Mercier, J.-L., 1968: Ann. géol. Pays hellén., 20:599-792.

Faugères, L. & Gillet, S., 1974: Ann. géol. Pays hellén.,25:407-438.
(27, 28, 29, 31) Gramann, F. & Kockel, F., 1969: Geol. Jb., 87:445-484.
(30) West, J., 1973: s. above.

Stratigraphy and Tectonics of Late Cenozoic Rocks in Western Peloponnesus

H. Streif

In the Bay of Elis, western Peloponnesus, late Cenozoic sediments unconformably overlie Mesocoic to Paleogene rocks. The lowest part of the Cenozoic sequence is the Vounargon Formation, a 500 m thick complex of sandy, silty, and clayey marine shallow-water sediments containing thin intercalations of brackish-lagoonal sediments and lignite. Rich marine mollusc faunas indicate a Late Pliocene age for these deposits (Mitzopoulos 1940; Streif 1964) and palynological data allow a correlation with the Akça assemblage in Turkey (Benda 1971; Benda & Meulenkamp 1972). Brackish and fluvial conditions predominated during the deposition of the younger, 100 m thick Khelidoni Formation. The lignite beds in the lower part of this formation contain a typical Akça sporomorph assemblage. The gradual change from shallow marine to fluvial facies in the Vounargon and Khelidoni Formations indicates that deposition kept pace with or slightly exceeded the tectonic subsidence.

The depositional environment changed considerably at the beginning of deposition of the Lala Formation. Conglomeratic sediments derived from the northeastern margin of the Bay of Elis were deposited by braided river systems. No fossils have been found in the Lala Formation but the overlying Tzoyia Formation contains *Equus* and *Cervus* indicating an early Pleistocene age (Sickenberg pers. comm. 1964) older than the Marathousa Stage of Megalopolis (Sickenberg 1975). It is highly probable that during the sedimentation of the Lala and Tzoyia Formation the tectonic subsidence of the basin changed to faulting and block movements and affected an alteration of the paleorelief.

The Katakolon Formation, comprising the Katakolon littoral limestone and the Miraka fluvial terrace deposits, overlies the older rocks with an angular unconformity of as much as 20 to 30°. The littoral limestone crops out at an average altitude of 20 to 40 m above present sea level and contains a rich but non-definitive assemblage of marine molluscs. The Miraka terrace deposits are normally 35 m above the level of the present river. Both lithologic units are cut by faulting which occurred during the development of the present geographic outline of the Bay of Elis.

Overlying and incised into the Katakolon Formation are the littoral limestone and fluvial terrace deposits of the Ayios Andreas Formation. Keraudren (1966) found *Strombus* in the littoral limestone and therefore correlated it with the "cycle eutyrrhénien".

Formed during the Holocene were two fluvial terraces and coastal bar-

rier and lagoonal systems which partly can be dated archeologically.

References

Benda, L., 1971: Beih. geol. Jb., 113:1-46.
Benda, L. & Meulenkamp, J.E., 1972: Z. dtsch. geol. Ges.,123:559-564.
Keraudren, B., 1966: Comptes rend. Acad.Sci. Paris, (D),262:2132-2134.
Mitzopoulos, M.K., 1940: Praktika Acad. Athènes, 15:429-436.
Sickenberg, O., 1975: Annales géol. Pays hellén., 46:4-73.
Streif, H., 1964: Unveröff. Diplomarbeit: 138 p.

Late Cenozoic Sediments on the Islands between Euboea and Turkey

H. Besenecker and D. Büttner

The study area (fig. 1) can be divided regionally and geodynamically into three parts:

the *western Aegean area* with predominantly continental Neogene deposits on the Magnesian Peninsula, the islands of Alonnisos and Piperi of the Northern Sporades, and parts of Euboea,

a predominantly *marine to brackish area* on the Island of Skyros with near-shore sandstone and marl of Tortonian age and travertine, intercalated in different niveaus,

the *eastern Aegean region* containing Lesbos, Chios, Karaburun Peninsula, Psara and Samos, with exclusively continental Late Cenozoic sediments and thick volcanic rocks.

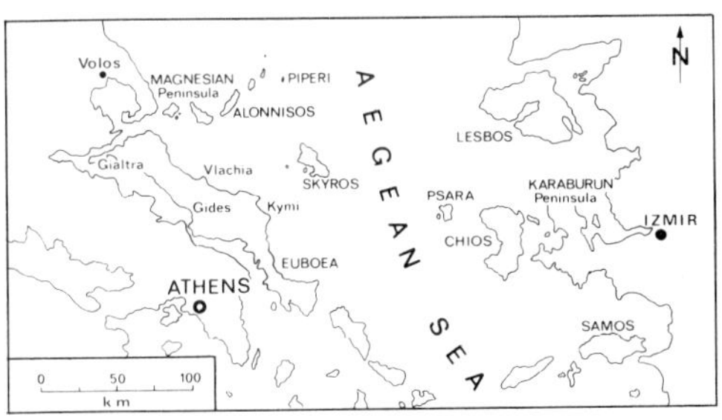

Fig. 1. Geographic situation of the study area.

Western Aegean Region

In the *Kymi basin of Euboea* thick limnic marls with lignites and layers rich in fossils were formerly dated as Eocene (Gorceix 1873) or Upper Oligocene (Saporta 1873). Melentis (1966) deduced an Early Miocene age (Burdigalian-Helvetian) by finding a lower jaw of *Brachyodus onoideus*

(Deperet 1895). Overlying conglomerates and sandstones with tuffitic layers are presumably of Pannonian age.

Limnic sediments in the *Gides basin* are overlain by coarser deposits with intercalated travertine. The Pliocene vertebrates (Pikermian faunas) described by Mitzopoulos (1947) apparently come from this "Upper Series". Other Pikermian faunas are known from Northern Euboea near Ellinika, Achladi and Limni from formations similar to those of Gides (Mitzopoulos 1947).

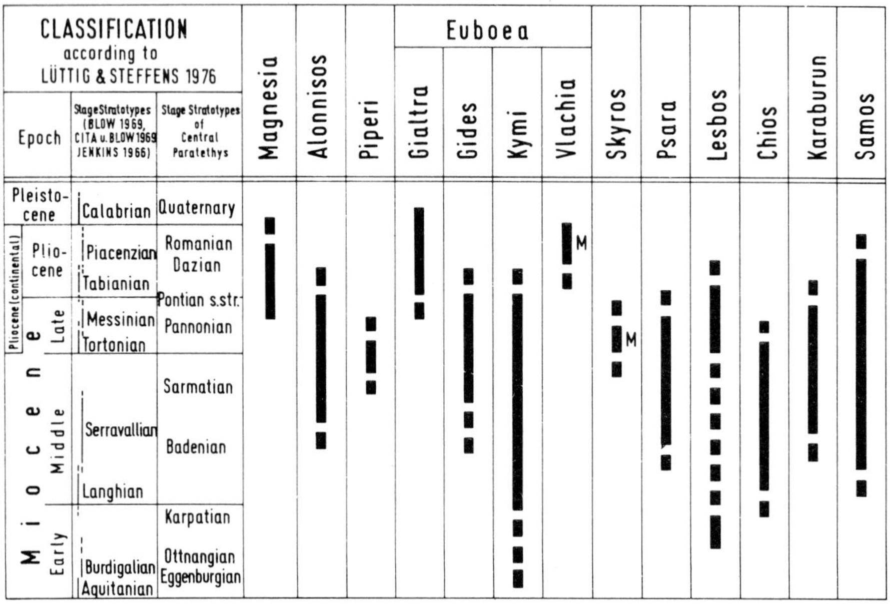

Fig. 2. Tentative Correlation of Neogene deposits on the Islands between Euboea and the Anatolian coast; M = marine sediments.

On the *Peninsula of Gialtra* (Northwestern Euboea), lacustrine marls with some larger lignite beds have been assigned to the Pleistocene (Psarianos & Thenius 1953) by means of *Elephas (Archidiscodon) meridionalis*.

Ostracods (det. M. Bassiouni & F. Gramann, Hannover) and pollen (det. L. Benda, Hannover) of the Akça assemblage (Benda 1971) suggest, however, a Pontian age.

On the *eastern coast of central Euboea*, marine shore deposits with rich molluscan and foraminiferal faunas have been dated as Pliocene (Astian) by Georgiades-Dikeoulia (1970). The ostracods (det. Gramann) confirm this age determination.

The Neogene of Alonnisos can be divided into a lacustrine limestone-marl formation below, travertine in the middle, and continental red conglomerates above. Ostracods from the marls suggest a Sarmatian to Pannonian age, whereas a Pannonian age is confirmed for the continental red formation by an *Hipparion* molar (Schneider 1973).

Small deposits, similar to the transition between the marl and the

travertine of Alonnisos crop out in the southern part of *Piperi Island*.

Small erosional relicts of a Neogene cover are conserved on the *Magnesian Peninsula*. The sequence consists of varicoloured conglomerates followed by marl and bedded limestone and topped by compact travertine. Plant debris, few gastropods, and ostracods (det. Bassiouni & Gramann) have been found, which indicate a Pliocene to Quaternary age. A time correlation with deposits of the Gialtra Peninsula and the coarse clastics of Alonnisos is probable.

Island of Skyros

Along the northeastern coast of Skyros, littoral marine to brackish sediments were dated by Papastamatiou (1961) as Pliocene and by Grekoff et al. (1967) as Upper Miocene. Our preliminary results confirm probable Tortonian age. The ostracode fauna (det. Bassiouni & Gramann) show relations to the Dafni formation in the Strimon basin (Gramann & Kockel 1969) and to the Apostoli and Barbara formations of Crete (Sissingh 1972). These relations are supported by the occurrence of otoliths (det. W. Schwarzhans, Berlin) of the species *Gaudiculus rumanus* (Weiler 1943) from the Sarmatian and Tortonian of the Paratethys.

Beach terraces, presumably belonging to the Tyrrhenian, unconformably overlie the dislocated Late Miocene beds.

Eastern Aegean Region

Thick ignimbritic blankets were deposited in Miocene time on *Lesbos Island* (Borsi et al. 1972). Overlying andesitic lavas and tuffs are interfingering with limnic sediments of Pannonian age (Hecht 1972).

The limnic and fluviatile deposits of *Chios*, attaining a thickness of 500 m, have been lithologically subdivided and correlated with sediments of Badenian and Sarmatian age of Turkey (Besenecker 1973). Their upper parts are dated by abundant vertebrate faunas as Sarmatian, and by pollen (Yeni-Eskihisar assemblage, Benda 1971) as Sarmatian to lowermost Pannonian.

Deposits on the adjacent Turkish *Karaburun Peninsula* can be correlated with those of Chios by their lithological appearance and vertebrate content.

On the islet of *Psara*, the lower part of the Neogene formation with vertebrate bearing beds (*Dinotherium bavaricum*; Besenecker & Symeonides 1974) were presumably deposited during the Badenian and Sarmatian (analogous to Chios) while the overlying thick pyroclastics (analogous to Lesbos) were deposited in Pannonian time.

The thick Neogene fill of the Mytilini and Karlovasi basins on *Samos Island* have been studied and subdivided by Meissner (1976). The famous *Hipparion* fauna from Samos must be regarded as of the same age as the Pikermi fauna, despite some differences, and belongs to the Pliocene (Pannonian) in continental terms. The Yatagan member (Becker-Platen 1970) of East Anatolia can be correlated with the *Hipparion* bearing beds of Samos.

Conclusions

The study area apparently formed a landmass until early Pannonian according to vertebrate stratigraphy or early Tortonian according to marine stratigraphy (fig. 2). During the Early and Middle Miocene (? Karpathian, Badenian, and earlier Sarmatian) fluviatile and terrestrial sedimentation prevailed in the whole area. During the Sarmatian and earliest Pannonian abundant multiform lakes dominated the landscape. Lacustrine beds, mainly of limestone and marl, from Kymi, Chios (Nenita member), Samos (Pythagorion and Hora sequence) and from Southwest Anatolia (Sekköy member) give evidence of this.

Probably in earliest Pannonian or earliest Tortonian time strong vertical tectonic displacements remodelled the depositional area. A depression between Euboea and Psara opened a seaway from the south over Skyros and presumably via the Dardanelles into the Paratethys. This seaway separated the western sedimentary area from the eastern one; but the subsequent depositional development in both regions was analogous because of similar climatic and morphotectonic influences. Since differential tectonic movements had created a high relief, clastic fluviatile and terrestrial sedimentation prevailed in the following time-span. Furthermore, an accentuated volcanic activity led to the deposition of thick pyroclastics. During the Late Pliocene and earliest Pleistocene, lakes accumulated limestone and marl.

Meanwhile, the marine connection between Tethys and Paratethys must have been interrupted, so that a direct interchange of land vertebrates was possible. During the Late Pliocene (Astian), however, the sea again invaded the region between Euboea and Skyros.

In the entire area Pleistocene beach terraces have only been found on Skyros - a fact that supports the assumption that greater land masses, particularly in the eastern Aegean region, submerged only in postglacial time.

References

Becker-Platen, J., 1970: Beih. Geol. Jb., 97:244 p.
Benda, L., 1971: Beih. Geol. Jb., 113:46 p.
Benda, L. & Meulenkamp, J.E., 1972: Z.dtsch. Geol. Ges., 123:559-564.
Besenecker, H., 1973: Diss. FU Berlin:196 p.
Besenecker, H. & Symeonides, N.K., 1974: Ann. Géol. Pays Hellên., 26:109-117.
Borsi, S. et al., 1972: Z. dtsch. Geol. Ges., 123:521.
Georgiades-Dikeoulia, E., 1970: Ann.Géol.Pays Hellên.,21:661-670.
Gorceix, M.H., 1873:Ann.scient.de l'École Norm.Sup.,2:317-322.
Gramann, F. & Kockel, F., 1969: Geol.Jb., 87:445-484.
Grekoff, N. et al., 1967: C.R. Acad.Sci. Paris, 265:1276-1277.
Guernet, C., 1971: Thèse Paris: 395 p.
Hecht, J., 1972: Z. dtsch. Geol. Ges., 123:423-432.
Lüttig, G. & Steffens, P., 1976: Paleogeographic Atlas of Turkey, Explanatory Notes: 64 p.
Meissner, B., 1976: N. Jb. Geol. Paläont. Abh., 152(2):161-176.
Melentis, J.K., 1966: Ann. Géol. Pays Hellên., 17:298-322.
Mitzopoulos, M.K., 1947: Ann. Géol. Pays Hellên., 1:209-215.
Papastamatiou, J., 1961: Bull. Geol. Soc. Greece, 6:219-237.

Psarianos, P. & Thenius, E., 1953: Prakt.Akad. Athenon, 28:413-424.
Saporta, G., 1873: Ann. Ec. Norm. Sup. 2.
Schneider, H.E., 1972: N.Jb. Geol. Paläont.Mh.:183-195.
Sissingh, W., 1972: Utrecht Micropaleont. Bull. 6:187 p.
Weiler, W., 1943: Senckenbergiana, 26(1/3):87-115.

Late Cenozoic Development of Kos, Aegean Sea*

H. Besenecker and O. Otte

From 1970 till 1974, investigations of the entire late Cenozoic successions of Kos were made in order to reconstruct the geodynamics of this area. The successions were subdivided into lithostratigraphic units (fig. 1). Their extension as well as their facies were the object of further investigations, which have supported the following outline of the sedimentary history and tectonic events.

Phase A

In the eastern part of the island and on the Kefalos peninsula, the alpine basement is overlain by transgressive biodetritic shallow-water limestone (Asklupi m., Pesia m.) which indicate continental influence by local intercalations of silt, sandstone and conglomerate. Tectonic contacts predominate above the basement. On the northern coast of Kefalos, however, basal red conglomerates, clay and siltstone (Sideropetra m.) show a sedimentary unconformable contact.

Desio (1931) dated these sediments as of Middle Miocene age because of a rich marine mollusc fauna. Foraminiferes, newly discovered in the calcareous parts of this series, prove an *Early Miocene* age.

Phase B

At the beginning of phase B a fundamental tectonic movement took place connecting the sedimentation area of Kos with the Aegean landmass until phase E. Probably in early Middle Miocene, vertical movements formed continental basins filled with coarse-grained red sediments, with sedimentological criteria suggesting terrestrial mass transport (Pili m.) as well as fluviatile sedimentation (Soccoro m.). They are overlain by limnic, thin-bedded calcilutites and marls in the eastern part of the island with a thick pyroclastic layer, intercalations of massive travertinous

* For the geographic position of Kos see Büttner et al. fig. 1, Nr. 24.

limestones, and thin lignite layers in several stratigraphic levels (Vasilios m.). The fresh water limestones outline a large limnic basin which was separated by a ridge in central Kos from another basin with analogous facies (Protocaravo m.) established in western Kos.

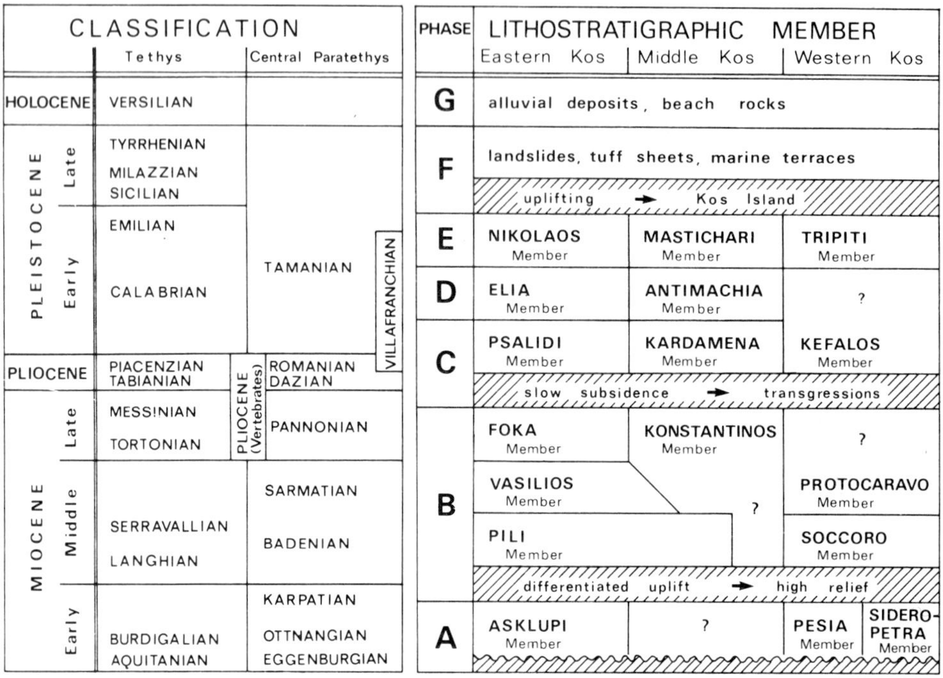

Fig. 1. Late Cenozoic lithostratigraphic units of Kos and their tentative correlation with classifications of Tethys and Central Paratethys.

In central Kos the limestone facies is substituted by an alternation of fluviatile, fine-grained and coarse-grained clastics with limnic limestones and marls (Konstantinos m.). Towards the end of phase B the fluviatile facies extends over the eastern part of the isle (Foka m.). The fluviatile sediments of the Konstantinos m. and the Foka m. were transported in southerly to easterly directions.

The volcanic deposits, locally more than 200 m thick, consist of ignimbrite layers, volcanic breccias and fine clastics with strongly varying thicknesses. Biotites out of the ignimbrite revealed a K-Ar age of 10.4 $\pm$ 0.2 m.y. (Altherr et al. 1976) which supports our former assumption of a Sarmatian age of the thin limestone flats.

After the fundamental change of the paleogeographical trend in the beginning of phase B a period of progressive tectonic quiescence followed, which is evident in a continuous still-water sedimentation. Towards the end of phase B, i.e. in the Pannonian, Kos was integrated in a predominantly fluviatile sedimentation area. At the end of the process, the relief was largely bevelled.

Phase C

In the beginning of phase C, a large shallow sedimentary basin was formed with only little relief towards the sea intruding from the south. Weak tectonic activity is indicated by synsedimentary tilting and subsidence; only occasionally it lead to local facies contrasts.

The predominant facies consists of an even alternation of varicoloured sediments containing non-compacted clays, silts and sands in beds of 1 or 2 m thickness disconformably overlying the sediments of phase B (Psalidi m., Kardamena m., Kefalos m.). Well sorted sands and gravels occur in minor quantities. Numerous thin layers of lignite and other organic contents, as well as sedimentary characteristics suggest a flat, temporarily desiccated area with lakes, swamps, meandering rivers and occasional ingressions of the sea.

The higher part of the Psalidi m. of east Kos is characterized by a calcareous facies with white silty, limnic marls containing an endemic gastropode fauna. This is strong evidence of a long and continuous isolation of a sea basin, from the surrounding sedimentation area.

The lower part of the Kardamena m. contains among others relicts of Hipparion and therefore probably belongs to the Upper Pliocene. The macro-mammal and micro-mammal assemblages of the major portion of the Kardamena m. and the Psalidi m. allow a correlation with Early Pleistocene (Middle *Villafranchian* and Villanyian, respectively). Pollen from samples of the Psalidi m. and Kardamena m. are indicating an "Akça" assemblage (Pliocene to Early Pleistocene, Benda 1971).

Phase D

Increased tectonic movements lead to a new paleogeographic pattern. The sedimentation area around Kos subsided below sea level. Sedimentary structures show continued tilt and a sediment transport in predominantly southerly direction.

Sediments consist mainly of neritic sands with a rich marine invertebrate fauna. Fine-grained clastic sediments with silts and clays (Antimachia m.) predominate locally. Clastic coarse-grained, mainly fluviatile, sedimentary successions (Elia m.) in Eastern Kos with little marine influence indicate local uplift and erosion of older Neogene strata.

Marine microfauna from the lower part of the Antimachia m. indicates an early Pleistocene age (Calabrian). The overlap of Villafranchian strata by Calabrian strata on the island of Kos is important in regard to the biostratigraphic correlation of Pleistocene strata of continental and marine origin.

Phase E

Continuous vertical differential movements in addition to the progressing subsidence of the Aegean landmass separated the sedimentation area during phase E, from the terrestrial supply areas. The sediments (Nikolaos m., Mastîichari m., Tripiti m.) consist of detritic shallow-water limestones with resedimentary components. Fauna and flora indicate a fully marine, turbulent clear-water biotype.

Phase E terminates the early Pleistocene sedimentary history and tectonics of this area.

Phase F

Phase F begins with the elevation of the island of Kos during intense fracturing with vertical fault displacement of more than 1000 m. The island became subdivided into several blocks with conspicuous displacements in the east as well as in the extreme west of the island. The newly developed relief is characterized by deep canyons and large terrestrial mass transports achieved by sheet floods and numerous landslides.

The intense fault dissection has most likely entailed the subsequent intense volcanism and a variety of related phenomena. Tuff sheets cover large parts of the island (Kefalos Tuff, Plateau Tuff) preserving the formerly developed relief, especially in central Kos. In coastal regions marine terraces of possibly Tyrrhenian age are intercalated with the Plateau Tuff.

Phase G

Phase G comprises the present development (*Holocene*) of the island with extended alluvia, beach rocks etc. and tectonic movements which entailed vertical shifts e.g. of the beach terraces by more than 20 m.

References

Altherr, R., Keller, J. & Kott, K., 1976: Bull. Soc. géol. France, 18,2:403-412.

Benda, L., 1971: Beih. Geol. Jb., 113:46 p.

Besenecker, H. & Otte, O., 1972: Z.dtsch. Geol. Ges., 123:527-539.

Charrier, G. & Giglio, A., 1969: Boll. Assoc. Mineraria Subalpina, 6/3: 482-516.

Desio, A., 1931: Mem.Descr.Carta Geol. Ital., 24:534 p.

Jongsma, D., Wissmann, G., Hinz, K. & Garde, S.: "Meteor"-Forsch.-Ergeb. Reihe C, 27:3-31.

Keller, J., 1969: Bull. Volcanologique, 23:942-959.

Lüttig, G. & Steffens, P., 1976: Paleogeographic Atlas of Turkey, Explanatory Notes: 64 p.

Sedimentary History and Tectonic Movements During the Late Neogene*

H. Böger

The northeastern part of Kos (Dodecanese) is covered by a Neogene sedimentary column up to 750 m thick (Böger et al. 1974). This mostly limnic section can be didived into several lithostratigraphic units, each representing different environmental settings. Some of them can be interpreted as reflecting special paleogeographic conditions and tectonic movements in connection with the Plio-/Pleistocene history of the Aegean basin, at least of its southeastern part. The stratigraphic data accurately reflect the intrusion of a monzonitic pluton at 10 to 12 m.y. BP, and the uplift from a depth of 10 km (Altherr et al. 1976). Also the presumed transport of the Lykian Nappes on top of the eroded monzonite is evident in the sedimentary history.

The Neogene history on Kos began with probable transgressive nearshore marine limestones, of supposed middle Miocene age (Asklupi beds). At about the same time, the monzonite intruded at great depth into schists considered to be of Paleozoic age. After this event, sedimentation was interrupted for at least 2. m.y. Within this period of time, uplift, erosion of the pluton and its schist mantle took place, and immediately afterwards, the transport of Lykian Nappes on top of the eroded surface.

Neogene sedimentation was resumed with transgressive conglomerates of varying thickness followed by a series of wellbedded limnic limestones and pyroclastic rocks (Palioskala and Vasilios beds, at least 200 m). Tectonic movements were confined to the downwarping of intracontinental sedimentary basins. The exact stratigraphic position of these beds has not yet been determined but they are probably of Messinian (late Miocene) age.

Once more sedimentation was interrupted. Uplift and faulting of the limnic basin fill preceded and accompanied a rather long period of erosion. The sandstones and conglomerates of the Phoka beds are regarded as reflecting the latest of these and other erosional events probably far beyond the Kos area.

The Phoka beds grade into a series of lacustrine marls with lignites and a rather abundant fauna of limnic Gastropods and Pelecypods (Sefto, Kos, Vokasia, and Irakli beds). Palynologic studies have shown an uppermost Pliocene age (Akça) according to the continental time scale.

These marls grade into sandy marls, sands and conglomerates (Elia beds), still with limnic fauna but also with occasional marine ingressions, indicating the beginning of subsidence of parts of the Aegean Sea and at the same time uplift and erosion of others.

At the top of the column, thick marine limestones with coralline biostromes, algal mats, and oyster beds occur. Presumably, they are of early Pleistocene age.

At the end of the evolution, local tectonic movements cause pre-Neogene parts of the alpine systems to emerge through their Neogene cover

* For the geographic position of Kos see Büttner et al., fig. 1, Nr. 24.

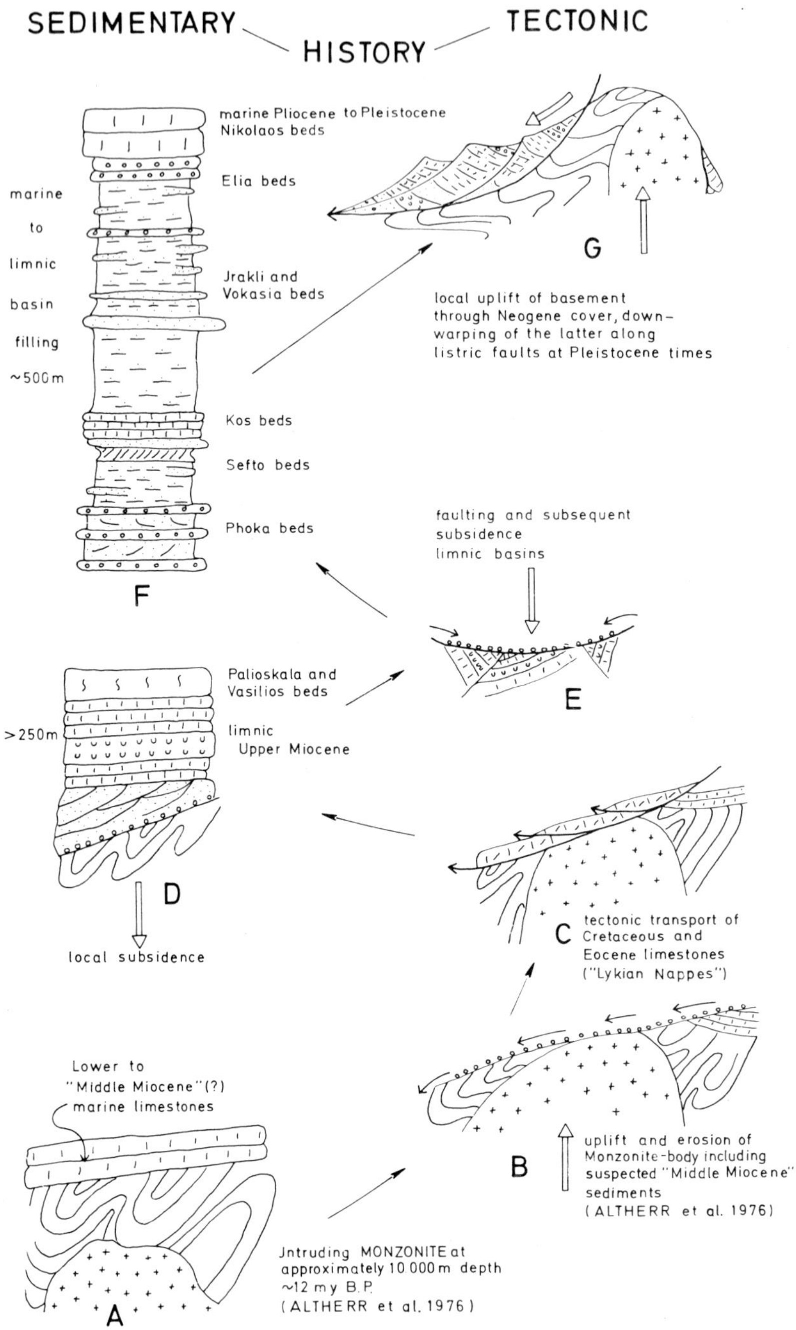

Fig. 1. Sedimentary history and tectonic movements on Kos.

which slides along listric faults, in the direction of the subsiding areas.

References

Altherr, R., Keller, J. & Kott, K., 1976: Bull. Soc. géol. France, 7, 18:403-412.
Böger, H., Gersonde, R. & Willmann, R., 1974: N. Jb. Geol. Paläont., Abh., 145:129-152.

Neotectonics in the Peloponnesian Coastal Regions

D. Kelletat, G. Kowalczyk, B. Schröder and K.-P. Winter

The division of the Peloponnesus into a mosaic of tilted fault blocks, horsts and grabens started after the last nappe movement, which ceased in the Tortonian at the latest and before the ingression of the sea in the Lower Pliocene. The fault tectonics is characterized, within the uppermost tectonic level, only by tensional features and can be demonstrated by the superposition or apposition of Pliocene beds on or above pre-existent horsts,grabens and faults (Kowalczyk et al. 1977). The single fault blocks are bordered by SSE-and ENE-striking faults in general. The regional distribution of areas with subsidence or uplift later controlled the distribution pattern of the marine Pliocene and the origin of the present gulfs.

However, the marine Pliocene only invaded in the coastal regions of the present Peloponnesus; therefore stratigraphic data about Neogene tectonics are restricted to this area but also concern the inland basins.

The ingression of the sea did not occur simultaneously over the whole region, it first began in the lower Pliocene in the western and southwestern part of the Peloponnesus and later, in the upper Pliocene, reached the northern and southeastern parts.

Synsedimentary normal faults and the filling of the inland basins of Megalopolis and Tripolis with continental upper Pliocene beds demonstrate continuing tensional fault tectonics during the Pliocene.

In the Early Pleistocene marine sedimentation persisted in the western and southern Peloponnesus, in the northern continental beds occurred, with the Gulf of Corinth disappearing since the latest Pliocene. In the Calabrian or just a little younger wide, pedimentlike erosional plains formed in the coastal areas of the Peloponnesus, probably indicating tectonic stagnation (Kelletat et al. 1976). Subsequently fault tectonics started again, in part reactivating pre-Pliocene fractures but also generating new fault patterns. This fault tectonics has affected the Pleistocene coastal terraces as well as holocene beds.

During the Late Neogene and the Quaternary, adjacent fault blocks have

formed bigger areas with the same tendency in tectonic development producing widespread regions of subsidence or uplift (resp. tilting).

The regional distribution of areas with subsidence or uplift has also formed the geomorphological structure of the present Peloponnesus. They strike in SE and are interrupted in the northern Peloponnesus by an E-W striking structure.

Areas with uplift	Areas with subsidence
- Westernmost Elis	- Plain of Elis, coast of Elis
- Eurymanthos, West-Messinian Peninsula	- Gulf of Messenia, Pamisos depression (Basin of Megalopolis)
- Mani Peninsula, Taygetos	- Gulf of Laconia, Evrotas depression (Basin of Tripolis)
- Peninsula of SE-Laconia, Parnon	- Eastern coast of Peloponnesus, Gulf of Argos, Argolis
- Northern Peloponnesus	- Gulf of Corinth, northern coast of the Gulf of Corinth

The evolution and amount of Pliocene and younger vertical movements are evident in the regional distribution, the present altitude and the stratigraphy of the marine Neogene, the pediments or uppermost erosional plains and the Pleistocene coastal terraces (s. fig. 1 and Kelletat et al. 1976).

It is significant that the Peloponnesian marine Neogene is widely distributed at high elevations in the western part and confined to remnants at low elevations in the eastern part.

This arrangement is repeated in the same direction at least on the two southeasternmost peninsulae. Here as elsewhere, the present extension of Neogene sediments has been determined by erosion, but the fundamental features of the original extension are still recognizable and can be determined by other evidences. Since the former level of the Neogene sea is unknown, the interpretation is restricted to the comparative study of the positions of isolated outcrops.

The marine Neogene transgresses only marginally into a continental area along the pre-existing tectonic framework. Consequently, the central mountain ranges - ridge areas of the peninsula - are free of coeval sediments. Extreme marine ingressions occurred at inland prolongations of gulf trenches, the coastal configuration of the southern peninsula was more pronounced during Neogene than it is today.

Two more features are conspicuous, and are doubtless consequences of post-Neogene tectonic dislocations:

a) Areas of most widespread and uplifted Neogene sediments occur in the northernmost parts of the peninsulae and from there decrease gradually with perpetual subsidence towards the capes.

b) A similar disparity is also recognized between the western and eastern parts of the peninsulae.

Pediments and pediment-like plains can be used in a similar way as Neogene sediments to reconstruct the subsequent tectonic evolution for the following reasons:

a) Their peripheral locations, the connection with Neogene marine sediments as well as with younger coastal terraces, indicate that these pediments are likely to have been adjusted to the same former marginal base level of erosion.

b) Their position above the oldest recognizable Quaternary coastal

forms and the accompanying morphological features moreover indicate that they date about the Pliocene/Quaternary boundary. As a result they appear to be the next younger element to date the tectonic events.

The absence of pediment-like plains in the east of the Peloponnesus is for instance again caused by subsequent submergence of the eastern part.

Vertical movements during the younger Quaternary can be estimated by means of the regional distribution and of the altitudes of marine terraces which are pre-Tyrrhenian but younger than the earliest Pleistocene. Several terraces of pre-Tyrrhenian age have developed in areas with strong uplift already indicated by emerged Pliocene sediments and degradational surfaces. Numbers and altitudes decrease in areas of minor uplift; in areas of subsidence they are either absent or have submerged below the recent sea level. Furthermore, the altitudes of terraces which can be correlated to each other and the changes of their vertical distances give information as to small scale movements during the entire Pleistocene.

Young Pleistocene terraces show the same tendency, except that the amounts of vertical displacements are smaller than in the older terraces.

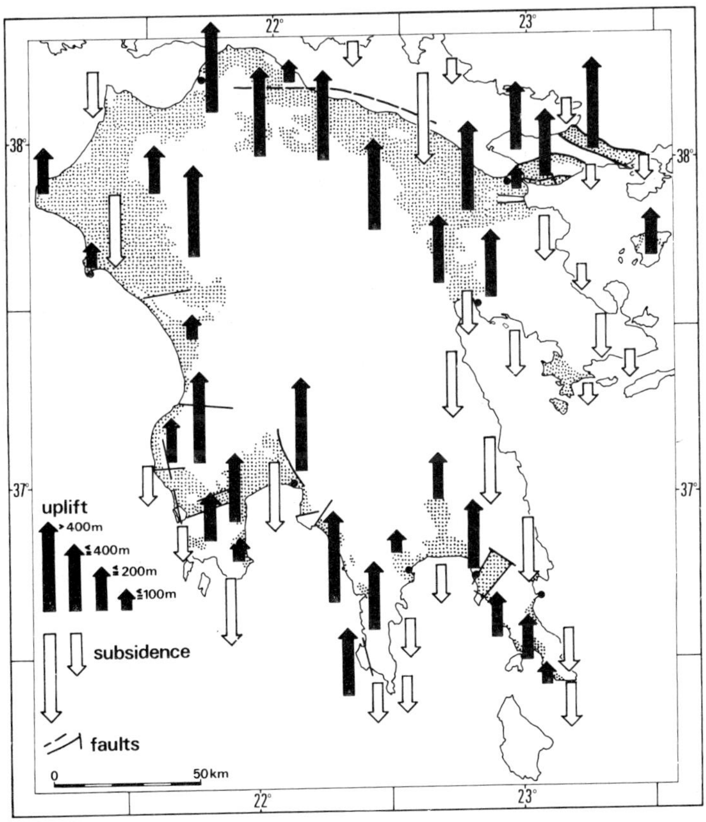

Fig. 1. Tendency of vertical movements since Pliocene: Black arrows = upward; Light arrows = downward (dimension only relative to each other); Dotted areas = Neogene sediments.

The succession of Quaternary coastal terraces in the Peloponnesus (as well as on the island arc of the Aegean sea) has many features in common with coastal terraces in areas of the circum-Pacific belt of earthquake activity (seismotectonic origin of terraces).

Achaia and Elis

Older Quaternary coastal terraces are almost absent but terraces of the Late Quaternary exist locally.

The westernmost part of Elis (Peninsula of Killini and Katakolon) contains two smaller districts with uplift during the Quaternary resp. since Miocene. Within a tilted fault block north of Katakolon Eu-Tyrrhenian deposits attain the 100 m alt. On the Peninsula of Killini marine beds of the Pleistocene have been uplifted up to 180 m alt. and only young Pleistocene terraces occur which are also tilted. Miocene sediments are lacking. This exceptional tectonic development is caused by an active diapiric culmination of Triassic gypsum. Several phases of salt-tectonic movements started during Miocene at the latest and continued with rather reduced activity during the Pliocene until the lower Quaternary. Compressional fault tectonics in the Late Quaternary reported from Elis by Mercier et al. (1972) are probably caused by this special geologic situation. Salt tectonics must also have complicated the pattern of structural evolution.

The plain adjoining to the east was an area of major subsidence since the Miocene. Marine Miocene and Pliocene deposits reach thicknesses of 600 m (Sosti, BP Co 1971) and even more than 1700 m (Pirgos, Mitzopoulos 1940). At several times, marine sediments have deposited in this area during the Quaternary as indicated by coastal terraces inland of the Peninsula of Killini and the superposition of Quaternary terraces must be taken into account (Philippson 1930).

West of the Eurymanthos mountains marine Pliocene beds have been uplifted to an altitude of 400 m (north of Olympia). Extensive erosional plains developed at 600 m and 750 m resp. overlie conglomerates of the late Pliocene or earliest Pleistocene (Streif 1964). They must be interpreted as pediment plains of fluvial gravel plains whose level was accordant to the sea level at the time of their origin. The uplift which took place since the late Pliocene must be calculated at a rate of at least 400 m or probably even of 750 m.

In the area between the Alphios river and Kiparissia marine terraces are almost completely lacking; Neogene deposits occur at relatively low levels. Refering to the recent sea level, Pliocene beds have been uplifted, but later on in the Quaternary this area must have subsided.

Messinia

In Messinia Miocene deposits are unknown. During the late Pliocene the west-Messianian Peninsula subsided as a whole. Marine sedimentation reached the area between Pylos and the Pamisos depression; the southern part of the peninsula became a separate island. This paleogeographic setting obviously persisted up to the earliest Pleistocene. During the Late Neogene and Quaternary only tensional fault tectonics are known from Messinia as in the other areas of the southern Peloponnesus. This fault

tectonics, however, was not continuous but interrupted by several phases of stagnation. Such a phase is probably testified by the Early Pleistocene erosional plain (pediment).

The uppermost erosional plain extending locally for many kilometers, either coincides with the distribution of the Pliocene/earliest Pleistocene or is orographically situated a little higher morphologically cutting the latter in many places. In the area north of Mesopotamos the distribution of Pleistocene beds and of the uppermost erosional plains indicate the strongest uplift since the Early Pleistocene (approx.400 m). The levels of the Pliocene base as well as those of the erosional plains dip to the south and submerge below the recent sea level; north of Gargaliani, however, they dip to the north. Therefore the West Messinian Peninsula as a whole gives the impression of a dome-like structural uplift but since the Upper Pliocene the tectonic evolution is characterized by block faulting only.

Old and middle Pleistocene coastal terraces also indicate the regional uplift and subsidence already previously demonstrated by the position of older erosional plains. Continuing fault tectonics are shown by the amounts of vertical displacements of single faults which have thrown these terraces increasing with growing ages of the terraces (e.g. near Gargaliani, where in terraces about 200 m alt. a throw of 30 m is attained while only a few meters are reached in those near to 60 m alt.).

The two youngest Pleistocene terraces (Eu- and Neo-Tyrrhenian) are also affected by faults with throws ranging about 1 m (exc. Isle of Sphakteria). Due to the grouping of smaller faults these amounts grow with increasing distances. Altitudes and distribution of these two terraces also indicate - though less significantly - continuous vertical movements of West Messinia, e.g. Isle of Sphakteria and the surroundings of Marathopolis.

Therefore, yet only with reservations, it can be deduced that post-Tyrrhenian southward tilt - varying in amount - has affected the entire western coast (and, different from the Pliocene/earliest Pleistocene beds, also the region north of Gargaliani).

On the contrary, Eu-Tyrrhenian terraces on the eastern coast of the West Messianian Peninsula indicate constant subsidence to the south.

The depression of the Pamisos river indicates in its southern part an area of subsidence since Pliocene (minimum age). In the northern part only Neogene sediments occur above recent sea level; possibly Miocene beds exist in the southern part of this depression.

East of Kalamata the base of the marine Pliocene has been uplifted to a range of 350 m alt. forming several fault blocks with NE-striking faults. It is often separated from the pre-Neogene basement by N/NW-striking border faults. The uppermost erosional plain (up to 900 m alt.) reach considerably higher than the Neogene sediments (Philippson 1930), lowering down to 400 m in the north.

In the area north of Kalamata there must have been a subsidence during the Pliocene after a preceding Miocene uplift, changing to strong uplift again at the Pliocene/Pleistocene boundary and persisting during the Quaternary. Several fault block units have been moved vertically at different rates and at different times as indicated by the regional distribution and altitudes of Pliocene sediments as well as by the uppermost erosional plain and by the youngest terraces.

Mani, SE-Laconia and Argolis

The neotectonic development of this region is characterized as follows:

Extreme marine ingressions of Neogene age coming from the inland prolongation of gulf trenches show that at that time the coastal configuration of the southern Peloponnesus was essentially more pronounced than it is today.

Especially in the northern parts of the smaller peninsulae post-Neogene tectonic dislocations were at first more or less strong uplift movements accompanied by erosion of the marginal Neogene sediments.

At the Pliocene/Pleistocene boundary probably a marked interruption of displacement activities took place shown by the development of large pediment forms, evidently belonging all to the same base level. In the older parts of the Quaternary dislocation revived and intensified mostly with the same pattern as before: uplifting dominates in the northern parts of the peninsulae, accompanied now by subsidence of the cape regions and the middle parts of the eastern coasts. These tectonic activities lasted till the end of the Pleistocene (or until today) as indicated by the position of the Eu- and Neo-Tyrrhenian coastal sediments.

The general picture is locally more complicated. So, in the SW part of the Mani Peninsula, an older Quaternary uplift changed to subsidence in the younger Pleistocene and Holocene and the Cape Grosso Peninsula has been uplifted 300 m during that time within an area of strong subsidence.

The western part of the SE-Laconian Peninsula shows interruption of uplifting in the Molai graben and this depression is blocked at the nearly coastal strip (Cape Xili) by a small part of uplifted Neogene sediments. The gradual south-dipping of the coastal region east of the Laconian Gulf is interrupted by upwharping in the Neapolis-Elaphonisos area striking SW-NE.

The longlasting subsidence of unknown amount north of Monemvasia to south of Argos is restricted to a small coastal strip while the Parnon mountains have been uplifted several hundred meters since the Pliocene.

The gulf trenches do not in every case separate blocks of different tectonic behaviour, e.g. at both sides of the northern part of the Argolian Gulf Tyrrhenian sediments hold the same altitude today.

Northern Peloponnesus

Two periods of tensional fault tectonics (during the Pliocene and after the Early Pleistocene) probably coincide with volcanic activity in the NW-Cyclades (Fytikas et al. 1976; Lippolt & Schröder not publ.). These periods were interrupted by a phase of tectonic stagnation or even compression (Dufaure et al. 1976).

During the *Pliocene* fault tectonics (often synsedimentary) with individual evolution of fault blocks (subsidence of basins and sedimentary fill up to 1.500 m) started at the Miocene/Pliocene boundary. The first volcanic period lasted about 4,5 - 2,2 m.y.

In the *Early Pleistocene* fluvial conglomerates up to 600 m in thickness (probably Early Pleistocene in age too after few paleontological evidence and radiometric age determinations of unconformably overlain volcanic rocks east of Corinth) have been cut by erosional plains. Indications of compressive tectonics have been described west and east of Corinth.

Middle and *Young Quaternary* coastal terraces (and even the Early Pleistocene beds) have been affected by tensional fault tectonics. At the southern coast of the Gulf of Corinth a flight of coastal terraces are clearly developed for more than 100 km and rise over 450 m. The second volcanic period in the NW-Cyclades lasted since 0.9 m.y. up till now.

Vertical deformation displacements probably reach up to 2.000 m during the entire Quaternary (Gulf of Corinth floor compared to the N-Peloponnesus), near to 100 m during the Late Quaternary (since the Tyrrhenian stage) and amount of a few meters during the Holocene. Variable vertical displacements since historic time reach close to 3 mm/y. The secular rate of vertical displacements varies between 0,2 - 1mm/y.

Tilts varied from 0.3 - 2 m/km for the lower prominent terraces (Holocene till Tyrrhenian stage) to 3 m/km for terraces of Chaudian age and up to 30 m/km for the older ones.

Quaternary tectonics with vertical deformation up to 500 m and even more since the Early Pleistocene can be studied in detail in the uplifted parts of the Neogene/Quaternary basin, especially in the well exposed region near Corinth. Here the area of uplifted coastal terraces south of the main fault at the southern border of the Gulf of Corinth is connected to the eastward area of subsidence.

The disentaglement of tectonically superimposed cycles of coastal terrace sediments affected by synsedimentary fault tectonics is still in progress.

References

BP Co., 1971: Geol. Greece (Inst. Geol. Subsurf. Res.),10:73 p.

Dufaure, J.J., Kadjar, M.-H., Keraudren, B., Mercier, J., Sauvage, J. & Sebrier, M., 1975: C.R. somm. Soc. géol. France: 18-20

Fytikas, M., Giuliani, O., Innocenti, F., Marinelle, G. & Mazzuoli, R., 1976: Tectonophysics, 31:T29-T34.

Kelletat, D., Kowalczyk, G., Schröder, B. & Winter, K.-P., 1976: Z. dtsch. geol. Ges., 127:447-465.

Kowalczyk, G., Richter, D., Risch, H. & Winter, K.-P., 1977: N.Jb. Geol. Paläont. Mh., in press.

Mitzopoulos, M.K., 1940: Prakt. Acad. Athènes, 15:429-436.

Mercier, J., Bousquet, B., Delibasis, N., Drakopoulos, J., Keraudren, B., Lemeille, F. & Sorel, D., 1972: C.R. Acad. Sci. Paris, D 275: 2307-2310.

Philippson, A., 1930: Geogr. Abh., III, 3:9-96.

Streif, H., 1964: Diplomarbeit, Freiburg i.Br.: 145 p., unpubl.

Postalpine Geodynamics of the Peloponnesus

H. Berckhemer and G. Kowalczyk

The Peloponnesus has fully participated in all major tectonic phases of the alpidic development of the Hellenides (Jacobshagen et al. this vol.). Although the allochthony of most of the sedimentary cover complicates the reconstruction of its origin, it can be regarded as a crustal block of continental type, perhaps underlain by pre-Apulian-Adriatic continental basement.

The Peloponnesus is tectonically disconnected from the Greek mainland by a pronounced fault zone along the Gulf of Corinth, possibly with right lateral strike motion.

A detailed investigation of Cenozoic vertical movements has been carried out by Kelletat et al. (1976 and this vol.), Schröder & Kelletat (1976), Kowalczyk et al. (1975).

After the end of the pre-Tortonian orogenesis (Kowalczyk et al. 1977) the southern and southeastern Peloponnesus was uplifted several kilometers, accompanied by erosion down to the autochthonous Plattenkalk series. Contemporary uplift of the northern Peloponnesus, where even the uppermost nappe is preserved, was less extensive. Subsequent subsidence indicated by marine ingression started in the Lower Pliocene and produced significantly higher sedimentation in the North and Northwest of the Peloponnesus than in the South. Disintegration into several fault blocks with individual movement had already been developed. These blocks are bordered by N or NNW and E or ENE striking faults which have controlled the regional distribution of the Neogene deposits.

Uplift of nearly the entire Peloponnesus increased at the end of the Pliocene. Up to 2 km were reached in the northern Peloponnesus. The regional vertical movements have been strongly modified by individual movements of the earlier developed fault blocks but they represent tensional fault tectonics everywhere.

The three southern peninsulae demonstrate the tectonic style of individual fault block movements since the Pliocene. They all show the eastward tilting connected with subsidence toward the south along E striking step faults (fig. 1), this movement can be understood as differential antithetic block rotations about N - S striking axes.

There is a remarkable correlation between high values of free air gravity anomalies (Makris et al. 1973) and zones of relative subsidence (fig. 1), suggesting a nonisostatic mass surplus under the eastern Peloponnesus which was produced by some process in the past and has since diminished. The block rotation as such, however, must be caused by a couple of horizontally directed forces. Available fault plane solutions of shallow shocks (Ritsema 1974; McKenzie 1972) show very little regularity in the crustal body of the Peloponnesus. Shocks occurring off shore in the region of the Hellenic trench, however, show thrust faulting and almost horizontal pressure axis normal to the strike of the arc. Precise hypocenter determinations have been carried out in the area for a period of two years using a specially installed seismograph net (Leydecker et al. this vol.). More than 90% of the hypocenters are confined to the uppermost 30 km of the earth but some shocks, even in the area of the trench, are 40 to 50 km deep. Below the central Peloponnesus east of

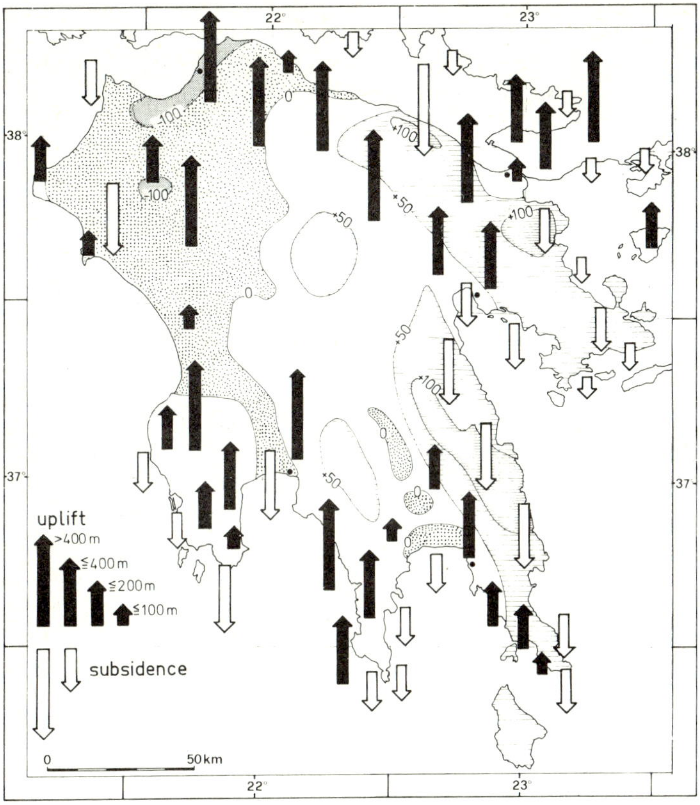

Fig. 1. Quaternary vertical movements (comp. Kelletat et al. this vol.) and isolines of free air gravity anomalies (in mgal) on the Peloponnesus.

20°E shocks of intermediate depth indicate a Benioff zone dipping down under 40° approximately towards the center of the Aegean Sea to a depth of 160 to 180 km (see also Richter & Strobach this vol.). As in other island arcs, the top of the Benioff zone coincides with the Bouguer gravity minimum, that is, with the maximum thickness of the sialic and sedimentary part of the crust.

A refraction seismic profile with shots in the Ionian Sea crossing the Peloponnesus in ENE direction was performed in 1971. A crustal thickness of 20 to 22 km with as much as 10 km of sediments was found in the Ionian Sea southwest of the Peloponnesus (Weigel 1974 and this vol.). Approaching the coast the main refractor of the Ionian crust dips down at an angle of 5 to 10 degrees. Below the western Peloponnesus the crustal structure changes completely and a crust-mantle boundary is found at a depth of 45 to 50 km indicating a possible superposition of two sialic slabs (Makris this vol.). Towards the east the crustal thickness decreases to 30 to 25 km in accordance with gravity.

The Hellenic trench is very well developed south of Crete but becomes irregular and flattens out along the Peloponnesus. Also, the free air gravity minimum associated with the trench diminishes. This may indicate

that the former trench area is now filled with young sediments that were scraped off and accumulated by low angle underthrusting of the Ionian crust. This is directly visible on airgun profiles (e.g. Finetti 1976) and supported by the seismic stress field and by the existence of the "hummocky and rolling landscape" in the Ionian Sea with regular folding aud faulting of the Pliocene-Quaternary sediment cover (Sigl et al.1973).

Quaternary volcanism occurred northeast of the Peloponnesus on the islands of Metana and Aegina. From their K_2O/SiO_2 ratio the volcanic rocks compare well with those from other island arc regions where a Benioff zone lies in a depth of 120 or 150 km respectively (Ninkovich et al. 1973).

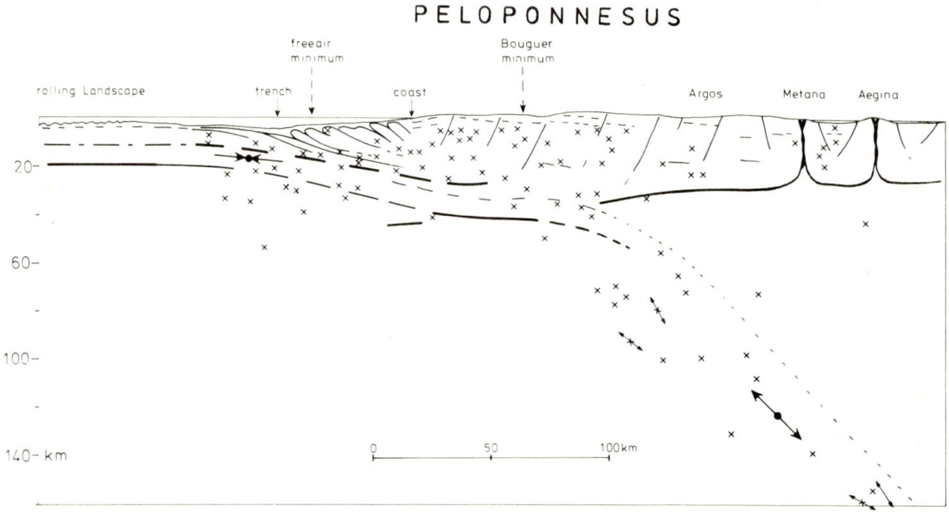

Fig. 2. Geodynamic section through the Peloponnesus area based on data of neotectonics, seismicity, deep seismic soundings and gravity. Heavy lines are boundaries deduced from deep seismic soundings.

We propose a geodynamic model of the Peloponnesus (fig. 2) by drawing a generalized W-E cross section. Low angle underthrusting of the Ionian crust is supported by a gently dipping refractor, the seismic stress field, the thickness of sediments in the trench region, the possible compressional features of the hummocky and rolling landscape and by the upward bending and exposure of the lower Hellenic nappes near the Ionian coast (Jacobshagen et al. this vol.). The antithetic block rotation within the Peloponnesus body is the consequence of underthrusting in the west. Towards the Aegean Sea the tectonic stile changes into predominant extension with normal faulting. In this zone Quaternary volcanism occurred above hypocenters at 120 to 150 km depth. In contrast to other island arcs the nonvolcanic outer arc is dominating here while the volcanic inner arc is relatively poorly developed. High heat flow, a low seismic velocity of the uppermost mantle ($v_p \approx 7.7$ km/s), and high attenuation of seismic wave energy in the region of the Aegean Sea is diagnostic for hot upwelling mantle material. The positive free air gra-

vity anomaly also indicates that isostatic equilibrium has not yet been restored. Upwelling mantle material is expected hydrodynamically from its displacement by the volume of the downward moving lithospheric slab (Berckhemer 1974; Andrews & Sleep 1975). Diapirism can be caused by melting of magma in the dehydration zone of the subducted oceanic crust or by gravity instability within the asthenosphere - lithosphere system. A downgoing slab of the Ionian lithosphere was only deduced from the trend of the Benioff zone but does not conflict with other data. Tension in the slab indicates high density caused perhaps by a high pressure phase transformation. The well established discontinuity at 45 km depth is assumed to correspond to the crust-mantle boundary of the subducted Ionian lithosphere. The distance from the trench to the Bouguer minimum and volcanic zone are similar in some of the Indonesian and Pacific island arcs.

References

Andrews, D.J. & Sleep, N.H., 1974: Geophys.J.Roy.astr.Soc.,38:237-251.
Finetti, J., 1976: Boll. geofiz. teor. appl., 19:31-65.
Kelletat, D., Kowalczyk, G., Schröder, B. & Winter, K.-P., 1976: Z. dtsch. geol. Ges., 127:91-109.
Kowalczyk, G., Winter, J. & Winter, K.-P., 1975: Bull. geol. Soc. Greece, 12:40-51.
Kowalczyk, G., Richter, D., Risch, H. & Winter, K.-P., 1977: N.Jb. Geol. Paläont. Mh.: 549-564.
Makris, J., Mavridis, L-N., Menzel, H., Stavrou, A. & Veis, G., 1973: Z. Geophys., 39:929-936.
McKenzie, D., 1972: Geophys. J. Roy. astr. Soc., 30:109-185.
Ninkovich, D. & Hays, D., 1972: Earth Planet.Sci.Letters, 16:331-345.
Ritsema, A.R., 1974: Netherl.Meteorol.Inst.De Bilt Sci.Report, 74,4.
Schröder, B. & Kelletat, D., 1976: N. Jb. Geol. Paläont. Mh.:174-186.
Sigl, W., Hinz, K. & Garde, S., 1973: Meteor Forsch.Ergeb.,14:51-54.
Weigel, W., 1974: J. Geophys., 40:137-140.

Crustal Fracture Pattern of the Aegean Region

P. Kronberg and R. Günther

LANDSAT imagery covering the Aegean region was mapped for surface traces of steep fractures. First, the individual LANDSAT scenes at a scale of 1 : 1 000 000 were mapped for fracture traces averaging in length from 5 to 20 km. Secondly, a LANDSAT mosaic covering the whole Aegean region was compiled and evaluated for linear geologic features up to several hundred kilometers in length. Finally, the bathymetric map of the Aegean Sea (Pfannenstiel 1960) was evaluated for indications of tectonic fea-

tures. The analysis and correlation of the above data with field observation and published data have not yet been completed.

The regional pattern of crustal fracturing (fig. 1) is characterized by predominating trends on the Greek and Turkish mainlands as well as within the area of the Aegean Sea. The ENE-striking fracture systems (with indications of equidistant spacing) e.g. can be traced from Turkey through the Aegean Sea into the Greek mainland and into the Ionian Sea. The NW/SE-, NNW/SSE- and SW/NE-, SSW/NNE-striking fracture systems occur throughout the region with similar continuity.

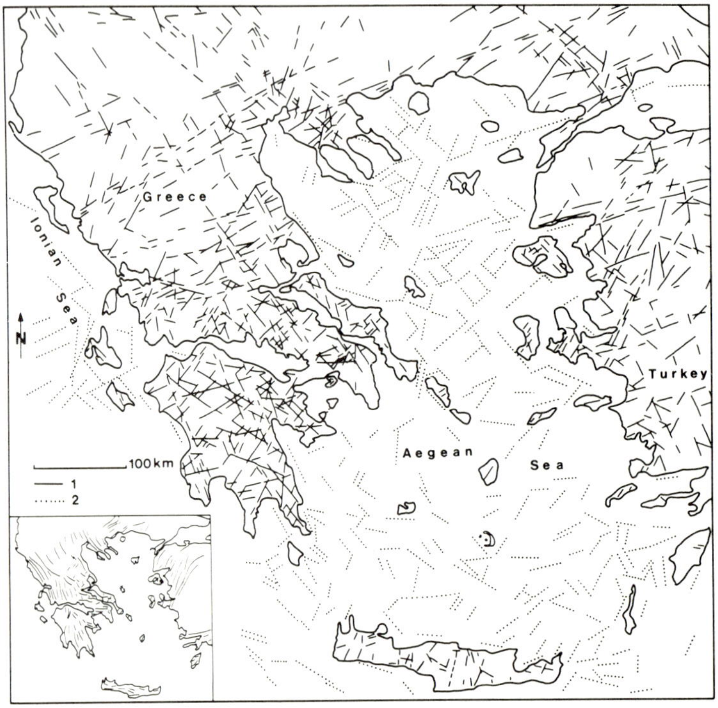

Fig. 1. The crustal fracture pattern of the Aegean region as mapped on a LANDSAT-mosaic and as deduced from bathymetric data. 1 = major linear structures on the LANDSAT-mosaic; 2 = structural trends of the seafloor morphology. The small map shows major structural trends of folding (after Jacobshagen 1976).

The individual fracture systems are very persistent in their orientation and continuation. They cut indifferently through the various Palaeozoic, Mesozoic and younger geological units or zones of Greece and Turkey of greatly different age, geological development, and style of deformation. This may indicate that the origin of the regional crustal fracture pattern of the Aegean region must be independent of the geological units in Greece and Turkey and their movements and strike directions (see also small map in fig. 1).

The regional fracture pattern (fig. 1) appears to be part of the late Tertiary to Recent intense block faulting, uplift, and subsidence within

the Aegean region. It may either be related to the plate-tectonic opening of the Aegean Sea basin, or to a rejuvenation of old zones of weakness within the basement.

Most of the major trends of fracturing shown in fig. 1 are known to have controlled the palaeogeographic and structural development of the regional units throughout geological time. For instance, the sigmoidal bending of the fold axes within the palaeozoic Serbo-Macedonian Massif from the general NW/SE-trend into an ENE-direction (Kockel & Walther 1968) may be influenced by deepseated structures. NW/SE-striking fractures controlled the development of the Hellenide zones during the Mesozoic and Tertiary, predetermining later ridges and basin areas, facies zones, thrusts, and nappe fronts.

Synsedimentary blockfaulting controlled facies changes and rates of sedimentation in the Neogene basins of the Aegean region (Meulenkamp & Zachariasse 1973; Fairbridge 1973; Jongsma 1975). Andesites of Miocene/Pliocene age extruded along NNE-trending fractures in the Karaburun-peninsula (Brinkmann et al. 1972) and the young Quaternary eruption of the Santorin volcano took place at the intersection of NW/SE- and NE/SW-striking fracture systems. Along the coastal areas of the Peloponnesus vertical movements along NW/SE-striking faults created offsets of up to 500 meters in subrecent to recent times (Kelletat, pers. comm.).

LANDSAT data, field observations, and published data for selected areas of Greece indicate that the regional fracture pattern of the Aegean region is controlled by old (Precambrian ?) lines of crustal weakness, rejuvenated repeatedly throughout geological history. The geomechanical function of the fracture systems may have changed in space and time as the regional stress fields varied.

This interpretation coincides with the conclusions drawn from LANDSAT-imagery covering Central Europe, parts of East Africa, and western Arabia

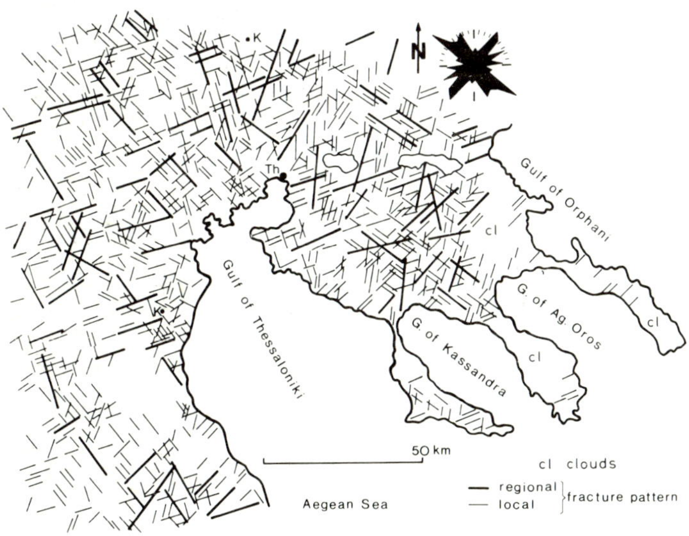

Fig. 2. A comparison of the orientation of regional and local fracture systems as mapped on a LANDSAT-mosaic and on individual LANDSAT images.

(vide Kronberg, Günther & Schönfeld 1976; Kronberg 1977; Günther 1977; Kronberg & Günther 1977).

Synoptical LANDSAT-imagery also provides information on the relations between regional and local fracture patterns. Local fracturing within blocks outlined by higher-order fractures is also controlled by the pattern of crustal fracturing as shown by a comparison between the maxima of the rose diagram included in fig. 2 and the predominating trends of fracturing in the map area. The diagram represents the trends of late- to post-Tertiary faulting in the area around Lake Volvi (Kockel & Walther 1968), which coincide with the predominating trends of regional and local fracturing within the entire region (fig. 2).

This area is crossed by several internal units of the Hellenide fold belt: The Palaeozoic Serbo-Macedonian massif, the Vardar zone with Palaeozoic and Mesozoic sediments and ultrabasics, the Pelagonian Zone with metamorphosed and non-metamorphic sequences of Palaeozoic and Mesozoic age, and depressions underlain by Tertiary and Quaternary deposits. Even though these zones have undergone several periods of folding and thrust faulting during the Mesozoic and early Tertiary, and block faulting in late to post-Tertiary, the same fracture patterns occur in all three units as well as in the areas of Tertiary and Quaternary sediments. This situation demonstrates that old fracture patterns persist through time.

If fracturing within the Aegean region is controlled by old lines of crustal weakness, one should expect that the seismotectonic pattern in the Aegean region should also be related to the old pattern of crustal fracturing. We therefore tried to correlate the position and strike directions of fracture traces mapped on LANDSAT imagery with the position of epicenters and the strike direction of fault planes of subrecent and recent earthquakes as calculated by Ritsema (1974).

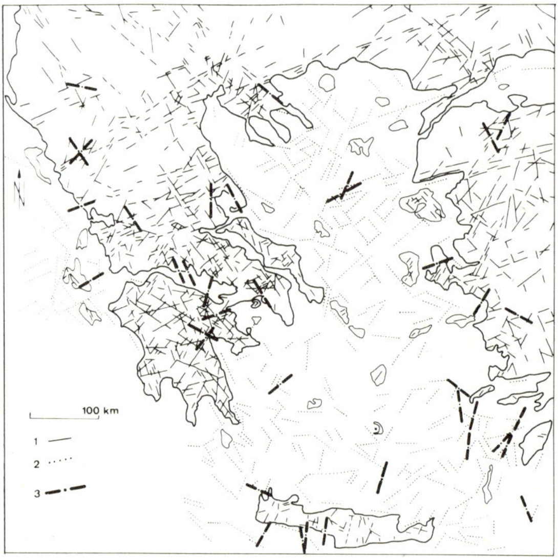

Fig. 3. The strike directions of seismic fault planes and the strike directions of fracture traces as mapped on the LANDSAT-mosaic. 1 = fotolineation; 2 = trend of seafloor morphology; 3 = seismic epicenter and strike of the fault plane.

In fig. 3 the orientation of the majority of the fault planes calculated by Ritsema corresponds evidently to the strike direction of the fracture systems of the various azimuths mapped on the LANDSAT mosaic.

Furthermore, the orientation of the fault planes appears to be related to the preferred trends of regional fracturing: WSW/ENE, NNW/SSE, SW/NE, WNW/ESE, NW/SE. These fracture systems occur through most of the Aegean region. This leads to the conclusion that the recent earthquake activity of the Aegean region follows the old regional crustal fracture pattern.

Data from the Peloponnesus present similar results. We correlated the position of earthquakehypocenters (Leydecker 1975), with the position and strike direction of larger fracture systems mapped on the LANDSAT mosaic. Only hypocenters with a positional error of less than 5 kilometers were selected for the correlation. 16 earthquakes range in the depth between 0-10 km, 30 from 10-30 km and 2 are deeper than 30 km.

Of these hypocenters, 41 coincide either with a fracture zone, its projection, the intersection of fracture zones or lie within 5 km of a mapped fracture zone. Thus the large fracture zones represent deep vertical zones of crustal fracturing which at least in part predetermine the earthquake distribution.

We conclude that old deep seated zones of crustal weakness are still active. In the Peloponnesus the earthquakes accompany recent blockfaulting (Kelletat et al. 1976) and perhaps tensional forces behind a subduction zone (vide Leydecker 1975; Makris 1975; Berckhemer & Kowalczyk this vol.). In any case, the young fracturing of our region does reflect the old pattern of crustal weakness.

References

Brinkmann et al., 1972: Geologica et Palaeontologica, 6:139-150.

Fairbridge, Rh.W., 1972: in: Stanley: The Mediterranean Sea:99-113.

Günther, R., 1977: Geotekt. Forsch.: in print.

Jongsma, D., 1975: Thesis, Cambridge.

Kelletat, D. et al., 1976: Geodynamics of the Mediterranean Region.

Kockel, F. & Walther, H.J., 1967:Bull.Geol.Soc. Greece, 7:1-16.

Kronberg, P. et al., 1976: Proceed. 2nd Intern.Confer. "New Basement Tectonics".

Kronberg, P., 1977: Geotekt. Forsch. 53: 7-41.

Kronberg, P. & Günther, R., 1977: Geotekt. Forsch. 53: 88-106.

Leydecker, G., 1975: Ber.Inst. Meteor.Geophys. Univ. Frankfurt, 29.

Makris, J., 1975: Geophys. 41:441-443.

Mercier, J. & Vergely, P., 1972: Z. dtsch. Geol. Ges., 123:469-489.

Meulenkamp, J.E. & Zachariasse, W.J., 1973:Messinian events in the Mediterranean.

Pfannenstiel, M., 1960: Bull. Inst. Océanogr. Monaco, 57.

Ritsema, A.R., 1974:Roy.Netherlands Meteor. Inst.De Bild.Sci.Rep.,74,4.

Some Aspects of the Evolution of Marginal Seas Deduced from Observations in the Aegean Region

H. Berckhemer

Physiographically the Aegean is a marginal sea bordering or even penetrating the European continent and being bounded toward the eastern Mediterranean sea by the Hellenic island arc. The question is to what degree this structure resembles Pacific, Indonesian or Central American marginal sea-island arc systems.

Referring for exemplification to the Peloponnesus area (Berckhemer & Kowalczyk this vol.) the following features typical for marginal sea-island arc structure are found in the Aegean region:

1. Nonvolcanic frontal arc (Hellenic island arc).
2. Deep-sea trench with a pronounced negative free-air gravity anomaly.
3. High seismicity with foci at normal and intermediate depth.
4. Benioff zone, dipping at 35-45° toward the central Aegean, at least in segments well recognizable.
5. Inner island arc with quaternary volcanism (Aegina, Methana, Milos, Santorin, Nysiras, Antoparos, Kos) above hypocenters at 120-150 km of depth. The K_2O/SiO_2 ratio is typical for andesitic volcanism related to the corresponding hypocenter depth (Ninkovich & Hays 1972; Puchelt this vol.).
6. Distance between trench and volcanic arc is about 200 km.
7. Tensional tectonics in the back-arc region (Jacobshagen et al.this vol.).
8. High heat flow, up to 2,7 HFU, in the central Aegean sea (Jongsma 1974).

Differences compared with many other island arcs are:

1. Small dimensions. Radius of curvature of the frontal arc only 300 km. Its development began, however, probably not before late Miocene.
2. Predominantly sialic crust of 18 to 30 km thickness and a somewhat reduced submoho velocity v_p = 7.7 - 7.9 km/s in the back arc area (Makris this vol.). No true sea-floor spreading.
3. Maximum focal depth is only 200 km with some scattered intermediate depth earthquakes outside the Benioff zone as defined above.
4. Nonvolcanic frontal arc is dominating compared with the poorly developed volcanic arc.
5. Shallow water depth and sialic islands behind the volcanic arc.
6. As a unique case: alpidic belts can be traced all through the Aegean from the Dinarides to the Taurides (Jacobshagen this vol.).
7. No typical oceanic crust (thickness approximately 20 km including some 10 km of Neogene) and a huge sedimentary chain (East Mediterranean ridge) in front of the Hellenic trench. Duplication of the trench system in its eastern part.

These differences seem to be more of a gradual than of a principal character. According to Karig's classification (Karig 1971) the Aegean region can be considered a young, small, inactive marginal basin with high heat flow in a somewhat hampered state of development.

If this is true, essentially the same physical and chemical processes should be responsible for the evolution of this island arc than for others and no ad hoc mechanism is needed. On the other hand, conclusions

drawn from observations in the Aegean area might be applicable also in other marginal seas.

The intimate connection of the alpidic orogeny and the succeeding island arc development in the same area is emphasized. The present frontal arc as well as the back arc islands were fully involved in the alpidic orogeny with phases of compression, obduction, overthrusting, flysch deposition, outward transport of nappes perhaps partly by gravity sliding. During the high-orogenic phase in Lower Miocene (Burdigalian) the median Aegean crystalline belt was rapidly rising and subsequently eroded 15 - 20 km deep (Dürr et al. this vol.). The existence of an Aegean mountain massif similar to the present Hellenides or Dianrides is postulated for that time. In middle to late Miocene already destructive tectonics commenced with intense block movements, subsidence, and simultaneous extension in the inner zone as well as raising and compression in the external zone. The Aegean archipelago proper developed in the late Pliocene and was tectonically reactivated when 3 m.y. ago the island-arc volcanism commenced (Kowalczyk this vol.).

An already existing mountain range disintegrated and was destroyed. Why just here and not somewhere else in the alpidic belt? The answer must be found in the fact that only this part of the Alpine chain is bordered at one side by a pseudo-oceanic crust, perhaps the remainder of a Tethyan ocean basin.

It can be shown (e.g. Artyushkov 1974) that a mountain range, even in isostatic equilibrium, exerts strong nonhydrostatic horizontal forces of the order of the weight of its topographic relief to the adjacent thinner crust. Stresses in the order of 0.3 - 1 kbar are to be expected. The mountain will flow apart if no sufficient counter-support is provided by adjacent continental blocks or if its internal strength is insufficient. Low strength, however, is expected inside a young mountain body because it is still under elevated temperature.

This gravity controlled disintegration is regarded (Berckhemer 1977) as the cause for the development of the island arc and its marginal sea. First the eastern Mediterranean crust was passively overridden. Subsequently, the suppressed lithosphere became actively pulling due to a density increase by phase transformation. The rate and direction of subduction is governed by the relative motion of Africa, Europe, and Anatolia and by the expansion of the Aegean sea. This expansion, in turn, is reinforced by upwelling mantle material displaced by the intruding downgoing lithosphere. In case of a preexisting gravitational instability in the mantle, the distribution in the lithosphere might cause an asthenospheric diapir to rise which could contribute as a third agent to the expansion. In principle this process may continue until equilibrium is achieved with the crust of the adjacent oceanic area. Because of the anomalously thick crust in the eastern Mediterranean the expansion and crustal thinning in the Aegean is limited and no true seafloor spreading has occurred.

By comparing the situation in the Aegean with true marginal seas such as the Andaman Sea, the Bering Sea, or the Eastern Caribbean it becomes evident that in all these cases the frontal island arc, even if only preserved in rudiments, merges continuously into continental orogenic belts of Tertiary age. Therefore the evolutionary model outlined above might be applicable in modified form also to other marginal seas.

References

Artyushkov, E.V., 1974: Journ. Geophys. Res. 79:741-752.
Berckhemer, H., 1977: In the Structural History of the Mediterranean Basins. Edit. B. Biju-Duval, L. Montadert, Edition Technip, Paris:303-313.
Jongsma, D., 1974: Geophys. J.R. astr. Soc., 37:337-346.
Karig, D.E., 1971: Journ. Geophys.Res. 76:2541-2561.
Ninkovich, D. & Hays, D., 1972: Earth and Planet. Sci. Letters, 16:331-345.

The Late Miocene Evaporitic Sequence on the Ionian Islands and Geodynamic Implications for the Eastern Mediterranean Sea

K. Braune, F. Fabricius and K.O. Heimann

Investigations of Upper Miocene sequences in Western Greece aimed at getting information about movements of the Eastern Mediterranean basin during Late Miocene and Early Pliocene. They constitute part of the German co-operation in the DSDP of JOIDES in the Mediterranean, Leg 42 A of R.V. Glomar Challenger.

Fig. 1. The Ionian Islands.

Our interest concentrated on the following questions:

1. Stratigraphical position and paleogeographical framework of the Ionian evaporite-bearing sequence and its correlation with analogous formations in the Mediterranean;
2. Sedimentological rhythms in evaporitic sequences;
3. Depth and conditions of deposition of these formation in a Messinian "Mediterranean".

The investigations are based on the following localities: Zakynthos,

Cephalonia, Corfu, Mathraki Island, North-western coast of the Greek mainland (fig. 1). Field trips in Sicily and Spain allowed to compare lithological and tectonic aspects. The investigations comprised 1. sedimentological studies (Thin sections, X-ray, carbonate- and C org.-analysis); 2. paleontological studies (pollens: L. Benda/Hannover, E.Schulz/Würzburg; nannofossils: C. Müller/Frankfurt/Main); diatoms: Schrader/Kiel; vegetable remains: W.Jung/München; foraminifera-taxonomy: J.Just/München.

The previously held opinion that during Late Miocene the area of the Ionian Islands might have been a local marginal basin has to be revised: foraminiferal assemblages from marly beds directly underlying evaporites as well as those covering them, indicate open marine conditions, with an impoverishment, however, in size and number of species towards the base of the evaporites. Erosional phenomena are lacking. Hence, ephemeral drying-out excepted, we can exclude one or more periods of long-time subaerial exposure.

It is certain that all of the following features present in the Ionian outcrops can be interpreted as part of the marginal zone of a large Messinian "Mediterranean" basin: - reef-like structures of shallow water carbonates with algae, corals, boring organisms and serpulae; lumachellic beds with neritic molluscs and gastropods; layers with well preserved vegetable remains of marine but more often terrestrial origin; intercalated conglomerates and calcarenites, the latter showing ripple-marks and cross- bedding; important admixtures of reworked Eocene to Middle Miocene marine fossils.

A paleobathymetrical attempt was made to compare Messinian foraminiferal assemblages with recent ones from the shelf off Cephalonia. This comparison suggests a water-depth of not more than 60-180 m. Sedimentary and paleontological features prove a continuous shoaling of the area since Helvetian time (Braune 1971); a tendency to deposition in greater water-depths is not observable until the end of the Early Pliocene.

The problem of exact stratigraphical boundaries, especially at the top of the evaporitic sequence, is extremely delicate. Due to hypersaline conditions the fully marine fauna deteriorated; in various outcrops the uppermost sediments of the Messinian are devoid of fossils or contain a brackish macrofauna. Hence, efforts to establish faunal correlations must be considered with certain reservations. In most places the boundary between the Messinian (defined by the evaporites as "index facies") and the "Early Pliocene" (often represented only by littoral sediments) is a facies boundary. There are arguments for considering this boundary as diachronous throughout the Mediterranean.

In order to correlate the rhythmic evaporite facies on the Ionian Islands with those in Sicily, Spain and North Africa and all of these with the deep sea cores, we have to assume that the former "Mediterranean" was not that deep as today. Evaporation of a water column in a desiccating deep-sea-basin would not have allowed (as present however) sedimentation of this series and that thick in marginal basins, rather quickly cut-off from the main, lowering water-body. According to the hypothesis of a dee-sea-basin (4000 - 5000 m deep) all marginal areas of the Mediterranean with Messinian-age rhythmic sedimentation should have been uplifted to their present position during post-Messinian time. This is contradicted by recent subsidence of the present basin centers of the Ionian Sea and the Eastern Mediterranean. Evidence of this subsidence includes step faults accompanying the margins of these basins, west of

the Greek shelf and east of the Sicily-Tunesia-Platform, shifting of the Nile delta towards the south, the collapse of the Aegean area and volcanic as well as seismic activity along the margins of the young basins (North Africa, Spain).

Critical analysis of all the Late Miocene evaporite-bearing outcrops known to us on land or from deep-sea cores leads us to conclude that the Mediterranean depression during the Late Meocene time cannot have been as deep as today. We concede 500-1000 m as a maximum in the centers, 0-200 m in marginal areas. This shallow basin hypothesis implies intra-Messinian subsidence of 1000-1500 m providing the necessary space for the sediments of this interval. The possibility of tectonic movements for the formation of rhythmic sediments cannot be excluded, the character of a relatively flat salt pan, however, persisted.

The main differences in altitude between Messinian land- and sea-sites thus have to be attributed to Pliocene-Quaternary tectonics.

References

Braune, K., 1973: Senckenbergiana marit., 5:99-133.
- , 1975: Rapp. Comm.int. Mer Médit., 23,4a:151-152.
Braune, K., Fabricius, F. & Heimann, K.O., 1973: in: Drogger: Messinian Events in the Mediterranean:192-201.
Braune, K., & Heimann, K.O., 1973: Rapp. Procès-Verb.Réunions CIESM, 22,2a:128-129.
- , 1973: Bull. Geol. Soc. Greece, 10,1:25-30.
Braune, K., Heimann, K.O. & Fabricius, F., 1972: Z.dtsch. Geol. Ges., 123:555-558.
Heimann, K.O., 1975: Rapp. Comm. int. Mer Médit., 23,4a:147.
- , 1977: Diss. Techn. Univ. München, 158 p.
Heimann, K.O., Jung, W. & Braune, K., 1975: Mitt. Bayer. Staatssamm. Paläont. hist. Geol., 15:169-177.
Heimann, K.O. & Jung, W., 1976: Mitt. Bayer. Staatssamml. Paläont. hist. Geol. 16:105-111.
Heimann, K.O. & Mascle, G.H., 1974: C.R.Acad.Sci.Paris,279,D:1987-1990.
Hsü, K.J., 1973: Messinian Events in the Mediterranean:60-67.
Hsü, K.J., Cita, M.B. & Ryan, W.B.F., 1974: Initial Reports of the DSDP, 13,2:1203-1232.
Hsü, K.J., Ryan, W.B.F. & Cita, M.B., 1973: Nature, 242:240-244.
Mascle, G.H. & Heimann, K.O., 1975: 1.Messinian Seminar, Erice: in press.
Ryan, W.B.F., Hsü, K.J. et al., 1970: Geotimes:12-15.
Schreiber, B.C., 1974: Ph. D. Thesis, Rensselaer, Polytechn. Inst. N.Y.
Sonnenfeld, P., 1974: Geol. Rdsch., 63,3:1133-1172.
- , 1975: Journ. Geol., 83,3:287-311.

Result of Reflection Seismic Surveys in the Southern Aegean Sea

D. Jongsma, G. Wissmann, K. Hinz and S. Gardé

Reflection seismic surveys were carried out in the Southern Aegean Sea by R.V. METEOR (cruise no. 33/1974) and by R.R.S. Shackleton in 1972 and 1974 (fig. 1). Altogether 2600 n.m. were measured using conventional air gun single trace profiling technique. The Southern Aegean Sea consists of a series of elongated depressions which in general follow the trend of the volcanic arc. The depth of the basins increases eastward with a maximum of 2525 meters attained west of Karpathos.

One of several unconformities is traceable on seismic records everywhere in the Cretan trough and is referred to as event A. It generally marks the transition between unconsolidated sediments and indurated rocks. The erosional and/or non-depositional surface represented by the event A has been dated by extrapolation of land geology as ranging from Tortonian (7.5 m.y.) to Lower Pliocene. In the deep basins event A is interpreted as the contact between Messinian evaporites and Pliocene marine sediments. DSDP site 378 which was proposed by us identified event A as the equivalent of horizon "M", e.g. top of the Messinian evaporite formation. The places where event A exhibits a similar seismic picture as at site 378 are shown by dots along the profiles in the track chart. In other parts of the Cretan Trough, event A is probably underlain by Palaeozoic or Mesozoic rocks. Near Karpathos we think that the event A can be correlated with an erosional unconformity between Oligocene and Upper Pliocene sediments.

A contour map of the depth to event A shows a similar configuration to that of the sea bottom but in detail this surface is more structured. This clearly points to a tectonic origin of the Southern Aegean Sea after event A time.

Thick sediments of up to 1.5 km have been deposited in local basins. Strong vertical movements on normal faults were responsible for the formation of the Cretan Trough and subsequently controlled the sedimentation. Sediments within the Cretan Trough have been deformed by Upper Pliocene/Pleistocene faulting and folding. Quaternary intrusions of volcanic? material are sporadically observed over the whole Cretan Trough but do not suggest active sea floor spreading. Extensional processes are seen to be the dominant feature in the tectonics of this area since ca. 10 m.y. It is not possible from the reflection seismic data to decide whether the origin of the Southern Aegean Sea is due to subduction or to rifting caused by mantle diapirism.

References

Jongsma, D., Wissmann, G., Hinz, K. & Gardé, S., 1977: "Meteor" Forsch. Ergeb., C. 27:3-31.

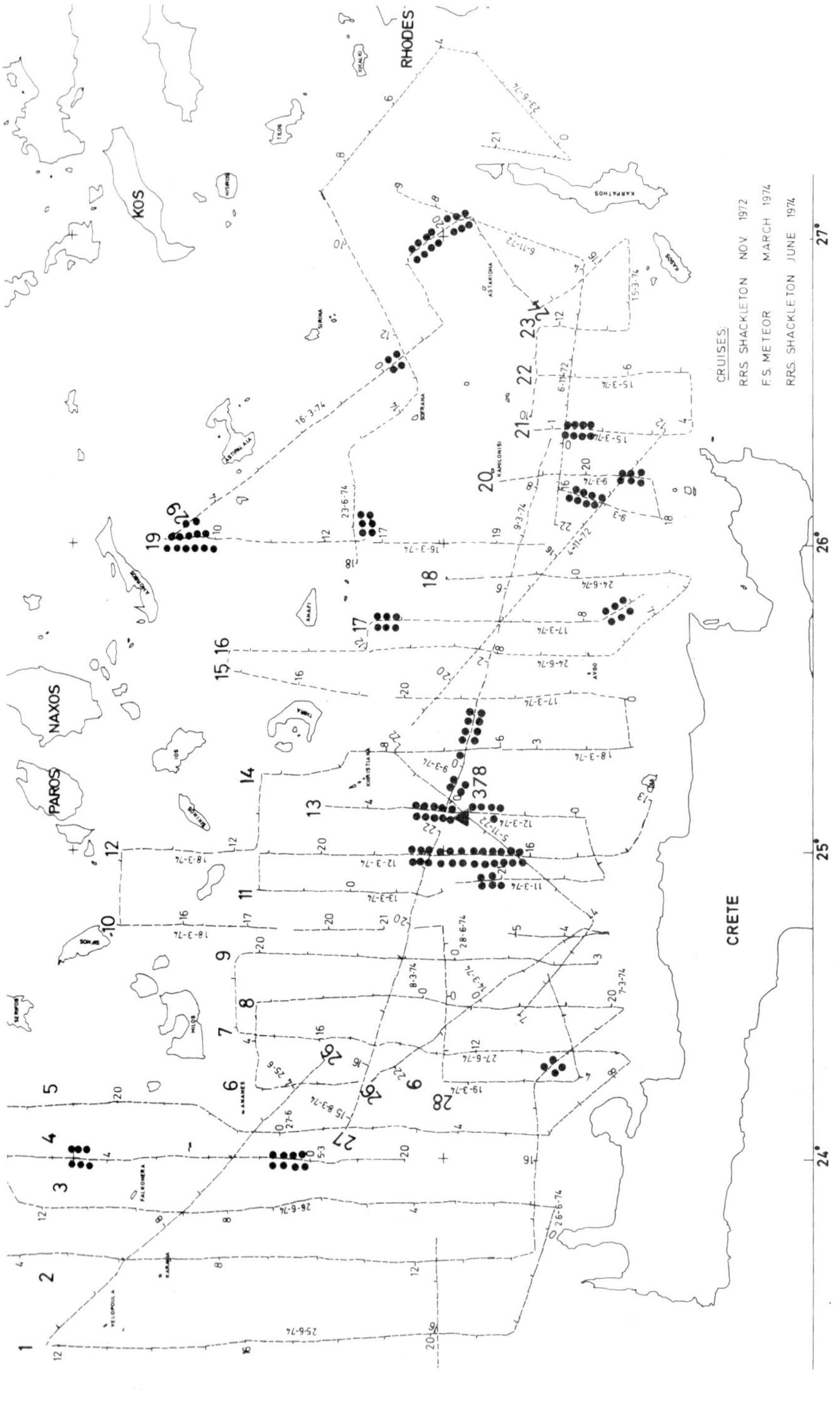

Fig. 1. Seismic reflection coverage. Profiles are numbered at the top; date and time with hour marks. Dots show seismic indications for horizon M (Messinian evaporites).

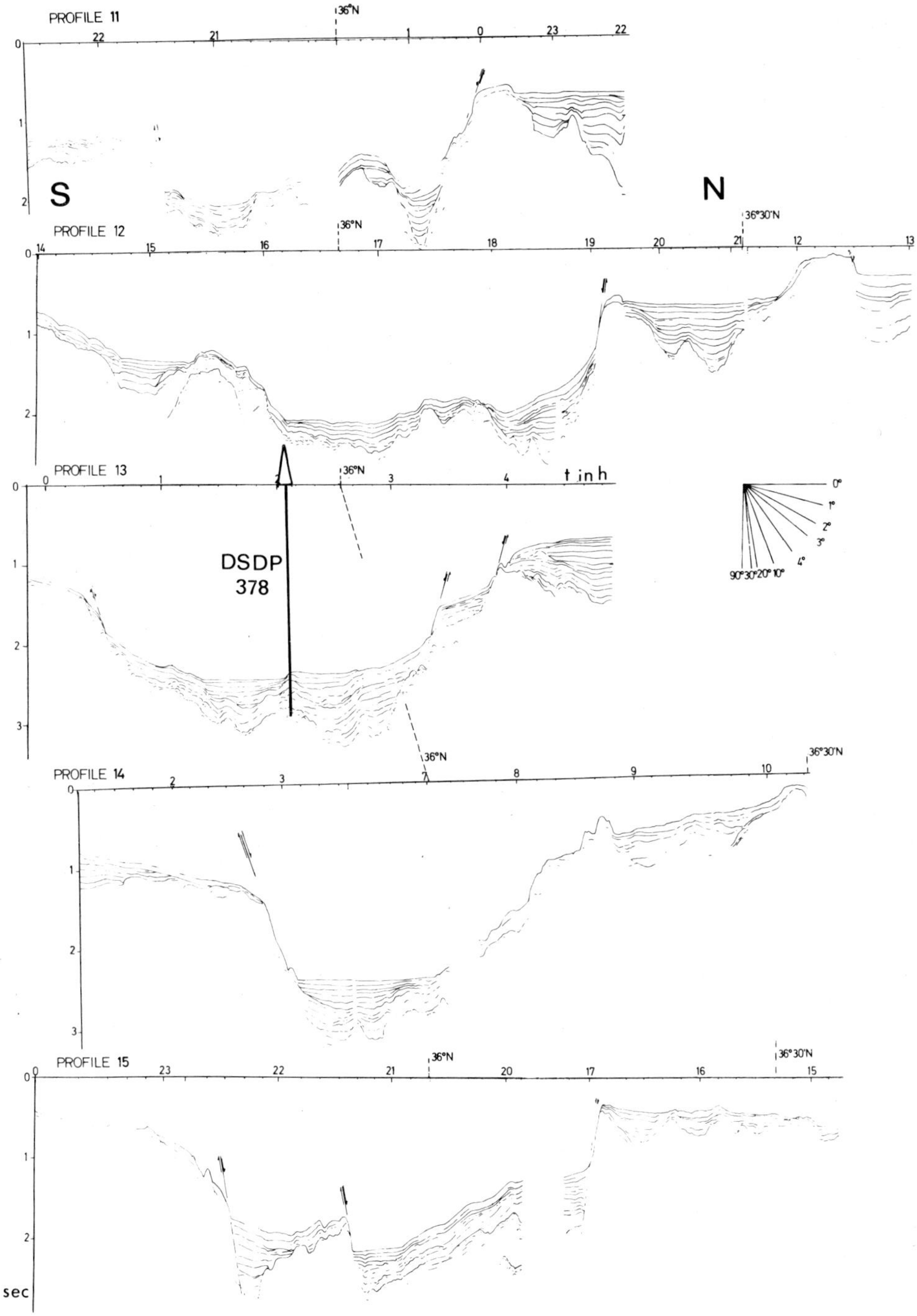

Fig. 2. Linedrawings of seismic reflection profiles in the vicinity of DSDP site 378. The horizontal scale shows hour marks, as in fig. 1, the vertical scale is the two-way travel time in seconds. The vertical exaggeration is roughly 1 : 12.

Control of Vertical Movements by a High-Precision Gravity Network in the Eastern Mediterranean

W. Torge

In the eastern Mediterranean, plate movements and related orogeny produce complex recent vertical displacements. Because the method of levelling is not applicable for the detection of vertical movements in large water areas, only repeated gravity measurements can give significant information about mass displacements in vertical direction. The accuracy of ± 0.015 mGal, obtainable even for large gravity differences and point distances (Torge 1971), will permit to control variations in elevation of ± 0.05 m, assuming a free-air relationship between gravity and elevation. This assumption is not valid everywhere, but current investigations for instance will probably improve our understanding of the correlation between elevation and gravity (e.g. Schleusener & Torge 1971).

To detect possible vertical movements and associated gravity variations in the Aegean plate region, a high-precision gravity network was established in 1974 by the Institute of Theoretical Geodesy, Technical University Hannover, in cooperation with the Department of Geodetic Astronomy, University of Thessaloniki (Torge et al. 1976). The net consists of 46 stations, covering the Greek mainland, the Aegean islands, and the island of Crete. The station sites were selected so that local mass displacements and site destructions are unlikely. Generally the observation points have been put in churches situated in probably stable regions.

The measurements have been carried out in spring 1974, using four LaCoste-Romberg gravity meters, the total gravity difference in the network being approximately 670 mGal. Altogether 348 gravity differences have been observed for the determination of 46 gravity values, 3 calibration factors (referring to a "reference" gravity meter) and 4 linear drift parameters. The accuracy of one observed Δg depends on the instrument and on the transportation (± 0.01 ... 0.04 mGal). From a free net adjustment (e.g. Mittermayer 1971), gravity values relative to a mean gravity level have been determined to ± 0.01 (0.004 ... 0.015)mGal. As the maximum vertical movements will probably not exceed the order of 1 cm/a, a reoccupation of the stations is not planned until 1984.

References

Schleusener, A. & Torge, W., 1971: Z. f. Geophysik, 37:679-701.
Torge, W., 1971: Deutsche Geod. Komm., B 183.
Torge, W., Mavridis, L.N., Drewes, H. & Arabelos, D., 1976: Zeitschr. f. Verm.wesen, 101:213-220.
Mittermayer, E., 1971: Zeitschr. f. Verm.wesen, 96:401-410.

Part 6

Synthesis

Structure and Geodynamic Evolution of the Aegean Region

V. Jacobshagen, St. Dürr, F. Kockel, K.-O. Kopp and G. Kowalczyk,
with Contributions of H. Berckhemer and D. Büttner

Contents

1. Introduction
(V.J., K.O.K.)

In this paper, we try to synthesize new facts and hypotheses concerning the structure and the evolution of the Hellenides in Mesozoic and Cenozoic times.

The classical structural scheme of the Hellenides (Philippson 1898 etc.; Renz 1940, 1955; Brunn 1956, 1960; Aubouin 1959; Aubouin et al. 1963) was a geosynclinal model with several basins and ridges based on the definition of isopic zones. According to the authors mentioned, in the course of several orogenic phases a pile of nappes originated in the internal zones of predominantly eugeosynclinal character and was thrust onto the autochthonous and miogeosynclinal external zones. Deformation, deposition of flysch at the front and of molasse on the back-side migrated outward.

More recently, Benioff zones (Caputo et al. 1970; Galanopoulos 1973; Papazachos 1973 etc.) and plate boundaries (Ritsema 1969; McKenzie 1970)

were drawn. Dercourt (1970), Bernoulli & Laubscher (1972) and Zimmermann (1972) attributed allochthonous ophiolite sheets of the internal zones to a Lower Cretaceous obduction of an oceanic plate. Problems of palaeo-oceanic troughs and their importance for the tectonic development of the Hellenides were discussed e.g. by Hynes et al. (1972), Aubouin (1973), Smith & Moores (1974), Mercier et al. (1975) and Jacobshagen et al. (1976a).

The German geodynamics program lead to a new structural scheme of the Hellenides which is briefly outlined below. Referring to this scheme, the major features of the geotectonic evolution will be drawn out under the scope of plate tectonics in a further chapter. A final discussion deals with geotectonic models.

2. A New Structural Pattern of the Hellenides
(St. D., V.J., F.K., K.O.K.)

Basal facts and ideas for the new scheme are the following:

- A new stacking order of nappes in the Peloponnesos (Jacobshagen et al. 1976b and this vol.).
- A more detailed knowledge on the geology of Crete (vide Kopp this vol.).
- A structural division of the Median Crystalline Belt of the Hellenides i.e. the metamorphic complexes of the Thessalian, the Attico-Cyclades and the Menderes "massifs" (Dürr 1975; Dürr et al. this vol.).
- The definition of the Pelagonian platform by Kauffmann (1976 a,b) which comprises the Pelagonian and sub-Pelagonian sedimentary zones and, moreover, also the Parnassus zone (Jacobshagen et al. 1976a).
- The separation of ophiolite complexes from the subjacent units (Bernoulli & Laubscher 1972, Zimmermann 1972; Jacobshagen et al. 1976a).
- The separation of a Cimmerian circum-Rhodope mountain belt from the inner Vardar zone (Kauffmann et al. 1976).
- Systematic observation and interpretation of high p/T metamorphism in the Hellenides (Schubert & Seidel 1972; Seidel & Okrusch this vol.).
- Re-investigation and geotectonic interpretation of the Hellenide flyschs (Richter 1976a; Richter & Mariolakos this vol.).

From these results the sequence of nappes shown in tab. 1 was deduced for the Hellenides outward the Vardar zone. Definitions of these units will be given below.

All the nappes are defined mainly on topological facts and delimitated by their basal thrust planes. They are, moreover, characterized by stratigraphical or petrological features, but only some of these units coincide exactly with the mentioned isopic zones. High pressure metamorphics with glaucophane parageneses are found at the basal thrust planes of all groups of nappes indicating former subductions and palinspastic gaps (lower West Hellenic Phyllite series, a series called herein Blueschist units, parts of the Inner Hellenic outliers). Within the Central Hellenic nappes, the sedimentary sequences are, furthermore, connected by lateral passages.

Finally, it may be stated that the boundary between the West Hellenic and the Central Hellenic nappes coincides with the limitation of the external and the internal flysch regimes:

According to Richter (1976a) the final flysch of the Central Hellenic nappes and the Vardar zone is of Upper Maestrichtian or Palaeogene to Priabonian age whereas the West Hellenic flysch was deposited between the Priabonian and the Burdigalian. On the base of these different ages

Tab. 1. Nappe Sequences of the Hellenides outward the Vardar Zone.

	Western and central continental Greece, Peloponnesus		Median Crystalline Belt
Inner Hellenic nappe		Ophiolitic outliers	
Central Hellenic nappe		Pelagonian nappes (including sub-Pelagonian areas)	
		Parnassus nappe	Blueschist units
		Pindus nappe	
West Hellenic nappe	Upper unit Ionian and Gavrovo/Tripolitza sediments)		Basal units
	Lower unit (Phyllite series s.str.).		
Autochthonous	pre-Apulian foreland	Talea Ori series (Plattenkalk)	

of the flysch we distinguish external zones (authochthonous, West Hellenic sedimentation areas) from internal ones (Pindos zone, Pelagonian platform, Vardar zone). The subdivision of Aubouin et al. (1963) is different as it refers to the age of the main deformations.

A new tectonic map of the Hellenides (fig. 1) and a cross-section through Northern Greece (fig. 2) are based on the above structural division. They illustrate the following definition of the several units and the discussion of some major tectonic problems. It is emphasized that our structural scheme refers in detail to the Greek mainland including the Peloponnesus. Within the Aegean Sea and in the Crete arc deviations of this scheme are obvious. They are outlined briefly below (2.7.).

2.1. Autochthonous

Although a subduction beneath western Greece and Crete is highly probable (Leydecker 1975; Leydecker et al. this vol.; Richter & Strobach this vol.) the pre-Apulian foreland and the areas of the Talea Ori series may be regarded as authochthonous.

The pre-Apulian foreland, not studied during the present program, is built up by a neritic carbonate sequence of Triassic to Eocene age covered by marine Neogene and shows a very moderate deformation predominantly by fractures (e.g. Bornovas 1964; I.F.P./I.G.S.R. 1966; B.P.Co 1972*).

* I.F.P. = Institut français du Pétrole.
I.G.S.R. = Institute of Geology and Subsurface Research.
B.P.Co = British Petroleum Co.

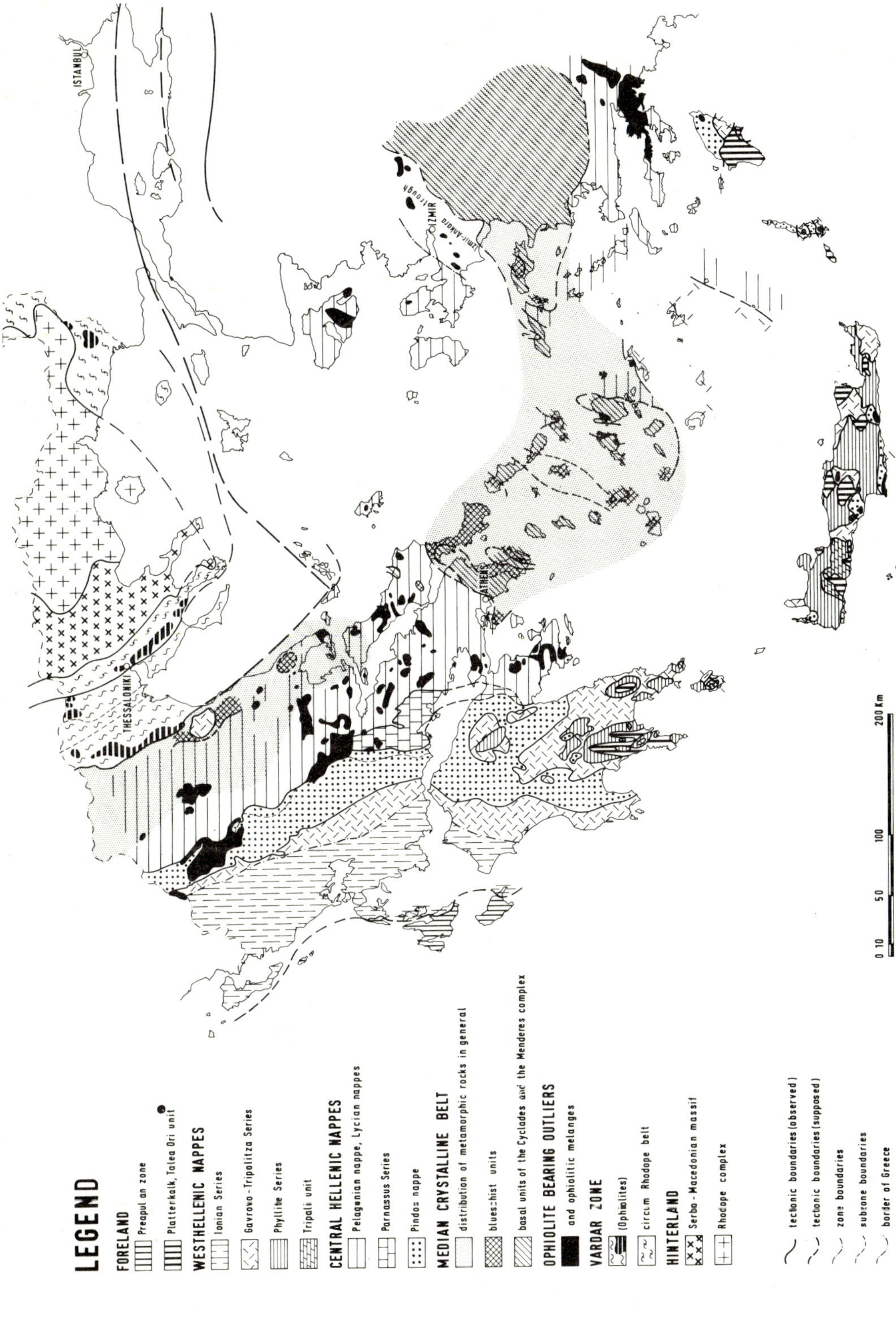
ISTANBUL
IZMIR
Izmir-Ankara trough
THESSALONIKI
ATHENS
0 10 50 100 200 Km
LEGEND
FORELAND
Preapulian zone
Plattenkalk, Talea Ori unit
WESTHELLENIC NAPPES
Ionian Series
Gavrovo-Tripolitza Series
Phyllite Series
Tripali unit
CENTRAL HELLENIC NAPPES
Pelagonian nappe, Lycian nappes
Parnassus Series
Pindos nappe
MEDIAN CRYSTALLINE BELT
distribution of metamorphic rocks in general
blueschist units
basal units of the Cyclades and the Menderes complex
OPHIOLITE BEARING OUTLIERS
and ophiolitic melanges
VARDAR ZONE
(Ophiolites)
circum Rhodope belt
HINTERLAND
Serbo-Macedonian massif
Rhodope complex
tectonic boundaries (observed)
tectonic boundaries (supposed)
zone boundaries
subzone boundaries
border of Greece

The Talea Ori series (Plattenkalk, Ida series) with a thick neritic sequence of Permian and Lower Mesozoic age followed by pelagic carbonate rocks (Jurassic to Eocene) and a thin Oligocene flysch (Bizon et al. 1976) was defined as a new palaeogeographic unit of the southern continental margin of the Tethys by Epting et al. (1972). Bonneau (1973), Kuss & Thorbecke (1974) and many others believed it to be a prolongation of the Ionian zone. Jacobshagen et al. (1976b and this vol.), however, refuted this with stratigraphic and tectonic arguments: The Ionian zone passes laterally to the Gavrovo zone in continental Greece, whereas the

Tab. 2. Nappe Sequence of the External Hellenides.

Bonneau (1973); Kuss & Thorbecke (1974) and others	Creutzburg & Seidel (1975)	Kopp & Ott (1977)	This paper, according to Jacobshagen et al. (1976b)
Gavrovo-Tripolitza Zone including Phyllite series	Tripolitza series	Tripolitza unit	Gavrovo-Tripolitza and Ionian series (Upper West Hellenic nappe)
	Phyllite-quartzite series, incl. the Tripali unit	Phyllite	Phyllite series (Lower West Hellenic nappe)
		Trypali unit	Trypali series
Talea Ori series (Ionian zone)	Talea Ori series (?Ionian zone)	Talea Ori (Ionian zone)	Talea Ori series and pre-Apulian foreland
pre - A p u l i a n foreland			

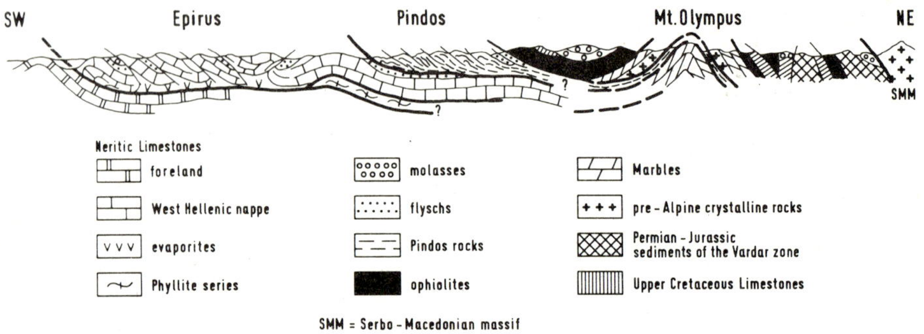

Fig. 2. Cross-section through the Hellenides of northern continental Greece. Re-interpretation after Aubouin (1973: fig.2B).

Fig. 1. Tectonic map of the Hellenides, compiled by D.Büttner and V.Jacobshagen, with contributions of R.Altherr, St.Dürr, G.Kauffmann, F.Kockel K.-O.Kopp & E. Seidel. Post-orogenetic basins of the Cenozoic neglected.

structurally much deeper Talea Ori series below the Phyllite series may be connected with pre-Apulian foreland. It is not excluded that a connection to the Ionian depocenter might have existed somewhere norht of the Peloponnesus (fig. 3 b).

The different geotectonic interpretations of the West Hellenic nappes and the Talea Ori series are summarized in tab. 2 and fig. 3a-b.

2.2. West Hellenic Nappes

The complex of the West Hellenic nappes must be divided structurally into two units. The lower one comprises the Phyllite series on the Peloponnesus and Crete and perhaps the Trypali unit of Crete. The upper one consists of the Mesozoic-Tertiary sediments of the Ionian and the Gavrovo/ Tripolitza zone as defined by Aubouin et al. (1963).

2.2.1. Lower unit (Phyllite series): The Phyllite series and its structural position pose serious problems: In eastern Crete it has been described as a tectonic mélange (Wachendorf et al. 1974); from inconsistencies in stratigraphy, deformation, and metamorphic grades it seems probable that this series is a complex of heterogenous elements (Creutzburg 1958). In eastern Crete micaschists and amphibolites possibly of pre-Alpidic age are enclosed in very low grade metamorphic schists (Wachendorf et al. 1975; Creutzburg & Seidel 1975; Seidel & Okrusch this vol.).

The Phyllite series is up to 2000 m thick and consists of phyllites with intercalations of (often neritic) carbonate rocks, quartzites, conglomerates, evaporites, and basic or intermediate volcanic rocks, Upper Palaeozoic and Triassic fossils are known locally (vide Creutzburg & Seidel 1975; Jacobshagen et al. 1976b and this vol.). Considerable parts of the series, on the Peloponnesus (Paraskevopoulos 1964; Thiébault 1975) as well as on Crete (Schubert & Seidel 1972) are affected by a post-Triassic high pressure/low temperature metamorphism. Since other weakly metamorphosed parts of the series are basal formations of the Tripolitza series (vide Kopp, first contr. this vol.; Thiébault 1975), the high pressure metamorphism could perhaps serve as a diagnostic feature of a Phyllite unit s.str.

The palinspastic position of the Phyllite series is enigmatic. Till now, two models are in discussion:

1. The Pyllite series is the original substratum of the Ionian and external parts of the Gavrovo-Tripolitza sediments (fig. 3a). Its carbonate roof was sheared off by Alpidic movements (e.g. Aubouin et al. 1971; Bizon et al. 1976; Jacobshagen et al. 1976b). This conjecture, however, would not explain the high p/T metamorphism of the Phyllite series which points to an Alpidic subduction.

2. Wunderlich (1971) and Kopp & Ott (1977) assumed that the Phyllite series was deposited in a Penninic type trough of Mesozoic age. If this is true an Alpidic subduction could explain its special metamorphism. One of the arguments against this idea is the lack of fossils younger than Triassic in the Phyllite series. Moreover, it is difficult to understand the close similarities between the Talea Ori series and the Ionian sequence (see below) which in this case would have been separated by the Phyllite trough. It is possible to escape this contradiction with the assumption that the two series were connected on the northwestern end of this speculative trough (fig. 3b) as traces of the Phyllite series cannot be persued beyond the Northern Peloponnesus.

2.2.2. Upper unit: In continental Greece, the upper West Hellenic unit consists of the Ionian and the Gavrovo-Tripolitza sediments. Stratigraphical and sedimentological analyses (I.F.P.(I.G.S.R. 1966; B.P.Co 1971 etc.) revealed that from Triassic to Middle Liassic the whole deposition area belonged to a uniform carbonate platform with an evaporitic base in its external parts. The external margin of this platform subsided from the late Jurassic to the Eocene testified by the pelagic Ionian sequence whereas the central and internal parts of the platform

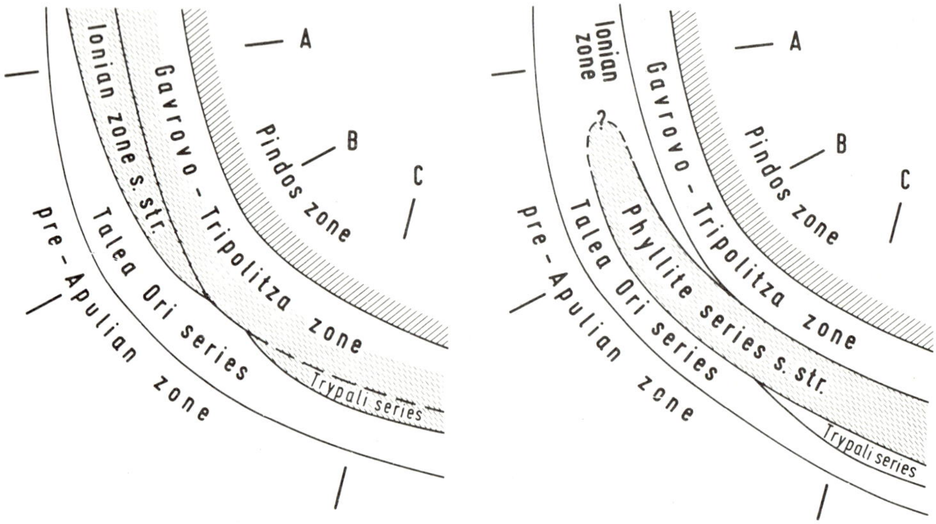

Fig. 3. Hypothetical distribution of the external Hellenide zones during the early Mesozoic. The marked cross-sections refer to northern continental Greece (A), the Peloponnesus (B), and Crete (C).
a) The Phyllite series (hatched) is interpreted to be the pre-Upper Triassic basement of the Ionian zone and of the external parts of the Gavrovo-Tripolitza zone which was sheared off and subducted during an Alpidic orogenesis.
b) The Phyllite series is thought to be the metamorphic filling of a special Penninic type trough. The neritic sequences vary laterally on both flanks. How far the Phyllite trough might have extended to the north, is open.

persisted. A lateral connection between both sequences is undoubted. Moreover, both are covered by a coherent area of external flysch which is not dissected by any major thrusts (B.P.Co 1971, sheet 2). Therefore a Gavrovo-Tripolitza nappe thrust onto the Ionian zone, as proposed by Dercourt et al. (1976), is not likely.

Differences in Mesozoic lithofacies determine the style of the internal structures of the Upper West Hellenic unit: In the Ionian zone, the evaporitic base caused large folds and upthrusts. In the Gavrovo-Tripolitza zone, without these evaporites, block faulting predominates, with movements beginning prior to the end of sedimentation (Richter 1976a).

The Parnon window in the eastern Peloponnesus shows a minimum amount of thrust transport of the West Hellenic nappes of 150 km. The age of this overthrust is between Middle or Upper Burdigalian and Lower Tortonian (Kowalczyk et al. 1977).

2.3. Central Hellenic Nappes

The Central Hellenic nappe complex comprises the well-known Pindos and Parnassus nappes and furthermore the Pelagonian nappes (defined below). Most of the distribution area of the Pelagonian nappes was affected by Palaeogene metamorphism (the Thessalian, the Attico-Cycladic and the Menderes "massifs" of former authors). These metamorphic areas were outlined summarily as the Median Crystalline Belt by Dürr et al. (this vol.). The following units are recognized within this belt, from top to bottom:

(D) Ophiolitic outliers, some with slices of metamorphic rocks at their base

(C) Pelagonian nappes with late Palaeozoic and mesozoic sediments locally overlying a Hercynian basement.

(B) Blueschist units, the stratigraphic and metamorphic ages of which are known very incompletely.

(A) Basal units: metamorphic platform carbonates of Triassic to Upper Cretaceous or even Palaeogene age.

The upper two nappes are usually covered by transgressive Cretaceous limestones and by the internal flysch of the Hellenides (Upper Maestrichtian to Eocene) both being weakly metamorphosed in large areas.

This sequence is similar to that of the Central Eastern Alps: The Pelagonian nappes correspond well to the Upper Austroalpine nappes, whereas the Blueschist units and the Basal units may be compared in position with the Penninic of the Tauern window consisting of Schieferhülle and Zentralgneis with sedimentary cover.

2.3.1. Pindos nappe: The Pindos nappe is characterized by a pelagic sequence of cherty limestones and radiolarites, interrupted by a first Pindic flysch of Barremian to Turonian age and ending with the internal Palaeogene flysch.

On the Peloponnesus, Kiskyras (1964) and Fytrolakis (1972) suggested that the Pindos sequence was deposited in two separate areas; this has been disproved by Doert (this vol.). Discussion is open whether the pelagic sedimentation may have taken place in a deep-sea trough (Aubouin 1965) or in a rather shallow basin (Caron 1975) and on oceanic or continental crust. At the external front, a connection to the Gavrovo-Tripolitza platform was supposed by Aubouin (1959) for continental Greece. Intermediate sequences are known from the Crete arc (Ethia series, Creutzburg & Papastamatiou 1969; Mangassa series, vide Baumann et al. 1977; Aubouin et al. 1976a).

On the internal side, the Pindos sequence is connected palaogeographically with the Central Hellnic Parnassus unit and parts of the Pelagonian platform called herein "sub-Pelagonian" by sedimentary passages (Papastamatiou & Tataris 1963; Koch & Nicolaus 1969). Nevertheless, Hynes et al. (1972) and Smith et al. (1975) supposed an Othrys ocean between the Pindos depocenter and the Pelagonian continental margin and believed that lithospheric remnants of this ocean were thrust to the east onto the Pelagonian platform during the Lower Cretaceous.

The inner parts of the Pindos zone were affected by the Eocene tec-

togenesis: Pindos clasts occur in the Lower Oligocene Molasse of the Mesohellenic trough (Brunn 1956) and in Oligocene olisthostromes of the Tripolitza flysch (Brunn 1956; Richter 1976a; Kowalczyk et al. 1977). In outcropping Pindos rocks, traces of this Eocene deformation were found only in the eastern Peloponnesus (Doert 1976 and this vol.). The overthrust of the Pindos nappe onto the external zones, however, and the internal deformation of the main parts of this unit originated in the Middle Miocene. This may be concluded from the age and the deformation of the Tripolitza flysch (vide Doert 1976 and this vol.). The nappe transport seems to be affected partly by gravitational gliding (Temple 1968) induced by the beginning uplift of the central Aegean area (Dürr et al. this vol.). The displacement of the Pindos nappe exceeds 100 km on the Peloponnesus and in northwestern Greece where the Mt. Olympus window exposes Gavrovo-Tripolitza rocks (Fleury & Godfriaux 1975).

2.3.2. Parnassus unit: According to Celet (1962), the Parnassus unit is a small platform with neritic carbonates of Upper Triassic to late Cretaceous age and several bauxite intercalations. There is no certain continuation of this platform north of the Spercheios river in Greece. To the west, it is tied to the Pindos zone by intermediate series (2.3.1.) and by the uniform cover of the internal flysch (Richter 1976a).

To the south and southeast, a primary end of the Parnassus zone (Dercourt 1964) is suggested by transitions into the "sub-Pelagonian" series of the Argolis (Bachmann & Risch this vol.). In continental Greece, however, the Parnassus unit, moderately folded corresponding to its material, overrides the Pindos zone as a nappe and is overridden by sub-Pelagonian outliers for more than 20 km (Celet 1962; 1976; Schwan 1976a and this vol.).

2.3.3. Pelagonian nappes: With the recognition of the Greek ophiolite complexes as obducted oceanic fragments, it became obvious that the Permian to Jurassic sedimentary sequence of the "sub-Pelagonian" and "Pelagonian" zones are quite similar and originally belonged to the same Pelagonian platform (Kauffmann 1976a,b) which developed on a thick Hercynian basement. Only some parts of this platform subsided early, in the Triassic (central Othrys, Ferriere 1976) or in the Liassic (Argolicum, Jacobshagen et al. 1976b). These still may be called "sub-Pelagonian". Regarding the Triassic and Jurassic sequence of the Parnassus zone as well as its lateral passages to sub-Pelagonian areas mentioned above it must be attributed to the Pelagonian platform, too.

At the end of the Jurassic, nearly the whole platform subsided and was covered first by pelagic sediments, then by olisthostromes and by the Eohellenic ophiolite nappes. In front of the latter, a marginal strip of the former platform remained free ("Parnassus zone"), while depressions were filled with the detritus of the Eohellenic cordillera (Boeotian flysch). The first Pindic flysch is interpreted as a distal continuation of the upper part of the Boeotian flysch.

After a phase of uplift and erosion the area of the future Pelagonian nappes which suffered the Eohellenic orogeny was flooded by a shallow sea from late Barremian to the Maestrichtian. In the Palaeogene a second flysch (final flysch of the internal zones) was sedimentated undiscriminatingly over the whole area of the Pelagonian platform and the Pindos depocenter. During the Eocene orogeny (see below), the whole area of the Pelagonian platform was dissected into thrust sheets of different size.

Referring to this history, the definition of the Pelagonian nappes comprises all units derived from that major part of the Pelagonian platform which suffered an Eohellenic orogenesis. The sedimentary bulk of these nappes is attached to the Hercynian basement (Thessaly, Chios) or sheared off (Euboea, Cyclades). As far as these sediments have been affected by Alpidic metamorphism (comp. 2.4. and 3.), the Pelagonian nappes form parts of the Median Crystalline Belt.

2.3.4. Blueschist units of the Median Crystalline Belt (=Variegated Series High Pressure Units, Dürr et al. this vol.): Windows in Thessaly expose a unit characterized by high p/T metamorphism beneath the Pelagonian basement (Derycke & Godfriaux 1976) containing metaclastics, metavolcanics, marbles, flyschoid sequences, and rocks of an ophiolite suite. A correlation of this unit with similar groups in Southern Euboea, Attica, the Cyclades, and farther to the east is likely (Dürr et al. this vol.). All these units contain variegated series of continental margin, oceanic trench, and oceanic origin. Their palinspastic position was probably inward the Pelagonian platform with respect to new observations of Derycke & Godfriaux (oral communication, Athens 1977), i.e. in the Vardar ocean. Subduction from a Pindos ocean seems unlikely because of palaeogeographical reasons. With respect to their probable Vardar origin, the Blueschist units were not included in the Central Hellenic nappes, at present.

The age of subduction and metamorphism of the Blueschist units is uncertain: In the respective units of southern Euboea, the oldest fold axes are believed to be of late Jurassic/early Cretaceous age by Katsikatsos et al. (1976). K/Ar age determination of the blueschists of Siphnos, however, point to a Palaeogene metamorphism (Altherr et al. 1977).

2.3.5. Basal units of the Median Crystalline Belt: The basal units of the Median Crystalline Belt are best represented by the Menderes massif s. str. in western Anatolia. It consists of an old continental basement (e. g. Menderes gneisses), covered by Palaeozoic schists and a marble envelope with metabauxites. The marbles are a metamorphosed Triassic - Upper Cretaceous platform sequence which once was covered probably by a Palaeogene flysch (Dürr 1975).

Similar marbles are known in the Cyclades, Attica and southern Euboea where their Mesozoic age, disputed since more than a century, has been proved by Marinos (1955), Katsikatsos (1970) and others. They all define basal windows of external appearance with stratigraphic sequences similar to the autochthonous/parautochthonous of the Western Taurides or with the Parnassus zone in Greece (Dürr 1975). A parallelization with the Gavrovo-Tripolitza window of the Thessalian Mt. Olympos is also possible. Structural features in Southern Euboea and Attica (Katsikatsos et al. 1976a,b; Argyriadis et al. 1976), however, and other observations on the Cyclades point to a more internal palinspastic position (Dürr et al. this vol.).

The basal units were overridden by the Blueschist units and the Pelagonian nappes presumably during the Eocene. The emplacement was accompanied by an initial medium to high pressure metamorphism. The maximum of pressure was attained in the Cycladic archipelago at the center of the Aegean arc with Middle to Late Eocene radiometric ages of the glaucophane bearing schists of Naxos (Andriessen et al. 1976). This orogenetic event might be interpreted as a continent/continent collision leading to excessive crustal thickness.

The first Alpidic stage of metamorphism was superseded by a slow heating culminating in the Middle Miocene, with local anatexis and acid plutonism, followed immediately by a rapid uplift in the same area where today the crust is rather thin (Dürr et al. this vol.).

To explain the presence of non-metamorphic outliers of Pelagonian ophiolites and molasse in the central and south Aegean area, a complicated late-orogenic history must be assumed for the (?early) Upper Miocene, before the beginning of break-down of the intra-arc basins (Cretan and North Aegean Seas) outlined by Dürr et al. (this vol.) and Roesler (this vol.).

2.4. Inner Hellenic Nappes

Several generations of ophiolitic outliers are known in Greece. As all are supposed by us to have originated inward the Pelagonian platform, these tectonic complexes are called Inner Hellenic Nappes. Often the ophiolites are underlain by ophiolitic mélanges sensu Gansser (1974) and covered by meso-autochthonous transgressive sediments. Thus, their first emplacement can be fixed stratigraphically. Furthermore, they are often accompanied by slices of metamorphic rocks representing different p/T regimes, such as high p/T glaucophane schists versus low p/T cordierite-andalusite/sillimanite gneisses. Radiometric ages for some of the ophiolitic rocks and for some of the metamorphic slivers dragged along with them are available.

Combining the evidences, two or three episodes of obduction were recognized: in the Tithonian to early Cretaceous, probably in the Upper Cretaceous, and in the Palaeogene

a)Tithonian/early Cretaceous (Eohellenic), ophiolite-bearing outliers rest upon the upper Central Hellenic units of continental Greece and the Peloponnesus. The most famous is the Vourinos massif in Northern Greece (Moores 1969; Zimmermann 1972). Its petrology and structure are closely comparable with recent oceanic lithosphere. A glaucophane-producing metamorphism in underlying crystalline rocks was related to this obduction by Mercier (1973b). Other groups of Eohellenic outliers were separated in the Othrys area (Hynes et al. 1972; Smith & Woodcock 1976), on top of the "sub-Pelagonic" Argolicum of the Peloponnesus (Jacobshagen et al.1976b, and this vol.) and above the Thessalian crystalline complex of the Pelion peninsula and of the Northern Sporades (Wallbrecher 1976, Jacobshagen & Skala 1977). The southernmost occurrences were detected on Gavdos (Vicente 1970), in central Crete (Seidel et al. 1976) and probably on Karpathos (Aubouin et al. 1976a).

b) In Anatolia, late Cretaceous emplacement of ophiolites is known from the Izmir-Eskisehir zone (Kaya 1972) as well as from the Diabase and Peridotite Nappes of the Lycian Taurus (Graciansky 1972; Poisson 1975). Probably, parts of the Cretan ophiolites with their accompanying metamorphic rocks, the Asteroussia nappe (Bonneau 1973) and the serpentinite-amphibolite association (Creutzburg & Seidel 1975), are correlated with the mentioned Lycian units (Bernoulli et al. 1974). A basalt, granitoids and various metamorphic rocks have furnished late Upper Cretaceous radiometric ages for the magmatic and metamorphic events (comp. Seidel et al. 1976). During the Miocene, these outliers were brought into their present position.

Upper Cretaceous metamorphism and extrusion of pillow diabases are also known from continental Greece (Hynes et al. 1972; Jacobshagen et al. 1976a). In these regions, however, traces of a synchronous tectogenesis

are lacking or at least doubtful.

c) Palaeogene ophiolite emplacement is evident in the southern Pelion and on the Northern Sporades and assumed for Euboea and the Peloponnesus (Jacobshagen et al. 1977).

2.5. Vardar Zone and Circum-Rhodope Belt

According to Kossmat (1924), the region between the Pelagonian nappe and the Serbomacedonian massif is regarded as Vardar zone s.l. It can be divided into two major units. The western one - here preliminarily called Vardar zone s.str. - was described in detail by Mercier (1973a,b).

During the Triassic and Jurassic up to the Kimmeridgian the western subzones of the Vardar zone (Almopias and Païkon subzones) obviously had not yet been differentiated into troughs and ridges. The Hercynian or pre-Hercynian basement in like manner is overlain by metamorphosed platform carbonates, intercalated with Kimmeridgian meta-alkali-rhyolites, meta-keratophyrs and metaspilites. In the Almopias subzone allochthonous slabs of metaophiolites are frequent, probably of Upper Kimmeridgian to "Portlandian" age which transgressively are covered by Apto-Albian conglomerates and Orbitolina-limestones. These pass continuously into neritic to pelagic carbonates of Cenomanian to Maestrichtian age. With the Upper Maestrichtian a genuine flysch develops within the trough.

The Païkon subzone remained a ridge. Lowermost Cretaceous clastic sediments unconformably overlying the platform are only locally preserved. The transgressive Upper Cretaceous shows a purely neritic development with only minor clastic intercalations. Ophiolites are absent.

Farther to the East, the Peonias subzone (unité de Guevguelije) only gives an incomplete record of the Mesozoic development. A metamorphic, Hercynian or pre-Hercynian basement is known as well as platform carbonates of Triassic to Kimmeridgian age. At its western border a thick meta-rhyolite sequence of upper Kimmeridgian age resembles the Païkon development. Most characteristic is a thick non-metamorphic ophiolite complex in the higher Jurassic, which is penetrated by granitic bodies of nearly the same age. The Portlandian, transgressively overlying the ophiolites as well as the pre-Mesozoic basement, shows a wide range of facial differentiation into molasse sediments, reef structures and platform carbonates, containing bauxites horizons. Only in Lower Cretaceous times a real flysch series occurs. Nothing is known about the development during the Upper Cretaceous. Priabonian molasse sediments cover large areas of this subzone forming an individual trough which continues below the Thermaic Gulf.

A small rim of pre-Hercynian crystalline rocks which in Greece is mostly covered by Neogene separates the Vardar zone s.str. from the early Alpidic (Cimmerian) circum-Rhodope Belt (Kauffmann et al. 1976). In this belt, Triassic platform carbonates are overlain by Lower and Middle Jurassic flysch deposits. The belt also contains a very thick sequence of intermediate to acid metamagmatites, as well as mafic and ultramafic complexes. Several slabs of diaphthoritic basement occur along major upthrust zones. Regional metamorphism during the upper Middle Jurassic reached high degrees of the greenschist facies and affected also parts of the pre-Hercynian basement. Syn- and postkinematic batholiths intruded preferably at the suture between the circum-Rhodope Belt and the Serbomacedonian massif. Transgressive Tithonian molasse sediments already contain pebbles of early Mesozoic metamorphic rocks. These deposits were compressively deformed during the Cretaceous (Eohellenic phase?). A

third orogenic phase of minor importance affected the circum-Rhodope Belt after the Priabonian. Several slabs of a diapththoritic basement occur along major upthrust zones.

To the N, the circum-Rhodope Belt extends into the Šumadija zone and the Veles series of Southern Yugoslavia and has a wide extension in the Transsylvanian Ore Mountains (Andelković & Lupu 1967). To the S and to the E, relics of the belt have been found at the tips of the Chalkidiki peninsula where its regional fold axes swing from the NNW direction towards E and finally towards NE.

In the region of Alexandroupolis and on Samothraki, Jurassic flysch series (Kopp 1969; Lebküchner 1972), meta-magmatites and ultramafites are exposed, with NNE strike directions.

This range of early Mesozoic metamorphic rocks borders the eastern Rhodope massif and continues to the Strançа Mountains and to the coasts of the Marmara Sea, thus attaining a width of at least 200 km.

The style of deformation of the whole Vardar zone s.l. is characterized by steep schuppen structures predominantly with western vergence.

2.6. Hinterland

The relative hinterland of the Hellenides in Greece at least comprises two different crystalline complexes: The Serbo-Macedonian and the Rhodope massifs.

The Serbo-Macedonian massif is the only piece of crust in the Aegean region which was approximately consolidated in pre-Alpidic times. It consists of very thick meta-sedimentary series containing marbles, meta-ophiolites and granitoide intrusions. The medium grade metamorphism (almandine-amphibolite facies) ascended regionally to anatexis and palingenetic granites. For these events, Yugoslavian authors assume a late Precambrian orogeny and Ordovician refolding. Hercynian K/Ar-ages (Borsi et al. 1965) may be due to rejuvenation.

A late Palaeozoic and early Mesozoic sedimentary cover was incorporated in this basement during the mid-Jurassic orogenesis which also effected a widespread diaphtoritic remobilization of the pre-Alpidic rocks and granitoide intrusions. Post-orogenetic porphyritic intrusions of Palaeogene age are known from the western margin of the Serbo-Macedonian massif.

East of the Strimon river, the Serbo-Macedonian massif overlies the adjacent Rhodope massif with a post-Oligocene overthrust. The latter consists of lower gneisses, amphibolites, synkinematic granitoids and an upper marble-greenschist sequence which are intruded by granitoids bodies (so-called"South Bulgarian Granites").

Bulgarian authors assume a Precambrian deformation of the schists wheras the granitoids are regarded of Hercynian or Laramide age, respectively (Zagorcev et al. 1971; Zidarov et al. 1971).

In Greek Macedonia, however, it was shown that at least the upper series has been deformed more or less synchronous with the intrusion of granites of the South Bulgarian type and is at least younger than Precambrian (Meyer 1968; Kronberg et al. 1970; Birk et al. 1971). K/Ar ages from biotites of these batholiths are nowhere older than Oligocene. Current investigations of Lundgren and Armour-Brown (not publ.) show that the Rhodope massif is subdivided by the important Nestos line into two complexes and that many of its granites are very young. The Eo-Oligocene molasse is deformed parallel to the regional B of the metamorphic rocks. Facing these facts, it seems impossible to regard the Rhodope massif as a part of a pre-Alpidic consolidated hinterland of the Hellenides. It is

believed to be an Alpine element of uncertain geotectonic character within the mountain belts of the Balkan peninsula.

2.7. Geological Relations between the Hellenides and the Orogenic Belts of Anatolia

The Aegean arc with its embryonal intra-arc and back-arc basins connects the Alpidic mountain chains of the Balkan peninsula and those of Anatolia with remarkable discontinuities and with tectonic units relaying each other along strike. With a few exceptions, the correlation between these units is more or less problematic.

Thus, the crystalline rocks of the North Anatolian belt have been tied to those of the Serbo-Macedonian massif by Brinkmann (1968). However, both are separated at least by the circum-Rhodope belt.

Equivalents of the Vardar zone units and of the Eohellenic ophiolite nappes, are known neither on the eastern and northeastern Aegean islands nor in western Anatolia. Their eastern continuation is an enigma.

Chios (Besenecker et al. 1968) and parts of northwestern Anatolia have been assigned tentatively to the Pelagonian realm (Kauffmann 1976b) or regarded as its eastern equivalent (Dürr 1975; Dürr et al. this vol.).

The very complex Izmir-Eskisehir zone (Brinkmann 1968, 1972) can hardly be the direct continuation of the Vardar zone. Its Upper Cretaceous to Palaeogene flyschs, its ophiolitic rocks (possibly obducted during the Upper Cretaceous) and its high pressure metamorphics (Brinkmann 1971; Kaya 1972, 1976) might be related to those of the Upper Lycian nappes of the Taurides (Brunn et al. 1971). Most of the western Lycian nappes (Graciansky 1972) continue to southeastern Aegean islands, the upper ones as far as Crete (Bernoulli et al. 1974).

In continental Greece, however, no analogs for the Izmir-Eskisehir zone and for the Lycian nappes have been identified despite several affinities to the Pindus zone. Both may correspond to Central Hellenic units, where locally Upper Cretaceous flysch formations are known (e.g. Athens schists: Marinos et al. 1971).

These correlations are acceptable, if the basal units of the Median Crystalline Belt are regarded as windows exposing a relatively external platform, if hence all the above named units palinspastically must be placed to the interior of the Median Belt.

Within this "internal realm", the Eohellenic phase seems to be confined to the west with the easternmost occurrence on Karpathos (Aubouin et al. 1976a), whereas Upper Cretaceous orogenic events clearly gain much in importance to the east; how exactly and why, is not understood at present.

The lower Central Hellenic nappes were not affected by the Mesozoic tectogeneses. Their prolongations to Western Anatolia are uncertain. Some of the Basal units of the Median Crystalline Belt may be equivalents of the Parnassus platform, at least in the Cyclades. The Pindus zone, however, cannot be connected directly neither with certain Lycian nappes nor with the Izmir-Eskisehir zone regarding considerable differences in their respective geological evolutions.

The external realm (West Hellenic Nappes and "Parautochthonous-Autochthonous" of the outer Hellenides and Taurides) offers hardly fewer complications. Obvious similarities in facies and stratigraphy lead to the opinion that all the pre-Apulian, Talea Ori/Plattenkalk, Ionian and Gavrovo-Tripolitza series with their possible equivalents originally belonged to a southern continental margin area (Bernoulli & Jenkyns 1974) later

incorporated into the orogen. This view is also supported by the uniformity of the external flysch concerning its age and stratigraphy.

The high pressure metamorphic Phyllite series on the Peloponnesus and Crete, however, suggests the former existence of a southern Tethys ocean (comp. 2.2.1.), which must have divided the sedimentation areas of the external zones.

Within the autochthonous, the Talea Ori series (confined to the Peloponnesus, Crete and Kasos) is supposed not to be a longitudinal continuation of the Ionian zone (2.1.2.). Correlations of the series called Ionian or Talea Ori/Plattenkalk on Karpathos, Rhodes, and in the southwestern Lycian Taurus are controversial amongst different authors (comp. e.g. Mutti et al. 1970; Bernoulli et al. 1974; Kuss & Thorbecke 1974; Aubouin et al. 1976; Brunn et al. 1976; Baumann et al. 1977) and highly questionable.

The southernmost outcrops of the Ionian zone are on the northwestern Peloponnesus; possibly this zone ends beneath the west Peloponnesian Neogene.

The Gavrovo-Tripolitza zone can be persued to eastern Crete. Further east and north appear still other series (Karpathos: Davidson Monett 1974; Rhodes: Mutti et al. 1970; Leboulenger & Matesco 1975; on some Carian islands/Southern Sporades and southeastern Cyclades: Dürr et al. this vol.). Some of them show stratigraphic and tectonic features of both the Tripolitza and the Parnassus units. Other more pelagic ones (e.g. on Karpathos) have been compared to the distal Pindos series of Crete (Davidson Monett 1974; Aubouin et al. 1976a). Relations to sequences of certain units within the Lycian nappes obviously do exist.

3. Geological Evolution of the Aegean Region (V.J., G.K., H.B.)

3.1. Palaeogeographic Pattern of the Hellenides

The palaeogeographic pattern of the Hellenide depositional area during the Triassic and Jurassic is summarized in the schematic cross-sections of fig. 4 d:

- The Mesozoic sediments of the West Hellenic nappes were parts of a broad mobile carbonate belt adjacent to the foreland. It must be seriously taken into account, however, that this belt was probably subdivided by the oceanic trough of the Phyllite series at least in the Hellenic arc south of the Gulf of Corinth (fig. 3b).
- In view of the lateral passages to the adjacent Parnassus and Subpelagonian sequences, an Othrys ocean between the Pindos area and the Pelagonian platform is improbable. If one speculates that the Pindos depocenter might have graded outward into an oceanic basin, any direct connections between the Ethia and the Mangassa series of Crete with the rocks of the Pindos nappe would be denied.
- The Pelagonian platform - subdivided into different sedimentation areas at times - bordered the mentioned Pindos depocenter to the east. It had a minimum original width of about 250 km. Toward the interior of the orogen it was flanked by an oceanic Vardar basin.
- All parts of the present Vardar zone with its basins and ridged had a sialic crust during the early Mesozoic which probably was thin considering the masses of basaltic lavas there. The westernmost Almopias subzone with

its ophiolites and pelagic sediments is supposed to have bordered onto an oceanic trough. If it is true that three times oceanic sheets have been obducted from this basin (2.4.), it must have been very large and is believed to have been the Tethys proper in accordance with Bijou-Duval et al. (1977). Moreover, it must be taken into account that the circum-Rhodope Belt is an individual early-Alpidic orogen (see below) the Jurassic flysch of which might have bordered a separate early Mesozoic ocean.

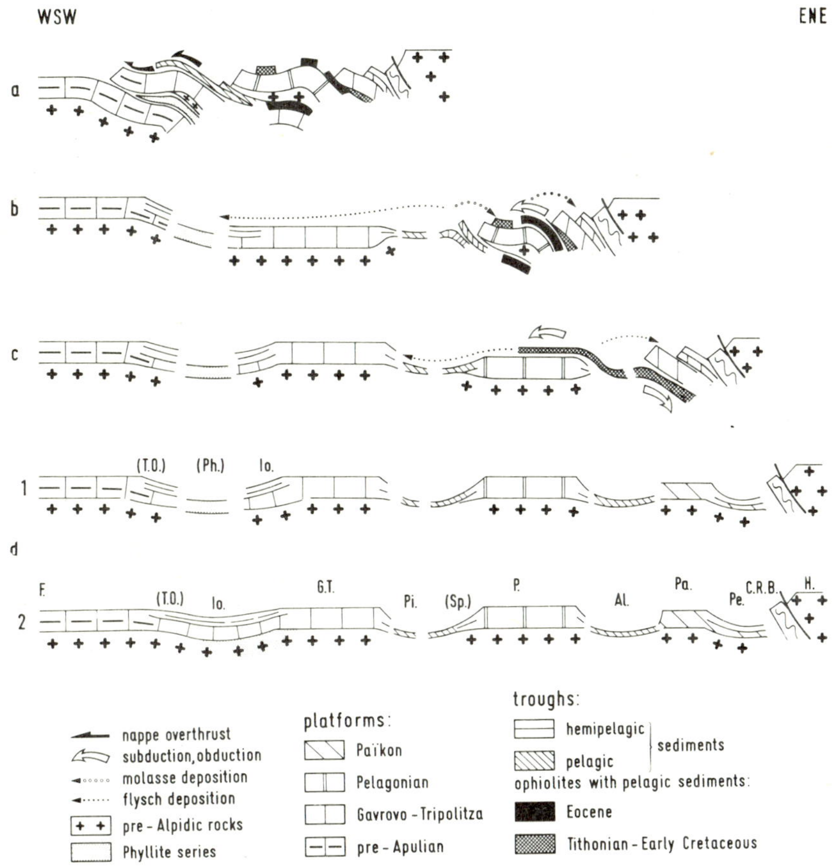

Fig. 4. Stages of the orogenic evolution of the Hellenides. The section crosses northern continental Greece.
a) Middle Miocene: Overthrust of the Central Hellenic nappes onto the Gavrovo-Tripolitza platform, transportation of the West Hellenic nappes over the foreland.
b) Middle to Upper Eocene: ophiolite obduction onto the Pelagonian platform from the Vardar ocean which is closed at that time; subduction of an oceanic slab (from the "Pindos" or the Vardar ocean) beneath the Pelagonian platform ("Blueschist units"). Folding of the internal Pindos trough. Sedimentation of the external flysch starts; molasse deposition in the Mesohellenic trough, in the Vardar trough, and in Thrace.
c) Tithonian-Lower Cretaceous: Subduction of the Vardar ocean beneath the

3.2. Orogenic Evolution of the Aegean Region

The orogenic system of the Hellenides was formed by five paroxyms between the Middle Jurassic and the Middle Miocene. Some major steps of the orogenic evolution are sketched in fig. 4. The present distribution of rocks deformed during the main tectonic phases is shown in figs. 5 and 6.

In the circum-Rhodope belt orogenetic deformation, granitic intrusions, and metamorphism with greenschist facies happened probably at the Dogger/Malm boundary (Orogenesis II a, Kockel et al. 1971). This paroxysm was followed by a Tithonian molasse. A late orogenetic volcanism was not developed. Details of the mechanism of this orogenesis are unknown.

During the Tithonian or early Cretaceous, ophiolite sheets with pelagic sediments were obducted from the Vardar ocean onto the Pelagonian platform. This event, called Eohellenic phase, appears restricted to the western branch of the Hellenic arc, but it profoundly altered the palaeogeographic pattern.

Locally, slices of rocks affected by a high p/T metamorphism were dragged along the base of ophiolites (Vicente 1970; Seidel et al. 1977). Flyschs with ophiolitic detritus derived from the rising cordillera were deposited between the uplifts of the Parnassus zone and later spread all over the Pindos trough. Eohellenic deformation also affected the whole Vardar zone (Orogenesis II b, Kockel et al. 1971) here, however, being accompanied by greenschist metamorphism. This points to a pattern of paired metamorphic belts for the Eohellenic phase. Aptian to Albian flysches with ophiolite debris were also sedimentated in the Almopias and the Païkon subzones (Mercier 1973a). Eohellenic molasses and subsequent volcanism are unknown. At the end of the Barremian a transgression started in the course of which the Eohellenic belt was flooded gradually by a flat sea: Rudist limestones finally covered the whole area.

Another orogeny is evident in the southern and eastern Aegean area and was dated as late Upper Cretaceous in analogy to western Anatolia (2.4.). Thrusts of this age were observed only on Chios (Besenecker et al.1968). Obduction probably played an important role. In continental Greece, an Upper Cretaceous metamorphism affected at least parts of the Pelagonian nappes (e.g. Hynes et al. 1972). Here, the Upper Cretaceous neritic carbonate sediments of the inner zones are conformably followed by the internal flysch (Upper Maestrichtian resp. Palaeocene to Eocene) which covered even the Pindos trough. This Palaeogene flysch of the Hellenides is believed to have come from late Cretaceous cordilleras. Volcanism and molasse deposition subsequent to the late Cretaceous paroxysm have not been observed.

Fig. 4 (cont.)
circum-Rhodope belt and the hinterland; obduction of the Eo-hellenic ophiolites onto the Pelagonian platform; sedimentation of ophiolite-bearing flyschs.
d) Lower Malm: Troughs and ridges after the circum-Rhodope orogenesis. 1) Phyllite series understood as substratum of the West Hellenic carbonate series (vide fig. 3a); 2) Phyllite series interpreted as filling of a special trough (vide fig. 3b).
Al = Almopias trough; C.R.B. = circum-Rhodope belt; F = foreland; G.T. = Gavrovo-Tripolitza platform; H = hinterland; Io = Ionian trough resp. zone; P = Pelagonian platform; Pa = Païkon platform; Pe = Peonias trough; Ph = Phyllite trough; Pi = Pindos trough; Sp = sub-Pelagonian zone; T.O.= Talea Ori series.

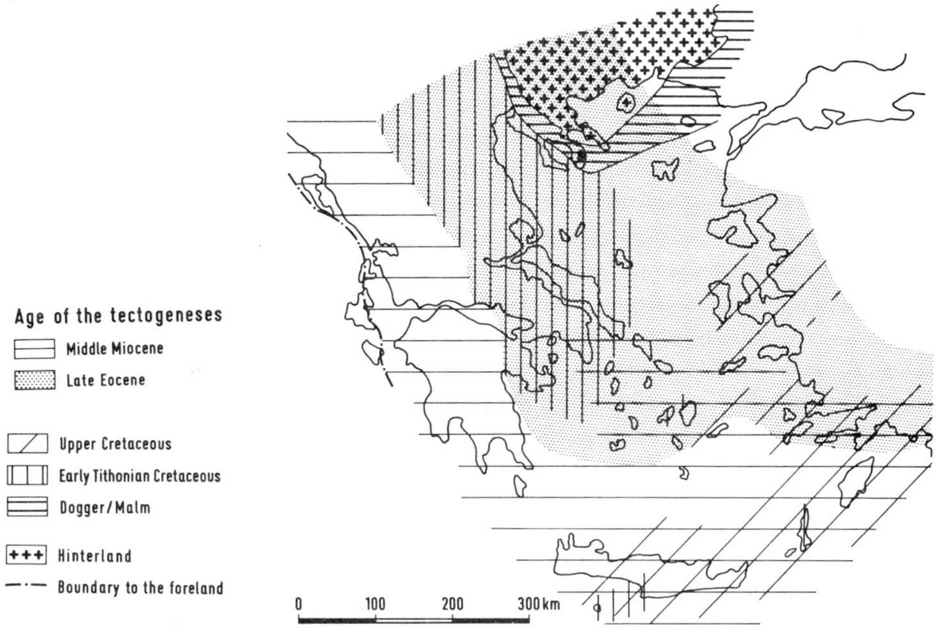

Fig. 5. Present distribution of rocks deformed during the main tectogenetic phases of the Hellenides.

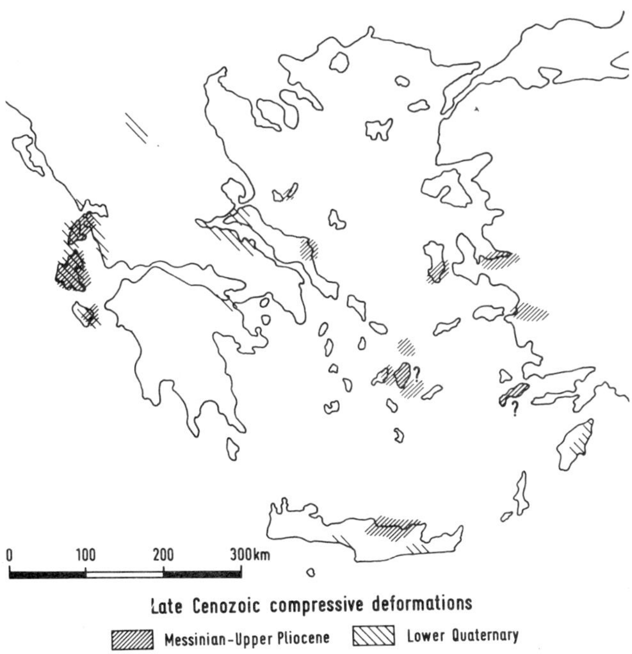

Fig. 6. Distribution of compressional structures younger than Middle Miocene in the Aegean region.

A fourth tectogenesis during the Eocene affected nearly the whole area of the internal zones, also the eastern parts of the later Pindus nappe in the Peloponnesus (Doert 1976). The Pelagonian nappes and the Blueschist units were emplaced, and a narrow fold arc originated, today intersected by the Median Crystalline Belt (Jacobshagen et al. 1975; Jacobshagen & Skala 1977). The deformations were accompanied by low to medium grade metamorphism within the Pelagonian nappes and the neighbouring Vardar zone. At the same time, a high p/T metamorphism affected the lower units of the Median Belt (Andriessen et al. 1976; Altherr et al. 1977). Thus, we observe again a paired belt pattern. Traces of Eocene ophiolite obduction may be seen e.g. in the Pelion and on the Northern Sporades. Flysch deposition ceased in the inner Hellenides and started in the external zones during the Upper Eocene (Richter 1976 and this vol.).

At the same time, molasse was already deposited on the hinterland and in basins within the new cordillera (Ivanov & Kopp 1969; Mercier 1973a; Brunn 1956; Roesler this vol.). The late Palaeogene volcanism of Thracia (Central Rhodope) with acid to intermediate calc-alcaline rocks (Ivanov & Kopp 1969) is subsequent to the Eocene tectogenesis.

At the end of the Lower Miocene, a fifth paroxysm deformed the external zones of the Hellenides. With the Burdigalian overthrust of the Pindos nappe associated flysch sedimentation ceased there. Features of gravity transport observed in the Pindos nappe of the Peloponnesus might be related to an incipient doming of the central and southern Aegean area which may be deduced from the known items of its deep-seated Miocene metamorphism and magmatism (vide Dürr et al. this vol.).

The large overthrust of the West Hellenic nappes onto the foreland during the Middle Miocene may have resulted from a wedge of the foreland crust which deeply split the West Hellenic crust (Jacobshagen et al. 1976b). This phenomenon could be caused by a collision of the West Hellenic arc and the Adriatic foreland during the Middle Miocene.

In the central Aegean area, this Middle Miocene paroxysm was superseded by very large vertical movements (Dürr et al. this vol.). Finally, limited nappe thrusting, possibly caused by gravitational slides, occurred, as evident on Naxos (Jansen 1973), Mykonos (Dürr et al. this vol.) and Kos (Altherr et al. 1976).

The tectonic events were accompanied by andesite eruptions with dates ranging from 22 to 13 m.y. in the Central and North Eastern Aegean region (Borsi et al. 1972; Fytikas et al. 1976).

With the end of the Miocene paroxysm, the character of deformation changed completely within the Aegean area proper to extension, fracturing and vertical movements. These processes were interrupted by short periods of weak compression in limited areas (fig. 6):

During the late Middle Miocene, a marine basin originated in the western and central Cretan Sea and was fully developed in the Upper Miocene with grabens striking parallel to the Crete arc and being crossed by transverse faults (fig. 7). The amounts of fractural displacement and the rates of sedimentation (above seismic event A, Jongsma et al. 1977) are high. A second marine basin subsided in the North Aegean region. Temporarily, both Aegean basins were connected (Büttner & Kowalczyk this vol.). Furthermore, many continental basins were created by Upper Miocene fractural tectonics which also traced out the configuration of the Lower Pliocene coasts.

At the Miocene/Pliocene boundary a compressional phase of minor inten-

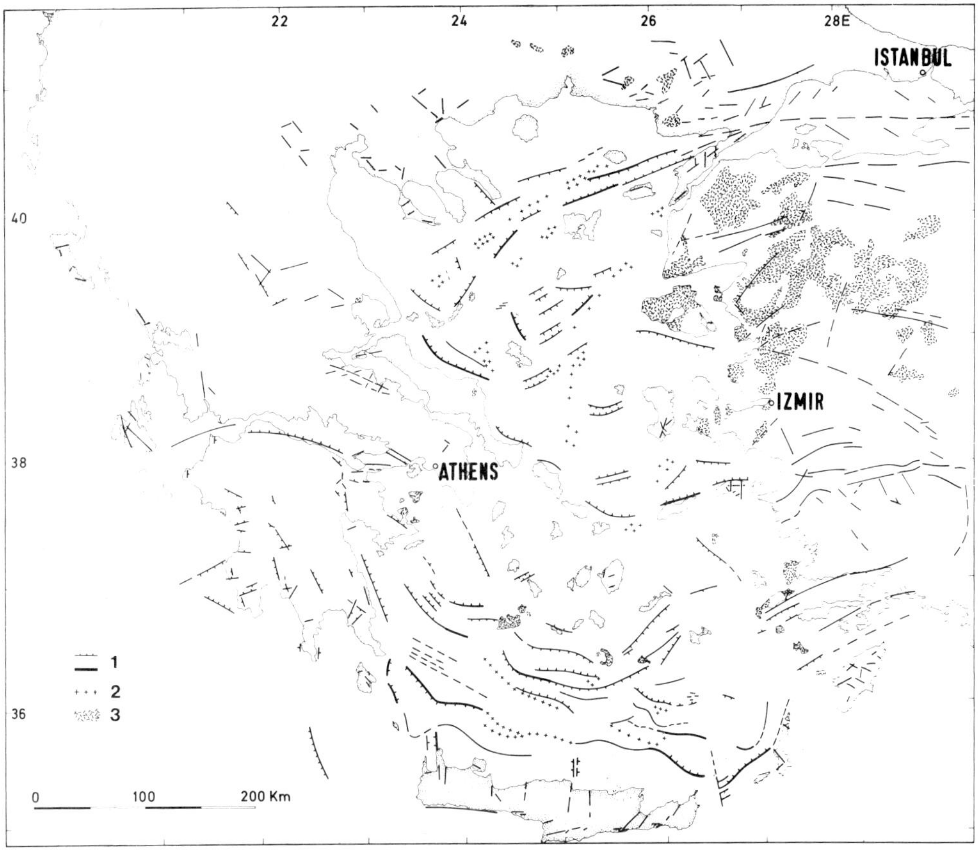

Fig. 7. Compilation of the post-Palaeogene fault pattern in the Aegean region. On land only faults are shown, which have affected Neogene rocks (taken from available publications and maps: in the southern Aegean Sea after Jongsma et al. 1977; in the northern Aegean Sea derived from steepest gradients of bathymetry).
1 = Faults (heavy lines: more than 500 m total throw)
2 = Magnetic anomalies as indications of magmatic intrusions
3 = Cenozoic volcanic rocks.

sity produced folds and upthrusts on external islands of the Hellenic arc (Levkas, Kephallinia: I.G.S.R./I.F.P. 1966; B.P.Co. 1971; Richter 1976b; Crete: Kuss & Thorbecke 1974).

During the Pliocene, tensional fracturing gave rise to new basins both with marine and continental sediments. These fault tectonics had a new climax at the Pliocene/Pleistocene boundary indicated by unconformities in the southern Aegean region and by the cease of marine sedimentation on the present mainland and on most of the islands. Compressional deformations of Late Pliocene/Early Pleistocene age were observed in the external parts of the Hellenic arc (Ionian Islands, Crete, Rhodos). In the Early Pleistocene compressive stress seems to have affected also central

parts of the Hellenides (Faugères & Vergely 1974; Mercier et al. 1976; Roesler this vol.).

The Neogene fault tectonics had produced an agglomerate of crustal blocks, each with an individual tectonic history. Local movements, however, were superimposed by regional tectonic trends which may have changed in the course of time (Kelletat et al. 1976). Tilting perpendicular to the arc and strike-slip faulting affected mainly the external parts of the Hellenides. Fractures in new directions were added to the older, still active faults (Meulenkamp & Zachariasse 1973; Kowalczyk et al. 1977), suggesting a change in the orientation of the stress field. We do not agree with Kronberg & Günther (this vol.) who concluded that the present fracture pattern of the Aegean region reflects the grain of a Precambrian basement. This fixistic view is disproved by the large data support of a complex polyphase orogenic evolution with extreme amounts of nappe transport and plate movements.

The present South Aegean volcanic arc originated at the Pliocene/Pleistocene boundary with calc-alcaline rocks ranging from basalts to rhyolites (Paraskevopoulos 1958; Davis 1967, 1968; Nicholls 1971a-c; Ninkovich & Hays 1972; Pichler & Kussmaul 1972; Keller 1969a,b; Di Paola 1976). Radiometric ages of its rocks range from 2,7 m.y. to present (Fytikas et al. 1976). Concerning the $K_{57.5}$ values sensu Dickinson (1975) and the strontium isotope ratios (Pe & Gladhill 1975) the lavas show similarities to continental margin suites. According to Ninkovich & Hays (1972) and Puchelt (1975 and this vol.) the contents of K_2O at constant SiO_2 are related to the depth of the underlying Benioff zone, for which 150-175 km were calculated by Richter & Strobach (this vol.). Rhyolitic magmas on Kos, Antiparos, Milos and Santorini have been related to anatectic processes within the crust (Burri & Davis 1969; Pichler & Stengelin 1968; Keller 1969a,b).

3.3. Present Geodynamic Situation

Late Neogene tectonics continues at present in the Aegean area. Geomorphological data show that the updoming of the Cyclades region (Schuiling 1969, 1973; Makris 1973, 1977 and this vol.) still goes on (Sabot & Papanikolaou 1976). This may also explain why north of the Cyclades deposits of the Tyrrhenian are below sea level in general while on south Aegean islands, on the Peloponnesus and on the Ionian islands they are elevated.

The North Aegean lineament indicated by the present isobaths (Giermann & Pfannenstiel 1960) extends the North Anatolian wrench fault from the Galipoli peninsula to the WSW. In the North Aegean Sea, this lineament is accompanied by active graben structures; at least from the Saros graben indications of a rift structure have been observed (Needham et al. 1973). The right-lateral North Aegean lineament may form a conjugate shear pair with the left-lateral wrench faults along the west side of the Vardar zone (Jacobshagen 1972, 1973). This fault system is accompanied by some curved transcurrent faults of the Chalkidiki with displacement rates up to 8 km (Kockel et al. 1971).

The present stress field obtained from seismic fault-plane solutions is shown in fig. 8. The Hellenic arc is surrounded by a belt with NE trending horizontal compression. Inside the Aegean and in western Anatolia, tension diverging from the central Aegean is prevailing. In the northern Aegean an almost EW striking belt of compression indicates thrust faulting, perhaps associated with the motion of the Anatolian block relative to Europe (Papazachos 1976).

This situation is in good harmony with other facts as the distribution of gravity anomalies and the structure of the crust (Makris 1977 and this vol.), the existence of the subrecent volcanic arc in the Aegean Sea, and the situation of an active Benioff zone beneath the Aegean area (Papazachos 1973; Leydecker 1975; Leydecker et al. this vol.; Richter & Strobach this vol.) which flattens to the exterior (Berckhemer & Kowalczyk this vol.) and reaches the sea floor between the Hellenic arc and the Mediterranean ridge. The compressional deformation of this ridge was dated as Pliocene by Mulder et al.(1975) and may be still active (Finetti 1976, Kenyon & Belderson 1977). On both flanks of this ridge flyschs are deposited (Stanley 1974).

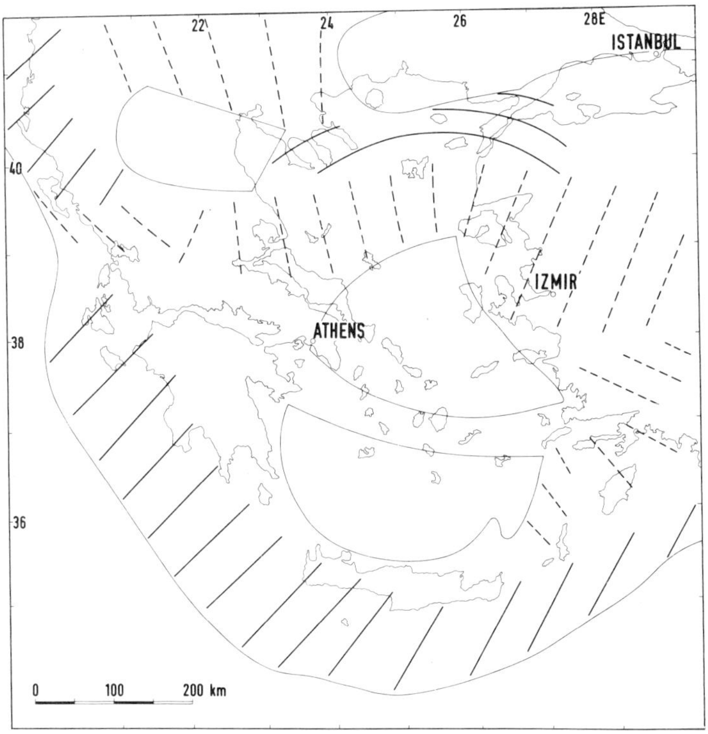

Fig. 8. Present stress field in the Aegean region derived from about 250 seismic fault plane solutions (compiled by Ritsema 1974). Solid lines = directions of horizontal compression; dashed lines = directions of horizontal tension; thin solid lines: boundaries of regions of normal deep seismicity.

All these facts testify that a new orogenic belt is just developing outside the Hellenic arc. The present geodynamic processes in the interior of this arc characterize the evolution of the youngest back-arc basin within a sequence of Alpidic orogenic events.

The neotectonic activities of today must be understood merely as one stage within the orogenetic development according to the definition of Aubouin (1973).

4. Discussion
(V.J.)

Within the last few years a great number of geotectonic models were proposed to explain the orogenetic evolution of the Hellenides on the whole or partially: On the one hand approaches were made in accordance with the ideas of plate tectonics (e.g. Bernoulli & Laubscher 1972; Dewey et al. 1973; Boccaletti et al. 1974; Smith & Moores 1974). On the other hand, thermodiapirism was believed to be the origin of orogenesis causing gravity transport of nappes to the foreland as well as deep erosion, crustal thinning and final break-down in the interior of the uplift (e.g. Wachendorf et al. 1975; Baumann et al. 1976, 1977 and this vol.). These conceptions extrapolate from the geophysical results of Makris (1973, 1975 and this vol.) which characterize the present geodynamic situation.

A third group of authors tried to reconcile the ideas of plate tectonics and thermodiapirism (Brunn 1976; Schwan 1976b, 1977). Brunn argued that changes of the movement direction of the Arabian plate relative to the European plate induced narrow convex arcs (e.g. the Aegean arc) in the interior of which the lithosphere was domed up and ejected radially over the foreland. From a general view, updoming within the back-arc regions is postulated also in plate tectonic models (e.g. Andrews & Sleep 1974; Dennis 1976).

From the foregoing review of the orogenic evolution of the Aegean area the following facts would support a plate tectonic model:

- If the huge amount of nappe overthrusts postulated above is true, large parts of the deeper lithosphere must have vanished. In a section through northern continental Greece, a shortening of at least 500 km must be taken into account even west of the Vardar zone. The mechanisms of subduction, obduction and collision would provide a good explanation.
- Oceanic crust seems to be repeatedly involved into the orogen, forming ophiolite nappes. These would be easily understood as features of obduction but may hardly be explained by a model of up-doming and gravity tectonics.
- The distribution of different types of metamorphism fits to the model of paired metamorphic belts at least for the Eohellenic and the Eocene paroxysm.

Some other observations cannot be explained by the model of thermodiapirism, such as:

- The disconformity of orogenic belts: the Eocene structures cross the Median Crystalline belt of Miocene origin in the central Aegean region according to Jacobshagen et al. (1975).
- The irregular distribution of the areas deformed during the different tectogeneses of the Hellenides (fig. 5). There is no general trend of the wandering of an orogenic wave which must be postulated in the case of the mantle dome model.
- The fact that the younger orogenic processes re-affected wide areas of earlier deformation.

It may not be excluded that these features could have originated between some minor plates moving in different directions in the course of time.

From this point of view, the updoming of the South Aegean area since the Miocene is interpreted as induced by foregoing subduction in accordance with Andrews & Sleep (1974) and Dennis (1976). Thermodiapirism is not believed to be the source of orogenesis.

It appears also doubtful whether the original conception of Berckhemer (1977 and this vol.) should be applied to the Aegean region. If it is true that the different tectogeneses within the orogenetic evolution of the Hellenides are characterized by subduction and/or obduction processes there is no urging necessity to introduce the new mechanism of gravity decay of the Neogene Hellenide ranges to explain the recent Benioff zone.

Regarding the palaeogeographical pattern of the Hellenides, that is, two or three troughs with oceanic crust between carbonate platforms with longitudinal extension, the irregular distribution of the different tectogeneses, a model of a fractured Apulian continental microcontinent lapping into the Tethys ocean with several peninsulas would supply a satisfying explanation. Orogeny, then, should have been caused by subduction and collision processes in Mesozoic and Cenozoic times. At present, observations are not yet sufficient, however, to reconstruct the history of plate movements within the Aegean area precisely. A better knowledge predominantly of the metamorphic zones (especially the Phyllite series and the "Blueschist units") and the Mesozoic volcanism as well as extended radiometric and palaeomagnetic investigations would essentially improve the base for a more detailed interpretation.

Acknowledgement: The authors are very indebted to all participants of the program who allowed them to refer to unpublished results, furthermore to R. Altherr, H. Closs, M. Okrusch, D. Richter, W. Schwan and E. Seidel for reading the manuscript and critical discussions. R. Altherr, G. Kauffmann and E. Seidel contributed also to the tectonic map. Technical assistance by R. Cousino, H. Graetsch, W. Jung and D. Reich is gratefully acknowledged.

References

Altherr, R., Keller, J. & Kott, K., 1976: Bull. Soc. géol. France (7), 18:403-412.

Altherr, R., Harre, W., Kreuzer, H., Okrusch, M. & Seidel, E., 1976: Internat. Sympos. structur.Hist.Mediterr.Basins, Split:315-316.

Andelkovic, M.Z. & Lupu, M., 1967: Rept. 8.Congr.Ass.géol.Carpato-Balcanique, sect. Géotectonique:15-28.

Andrews, D.J. & Sleep, N.H., 1974: Geophys.J.roy.astr.Soc.,38:237-251.

Andriessen, P.A.M., Boelrijk, N.A.J.M., Hebeda, E.H., Priem, H.N.A., Verdurmen, E.A.Th. & Verschure, R.H., 1976: 4th Europ.Coll. Geochronol. Cosmochronol. Isotope Geol.

Argyriadis, I., Mercier, J.-L. & Vergely, P., 1976: C.R. Acad. Sci. Paris, (D),283:459-467.

Aubouin, J., 1959: Ann. géol. Pays hellén., 10:1-484.

- , 1965: Developments in geotectonics, 1:1-359.

- , 1973: Bull.Soc. géol. France, (VII), 15:426-460.

Aubouin, J., Brunn, J.H., Celet, P., Dercourt, J., Godfriaux, I. & Mercier, J., 1963: Livre Mém. P. Fallot, 2:583-610.

Aubouin, J., Bonneau, M., Celet, P. et al., 1971: Ann. Soc. géol. Nord, 90:277-306.

Aubouin, J., Bonneau, M. & Davidson, J., 1976: Bull. Soc. géol. France (VII), 18:385-401.

Aubouin, J., Bonneau, M., Davidson, J., Leboulenger, P., Mattesco, S. &

Zambetakis, A., 1976b: Bull.Soc.géol.France (VII),18:327-336.
Baumann, A., Best, A., Gwosdz, W. & Wachendorf, H., 1976: Tectonophys. 30:T33-T40.
Baumann, A., Best, G., Wachendorf, H., 1977: Geol.Rdsch.,66:492-522.
Berckhemer, H., 1977: Internat.Sympos.structur.Hist.Mediterr.Basins, Split:303-314, Paris.
Bernoulli, D. & Jenkyns, H., 1974: in Dott, R.H.jr. & Shaver, R.H.(eds.): Spec.Publ.Soc. econ.Paleontologists Mineralogists, 19:129-160.
Bernoulli, D. & Laubscher, H., 1972:Eclogae geol. Helv., 65:107-118.
Bernoulli, D., Graciansky, P.C.de & Monod, O., 1974: Eclogae geol. Helv., 67:39-90.
Besenecker, H., Dürr, St., Herget, G. et 5 al., 1968: Geologica et Palaeontologica, 2:121-150, Marburg.
Biju-Duval, B., Dercourt, J. & Le Pichon, X., 1977: in Biju-Duval, B. & Montadert, L. (eds.): Structural history of the Mediterranean basins: 143-164, Paris.
Birk, F., Boer, H.U.de, Kronberg, P. et al., 1971: Beih.geol.Jb., 88:1-179.
Bizon, G. & Thiébault, F., 1974: C.R. Acad.Sci.Paris, (D),278:9-12.
Bizon, G., Bonneau, M., Leboulenger, P., Mattesco, S. & Thiébault, F., 1976: Bull.Soc.géol. France (VII),18:337-345.
Bocaletti, M., Manetti, P. & Pecerillo, A., 1974: Geol. Soc. Amer. Bull., 85:1077-1084.
Bonneau, M., 1973: C.R. Acad. Sci. Paris, (D), 277:2453-2456.
Bornovas, J., 1964: Inst. Geol.Subsurface Research, 10:1-142, Athens.
Borsi, S., Ferrara, G. & Mercier, J., 1965:Ann.Soc.géol. Nord,84:223-225.
Borsi, S., Ferrara, G., Innocenti, F. & Mazzuoli, R., 1973: Bull.volcanol., 36:473-496.
B.P. Co., 1971: Geol. Greece, 10:1-73, Athens.
Brinkmann, R., 1968: Geologica et Palaeontologica, 2:111-119, Marburg.
- , 1971: in Campbell, S. (ed.): Geology and history of Turkey:171-190.
Brunn, J.H., 1956: Ann. géol. Pays hellén., 7:1-358.
- , 1960: Bull. Soc. géol. France (VII), 2:470-526.
- , 1976: Z.dt. geol. Ges., 127:323-335.
Brunn, J.H., Dumont, J.F. et 7 al., 1971: in Campbell, S.: Geology and history of Turkey:225-255.
Brunn, J.H., Argyriadis, I. et 4 al., 1976: Bull. Soc. géol.France (VII), 18:481-497.
Burri, C. & Davis, E., 1969:Prakt.Akad.Athinon, 43:175-181.
Caputo, M., Panza, G.F. & Postpischl, D., 1970: J.geophys.Research, 75: 4919-4923.
Caron, D., 1975: Thèse 3me cycle Univ. P. et M. Curie:1-119, Paris.
Celet, P., 1962: Ann. géol. Pays hellén., 13:1-446.
- , 1976: Bull.Soc.géol. France, (VII), 18:299-307.
Creutzburg, N. & Papastamatiou, J., 1969: Sitz.-Ber. Heidelberger Akad. Wiss., math.-naturwiss. Kl.:3-63.
Creutzburg, N. & Seidel, E., 1975:N.Jb. Geol. Paläont.,Abh.149:363-383.
Davidson Monett, J., 1974: Thèse 3me cycle Univ.Paris VI, 157 p.
Davis, E.N., 1967: Prakt. Akad. Athinon, 42:235-252.
- , 1968: Geol. Rdsch., 57:811-821.
Dennis, J.G., 1976: Z.dtsch. geol. Ges., 127:73-85.
Dercourt, J., 1964: Ann. géol. Pays hellén., 15:1-418.
- , 1970: Bull. Soc. géol. France (VII), 12:261-317.
Dercourt, J., Wever, P. de & Fleury, J.-J., 1976: Bull. géol. France

(VII), 18:317-326.
Derycke, F. & Godfriaux, I., 1976: Bull.Soc.géol.France (VII),18:252.
Dewey, J.F., Pitman, W.C., Ryan, W.B. & Bonnin, J., 1973: Geol. Surv. Amer. Bull., 84:3137-3180.
Dickinson, W.R., 1975: Geology, 3:53-56.
Doert, U., 1976: Z. dt. geol. Ges., 127:417-428.
Dürr, S., 1975: Unveröff.Habil.Schr. Univ.Marburg, 106 p.
Epting, M., Kudrass, H.R., Leppig, U. & Schäfer, A., 1972:N.Jb.Geol. Paläont., Abh., 141:259-285.
Faugères, L. & Vergely, P., 1974: C.R. Acad.Sci. Paris, (D),278:1313-1316.
Ferrière, J., 1976: Ann. Soc. géol. Nord., 96:121-134.
Finetti, I., 1976: Boll. geofis. teor. appl., 19:31-65 , Trieste.
Fleury, J.-J. & Godfriaux, I., 1976: Ann.Soc. géol. Nord, 94:149-156.
Fytikas, M., Giuliani, O., Innocenti, F., Marinelli, G. & Mazzuoli, R., 1976: Tectonophysics, 31:T29-T34.
Fytrolakis, N., 1972: Bull. geol. Soc. Greece, 9(2):122-132.
Galanopoulos, A.G., 1973: Bull. geol. Soc. Greece, 10:67-69.
Gansser, A., 1974: Eclogae geol. Helv., 67:479-508.
Giermann, G. & Pfannenstiel, M., 1960: Bull. Inst. Océanogr. Monaco, 1192: map C.
Graciansky, P.C. de, 1972: Thèse Univ. Paris-Sud, 762 p.
Guernet, C., 1971: Thèse Sorbonne, 319 p.
Hynes, A.J., Nisbet, E.G., Smith, A.G., Welland, M.S.P. & Rex, D.C., 1972: Z.dt.geol. Ges., 123:455-468.
I.F.P./I.G.S.R., 1966: Etude géologique de l'Epire (Grèce nord-occidentale):306 p.
Ivanov, R. & Kopp, K.-O., 1969: Geologica et Palaeontologica, 3:123-151.
Jacobshagen, V., 1972: Z.dt.geol. Ges., 123:445-454.
- , 1973: Bull. geol. Soc. Greece, 10:90-93.
Jacobshagen, V. & Skala, W., 1977: Ann. géol. Pays hellén. (in print).
Jacobshagen, V., Skala, W. & Wallbrecher, E., 1975: Rapp. Comm. int. Mer Médit., 23 (4a):173 p., Monaco.
Jacobshagen, V., Risch, H. & Roeder, D., 1976a:Z.dt.geol.Ges.,127:133-145.
Jacobshagen, V., Makris, J., Richter, D. et al., 1976b: Z.dt.geol. Ges., 127:337-363.
Jacobshagen, V., Märtz, J. & Reinhardt, R., 1977: N. Jb. Geol. Paläont. Mh.:613-620.
Jansen, J.B.H., 1973: Geological map of Naxos, 1:50.000.
Jongsma, D., 1975: Thesis Univ. Cambridge, 85 p.
Jongsma, D., Wissmann, G., Hinz, K. & Garde, S., 1977: "Meteor"-Forsch.-Ergebn., (C), 27:3-30, Stuttgart.
Katsikatsos, G., 1970: Ann.géol.Pays hellén., 22:62-76.
Katsikatsos, G., Mercier, J.-L. & Vergely, P., 1976a: C.R. Acad. Sci. Paris, (D), 283:459-462.
- , 1976b: C.R. Acad. Sci. Paris, (D), 283:1613-1616.
Kauffmann, G., 1976a: Bull.géol.Soc.France (VII), 18:308.
- , 1976b: Z.dt.geol.Ges., 127:387-398.
Kauffmann, G., Kockel, F. & Mollat, H., 1976: Bull. Soc. géol. France (VII), 18:225-230.
Kaya, O., 1972: Z.dt.geol. Ges., 123:491-501.
- , 1976: Bull. Soc. géol. France (VII), 18:479-480.
Keller, J., 1969: Bull. volcanol., 33:942-959.
- , 1971: Acta 1st int.sci.Congr.Volcano Thera:152-169.
Kelletat, D., Kowalczyk, G. et al., 1976:Z.dt.geol.Ges.,127:447-465.

Kenyon, N.H. & Belderson, R.H., 1977: in Biju-Duval, B. & Montadert, L. (eds.): Structural history of the Mediterranean basins:233-240 Paris.
Kiskyras, D., 1964: Bull. geol. Soc. Greece, 5 (2):1-21.
Koch, K.E. & Nicolaus, H.J., 1969: Geol. Greece, 9:190 p., Athens.
Kockel, F., Mollat, H. & Walther, H.W., 1971: Geol. Jb.,89:529-551.
Kopp, K.-O., 1969:Geotekton.Forsch.,31:97-116, Stuttgart.
Kopp, K.-O. & Ott, E., 1977: N.Jb. Geol.Paläont., Mh.:217-238.
Kossmat, F., 1924:Geologie der zentralen Balkanhalbinsel, 198 p.
Kowalczyk, G., Richter, D., Risch, M. & Winter, K.P., 1977: N. Jb. Geol. Paläont., Mh.:541-564.
Kronberg, P., Meyer, W. & Pilger, A., 1970: Beih.geol.Jb., 88:133-180.
Kuss, S. & Thorbecke, G., 1974: Ber. naturf. Ges. Freiburg i.Br.,64: 39-75.
Lapièrre, H. & Rocci, G., 1976: Tectonophysics, 30:299-313.
Lebküchner, H., 1972: Geological map of Greece 1:50.000, Sheet Samothraki. Athens (N.I.G.M.R.).
Leboulenger, P. & Mattesco, S., 1975: Trav. Dept. Géol. struct. Univ. P. et M.Curie, Thèse 3me cycle 5 & 6, 217 + 23 p., Paris.
Leydecker, G., 1975: Diss. Univ. Frankfurt, 75 p.
Makris, J., 1973: Bull. geol. Soc. Greece, 10:206-213.
- , 1977: Proceed. 15th gen.Ass. europ. seismol. Comm.Krakow 1976, (in print).
Marinos, G., 1955: Bull. geol. Soc.Greece, 2 (1):121-123.
Marinos, G., Katsikatsos, G., Georgiadhis-Dikeoula, E. & Mirkou, R., 1971: Ann. géol. Pays hellén., 23:183-216.
McKenzie, D.P., 1970: Nature, 226:239-243.
Mercier, J., 1973a: Ann. géol. Pays hellén., 20:1-596.
- , 1973b: Ann. géol. Pays hellén., 20:599-792.
Mercier, J.-L., Vergely, P. & Bebien, J., 1975: C.R. Soc. géol. France, (D), 282:108-112.
Mercier, J.-L., Carey, E. et al., 1976: Bull. Soc. géol. France (VII), 18:355-372.
Meulenkamp, J.E. & Zachariasse, W.J., 1973: in Drooger, C.W. (ed.): Messinian events in the Mediterranean:202-205, Amsterdam.
Meyer, W., 1968: Geologica et Palaeontologica, 2:173-192, Marburg.
Moores, E.M., 1969: Geol. Soc. Amer. Spec. Pap., 118:74 p.
Mulder, C.J., Lehner, P. & Allen, D.C.K., 1975: Geol. en Mijnbouw, 54:208-221.
Mutti, E., Orombelli, G. & Pozzi, R., 1970: Ann. géol. Pays hellén., 22:77-226.
Needham, H.D., Le Pichon, X. et 5 al., 1973: Bull. geol. Soc. Greece, 10:152-153.
Nicholls, I.A., 1971a: Tectonophysics, 11:377-385.
- , 1971b: J. Petrol. 12:67-119.
- , 1971c: Contrib. Miner. Petrol., 30:261-276.
Ninkovich, D. & Hays, J.D., 1972: Earth planet. Sci.Lett.,16:331-345.
Paola, G.M. di, 1974: Bull. volcanol., 38:944-987.
Papastamatiou, J. & Tataris, A.A., 1963: Bull. geol. Soc. Greece, 5 (1): 83-88.
Papazachos, B.C., 1973: Bull. geol. Soc. Greece, 10:165-168.
- , 1976: Tectonophysics, 33:199-209.
Paraskevopoulos, G.M., 1958: Tschermaks miner.petrogr.Mitt., 3 F., 6:13-72.
Paraskevopoulos, G.M., 1964: N.Jb. Miner., Abh., 101:195-209.

Pe, G.G. & Gladhill, A., 1975: Lithos, 8:209-214.
Philippson, A., 1898: Ann. Géogr., 7:112-141.
Pichler, H. & Kussmaul, S., 1972: N.Jb. Miner., Abh., 116:268-307.
Pichler, H. & Stengelin, R., 1968: Geol.Rdsch.,57:795-810.
Poisson, A., 1975: Rapp. Comm. int. Mer Médit., 23 (4a):207-209, Monaco.
Puchelt, H., 1975: Rapp.Comm.int. Mer Médit., 23 (4a):193-194.
Renz, C., 1940: Pragm. Akad. Athinon, 8:171 p.
- , 1955: Die vorneogene Stratigraphie der normalsedimentären Formationen Griechenlands. 637 p., Athens.
Richter, D., 1976a: Z. dt. geol. Ges., 127:467-483.
- , 1976b: N.Jb. Geol. Paläont., Abh., 151:224-252.
Ritsema, A.R., 1969: Geol. Rdsch., 59:36-56.
- , 1974: Kon.Nedelands meteor.Inst. De Bilt,Wet.Rapt., 74-4:36 p.
Sabot, V. & Papanikolaou, D., 1976: Praktika Akad. Athinon, 51:86-96, Athens.
Sannemann, W. & Seidel, E., 1976: N.Jb. Geol. Paläont.,Mh.:221-228.
Schubert, W. & Seidel, E., 1972: Z.dt. geol. Ges., 123:371-384.
Schuiling, R.D., 1969: Verh. kon.geol.Mijnbouw Gen.,26:143-148.
Schuiling, R.D., 1973: in Jong, K.A. de & Scholten, R.: Gravity and tectonics:35-47.
Schwan, W., 1976a: Z.dt.geol. Ges., 127:373-386.
- , 1976b: Z. dt. geol. Ges., 127:105-124.
- , 1977: Z. dt. geol. Ges., 128:143-152.
Seidel, E., Okrusch, M., Kreuzer, H., Raschka, H. & Harre, W., 1976: Contrib. Miner. Petrol., 57:259-275.
Seidel, E., Schliestedt, M. et al., 1977: Geol. Jb. (in print).
Smith, A.G. & Moores, E.M., 1974: Geol. Soc. London, Spec. Publ., 4:159-185.
Smith, A.G. & Woodcock, N.H., 1976: Geology, 4:653-656.
Smith, A.G., Hynes, A.J. et al., 1975: Eclogae geol Helv.,68:263-281.
Stanley, D.J., 1974: Soc.econ.Paleontol.Mineral.,Spec.Publ.,19:240-259.
Temple, P.G., 1968: Geol. Soc. Amer. Bull, 79:687-700.
Thiébault, F., 1975: C.R. Acad. Sci. Paris, (D), 280:947-949.
Vicente, J.C., 1970: Bull. Soc. géol. France (VII), 12:481-495.
Wachendorf, H., Baumann, A., Gwosdz, W. & Schneider, W., 1974: Z. dt. geol. Ges., 125:237-252.
Wachendorf, H., Best, G. & Gwosdz, W., 1975: Geol. Rdsch., 64:728-750.
Wallbrecher, E., 1976: Z. dt. geol. Ges., 127:365-371.
Wunderlich, H.G., 1971: N.Jb. Geol. Paläont., Mh.:566-579.
Zagorčev, I. et al., 1971: Minist. Chemie, Metallurgie, Komm. Geologie, Ann. jubilaire, 18:137-156.
Zidarov, N., Zagorčev, I. et al., 1971: Jb. staatl. Wirtschaftsver. "Geol. Erkundungen", 20:261-278.
Zimmermann, J. jr., 1969: Geol. Soc. Amer. Abs. with Progr., 245 p.
- , 1972: Geol. Soc. Amer. Mem., 132:225-239.
Zimmermann, J. jr. & Ross, J.V., 1976: Bull. geol. Soc. Amer., 87: 1547-1550.

Chapter 6

The Crusts of the Alps, Apennines, Hellenides: Comparisons and Correlations

Crustal and Structural Features of the Margins of the Adria Microplate

P. Giese and K.-J. Reutter

Contents

Introduction

Since the first systematic geological investigations in the Mediterranean orogens, attention has always been drawn to the strange loop with which Apennines, Alps, Dinarides, and Hellenides encircle the seemingly unfolded area of the Po plain, the Adriatic Sea and the Apulian platform. The contrast between this more than 1300 km long and 100-200 km wide area, which evidently behaved as a relatively rigid element, and the strongly compressed and plastically deformed surroundings induced E. Argand as early as 1916 to consider it a promontory of the African continent pushed into the European continent.

Thus, when plate tectonics provided an essential key for the understanding of orogenic processes it seemed obvious that this rigid area could be considered as a small continental plate surrounded by previously active convergent margins, except for its southern border. The investigation of the magnetic anomalies in the Atlantic Ocean showed that the opening of areas with oceanic crustal and sedimentary conditions in the surroundings of this Adria microplate, as well as their closure in the course of the orogeny of the Mediterranean mountain chain must be seen as a consequence of the differential and finally converging motions of the African continent and the pre-Alpine European continent. It could also be demonstrated paleomagnetically that the Adria microplate, although not directly connected with the African continent, behaved as if it really was a promontory of Africa, except for some marginal parts. According to plate tectonics the circum-Adriatic mountain chains are supposed to be produced by the collision with the European continent or fragments of it after the complete consumption of the separating oceanic areas (eugeosynclines) which supplied the ophiolites and related sediments to these orogens. Although the general plate tectonic concept is widely accepted, the enormous amount of geological information about these chains still does not allow unequivocal interpretations. The great differences of

structural settings in the chains surrounding the Adria microplate prove that the orogenic processes were extremely complex and need sophisticated models of subduction and collision, because they did not follow the same scheme everywhere.

One of the tasks in the study of the nature, shape, and behaviour of the Adria microplate must consist of revealing the structures of its collision zones with Europe and other minor continents. When trying a tectonic interpretation of crustal profiles, the geological development of the crustal elements, the different geotectonic roles of foreland and hinterland areas, and the polarity in the evolution of the orogen have to be taken into account. The clear asymmetric nature of the orogens should also be reflected in their crustal structure. The question arises as to how the crustal setting is related to the asymmetry which is inherent in convergent plate margins. Another question is whether the nappe thrust planes can be considered as synthetic branches of crustal thrusts and the pre-collisional subduction plane.

Under these aspects the circum-Adriatic mountain chains shall be treated in the following. The discussion is based on a number of crustal profiles shown in fig. 3, a foldout at the end of this paper.

Western Carpathians

The European foreland which is covered with thick Neogene molasse sediments dips southward under the nappes of the Outer Carpathians (or Flysch Carpathians). Their very thick and mostly clastic sediments were deposited in several basins of the mobile shelf and continental slope of the pre-Alpine European continent. The narrow Pieniny Klippen belt separates the Outer Carpathians from the Inner Carpathians. The facies of the Pieniny sediments can be compared with the Ultrahelvetic zone of the Alps (Tollmann 1969). The Inner Carpathians consist, to a great extent, of thrust complexes of basement rocks and sediments. At least their internal units correspond to the Austroalpine of the Eastern Alps (Tollmann 1969) and, hence, to the Adria microplate. In contrast to the Alps and the other peri-Adriatic chains, no ophiolites occur in the Western Carpathians and, therefore, no typical eugeosynclinal sediments deposited on an oceanic basement can be recognized.

Inner and Outer Carpathians show intense folding and far reaching northward transported nappes. However, the deformation of the Inner Carpathians is of Upper Cretaceous age (paleo-Alpine) and that of the Outer Carpathians of Oligocene and Miocene age. For these younger events the northward directed orogenetic evolution (polarity) expressed in the direction of nappe movement and migration of depositional troughs seems to be clear. The preceding thrust movements of the Inner Carpathians had the same direction, but from some places the Pieniny Klippen belt southward thrusting of late Cretaceous age is also known (Andrusov 1968; Birkenmajer 1974).

During the Tertiary orogenic movements the Cretaceous tectonized block of the Inner Carpathians acted as a rigid hinterland. Thrusting of at least 100 km can only be observed in the allochthonous Outer Carpathians. The autochthonous substratum dips southward continuously under the imbricated nappes and does not reappear in fensters far behind the front of thrusting.

In the course of extensive seismic crustal investigations in Southeast

Europe, some profiles through the Eastern and Western Carpathians were recorded (Geophys. Transact. 1972). The three international profiles III, V, and VII all reveal one common feature: From the European foreland towards the internal side of the Carpathians, the crust/mantle boundary dips down to 40-50 km beneath the Pieniny Klippen belt. At this suture the crust/mantle boundary appears abruptly at a shallow depth of only 25-30 km. This can be followed through the Inner Carpathians into the Pannonian basin.

Thus a clear correlation between surface geology and crustal structure exists. The Inner Carpathian-Pannonian hinterland, together with the wedge of Outer Carpathian nappes, appears to be thrust upon the foreland during Tertiary. The configuration resembles that of converging plates in which the substratum of the Pieniny and Outer Carpathian sequences was subducted synthetically. Neogene volcanism and tensional tectonics which confer back-arc characteristics to the hinterland, confirm such an orogenetic model.

During the Upper Cretaceous and early Tertiary, however, mainly the Inner Carpathians and the Pannonian area suffered strong deformation. The nature and the position of hinterland, foreland, and subduction zone of that orogenic period were quite different. These earlier processes formed the hinterland plate from very different crustal fragments (Channel & Horvath 1976). Thus at least one reorganization of the subductory system must have taken place during the orogeny of the Carpathian mountains, probably in the Lower Tertiary.

Border Region Carpathian - Eastern Alps

The transition between the Carpathians and the Eastern Alps is hidden by the young sediments of the Inner Vienna Basin. From structural and lithological comparisons, however, it can be concluded (Tollmann 1969) that the Inner Carpathians are directly connected with the Austroalpine units of the Eastern Alps. The Pieniny Klippen belt, which plays such an important role in the Carpathians, can be followed into the Flysch zone of the Eastern Alps where it splits up and disappears towards west.

In the transition from the Carpathians into the Eastern Alps, crustal structure must have undergone a distinct change, as the Bouguer gravity shows. While in the Carpathians the axis of the gravity minimum is situated in the Outer Carpathians it is to be found in the Eastern Alps under the main crest. In the Carpathians as well as in the Alps, the isanomalies are generally parallel to the strike of the main geologic units. But at the eastern border of the Alps, where the Graz basin begins, they run perpendicularly to the geologic strike.

Seismic data for this interesting transition zone provide the Alpine Longitudinal Profile with the recordings between the shotpoints D, E, and F (Alpine Explosion Seismology Group 1976, Aric et al. 1978).

In the Graz basin, a thin crust of 25-27 km has to be expected up to its western border (Aric et al. 1978; Gutdeutsch pers. comm.). From the west, a thick crust of the Eastern Alps can be followed up to at least 50 km beyond the eastern border of the Tauern fenster. The seismic data do not allow to prove that the deep crust/mantle boundary of the Eastern Alps is a direct continuation of the shallow one of the Graz basin. The situation may be more complex because in the transition zone at a depth of 30-33 km an additional discontinuity exists at which the velocity in-

creases to about 7 km/s. Whether this interface has to be directly connected with the shallow crust/mantle boundary under the Graz basin or not is an open question and shall be shortly discussed.

In spite of the geological connection between the Carpathians and the Austroalpine zone of the Eastern Alps there is a fundamental difference in the geotectonic setting. While the Inner Carpathians during the Tertiary orogenic movements represented a rigid crustal block, the Austroalpine represents a wedge-like thrust sheet whose thickness reaches that of an entire continental crust only South of the Periadriatic lineament (Gailtal-line). Its allochthonous position becomes evident especially in the huge tectonic fenster of the "Tauern", where it is underlain by metamorphic sediments and ophiolites of the Penninic eugeosyncline and Palaeozoic as well as basement rocks which belong to the tectonically mobilized border of the pre-Alpidic European plate. The easternmost fensters in which the Penninic substratum appears, are found where the Eastern Alps are truncated by the Pannonian plain (Wechsel and Rechnitz windows).

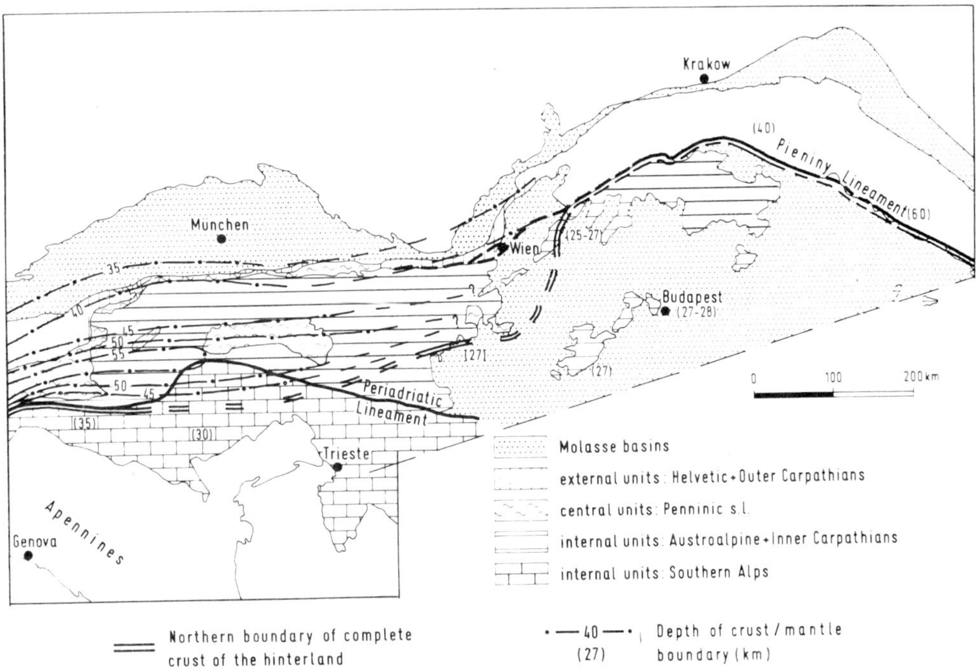

Fig. 1. Map showing the main tectonic units of the Alps and the Western Carpathians and contours of the crust/mantle boundary. The hatched parts indicate the areas where the Austroalpine domain including the Inner Western Carpathians and the Southern Alps are underlain by their primary upper mantle.

The crust/mantle boundaries mentioned above deduced from the Alpine Longitudinal Profile thus belong to different crustal units: near the Tauern window it is the base of the pre-Alpidic European plate, deeply pressed down by its tectonic overburden. In the Graz basin, however, the base of the rigid Pannonian crustal block is present. We conclude that

this block looses its basal parts and becomes thinner towards the Alps. Therefore, it may be possible that both crust/mantle boundaries overlap to some degree. The interface on top of the 7 km/s-layer encountered in the transitional zone between the two crustal elements may belong to an intermediate crustal fragment which appears to be jammed in between the European plate and the Pannonian hinterland.

Eastern Alps

The geology of the Eastern Alps is determined by the Austro-Alpine and South Alpine basement and sedimentary rocks which occupy the whole width of the Eastern Alps from the Po plain to the external molasse basin. According to their stratigraphic evolution the Austro-Alpine and South Alpine realms are related but they are tectonically separated by the Gailtal-Tonale section of the Periadriatic line. While the Austro-Alpine to the north of this lineament is directly connected with the Inner Carpathians, the Southern Alps, south of the Periadriatic line, pass into the Dinarides.

The Austro-Alpine represents a huge overthrust mass which consists of a complex system of nappes. Under it a stack of very differently developed tectonic units is hidden. It has to be derived from the palaeogeographic zones originally situated between the pre-Alpidic European foreland and the Austro- and South Alpine hinterland. In this area a system of basins and ridges was situated corresponding to the continental slopes and the Penninic realm which, during the Lower Cretaceous, was subdivided into a northern oceanic basin, a central ridge of continental crust and a southern oceanic basin. These sediments and basement rocks appear in a series of fensters (Unterengadin, Tauern, Wechsel, Rechnitz) affected by an Alpidic metamorphism, or as sheared-off unmetamorphic sediments along the external front of the Austro-Alpine "Kalkalpen". The enormous compression of the underlying units is best revealed along the Western border of the Eastern Alps, where the Austroalpine units bend back towards the Insubric line.

A first phase of compressional tectonics and metamorphism is of early Upper Cretaceous age (Trümpy 1973, 1975: palaeo-Alpine event). Another important period of thrusting and subsequent metamorphism falls into the Upper Eocene (meso-Alpine event). During the Lower Miocene (neo-Alpine event) strong pressure moved the Alps again towards north. On that occasion occurred the overthrust of the Austro-alpine units over the molasse basin of the foreland.

Across the Eastern Alps a classical profile exists between Eschenlohe and Lago Lagorai which had been interpreted as a symmetric crustal structure (Prodehl 1965; Giese 1968a; Angenheister et al. 1972). It should be reconsidered in the light of new data and experience.

The continuously increasing crustal thickness is clearly evidenced by the data of the P^M group when going from the northern Alpine margin up to the Periadriatic line near Merano (profile Eschenlohe-S, profiles from shotpoint D to W, E, and S Miller 1976, profile Lago Lagorai-N). The profile Lago Lagorai-E also shows overcritical reflections from the crust/mantle boundary resulting in a depth of 50 km for this interface. This value, projected parallel to the isogals onto the profile Lago Lagorai, gives a depth of 50 km near Lago Lagorai. Thus the course of the crust/mantle boundary can be delineated from the northern foreland of the

Alps up to the region near Bolzano (Behnke et al. 1962; Giese 1968a; Miller 1976).

A new refraction profile from Padova to Innsbruck reveals a crustal thickness of about 30 km under the Colli Euganei and for the southern border of the Alps near Vicenza (Polizzi 1976). This thickness can be regarded as normal for areas not touched by young orogenic movements.

For the part of the profile between Lago Lagorai and the southern border of the Alps, the crust/mantle boundary had formerly been traced by simple interpolation between the two depth values. The result was a southward updipping interface which corresponds to the gravity data. Nevertheless we are not forced to accept this model. Another solution seems to be more probable, and it has been presumed by Closs (1975).

On the record section of the profile Lago Lagorai-W (Behnke et al. 1962) we can identify clear later arrivals at a distance of 70 to 100 km from the shotpoint. These can be well interpreted as a reflection group. The depth of the reflecting surface can be calculated with 20 ± 2 km. In view of this interface we propose the following new interpretation of the crustal structure in the Southern Alps: This interface is considered to represent the lower part of the crust of the Adriatic plate which has, in the S, a normal thickness of about 30 km. This model implies a wedge-like Adria microplate getting thinner towards the Periadriatic line of the Alps by a reduction of the uppermost crust, including its sedimentary cover, as well as by a loss of parts of its upper mantle and of the lower crust.

The northern pre-Alpidic European crust/mantle boundary on the other side dips southward down under the crust and the upper mantle of the Adria microplate. It disappears here and is no longer detectable. Hence, in terms of plate tectonics, the European continent seems to represent the lower plate, which is marginally subducted beneath the Adria microplate.

There are, however, several facts which should be kept in mind. Dating of metamorphism proves that the southern Penninic oceanic basin was already closed in the Upper Cretaceous. According to Frisch (this vol.), a new subductory system developed thereafter which caused the closure of the northern oceanic basin in the Upper Eocene and the collision of the downgoing European plate with the Adria microplate. The neo-Alpine compressional tectonics must be considered as a post-collisional process in the course of further subduction of lithosphere. They may be due to the phenomenon of suture progradation (Roeder this vol.) rather than to a new subduction system.

The orogenic history implies that after each tectonic event a reorganization of the subductory system took place. As all these different kinematic patterns acted in more or less the same direction their effects were superimposed without obliterating the plate collisional model which is revealed by the seismic refraction profiles.

It is remarkable that in contrast to the Carpathians, the Apennines, the Hellenides and the Dinarides the Alpine hinterland was not subjected to strong tensional tectonics. Only the magmatism along the Periadriatic line and the volcanism of the Colli Euganei point to this geotectonic function of the Southern Alps.

Border Region between Eastern and Western Alps

Geologically the border between Eastern and Western Alps is marked by the ending of the broad Austroalpine zone near the Rhine valley. To the west the Alps are essentially built up of tectonic units of the foreland (Helvetic nappes) and of the complex Penninic eugeosyncline. The Austroalpine realm is only represented by a small strip of metamorphic basement rocks, which can be followed immediately north of the Insubric line (westernmost part of the Periadriatic line) into the Sesia-Lanzo zone. The Southern Alps, situated to the south of the Insubric line, do not show any structural changes from east to west. The Insubric line forms a sharp boundary between units affected by strong Alpine metamorphism in the north, and rocks without metamorphism of this age in the south.

The Periadriatic line distinctly changes its position with respect to the contours of the Bouguer gravity from the Eastern Alps into the Swiss Alps. It starts from the gravity minimum in the Eastern Alps and shifts towards the axis of the gravity maximum in the Western Alps.

From a number of seismic-refraction profiles, a cross-section between Chur and Bergamo has been constructed (Choudhury et al. 1971). It shows a markedly asymmetric structure. South of the Insubric line the crustal thickness is distinctly smaller than north of it in the central zone of the Alps. An overlapping of two crust/mantle boundaries could not be proved by seismic data.

Geological considerations, too, suggest an overlapping which would correspond to a crustal overriding of the Southern Alps resp. of the Adria microplate. It is indicated by the strong contrast of the highly compressed and metamorphic rocks north of the Insubric line and the slightly folded non-metamorphic rocks south of it, and also by the presence of Tertiary plutonic rocks near this line (Bergell, Adamello). As in the case of the Eastern Alps, the thick crust to the north of the Insubric line can be interpreted as the rim of the pre-Alpine European continent which was loaded by allochthonous (obducted) material and thickened by the orogenic compression.

Western Alps

Important structural elements of the Western Alps are the autochthonous crystalline massifs (Aare, Mt. Blanc, Belledonne, Pelvoux, Argentera), which represent upwarpings of the basement of the northwestern and western foreland. They are surrounded by autochthonous sediments (Helvetic and Dauphinée) and nappes from the zones at their internal side (allochthonous Helvetic and Penninic units). According to the sedimentary record, the Penninic realm was composed of very heterogeneous zones during the Jurassic and Lower Cretaceous. The eugeosynclinal, probably oceanic, trough, which terminated towards the south, was limited by a broad intermediate continental ridge (Briançonnais). At the internal side of this ridge, the Piedmont or Penninic-Ligurian eugeosynclinal trough with its partly oceanic basement was situated. As almost everywhere in the circum-Adriatic chains, in the Western Alps the former border of the Adria microplate cannot be easily determined (Trümpy 1975). Most geologists consider the strongly metamorphosed Sesia-Lanzo zone as an Austroalpine crustal element, but it is separated from the Southern Alps near Ivrea by the narrow Canavese zone which shows eugeosynclinal features.

The Upper Cretaceous ages of the HP/LT metamorphism of the Sesio-Lanzo Zone and parts of the Piedmont zone indicate paleo-Alpine compressional movements and subduction. Dal Piaz et al. (1972) and Compagnoni et al. (1975) even suppose that continental collision had already occurred during this period. However, as locally in the Piedmont zone, deep water deposition continued until the Lower Tertiary and its final closure, together with that of the Valais trough, was probably achieved only with the meso-Alpine orogenetic event (Trümpy 1973; Laubscher 1975). During this phase Penninic sediments and basement rocks were thrust as huge nappe complexes towards the external zones. While tectonics with the same polarity continued until the Miocene in the external zones, the internal zones were subjected to a retrocharriage during the Oligocene which led, in an Apenninic sense, to fold and thrust towards the internal side of the Western Alps.

From a gravimetric point of view, the crustal structure of the Western Alps is governed by the gravity high of the Ivrea zone, situated between Locarno in the north and Cuneo in the south (Niggli 1946). In the past, its crustal structure was regarded as quite anomalous (Berckhemer 1968, 1969; Giese 1968b). Considering its asymmetry, however, the Ivrea structure can be understood as an extreme modification of this general type. The overlapping is clearly evidenced here by numerous profiles. Giese, Günther & Reutter (1971) and Giese & Prodehl (1976) have demonstrated the regional distribution of the overlapping of the Adria microplate onto the European plate. The result is an eastward extension of the European crust at least as far as Torino.

The structural and crustal disposition of the Western Alps corresponds to a marginal overriding of the hinterland complex over the pre-Alpidic European foreland plate as a consequence of the total consumption of the oceanic crust of the Penninic-Ligurian and the Valais basins. However, as in the Eastern Alps, the structure cannot correspond to such a simple model because of the repeated reorganizations of the subduction system. Thus, the thick sialic root beneath the central zone of the Western Alps may represent a younger meso-Alpine or neo-Alpine collisional structure along a more external, also southeast dipping, subduction plane. The polarity of the orogenetic movement was always the same; the model of synthetic thrusting and subduction has been preserved; the crustal effects of the retrocharriage along the Ivrea cross section were evidently too weak to obliterate it.

The granulitic and ultramafic rocks in the Ivrea zone, which geologically belong to the Southern Alps, indicate that a basal part of the crust of the hinterland and even its transition into the upper mantle is exposed. This quite special situation may be due to the steepening and even overturning of the internal zones in the course of the retrocharriage.

Transition from the Western Alps to the Northern Apennines

The small mountain chain north of Savona and Genova which orographically connects the Alps and the Apennines represents a transitional area. It was subjected to Alpine as well as to Apenninic movements (Günther & Reutter this vol.).

Western Alps and Apennines are parts of the same orogenetic system. Thus, eugeosynclinal rocks which are found in an internal allochthonous

position in the Alps as well as in the Northern Apennines can be derived from the same oceanic Penninic-Ligurian eugeosyncline. However, the transport directions of the Western Alps (W to SW) are almost opposed to that of the Northern Apennines (N to NE) (Reutter, Günther & Groscurth this vol.).

The main orogenetic phase of the Western Alps (Upper Eocene, meso-Alpine event or Ligurian phase) preceded that of the Northern Apennines (U-Miocene, Tuscan phase). Therefore, in the transitional area, considerable parts of the Western Alps were affected by the two orogenetic phases to such a degree that the divergent transport directions were superimposed. During the Ligurian phase the basement of the Po plain and the autochthonous basement of the Apennines acted as a hinterland, and the European plate as a foreland of the Western Alps. However, during the Apenninic orogeny, the Po plain and the autochthon of the Northern Apennines, as parts of the Adria microplate, assumed the role of the foreland, whereas the Western Alps and their prolongation to Corsica became the hinterland.

In the orographic transitional zone between the Western Alps and the Northern Apennines, the region of Genova - Savona, a crust/mantle boundary could be found at a depth of 25-30 km. On the other hand, a further discontinuity at a depth of about 50 km was revealed which can be interpreted as a second crust/mantle boundary. While the upper boundary must be correlated with the crust/mantle boundary of the Adria microplate, the deeper one has to be associated with that of the pre-Alpine European continent. Thus, this crustal overlapping is in principle the same as that in the Ivrea zone. The shallow crust/mantle boundary continues eastwards along the Ligurian coast at the same depth of 25-30 km, as it is situated W of Genova.

In contrast to the well-developed asymmetry of the crustal structure in the Carpathians and the Alps, which is evidently due to the polar development of the orogen, in the Alps/Apennines transitional zone with its changing polarities, a more complicated structure has to be expected. On the one hand, during the Apenninic orogeny the margin of the Adria microplate was loaded with a thrust nappe consisting of tectonic units which clearly belong to the Western Alps and which are still connected with them. On the other hand, deeper parts of the Alpine crust and mantle which may at least partially correspond to the pre-Alpidic European plate are pushed beneath the Adria microplate. These contradictions had been presumed in previous plate tectonic theories (Boccaletti, Elter & Guazzone 1972).

The question arises as to whether continental collision was already completed during the Ligurian phase which there led to an asymmetric structure similar to that near Ivrea, or whether there was still a suture with oceanic crust left which was only closed during the Apenninic Tuscan phase. In any case this kind of crustal doubling must be related to the preceding orogeny of the Western Alps.

The retrocharriage of the Western Alps, which postdates the Ligurian phase, can be considered to be the beginning of the Apenninic orogeny. The deformation then slowly proceeded towards the external Apenninic zone. This, finally, means that the almost vertical setting of the different units and especially of the Southern Alps near Ivrea were due to the Apenninic events. This is also supported by the gravity high and the steep or probably overturned margin of the Adria microplate which was hidden beneath the central zone of the Western Alps (Dora Maira Massif)

in the course of the retrocharriage.

No matter how the complex crustal structure in the transitional zone was generated, it now resembles the model of flake tectonics, once proposed by Oxburgh for the Eastern Alps.

Northern Apennines - Corsica

As pointed out before, the Northern Apennines on one side and the southern part of the Western Alps and their prolongation to Corsica on the other form an orogen (Reutter, Günther & Groscurth this vol.: "double orogen") with divergent, partly superimposed structures. The Apenninic part of it becomes broader from its onset near Torino towards the southeast, whereas the Alpine part seems to become less important in this direction.

In northeastern Corsica a tectonic stack of Penninic-Ligurian eugeosynclinal sediments containing ophiolites was thrust (obducted) upon these parautochthonous units and the autochthon, which consists of basement rocks without any mesozoic sedimentary cover (Seidl & Zacher this vol.). HP/LT metamorphism in the basal part of the stack of eugeosynclinal meterial (Ohnestetter et al. 1976) indicates an eastward subduction of Lower Tertiary age. The same conditions are present in the ophiolitic complex north of Genova (Sestri-Voltaggio Zone) and in the internal zones of the Western Alps.

In the Northern Apennines several nappes of eugeosynclinal sediments and ophiolites (Ligurides) exist which could have been derived paleogeographically from the Penninic-Ligurian eugeosyncline. They overlie allochthonous and autochthonous sequences of Mesozoic and Cenozoic pelagic sediments and Neogene graywacke flysches (Tuscanides and Umbro-Marches). The internal parts of the Tuscanides were affected by strong deformation and low-grade metamorphism.

The nappe tectonics have been interpreted as a consequence of obduction of the eugeosynclinal material from the southwest upon the internal part of the Tuscanides (which correspond to the margin of the Adria microplate) and subsequent spreading of the allochthon by gravity towards the northeast and east. Gravity sliding was facilitated by shifting of the axis of the depositional trough of the Neogene flysches from the southwest towards the northeast. A belt of compressional tectonics and upwarping follows in the same direction from the uppermost Oligocene until the Upper Miocene. During the Pliocene, the depositional trough migrated into the Po basin. Behind the compressional zone, tensional tectonics and magmatic activity developed.

The shallow crust/mantle boundary found in the transitional zone from the Western Alps to the Northern Apennines continues eastwards into the shelf region of the Ligurian Sea. Here the crust shows a thickness of about 20-25 km. Towards the Northern Apennines this crust thickens continuously to 35 km under the crest of the Apennines (Letz et al. this vol.). It continues probably without large displacements into the crust of the Po plain having here about the same thickness (Giese & Morelli 1975; Stein this vol.). This result, as well as geological evidence, leads to the conclusion that the crust of the Ligurian shelf region can be seen as the present margin of the Adria microplate.

Further, in the Ligurian Sea, between La Spezia and Elba, a second deeper crust/mantle boundary could be revealed. This deeper crust seems

to be connected with the crust of the Corso-Sardinia continental fragment, although in the Corsica basin, the upper mantle is somewhat updömed.

While the structural and crustal conditions of Corsica are compatible with a subductory process similar to that supposed for the Alps and Carpathians in as much as the crust of the foreland grows thicker towards the internal side of the hinterland, the crustal structure of the Apennines is quite different. Although the allochthonous material has to be derived from the southwest (Penninic-Ligurian trough and internal Tuscan zone) and its advancement onto the foreland can be well dated, the crust of the foreland does not get thicker towards the hinterland. This is also evidenced by the gravity anomalies, which show a marked minimum at the northeastern (external) edge of the Northern Apennines. There the top of the basement has to be expected at a depth of more than 10 km. But already in the central zone of the Apennines, it was found to be only 3-4 km dep, and in all the seismic profiles through the Ligurian Sea (Letz et al. this vol.) it appeared on the shelf and its outer edge at 5 or less km of depth. Therefore, the Apenninic thrust sheets are relatively thin and not wedge-shaped. This configuration may be explained by marginal subductory downbending of the margins of the Adriatic plate beneath an accretionary wedge of Penninic-Ligurian oceanic sediments and ophiolites (Reutter & Groscurth this vol.) and a subsequent uplifting in the course of the continental collision and underthrusting of the margin of the Corso-Sardinian continental fragment.

Disregarding the special conditions at the southwestern border of the Adriatic plate, the greatest crustal thickness of 35 km or even more is found in the very external zone of the Apennines at the border of the Po plain. It corresponds to the belt of Pliocene compressional tectonics as pointed out by Elter et al. (1975). At its internal side, in Tuscany, the crust is considerably thinner. Thus, the tensional tectonics, young magmatism, and the high heat-flow values confer the characteristics of a hinterland to this region even though this configuration is established in the Adriatic plate. It may represent the onset of a new subductory system whose importance increases towards the Southern Apennines.

Southern Apennines, Calabria, and Sicily

The mountain ranges bordering the Tyrrhenian Sea to the east and south show, once more, the effects of strong compressional tectonics. The internal chains of the Southern Apennines as well as of Sicily, consisting of Mesozoic carbonate sediments and Neogene flysches, form allochthonous thrust sheets. They override similar sediments of the external zones and of the respective forelands represented by the Apulian and the Iblean platforms (Andreieff et al.1974; Ogniben 1969, 1975; Catalano et al. 1976; Amodio-Morelli et al. 1976). In Calabria, where the structures are sharply bent from the Apenninic NW-SE direction into the Sicilian E-W direction the appearance of crystalline basement rocks emphasizes the special geotectonic situation of this region.

An unravelment of the nappes according to their stacking order reveals that towards the internal parts there was an array of longitudinal zones with very thick shallow-water carbonates and basins with thin deep-water deposits. The most internal elements are partly metamorphic flysches of Cretaceous age and, in Northern Calabria, metamorphic ophiolites associated with Jurassic and Lower Cretaceous sediments. Metamorphism

is of the HP/LT type. Because of the oceanic character of these internal formations and their stratigraphic and lithological affinities, they are considered to be a southward prolongation of the Penninic-Ligurian eugeosyncline.

A problem is posed by the huge mass of crystalline basement rocks of Calabria and northeastern Sicily, which consists of several tectonic units of metamorphic, magmatic, and sedimentary rocks of Palaeozoic age. In Sicily these basement rocks hold an internal position, whereas in Northern Calabria they are relatively external. There, they overlie internal carbonate units and the Penninic-Ligurian series, and many authors therefore consider the basement rocks as allochthonous thrust sheets of a very internal origin. However, the relationships of the eastern basement units to the Neogene sediments and olistostromes along the Ionian coast and some geophysical data may be used as an argument for a more autochthonous, i.e. external position (Görler & Giese this vol.).

The tectonic setting of the internal zone may have been achieved by polyphase movements. First, northwestward directed movements of probably Upper Cretaceous and Lower Tertiary age, culminating in the Upper Eocene "Ligurian Phase" probably led to an "Alpine chain" (Amodio-Morelli et al. 1976). Just as in the Corsica/Northern Apennines segment, during the Oligocene movements inverted their directions and Penninic-Ligurian material was thrust upon the bordering Apenninic-Sicilian internal carbonate zones. It is not clear if the HP/LT metamorphism of the eugeosynclinal ophiolites and sediments is due to the first or second phase. During the Upper Miocene and Lower Pliocene the internal zones proceeded towards their foreland. Only in the Calabrian arc do the orogenic processes not yet come to an end (Görler & Giese this vol.).

Southern Italy has been studied by an extensive seismic-refraction program thus giving a detailed picture of crustal structure in this region (Cassinis et al. 1969; Colombi et al. 1975; Morelli et al. 1975; Schütte this vol.). In the Southern Apennines, Northern Calabria, and Sicily a common type of crustal structure could be proven: The Tyrrhenian shelf and its coastal region show a crustal thickness of only 20-25 km. Going inland crustal thickness abruptly increases within 10-20 km up to 40-45 km. Although for Southern Calabria this change in crustal thickness has not been ascertained since seismic data are available only from the Ionian coast, the strong gravimetric gradient along the western shore implies a similar structure. Towards the Apulian foreland as well as to the southern coast of Sicily the crustal thickness is reduced to 30 km. The structure of the transition to the Ionian Sea is unknown due to the lack of data.

In addition to this general picture which supports this asymmetry structure, two further details could be detected (Schütte this vol.). The seismic-refraction profile Palermo-Peastum-Gargano reveals a discontinuity at a depth of 60 km offshore in the Tyrrhenian Sea. It is correlated with the deep crust/mantle boundary under the crust of the Southern Apennines. Thus the crust/mantle boundary of the Adria microplate dips down towards the Tyrrhenian Sea and a crustal doubling exists under the Tyrrhenian shelf. Due to the principal similarity of crustal structure along the Tyrrhenian shelf, it is probable that the overlapping of two crusts is a general feature.

Besides this crustal doubling, a further special structure could be revealed in the Sila east of Cosenza. A high-velocity layer of 7.5-8.0 km/s in a depth of about 20 km was found to be underlain by a velocity

inversion (Schütte this vol.). There is no proof, however, that the shallow crust/mantle boundary under Sila is connected with that of the Tyrrhenian shelf. It may indicate a special geotectonic setting of the Sila block due to later movements of the Calabrian arc (Görler & Giese this vol.).

The same crustal structure as in the Southern Apennines is met below the northern coast of Sicily. Again the first crust/mantle boundary of the Tyrrhenian shelf area, 20 km deep, is underlain by a second crust with its lower boundary at about 40 km. Towards the southern Sicilian coast this latter crust climbs up to 30 km and to 20 km in the Sicilian Sea.

The crustal structure along the southeastern Tyrrhenian coast corresponds to the type present in the Carpathian and the Eastern and Western Alps inasmuch as the crust of the hinterland represented by the Tyrrhenian Sea and its coastal zone seems to be thrust upon the crust of the foreland.

Just as in the Carpathians, the crust of the hinterland is a composite one which was assembled to the present state by previous subductory processes. During the Upper Cretaceous and Lower Tertiary the consumption of the southern prolongation of the Penninic-Ligurian eugeosyncline took place by southeastwardly dipping subduction beneath the internal zones of the South Apenninic-North Sicilian carbonate platform. It has not yet been ascertained if thereby continental collision was already completed as suggested by Amodio-Morelli et al. (1976), or if an eastern part of the eugeosyncline was preserved (Görler & Giese this vol.). In the latter case the conditions would have been the same as in the Northern Apennine/Corsica segment. In either case, starting in the Oligocene just as in the north, the whole subductory system was reorganized and consumption of the remaining eugeosyncline or the area bordering the Alpine suture to the east took place along a northwest dipping subduction plane. Thus Penninic-Ligurian sediments and ophiolites, in Calabria possibly together with some of the basement units, were thrust upon the internal (Campanian-Lucanian and Panormide) carbonate zone of the Apennines and southern Sicily.

A renewed reorganization took place in the Middle Miocene. Subduction occurred again along north (Sicily) and west dipping planes, but now the thin continental crust of the basins within the carbonate zones (Lagonegro and Molise basin in the Southern Apennines, and Imerese and Sicano basin in Sicily), was consumed, which created the crustal structure as revealed in the seismic profiles. These show that the Iblean foreland, as part of the African plate, dips beneath the crust of the north coast of Siclily and the Tyrrhenian shelf. The same phenomena are found in the Apulian foreland which, as part of the Adria microplate, dips beneath the crust of the internal carbonate zone of the Southern Apennines.

During this process the Tyrrhenian Sea opened and, together with the already deformed internal parts of the Southern Apennines and Sicily, as a hinterland. These parts, represented by the Campanian-Lucanian zone (Apennines) and the Panormide zone (Sicily), however, must be considered as a mobilized branch of the Adria microplate which extends into the North African Maghrebides. It lost its stable-plate behaviour only during the Miocene. The transition from stable-plate to mobile-arc behaviour lies in Tuscany, where it is indicated by the increasing thickness of the crust towards the external side of the Northern Apennines.

Subduction in Sicily and in the Southern Apennines stopped in the Lower Pliocene. Off the Ionian coast, however, it is still going on. The

oceanic area of the Ionian Sea, which separates the African plate from the Adria microplate, offered no possibility of collision and could not stop the southeastward movement of the Calabrian arc (Görler & Giese this vol.).

Hellenides

The broad Hellenide-Dinaride segment of the Mediterranean mountain belt follows the contours of the eastern margin of the Adria microplate. Tectonic transport and structures are directed towards the Adria microplate ("Africa-vergent"), and hence the Hellenides and Dinarides are diametrically opposed in a convergent sense to the Apennines. In a divergent sense they are opposed to the ("Europe-vergent") Alpine-Carpathian segment.

Just as in the other chains which surround the Adria microplate, the orogenetic compressional movements, expressed in thrusts and nappes, are of ages varying from the early Cretaceous to the Middle Miocene. According to Jacobshagen, Dürr, Kockel, Kopp & Kowalczyk (this vol.), the Hellenides can be subdivided into the pre-Apulian foreland, West Hellenic nappes, Central Hellenic nappes, Inner Hellenic nappes, Vardar zone, and the Serbo-Macedonian hinterland.

The pre-Alpine and the Talea Ori zones of the Adria microplate represent the foreland of the Hellenides. As several fensters prove, especially on the Peloponnesus, these zones extend far beneath the West Hellenic and Central Hellenic nappes at a shallow depth. From the foreland to the internal parts several zones of Mesozoic deep-water and ridge sediments are found. Thus, it is not clear whether the former border of the Adria microplate corresponds to the basin of the West-Hellenic Ionian zone, to that of the Central-Hellenic Pindos zone, or to the ophiolitic belt of the Vardar zone.

A great paleogeographic problem is posed by the ophiolites and oceanic sediments which are intimately connected with Triassic and Jurassic sediments of the external Pelagonian zone. According to Hynes et al. (1972), Smith & Moore (1974), and Channel & Horvath (1976), they have to be derived from an "Othris ocean" which separated the Pelagonian crustal element from the Adria microplate during the Jurassic and Lower Cretaceous. However, Bernoulli & Laubscher (1972), Aubouin (1973), and Jacobshagen, Dürr et al. (this vol.) consider these oceanic rocks as a huge nappe which was thrust from the Vardar zone across the Pelagonian platform in the Lower Cretaceous. In the first case, these "eo-Hellenic" movements would show a vergence opposed to that of the later orogenic transport directions.

The Vardar zone is a strongly deformed belt which has the structural and metamorphic characteristics of a root zone (Zimmermann & Ross 1976). It represents the suture of an originally oceanic area from which not only the "eo-Hellenic" ophiolites, according to the hypothesis of Bernoulli & Laubscher (1972) but also the Upper Cretaceous and early Tertiary ophiolitic nappes of the Inner Hellenic zones must have originated. The Vardar zone separates the Hellenic nappe system from its hinterland, the Serbo-Macedonian and Rhodopian massifs. In a similar way as in the Carpathians, this hinterland area was not a rigid continental plate: it was subjected to deformations in the Late Cretaceous and to intrusions during the Tertiary.

The internal zones were affected by several phases of tectonic movement and metamorphism. The deformation started as early as in the Jurassic in the circum-Rhodope belt (cf. Jacobshagen, Dürr et al. this vol.). A very important phase occurred during the Lower Cretaceous which led to the above mentioned emplacement of ophiolitic nappes into the Central Hellenides. Upper Cretaceous and Paleocene events are also documented. Of special importance is the Upper Eocene phase, to which the emplacement of the Inner Hellenic nappes and internal thrusting within the Central Hellenic nappes are attributed. In the course of the Miocene, the central Hellenic nappes were thrust upon the West Hellenic ones and both upon the foreland.

The Aegean and the East-Mediterranean arc systems are still tectonically active.

Each period of tectonic activity was accompanied or followed by flysch or molasse deposition. From the Paleocene until the Upper Miocene the basin of flysch deposition migrated from the central (Pelagonian) to the external (pre-Apulian) parts of the Hellenides. As in the Carpathians, in the Northern and Southern Apennines flysch deposition was followed first by compressional and then by tensional tectonics. In the central and internal parts of the orogen, volcanic and plutonic activities of magmas with crustal affinities are connected with the tensional tectonics.

Structurally the Western and Central Hellenides are characterized by flat-lying thin nappes, as evidenced by many fensters. The Olympos fenster shows the Pelagonian nappe to be overthrust over West Hellenic sediments and HP/LT metamorphic ophiolites of uncertain attribution. The basal parts of the huge Median Crystalline Belt (Jacobshagen, Dürr et al. this vol.) also represent similar fensters. This implies that nappe thrusting exceeded 100 km. Because of this far-reaching nappe transport and the small thickness of the thrust sheets gravitation must have been the main agency of this transport.

The Ionian Sea at the external side of the Hellenides and the Aegean Sea at the internal side are marked by positive Bouguer anomalies. As the internal positive anomalies extend into the Vardar and the Pelagonian zone, the intervening zone of strong negative Bouguer anomalies lies in a relatively external position of the mountain chain (Makris 1977; Makris et al. 1973). Its axis is situated near the present thrust zone between the West and the Central Hellenic nappes. The strongest anomaly of -140 mgal could be detected in the Pindos mountains.

Some seismic-refraction profiles helped to reveal further details of crustal structure. Along a profile through the Peloponnesus (Jacobshagen, Makris et al. 1976) in the pre-Apulian foreland near the coast of the Ionian Sea, the crust was found to be only 20-22 km thick. This thin foreland crust continues inland for a distance of about 20-30 km under the area of the West Hellenic nappes. There the crustal situation changes abruptly. The shallow crust/mantle boundary of the foreland disappears completely eastwards and in its place a "separate" crust/mantle interface at a depth of 45-50 km is indicated by reflected waves. It seems that the two boundaries overlap each other. The relatively low average crustal velocity, derived from the deep crust/mantle boundary reflection, shows that a considerable portion of sialic material must exist in this thick crust.

The crustal thickness under the western and central Peloponnesus decreases distinctly towards the internal zones according to the gravimetric data. Since no seismic observations are available from east to west at the

eastern end of the profile, details of the eastward up-dip of the crust/mantle boundary are unknown. A strong gradient in the Parnon mountains of the eastern Peloponnesus, however, indicates further complications of crustal structure. The central zones of the Hellenides are now occupied by the Aegean Sea, which presently acts as a hinterland of the East Mediterranean arc, a crustal thickness between 25 and 30 km has been measured. This value is adequate to such a geotectonic situation.

Although the orogenetic polarity of the Hellenides and, hence, the location of hinterland and foreland is apparently clear, this orogen does not correspond to the model of the Carpathians and the Alps where the hinterland is in the position of an upper plate that marginally overrides the down-dipping foreland. Perhaps, as in the case of the Northern Apennines and Calabria, the deviation from a simple plate tectonic model may be due to the successive superposition of different stress and movement patterns.

If the Pelagonian crustal element was a continental fragment surrounded by oceanic areas (Channel & Horvath 1976: "Drina Ivanjica element" and "Pelagonian plate"), at least two continental collisions must have occurred. According to Hynes et al. (1972), consumption of the hypothetical Othris ocean led to the eastward obduction of the Central Hellenic ophiolites during the Lower Cretaceous. This assumption would imply an inversion of the polartiy of the following orogenetic processes.

The Vardar ocean, from which, according to Bernoulli & Laubscher (1972) these ophiolites were obducted in a southwestern direction, was closed in the Eocene after further obduction of oceanic material during the Upper Cretaceous and Lower Tertiary (Jacobshagen et al. this vol.). Zimmermann & Ross (1976) argue that antithetically to the foregoing obduction, the Rhodopian hinterland was thrust to the southwest far beneath the Pelagonian platform. This last situation would be similar to that in the Northern Apennines, where obduction of ophiolites and sediments upon the margin of the Adria microplate was followed by underthrusting of the hinterland in the same direction.

In whatever manner these obduction and collision processes took place, after each of these events readjustments must have occurred since the tectonically emplaced overburden, lying on the glaucophane bearing metamorphic zones must have been reduced by further tectonic (gravitational?) processes. No seismic refraction data are available to elucidate the crustal structure of the internal zones.

The gravity minimum of the Hellenides reflects another crustal event outside the Vardar root zone. In the profile through the central and external zones of the Hellenides in the Peloponnesus an underthrusting of an eastern crustal portion beneath the crust of the foreland is suggested by seismic data (Jacobshagen, Makris et al. 1976). If this interpretation is correct, the model of the internal zones of the Northern Apennines may again be applied. This would mean that the crust of the pre-Apulian foreland which, according to the presence of a series of fensters, underlies most of the Peloponnesus, must first have bent down in order to receive the load of the Western and Central Hellenic nappes. Subsequently it was antithetically underthrust by an eastern continental crust. The HP/LT metamorphism (glaucophane) of the Phyllite series at the base of the Tripolitza nappe indicates that initially there was a great tectonically emplaced overburden. In the course of uplifting in connection with the crustal underthrusting towards the southwest the allochthonous sediment sequences slid in the same direction. This movement was controlled by

gravity which acted on the slope of the shifting configuration of an external flysch trough and an internal upwarping compressional belt.

The suture of the crustal underthrusting, revelaed in the seismic refraction profile, may correspond to the basement of the Ionian zone or to that of the Pindos zone. As already pointed out, it must not have been oceanic; it could have been only thin continental. In this case it would be within the Adria microplate and not at its margin, and the structure would represent intraplate thrusting rather than continental collision. Such internal tectonics along zones of weakness within the Adria microplate in the form of overthrusts resemble the situation in the Southern Apennines and Sicily. The crust of the eastern side of this suture is probably a composite one which originated from other crustal events during the Cretaceous and Lower Tertiary. Its internal structure may have been modified during the Miocene phases.

A last reorganization of the Hellenides obviously took place during the Pliocene when the still active Aegean arc system came into being and overprinted the former crustal conditions. Thus, on the island of Crete, the foreland units exist together with the West Hellenic units and the Pindos nappe, indicating that the Adria miroplate had an extension into this southeastern part of the Mediterranean. However, now the foreland crust and the nappes act together as the front of a new Aegean hinterland. With a belt of active tectonics in form of the Hellenic trench and the Mediterranean ridge at its external side now it is moving onto the continental rise of Africa.

Indeed, according to Makris (1977), the island of Crete shows a thick crust of 35-40 km. Immediately off the north coast gravimetric and seismic data indicate a crust of only 20-22 km which corresponds to the backarc conditions of the hinterland. This abrupt change of crustal thickness suggests a crustal structure, which is synthetic to the orogenic evolution.

Dinarides

The paleogeographic and structural zones of the Hellenides run into the Dinarides with only slight modifications (Aubouin 1973). The Karst zone of Yougoslavia represents a dominant element whose neritic sediments were deposited on a portion of continental crust which obviously belongs to the Adria microplate. As the Dinarides have their equivalent in the Central Hellenides (Parnassos zone), the Ionian as well as the Pindos deepwater basins gradually loose their paleogeographic and structural significance towards the northwest.

The problem of the proveniance of the Serbian ophiolites is the same as in the Hellenides: Either they were thrust out of a hypothetical Othris ocean which should have been separated from the Vardar ocean by a continental fragment ("Drina-Ivanjica element", Channel & Horvath 1976) or they had their origin in the Vardar ocean and were overthrust towards the southwest once during the Lower Cretaceous and again during the Lower Tertiary together with their extraneous substratum. As already mentioned this latter hypothesis would not require a polarity reversal during the orogenic development. In any case, at present the Serbian ophiolites are separated from the Vardar ophiolites by a non-ophiolitic zone which corresponds to the Pelagonian platform of the Hellenides.

The Vardar root zone and the Serbo-Macedonian crystalline hinterland

disappear near the river Save beneath the Tertiary mostly molasse type sediments of the Pannonian basin. Therefore, the important questions are left open to discussion whether and what way the eugeosynclinal, formerly oceanic, zones of the Alps (Penninic), of the Southern Carpathians and Apuseni Mountains and of the Dinarides are connected. The solution of the problem would also explain of what crustal elements the Pannonic basin is composed.

The timing of orogenic events is the same as in the Hellenides. The width of early-Tertiary overthrusting of the internal zone over the Karst zone, which again follows a shifting of flysch deposition, is at least 80 km for the Bosnian zone and 80 km for the Serbian ophiolites (Aubouin et al. 1970). Perhaps as a consequence of the termination of the Ionian and Pindos deep-water basins towards the northwest, the amount of overthrusting of the Karst zone over more external units is much less than that of the corresponding zones in the Hellenides; it dies out towards the junction of the Dinarides with the Southern Alps.

Under the auspices of the international program aiming to investigate the crustal structure of SE-Europe, two seismic-refraction profiles transversing the Dinarides (Geophys. Transactions 1972) have been investigated. Although the data are scanty, the main features are visible. The Inner Dinarides show a crustal thickness of about 25 km, a value which corresponds to that of the Inner Carpathians. The external Dinarides are characterized by a maximum depth of the crust/mantle boundary of 50 km beneath the Karst zone. Structural details of the transition from internal to the external Dinarides are lacking. The same must be said for the transition from external Dinarides to the Adriatic foreland where, according to the known values of Apulia, a crust of 30 km or somewhat less (3-5 km) can be expected. Therefore, the question as whether the downdip of the crust/mantle boundary from the Adriatic Sea to the Dinarides is continuous or if a crustal break exists as in the western Peloponnesus.

As the zone of crustal thickening holds the same position in relation to the geological structures as in the Hellenides, one may suppose that in the Dinarides an internal continental crust was also underthrust antithetically to the sense of the overthrusting of the nappes. The surface structures neither support (lack of fensters) nor contradict (absence of thick crystalline nappes) such an interpretation. A comparison with the crustal imbrication of the Eastern Alps would favour an upper prosition of the crust of the Adria microplate in the Dinarides in the same way as the Alpine crustal structure maintains its setting in the Apennines. However, little is known about the connection of the Dinarides with the Eastern Alps. The width of thrusting is clearly reduced towards the Alps and the gravity minimum beneath the external Dinarides seems to die out between the maxima of the Verona region and the Pannonian basin.

Conclusions

Geotectonic processes taking place in the Mediterranean were the generation of oceanic basins, the consumption of these basins, and the collision of great or small elements of continental crust. The opening of the Atlantic ocean and the motions of the African plate relative to the pre-Alpine European plate controlled these processes. The loop of the circum-Adriatic orogens is due to the irregular (promontory-like) shape and in-

 to p. 582

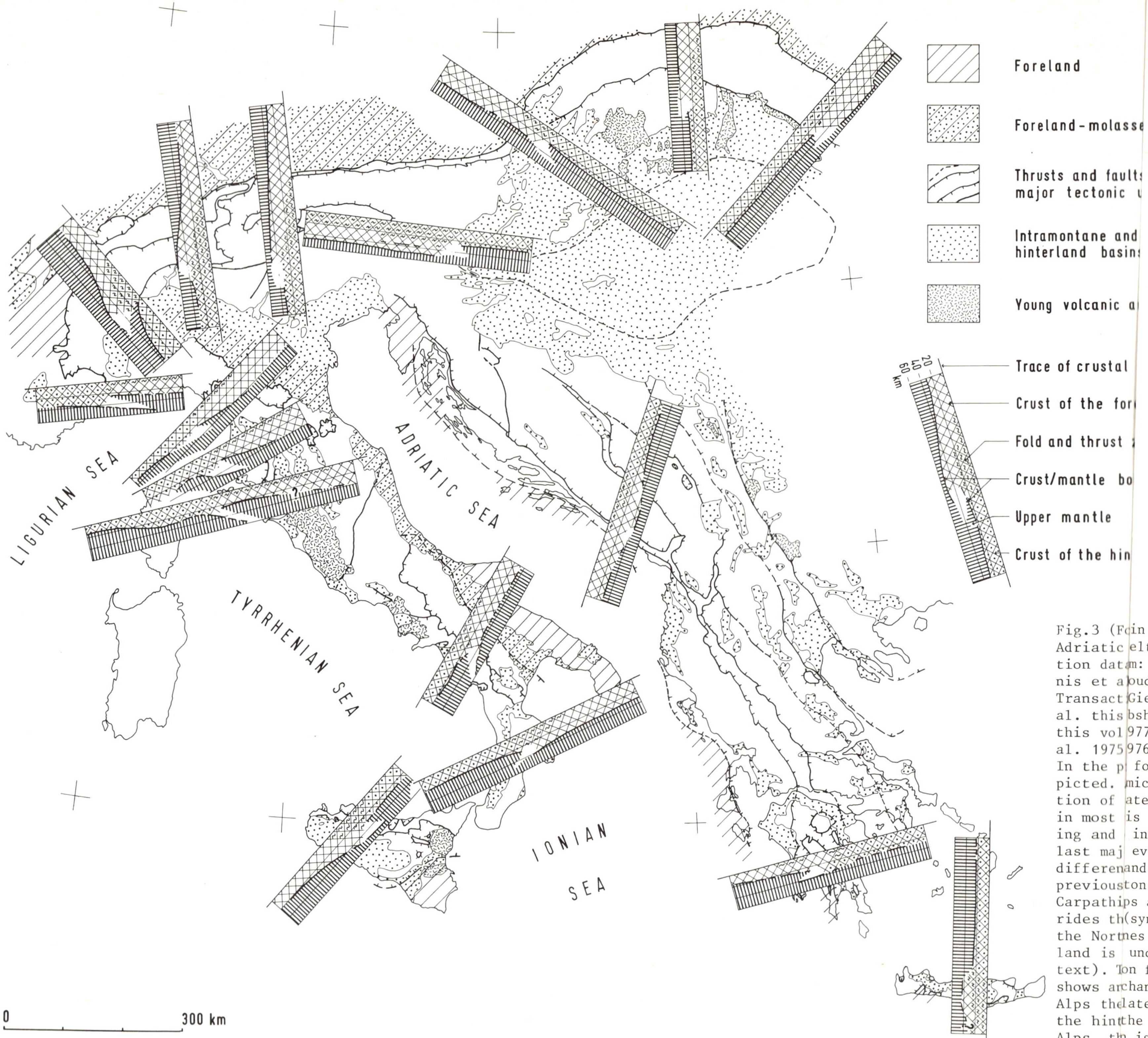

Fig.3 (Fc in crustal and structural features of the circum-Adriatic elts. The crustal profiles are based on seismic-refraction dat m: Angenheister et al. 1972; Aric et al. 1978; Cassinis et a budhury et al. 1971; Colombi et al. 1973; Geophysical Transact Giese & Morelli 1975; Giese & Prodehl 1976; Giese et al. this bshagen et al. 1976; Letz et al. this vol.; Makris this vol 977; Miller 1976; Miller et al. this vol.; Morelli et al. 1975 976; Schütte this vol.
In the p foreland, the thrust belt and the hinterland are depicted. mic-refraction method does not allow the discrimination of ated structures within the zone of crustal overlapping, in most is critical area had to be left blanc. Crustal thrusting and indicated in the sections are essentially due to the last maj event. In many cases the hinterlands are composed of differen and paleogeographic units which were assembled during previous tonic events (composite hinterland). Note that in the Carpathi ps and the Southern Apennines the hinterland overrides th (synthetic collisional model, see text), whereas in the Nort nes and in the Hellenides (Peloponnesus) the hinterland is under the foreland (antithetic collisional model, see text). T on from the Western Alps to the Northern Apennines shows a change of the geotectonic situation. In the Western Alps th late represents the foreland and the Adria microplate the hin the Northern Apennines, which are younger than the Alps, th is reversed due to a flip of subduction which predated t c continental collision (see text; for more details: Reutter Groscurth this vol.).

homogenous crustal structure of the Adria microplate which interacted with the pre-Alpine European plate.

All circum-Adriatic mountain chains represent zones of strong compression. They contain an excessive quantity of strongly folded and thrusted sedimentary rocks and slices of basement rocks. The units preserved in the compressional zones were evidently stripped off from their substratum which was consumed in the course of orogeny. As mainly oceanic crust or thin continental crust was destroyed, the mechanism of consumption is thought to be subduction. Subduction was followed by collision of the pre-Alpine continental crust with the crust of the Adria microplate or thick crustal elements of the marginal zones of these continents.

Without exception the circum-Adriatic orogens show a clear polarity in their evolution which is expressed in the migration of the depositional troughs of flysch sediments and a following belt of compressional tectonics. It may be explained as follows: Subduction starts with strong subsidence of the lithosphere in the areas of oceanic crust (eugeosyncline) which separates the continental areas. Lateral shortening causes the characteristic asymmetrical disposition of active plate margins with an upper and a marginally down-going plate and a subduction plane between them. As subduction proceeds, the rim of the continent attached to the down-going lithosphere also has to subside. When the accretionary wedge at the front of the upper plate reaches the submerging continent, the trench-like depression between the plates is filled with detritus (flysch) and with allochthonous materials (nappes). As this process moves from the margin of the down-going continent towards its interior, it causes the migration of the flysch depression and the advancement of the allochthonous units at its internal side. Introduction of heat into the marginally subducted continental crust confers a greater tectonic mobility to it and may be the source of its buoyant uprise. The foreland is situated in front of the polar process and the hinterland behind it.

The complete consumption of the oceanic crust between the continental foreland and hinterland is convincingly proved not only by surface geology but also by seismic investigation of the crustal structure in all cross-sections through the circum-Adriatic chains. This implies that, at a certain moment of orogeny, continental collision must have occurred. Analysis of the seismic data shows that there are two types of crustal collisional structures: either the crusts overlie each other synthetically to the thrusting of nappes and preceding subduction or they overlie each other antithetically to these directions (fig. 2).

In the case of synthetic collisional structure, the hinterland with its entire crust and even portions of the upper mantle overrides the foreland and intermediate units as the most internal unit. The crustal structure depicts a model of plate convergence with the hinterland in the position of the upper plate, the foreland in that of the lower and partially subducted plate. Examples are the Alps (Closs 1975), the Carpathians, and the Southern Apennines.

In the antithetic case, the crust of the hinterland is underthrust beneath the crust of the foreland and only a relatively thin upper layer of sediments or uppermost basement overrides the foreland and the intermediate allochthonous units as most internal nappe. Such a splitting of the hinterland resembles the model of flake tectonics (Oxburgh 1972) to a certain degree. Underthrusting of the crust of the hinterland causes the continental border of the foreland to rise up and the previously obducted accretionary wedge of allochthonous units to be distributed by

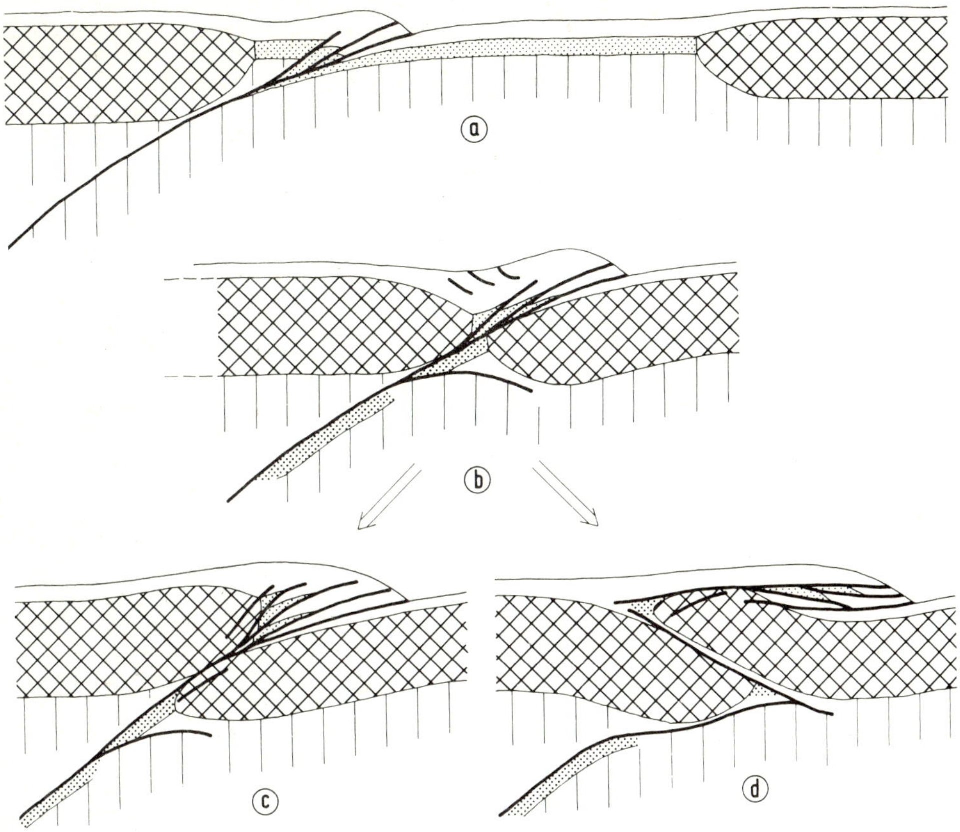

Fig. 2. Sketch of the development (a, b) of a synthetical (c) and an antithetical (d) continental collision.

gravity. Geologically this situation is indicated by the thin thrust sheets of allochthonous units and by the many fensters in which the foreland is exposed even in internal parts of the orogen. Examples are the Northern Apennines and the Hellenides.

We don't consider that continental underthrusting which leads to an antithetical collisional structure was generated by the flip of the subduction plane into a dip opposite to that which preceded the collision. This is evidenced in the Northern Apennines by the uninterrupted polar orogenic evolution which outlasted the collision during the entire Miocene. We assume that heating of the border of the continental margin during initial marginal subduction allowed the separation of the crust of the foreland from the underlying lithosphere and the underthrusting of the crust of the hinterland.

Why did such fundamentally different collisional structures develop? As outlined in the chapter Corsica/Northern Apennines, the evolution of this double orogen was determined by a true flip of the subduction system which consumed the oceanic Penninic-Ligurian eugeosyncline. The Alpine

synthetic collisional process, which preceded the Apenninic orogeny, caused the margin of the European plate to dip southeastwards under the load of the Alpine thrust masses. This inclination seems to have survived the collision of the Corso-Sardinian continental fragment with the thin crust of the border of the Adria microplate which occurred after the consumption of the residual oceanic crust along the flipped subduction (Reutter, Günther & Groscurth this vol.).

It is questionable whether this model can also be applied to other cases of antithetic crustal collisional structures. As already pointed out, in the Hellenides polarity reversals in the orogenic history have been proposed by Smith & Moores (1974), however, this hypothesis is not generally accepted.

Tectonic settings, radiometric ages, and crustal structures show that in all circum-Adriatic chains the orogeny was not confined to a single process of subduction and collision due to a complex paleogeographic starting situation. As demonstrated in most palinspastic reconstructions, the Adria microplate, as well as the border of the European continent, did not possess crusts of homogeneous thickness. They were lacerated to strips of either compact, reduced, or even oceanized continental crust during phases of crustal spreading (Görler & Giese this vol.). The Briançonnais ridge of the Western Alps, which separated the Valais basin from the Piedmont basin, is an example of a continental fragment situated within an eugeosynclinal oceanic area. It seems that, if such an intrageosynclinal compact crustal fragment was affected by a subduction process, the subduction plane jumped from its former site to a next and more external zone of minor resistance to crustal shortening (Roeder this vol.: suture progradation). It is also possible that the subduction plane branched into two contemporaneously active parallel sutures.

The main causes of the variations in shape and structure of the forelands and hinterlands during the orogeny are the inhomogeneity of the crust involved in the orogenic process and shifting of the suture of the subduction. Thus former marginal parts of the forelands which probably had a compact crust were detached from the foreland and were welded to the front of the hinterland (composite hinterland). Variations of this kind probably occurred during each major orogenic event. Therefore, the crustal structures of the circum-Adriatic orogens have to be considered as the result of multiple crustal collision and thrusting. Often this may be hidden in an apparent general crustal thickening (e.g. Eastern Alps).

The incorporation of parts of the foreland into the hinterland can easily be observed in the Cretan arc. Parts of the former (Miocene) Hellenic foreland together with the Hellenic nappes on its back have assumed the role of the hinterland. There is another good example in the Northern Apennines: In Tuscany the crustal thickness, the heat flow, and the magmatism are indicative of a hinterland although this region is a marginal part of the Adria microplate. It had acted as a typical foreland until the Upper Miocene. This is all the more remarkable since the new hinterland developed from a preexisting antithetic collision zone.

The obvious contrast of the antithetic crustal structure of the Northern Apennines to the synthetic structure of the Southern Apennines originated from such a further development of the latter. After passing through a stage similar to that of the Northern Apennines the subduction jumped into the deep basins situated within the Adria microplate (Lagonegro basin, Molise basin). The new collisional structure, which was su-

perimposed on the older ones, corresponds to a synthetic overthrusting of a composite hinterland, which had crustal elements of the Adriatic microplate at its front, over its inner parts. This configuration seems to have a weakly developed equivalent in the synthetic asymmetry of the crustal structure in the external Northern Apennines (Elter et al. 1975).

The foregoing considerations need some specification of the term hinterland. As the hinterland can be roughly compared with the upper plate of a converging plate system, it should have a complete crust and its own crust/mantle boundary. Structurally it must be free of compressional tectonics which are contemporaneous with those at its frontal border. Its crustal structure may be composite, i.e. the crust was assembled during earlier tectonic events (e.g. Inner Carpathians, Dinarides, Hellenides, Southern and Northern Apennines), or it is more or less unchanged from the beginning of the compressional phases of the orogenic cycle (Southern Alps). In both cases it may be normal continental (e.g. hinterland of the Alps), reduced continental with transition to back-arc conditions (e.g. Aegean Sea, Pannonian basin), or oceanized with fragments of continental crust (e.g. Ligurian Sea, Tyrrhenian Sea).

At the front of the hinterland, there is a relatively stable wedge of varied material which gradually thins out towards the foreland. It overrides the foreland and intermediate units. In the Eastern Alps the Austroalpine represents such a wedge fo the hinterland. Lithologically it corresponds to the South Alpine hinterland, but essential parts of its crust were removed and have disappeared in the suture zone. In contrast with the Alps, in the Carpathians and Southern Apennines the frontal wedge of the hinterland consists of stacked sediments which were scraped off from the foreland and added to the hinterland in a similar way as an accretionary wedge at the front of the upper plate. It is a curious fact that the boundary between the true hinterland and its frontal wedge in the Carpathians is marked by the Pieniny lineament, whereas in the Eastern Alps it is marked by the Periadriatic lineament. In the border region, the two lineaments alternate and the Inner Carpathian hinterland passes laterally into the frontal wedge of the hinterland of the Eastern Alps.

The causes for the variations of crustal development of the hinterlands (normal continental or back-arc type) are unknown. The difference between the down-bending velocity of the lithosphere of the foreland and the velocity of the advancement of a stable upper plate as well as variations of the dip of the subduction plane may be suggested as possible causes.

A comparative seismic-refraction and geological study of the circum-Adriatic chains illustrates that even in the case of the relative small Adria microplate, the model of plate tectonics is a very useful vehicle for the understanding of orogenetic processes. The dimensions of the mountain belts of the area studied are very small compared to those of the Pacific active margins. However, it permits the study of boundary conditions in order to test the validity of the model. The Adria microplate does not correspond to the conventional idea of rigid plates. The instable forelands and hinterlands whose configurations vary with each important tectonic event, cannot be fully described with the basic terms upper and lower plates. We found that the application of the plate tectonics model must be modified according to local conditions. Gravity tectonics and other generally known tectonic principles have to be combined with it. The plate tectonics model then offers the clue to the interpretation of very numerous partly isolated data and to an improved geodynamic concept of the geolo-

gical history of the central Mediterranean area. We are convinced that the structural and crustal models of the circum-Adriatic orogens will prove useful tools for the study of other parts of the Tethyan system and of similar orogens elsewhere.

References

Alpine Explosion Seismology Group, 1976: Pageophy.,114:1109-1130.

Amodio-Morelli, S., Bonardi, G., Colonna, V., Dietrich, D., Giunta, G., Ippolito, F., Liguori, V., Lorenzoni, S., Pglionico, A., Perrone, V., Piccarreta, G., Russo, M., Scandone, P., Zenettin-Lorenzoni, E.& Zupetta, A., 1976: Mem. Soc. geol. Ital. in press.

Angenheister, G., Bögel, H., Gebrande, H., Giese, P., Schmidt-Thomé, P. & Zeil, W., 1972: Geol. Rdsch., 61:349-395.

Andreieff, P., Broquet, P., Duée, G. & Mascle, G., 1974: B.S.G.F. XVI: 295-302.

Andrusov, D., 1968: Verlag der Slowakischen Akademie der Wissenschaften, Bratislava: 188 p.

Aric, K., Gutdeutsch, R., Albu, I., Mituch, E.& Posgay, K., 1978: Geophys. Transactions, in press.

Aubouin, J., Blanchet, R., Cadet, J.-P., Celet, P., Charvet, J., Chorowicz, J., Cousin, M. & Rampnoux, J.-P., 1970: Bull.Soc.géol. France, XII: 1060-1095.

Aubouin, J., 1973: Bull. Soc. géol. France, XV: 401-644.

Behnke, C., Giese, P., Prodehl, C. & Visintini, G. de, 1962: Boll.Geofis. teor. ed appl., 4:110-132.

Berckhemer, H. (German Research Group for Explosion Seismology), 1969: Schweiz. Min. und Petr. Mitt., 48:235-246.

– , 1969: Tectonophysics, 8:97-105.

Bernoulli, D. & Laubscher, H., 1972: Eclogae geol. Helv.,65:107-118.

Birkenmajer, K., 1974: in: M. Spencer (ed.): Mesozoic-Cenozoic Orogenic Belts, The Geological Soc.:Spec. Publ. No.4:125-157.

Boccaletti, M., Elter, P. & Guazzone, G., 1972: Nature Physical Science, 234:108-111.

Cassinis, R., Finetti, J., Giese, P., Morelli, C., Steinmetz, L. & Vecchia, O., 1969:Boll.Geofis.teor. ed app., 11:140-160.

Catalano, R., Channel, J.E.T.& d'Argenio, B., 1976: in: G.Pialli (ed.): Paleomagnetic Stratigraphy of Pelagic carbonate sediments, Mem. Soc. geol. Ital., 15:95-114, Roma.

Channel, J.E.T. & Horvath, P., 1976: Tectonophysics, 35:71-103.

Choudhury, M., Giese, P.& Visintini, G. de, 1971: Boll. Geofis. teor. ed appl., 13:210-240.

Closs, H., 1975: Rapp. Comm. int. Médit., 23:55-57.

Colombi, G., Giese, P., Luongo, G., Morelli, C., Riuscetti, M., Scarascia, S., Schütte, K.-G., Strowald, J.& Visintini, G. de, 1973: Boll. Geofis. teor. ed appl., 59:225-254.

Compagnoni, R., Dal Piaz, G.V., Hunziker, J.C., Gosso, G., Lombardo, B. & Williams, P.F., 1975: Italy-USA Cooperative Project on High Pressure-Low Temperature Metamorphism in the Western Alps. Internal Report 2,54 p.

Dal Piaz, G.V., Hunziker, J. & Martinotti, G., 1972: Mem. Soc. geol. Ital. 11:433-466.

Elter, P., Giglia, G., Tongiogi, M., Trevisan, L., 1975: Boll. Geofis. teor. ed appl., 17:3-18.

Geophysical Transactions, 1972: 172 p., Budapest.
Giese, P., 1968a: Geophys. Abh.Inst.Meterol.u.Geophys. FU Berlin 1 (2): 202 p.
- , 1968b: Schweizer. Min. und Petr. Mitt., 48:261-284.
Giese, P., Günther, K.& Reutter, K.-J., 1970 (1968): Zeitschr. Dtsch. Geol. Ges., 120:151-195.
Giese, P. & Morelli, C.: in: Ogniben et al. (ed.): Structural model of Italy, Consiglio Naz. Ricerche Quaderni, 90,4:4, Roma.
Giese, P. & Prodehl, C., 1976: in: Giese, P., Prodehl, C. & Stein, A. (eds.): Explosion Seismology in Central Europe:347-376.
Hynes, A.J., Nisbet, E.G., Smith, A.Gilbert & Welland, M.J.P., 1972: Zeitschr. Dtsch. Geol. Ges., 123:455-468.
Jacobshagen, V., Makris, J., Richter, D., Bachmann, G.H., Doert, U., Giese, P. & Risch, H., 1976: Zeitschr. Dtsch. Geol. Ges., 127:337-363.
Laubscher, H.P., 1975: Amer. J. of Sci., 275:865-876.
Makris, J., 1977: Hamburger Geophys. Einzelschriften, 27:1-98.
Makris, J., Mavridis, L.N., Menzel, H., Stavrou, A. & Veis, G., 1973: Z. Geophys., 39:929-936.
Miller, H., 1976: Pageoph., 114:1109-1130.
Morelli, C., 1975: in: Ogniben et al. (ed.): Structural model of Italy, Consiglio Naz. Ricerche, Quaderni, 90, 427-447, Roma.
Morelli, C., Giese, P., Cassinis, R., Colombi, B., Guerra, I., Luongo, G., Scarascia, S. & Schütte, K.G., 1975: Boll. Geofis. teor. ed appl., 67:183-210.
Niggli, E., 1946: Eclogae geol. Helv., 29:211-220.
Ogniben, L., 1969: Mem. Soc. geol. Ital., 8(4): 453-763.
- , 1975: in: Ogniben et al. (ed.): Structural model of Italy, Consiglio Naz. Ricerche, Quaderni, 90, Roma.
Ohnestetter, M., Ohnestetter, D. & Rocci, G., 1976: Bull. Soc. géol. France, 18:115-134.
Oxburgh, E.R., 1972: Nature, 239:202-204.
Polizzi, D., 1976: Unpublished Report, Universitá degli Studi di Milano, Istituto di Geologia, 92 p.
Prodehl, C., 1965: Boll. Geofis. teor. ed appl., 7:35-88.
Smith, A.G. & Moores, E.M. 1974: in: M. Spencer (ed.): Mesozoic-Cenozoic Belts, The Geological Society: Spec. Publ. No. 4:159-185.
Tollmann, A., 1969: Sonderdruck aus Geologie, 18: H. 10.
Trümpy, R., 1973: in: Gravity and Tectonics, De Jong, K.A. & Scholten, R. (eds.):229-251.
- , 1975: Amer. J. of Sci., 275A:209-238.
Zimmermann, J. & Ross, J.V., 1976: Geol. Soc. of America Bull., 87:1547-1550.

Chapter 7

Plate Tectonic Conclusions

Three Central Mediterranean Orogens - A Geodynamic Synthesis

Dietrich Roeder

Contents

1 Introduction

The present article is a review of a major collection of data and ideas achieved by an international effort. Although the contributions differ widely in scope, topic, and point of departure, they are rather successful in bridging the gap between the plate tectonic model as understood in the classical summaries (Dewey & Bird 1970; Le Pichon et al. 1973) and the descriptive geology of mountain ranges as in Spencer (1974). Although the present article is based on hundreds of references, it is by agreement between all contributors and editors that contributors to the present volume are not quoted in the present article.

At the beginning of this article I express my thanks for the opportunity to collaborate with European colleagues during the last phase of this important research project. Funding by the German Research Society (DFG) and the Technical University of Munich enabled me to enlarge on previous field experiences in Greece and in the Alps. The coordinators of the geotraverses and many of the contributors have freely shared with me their published and unpublished data and ideas, in the field, in meetings, in innumerable letter communications, and in patient and critical reviews of the present manuscript. Their unselfish help is thankfully acknowledged. I also acknowledge the continuous and beneficial exchange of ideas with my friends and co-editors, in particular with Dr. Hans Closs.

The most elaborate way to analyse a structurally complex area is a rigorous plate reconstruction such as the ones made by Smith (1971), Dewey et al. (1973), Biju-Duval & Le Pichon (1977). We did not attempt this approach here, because little of the required analytical work was done in the German research groups during the research period reported. However, our work organization in the form of geotraverses across orogenic belts has evolved into another rigorous and rather successful approach. It is recommended wherever it is necessary to apply geodynamic results or ideas to more local problems.

The present research project has generated new data and ideas from all geoscience specialties. We have used these ideas to develop a new view on orogeny as a geodynamic process. Essential parts of this new view are derived from the plate tectonic earth model which has been adopted rather than developed locally. However, the present research project has filled in many details which are only vaguely or wrongly covered in previous plate models of orogeny. Thus, the emerging new view on orogeny is a collective achievement of collaborating and interacting workers, rather than a personal individual view of the type many of the older orogenic theories have been.

In confirmation of one of the oldest geotectonic observations (Dana 1873; Suess 1885; Argand 1924), we have found that our orogenic belts grade and merge into one another, but show strong differences in size, style, and orogenic timing. This is most elegantly explained by the geometrically rigorous plate tectonic earth model, in which the orogenic belts function as geometrically and kinematically critical parts of continuous plate boundaries. We have also found that substantial parts of the overall tectonism represent intraplate strain and not plate boundary effects, and we recognize that the typical orogenic architecture of a tightly folded and plutonized orogenic core flanked by one or two marginal thrust belts is the product of crustal convergence combined with topographic upbulge and lateral spreading under the influence of gravity. Hence the new view on orogeny combines conventional mobilistic and fixistic arguments to confirm a basic mechanism of plate tectonics.

The organization of this article follows only in part the work organization by geotraverses. In a first part, comprising sections 2 through 5, the impact of the plate tectonic model on the present understanding of the central Mediterranean mountain ranges is studied, in the form of a review of the geotraverses. The largest part, from section 6 through 21, present the results of the research , and the smallest part, section 22, contains a diagnosis of Central Mediterranean orogeny.

2 Alps

In this classical orogenic belt, a north-vergent continent-continent collision with weak arc magmatism is followed by strong intracratonic deformation and uplift (fig. 1). In this collision, the Adriatic microcontinent overrode the Eurasian craton, with an upper and younger of two sutures at the base of the Austro-Alpine units. The collision was preceded by an arc-continent collision of Tethyan oceanic material obducted over a fragmented Eurasian borderland, with a Cretaceous suture at the base of the South Penninic ophiolites. Interplate wedges of fine clastics and olistostromal material are associated with both sutures, constituting the eugeosynclinal fill of the Mesozoic Alpine trough.

Late Cretaceous high p/t metamorphism, associated with the Penninic obduction, requires burial to at least twice the thickness of the Austro-Alpine plate. This implies an event of root-zone kinking with resulting burial of the accretionary wedge to the required depth and re-emergence after collision. Presumably this kink did not survive the Eocene continent-continent collision.

A kinematic sequence of megatectonic events in the Western Alps (Milnes 1974) shows thick-skinned folding after thrust emplacement of the ophiolites but before its polyphase redeformation and nappe transport. This sequence suggests progradation of the late Cretaceous obduction into the foreland, causing thick-skinned upright folds and part of the Eo-Alpine metamorphism in the middle Penninic complex, followed by the second collision and the progradation of the Austro-Alpine suture into the same foreland. This caused recumbent redeformation of the late Cretaceous folds and was followed by updoming of the gneiss-cored recumbent nappes during the Meso- and Neo-Alpine events (fig. 2).

The double collision in the Alps explains the absence of Cretaceous arc vulcanism. It implies a backarc basin between the north-facing Penninic arc and the then extinct north edge of the Adriatic microcontinent.

The Alpine lower plate is the site of a post-collisional event of low p/t metamorphism attaining medium grade or amphibolite facies with kyanite requiring more heat than available beneath the sialic cover of the Austro-Alpine and in the time span between the collision and the radiometric closing dates. This event, therefore, cannot be explained by subductive burial alone. It requires an additional heat contribution, either a center of arc magmatism related to a deeper subduction dipping beneath Europe or strain heat from folding and intracrustal decollement. Subduction beneath the Helvetic shelf would explain the origin of the thick-skinned Helvetic thrust-fold belt in a position antithetic to subduction, but not conclusively. Also, the stratigraphic record in the Helvetic facies belt leaves little time for subduction. Strain heat in addition to burial under material with a normal geothermal gradient is a plausible explanation for the intracrustal low-velocity channels but its possible quantity is unknown.

3 Corsica and Northern Apennines

This geotraverse deals with a double orogen that resulted from a polarity reversal, with crustal data suggesting further complications (fig. 3). In the simplest model, Jurassic spreading generated the Penninic-Ligurian oceanic trough within the subsiding Tethyan borderland. A Europe-facing arc developed at an unknown time within this ocean, probably by convergence across one of the Atlantic-Tethys transform faults active before the break-away between Spain and the Grand Bank (Pitman & Talwani 1970). This arc collided with the mid-Penninic part of the Eurasian craton in late Cretaceous time and with the Iberian-Corsican craton during the Eocene. This collision lead to the obductive preservation of ophiolite nappes and accretionary wedges in the Western and Piemontese Alps, the Ligurides, and in Corsica, although arc material has not been formally identified.

Convergence east of the collided arc may have started across a branch of the Atlantic Tethyan transform system diagonally across the remaining ocean basin. In the Carpathian-Alpine segment, the new subduction system

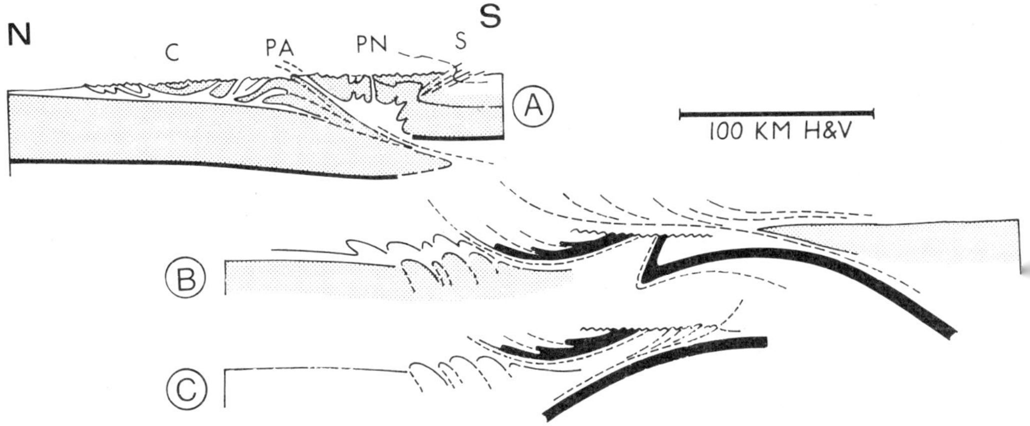

Fig. 1. Alps, crustal section and evolutionary details. Shaded: continental crust and related sediments. Massive black: oceanic crust and ophiolites. Dashed lines: thrust faults. (A): present state after Gwinner (1971) with positions of M discontinuity (thick black lines) after Closs (1975) and Giese (pers.comm.). C, Austro-Alpine. PA, Tauern window, PN, Periadriatic line. S, South Alpine root zone. (B) Penninic arc-continent collision and root-zone kinking, schematic. (C) Alternative interpretation of Penninic event, assuming flip and north-dipping subduction subsequent to arc-continent collision.

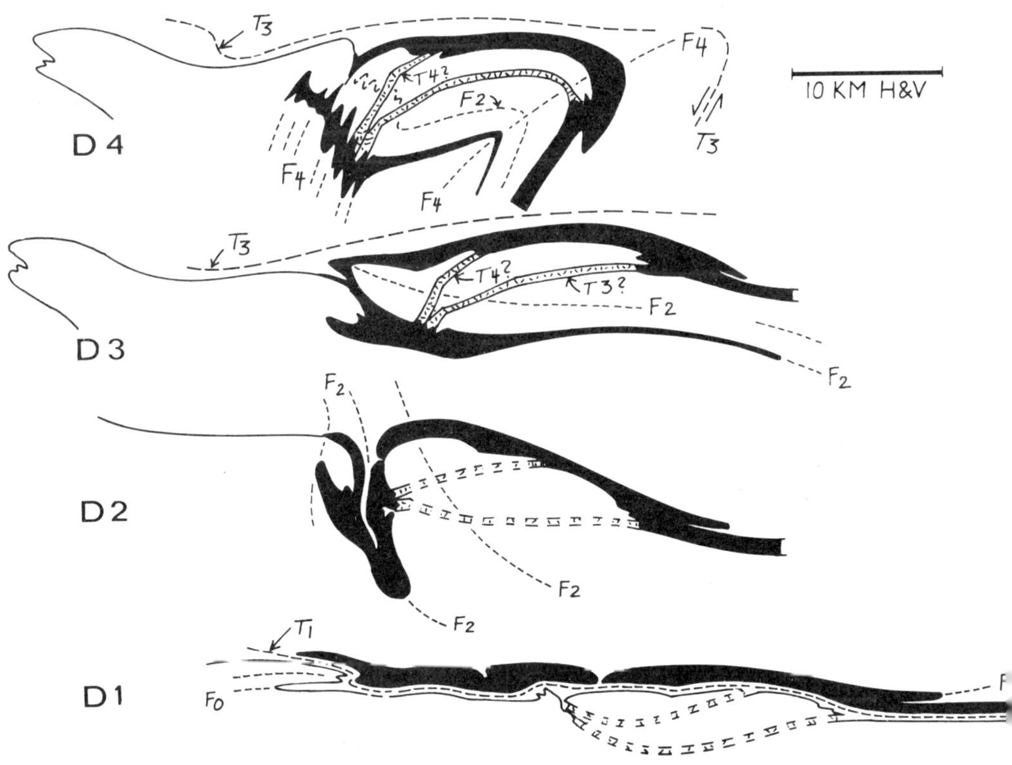

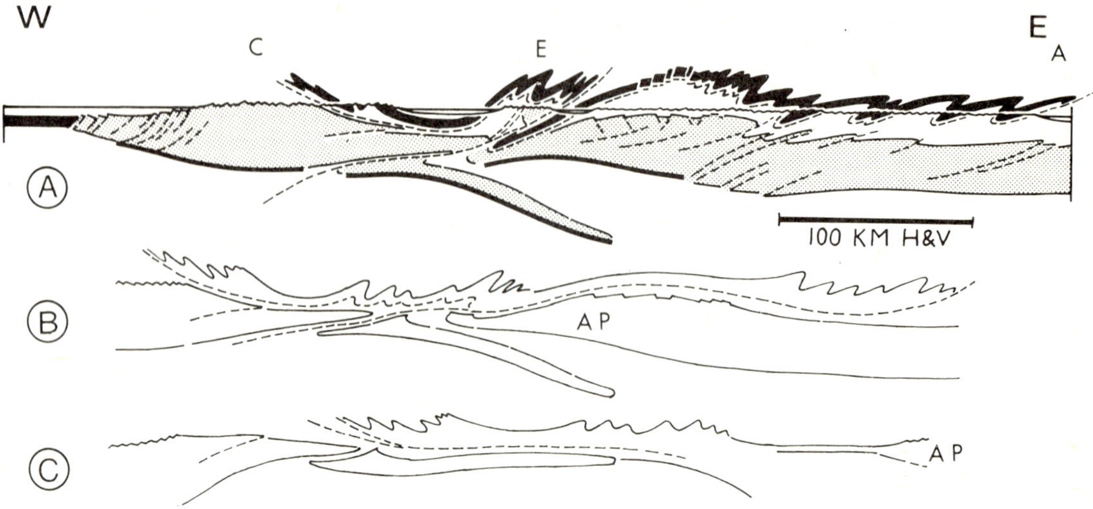

Fig. 3. Crustal section and evolutionary details of Corsica (C), Elba (E), and Northern Apennines. Shaded: continental crust. Black: oceanic crust and ophiolites. Dashed lines: faults. A: Adriatic sea. Interpretation of crustal data (thick black lines) as flip with intra-crustal shearing. Section A, present state (post-flip), with main Tuscan suture assumed east or west of Elba merging with a shallow-west dipping intracrustal shear. Section B, late Neogene state (post-flip), with crustal shear and some late Tuscan deformation and thrusting restored, but eastward obduction upon Adriatic plate (AP) completed. Section C: End of Penninic-Ligurian obduction (pre-flip, late Eocene).

Fig. 2. Four interpretive structural cross sections of the Monte Rosa and Bernhard nappes, south and Central Penninic zones, Swiss Alps, after data in Milnes (1974) and Gwinner (1971). Black: ophiolites. Dot pattern: blastomylonites of unknown age. The sections represent a kinematic sequence which can be correlated with the major plate tectonic events. D1, late Cretaceous obduction of South Penninic-Liguride ophiolites onto gneissic basement of Monte Rosa nappe, forming thrust T_1 preceded by undated folds F_0. D2, latest Cretaceous and Eocene thick-skinned folding F_2 either antithetic to a flipped Penninic subduction or as crustal progradation of the 0 Penninic-Liguride suture. D3, Eocene flattening beneath overriding (T_3) Dent Blanche nappe of Austro-Alpine complex. T_3 is the continent-continent collision suture. Blastomylonites may be in part of this age. D4, Neogene crustal compression subsequent to collision, forming F4 and south Alpine root zone. Blastomylonites may be in part of this age.

was located at the north-west front of the Adriatic microcontinent and was dipping toward the southeast. In the Ligurian-Apennine segment it was located on the eastern or back slope of the collided Penninic arc and was dipping to the west. A middle section, not paleogeographically identified, must have remained without convergence, possibly shifting its position along the plate boundary. This second convergence eliminated the remaining oceanic crust and led to a continent-continent collision. Beginning and collisional end of this subduction are strongly diachronous. In the Alps, the Austro-Alpine front collided in late Eocene time and therefore must have started subduction during the Paleocene or Eocene. In the Corsica-Apennines section, the convergence is dated as Miocene, with a collision in late Miocene time.

In the Apennines system, the post-flip subduction reversed the transport of the northward obducted Ligurian nappes and carried them eastward or northward. Convergence persisted after collision and began to involve the Mesozoic platform sediments and the basement of the Adriatic microcontinent. Thrust processes during the post-flip stage added an accretionary wedge and the fill of a forearc basin to the previously assembled nappe stack; redistribution of sedimentary material was powered by gravity spreading and, at the front, by extensive gravity gliding into the eastward and northward shifting trench or supracratonic depocenter.

Crustal data suggest yet another, northeast-dipping thrust or subduction event located approximately where one would expect the Penninic-Ligurian collision trace, but it appears uninterrupted by the southwest-dipping Tuscan subduction. It is interpreted either as an example of flake tectonics (Oxburgh 1972) or as an arc-arc collision.

4 Southern Apennines in Calabria

This system is a complete east-facing arc-trench system with a west-dipping Benioff zone (fig.4B). It is in the process of colliding with the west-facing Hellenic arc. From its external front on toward the west, it contains an accretionary wedge with a sediment ridge at the site of collision in the Gulf of Taranto. The outer non-volcanic arc forms the peninsula of Calabria and the north-Sicilian mountain ranges and consists of predeformed cratonic basement with Mesozoic cover sediments. The forearc basin between Calabria and the Eolian islands is deep enough to be devoid of sediments and to represent thinned,stretched,or faulted continental crust. The cluster-like Eolian volcanic arc is not yet consolidated into a coherent magmatic body. The backarc region is formed by the south part of the Tyrrhenian sea. Its bottom topography of ridges and troughs suggests tensional neotectonic activity at the beginning of sea floor spreading.

Four features of this system are of general geodynamic significance:

1) The crustal east front of the upper plate, located beneath the eastern slope of Calabria and dated by its trench fill as late Oligocene to Recent, is the only suture of west-dipping subduction. In the Calabrian system it is an east-facing continental margin, whereas to the north it becomes the east front of Apennine structure, located within continental crust. In Calabria it has been active for a longer geological life span than the eastward thrusting in the rest of the Apennines, and in Calabria it is not preceded by earlier, more westerly positions of crustal suturing. The origin of this suture, therefore, may be one of the Mesozoic transform faults connecting the spreading North Central Atlantic

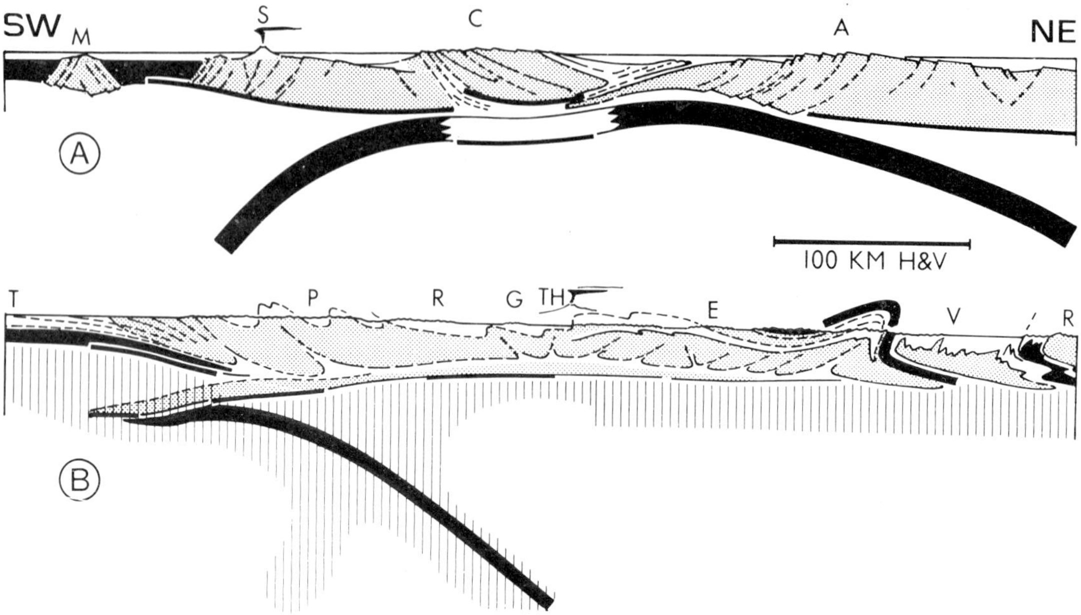

Fig. 4. Two interpretive crustal sections, (A) and (B). Shaded, continental crust. Black, oceanic crust. Thick lines, M discontinuity from refraction seismic data. (A), Calabrian arc adjacent part of Apulian-Hellenic arc. M, Marsili sea mount, S, Stromboli. C, Calabria. A, Apulia. The west-dipping subduction zone is documented by deep seismicity and arc vulcanism. The east-dipping subduction is not documented, but inferred from regional plate mapping. (B), Hellenides. T, Hellenic trench. P, Peloponnes. R, Argolis. G, Aegina. TH, Hellenic arc vulcanism, projected. E, Evea Island. V, Vardar zone. R, Rhodope. Crustal data interpreted as Neogene east-vergent thrusting absorbed within crustal body of orogen. For interpretation and identification of structural details, see fig. 5.

with Tethys across fields of small plates and mixed crust in the west Tethyan borderland.

2) The Calabrian outer arc shows an unusual structure with eastward tilted panels of westward thrust-faulted continental crust and Mesozoic shelf-type or pelagic sedimentary cover. Thrusting and tilting are mechanically compatible. The tilting may or may not date the thrusting; it is dated as late Oligocene through Recent. Part of this tilting probably reflects the rotation of normal fault blocks in conjunction with Tyrrhenian rifting. This architecture makes the Calabrian outer arc very similar to some cross sections of the Ryukiu arc. The westward thrusting of the central panels, however, is not conclusively explained. It may be a retrocharriage or antithetic thrust developed during the collision of the Apennine orogen with the Adriatic plate, or it may reflect the bending of the arc system in a horizontal plane.

3) The trench slope profile of the Calabrian arc shows a bilateral ridge instead of the more common accretionary wedge. This ridge is underlain by thick and strongly disturbed sediments, and it serves as point of origin of olistostromes in both directions. It may be a principal site of dewatering for the geopressured trench fill of the Calabrian as well as the Hellenic arc, similar in structure and function to the lines of shale diapirs in the vicinity of the Barbados ridge and to the olistostromal ridge of the Molucca Strait (Roeder 1977a; Silver & Moore 1977).

4) Strongly evident only southwest of the Calabrian geotraverse in Sicily, the foreland of the Calabrian arc is fragmented and produces in Mt. Etna, upper-mantle derived mafic volcanic rocks. Upon further subduction, these volcanics will line the collision suture as a basal mafite body as in several other sites worldwide.

5 Hellenides

By far the most complex of our orogenic systems, the Hellenides (fig. 4B) represent a south-facing or Africa-vergent stack of thrust sheets in the upper plate of an active Africa-vergent or NE dipping subduction zone, resulting from 150 m.y. of polyphase plate tectonic history. In complexity and finite distance of plate movements, the events recorded in this geotraverse are of the same order of magnitude as anywhere in the great planetary convergence systems. Although its untangling is not yet conclusive, the present research effort has produced a major re-organization of the regional geological data (converting the old facies zones into a systematics of stacked thrust sheets) and two somewhat different plate-kinematic interpretations.

Not all events known by their structure, their stratigraphic effects, and their radiometric ages have as yet been classified and identified plate tectonically, but the list of events includes Andean-type subduction, ophiolite obduction, root-zone kinking and flip, suture progradation, Atlantic-type rifting, foreland plate fragmentation, backarc- and forearc sprading. Associated secondary tectonic features include arc magmatism, metamorphism, and vulcanism, pressure-metamorphosed accretionary wedges, tectonic melanges, mafic and felsic intraplate vulcanism, block rotation, and gravity spreading of thrust sheets. The topology of the structural units and the tectonic events are further complicated by the crustal architecture suggesting a flip or arc-arc collision not yet conclusively identified at surface. Regardless of the as yet uncertain details of plate reconstructions, also intracontinental plate movements must have contributed to the orogenic history, including early Mesozoic transform faulting between the Atlantic and Tethys, major arc-associated or intra-Tethyan wrench faulting, and oroflexing or map-view bending of the subductive fronts.

In detail, the Hellenic orogenies represent the continuous consumption of Tethys and its southern marginal seas, and the foreshortening, by suture progradation or intraplate convergence, of the fragmented southern Tethyan borderland (fig. 4 B). Depending on the assumed model, five or six sites of subduction-like processes leading to northeast-dipping plate overlaps have been identified. Subduction of the Tethyan Paleozoic and Mesozoic oceanic crust took place at the internal edge of the Vardar zone, perhaps within the Vardar zone, and below the Eohellenic obductites. Progradation of the Eohellenic suture into the Median Crystalline Belt created a second (and third) subduction zone below the Pelagonian basement nappes and below

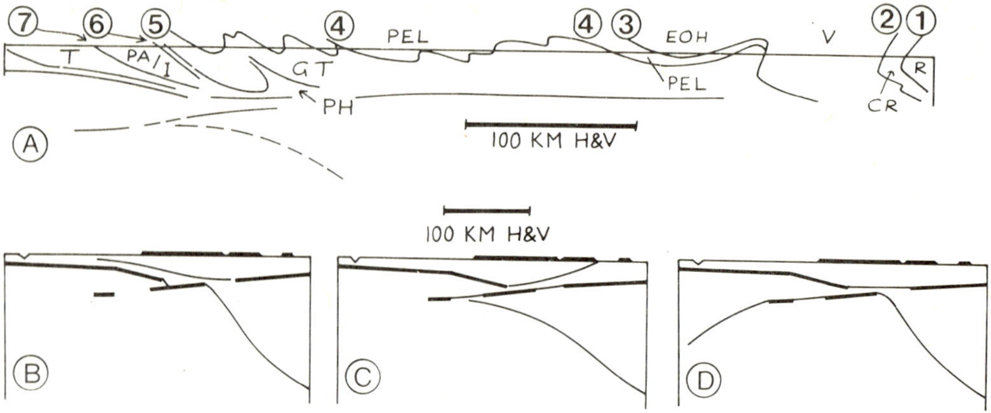

Fig. 5. Schematic structure section across the Hellenides (A) and crustal section interpreted in three different plate tectonic ways (B, C, D). (A) contains major sutures and crustal thrusts in the Hellenides (encircled numbers) and some thrust units (letters). (1) early Mesozoic eastward obduction onto Rhodope (R). (2) circum-Rhodope suture (CR), possibly identical with the main Tethyan suture. V, Vardar zone. (3) Eohellenic (EOH) obduction, possibly main Tethyan suture. (4) thrust below Pelagonian unit (PEL); interpreted as major Eocene subduction. (5) Pindos thrust, either scar of closed local basin or prograded extension of subduction (4). (6), intracrustal shear at Phyllite zone (PH) and Gavrovo-Tripolitza (GT) zone. (7), presumed thrust at base of Pre-Apulian-Ionian zone (PA/I). B, C, D, crustal section from Hellenic trench across Peloponnes and across island of Aegina. Thick lines, observed positions of M discontinuity. Thin lines, interpretive connections. B, interpretation as northeast-dipping subduction. C, interpretation as Plio-Pleistocene flip-like dissection of Benioff zone by gently west-dipping intrecrustal shear. D, interpretation as arc-arc collision.

the Pindos nappe. This double event may represent either one or two separate ocean basins and their subduction. Further progradation closed an intra-cratonic trough containing the Phyllite zone and little or no oceanic crust. Present subduction contains mixed crust in the lower plate, perhaps similar to pre-orogenic crustal conditions in the Phyllite zone.

6 Orography and Plate Tectonics

The Alpine-Mediterranean orogens and their complex loops have always been explained as effects of crustal convergence between Africa and Eurasia, but it was not until the mapping of Mesozoic to Recent sea floor spreading in the Atlantic that the pattern of this convergence could be charted and related quantitatively to the Alpine-Mediterranean orogenic history.

Pitman & Talwani (1970) and more recent authors have shown that this pattern was governed by persistant vectorial differences of spreading north and south of the Azores triple junction. The history of Atlantic sea-floor spreading thus furnished a history of trans-Tethyan plate interaction, including about 1000 km of left-lateral offset in an east-west direction and 800 km of north-south convergence, and also including a Recent instantaneous north-northeasterly convergence of 2.3 cm/y in Sicily and of 2.6 cm/y in Crete (Le Pichon et al. 1973).

In addition to these mappable movements of the tabular framework, complications are suggested by the loop pattern and the orogenic history of the mountain belts, by the paleogeography of ocean basins and platforms, and by paleomagnetic data. These complications are not only caused by the outlines of the crustal fragments as suggested by Argand (1924) but also by individual rotations of small plates and microcontinents (Smith 1972; Dewey et al. 1973). Since only one of these two categories of movement is accessible to quantitative reconstruction, the remainder of the movements must be obtained from the map-view patterns of orogenic loops and from the crustal history of each individual mountain belt.

Jammed into the convergent system of plates framing the Mediterranean area is a northward convex loop of Mesozoic to Recent orogenic belts in three subparallel and crosscutting or overprinting and/or colliding and discordant tiers. The southernmost tier is active and includes the southeast-facing (Africa-vergent) Calabrian-Sicilian arc and the southwest-facing (Africa-vergent) Hellenic arc, both in collision across the Gulf of Taranto. The active belts merge along strike with extinct or significantly less active thrust belts of the middle and northern tiers. They loop around the Adriatic plate comprising Italy and the Adriatic sea. At its northern end, the Adriatic plate forms the north-facing (Europe-vergent) upper plate of the Eocene Alpine continent-continent collision, whereas on the east and west flanks it forms the lower plate of collided and prograded segments of the middle tier of orogens, the quiescent west-facing (Africa-vergent) external Dinarides to the east and the active, east-facing (Africa-vergent) Apennine foothills belt to the west.

The outer orogenic tier consists of an outward facing (Europe-vergent) arc-continent collision of Cenomanian to late Eocene age. Exposed in its original collisional position in Eastern Corsica, the belt crosses the Ligurian Sea and ties with the Ligurian and Piemontese ophiolites into the South Penninic units of the western Alps and the Tauern window.

The Tyrrhenian Sea is an active backarc basin in the late rifting or early spreading stage extending between the Calabrian arc and a remnant arc composed of Corsica and Sardinia. The site of this spreading is in the midst of the Tertiary orogenic belt and leads to indiscriminate separation of orogenic fragments on both shores, but mostly offshore. An earlier stage of backarc basining in Oligocene to Miocene time opened the Provencal basin between the Corsic-Sardinian remnant arc or microcontinent and the Iberic block (Auzende et al. 1973).

The Aegean sea is a vaguely circular or domal spreading center in the late rifting stage with Pliocene to Recent block tilting and faulting crosscutting slightly older nappe soles. Its shape, suggestive of a mantle diapir, can also be interpreted as reflecting backarc spreading that is localized by the curvature of the Hellenic arc.

7 Crustal Structure

Refraction seismic data not only support surface geological interpretations, but also have revealed major tectonic features not recognized in surface data. Regional mapping of the M discontinuity has shown in several orogenic belts, overlaps of entire segments of continental and/or mixed crust. In some profiles, their polarity is opposed to the polarity of the thrust belts. Refraction seismic data, therefore, have become a major tool for charting polarity changes or flips of subduction zones.

Three types of crust have been identified in the Central Mediterranean area, oceanic crust with the M horizon at less than 15 km of depth, mixed crust with the M horizon between 15 and 25 km, and continental crust with M deeper than 25 km. For some oceanic areas, the crustal thicknesses confirm the bathymetric state of evolution. The crust of the Ionian Sea, for example, is 17 to 18 km thick (Hinz 1972) but has a late Jurassic spreading age (Klitzsch 1970). On the curve of oceanic subsidence (Sclater and Francheteau 1970), oceanic crust of this age should be deeper than 5 km, but for sites adjacent to the continent and likely containing a percentage of continental crustal material, the water depth is compatible with the geologically suggested age of the spreading. The crust of the Tyrrhenian sea resembles that of the mid-Atlantic ridge (Allan 1970 unpubl. in Lort 1971) with rugged relief of tensional ridges and grabens overlying a very shallow layer 3 (7.0 to 7.3 km/sec at 2.9 to 3.2 km of depth below water bottom). The Recent normal faults observed in southwest Italy, as well as the chemistry of the Tyrrhenian volcanic province, are without doubt related to the Tyrrhenian spreading and support the assumption of mixed crust in that basin.

Beneath the orogenic belts, the base of continental crust is encountered at a great variety of depths and at dip angles of up to 20 degrees. Some of these slopes in the M horizon may represent wedge-outs of continental crust due to extensional deformation during an event of rifting such as beneath an Atlantic-type margin. Other such M slopes are associated with overlaps or repetitions of M segments; they represent crustal thrusting and/or subduction that can be related or juxtaposed with surface evidence of orogeny.

Plate tectonic interpretation of surface structural features suggests that the crustal data may not always present the complete record. In some profiles, some of the flip interpretations are not conclusive. Crustal overlaps in senses opposed to that of thrusting may represent arc-arc collisions in which only fragments of the three plates remain detectable. The preservation of an M segment is dependent upon the basalt-eclogite transformation, because the p-wave velocity of eclogite is the same as that of mantle rocks (Yoder 1976). The transformation takes place at pressures equivalent to 35 km and more of overburden at temperatures at and above 600° (Winkler 1967). Such temperatures can be expected in lower plates of subduction zones under a variety of kinematic conditions (Toksöz et al. 1971).

Segments of the M horizon below 35 or 40 km, therefore, are probably underlying sialic crust. Segments of the M horizon, required in plate reconstructions, could have disappeared if they underlay basaltic oceanic crust. Such models, therefore, can only be tentative.

8 Plate Motions and Neotectonics

Present crustal movements of the Mediterranean orogens and their tabular surroundings are not only an important environmental factor but also are key elements in relating structural data to the plate tectonic framework and in interpreting older orogens with the use of data from active belts. Seismotectonic and geomorphological research, therefore, was a major effort in the overall program, in particular to outline fault zones not yet apparent in the seismicity data.

There are four categories of neotectonic movements: 1) movements on the major plate boundaries. 2) movements between small plates, on arcs between subduction zones, at centers of backarc spreading, and at the boundaries of rotating microcontinents or oceanic areas. 3) Intraplate movements show regionally persistent directions of σ_1 and σ_3, but are not tied to plate-boundary-like narrow belts. 4) Dislocative and compressional or extensional movements may be related to gravity spreading of topographic highs with or without crustal convergence.

All of these movements are capable of producing earthquakes in the sediments as well as in the crust. All of them can occur in zones of regional to continental extent, and all can achieve strain rates measurable in cm/y. All of these movements are instantaneous, whereas significant geological structures constitute large amounts of finite strain. Therefore, diagnostic differences between some of these categories are only in the duration of movements through geologically significant time spans. Sublithospheric earthquakes are considered to be diagnostic of convergence across plate boundaries. They constitute Benioff zones, regardless of the cluster shape of the earthquake distribution. This is a well founded conclusion of modern geodynamics (Isacks et al. 1968).

Movements of the first category include compression at rates of 1.9 to 2.6 cm/y and at nearly northerly azimuths (Le Pichon et al. 1973) between the African and Eurasian plates. A segment of this plate boundary extends from the Azores triple junction, with compressional first arrivals through the Strait of Gibraltar to the Calabrian arc, the Hellenic and Tauride arcs to the Zagros front.

There are small plates with movements of the second category on both sides of the main plate boundary. Spreading in the Strait of Tunis at an azimuth of 45° based on data from the Tunisian Atlas (Castany 1951) and from the Malta graben (Biju-Duval et al. 1974) is copolar with the Gulf of Suez spreading. Although the connection between these two areas is not established, it is assumed to form a separate Levantine plate west of the Dead Sea-Aquaba left-lateral transform (Freund et al. 1968; McKenzie et al. 1970) and south of the Calabrian, Hellenic, and Tauride arcs.

North of the main boundary, backarc spreading, oroflex-like loop tightening, gravity spreading of rising orogenic masses outline the geometry of the Calabrian and Hellenide arc systems as they simultaneously collide and tighten their loops.

Movement of the Levantine plate relative to Africa causes a southwesterly convergence of an estimated 3.5 cm/y across the Hellenic trench which fits well with the dimensions of the Hellenic Benioff zone. The Calabrian arc and its conically curved Benioff zone suggest a southeasterly convergence vector and imply an easterly or northeasterly movement of the arc relative to the European plate, at an estimated rate of 4 to 5 cm/y which can only be expressed as divergence across the rifting and/or spreading Tyrrhenian sea. The triangular shape of this basin may suggest high-lati-

tude spreading around a pole in the Ligurian sea. Such a movement would cause a counterclockwise rotation of the Adriatic plate relative to Europe and the Balearic Sea, with right-lateral strike slip in the northern Alps or at the Peri-Adriatic line, and with convergence across the west Alpine front.

Alternatively, this movement may belong to a large system of low-altitude spreading around a pole in the North Atlantic and may extend across continental Europe connecting the Rhone and Bresse grabens with the Rhine and North Sea grabens (Illies 1975) and a triple junction to the left lateral Languedoc wrench faults (Mattauer et al. 1972) near Valence.

Diffuse fields of stress with horizontal planes are associated with post-collisional tectonics in orogenic belts and with intraplate deformation described as slip-line tectonics (Molnar & Taponnier 1976). Such deformation is likely an evolutionary stage prior to plate fragmentation and sea floor spreading.

9 Shelf Deposits and Plate Fragmentation

Patterns of crustal subsidence are expressed in finite sedimentation rates of shallow-water shelf rocks. Such patterns reflect lithospheric fragmentation and its timing, because the subsidence of cooling oceanic crust takes place predictably along a time-depth function known as the Sclater curve (Sclater & Francheteau 1970) and is transmitted into adjacent continental or mixed crust. Faulting along the continent-oceanic interface of Atlantic-type margins normally ceases after initial rifting (Beck & Lehner 1974). On the shelves, therefore, the oceanic subsidence effect is closely reproduced by the cancelling effects of sialic buoyancy and of sediment load. The effect diminishes inland along a function determined by the flexural rigidity of the continental lithosphere (Walcott 1970), but is detectable on many ancient shelves 300 to 500 km inland from the continent-ocean interface.

A major subsidence cycle is recognizable in the eastern parts of the East Mediterranean area and along the northweast flank of the Adriatic plate. It had its rifting stage during the Permian and early Triassic, setting the scenario for the great circum-Tethyan carbonate platforms (Bernoulli & Jenkyns 1972). The shelf area affected by this subsidence fringed the triangular west-pointing Tethys.

Jurassic spreading in the Central Atlantic (Le Pichon & Fox 1971; Pitman & Talwani 1972) ended to the north at the Grand Banks-Iberia transform fault. Easterly continuation of this spreading developed within the circum-Tethyan shelf along a complex system of transform splays and offset oceanic-floored rifts. The southernmost of the fractures extended through the Saharan Atlas and split the Adriatic plate off the Libyan and Egyptian shelf. More northerly branches of transforms produced the mixed crust and succession of platforms and basins evident in the Hellenides and Dinarides and shaped the paleogeographic subdivision of the Penninic mixed-crust area into south, middle, and north Penninic belts. Westerly extensions of rifts into the European continent, such as the Dauphiné trough (Gignoux 1958; Dietrich 1976) and the south-facing North Pyrenean shelf edge or trough (Mattauer & Henry 1974) may represent failed arms of triple junctions (Dewey & Burke 1974).

In the Penninic trough of the Alps and the Apennines, rapid succes-

sion of rifting, subsidence, and renewed rifting leads to complex and not always interpretable curves of subsidence. The date of the changeover from initial rifting and block faulting to sea-floor spreading and epeirogenic subsidence, critical to the fitting of the subsidence curve, is difficult to recognize in marine sedimentation before and after the changeover. Available thickness data are cumulative maximum figures not always derived from measured sections.

10 Interplate Units and Accretionary Wedges

Clastic sedimentary units, comprising eugeosynclinal fill, flysch, and molasse, are found sandwiched between the upper and lower plates of subduction zones or between orogenic forelands and hinterlands. During the life of an orogenic belt, these units move intermittently with either plate and therefore, they have been termed *interplate sediments*. Their plate tectonic interpretation, not significantly different from the classic views on synorogenic sediments, is based on arc-trench models by Seely et al. (1974), Karig & Sharman (1975), and Hamilton (1976). In terms of these models, most interplate sediments in the eastern Mediterranean area are parts of accretionary wedges deposited on the lower plate.

In the Hellenides, flysch wedges record several convergence events. Melanges and a wide belt of Cretaceous flysch characterize not only the west-facing (Africa-vergent) obduction of the Eohellenic ophiolites, but also renewed transport of the obductive thrust sheets during subsequent events of suture progradation westward. A Paleocene to early Eocene flysch covers both obductites and their foreland; it records the suturing of a Tethyan basin and the emplacement of the internal Hellenic thrust units. The external Ionian flysch unconformably covers a Karst topography on a Mesozoic carbonate platform which became involved in thrust transport as the suture shifted from a position above the flysch wedge into the substratum of the flysch. The pre-Apulian foreland of the Hellenic thrust complex is at present overriding a recent accretionary wedge in the Ionic sea.

In the Rheno-Danubian flysch at the north front of the Alps, few if any of the complex stages of orogenic evolution are recorded, although it is sandwiched between both plates and although it comprises all stages of Cretaceous and Paleogene. To explain this wedge, three geodynamic interpretations are being discussed:

1) The flysch does not represent an accretionary wedge. It was deposited in a separate intracratonic rift overrun by the Adriatic plate after the continent-continent collision.

2) The flysch forms the accretionary wedge of the intra-Tethyan, Penninic, north-facing (Europe-vergent) arc system and was overrun by the Adriatic plate after the continent-continent collision.

3) The flysch accumulated in the backarc basin between the Penninic arc system and the Adria microcontinent and became an accretionary wedge only after the beginning of subduction at the north front of the Adriatic plate, probably in the late Cretaceous. The second of theses choices is the most similar to the recent arc-trench models, but the third choice is preferred because the Rheno-Danubian flysch probably extends into the Ligurian basin southwest of the present Alps.

The imbricate thrust plates of the Apenninies largely consist of interplate material overlying lower-plate material. Ophiolite suites at

less than thrust-plate size for melange fragments together with deep-ocean sediments, flysch, and bouldery marine units.

Apennine interplate clastics, occupying a rather high percentage of the total orogenic volume, consist of ophiolite fragments, deep-ocean sediments, flysch, and bouldery marine units, that is, olistostromes and melanges. Their interpretation is based on two elements of a plate model:

1) the interplate clastics were deposited behind a west-facing (Europe-vergent) trench,

2) the present stacking order was achieved after a flip of the subduction into a northeast-facing (Africa-vergent) polarity.

Remnants of the accretionary wedge of the pre-flip stage are exposed, as Schistes Lustrés or Bündnerschiefer in the Piemontese Alps and in Corsica,overlain by an ophiolite slab derived from the median high of a west-facing (Europe-vergent) oceanic arc. The post-flip accretionary wedge can be recognized in the Oroco, Caio, and Canetolo nappes and their sediments.

Some of the bouldery deposits included in interplate units can be interpreted to reflect accelerated rates of uplift and erosion due to structural changes in the subduction system. In the Alps, such changes can be related to the mid-Cretaceous arc-continent collision within the Penninic system. In the Hellenides, similar structural events may be required by the abundant glaucophane schists but have not yet been identified by mapping.

The forelands of orogenic systems have also furnished olistostromes. During the Paleogene the foreland of the Hellenides consisted of the rifted carbonate banks of the greater Tripolitza platform. Its downbending renewed its fragmentation, touching olistostromes off its steep cliffs and fault scarps into the more remotely sourced flysch trench fill.

Shale diapirs activated by the dewatering of pelitic Neogene to Recent trench fill line the belt of arc-arc collision between the Calabrian and Hellenic systems onshore in Lucania and, by inference from the bottom topography, offshore in the Gulf of Taranto and the southern Messina Strait. A median ridge between the converging arc systems is the source of olistostromes shed symmetrically over the footslopes of both arc systems. This observation may have a bearing on the interpretation of trench fill worldwide. Subduction-related shale diapirs or local sources of olistostrome masses have been reported from the Makran ranges (Ahmed 1969), from Trinidad (Saunders 1974), Sarawak (Haile 1974), and the Molucca Strait (Silver & Moore 1977). They may be much more common in the fossil record than recognized by present geological methods.

Two geodynamic types of Molasse deposits occur in the Central Mediterranean area. Foreland molasse covers the lower plates of continent-continent-collisions, such as north and west of the Alps, in the Po Valley basin between the Northern Apennines and the Alps, offshore from the central Apennines in the Adriatic Sea, and in the Wallachian basin in front of the southeastern Carpathians. The front of the Southern Apennines is the Lucanian flysch trough, presently rising, possibly because of the collision of the Calabrian and Hellenic arc systems. Internal molasse of Oligocene to early Miocene age fills rifts and extensional basins behind active thrust fronts in the Hellenides. Similar deposits of Miocene and Pliocene age fill rift basins in the Eastern Alps, such as the Vienna Basin and the Styrian Tertiary basins.

11 Mechanics of Thrusting

The mechanics of thrusting is closely related to modern geodynamic interpretation and the confrontation of fixistic and mobilistic views. Assumed driving mechanisms of thrusting and nappe transport show the same confrontation; they include gravitational body forces as in gravity gliding and gravity spreading, and compressional surface forces as in rear-end push.

The origin of compressive stress in mountain belts at convergent plate boundaries is unknown, because subduction due to the gravitational sinking of the lithosphere is conceivable without any horizontal compressive stress. First-motion studies in the vicinity of trenches and in accretionary wedges show vertical or at nearly 45° inclined P-axes. The extensive and mobile slopes of accretionary wedges suggest a mechanism of gravity spreading in the form developed by Elliott (1976) after Bucher (1956) and Price & Mountjoy (1970). In this model, shear stress at the bottom of a moving thrust sheet or nappe complex depends only on the lithostatic stress and on the angle of the surface slope. Since this stress is in equilibrium, continuous incremental movement on thrust sheets and flattening of the slope may result from the convergence and its effect of resteepening the slope.

In the Alps, where many of the mappable thrust surfaces originated during the post-collisional stage, the gravity-spreading model is particularly successful. After collision, strong uplift in the central Alps was accompanied by strong transport on low-angle thrusts in the external parts. The rear portions of the older thrusts are strongly uplifted and show outward dips. However, gravitational downhill gliding is clearly not the mechanism of thrust transport, because the older thrusts predate the uplift and are crosscut by younger thrusts soling out at depth below the central uplift.

Comparisons of reflection-seismic profiles across thrust belts beneath trench innerwalls (Beck & Lehner 1974) with similar profiles across marginal thrust belts on the cratonic flanks of orogenic belts (Bally et al. 1966) show significant similarities and differences.

The similiarities include transport over a flat, blockfaulted, or warped, but uninvolved basement with sledrunner-shaped thrusts branching off upward. They also include two pressure-dependent layers of style, a superficial frontal zone and a deeper transport zone. Thrusting during synorogenic erosion is recognized in successions of older transport fabrics and younger frontal fabrics (Roeder 1973; Elliott 1976). As a corollary, thrust transport during increasing sediment load or imbrication, such as in submarine accretionary wedges, should show older frontal fabrics superposed by younger transport fabrics.

Thrust sheets of the Central Mediterranean area show some of the diagnostic features. Detachment planes are bedding-parallel or cutting up-section in the direction of movement. Frontal imbricate splays and brow folds affect entire subunits such as the Allgäu nappe of the Austro-Alpine unit (Gwinner 1971) or the Olonos chain at the west front of the Pindos nappe (Temple 1968). Folds, warps, and crosscutting steeper thrusts may record transitions to frontal-zone conditions due to synorogenic erosion. Other such secondary structures may be caused by deeper-seated disturbances. Polyphase thrusting is being recognized as the cause of complex juxtapositions of facies belts and nappe elements in the Alps (Tollmann 1973) and in the Apennines.

In analogy to the arc-trench model by Karig & Sharman (1975) the source area of many melanges is the median high, a thrust sheet emerging between the forearc region and the accretionary wedge. This is the simplest explanation of the ophiolitiferous melanges in front of ophiolite slabs, such as the Aroser zone in front of the South Penninic ophiolites in the Alps, and the flysch-sandwiched melanges in front of the coherent Eohellenic ophiolites. In the Apennines, the disintegration of several nappes into olistostromes suggests another model of a trench innerwall, possibly a stairway of several thrust scarps shedding olistostromes simultaneously.

12 Flips and Collisions

Flips or polarity reversals of subduction zones are fundamental changes in the architecture of subduction-related orogenic systems, comparable in importance only to collisions between continent and/or arcs (McKenzie 1969; Dewey & Bird 1970; Karig 1971; Roeder 1973). Some of the diagnostic features for flips are the same as for collisions, and only additional plate kinematic or topological data can decide.

A flip is the adjustment of plate convergence to stable conditions of buoyancy and gravitational sinking. In an arc-continent collision with the oceanic arc obducted upon continental crust, a flip restores stable plate convergence by placing the continental crust back into the upper plate. This process may take 1 to 10 m.y. to evolve, and it is recognized at the surface by obducted slices of oceanic crust on the back side of younger trench-associated thrusts, by a younger thrust belt on one flank of an orogen with an older thrust belt on the other flank, or by younger thrusts cutting an older thrust plate in a sense opposite to its transport. A second type of flip (Karig 1971) affects oceanic arcs without any buoyancy difference between both plates and is therefore ascribed to differential flow within the mantle overturning the downgoing slab beneath the flipping oceanic arc (Roeder 1975). Arcs reveal this evolution by showing an extinct and filled trench on their present back side. So far, flipped oceanic arcs have not been recognized in older orogenic belts.

Head-on arc-arc collisions may be mistaken for flip events where the geological record is limited to an Andean arc complex overridden by an obduction complex (Roeder 1973). Critical evidence for an arc-arc collision includes the sequence and the dynamics of events. A flip of an Andean system would place continental crust into the lower plate; it would, therefore, be unstable, and the assumption of an arc-arc collision is more reasonable. Critical evidence for arc-arc collisions also includes the topology of adjacent areas. The arc-arc collision between the Calabrian and the Hellenic arc, for example, is an argument for assuming the same interpretation in cross sections along strike to the north, instead of a flip event.

The model of flake tectonics (Oxburgh 1972) may provide a second alternative to the flip. In this model, an Andean margin is simultaneously overridden and underthrust by the incoming plate split into a downgoing slab and an overriding top portion.

New support for charting flips or collisions comes from the refraction-seismic exploration of crustal structure. The evidence used consists of repeated and overlapping velocity discontinuities that can be identified as the M horizon. The repetitions are interpreted as preserved crustal parts of fossil or presently active subduction zones which must

be reached by burial beneath sediment accumulations or by subduction beneath stacks of thrust plates. Furthermore, many high p/t metamorphic rocks show no tectonic fabrics as would be expected within an accumulating stack of trench fill and subduction-related thrusts.

Timing and palinspastics of many blueschist belts in the Central Mediterranean are consistent with collision events assumed to be the cause of root-zone kinking. However, few or none of these belts have the structural attitude of root zones. One must assume, therefore, that they formed, rose, and were eroded prior to arrival of younger thrust systems, and that they were flattened out during subduction beneath these younger systems. Structures of this type have been assumed in the South Penninic accretionary wedge in the Alps and Apennines and could have been present in several belts of the Hellenides.

14 Metamorphism

One of the geodynamic aims of studies of metamorphism is to furnish pressure-temperature data which can be used in reconstructions of orogenic belts and their evolution. Lithospheric mass movements displace rocks against the geothermal field by erosion, by burial, by subduction, and by diapiric intrusion of magma. Mass movements also distort the geothermal field by introducing volatiles and by developing strain heat. The mineral products resulting from these processes are described in terms of isoreaction-grads (Winkler 1976) which bear some relationship to the structural evolution of their crustal habitat.

Based on the geodynamic situation of the occurrence, metamorphic provinces can also be classified in four plate tectonic groups. These groups are only in part characterized by diagnostic mineral facies, but their distinction is useful in analysis of plate motions and plate configuration. The groups include:

1) Oceanic metamorphism in newly formed oceanic crust and upper mantle results from mantle-derived heat interacting with meteoric fluids. It predates orogeny but affects the future ophiolite.

2) Arc-associated metamorphism surrounds or overlies the plutons in the magmatic arc of a subduction system. This type forms the low p/t facies series in the trench-facing element of paired metamorphic belts (Miyashiro 1973).

3) Trench-associated metamorphism with high p/t facies series occurs in in geosynclinal slate belts, melanges, and ophiolites, in elements of accretionary wedges or interplate units of convergent plate boundaries. This metamorphism arises from deep burial and subduction.

4) Post-collisional metamorphism records the re-exposure by uplift and erosion of once deeply buried terranes. Additional heat contributions, such as by friction on thrust contacts and in fold belts, are suggested by some of the available data. This type shows a low p/t metamorphic facies series and may grade internally into arc-associated metamorphism. Its occurrence subsequent to paired-belt metamorphism is the principal agent in retrograding the high p/t parageneses in the sedimentary interplate complexes.

All four groups have been identified in the Eastern Mediterranean area, although not every known occurrence of metamorphism can be classified within this system. Metamorphism does not support the view that Eastern Mediterranean orogens differ from the normal plate tectonic types of mountain belts.

15 Oceanic Metamorphism

This type has been identified in a very careful study of ophiolites in the Northern Apennines; it is of great economic importance, but its significance to the plate tectonic model is limited to providing an argument that the ophiolites affected by it are fragments of oceanic crust and upper mantle.

16 Arc-associated Metamorphism

The range of p/t conditions found in arc-associated metamorphic terranes is enclosed by the Barrovian and Abukuma facies series, or between estimated geothermal gradients of 20 to 80°C/km. Since the world average geothermal gradient of stable cratons is 20°C/km (Winkler 1976), some arc metamorphism requires an additional heat contribution, but some metamorphic rocks are best explained by uplift by folding and buoyant rise of sialic crust as in the Andean orogenic model (Dewey & Bird 1970). At the realized pressures, anatexis starts between 650 and 700°C or at depths of 8 to 20 km, but in most arc-associated belts, metamorphic isograds are much lower and require smaller amounts of uplift.

Arc-associated metamorphism has been identified only in two of the Hellenide belts and, with remaining questions, in the Alps. A Jurassic event is associated with granite plutons in the circum-Rhodope belts, and Upper Cretaceous to Miocene events occur in the median crystalline belt of the Central Hellenic nappes. A late Cretaceous metamorphic event in the upper plate of the Alps implies a geothermal gradient of 60 to 100°C/km and suggests an additional contribution. Two views are possible: the contribution could be associated with the beginning of convergence and arc activity, or less likely it could have accompanied an event of rifting or incipient sea-floor spreading. Scarce traces of metamorphism of Triassic age in the Western Alps, not sufficient to establish entire orogenic belts but assumed to record the subduction of parts of Paleotethys, are more likely arc-associated than of another type.

Arc-associated metamorphism thus is documented for several of the Central Mediterranean arc-trench systems, but not in solid belts. This may generally reflect the short life spans of these arc-trench systems, but it is also possible in several examples that subsequent plate events, such as collisions with overriding units or the formation of extensional basins with thick sediments, or major events of sea-floor spreading, destroyed or disrupted the more obvious evidence.

17 Trench-associated Metamorphism

Glaucophane-bearing blueschists with pervasive cleavage systems or without any tectonic fabrics and occurring alone or associated with eclogites, are common in the Central Mediterranean arc-trench systems. While eclogites are stable in dry environments under a very wide range of conditions, parageneses of glaucophane without lawsonite suggest pressure ranges above 8 kb (Winkler 1976). These pressures are interpreted as indicators of subduction and rapid burial followed by emergence, in rocks belonging to the accretionary wedge external to the magmatic arc.

Burial under an accretionary wedge is possible only to a depth equal to

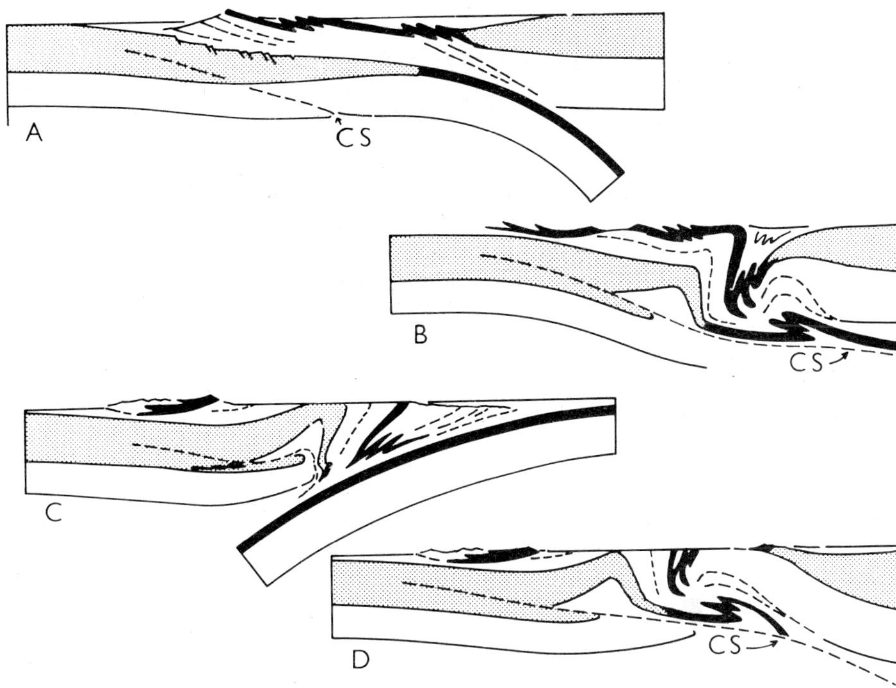

Fig. 6. Four interpretive crustal cross sections, showing plate mechanism of generating and preserving high p/t metamorphism. Shaded, continental crust. CS, crustal shear. Black: oceanic crust. (A), incipient arc-continent collision with continent trailing behind arc, possibly representing Penninic-Ligurian ocean in middle Cretaceous time. (B), advanced collision with kink of both plates above intracrustal shear. Kink provides rapid deep burial of accretionary wedge. (C), flip and continued subduction as one of two ways to rapid erosion and re-emergence of accretionary wedge. (D), Alternatively, post-collisional buoyant rise and continued decollement on crustal shear provides re-emergence.

the thickness of the wedge, typically 5 to 7 km. Ernst (1970) and Dal Piaz (1972), therefore, propose burial by subduction and re-emergence by a reversal of the subductive movement. A third mechanism (fig. 6) proposed in this volume for the Alpine blueschists, achieves the extreme burial by root-zone kinking of the arc-trench system as in a flip. The rapid emergence is achieved by buoyant uplift and erosion after the root-zone kinking.

18 Post-collisional Metamorphism

This type of low p/t metamorphism is petrologically identical with arc-associated metamorphism, but it occurs in the lower plates of arc-continent or continent-continent collisions, in locations that are pronouncedly negative thermal anomalies in the subduction model.

There are three principal sources of heat for metamorphism in this plate tectonic position: burial at the regional geothermal gradient, arc-

associated heat flow, and strain heat due to intracrustal thrust transport, folding, and cleavage transposition. In each instance, at least two of these heat sources must be considered. Burial beneath an upper plate, not much of a problem in subduction-related orogens, requires a fixed time span between the collision date and the radiometric closing date (Oxburgh & Turcotte 1974). The applicability of this heat source, therefore, depends on the plate tectonic history of the orogen. Arc-associated heat flow may occur in the lower plate if it is the upper plate of a second overridden subduction zone, as in a three-plate model of arc-arc collision.

Strain heat, not systematically studied during the present research project, may be a substantial heat contribution in collision orogens and underneath their forelands, if the results by Bird et al. (1975) are generally applicable. These authors show that at convergence velocities of 5 cm/y the temperature of dislocated rocks at 15 to 25 km of depth reaches the solidus of dry granite, with friction at the crustal decollement as a heat contribution in addition to the geotherm.

In a subducted lower plate, the geothermal gradient is anomalously low at 10°C/km as shown by the high p/t parageneses. In the models by Bird et al. (1975), the average gradient to a depth of 20 km and after subduction of 400 km of mixed and continental crust is only 8°C/km. Such low values during and immediately after continental convergence necessitate strong alternative contributions such as strain heat or, if the plate configuration allows for an arc-arc collision, arc-associated heat flow.

In the Alps, the Meso- and Neo-Alpine metamorphic events attain amphibolite grade with kyanite and some anatexis and are inferred to be post-collisional by the stratigraphy of contemporaneous sediments in the overridden foreland. The minimum pressure of 5 kb or 19 km is in tolerable agreement with the structural location of the metamorphic rocks beneath 5 km of eroded lower-plate material and 10 km of restored thickness of the upper plate. The estimated temperatures imply finite gradients of 33 to 43°C/km which can only be attained by an additional heat contribution and after a time span of thermal adjustment after subduction.

The proportions of heat by burial and additional heat depend on the relative timing of collision and cooling dates (Oxburgh & Turcotte 1974; Roeder 1977). Overlapping ranges of these figures and lack of adequate specialized studies are the reasons for two opposed views in the Alps:

1) The post-collisional metamorphic events may represent a geothermal normalization of the gradients after an intense event of subduction and depressed isotherms; the post-collisional uplift makes formerly deeply buried zones accessible at surface (Closs 1975).

2) additional heat is required because of the anomalously high gradients and because of the short time span between the emplacement of the Austro-Alpine plate and the thermal peaks (Oxburgh & Turcotte 1974; Roeder 1977).

Post-collisional metamorphism characterizes several other lower-plate areas. In the Apennines, greenschist-grade metamorphism in the Alpe Apuane , an anticlinal uplift of the shelfal (Tuscan) foreland zones, and more subtle thermal effects upon carbonaceous matter require a heat source in addition to burial beneath the internal Apennine nappes.

Metamorphic complexes in an interplate position suggest mixed heat sources or an undiagnostic geodynamic situation. This group of occurrences includes the greenschist facies found in the South Penninic trough fill as Schistes Lustrés in the Western Alps, the Piemont province of

the Italian Alps, and in Corsica. In some areas, this metamorphism may be retrogressive after a high p/t event associated with subduction.

In the low p/t series known from the Hellenides, arc-associated metamorphism and post-collisional metamorphism are not clearly distinguished. Greenschist-facies metamorphism on both sides of the Phyllite-zone sole fault as exposed in the Peloponnes may represent the effect of strain heat, whereas some of the thermal peaks of low p/t metamorphism in the Pelagonian complex of the Central Hellenic crystalline belt are more likely arc phenomena.

19 Magmatic Rocks

Volcanic rocks, dikes, plutons, and ophiolitic complexes follow closely a pattern of distribution set by the meatmorphic rocks, suggesting similar mechanisms and similar classifications for both.

Plate fragmentation and initial rifting are evident in volcanic suites and dike rocks. Felsic and alkalic rocks are of Permian age along the northwest edge of the Adriatic plate (southern Alps), the east edge of the Iberian plate (Corsica) and parts of the Eurasian plate (Maures-Esterel massifs of the French Alps and their foreland). They are of Middle Triassic age in the Peloponnes. They interfinger with non-marine transgressive miogeosynclinal sediments on Hercynian and older sialic basement, and their composition suggests a substantial admixture of cratonic anatexites.

Basaltic dikes and flows associated with rifting of Triassic to Pleistocene age occur as bedding-parallel ophite flows in the late Triassic salt domes in North Africa, Iberia, and France, in many areas of the Alps, Apennines, and Hellenides, and in the cratonic areas of northeast Africa.

Ophiolites are combined metamorphic and plutonic complexes formed hot on divergent plate boundaries and emplaced cold by obduction within collisional orogenic belts. In East Mediterranean orogens, ophiolites occur as large coherent slabs with thrust contact at the base and sedimentary contact at the top, or as smaller olistolithic fragments with tectonic or slump contacts on all sides.

The Central Mediterranean ophiolites may represent oceanic basement generated by sea-floor spreading in a normal mid-oceanic environment, transported beneath arc-trench gaps, and obducted during arc-trench collisions. However, ophiolites could also have been obducted out of backarc basins and onto the craton by thrusting antithetic to subduction (Dewey 1977). Reconstruction of the mechanism of emplacement, therefore, depends on the entire plate tectonic evolution of the system.

Complete suites contain, at their base, an ultramafic complex of lherzolitic or nearly pyrolitic composition with a metamorphic and strongly tectonized grain fabric. This tectonic layer is overlain by a plutonic series of ultramafic and mafic cumulates, and further by fine grained mafites in sheeted dikes and in flows including pillow lavas.

The base of an obducted ophiolite complex is a thrust fault post-kinematic to the tectonite fabric, at an angle to the lithological layering and cutting up-section toward the front or toe of the ophiolitic thrust sheet. Its stratigraphic location involves varying thicknesses of the tectonite layer at the base. The thickness of the plutonic complex, between the petrological Moho and the diagnostic sea-floor basalts, is

also variable but was determined by processes during the generation of the sea floor. This thickness, therefore, should reflect the p/t conditions in the crust and lithosphere during plate-boundary divergence with production of cumulates predominating at mid-ocean ridges, and with uplift and erosion of primordial mantle predominating near transform faults and fracture zones. Recognition of two types of ophiolite composition in the western and eastern Mediterranean (Nicolas & Jackson 1972) may demonstrate a regional difference in tectonic style of the Tethyan oceans. In the narrow pericontinental ocean basins of the western Tethys, divergence involved considerably more transform faulting than mid-ocean ridge spreading, and that therefore, the upper mantle rose more highly instead of becoming buried under the normal amount of plutonic products of sea-floor spreading. This explanation presents problems.

In pericontinental or cratonic areas, the mantle can rise from its subsialic depth of 35 to 40 km, or from its pericontinental depth of 15 to 35 km, only if most of the overburden is removed by erosion. There is as yet no evidence of such enormous erosion of sialic pre-Jurassic material.

Arc magmatism in the Central Mediterranean area is as scarce as arc-associated metamorphism. In several subduction systems, the magmatic arc should be located 50 to 200 km behind the subductive front, but no diagnostic arc rocks are known at the surface. Of the late Createaceous collision of the South Penninic system with Eurasia, the magmatic arc is assumed to be lost in a subsequent subduction beneath the Austro-Alpine north front of the Adriatic plate. In the Corsica-Apennines profile, the south Penninic arc became the lower plate after the pre-Tuscan flip without leaving an acceptable record of arc magmatism.

Tyrrhenian arc vulcanism of Miocene to Recent age extends from the Eolian islands north of Sicily to the Apennine volcanoes and to the islands of Capraia north of Elba. The magma produced by these volcanoes is basic, undersaturated, and Na-enriched (Marinelli 1967) and contains a clearly continental $^{87}Sr/^{86}Sr$ ratio of 0.707 to 0.711 (Hurley et al.1966). It suggests that the province is related to the east-facing subduction along the Apennines and the Calabrian arc, but also that the arc is in the earliest stage of extension, producing anatexites of mixed oceanic and continental origin.

Arc-associated vulcanism is also recognized in several complexes of calc-alkaline and alkaline eruptiva. The Eocene Austro-Alpine arc contains the Colli Euganei volcanic center and may also be the origin of a high volume of detrital andesite in the Tavayannaz sandstone within the north-Helvetic flysch complex of the Western Alps (Dietrich & Franz 1976).

In the Rhodope and Pelagonian massifs of the Hellenides, Jurassic to Miocene granitoid plutons (Altherr & Seidel 1978) are attributed to arc magmatism, associated with at least two consecutive south-facing subduction zones.

Tholeiitic effusives are associated with intraplate stress in several oceanic and cratonic areas. In the immediate foreland of a subduction, such mafics may record the tensional downbending of the lower plate. A magnetic anomaly beneath the Northern Calcareous Alps near Berchtesgaden could represent a mafite body extruded in late Eocene time in the foreland of the North Alpine collision. In ophiolite obductions, basaltic extrusives in flow contact with foreland rocks are found incorporated in

the ophiolite complex as basal mafites, such as in the Eohellenic obduction on the Peloponnes.

Intracrustal anatexis resulting from strain heat of collision or intraplate convergence may have caused plutonism such as the south Alpine plutons of Adamello, Bergell, and of the eastern Periadriatic area.

20 Suture Progradation

Most orogenic belts grow outward by incorporating their forelands. During the subduction of oceanic crust, this outward growth is confined to selective subduction (Moore 1975), sediment accretion (Karig & Sharman 1975), and an occasional incorporation of an ophiolitic lower-plate fragment (Beck & Lehner 1974). These processes leave little structural record in the evolving orogenic belt. However, a collision with a craton produces a chain of events related to major outward growth which is referred to as suture progradation.

In the Central Mediterranean orogens, prograding collision sutures have been causing many geological features, including foreland thrust belt, thrusts crosscutting former sutures, root-zone kinks or rotation zones involving parts or the entire thickness of the orogenic belt, dome-shaped centers of low p/t metamorphism dated as post-collisional, anatectic plutons in the foreland of the magmatic arc, and intracrustal layers of low P-wave velocity interpreted as layers of anatexis.

The arrival of a high-riding craton in the foreland of a subduction zone leads to the buoyant elimination of the deep-sea trench and to the loss of topographic slope at the surface of the thrust belt in the trench innerwall. This causes the tectonic style of the thrust belt to change from gravity spreading and low-angle transport of sledrunner thrust sheets to steep crosscutting reverse faults, to symmetrical folds above a basal decollement, and to kinking of root zones or rotation zones. Within the crust and lithosphere, the collision implies the loss of the gravitational component of subduction. If plate convergence continues, both overlapping plates may buckle or get kinked. The suture of convergence may re-establish itself as a boundary between an upper buckled part and a lower undeformed part, shaving off the rotation-zone kink in the lower plate. This intracrustal decollement builds up strain heat leading to anatexis along a seismic low-velocity layer, and to metamorphism and anatectic plutons rising from the decollement. Intracrustal kinking and decollement cause thickening of sialic material and contribute isostatically to the tectonic rise of the top surface of the orogen. The isostatic rise may outlast the convergence, building up a storage of orogenic energy which is released in continued marginal thrusting.

Two modifications of this model must be considered in the Tethyan realm:

1) In the Zagros foothills of the Iranides, convergence is expressed in folding of supracrustal sediments and in intracrustal seismicity confined to the foreland an reaching down to 60 km (Nowroozi 1971). This belt is not consistent with a flat intracrustal decollement beneath a system of gravity spreading. Rather, it suggests that the entire locus of convergence has shifted.

2) Mixed oceanic and continental crust in a lower plate causes subduction to evolve and to break up the lower plate, either in a series of interconnected branches as proposed in the model of prograding sutures,

or in the form of independent progressively younger subductions as proposed for the Alps in models emphasizing reduced plate mobility or limited amounts of finite subduction (Trümpy 1973; Bögel 1974).

A third modification of the basic model would include two or more branches of parallel Benioff zones as proposed for the Aegean sea (Papazachos & Comniakis 1977).

In the Alps and in the Hellenides, subduction of fragmented or mixed oceanic and continental crust is suggested by geological data. In the Alps, these units are stacked up beneath the principal suture, whereas in the Hellenides, they have agglomerated to form the upper plate of the present subduction.

In most Central Mediterranean belts, the documented orogenic time span is a multiple of the time required to cover palinspastic distances at even the minimum rate of oceanic subduction of 2 cm/y (Le Pichon et al. 1973). Therefore, collisions with continental fragments and the subduction of mixed crust probably slow down subduction.

This slowdown may be achieved by the elimination of the gravitational component of subduction, and its effectiveness would depend on the volume of cratonic fragments contained in the mixed crust. Based on the subsidence patterns of such fragments, a crude estimate would set 50 km as the minimum width of a cratonic block that could develop its own shelf subsidence pattern and could retard a prograding subduction. Bird et al. (1975) assume a width of 200 km for the transitional belt between oceanic crust and the craton with a full thickness of continental crust.

For the post-collisional tectonics of the Alps, a data body sufficient for some quantitative estimates is available. This tectonism includes the kinking of the South Alpine root zone, with uplift, erosion, and re-exposure of metamorphic belts, and with outward transport at low-angle thrusts. Thrusting since 20 m.y. amounted to 20 to 50 km and was probably caused by gravity spreading of the central Alpine uplift. However, an equivalent amount of crustal convergence is required to achieve the uplift.

Based on petrological data, the uplift of the Zentralgneis zone since the metamorphic peak at 20 m.y. amounted to 13 to 20 km. Of this uplift, only 3 km remain as present-day topography, while the remainder was removed by flattening and gravity spreading and by erosion. The erosion can be estimated at 7 to 12 km depending on thickness reconstructions (Bögel, personal comm., Oxburgh & Turcotte 1974), leaving 3 to 5 km of height reduction as a possible component of gravity spreading.

If only rock volume above sea level is assumed to have participated in the thrusting due to gravity spreading, the change in width of the thrust belt would have been 54 to 100 km, or an arithmetic mean of 78 km. However, if the seismic low-velocity layers are interpreted as intracrustal decollements, the spreading affected a wedge of 10 km of thickness below sea level. If this wedge is included, the change in width will be 0 to 63 km, or an arithmetic mean of 34 km.

Structurally, the change in width is expressed as foreshortening or thrust overlap and it may or may not be accompanied by an equal amount of crustal convergence. In order to estimate the amount of crustal convergence, one may consider the kinking of the South Alpine root zone which is the prime cause of central Alpine uplift. During this kinking, a 30 km long middle limb, that is, the present South Alpine root zone, became rotated into a vertical position. This movement amounts to about 30 km of horizontal crustal shortening. Where the South Alpine root zone is not developed, such as in the easternmost Alps, post-collisional crustal convergence may have been converted directly into thrust overlap.

21 Small-plate Rotations

The Mediterranean map pattern of shelves, ocean basins, and orogenic belts suggests the action of small plates in the size range between 100 and 1000 km of diameter. During the Mesozoic to Recent closing of Tethys, these small plates were located between Africa and Eurasia and performed rotations and translations in addition to the major plate movements across Tethys. Some of these small-plate rotations are consistent with paleomagnetic data.

The finite rotation of the Mediterranean small plates relative to either of the major Tethys border plates cannot be explained by convergence across the jagged outlines of the other plate, but requires individual and composite sets of finite rotations. Analysis of the orogenic belts and their complex kinematics as presented in this volume largely confirms the results of Smith (1971), but the driving mechanism of the small-plate rotations does not follow from the Tethyan convergence.

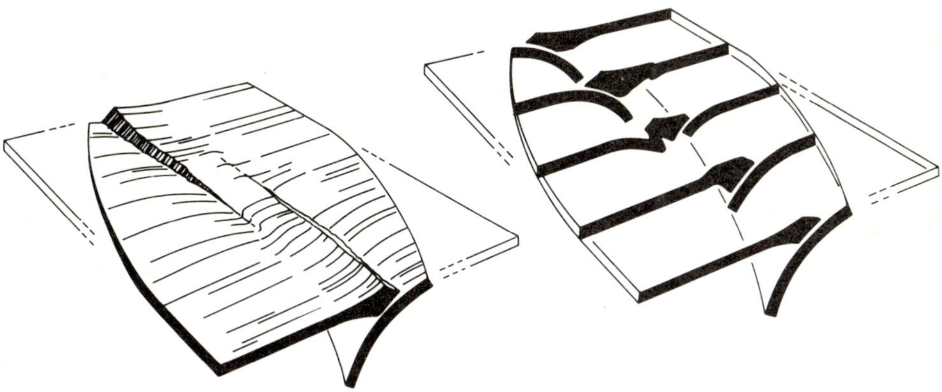

Fig. 7. Two schematic block diagrams showing interaction of offset trenches as a cause of small-plate rotation. Cumulative subduction on both trenches is shown. Left, view of surface of plate complex with a left-dipping Benioff zone on the near side replaced by an offset right-dipping Benioff zone on the far side. Right, subsurface structure of the same situation shown in three dimensions by seriated cross sections.

A partial mechanism for small-plate rotations may be provided by the fringing subduction zones themselves (fig. 7). Subsequent to arc-arc collisions, the continuing convergence may shift, by progradation and retrogradation, to new positions marginal to the collision orogen, in analogy to the arc-arc collision across the Molucca Strait and the southern Philippines (Roeder 1977b). If the main surface of convergence progrades into continental crust, convergence ceases. However, if oceanic crust on the internal flank of the collision orogen is reached by the convergence, subduction can continue at a new site offset from the site of collision. Subsequent plate tectonic events may follow a variety of evolutionary paths depending on the distribution of oceanic and continental crust.

After a collision of two arcs with oceanic crust on their internal or back sides, renewed subduction may generate two Benioff zones back to back, and offset along strike in the fashion of fig. 7 . The orogenic belt enclosed by the trenches will rotate relative to the surrounding plates. Each trench pinches out at an eulerian pole.

The lower plate of each trench is conterminous with the upper plate of the opposite trench, and both eulerian poles may be connected by an area of intraplate deformation. The amount of rotation of the orogenic belt and its offset marginal trenches relative to its forelands is equal to subduction. On the free end of a 1000 km long trench rotating by 30°, the amount of post-collisional subduction, therefore, is 500 km.

Two groups of orogenic belts and forelands in the Mediterranean area can be interpreted as having gone through stages of helical trench rotation. One is the sigmoidal arc of the Jura and Western Alps, the Apennines, and the Calabrian-Tellian system. The area common to two eulerian poles of convergence is located in the northwestern Apennines and in the maritime or Piemontese Alps.

With the Jura arc pushing westward and the Calabrian arc pushing eastward, the sense of rotation of the collision orogen relative to its forelands is counterclockwise. Nearly everywhere, except at the Calabrian-Sicilian arc, the frontal thrust belts are in late stages of intracrustal convergence. Therefore, the forelands are increasingly locked with the thrust belt and participate in the rotation. This is confirmed by paleomagnetic data for the Adriatic plate.

The preceding speculation is consistent with the paleomagnetic data, but not entirely with their interpretation. The counterclockwise rotation of the Adriatic plate relative to Europe very closely reproduces the rotation of Africa relative to Europe. However, only from a paleomagnetic point of view, is the Adriatic plate part of the African plate. There are numerous geological arguments for plate boundaries between Africa and the Adriatic plate during much of the Mesozoic era.

22 Mountain Belts in Retrospective

The foregoing has clearly shown that there is no principal difference in machanism, material, or style between specifically Central Mediterranean orogens and the prototype of the plate tectonic models of orogeny taken from Tethys and the circum-Pacific convergence belt. Caribbean orogenic belts, local convergence zones in the east part of the Sunda system, and many of the constituents of the polyphase and complex North American Cordillera are quite similar to the Central Mediterranean orogens.

In detail and in some of the dimensions, however, there is a distinct tendency toward smaller, shorter lived, less plutonized and more complex orogens produced by smaller distances of subduction. Complications, such as flips, collisions, and suture progradations are much more common occurrences than admitted during the development of the classical plate models. Specifically Mediterranean features include the effects of small-plate rotations and the subduction of mixed crust resulting in facies-nappe belts. These features may be also recognizable in other global convergence provinces.

In confronting our results with the classical problems of orogeny, namely the geosynclinal model, the geotectonic magma cycle, and the orogenic phase theorem, it is evident that most of the classical observa-

tions fit the plate tectonic model.

Miogeosyncline and eugeosyncline in their original simple definitions (Stille 1924) are rather useful terms for describing the facies of their plate tectonic equivalents, but extremely fixistic geosynclinal models (Aubouin 1965) are not compatible with present data.

The geotectonic magma cycle, developed within a fixistic earth model (Stille 1924), is easily translated into plate tectonic terms if the geochemically obsolete magmatic kinships are disregarded and if the causal relationship between stages is reduced to a coincidence in location. As already pointed out by Wilson (1968), Coney (1970), and Dewey & Burke (1974), Stille's stage of initial magmatism describes conditions at divergent plate boundaries, whereas synorogenic and subsequent plutonism are precise equivalents of arc activity, and final vulcanism accompanies intracontinental rifting prior to plate divergence. All of these forms of diastrophism are amply documented and relatable to plate events in the Central Mediterranean area.

Many of the paroxysmal events on the convergent plate boundaries in the study area are appropriately called phases and can be correlated into adjacent orogenic belts. The study area is small enough to demonstrate universally present orogenic phases although they cannot be correlated globally as was postulated in the model by Stille (1924). As an example of modern usage of the phase concept, the Eohellenic phase (Jacobshagen et al. 1976a) was defined as a plate tectonic event probably confined to the Hellenide segment of the Tethyan south margin.

The data presented in this volume suggest several mechanisms leading to a consistent relationship between the plate model and the phase theorem, as follows:

Convergence on the major plate boundaries is continuous but changes its rates and its azimuths, while movement on the minor plate boundaries is of shorter duration. Continuous subduction is absorbed in arc-trench systems in the form of progressive shearing in the accretionary wedge and in continuous addition to the magmatic arc. Only changes in the mode of plate convergence leave datable records of an orogenic phase. Most of the datable paroxysms can be identified as collisions of the arc with continental crust or crust fragments, some as arc-arc collisions, some as flips, and some as root zone kinks. Orogenic phases attributable to changes of subduction rate or azimuth have not yet been identified.

Our results show that some of the classical geotectonic problems are solved or that the avenues toward their solution are clear. The new limits of our knowledge are evident in quantitative and dynamic aspects of orogeny, in the thermal processes affecting the rocks during and after subduction, and in particular in the post-collisional metamorphism affecting the lower plates of quiescent subduction zones. We cannot distinguish with sufficient accuracy, the effects of burial by subduction, the heat contributions of overridden arc systems, and the contributions of strain heat due to intracrustal folding and low-angle thrusting. We ignore the causes of backarc spreading, and we cannot establish from the fragmentary geological record, topologically conclusive kinematic sequences of megatectonic events, such as flips and collisions.

The foregoing has also clearly shown that by adopting a plate tectonic view, previous achievements in the understanding of terrestrial processes can be preserved without blocking the avenues to further geodynamic insight.

References

Ahmed, S.S., 1969: Bull. Amer. Assoc. Petrol. Geol., 53:1480-1499.
Altherr, R. & Seidel, E., 1978: N. Jahrb. Geol. Pal. N.F. (in press).
Argand, E., 1924: Congr. Geol. Int. C.R. 13e Ser. I: 171-372.
Bally, A.W., Gordy, P.L. & Stewart, G.A., 1966: Bull.Canad.Petrol.Geol., 4,3:337-381.
Beck, R.H. & Lehner, P., 1974: Am. Assoc. Petrol. Geol. Bull.,58,3:376-395.
Berckhemer, H., 1977: in: Structural history of the Mediterranean basins: 303-314.
Bernoulli, D. & Jenkyns, H.C., 1974: Soc. Econ. Pal. Mineral., Spec. Publ., 19:129-160.
Biju-Duval, B., Letouzey, J., Montadert, L., Courrier, P., Mugniot, J.F. & Sancho, J., 1974: The Geology of Continental Margins:695-721.
Bijou-Duval, B., Dercourt, J. & Le Pichon, X., 1977: in: Structural history of the Mediterranean basins:143-164.
Bird, P., Toksöz, M.N. & Sleep, N.H., 1975: J. Geophys. Res.80:4405-4416.
Boccaletti, M., Elter, P. & Guazzone, G., 1971: Nature Phys. Sci., 234 (49):108-111.
Bucher, W.H., 1956: Geol. Soc. America Bull., 67:624-649.
Castany, G., 1951: Tunis Ann.Mines Geol., 8:632 p., Besançon.
Closs, H., 1975: Rapp. Comm. int. Mer Médit., 23,4a:55-57.
Dal Piaz, G.V., Hunziker, J.C. & Martinotti, G., 1972: Mem. Soc. Geol. Ital., 77,4:433-466.
Dana, J.D., 1873: Am. J. Sci., 5:347-350.
Dewey, J.F., 1976: Tectonophysics, 31:93-124.
Dewey, J.F. & Bird, J.M., 1970: J. Geophys. Res., 75:2625-2647.
Dewey, J.F. & Burke, K.C.A., 1974: Geology, 2:57-60.
Dewey, J.F., Pitman III, W.C. & Bonnin, J., 1973: Bull. geol. Soc. Amer. 84:3137-3180.
Dietrich, V.J. & Franz, U., 1976: Geol. Rdsch., 65,2:361-374.
Elliott, D., 1976: J. Geophys. Res., 81:949-963.
Ernst, W.G., 1970: J. Geophys. Res., 75:886-901.
Freund, R., Garfunkel, Z., Zak, I., Goldberg, M., Weissbrod, T. & Derin, B., 1970: Phil. Trans. Roy. Soc. London, A, 267:107-130.
Gansser, A., 1974: Eclogae geol. Helv., 67:479-508.
Gignoux, M., 1958: Géologie stratigraphique, 735 p., Paris.
Gwinner, M.P., 1971: Geologie der Alpen, 394 p., Stuttgart.
Haile, N.S., 1974: Geol. Soc. London, Spec. Publ., 4:333-347.
Hamilton, W., 1977: Maurice Ewing Series 1:15-32.
Hinz, K., 1972: Unternehmen Erdmantel DFG Forschungsbericht:174-175.
Hoeppener, R., 1956: Geol. Rdsch., 45:247-283.
Hurley, P.M., Faibairn, H.W. & Pinson, W.H., 1966: Earth Planet. Sci. Lett., 1.5:301-306.
Illies, J.H., 1975: Geol. Rdsch., 64,3:677-699.
Isacks, B.L., Oliver, J. & Sykes, L.R., 1968: J. Geophys. Res., 73: 5855-5899.
Jacobshagen, V., Risch, H. & Roeder, D., 1976: Z. dt. geol. Ges., 127: 133-145.
Karig, D.E., 1971: J. Geophys. Res. 76:2542-2561.
Karig, D.E. & Sharman III, G.F., 1975: Geol. Soc. Amer. Bull., 86: 377-389.
Klitzsch, E., 1970: Geol. Rdsch. 59(2):459-527.